Nondestructive Characterization of Materials II

Nondestructive Characterization of Materials II

Edited by

Jean F. Bussière

Industrial Materials Research Institute
National Research Council Canada
Boucherville, Quebec, Canada

Jean-Pierre Monchalin

CANMET, Energy, Mines, and Resources Canada
Ottawa, Ontario, Canada

Clayton O. Ruud

Pennsylvania State University
University Park, Pennsylvania

and

Robert E. Green, Jr.

The Johns Hopkins University
Baltimore, Maryland

PLENUM PRESS • NEW YORK AND LONDON

Library of Congress Cataloging in Publication Data

International Symposium on Nondestructive Characterization of Materials (2nd: 1986: Montréal, Québec)
Nondestructive characterization of materials II / edited by Jean F. Bussière...[et al.].
p. c.m.

Proceedings of the Second International Symposium on Nondestructive Characterization of Materials, held July 21–23, 1986, in Montreal, Canada, sponsored by the Industrial Materials Research Institute of the National Research Council Canada and other organizations.
Bibliography: p.
Includes indexes.
ISBN 0-306-42610-2
1. Non-destructive testing—Congresses. 2. Materials—Testing—Congresses. I. Bussière, Jean F. II. Industrial Materials Research Institute (Canada) III. Title.
TA417.2.I56 1986
620.1′127—dc19 87-16609
CIP

Proceedings of the Second International Symposium on Nondestructive Characterization of Materials, held July 21–23, 1986, in Montreal, Canada

A Division of Plenum Publishing Corporation
233 Spring Street, New York, N.Y. 10013

Printed in the United States of America

PREFACE

The possibility of nondestructively characterizing the microstructure, morphology or mechanical properties of materials is certainly a fascinating subject. In principle, such techniques can be used at all stages of a material's life - from the early stages of processing, to the end of a structural component's useful life. Interest in the subject thus arises not only from a purely scientific point of view but is also strongly motivated by economic pressures to improve productivity and quality in manufacturing, to insure the reliability and extend the life of existing structures.

The present volume represents the edited papers presented at the Second International Symposium on the Nondestructive Characterization of Materials, held in Montreal, Canada, July 21-23, 1986.

The Proceedings are divided into eight sections, which reflect the multidisciplinary nature of characterizing materials nondestructively: Polymers and Composites, Ceramics and Powder Metallurgy, Metals, Layered Structures/Adhesive Bonds/Welding, Degradation/Aging, Texture/ Anisotropy, Stress, and New Techniques.

Invited papers by R. Hadcock of Grumman Aircraft Systems, R. Cannon of Rutgers University, H. Yada of Nippon Steel and R. Bridenbaugh of Alcoa review respectively the processing of polymer matrix composites, ceramics, steel and aluminum, emphasizing the need for material property sensors to improve process and quality control. Two other invited papers, one by A. Wedgwood of Harwell and the other by P. Höller of the IzFP in Saarbrücken review state of the art techniques to characterize particulate matter and metals respectively.

The editors would like to express their gratitude to many individuals and several organizations. Foremost, our thanks to all the authors for their excellent contributions and cooperation in providing manuscripts. The symposium was sponsored by the IMRI, NRC Canada, the Physical Metallurgy Research Laboratory, CANMET Canada, and The Ministry of Trade and Commerce of the Province of Québec, who provided generous support. The cooperation of several societies in publicizing the meeting was greatly appreciated; these are: ASM, ASME, TMS-AIME, CSNDT, ASNT and SPI. Members of the international advisory committee provided useful guidance in the choice of speakers; these were: L. Adler, Ohio State University, H. Berger, Industrial Quality Inc., L. Bertrand, École Polytechnique de Montréal, P. Höller, IzFP, T. Kishi, University of Tokyo, G. Labbe, IRSID, R. Sharpe, AERE Harwell, W. Sturrock, Dept. of National Defence, D.O. Thompson, Iowa State University, and B.A. Wilcox, DARPA.

Finally, the editors would like to thank the conference secretary, Mrs. Gladys Cyr, for her devotion throughout the symposium and preparation of the Proceedings.

J.F. Bussière
R.E. Green, Jr.
J.P. Monchalin
C.O. Ruud

CONTENTS

I - POLYMERS AND COMPOSITES

III- METALS (continued)

IV - LAYERED STRUCTURES/ADHESIVE BONDS/WELDING

V - DEGRADATION/AGING (continued)

VI - TEXTURE/ANISOTROPY

VII- STRESS

VII - STRESS (continued)

VIII-NEW TECHNIQUES

VIII- NEW TECHNIQUES (continued)

PROCESS CONTROLS AND NONDESTRUCTIVE EVALUATION OF STRUCTURAL COMPOSITES

Richard N. Hadcock*, Peter J. Donohue**, and Richard F. Chance***

Grumman Aircraft Systems
Bethpage, New York

FOREWORD

Organic matrix structural components require the use of extensive and costly process control and quality assurance procedures to provide the quality and integrity required for aerospace applications.

This paper describes critical manufacturing operations and requirements. The current status of associated process controls and nondestructive evaluation techniques is addressed. Potential advanced techniques are identified which should reduce the high costs of these operations and, at the same time, improve quality and dimensional consistency of the finished parts.

INTRODUCTION

During the past 25 years, structural composites have been used for more and more military and commercial aircraft and helicopter structures as well as for spacecraft structures.

Of the many reinforcement and matrix combinations available, graphite fiber reinforced epoxy matrix composites have found the greatest utilization. Other fiber and organic or metal matrix systems have or are being developed; however, this paper addresses process controls and nondestructive evaluation of only epoxy matrix composites reinforced by graphite or boron fibers.

There will always be a need for structure weight savings since these can be directly transformed into increased payload, reduced acquisition and operating costs, or increased performance. Significant weight savings have been demonstrated by composite structures. In addition, the nature of composites allows parts to be tailored to provide the strengths, stiffnesses, and deflections required for the part being designed.

During the past 25 years, advanced composites have matured from a laboratory curiosity to materials which are being used to produce thousands of parts on more than 20 different aircraft. The successful application of these materials has, to a large extent, been accomplished by establishing well-

* Deputy Director, Technology Development
** Manager, Advanced Materials and Manufacturing Development
*** Manager, Quality Control Advanced Development

designed quality assurance programs with close quality assurance coordination throughout the design effort to establish inspectability requirements. This initial coordination and review continues into the tooling and manufacturing activities. A typical program sequence is depicted in Figure 1.

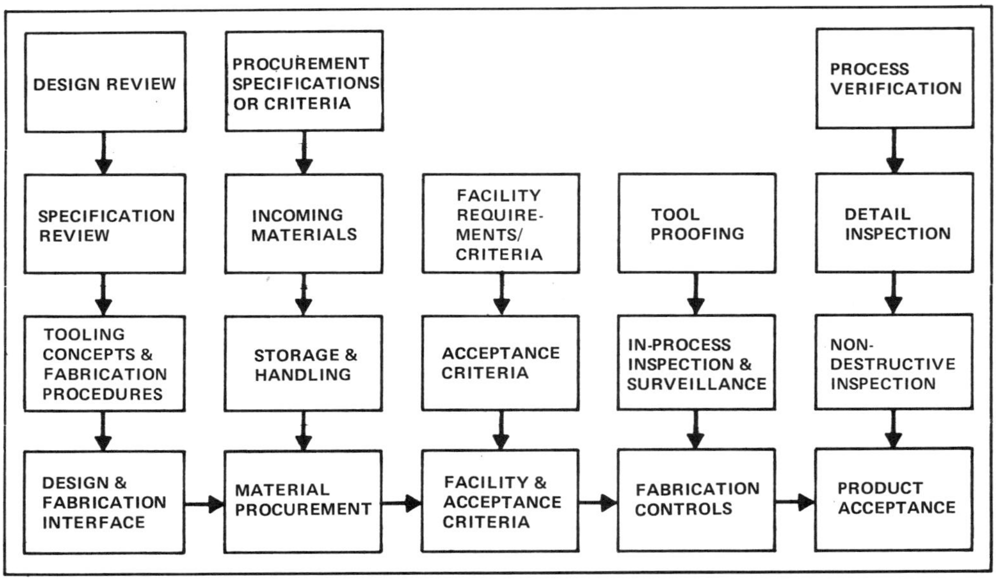

Figure 1 Quality Assurance Requirements for Advanced Composites

Composite laminates are manufactured from various intermediate production forms (e.g., prepregs) which change state through processing. Inspection operations must be added to the Quality Assurance Plan to ensure part integrity since fabrication deviations such as the number and orientation of plies, variations in the time-temperature-pressure cycle involved in autoclave curing, accidental inclusion of foreign matter, etc, significantly affect the integrity of the finished part.

In summary, incoming materials must be inspected for conformance to material specifications; layup tools must be inspected to determine their ability to meet heatup requirements and withstand specified thermal/pressure cycling; in-process quality control must be established for conformance to processing specifications in the fabrication of the composite material; and, as a final check, each composite part must be nondestructively inspected for defects. Further inspections and tests are then required for the various stages of assembly into the final end item. Figure 2 presents the general inspection requirements for an advanced composite component.

DESIGN AND SPECIFICATION REVIEW

Quality assurance personnel must participate in design review activities to ensure compatibility of requirements with related inspection techniques and their associated defect detection capabilities. The design and each material combination must be thoroughly evaluated.

Adequate quality control provisions are incorporated into materials procurement, processing and assembly specifications. Materials procurement specifications must list the constituents, component (prepreg), and storage requirements (see Tables 1 and 2). Process specifications must include detailed requirements for bleeder systems and the cure cycle to be used, as well as descriptions of the procedures to be used for assuring compliance with these requirements (see Table 3).

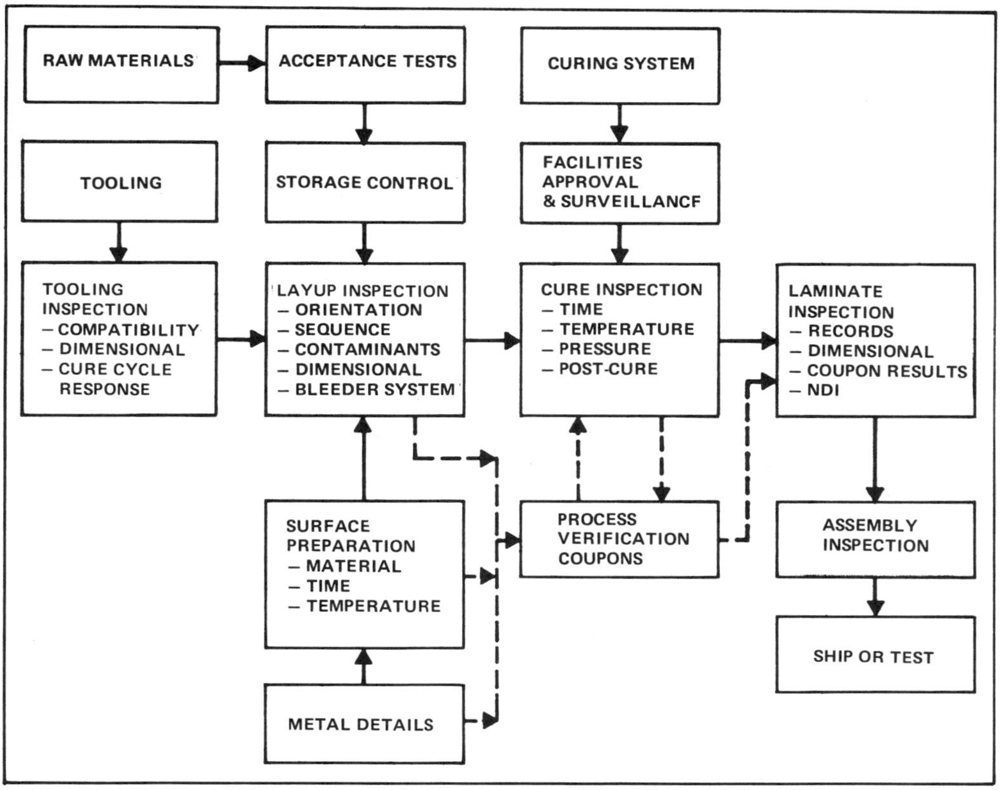

Figure 2 Inspection Requirements for Advanced Composites

TABLE 1 CONSTITUENT MATERIALS REQUIREMENTS, COMPOSITE MATERIALS PROCUREMENT SPECIFICATIONS

	RESIN	
• FORM • SPECIFIC GRAVITY • VISCOSITY	• MECHANICAL & ELECTRICAL PROPERTIES • SHELF LIFE • VOLATILES	• FLOW • TACK • TACK RETENTION
	FIBER REINFORCEMENT	
• TENSILE STRENGTH • TENSION MODULUS • DIAMETER	• DENSITY • FIBERS PER TOW • TWO TWIST	• SPACING BETWEEN SPLICES • FABRIC WEAVE (WOVEN MATERIALS)

RECEIVING INSPECTION, STORAGE, HANDLING, AND RECERTIFICATION

Once a composite material is received, Quality Assurance testing is performed to establish its conformance to the procurement requirements. These tests must take into consideration inter- and intra-batch variations. Typically, advanced composite prepreg materials should be tested or examined to determine uncured physical properties and tape quality, and panels should be cured and tested to determine mechanical properties (see Table 2). As a minimum, tension, flexure and horizontal shear properties should be determined at room temperature and the intended maximum service temperature.

TABLE 2 PREPREG REQUIREMENTS, COMPOSITE MATERIALS PROCUREMENT SPECIFICATIONS

UNCURED PHYSICAL PROPERTIES		
• FILAMENTS OR TOWS PER UNIT WIDTH	• TACK RETENTION	• LENGTH
• VOLATILES	• FLOW	• UNIFORMITY
• RESIN CONTENT	• ALIGNMENT	• STORAGE
• TACK	• GAPS/SPACING	• WORKMANSHIP
	• WIDTH	

CURED (LAMINATED) MECHANICAL PROPERTIES*	
LONGITUDINAL (0°)	TRANSVERSE (90°)
• TENSILE STRENGTH & MODULUS	• TENSILE STRENGTH & MODULUS
• COMPRESSION STRENGTH AND MODULUS	• FLEXURAL STRENGTH & MODULUS
• FLEXURAL STRENGTH & MODULUS	
• HORIZONTAL SHEAR STRENGTH	

*AT BOTH ROOM & SERVICE TEMPERATURE

TABLE 3 COMPOSITE MATERIALS PROCESS SPECIFICATION REQUIREMENTS

• HUMIDITY AND TEMPERATURE CONTROL	• QUALITY ASSURANCE
• MANUFACTURING DOCUMENTS	– SURVEILLANCE
• LAYUP	– IN-PROCESS INSPECTION
• BLEEDER MATERIALS AND PROCEDURES	– PROCESS CONTROL* FLEXURAL STRENGTH & MODULUS HORIZONTAL SHEAR STRENGTH ADHESIVE BOND STRENGTH
• BAGGING MATERIALS AND PROCEDURES	
• CURE CYCLES	
• POST CURE CYCLES	
• FINISHED PARTS	– VISUAL INSPECTION
• ADHESIVE BONDING	– NDI
• REPAIR PROCEDURES	
• MATERIAL REVIEW BOARD (MRR ACTION)	

* AT BOTH ROOM AND SERVICE TEMPERATURE

Because prepreg properties are known to change with time and temperature, even when stored under refrigerated conditions (0°F), these materials must be placed in quasi-bonded refrigerated storage and out time must be controlled. A specific storage life (shelf life) is established for each material and the material recertified when approximately two-thirds of the life has expired. Generally, recertification requires repetition of all or part of the original receiving inspection tests.

CONSTITUENT PROPERTIES

Composite parts currently being produced for aircraft must provide structural performance and integrity for the entire service life of the aircraft. Structural integrity is dependent upon the properties of the constituents of the material systems.

Chemical Composition

Fiber reinforcements have been fairly well characterized. Because most state-of-the-art composite resin systems are proprietary and their chemical composition is not identified, users depend solely on performance tests as a means of maintaining batch-to-batch quality control. While these tests are acceptable during introduction and development of composite materials, production use requires additional quality assurance tests to ensure consistency.

Variations in matrix systems have adversely affected the producibility of composite detail parts. J. Carpenter, H. Borstell, and others have studied the effects of resin system variations on integrity, reproducibility, and durability of composite structures. The variations investigated included type, purity, and concentration of chemical constituents, uniformity of mixing, and resin processing. Their approaches used a variety of available analytical methods to assure reproducibility of matrix systems between carefully established limits based on mechanical test data and processing experience under production conditions. Improved material reliability and durability were realized by characterizing approved resin systems and subsequently expanding existing specifications to include chemical composition and processing requirements.

The most promising chemical material characterization analysis methods are described:

- Infrared Spectroscopy - IR is used to describe the basic chemical structure of organic polymeric materials. The spectra obtained yield characteristic fingerprints which serve to identify the nonmetallic system as well as individual constituents

- Differential Scanning Calorimetry - DSC is a technique for evaluating the curing characteristics and other thermal properties of composite materials
- Rheometrics Dynamic Spectrometry - RDS is an indispensible technique for determining the flow characteristics of curing resins utilized in advanced composite prepreg systems. The instrumentation monitors the viscosity changes of the neat resin as temperature is increased. Farber[2] has described a resin extraction technique which can isolate resin from prepregged materials so that rheometrics analysis can be performed without resin alteration
- High Performance Liquid Chromatography - HPLC is a technique used to separate and quantitate the constituents of advanced composite prepreg resins. The resins are dissolved in a suitable solvent and introduced into the HPLC instrument, where separation and detection occurs. The data generated are used to verify proper resin formulation and advancement.

TOOLING CONCEPTS AND FABRICATION PROCEDURES

Ideally, process control specimens for destructive test should be cut from trim areas or protrusions on the part periphery to verify that the required mechanical strength has been obtained as a result of the specific processing employed. However, from cost considerations, most users net-mold their parts to near-finished size. Tool designs should be reviewed to ensure that adequate provision is made for separate quality assurance test coupons located at areas of the tool which will duplicate the cure parameters experienced by the part. If multiple lots of material are used in making the part, allowances should be made for additional coupons representing each lot.

Tool design should also be reviewed for selection of compatible tooling materials, fabrication, and processing parameters. Of particular importance are the thermal expansion and thermal conductivity of the basic tooling materials.

Tool Proofing

Prior to accepting a mold form (layup tool) for use, it must be proofed to verify its ability to meet the process heatup, pressure, and thermal requirements. The preferred procedure is to fabricate a preproduction part that can be thoroughly instrumented to provide its complete temperature response to the imposed cure cycle. The part is then evaluated by nondestructive testing followed by destructive testing through proof loading or evaluation of test coupons removed from strategic locations of the part. However, if raw material and fabrication costs preclude this approach it is essential that the layup tool be subjected to a tool-proof cycle that duplicates as closely as possible the conditions encountered during part cure.

Problem Areas

Many components with complex configurations require the use of novel tool concepts. These may have an adverse effect on nondestructive testability of the cured component. Intermediate septums for thick co-cured bonded assemblies can result in too much attenuation for conventional ultrasonic nondestructive testing (NDT) techniques. Foam tooling, removed by abrasive blasting, can cause irregular surfaces that do not adequately reflect sound waves normally used to test laminates. In these cases, the tool designer, manufacturing engineer, nondestructive test engineer, and design engineer must work very closely to minimize and identify "uninspectable" areas. Lower design allowables should then be be established for these areas. Potential improvements being investigated include the use of low-frequency ultrasonic NDT techniques.

LAMINATE LAYUP

Composite layup is accomplished either manually or by the use of automated tape laying machines. The prepreg tape is applied onto mylar templates (referred to as ply-on-mylar) or directly onto a layup tool in successive plies (ply-on-ply). Adequate control at this stage is important both in terms of part reliability and the significant investment in materials and labor.

Manual Layup

Ply-on-mylar layup is the more reliable procedure for manual layup. This approach offers the most opportunity for control and visual inspection of each individual ply as well as control of stacking sequence (see Table 4).

Excessive gaps between tape strips are very difficult to detect using ply-on-ply procedures. Quality assurance procedures for this method include a pre-layup inspection of the tape by viewing it on a light table for defects.

TABLE 4 COMPOSITE LAYUP INSPECTION REQUIREMENTS

• PLY MATERIALS	• TAPE GAPS	• FOREIGN MATERIALS
• PLY ORIENTATION	• TAPE OVERLAPS	• NUMBER OF PLIES
• PLY STAGGER INDEX	• BROKEN FILAMENTS	• PEEL PLY PLACEMENT
• PLY NESTING	• CURED RESIN	

Automated Composite Tape Laying

In the fully computerized factory, the end product is defined by a digital set of data residing in a central data base rather than paper drawings. The data contains all information, including engineering, procurement, tooling, manufacturing, and quality assurance, required for a predetermined form of manufacture. The data supports automatic operations for tool path and nesting data for least material waste. The digital design data will be used to produce an NC tape to directly fabricate the part by machining, routing, and drilling sheetmetal details, forging, casting, or by layup and cure of advanced composites. In the case of assemblies, parts lists, application lists, and the automated assembly operations themselves will be provided and controlled from the digital engineering data.

Both flat and contoured automated laminating machines are available. The flat laying equipment is simpler and useful on smaller components having shapes which may be obtained through subsequent operations such as draping in shallow molds or creep forming in special dies. Larger, more complex, components can be fabricated using 9-axis, computer-controlled composite tape laying machines (Fig. 3). These machines can automatically dispense unindirectional composite tape at speeds of 100 ft/min, while maintaining machine accuracy within a few thousandths of an inch. The tape can automatically be laid up in widths to 12 in. debulked, cut, and then overlaid, ply-on-ply, in several directions to achieve optimum part strength and flexibility. Perimeter programming for flat layup and natural path programming for contour layup reduces the programming task.

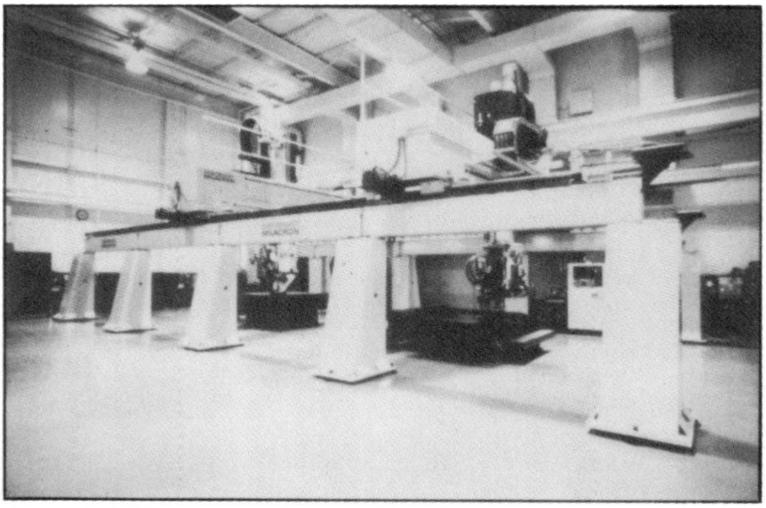

Figure 3 Composite Tape Laying/Placement Machine

Problem Areas

Recent trends towards the increasing use of automated tape laying machines and thick laminate designs are placing added importance on prepreg material quality. Manual layup permits the "man-in-the-loop" to readily detect and correct for prepreg anomalies such as tape width variations and inclusions. In the future, the prepreg manufacturers may be required to exercise additional controls over their processing operations. Potential improvements include automated in-line optical inspection for tape variations and other defects, and various closed-loop sensor and feedback techniques for tape processing and closer dimensional control.

Specific problems associated with automated tape laying are listed in Table 5.

TABLE 5 PROBLEMS ASSOCIATED WITH AUTOMATED TAPE LAYING

FLAT	• PLY-ON-PLY – CONTOUR ISSUES ("FLAT" IS GENERALLY A GENTLE CONTOUR) – PLY INSPECTION • INCOMING MATERIAL QUALITY – WIDTH CONTROL – TELESCOPING – FLAGS – ON-LINE INSPECTION • TAPE WIDTH – ECONOMICS OF SPEED – SCRAP LOSSES – PLIES AT SHALLOW ANGLES • TRIMMING/STACKING • INSTALLATION OF LOCAL BUILDUPS • MAINTENANCE; MANUAL PLY REPAIR
CONTOUR	• LAYING TO CONTOUR VS. FORMING TO CONTOUR • ROBOTIC PLACEMENT OF INTERNAL AND EDGE PLY BUILDUPS; COMPACTION, HANDLING • HANDLING FOR INTEGRAL STRUCTURE

LAMINATE PROCESS VERIFICATION COUPONS

Process verification coupons are normally processed for each operation in the manufacturing cycle involving a change in state of the materials used. The coupons are destructively tested to verify that the required mechanical strength has been obtained. As minimum, flexural strength and modulus, and horizontal shear strength, should be determined at room temperature. When more than one lot of prepreg raw material is used in a part, duplicate coupons should be processed because the resins may react differently to the cure cycle.

CURE

The autoclave cure process is most critical and must be very closely controlled. Current control procedures typically involve the following:

- Visual inspection verification of proper bleeding and bagging procedures (types and amounts of material)
- Vacuum leak check of bagged components prior to autoclave insertion
- Preset temperature controllers with upper limit alarms to control heat-up
- Loss of vacuum alarm to identify bag failures
- Strip chart recordings of part temperature, pressure, and vacuum to permit verification of the prescribed cure cycle.

These procedures have been adequate for control but are very labor-and-skill-intensive. They typically require the services of a manufacturing engineer to custom design the cure cycle based on the component mix in the load. A full-time autoclave operator is needed to activate the heat and pressure cycles at the appropriate times. Computer control of the autoclave operations is readily achievable and is currently being implemented by several companies. However, the proper cure cycle cannot yet be adequately performed automatically. The following is a discussion of the complexities and technologies being investigated relating to the total automation goal.

Computer-Aided Autoclave Curing

Material discrepancy, rework, or scrap is becoming an expensive problem. The value of an autoclave load is typically $30,000 to $200,000 per run (including materials, fabrication labor, overhead, etc). Added to this value is the expense of schedule slippage and disposition costs. It is evident that even a low scrap rate (5%) is unacceptable. Composite process equipment (autoclaves, presses, ovens, chemical reactors, etc) are constructed only to create a thermal environment proportional to a "set point" device. The actual part behavior in these vessels, however, varies significantly. The tooling mass has an associated nonuniform thermal conductivity. The part location within the vessel and the airflow dynamics all contribute to significant thermal history excursions on the actual part. It is virtually impossible to solve this problem, which results in process discrepancy and/or scrap situations, using manual or electronic cam controllers.

Mechanical cams and their electronic thermal programmable counterparts were the first attempts to "automate" and help reduce these manual limitations. These controllers, however, only ramp air temperature in the vessel or use fixed cascade "go/no-go" responses. They are unable to correct and optimize simultaneously the multiple variations which actually occur. In fact, their fixed response can actually create errors.

Conventional devices are not capable of identifying process deviation and documenting them for Quality Assurance. Therefore, the new generation control system must include full process data tracking, appropriate graphics information displays and error identification concepts to provide on-line QA monitoring of the process.

New manufacturing techniques are constantly being developed. Closed-loop material sensor feedback logic, zone thermal control, integrally heated tooling, and non-autoclave production techniques are areas currently being explored. Material science records will result in new sophisticated control algorithms. A flexible system which can operate any vessel type and control configuration would represent a major advance.

Available State-of-the-Art, Computer-Aided Curing Systems

Three automation systems utilizing temperature, pressure, and vacuum as control variables are currently available. These are Applied Polymer Technology's Composite Autoclave Process System (CAPS), Thermal Equipment Corporation's Autoclave Computer Control System (ACCS), and Research, Inc., (RI) systems:

- CAPS is available in three models. Model 110 is not suitable for use in an aerospace production environment because it has no error or data-tracking capability. Model 210 is both a complete manufacturing system and a materials-and-processes development system which includes data tracking, special graphic functions, and utility support routines. It can control a single autoclave. Model 310 is similar but controls up to four autoclaves simultaneously by utilizing a high-speed processor
- ACCS is available in two models. Model 216 can control one autoclave, while Model 1000 can control several autoclaves. An important feature of the ACCS models is automatic checking or thermocouple validity and removal of faulty sensing devices from the control algorithm during cure cycles. These models also incorporate provisions for recovery from power failures
- RI has several standard systems available and can also construct custom-made systems. The STAR SYSTEM series is designed to control various processing operations. The MICRICON unit can control the temperature profiles of multiple parts and the autoclave pressure

profile. However, it must be integrated with a MICRIHOST central command station for multiple autoclave operation.

Emerging Cure-Monitoring Technologies

Emerging technologies which may provide solutions to the many problems associated with computer-aided autoclave cure include dielectrometry, acoustic emission monitoring, and new modeling software.

Dielectrometry. Resin viscosity is an important process variable that should be monitored and controlled during autoclave/vacuum bag molding of composite parts. Since there is no currently available technique that directly measures resin viscosity during the cure cycle, electrical properties, correlate quite well with resin viscosity, could be monitored. Dynamic dielectric analysis (DDA) is a useful tool for developing and controlling cure cycles, since the dielectric medium (resin) is constantly changing during the cure cycle.

Micromet Industries' Eumetric System uses dielectric sensors to monitor the curing process. This equipment can be used in both the manufacturing and quality control areas and it should be available by late 1986.

Acoustic Emission. Monitoring of acoustic emission is an emerging technology for composite processing control. Several techniques are under investigation.

Modeling Software. Three IBM-compatible modeling software programs are available. These include Technology Pipeline (being used by Northrop), ROAST (developed and used by Lockheed-Georgia), and CUREL (developed by the University of Dayton). Since an automated autoclave control system would probably result in combining cure cycles, further evaluation of modeling software is warranted to determine both the compatibility of various cure cycles and the effects of particular loads and programmed cycles.

LAMINATE PHYSICAL PROPERTIES AND COMPOSITION TESTS

The cured composite must be characterized with respect to its density and fiber, resin, and void content. Fiber volume content establishes the composite's physical and laminate properties. Generally, higher fiber contents produce higher density and mechanical properties. The presence of voids significantly affects some of the mechanical properties. High void content increases the variation or scatter in strength and modulus properties accompanied by significant reductions in interlaminar shear, compression strength, and fatigue resistance. In addition, high void content increases the composite's susceptibility to degradation due to environmental effects such as humidity and temperature.

Density

Composite density measurements are made by the classical pycnometer method wherein the weight and volume of a composite specimen is compared to the weight and volume of mercury that the specimen displaces. The method determines an apparent density rather than the absolute density because the mercury is unable to penetrate the nonsurface-connected pores of the specimen.

Thermal Expansion

Thermal expansion characteristics of filamentary compositesare often an important factor in their design and use. Thermal deformations can be grossly affected by fiber orientation, laminate stacking sequence, and the thermal expansion characteristics of the constituent materials. Recording dilatometry using quartz dilatometers has been successfully used to measure the thermal expansion of undirectional and multidirectional laminates.

Composite Constituents

The composite constituents of general interest are the resin, fiber, and void contents. Nitric acid digestion is the recommended procedure for determining the resin content of a cured composite.

LAMINATE INSPECTION

Detail Inspection

Detail part inspection consists of laminate thickness and dimensional check of the periphery and moldline, and visual examination for defects. Although simple and routine, this inspection can provide valuable information of indications that unacceptable changes have taken place during the fabrication process. Changes in laminate thickness can indicate improper resin flow with resultant degradation of mechanical properties.

Nondestructive Inspection (NDI)

The primary purpose of NDI is to determine the existing state or quality of a material or part with a view towards acceptance or rejection. It is a comparative rather than an absolute method of inspection. The part is adjudged defect-free or defective by comparison to appropriate test standards which contain typical processing and fabrication anomalies representative of the actual parts.

Essential components of any NDI method are listed in Table 6. Interpretation is most important as NDI results are generally indirect indications such as dark or light areas on an X-ray film, the appearance of a pip or change in pattern on an oscilloscope screen, or a change in the color of a chemical. The interpreter must be highly trained and experienced, since interpretation of the result determines the parts' acceptance or rejection.

TABLE 6 ESSENTIAL COMPONENTS OF NDI METHODS

• *APPLICATION* OF A TESTING OR INSPECTION MEDIUM
• *MODIFICATION* OF THE TESTING OR INSPECTION MEDIUM DUE TO DEFECTS OR VARIATIONS IN THE STRUCTURE OR PROPERTIES OF THE MATERIAL
• *DETECTION* OF THIS CHANGE BY A SUITABLE DETECTOR
• *CONVERSION* OF THIS CHANGE INTO A FORM SUITABLE FOR INTERPRETATION
• *INTERPRETATION* OF THE INFORMATION OBTAINED

Selection of an NDI method for a specific part is equally important. Over-inspection results in excessive costs whereas underinspection might result in its failure if a critical defect goes unnoticed. Final selection of an NDI method must be based on engineering assessments of acceptable defects as well as thorough knowledge of the capabilities and limitations of the available methods see Table 7. Typical defects and detection method are shown in Figure 4.

NDI Standards

Test standards serve the purposes described in Table 8 and the criteria for their establishment are summarized in Table 9. The standards must closely represent the configurations important characteristics and properties of

TABLE 7 NDI METHOD SELECTION CRITERIA

- RELIABILITY
- SENSITIVITY
- EQUIPMENT REQUIREMENTS AND COST
- REQUIRED OPERATOR PROFICIENCY TO PERFORM TESTS AND INTERPRET DATA
- DEVELOPMENT NECESSARY TO ADAPT TO PRODUCTION LINE
- USEFULNESS TO OTHER APPLICATIONS
- SAFETY ASPECTS

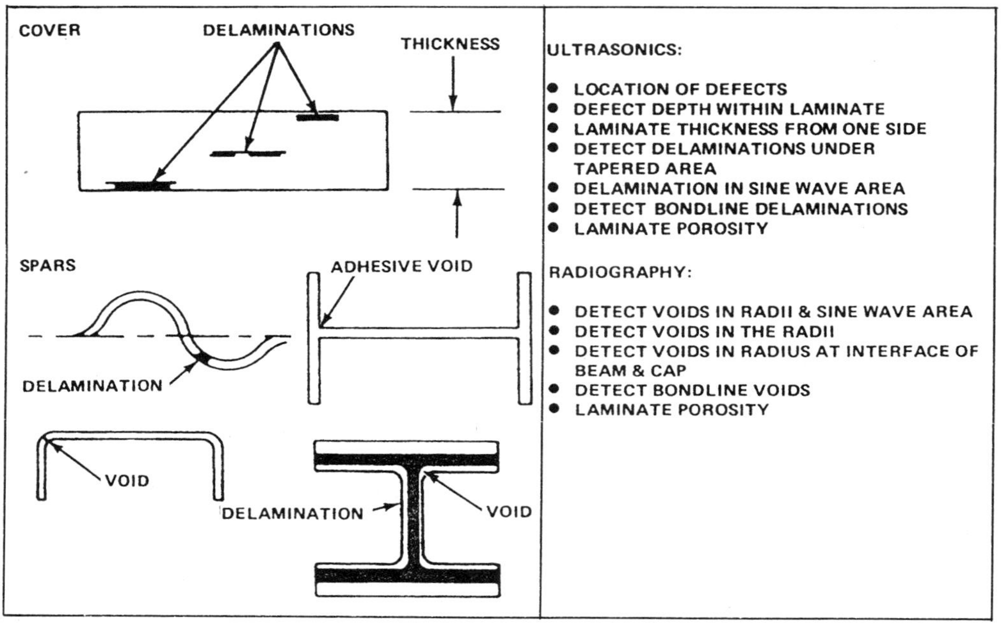

Figure 4 Structural Defects to be Found in Nondestructive Test

TABLE 8 NDI STANDARDS

- ESTABLISH INSPECTION PROCEDURES
- DETERMINE TEST SENSITIVITY LEVEL
- VERIFY EQUIPMENT PERFORMANCE
- ESTABLISH ACCEPT/REJECT CRITERIA
- ESTABLISH REPEATABILITY OF RESULTS

TABLE 9 CRITERIA ESTABLISHING NDI STANDARDS

- REPRESENTATION OF PART CONFIGURATION
- DEFECT TYPE AND SIMULATION
- DEFECT CRITICAL SIZE
- REPRESENTATION OF MATERIALS AND PROCESS VARIABLES
- DEFECT LOCATION AND DEPTH
- APPLICATION OF STANDARDS TO PART SERVICE AND ENVIRONMENT

the final part. They must also contain typical process and fabrication errors found in production. The type of simulated defect and its size, orientation, and location with respect to an inspection surface must be chosen to represent material and fabrication variables and not the capability of state-of-the-art NDI equipment.

The utilization NDI for advanced composites is summarized in Table 10.

TABLE 10 NDI OF ADVANCED COMPOSITES

DEFECT	METHOD OF INSPECTION
FIBER VARIATIONS	RADIOGRAPHY (X-RAY OR GAMMA-RAY) ULTRASONIC VELOCITY
DEBONDS & DELAMINATIONS	ULTRASONICS SONIC THERMAL
INCLUSIONS AND/OR VOIDS	RADIOGRAPHY ULTRASONICS

Radiography

Radiography has been applied extensively to determine fiber variations, voids, and inclusions. In this method the differences in the relative attenuation of a radiative energy beam by the various constituents of the composite are measured. For conventional X-rays, the radiative energy is produced by the impingement of a beam of electrons on suitable target. Gamma-rays are produced by radioactive isotopes, and neutron beams by various neutron generators. In the case of gamma and X-rays, inspection results can be recorded directly. Improvements in energy converter screens and image intensifiers have led to the development of means of recording transmitted images either for direct visual inspection or for online computer analysis. In the computer technique, filtering devices are employed to enhance the anomaly sought. In the case of boron-epoxy composites, single filaments in multilayer laminates can generally be observed under routine conditions by radiography due to the high opacity of the tungsten core of the filament to X-rays. For glass and graphite fibers in an organic matrix, the differences in opacity of the constituents are very much less. Nevertheless, the numerous radiographs made from these materials indicate relatively gross imperfections in filament alignment, as well as other imperfections, such as voids. In general, the lower the radiation energy employed, the better the result obtained. To optimize those radiographic parameters necessary to obtain maximum resolution or either filaments or voids, the maximum difference between the filament and matrix absorption must be obtained.

Radiographic methods are generally the most widely used for detecting inclusions. The existence of singular inclusions down to the resolution limits of the system and recording method can readily be observed. Aggregations of inclusions below this size are also observed as changes in the transmitted intensity.

The use of radiography for detecting delaminations in materials is generally unfeasible since the material cross-sectional thickness in unaffected and does not cause a change in X-ray absorption. However, when the x-radiation is directed parallel to the width of the delamination/void (such as tangent to a part radius), the defect does cause a reduction in absorption and is therefore detectable (see Figure 5). The technique also uncovers voids due to a lack of filler adhesive and improper ply compaction in the radius area of channels and sine wave spars.

Ultrasonics

The capabilities of currently available ultrasonic techniques are affected by the detection level flaw size and number of plies. The ultrasonic techniques normally employed can typically detect a 1/2 in. by 1/2 in. void

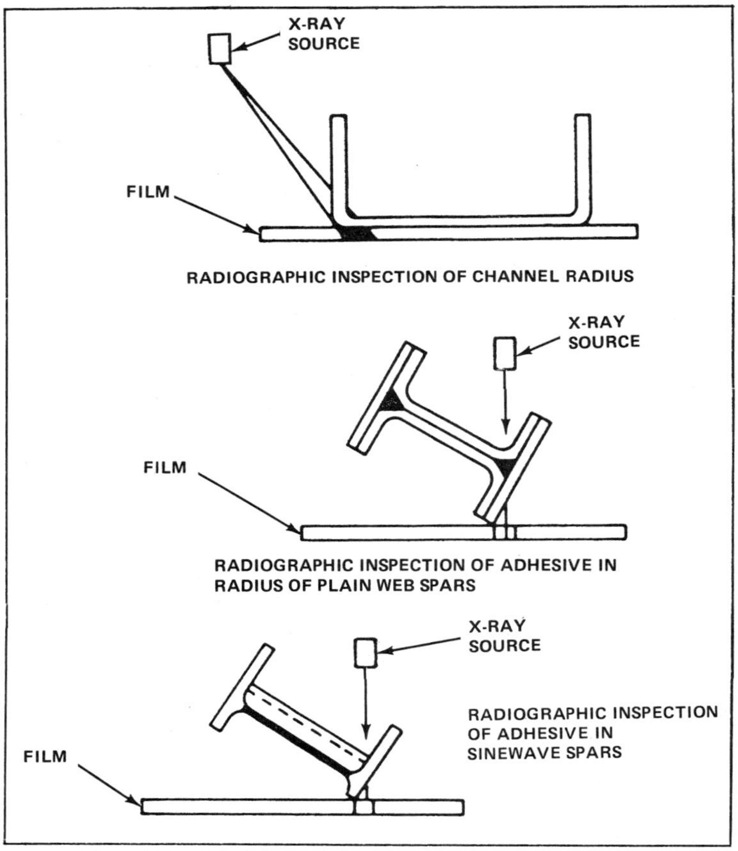

Figure 5 Radiographic Examination of Radius of Sinewave Spars

or delamination over a large range of ply thickness. Voids as small as 1/64 in. by 1/64 in. have been detected using techniques that cannot tolerate ply variations. Specialized techniques are much more costly to implement and should be avoided whenever possible.

The use of audible sound as a means of testing the quality of manufactured parts is not new. It is used extensively to reveal, within limits, the size and depth of subsurface flaws. The advantages and disadvantages of this NDI method are presented in Table 11.

TABLE 11 ADVANTAGES AND DISADVANTAGES OF ULTRASONIC NDI

ADVANTAGES
• ABILITY TO PENETRATE TO SUBSTANTIAL DEPTHS
• ABILITY TO TEST FROM ONE SURFACE ONLY
• SENSITIVE ENOUGH TO DETECT MINUTE FLAWS
• COMPARATIVE ACCURACY TO DETERMINE FLAW SIZE DEPTH
• ELECTRONIC OPERATION ENABLING RAPID & SUBSTANTIALLY AUTOMATED INSPECTION
DISADVANTAGES
• MANUAL USE REQUIRES SKILLS & MOTIVATION
• INTRINSICALLY FOR SMALL AREA COVERAGE – LARGE AREA COVERAGE REQUIRES COMPLEX SCANNING & NUMEROUS TRANSDUCERS IN AN ARRAY
• DIRECT COUPLING TO TEST ARTICLE REQUIRED

The most common technique is the ultrasonic pulse-echo method in which a pulsed beam of ultrasonic energy is directed into the material at right angles to the surface. Debonds, delaminations, and voids produce an echo which can be recorded. Analysis of the time difference between the far side and the imperfection allows the depth of the defect to be estimated. Alternately, a complete two-dimensional C-scan plot of the surface indicating the projected area and location of the imperfection can be recorded. Both one-sided and through-transmission methods are available using either direct contact or immersion techniques of coupling. When using the immersion technique, both the transducer and the test article are submersed in water or some other suitable liquid couplant. An additional benefit of the immersion technique is the ability to use higher frequency transducers that are extremely sensitive to the detection of small discontinuities. The ultrasonic attenuation monitoring method has proven to be very successful for detecting delaminators, voids, and porosity in laminated structures. Immersion or non-immersion techniques can be used. Many non-immersion systems utilize water coupling (squirter nozzles) and electronic data recording, and are completely computer-controlled.

Sonics

The surface of the material to be tested can be tapped with a coin or hammer; the operator then listens for a change in pitch of the sound generated from the material. The accuracy of such a test depends upon the acuteness of the operator's hearing and such variables as the amount of force applied to each tap, as well as the level of ambient noise in the vicinity of the test area.

A new NDI instrument, the Instrument Corporation Impactoscope™, eliminates the problem of operator subjectiveness associated with coin or hammer tap testing.

Thermal

When a reinforced plastic part is heated, heat transfer and dissipation occur much more readily if there are no defects such as delaminations or debonded areas. Consequently, the portion of the surface nearest a defect has the highest temperature. This phenomenon has been used as a method of inspecting composite laminates. The part is heated. During heating or as the part cools, the surface temperature is observed either by a sensitive infrared measuring device (radiometer) or by visual observation of the color change of a previously applied coating of temperature-sensitive, cholesterol-containing liquid crystals. Although colorless as isotropic liquids, cholesteric substances pass through a series of bright colors as they cool through their liquid-crystal phase. The color changes occur over a temperature range of 2° to 4°C.

Most of these NDI techniques are described in more detail by G. Epstein[3].

New Methods with NDI Potential

Some recent NDI methods holding promise for inspection of composites include optical and acoustic holography, and stress wave acoustic emission analysis. Holography is a method of recording the amplitude and phase of the optical wavefronts reflected from an object such that, when reconstructed, these wavefronts have the relative amplitude and phase of the original wavefronts. Acoustic holography is a means by which scattered or reflected sound waves are converted into optical wavefronts to obtain the same properties. The light waves do not suffer the reflection and mode conversion problems associated with acoustic energy.

In theory, the vibrational "signature" of sound material differs from that of defective material. Holograms provide a means for comparing the signatures of relatively large areas of materials of structures against known acceptable standards.

Ultrasonic holography has been applied to detect various interactions of sound waves within a material in the form of attenuation and/or interference. Techniques for performing coherent mapping of the data are based on a holographic recording of coherently pulsed ultrasonic wavefronts over a line, along a plan, or throughout a volume. While the method promises improved sensitivity and resolution over conventional ultrasonic imaging, considerable development is required to make the method practical.

Stress wave acoustic emission analysis has been successfully applied to measure certain mechanical properties of fiber-reinforced structures. Materials emit sound as they are strained. Some of the low-intensity, inaudible sounds are incipient indicators of failure. Microphones or accelerameters can be used to detect these signals. Both the frequency of emission and amplitude of the sound can be analyzed in real-time or recorded (see Figure 6).

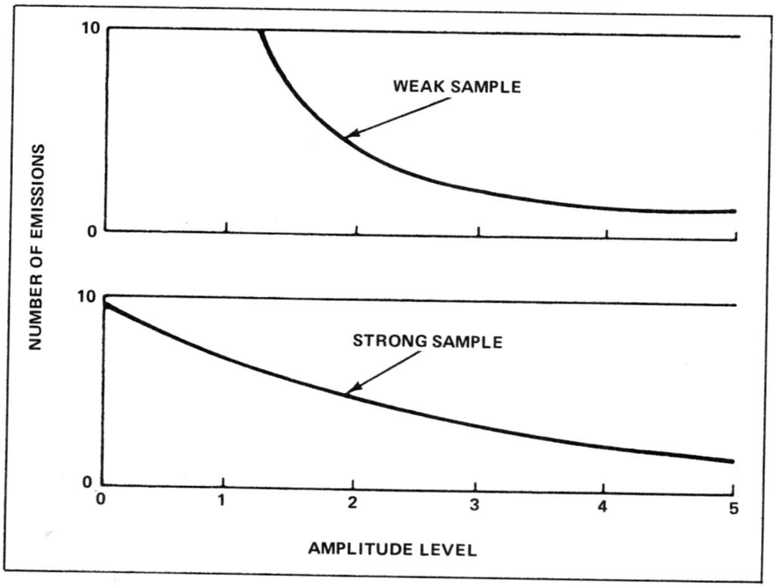

Figure 6 Differences in the Acoustic Emission Signal Amplitudes

<u>Problem Areas</u>

The major problems associated with laminate inspection are related to cost and sensitivity. Many of the dimensional and NDT inspections are labor and skill-intensive. Very precise manipulation over complex configurations has restricted the use of automated systems. Also, thick laminates have a high potential for developing detrimental porosity levels which must be considered when doing NDI.

Recent developments involving automated contour-following ultrasonic systems (see Figure 7) and micro-focus radiography (see Figure 8) show promise in meeting the sensitivity and alignment requirements for automation. In addition, recent studies shown good correlation between ultrasonic attenuation and various porosity levels (see Figure 9). Implementation of these techniques is expected to provide significant cost and quality improvements.

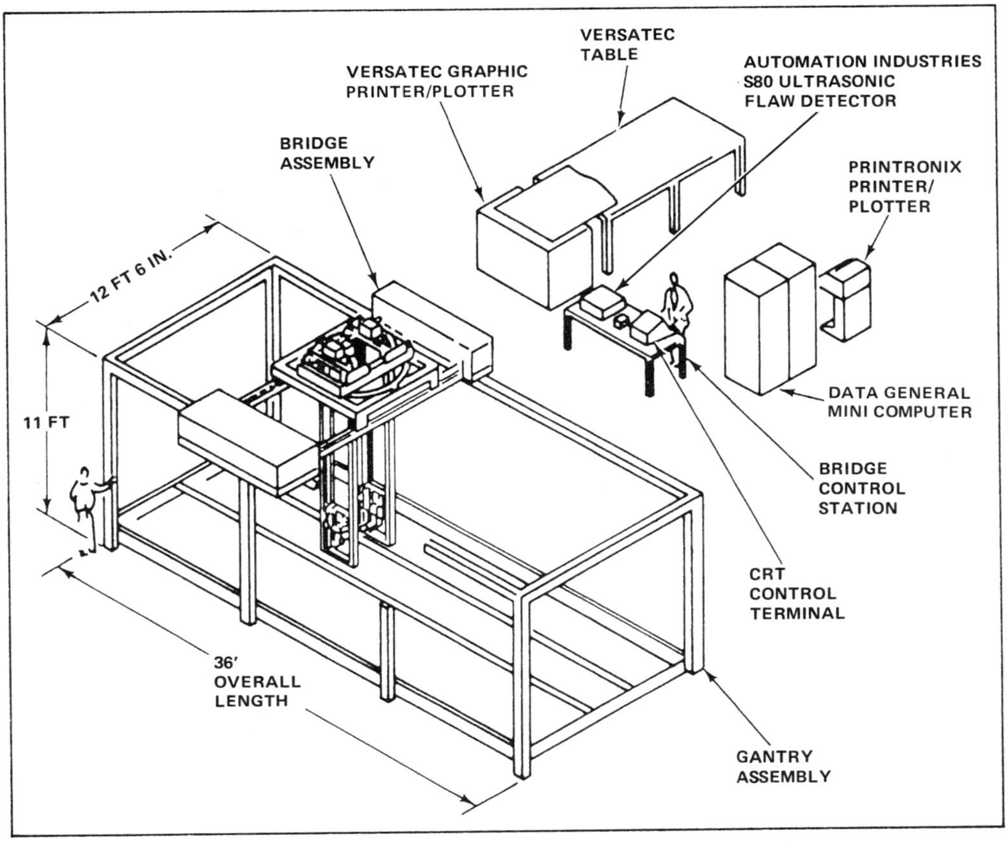

Figure 7 Automation Industries Contour-Following Ultrasonic System

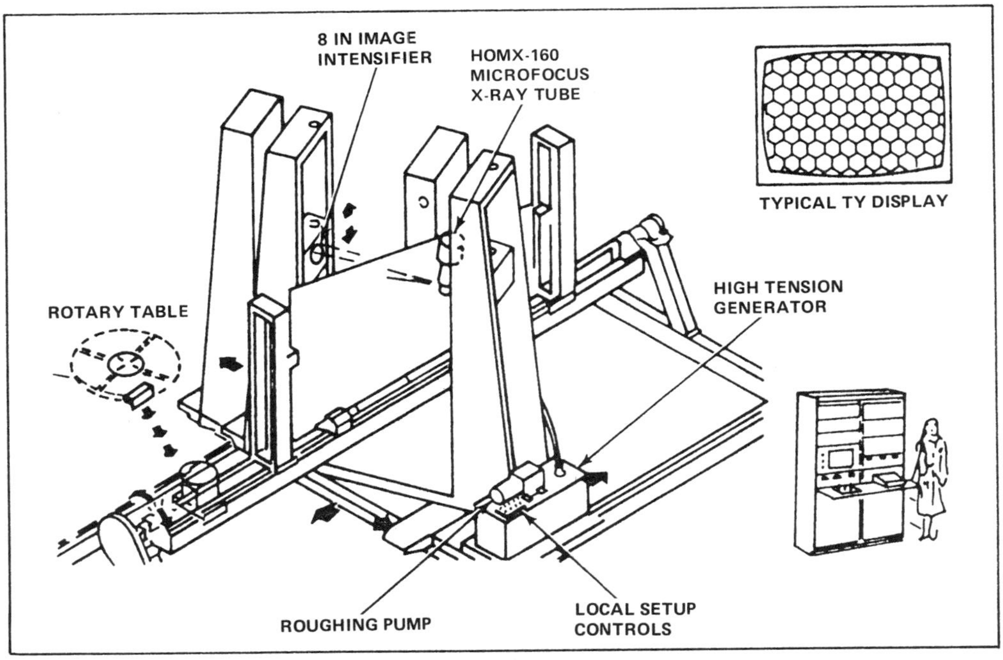

Figure 8 Ridge Micro-Focus Radiographic System

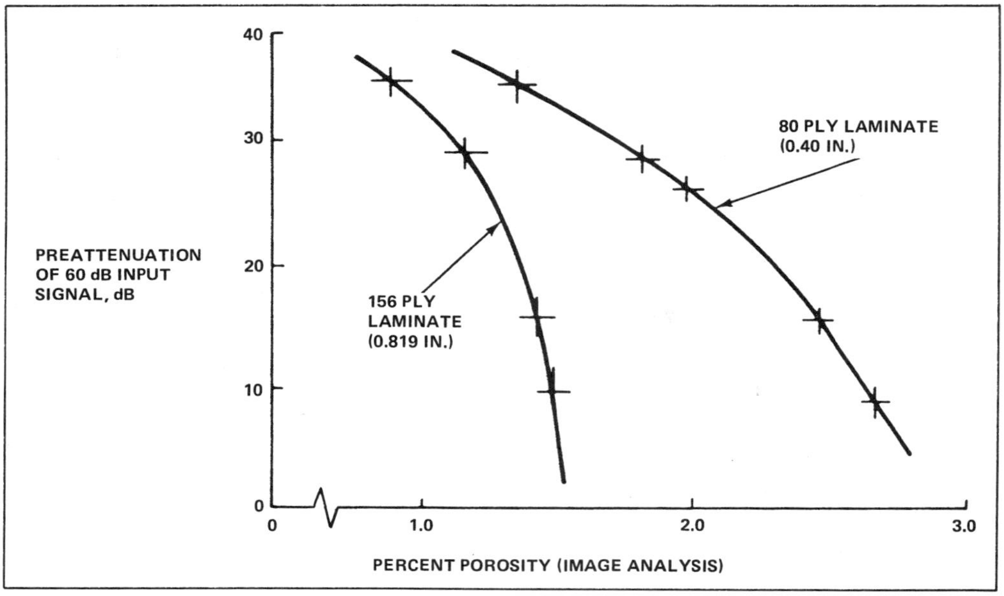

Figure 9 Correlation Between Ultrasonic Attenuation and Porosity

CONCLUSIONS

Current process controls and nondestructive evaluation techniques are adequate for assuring the quality of most of today's structural composite parts. However, in many cases these procedures are very labor and skill-intensive, resulting in high quality control costs. Quality Control procedures must be integrated and automated wherever possible to improve quality and minimize the costs. This will be especially important as most complex structures are built utilizing automated manufacturing techniques.

ACKNOWLEDGEMENT

The authors are grateful to Richard M. Collins, Robert J. Holden, Ronald P. Roberts, and others at Grumman for their assistance in preparing this paper.

REFERENCES

1. Carpenter, J.F., "Characterization and Control of Composite Prepregs and Adhesives," 21st National SAMPE Symposium and Exhibition, Breakdown of Material Progress, Vol. 21, April 1976.

2. Farber, Susanne F., "Resin Extraction of Epoxy Composities." SAMPE Journal, Volume 22, No. 2, March/April 1986.

3. Epstein, George, "Nondestructive Test Methods for Reinforced Plastics." Handbook of Composites, edited by G. Lubin, Van Nostrand Reinhold Company, New York, 1982.

IN-PROCESS ULTRASONIC NONDESTRUCTIVE EVALUATION

S. I. Rokhlin and Laszlo Adler

Department of Welding Engineering
The Ohio State University
Columbus, Ohio 43210

INTRODUCTION

Ultrasonic waves have great potential for use in in-process nondestructive evaluation. First, the ultrasonic wave velocity and attenuation may give important information on material property changes during processing. Second, they may give indirect information on other process parameters, for example, temperature or pressure. And third, ultrasonic waves may be used on-line for inspection of part quality. Measurements of the process parameters and material properties make possible the control of process variables to achieve the required material properties[1].

In this paper we describe the application of ultrasonic waves to in-process evaluation of spot welds and in-process monitoring of curing of polymers.

In-process testing of spot welds provides good illustrations of several advantages of in-process NDE over conventional testing after welding. The first aspect is cost: conventional NDT has a low speed of testing due to the need to position the transducer on the spot and make measurements for each single spot weld, so testing may be more time-consuming than manufacturing. This is especially important for parts with large numbers of spot welds. In-process techniques give weld quality determinations during the act of welding, in fractions of a second. The second aspect is of technical character: both conventional ultrasonic and electrical resistivity methods fail in identification of the stick weld (cold weld). This type of weld has very low strength but makes good electrical or acoustical contact and therefore cannot be identified by the above techniques. This is especially important for coated steel. The coating melts at a lower temperature than the base material. The melted coating will join sheets with excellent electrical and acoustical contact but this joint will be very weak. The in-process method resolves this problem as will be discussed below. Preliminary results on in-process evaluation of spot welds by Lamb waves have been published elsewhere[2]. The technique used was a further development of the Lamb wave technique described by Rokhlin and Bendec[3] and by Rokhlin and Adler[4] and used by them for post-service evaluation of spot welds.

IN-PROCESS EVALUATION OF SPOT WELDS

The principle of utilization of ultrasonic waves for in-process evaluation of spot welds is shown schematically in Fig. 1. Lamb waves and bulk ultrasonic waves are transmitted through the melted area during welding. Two independent ultrasonic channels are used in this system: one for ultrasonic Lamb waves with transducers placed on sheets at some distance from the electrodes and the second for ultrasonic longitudinal waves with transducers placed inside the electrodes. Both transducers were designed and built in our laboratory. To eliminate the effect of an acoustical coupling of the channels through the welded plates both ultrasonic generators were synchronized, separating the received signals by different time delays due to different acoustic paths. Both the received ultrasonic signals after peak-detection are digitized and fed to the computer for processing. Afterwards, the changes of the ultrasonic signals as a function of time are plotted together with the welding current on a chart recorder.

Plain carbon cold-rolled 0.045 in. thick steel sheets were used in this experiment. The weld current was the same (12 KA) for all specimens but the weld time varied from 2 to 15 cycles. Samples were in the form of two strips, each approximately 20 cm in length and 6 cm in width (length of overlap about 2 cm). Samples were prepared using an automated Taylor Winfield resistance welder (75 kVA, 60 Hz) with a programmable controller.

The spot welder was equipped with a digital data acquisition system. This system operates with three analog sensors to monitor electrode voltage, current and displacement and with ultrasonic transducers. The outputs of the sensors are digitized by a multichannel A/D converter and then stored in the microcomputer. The shear failure load of the welded samples were measured by tensile tests. Most of the samples failed through the nugget while some of them failed through the base material (welded sheets were torn in the vicinity of the weld). The nugget diameter was estimated under the microscope after fracture surface study.

The Lamb and longitudinal ultrasonic waves are transmitted through the nugget during welding. The behavior of the pool (its melting and solidification) affects the ultrasonic signal and therefore, can be related to the different stages of the weld. Ultrasonic behavior reflects temperature changes in the pool and therefore, the ultrasonic sensor plays the role of a temperature sensor. This makes it possible to relate indirectly the changes of the ultrasonic signal to the time of cooling of the pool (nugget) to some specific temperature.

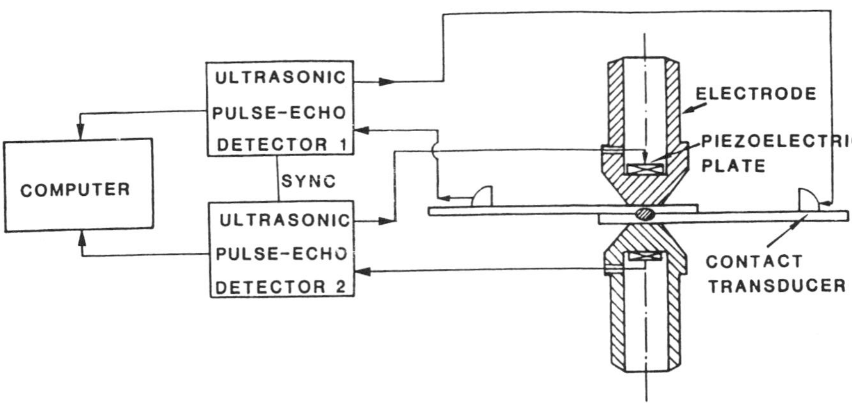

Fig. 1. Schematic block diagram of the spot weld monitoring.

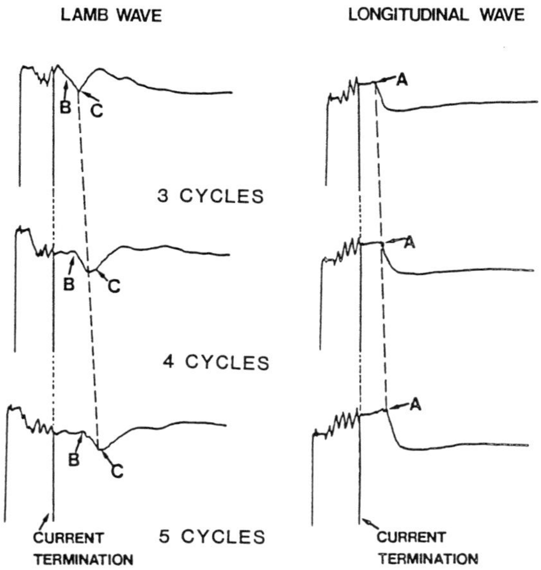

Fig. 2. Ultrasonic signal amplitudes versus time for different number of weld cycles. The plottings are shifted so that the moment of current termination for each recording corresponds to the same point on the time scale.

To clarify, typical ultrasonic signals for three, four and five current cycles are compared in Fig. 2. The figures are slightly shifted on the time axis in such a way that they correspond to the same moment of current termination. This moment is shown by an arrow in Fig. 2. Two sets of ultrasonic data are indicated: one for transmitted longitudinal wave, the second for transmitted Lamb wave. Sharp changes of the ultrasonic longitudinal wave trace are labelled as A. At the same time moment on the Lamb wave, trace indications (labelled B) can also be seen. Sharp minima of the ultrasonic Lamb wave signal are labelled as C. Comparison of the results for different cycles (Fig. 2) shows that the position of the minimum of the Lamb wave signal is strongly affected by the number of cycles and by the degree of development of the welding pool. To understand the signal behavior we must take into account the metallurgical and welding process factors affecting the ultrasonic wave propagation.

During the welding process the temperature of the welded region changes from room temperature to above the melting point of steel (1500°C). This results in strong changes of the ultrasonic wave attenuation. It has been established that the attenuation versus temperature has strong maxima at about 700° ÷ 800°C, corresponding to the α-Fe to γ-Fe (austenite) transformation[5]. Another maximum corresponds to the solid-to-liquid transformation (melting).

The velocity changes also affect the behavior of the transmitted Lamb wave signal. When the temperature changes from room temperature to the melting point, the shear wave velocity changes by a factor of 1.5. Therefore, the working point on the dispersion curve for Lamb waves shifts. This looks as though the thickness of the welded plate changes by a factor of 1.5. This means that the incident mode can no longer propagate and converts to other modes before it reaches the melting region. It converts to the modes existing at the new value of the parameter $K_t h$.

Elastic waves are also converted on the boundary between the liquid pool and the base material. The minimum of the transmitted signal corresponds to the austenite-ferrite phase transformation in the nugget which results in the maximum attenuation of the ultrasonic wave when it propagates through the nugget. The minimum of the transmitted signal is related to the temperature of this phase transformation. The time period from weld current termination to attainment of this temperature matches well the theoretical estimates and experimental results of different authors[6,7].

The minimum of this transmitted Lamb wave signal corresponds to the same temperature for the different samples. The time of cooling from the point of current termination to this minimum corresponds to the time of welding region cooling to this temperature. It is clear that the length of this time interval is connected to the volume of the formed liquid pool (its thermal mass), that is with the volume of the formed nugget. The larger the nugget formed, the longer the time interval. One can propose that the length of this time interval can be correlated with the weld quality (weld strength). Such correlation is shown in Fig. 3. The failure load of the welds is plotted versus the length of the time interval to the minimum (τ_{min}).

The shear fracture load data for different samples are plotted as a function of the spot diameter squared in Fig. 4. The samples which failed through the based material are shown by triangles. The stick welds have very low fracture load. For these welds, the experimental points are indicated by

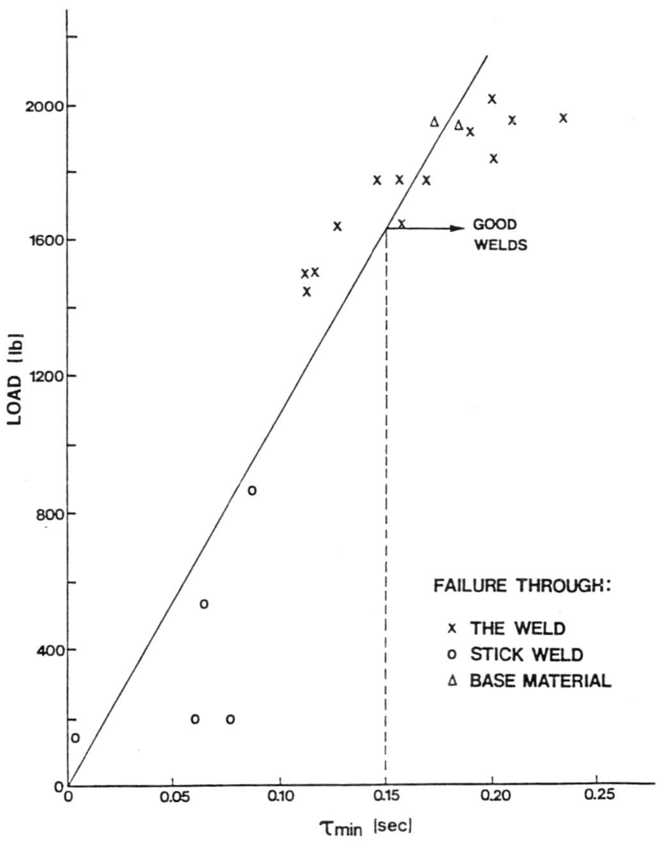

Fig. 3. Shear fracture load versus time shift of the minimum of the Lamb wave signal.

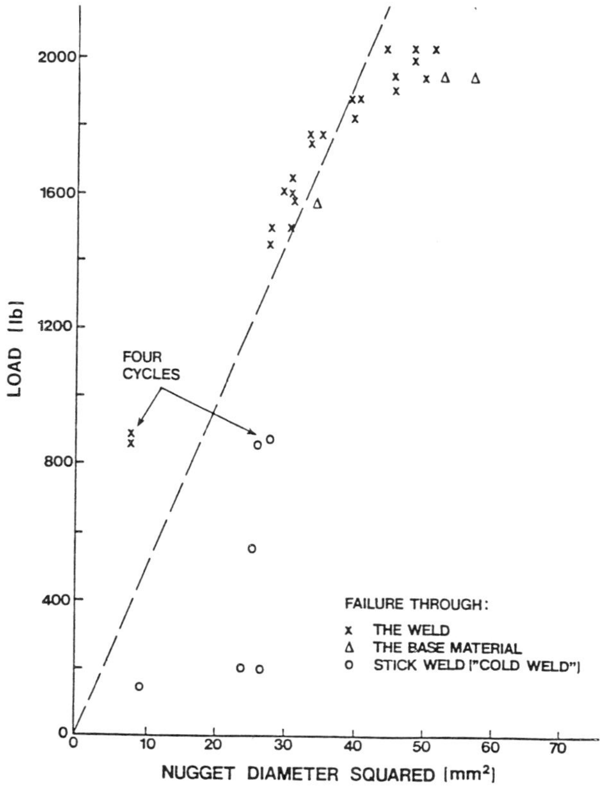

Fig. 4. Shear fracture load versus the spot diameter squared.

open circles. Such welds correspond to a sticking of the bonded sheets without metal melting. The welds for four cycles have both stick and melted areas so both diameters are shown on this figure. It is seen from Fig. 3 that this weld can be easily identified by the ultrasonic method discussed here. This is a very important advantage of this mehtod because the cold weld forms good acoustic contact and therefore cannot be identified by other ultrasonic methods.

IN-PROCESS MONITORING OF THE CURING OF COMPOSITE MATERIALS AND STRUCTURAL ADHESIVES

Thermosetting resins are widely used as matrices for advanced composite materials and structural adhesives. Fabrication of composite materials or adhesive joining consists of curing (polymerization) under proper thermal and pressure conditions of resin-impregnated fiber fabrics. During the curing reaction, the thermoset resin transforms from the visco-liquid state to the gel and then vitrifies to the gelled glass. This transformation is accompanied by strong changes of the viscoelastic properties of the material which are related to the molecular network. Therefore measurements of velocity and attenuation of ultrasonic waves during cure may give important information on the extent of the thermoset curing reaction and the mechanical properties of the material.

The application of ultrasonic waves for polymer cure monitoring was first described by Sofer and Hauser[8] and then by Papadakis[9]. Lindrose[10] shows that velocity and attenuation for longitudinal and shear waves change simultaneously during cure. The application of interface waves for insitu study of curing of thin epoxy films has been described by Rokhlin et al.[11].

For successful application of ultrasonic waves to in-process cure monitoring two questions have to be answered. First, how do the frequency of measurements and temperature of cure affect ultrasonic data? Second, how are ultrasonic data related to static polymer properties, such as resin viscosity in a liquid stage and its mechanical properties in a solid stage, which are extremely important for composite material production? It was recently shown by Elsley et al.[12] that in the initial stages of cure ultrasonic velocity changes simultaneously with viscosity. The effect of temperature on the velocity and attenuation of a longitudinal ultrasonic wave is discussed by Rokhlin[13] in the accompanying paper in these Proceedings. The relation between ultrasonic parameters and polymer strength, and the effect of temperature on cure, was studied by Rokhlin[14] and Rokhlin and Segal[15] for FM-73 structural adhesive, which is supplied as prepreg with resin and fibers in the same way as prepregs of composite materials.

Fig. 5 (Ref. 14) shows the variations in the phase velocity of an ultrasonic interface wave during a typical cure cycle at 120°C for the FM-73 adhesive. The same figure shows, at the top, the temperature variation. The ultrasonic data are represented for the phase shift $\Delta\tau$, from which one can easily calculate the variations in the velocity of the interface wave. An increase in phase shift corresponds to increasing phase velocity and shear modulus.

Characteristic points of the cure cycle are noted on the phase-shift curve. The adhesive prepreg starts softening upon heating, and hence the velocity of the interface wave (shear modulus) starts dropping. Near Point A the prepreg becomes liquid (between 60 to 70°C). This figure also presents the relative change in the thickness of the cemented specimen due to the thermal expansion. The measurements were performed by means of a mechanical indicator. The open triangles are for a specimen without the adhesive, whereas the darkened are for one with an adhesive.

When the adhesive becomes liquid due to heating, its thickness decreases. Then this happens (with the mechanical properties remaining un-

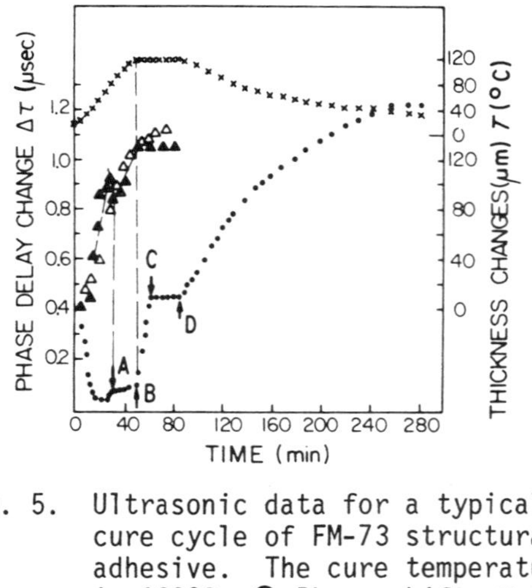

Fig. 5. Ultrasonic data for a typical cure cycle of FM-73 structural adhesive. The cure temperature is 120°C. ● Phase shift, X temperature. Thickness changes: ▲ for specimen with adhesive, △ for specimen without adhesive.

changed) the velocity of the interface wave should increase. When a temperature of 120°C is attained, the velocity of the interface wave starts increasing steeply (Point B on the phase variation curve). This is due to the increase in the cross-linking reaction in the adhesive and consequently, to the increase in its shear modulus. The temperature in this region is maintained constant (at 120°C). Approximately after 10 minutes of holding at this temperature the cross-linking reaction is completed (Point C on the velocity curve). Point D on the phase shift curve corresponds to the start of cooling of the specimen. The increase in the phase shift in this region corresponds to an increase in the shear modulus of FM-73 adhesive with cooling.

In order to establish that segment B-C in Fig. 5 corresponds to the high-rate cross-linking reaction, the ultrasonic data were compared with measurements of the variation in the shear strength of the bond for different time points of the cure cycle. The strength values thus obtained[14] are shown in Fig. 6. It is seen that a rise in the velocity of the interface wave (rise in the shear modulus of the adhesive) corresponds precisely to the time interval of the bond-strength growth.

The gel point is seen on the velocity curve, as well as on the curve of variation of the bond strength; it is indicated by an arrow. The measurements were also performed at different temperatures. Reducing the cure temperature decreases the rate of reaction and the gel point is shifted on the axis. The percent of crosslink connections in the gel point is a material property and is not affected by temperature. Therefore a shift of the gel point on the time axis characterizes quantitatively the variation in the reaction rate as a function of the temperature.

The curing data can be summarized in the form of the Arrhenius plot, from which the curing activation energy for the FM-73 adhesive of E = 9.3 kcal/mole was found[15]. It is in agreement with published data for different epoxy-resin systems.

The effect of frequency on the measured velocity and attenuation of an ultrasonic wave was recently studied by Rokhlin et al.[16] by a spectral analysis technique. The measurements were done in real-time during curing of the epoxy resin. As an example several cross sections of the attenuation surface for different stages of the curing reaction (attenuation as function of frequency) are shown in Fig. 7 (here are shown only several curves; data were taken once a minute[16]).

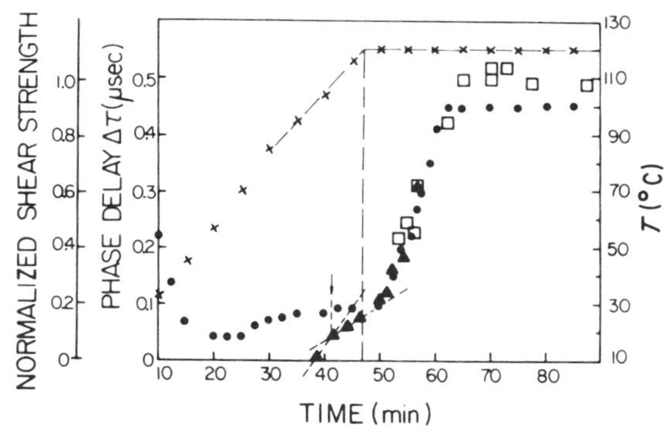

Fig. 6. Comparison of the ultrasonic data with the variation in relative bond strength in the course of the cure. ● Phase shift, X temperature. Strength: ▲ rapid cooling, △ slow cooling (after shift).

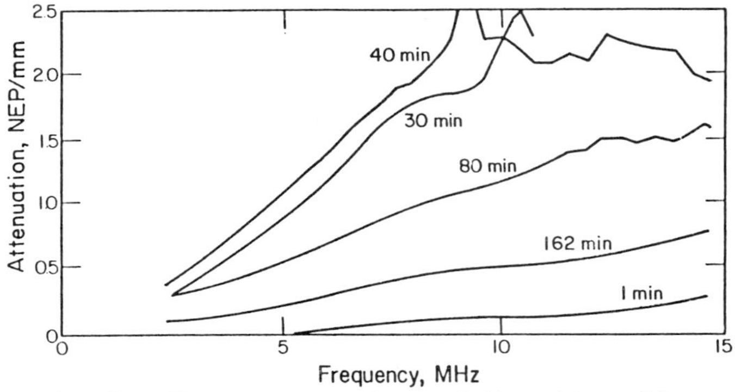

Fig. 7. The cross sections of the attenuation surface for different times of cure.

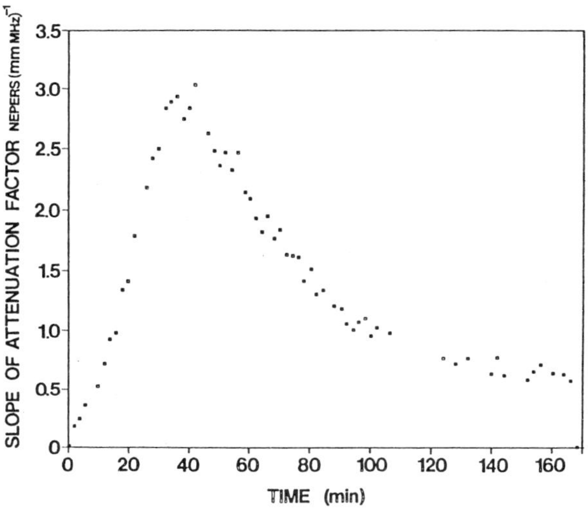

Fig. 8. The slope of frequency of the attenuation factor versus time of curing.

From these data we come to the very important conclusion: the attenuation coefficient behaves nearly linearly with frequency at all stages of the curing reaction from viscous liquid to solid state. The slope of the attenuation factor as function of frequency is strongly dependent on time of cure (on degree of cure) and passes through its maximum in the middle stage of the reaction. The slope of the attenuation factor is plotted as a function of time in Fig. 8. The linear behavior of attenuation versus frequency suggests that the attenuation effect cannot be explained by classical viscothermal absorption or relaxation theory. Such type of behaivor was found previously for some highly viscous liquids, for solid polymers and for biological tissue.

The analyses of the cross section of the attenuation surface for different frequencies[16] (attenuation as function of time) show that there is no observable shift of the attenuation maximum on the time scale due to changes of frequency (in the frequency range under study). The maximum for the slope of the attenuation factor (Fig.8) is close, in the time scale, to the maximum of the attenuation coefficient at different frequencies.

The phase velocity data were evaluated from the phase spectrum of the transmitted signal. Moderate ultrasonic velocity dispersion was found in any stage of the reaction[16]. No dispersion was observed in this frequency range for pure epoxy and hardener agents before mixing. In the first stage of the curing reaction the phase velocity of the mixture of epoxy and hardener is less than in the pure component.

These results lead to an important practical conclusion: that in the frequency range 1-20 MHz there is no observable effect of the frequency of ultrasonic measurements on kinetic curves such as dependence of velocity and attenuation on time of cure. Therefore the selection of frequency may be dictated by experimental convenience.

REFERENCES

1. R. Mehrabian and M. N. G. Wadley, Needs for process control in advanced processing of materials, J. of Metals Feb:51 (1985).
2. S. I. Rokhlin, R. J. Mayhan, and L. Adler, On-line ultrasonic Lamb wave monitoring of spot welds, Mater. Eval. 43:879 (1985).
3. S. I. Rokhlin and F. Bendec, Coupling of Lamb waves with the aperture between two elastic sheets, J. Acoust. Soc. Am. 73:55 (1983).
4. S. I. Rokhlin and L. Adler, Ultrasonic method for shear strength prediction of spot welds, J. Appl. Phys. 56:726 (1984).
5. L. C. Lynnworth and E. P. Papadakis, eds., "Measurement of Elasitc Moduli of Materials at Elevated Temperature," Report AD780231 for the Office of Naval Research, Panametrics, Inc., Waltham, MA (1974).
6. B. K. Bentley, J. A. Greenwood, P. McKnowlson, and R. G. Baker, Temperature distributions in spot welds, Brit. Weld. J. 10:613 (1963).
7. J. E. Gould and P. H. Chang, Thermal modeling of spot welds, J. Metals. May:39 (1986).
8. G. A. Sofer and E. A. Houser, A new tool for determination of the stage of polymerization of thermosetting polymers, J. Polym. Sci. 6:611 (1952).
9. E. P. Papadakis, Monitoring the moduli of polymers with ultrasound, J. Appl. Phys. 45:1218 (1974).
10. A. M. Lindrose, Ultrasonic wave and moduli changes in a curing epoxy resin, Exper. mech. 18:227 (1978).
11. S. I. Rokhlin, M. Hefets, and M. Rosen, An elastic interface wave guided by a thin film between two solids, J. Appl. Phys. 51:3579 (1980).
12. R. K. Elsley, W. J. Pardee, M. J. Buckley, F. Cohen-Tenoudji, K. W. Ferting, "Advanced concepts in the control of autoclave cure of composites, in: "Review of Progress in Quantitative Nondestructive Evaluation," vol. 5B, D. O. Thompson and D. E. Chimenti, eds., Plenum, New York (1986).
13. S. I. Rokhlin, Characterization of composite and adhesive cure by ultrasonic waves, these Proceedings.
14. S. I. Rokhlin, Evaluation of the curing of structural adhesives by ultrasonic interface waves. Correlation with strength, J. Comp. Mater. 17:15 (1983).
15. S. I. Rokhlin, Study of the cure kinetics of structural adhesives by ultrasonic interface waves, J. Mater. Sci. 20:3300 (1985).
16. S. I. Rokhlin, D. K. Lewis, K. F. Graff, and L. Adler, Real-time study of frequency dependence of attenuation and velocity of ultrasonic waves during the curing reaction of epoxy resin, J. Acoust. Soc. Am. 79:1786 (1986).

NONDESTRUCTIVE CHARACTERIZATION OF KEVLAR COMPOSITES USING PULSED NMR

George A. Matzkanin and Armando De Los Santos

Southwest Research Institute
San Antonio, Texas

INTRODUCTION

Nuclear magnetic resonance (NMR) is a branch of spectroscopy based on the interaction between nuclear magnetic dipole moments and a magnetic field. The term "resonance" is used because a natural frequency of the magnetic system, namely, the frequency of gyroscopic precession of the magnetic moment in an applied static magnetic field, is the quantity detected. Typically, the resonance frequency falls in the radiofrequency (RF) region of the electromagnetic spectrum. To implement the NMR method, a static magnetic field is applied to the specimen to polarize the magnetic dipole system and resonance detection is accomplished by coupling a suitable electromagnetic field to the specimen by means of an RF induction coil. Although NMR has been utilized as an investigative tool for many years in physics and chemistry laboratories, only recently has it been seriously considered for NDE of materials.[1]

Application of NMR to organic matrix composites involves measurement of the hydrogen nucleus (proton) NMR signal. Fortunately, the proton NMR signal is very strong and easily measured. Certain characteristics of the NMR signals can provide a variety of information about the specimen, including (1) the presence of atoms in different physical states (e.g. liquid or solid, crystal and/or amorphous); (2) the type and strength of interactions between nuclei of both the same species and different species; (3) changes in molecular structure and intermolecular binding; and (4) transport (diffusion) phenomena. Much of the physical and chemical information available through the use of NMR is associated with the relaxation characteristics of the nuclear magnetic moments, which can be measured using pulsed NMR techniques. The energy exchange between nuclear moments and the surrounding lattice is characterized by the spin-lattice relaxation time, T_1, while the energy exchange among nuclear magnetic moments is described by the spin-spin relaxation time, T_2. These relaxation times are very sensitive to molecular motions and structural changes and can be used to provide both qualitative and quantitative information on the dynamic environment in which the nuclei are located.

It is known that moisture intrusion can have an adverse effect on the structural integrity of organic matrix composite materials.[2] Not only does moisture migration along the fiber-matrix interface weaken the interface bond, but moisture also reacts with the organic matrix leading to degradation of the polymeric structure. In the case of Kevlar composite, this potential problem is complicated by the fact that the Kevlar fiber itself is hygroscopic and absorbs moisture. The presence of moisture in the constituent materials used in the fabrication of Kevlar composite can lead to improper cure of the composite. It is suspected that moisture in the constituent materials can migrate between the resin and fiber during cure and degrade the resin by altering the polymer cross-link structure.

The work reported in this paper was directed at using pulsed NMR to characterize the moisture in Kevlar fiber, neat resin (both cured and uncured), Kevlar prepreg, and Kevlar reinforced composite. The hydrogen NMR signals were analyzed to obtain information on the characteristic states of moisture in these constituent materials and to determine the capabilities of NMR for determining degraded resin.

EXPERIMENTAL APPROACH

Specimens

Kevlar 49 fiber roving was subjected to various drying times to obtain dry weight moisture levels of 0.75%, 0.22%, and 0.15%. Some of the fiber was conditioned in a humid environment to obtain a moisture percentage of 2.8%. Neat resin samples for NMR were prepared by adding water directly to the resin to produce moisture percentage levels ranging from 1% to 10%. NMR measurements were made on the resin samples both before and after cure. The prepreg samples measured were composed of resin impregnated Kevlar 49 fiber roving. The resin used for the prepreg had no added water and the Kevlar fibers used contained the moisture percentages noted above for the Kevlar roving. Measurements were made on two types of composite specimens. In one case, the composites were fabricated from resin to which various percentages of water had been added before fiber impregnation and cure. In the other case, cured composite specimens were subjected to an environmental chamber at 125°F and 95% RH for various periods of time to produce a range of moisture absorption.

NMR Measurements

The NMR measurements were made using a laboratory NMR system operating at a frequency of 27 MHz. The samples were put into 5 cm (2 in.)-diameter glass vials and placed into a detection coil in the gap of an electromagnet. Parameters measured for the various samples included the free induction decay (FID), the spin-lattice relaxation time, T_1, and the spin-spin relaxation time, T_2. For most of the samples, multiple relaxation times were observed. Processing of the data to obtain the various time constant components and other parameters of interest was done on an HP1000 computer.

RESULTS

Kevlar Fiber

Kevlar fiber was found to have a very distinctive NMR free induction decay signal with an easily identified moisture component as noted in the oscilloscope photograph in Figure 1. The characteristic beat effect noted for the moisture in Kevlar fiber was seen in all samples containing the fiber including the prepreg and the composites. This beat effect implies that the moisture in the fiber is apparently isolated from the more tightly bound hydrogen in the solid matrix and may be associated with the morphology of the Kevlar fiber. It has been noted in the literature that there is a tendency for water to form hydrogen bridges and thus coagulate into clusters inside the Kevlar matrix.[3] The pulsed NMR results appear to support this hypotheses. The amount of moisture in Kevlar fiber is easily determined using pulsed NMR by measuring the signal amplitude at two points on the free induction decay. This is because the initial portion of the signal (at 24.5 μs) is associated with the total hydrogen in both the solid and liquid phases, whereas the signal amplitude at a later time (80 μs) is associated with hydrogen only in the liquid phase. Thus a ratio of the NMR signal amplitude at these two measuring points should be correlated with the moisture percentage. These results for the Kevlar fiber samples are shown in Figure 2.

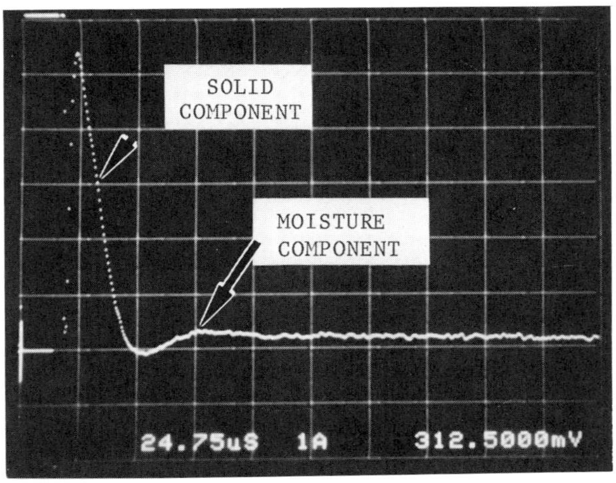

Fig. 1. Free Induction Decay from Kevlar Fiber Containing 0.75% Moisture

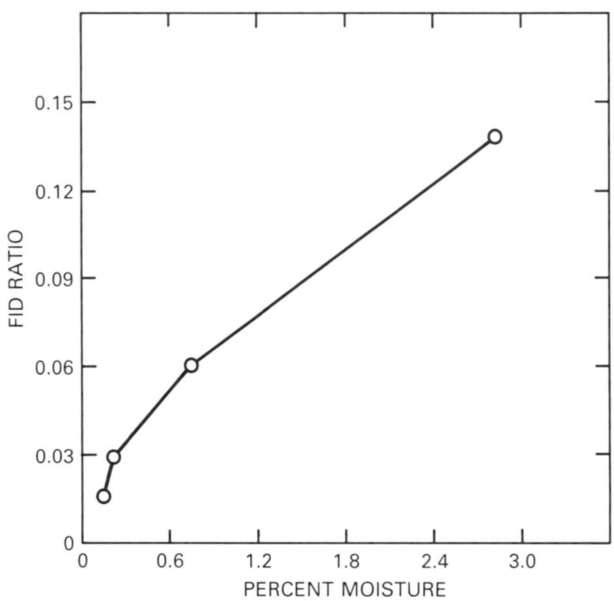

Fig. 2. Free Induction Decay Ratio vs Percent Moisture for Kevlar Fiber

Resin

In the case of the uncured resin, a distinct moisture component is not observed in the free induction decay because of the long spin-spin relaxation time for this viscous material. However, it was found that the spin-lattice relaxation time, T_1, for the neat resin with no added water was about 50 ms compared with resin containing added water which had a T_1 component of about 50 ms for the resin plus a T_1 component of approximately 1-2 s for the water in the resin. Therefore, the free induction decay difference signal obtained for two different pulse repetition rates contains a substantial water signal component whereas the resin contribution cancels out. A plot of the FID difference signal versus percent water in uncured resin is given in Figure 3.

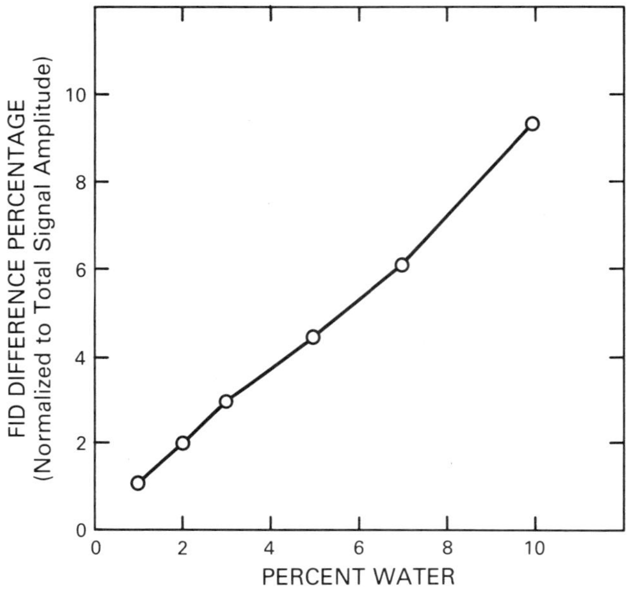

Fig. 3. Free Induction Decay Difference vs Percent Water in Uncured Resin

After the resin samples were cured, it was noticed that the resin with increased water added before cure became considerably more brittle. Although the NMR free induction decay signal did not show a distinct moisture component for the cured resin, relative measurement of the signal amplitude at various times during the free induction decay can be used as a determination of the amount of amorphous to crystalline character. The free induction decay ratio measured at times of 23.5 μs and 70 μs vs percent water added before cure is shown in Figure 4. The decrease in this ratio which implies decreasing amorphous character correlates well with the increased brittleness observed for the cured resin samples with greater amounts of added water.

Prepreg

The prepreg samples were composed of Kevlar 49 fiber containing various amounts of moisture impregnated with uncured epoxy resin with no added water. The NMR free induction decay signals observed were characteristic of both the uncured resin and Kevlar fiber NMR signals. Since the T_1 relaxation time component associated with the resin is considerably shorter than that associated with either the fiber or the moisture, the difference signal obtained for two different pulse repetition rates contains components only from the Kevlar and moisture as illustrated in Figure 5. Note the similarity of this difference signal to that obtained from Kevlar fiber alone shown previously in Figure 1. The moisture content of the prepreg (specifically the moisture in the fiber in the prepreg) can then be determined by measuring the ratio of the FID difference signal at 24.5 μs and 80 μs. A correlation between the FID difference ratio and the percent moisture for Kevlar prepreg is shown in Figure 6 (note the general similarity of this correlation to that obtained for the Kevlar fiber alone shown in Figure 2).

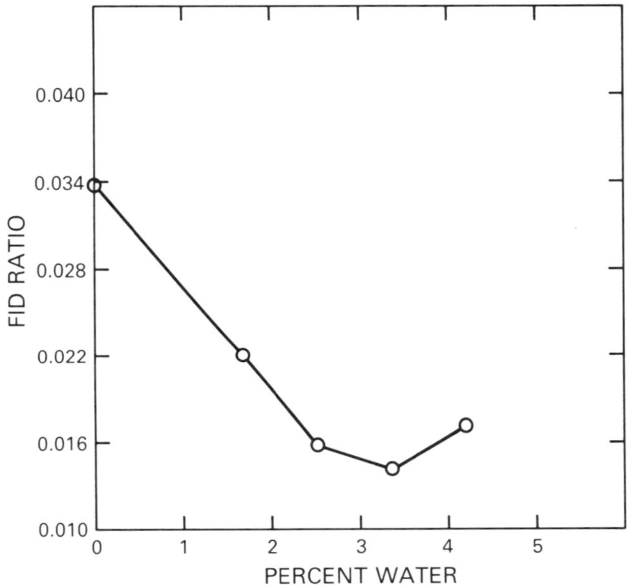

Fig. 4. Free Induction Decay Ratio vs Percent Water Added Before Cure for Cured Resin

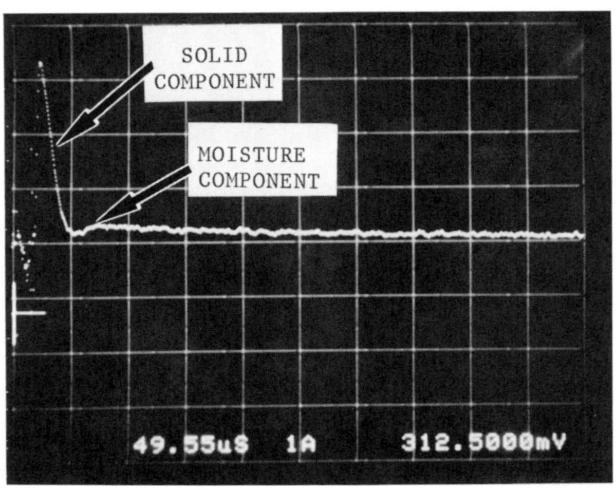

Fig. 5. FID Difference Signal for Kevlar Prepreg Obtained for Pulse Repetition Rates of 3 Seconds and 0.3 Seconds

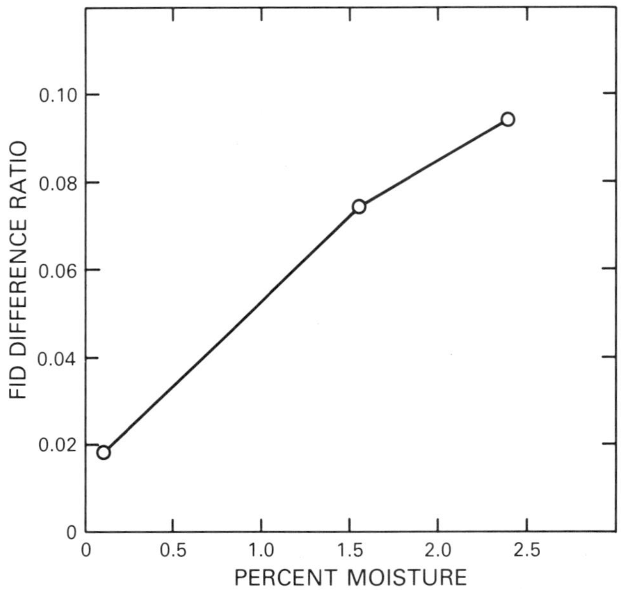

Fig. 6. Free Induction Decay Difference Ratio vs Percent Moisture for Kevlar Prepreg

Kevlar Composite

In earlier work performed on Kevlar composite samples, exposed to long term environmental conditioning, the free induction decay NMR signal was found to contain a distinct absorbed moisture component similar to that shown previously in Figure 1 for the Kevlar fiber.[4] The amplitude of the absorbed moisture component correlated extremely well with the percentage of moisture absorbed by the Kevlar composite as shown in Figure 7. However, for composite specimens in which controlled percentages of water had been added to the resin before fiber impregnation and cure of the resulting prepreg, the free induction decays showed no notable differences. However, the spin-lattice relaxation time could be resolved into three components as shown in Table 1. In comparison with measurements on the constituent materials it can be concluded that the shortest T_1 component, T_{11}, is dominated by the cured resin, the middle component, T_{12}, contains contributions from both the resin and the Kevlar fiber, whereas the longest component, T_{13}, is entirely attributable to the solid Kevlar fiber and the corresponding moisture in the fiber. Measurements of the long spin-lattice relaxation time constant, T_{13}, showed that this parameter decreased as the amount of moisture added to the resin before cure increased (see Figure 8 and Table 1). In measurements on Kevlar 49 fiber above, the long spin-lattice relaxation time also decreased with increased moisture content. The implication of this result is (1) moisture migrated out of the resin and entered the Kevlar fiber during cure, and/or (2) more moisture migrated out of the Kevlar fiber when less moisture was available in the resin. In either case, the indication is that more moisture is present in the Kevlar fiber (that is, shorter T_{13}) as the amount of water in the resin before cure is increased.

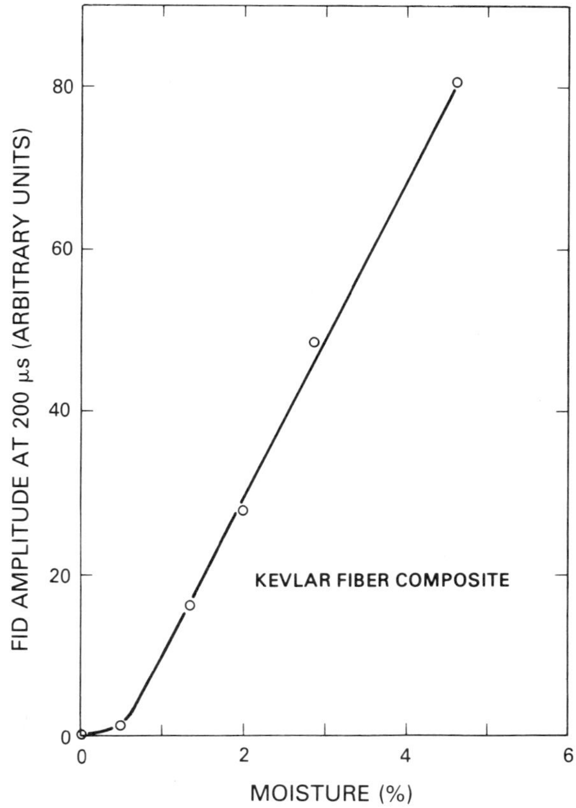

Fig. 7. Free Induction Decay Amplitude vs Absorbed Moisture for Kevlar Composite

Table 1. Spin-Lattice Relaxation Times for Kevlar Composite and Constituent Materials

Material	T_{11} (ms)	T_{12} (ms)	T_{13} (ms) (Wet - Dry)
Kevlar Fiber	--	100 - 150	450 - 1900
Uncured Neat Resin	--	51 - 59	--
Cured Neat Resin	6	33	100 - 150
Prepreg	--	50	500 - 1000
Kevlar Composite	6	58 - 150	450 - 1900

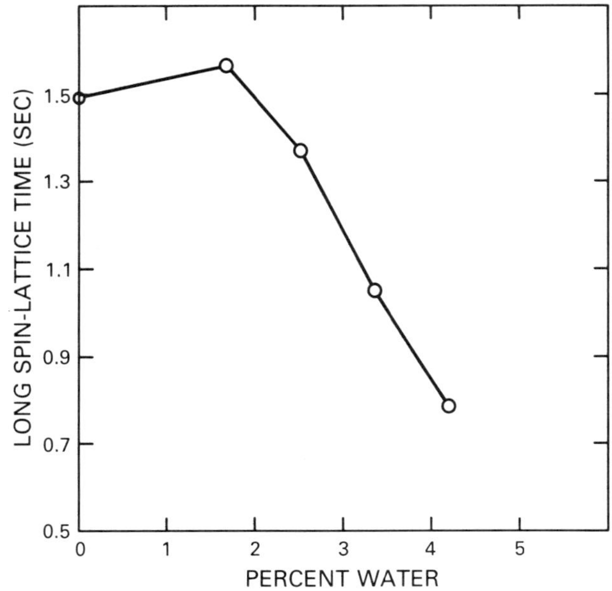

Fig. 8. Long Spin-Lattice Relaxation Time (T_{13}) vs Percent Water Added to Resin Before Cure for Kevlar Composite

DISCUSSION

Pulsed NMR studies of Kevlar fiber indicate that water molecules in the fiber exist in a phase distinct from other hydrogen atoms in the fiber molecular structure. The physical model suggested for Kevlar 49 is a crystalline core surrounded by an amorphous skin.[5] The core consists of extended chain, rod-like macro-molecules approximately 200 nm long aligned in the fiber direction. The chain ends distribute themselves along periodic transverse defect planes spaced approximately 200 nm along the fibers. The transverse planes in the core would appear to provide interstices in which water molecules may reside, thus producing the observed two-phase free-induction decay signal. There indeed is evidence that the macro-molecules may be bridged together by tightly-bound clusters of water molecules.[6] The observed decrease in Kevlar fiber relaxation time with increasing moisture content may be explained on the basis that cluster formation is favored with increasing concentration of water thus reducing the mobility of the diffusant[3] and producing faster energy exchange with the lattice.

The distinctive NMR free induction decay signal from moisture in the Kevlar fiber provides the opportunity of characterizing and measuring the moisture in the fiber even in the presence of uncured resin as in prepreg, or in a cured composite. In the case of the prepreg, the relatively short spin-lattice relaxation time of the uncured resin ($\approx$55 ms) compared with the relaxation time of the fiber and moisture in the fiber ($\approx$500-1900 ms) allows an FID difference signal to be obtained which is characteristic of the fiber alone. Results of spin-lattice relaxation time measurements on the cured composite indicate that moisture migrates between the resin and the fiber during cure with the direction of the migration depending on the relative amounts of moisture in the fiber and the resin. As the amount of water added to the resin before cure is increased, the spin-lattice relaxation time associated with the fiber decreases implying that the moisture is migrating from the resin into the fiber. Measurements on carefully prepared composite specimens containing known amounts of moisture in the resin and fiber, respectively, will be needed to verify this conclusion.

REFERENCES

1. J. D. King, W. L. Rollwitz, and G. A. Matzkanin, Magnetic Resonance Methods for NDE, Proc. 12th Symp. on NDE, 138-149 (1979).
2. "Advanced Composite Materials-Environmental Effects," ASTM STP 658, J. R. Vinson, ed., American Society for Testing and Materials, Philadelphia (1978).
3. J. M. Augl, Moisture Sorption and Diffusion in Kevlar 49 Aramid Fiber, NSWC/TR-79-51, 1-48 (1979).
4. G. A. Matzkanin, Applications of Nuclear Magnetic Resonance to the NDE of Composites, Proc. 14th Symp. on NDE, 270-286 (1983).
5. C. O. Pruneda, W. J. Steele, R. P. Kershaw, and R. J. Morgan, Structure-Property Relations of Kevlar 49 Fibers, Polymer Preprints, American Chemical Society, 22: No. 2 (1981).
6. R. G. Garza, C. O. Pruneda, and R. J. Morgan, Polymer Preprints, American Chemical Society, 22: No. 2 (1981).

CURE MONITORING OF COMPOSITE STRUCTURE

USING MECHANICAL IMPEDANCE ANALYSIS (MIA)

Bor Z. Jang and M. Rao

Materials Engineering Program
Department of Mechanical Engineering
Auburn University, Alabama 36849

INTRODUCTION

Very few techniques are sufficiently sensitive and reproducible to be used effectively as an in-process, real-time tool for cure monitoring of composite structures. In conventional cure cycles a fixed cure schedule with temperature monitoring, usually according to the prepreg manufacturer's instruction, is followed. But these fixed cure cycles do not precisely control the cure conditions of complex integrated structures of which the dimensions are large and variable. This is especially true if the structure contains various prepreg properties and forms. A closed loop feedback control of cure and automation of composite structure production have yet to be realized.

In the present investigation a new methodology for cure cycle design, in-process monitoring, and feedback control of composite structure is proposed. This technique directly gauges the rheological properties of the thermosetting resin in an integrated composite structure. The method of mechanical impedance analysis (MIA) involves measuring the relation between a force input and a motion output between various points on the structure. In the simplest form, the technique measures the excitation and response at a single point and the values of mechanical admittance (velocity/force) are plotted as a function of frequency. Rapid measurements and analysis of the dynamic mechanical properties of a structure can be obtained real-time by a spectrum analyzer assisted by a microcomputer. In a given cure process this technique provides information on the rheological characteristics that would define where a matrix starts to flow, the regions of minimum viscosity, and the onset of gelation and vitrification. In the solid state MIA can be used to measure post cured properties such as the glass transition temperature or modulus as a function of temperature.

REVIEW OF COMPOSITE CURE MONITORING AND CONTROL TECHNOLOGY

The most rudimentary cure monitoring methodology is to carefully follow the temperature/pressure/vacuum of the system. This may be augmented by the other four techniques: AC Dielectrometry [4-9], DC Resistance [4,9], Ultrasonic Waves [10-16], Thermography [22] and Acoustic Emission (AE) [1]. The AC dielectric techniques have the advantages that relations exist between dielectrometry and dynamic mechanical analysis (DMA) response and that commercial instrumentation is available for measurement. Limitations of this method are (a) changes in spacing between electrodes can modify the dielectric output, and (b) increasing the number of prepreg plies between electrodes lowers measurement reliability. The DC resistance measurements only require very simple electrical circuit and the resistance variations do correlate with viscosity [4]. However, no physical model correlating DC resistance and rheology is available. Acoustic emission sensors provide direct information on microcracking processes. This method is new and needs further development. No data seem to exist that correlate AE signals with rheological properties. Presumably, AE could be used to detect the possible microcracking process during the cooling stage of composite processing. These composite characterization techniques have been recently reviewed by Kaelble and Shuford [1], Kaelble [19], and Koenig [2]. The use of ultrasonic techniques for the measurement of dynamic mechanical properties of polymers has received moderate attention in the literature [10-16,19]. Although these techniques have been used sporadically in the cure study of thermosetting resins [10,12,16], they have yet to be applied successfully for the in-situ cure monitoring and control of integrated composite structures.

METHODOLOGY--MECHANICAL IMPEDANCE ANALYSIS (MIA)

The details of MIA theories and the general feasibility of utilizing these methods in monitoring the viscoelastic and rheological behavior of thermosetting resins and composite materials during cure have been established by the present authors [20,21]. The theories were developed on the basis that dynamic characteristics of a beam can be related to Z* (impedance), defined to be the ratio of force/velocity or force/acceleration. The Z*, or its reciprocal M* (admittance), of a specimen can be measured by various methods, including sinusoidal sweeping, random excitation, and transient excitation test.

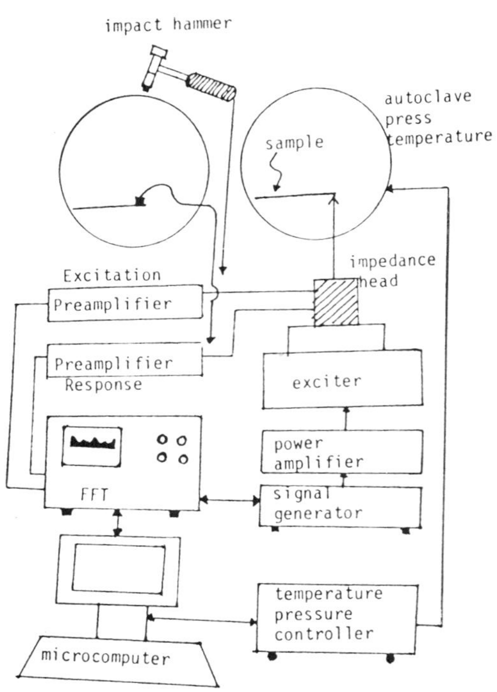

Fig. 1. Schematic of MIA set-up.

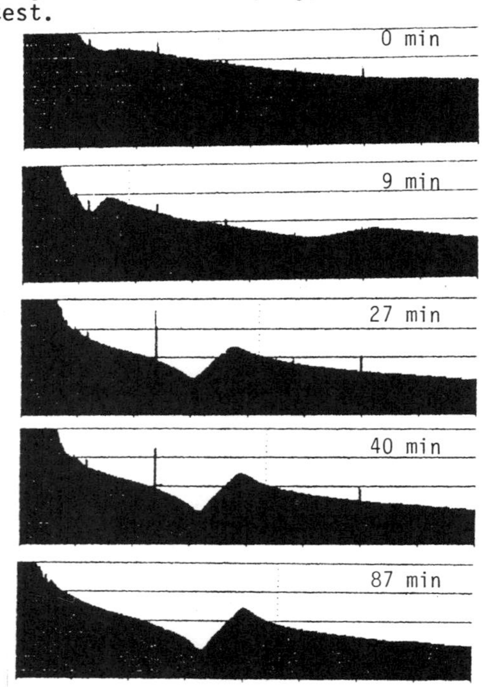

Fig. 2. Admittance spectrum of a beam as a function of epoxy curing time.

Some algorithmic approaches can be used to identify dynamic parameters such as resonant frequency, mode shape, and damping from the admittance data. By measuring the admittance continuously, the variation of the dynamic characteristics of thermosetting system with time can be obtained.

In a typical MIA apparatus (Fig.1) a sample is excited with its force and velocity response measured and analyzed real-time through a fast fourier transform (FFT) spectrum analyzer and a microcomputer. The mechanical admittance data after a selected number of scans can be displayed on the FFT screen. An impedance head consisting of two built-in transducers is used to measure the force and velocity of the sample. Alternatively, force and acceleration transducers may be placed at different locations, depending upon sample configurations and specific MIA techniques used. Signals are then amplified by two charge amplifiers and analyzed by the FFT. The driving force (excitation) comes from a shaker (exciter) driven directly by the FFT with an internal signal generating device. A separate signal generator may also be used to directly drive the shaker for improved accuracy. In the case of a transient test, the excitation is created by the impact hammer.

The result of a normal FFT analysis shows a distribution of frequency (e.g., Fig.2) from zero up to the Nyquist frequency fN, while the frequency resolution is determined by the number of frequency lines up to fN. The advent of Zoom-FFT provides a method by which increased resolution can be obtained within a smaller part of the frequency range. Typical outputs of the mechanical admittance (velocity/force) data would have the ordinate representing the amplitude of the admittance while the abscissa the frequency spectrum. The location, shape, and size of each peak can be used to determine the dynamic parameters. The storage modulus E' can be calculated if the natural frequency is identified from the location of each peak [20]. The values of loss tangent can be read off directly from each peak of the graph using the half-power points method or curve-fitting technique. Once the values of E' and tan (delta) are determined, then so is E". The numerical values of these parameters can be obtained on a real-time basis with a microcomputer.

EXPERIMENTAL

Several groups of experiments have been designed and carried out to determine the sensitivity of MIA techniques for cure monitoring of composite structure. The first group involved the measurements of dynamic response (in particular, mechanical admittance spectrum) of rectangular beams centrally loaded with both ends free. Analytical solutions to the dynamic equations describing the vibrations of samples of this geometry have been solved [20]. Each specimen is composed of three layers of glass fiber-epoxy prepreg, which are packed and sealed with thin aluminum foil to prevent epoxy leakage. The dynamic behavior of each specimen is monitored as a function of cure time and temperature. Vibrational mode shape, natural frequency, and damping can all be obtained in real-time as curing of epoxy proceeds.

In order to demonstrate that the techniques of MIA are appropriate for cure monitoring of a real composite structure a simple compression mould was constructed to accommodate the laminate being cured. This configuration is used to simulate press lamination of composite materials. The mould is composed of two mating pieces of aluminum that can be bolted together to form a cavity for the laminate. The mould also has the provisions to allow for possible screw fastening to the heating chamber. The data presented herein were obtained without fastening the mould to the chamber. The mould was allowed to naturally sit on the bottom plate of the heating chamber.

The system (mould & specimen) was then loaded at the geometric center with all four corners free. The dynamic response of the empty mould alone and of mould plus specimen with varying degree of curing was monitored continuously.

The third group of tests involved the monitoring of epoxy curing using transient excitation technique. An impact hammer provides a measurable input of mechanical force into the structure which sets it into vibratory motion. The input time history is registered by means of a force transducer built in the hammer. The force signal and the corresponding response signal picked up by an accelerometer are processed simultaneously in the spectrum analyzer. In this preliminary study, a room-temperature curing epoxy was poured into an open slot on an aluminum plate which was tightly clamped to a sturdy steel stand. The aluminum plate was excited on one rim while the response picked up on another. The modal parameters of the system were then monitored as a function of epoxy curing time.

RESULTS AND DISCUSSION

1. Time-Temperature-Transformation (TTT) Diagram of Thermosetting Resins (Cure Cycle Design and Optimization)

The ultimate criterion for the MIA technique is the capability of establishing TTT-diagrams of thermosetting resins to assist in selecting the optimal cure cycle. The cure cycle should be carefully designed and the curing process well monitored if the structure and properties of the resin are to be controlled. In the present investigation the phase transformation behavior of a selected group of epoxy resins were studied. Several specimens from the same batch of epoxy resin were allowed to isothermally cure at various temperatures, ranging from 110 to 190°C. During cure tan delta, E', and E" as a function of cure temperature and time were followed.

Typical outputs of the admittance data are shown in Figure 2 where the prepreg specimens of glassfiber-epoxy were cured isothermally at 169°C. At time t=0 (before curing), the curve hardly exhibts any peak where maximum admittance exists. This sort of curve is characteristic of any viscoelastic liquids with very small rigidity and great damping capability. As curing proceeds the rigidity increases while the damping ability decreases. Figure 2 only covers the admittance behavior over a narrow frequency range of 0-400 Hz. A wider coverage (e.g. up to 200 KHz) would include several peaks corresponding to the 1st, 3rd, 5th, 7th,... normal modes of vibration, respectively [20]. The natural frequency (Wn), damping (tan δ), and mode shape of each peak all vary with varying degree of cure. The values of damping coefficient can be calculated using either half-power-point method or curve fitting technique. A software package compatible with HP 9817 computer has been developed for this purpose. Natural frequency and damping coefficient of the first peak has been followed as a function of cure time (Fig.3). Same study may be repeated for several different cure temperatures. The results of a similar study for Kevlar fiber-epoxy prepreg sample are shown in Fig.4.

The times of resin gel and vitrification can be identified for each cure temperature. These data could then be integrated for the construction of TTT diagrams [3] to provide a framework within which an understanding of the physical properties of thermosetting matrices may be achieved. The study of cure TTT diagrams will be extended to include two-phase systems such as rubber-toughened epoxy. The curing of such rubber-modified systems may involve a change from an initially homogeneous solution to a heterogeneous multiphase morphology. Gelation may arrest the development

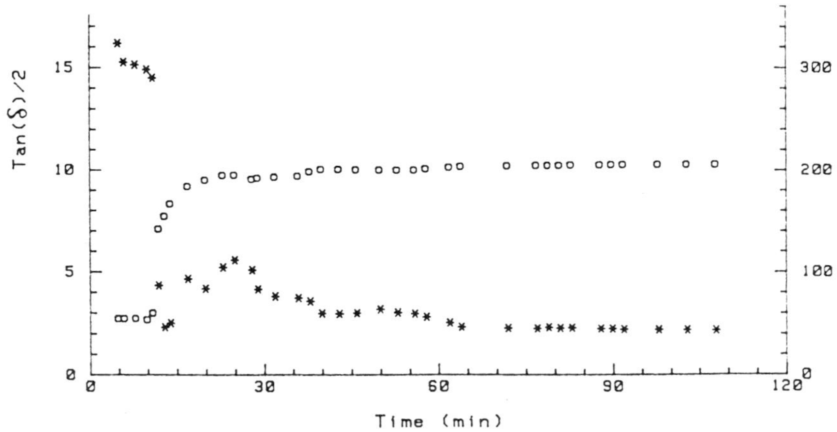

Fig. 3. Damping and natural frequency of a prepreg specimen plotted against cure time.

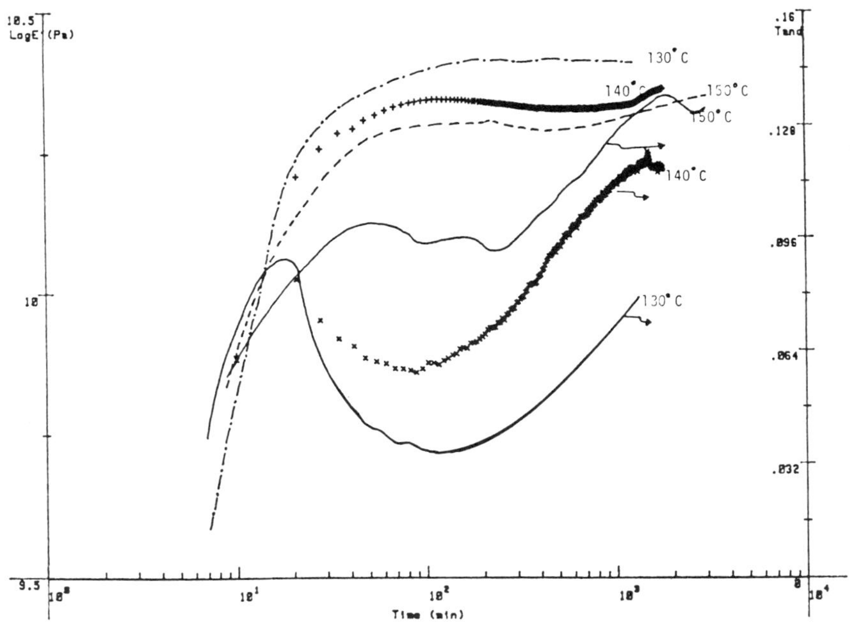

Fig. 4. Kevlar-epoxy prepreg cured at various temperatures.

of the rubbery phase and therefore procedures which alter the time and temperatures to gelation can be used to control the material properties. Control of the time-temperature history of the material during cure is a method of achieving desired degrees of phase separation, but a knowledge of the TTT diagram is prerequisite for such a procedure.

2. Cure Monitoring of a Model Composite Structure

Preliminary studies [20] have attested to the feasibility of monitoring the cure of a simplified, model composite structure. One possible methodology for cure monitoring involves the comparison of vibration spectra of a moulding press (or an autoclave) without prepreg and with prepreg in place for subsequent curing. Damping characteristics associated with epoxy reaction can then be identified and analyzed. Results of a similar but qualitative study are shown in Fig.5, where the dynamic response of a fiberglass-epoxy sample embedded in a mould is plotted over a frequency range of 0-1600 Hz. Again, a plot covering a wider frequency scope would have included several additional modes of vibration. Heating began at t=0 and T=44°C (1st plot) while reached a steady state of 158°C at @ t=35 min.. Two competing effects are expected to exist during heat-up from 44°C to 158°C. An increasing temperature should increase molecular mobility and therefore decrease resin viscosity. As temperature continues to increase the polymerization/crosslinking reaction would proceed to an ever increasing extent. These latter events would dramatically increase the molecular sizes and resin viscosity. These effects tend to dominate after the desired isothermal cure temperature is reached. Consider the first major peak, "Peak A", which could hardly be identified in the first plot (t=0, T=44°C). The location (natural frequency), damping, and mode shape of this peak shift back and forth during heat-up. At t=34 min., this peak becomes flattened, reflecting the fact that system vibration is totally dominated by the liquid resin. As curing proceeds, the peak reappears and, although shifting in location, becomes sharper and more clearly defined. Information of this nature could serve as a fingerprint for the cure state of a composite.

The early stage of this work has been primarily concerned with the driving-point impedance analysis (excited and measured at the same point). The experimental approach to the mechanical impedance analysis was based on the assumption that, by performing the test on an existing object, all of the dynamic properties were automatically taken into account. In general, the dynamic response can be analyzed as the summation of a number of normal modes. Each of the modes is characterized by unique modal parameters: natural frequency, damping, and mode shape. For the simpler geometries, analytical mechanical impedance analysis could be performed to support the experimental work. For complex structures, analysis must be performed using computer models based on finite element techniques.

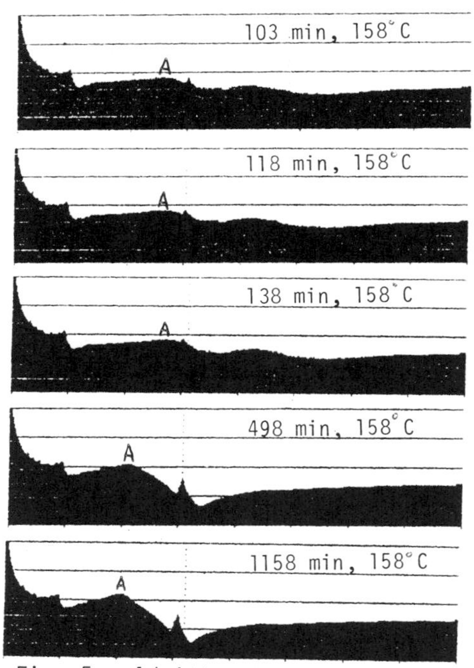

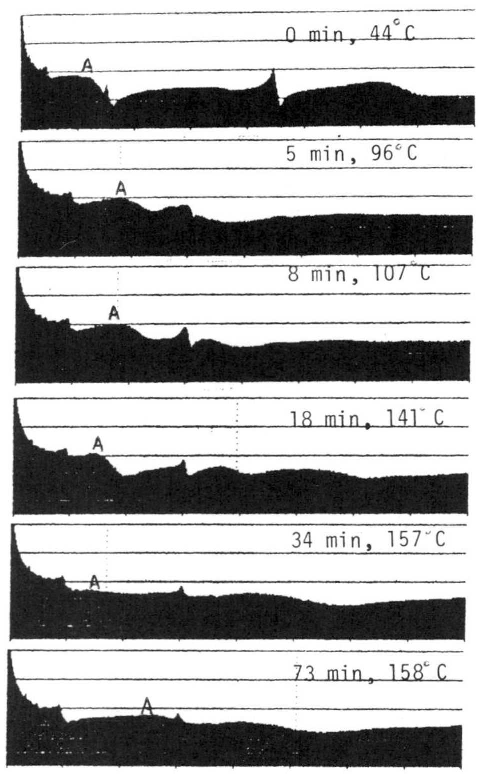

Fig. 5. Admittance spectra of a model composite structure recorded during heat-up and curing stages.

3. Impulse Techniques

The impulse techniques involve transient excitation, usually using a hammer with a built-in force transducer. The response signal is normally picked up at other locations via response transducers. This is the simplest and fastest technique for obtaining good estimates of the required frequency response data. The first plot in Fig.6 represents the admittance function (0-400 Hz) of a mould clamped on the surface of a sturdy testing stand. A corresponding function after the epoxy is poured into the mould cavity is shown in the second plot. The difference between these two spectra reflects the effects of epoxy resin. To include more information a broader frequency range (0-800 Hz) is scanned in the subsequent plots. As curing proceeds at room temperature, the peak values of the lower-order modes ascend while those of the higher-order ones descend. This is consistent with the results of a previous study [20] using rectangular beams centrally-loaded with both ends free. Vibration modes become increasingly better defined as epoxy curing becomes matured. A more quantitative study is now in progress.

Future cure study of composite structure will include automatic data acquisition and reduction of impedance mode characteristics as a function of cure temperature and time. Specifically, the rheological properties E' (storage modulus) and tan delta (damping) will be obtained real-time and shown in the computer screen, along with T-P-V parameters, as curing proceeds. In this phase of study the temperature history will be preprogrammed. The points where epoxy gels or vitrifies will be identified from the curves of E' or tan delta versus time. With this approach it would be possible to control composite curing processes using computer based monitoring of the process rheology. Thus all process decisions such

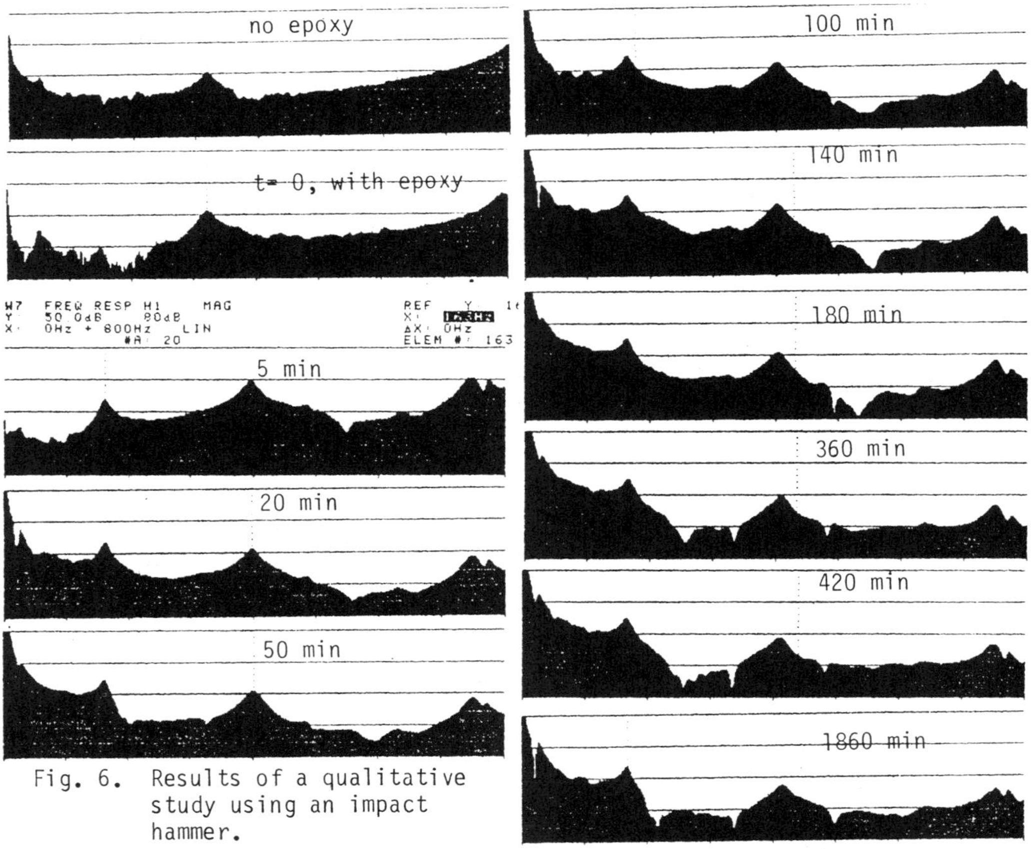

Fig. 6. Results of a qualitative study using an impact hammer.

as the control of heating rates and pressure application timing would be freed from the inherent errors of human judgment. If a previous study indicated, for instance, that a given value of E' or tan delta was the ideal rheological condition for lamination consolidation through the application of pressure, this point of time has to be clearly identified during cure. Otherwise, the end result could be a faulty product due to insufficient or excessive resin squeeze-out.

The proposed MIA techniques can directly measure the storage and loss components of rheological response and are highly sensitive to both flow and glass transitions at all states of cure. Such properties are better indices of the cure state than the more indirect dielectric signals. An operator independent, closed loop cure cycle based on the rheological property change appears to be probable and feasible. All that would be required is the development of the proper mathematical algorithm on the data file of rheological properties that would dictate the T-P-time parameters for subsequent curing process. Composite fabrication then could be based solely on the chemistry and physics of the curing process (as monitored by MIA) and independent of any operator decisions.

CONCLUSION

A potentially versatile and reliable methodology of sensing the dynamic mechanical properties of materials has been proposed and studied. The techniques of MIA are expected to be very powerful nondestructive evaluation (NDE) tools in the cure cycle design, optimization, monitoring, and control of composite fabrication. The maturization of this methodology would help to ensure more consistent composite quality, promote possible automation of composite production, and therefore increase the productivity. The techniques of MIA for cure monitoring appear to have the following advantages:

1. The MIA method affords a great deal of flexibility in selecting sample geometry and dimensions and curing apparatus configuration.
2. A wide scope of materials and physical states may be studied by MIA. These include thermoplastics and thermosetting polymers, neat resins and composites, and states of materials such as liquid, rubber, gel, glass, and crystalline. An inert substrate or container may be used when studying a high-flow liquid which cannot support its own weight.
3. The MIA method may be used to define the kinetics of curing and the optimum cure cycle. For this purpose, the techniques of MIA may have the following advantages: (a) Isolation of the storage and loss components of rheological response; (b) Applicable to unsupported and supported polymers; and (c) Sensitive to both flow and glass transitions at all states of cure. The time-temperature-transformation (TTT) diagram for each formulation can be established.
4. Since the data acquisition and reduction can be handled very rapidly and accurately, the technique may serve as an effective, real-time monitoring tool for polymer and composite processing. The technique can also be used in a closed-loop feedback control of composite curing.

ACKNOWLEDGMENTS

Financial support by NSF (DMS-8601416) and dynamic equipment grant by NSF (MSM-8604641) are gratefully acknowledged.

REFERENCES

1. D.H.Kaelble and R.J.Shuford, "Composite Characterization Techniques: Overview", US Army Mantech J. 10(2) (1985) 17.

2. J.L.Koenig, "Composite Characterization Techniques: Physiochemical" US Army Mantech J. 10(2) (1985)27.
3. J.K.Gillham, Formation and Properties of Network Polymeric Materials, Polym. Eng. Sci. 19 (1979) 676.
4. Y.A.Tajima, "Monitoring Cure Viscosity of Epoxy Comp.", Polym. Comp. 3 (1982) 162.
5. J.Chottiner, Z.N.Sanjana, M.R.Kodani, K.W.Lengel, and G.B.Rosenblatt, "Monitoring Cure of autoclave-Molded Parts by Dielectric Analysis", Polym. Comp. 3 (1982) 59.
6. D.E.Kranbuehl, S.E.Delos, and P.K.Jue, "Dynamic Dielectric Characterization of the Cure Process: LARC-160", 28th National SAMPE Symp. April 12-14, 1983, p. 608.
7. S.D.Senturia, N.F.Sheppard, Jr., H.L.Lee, and S.B.Marshall,"Cure Monitoring and Control with Combined Dielectric/Temp. Probes", 28th National SAMPE Symp. April 12-14, 1983. p.851.
8. C.A.May, "The Chem. Charac. and Processing Sci. at Comp.", Pure and Appl. Chem. 33 (1983) 811.
9. Y.Minoda, Y. Sakatani, Y. Yamaguchi, M. Niizeki, and H.Saigoku, "Toward Process Optimization by Monitoring the Electrical Properties during the Cure Cycle for CFRP" in Recent Adv. in Comp. in the US and Japan, ASTM STP 864, ASTM, Phila., 1985, pp. 489-501.
10. G. A. Sofer, A. G. H. Dietz, and E. A. Hauser, Ind. Eng. Chem. 45 (1953) 2743.
11. W. Roth and S. R. Rich, J. Appl. Phys., 24 (1953) 940.
12. J. H. Speake, R. G. C. Arridge, and G. J. Crutis,J. Phys. D.7 (1974) 412.
13. J. F. Bell, Ultrasonics, 6 (1968) 11.
14. C. G. Delides and T. A. King, J. Chem. Soc., Faraday Trans. 75 (1979) 359.
15. T. Wright and D. D. Campbell, J. Phys. E-10 (1977) 1241.
16. N. F. Sheppard, S. L. Garverick, D. R. Ray, and S. D. Senturia, "Microdielectrometry: A New Method for In Situ Cure Monitoring," Proc. 26th National SAMPE Symp. 26 (April, 1981) pp. 65-76.
17. R. Hinricks, "Interactive Computer Process System for Composite Autoclave Fabrication," Critical Review: Techniques for the Characterization of Composite Materials," Mass. Inst. Tech. (June 1981).
18. D. J. P. Harrison, W. R. Yates, and J. F. Johnson, "Techniques for the Analysis of Crosslinked Polymers,"J. Macromol. Sci., Rev. Macromol. Chem. Phys. C25 (1985) 481.
19. D. H. Kaeble, "Computer Aided Design of Polymers and Composites," Marcel Dekker, Inc. New York, 1985.
20. B. Z. Jang and G. H. Zhu "Monitoring of Dynamic Mechanical Behavior of Polymers and Composites Using MIA", Accepted by J. Appl. Polymer Sci.. 1986.
21. B. Z. Jang, G.H.Zhu, and Y.S.Chang, "Cure Monitoring and Control of Thermosetting Resins and Composites", to appear in Polymer Composites.
22. E. G. Henneke, K. L. Reifsnider, R. J. Schuford, Y. L. Hinton, and B. R. Markert, "Thermography-Application to the Manufacture and NDE of Composites" SPIE vol. 371, Thermosense V (1982) 98.

ULTRASONIC CHARACTERIZATION OF POLYMERS IN THEIR EVOLUTION FROM SOLID TO LIQUID STATE

F. Massines and L. Piché

Industrial Materials Research Institute
National Research Council Canada
75 De Mortagne Blvd, Boucherville, Québec, J4B 6Y4

INTRODUCTION

A polymer can be defined as a large molecule made of a number of repeating units. Actually many natural organic substances, such as hevea rubber, cellulose, starch, leather, proteins... are polymers. The hypothesis, that these materials are made of very long chains and that they owe their specific properties to this peculiar structure, was first put forth by Kekule in 1877. Since then, the macromolecular theory has been substantiated by its numerous success, both scientific - one of the most exhilarating being the discovery of the double helix structure for the DNA molecule by Crick, Watson and Wilkins (1953) - and industrial, with the advent of totally synthetic polymers which today represent an important portion of industrial materials.

From the point of view of processing and fabrication, it is useful to distinguish two classes of synthetic polymers. The thermosets are those where a crosslinking agent (hardener) is added to a low molecular weight liquid (resin); polymerization is made to occur in the mold and yields a stable end product. The process is irreversible. The second class, which is certainly the most important commercially, is that of thermoplastics, more commonly referred to as plastics. In fact, the name "plastics" comes from that the resin material is heated until pliable and soft enough to be formed into the desired shape. In contradistinction to thermosets, thermoplastics are high molecular weight polymers which preserve their chemical and structural identity during processing.

The art of plastic making has allowed for innumerable applications of polymers, with new ones constantly being found. This success can be attributed to the introduction of more advanced processing technologies, but also to a better knowledge and understanding of the basic properties of the material. This last factor can be expected to become even more important, since it is usually recognized that most of the "great" polymers have already been discovered, and that future developments will come from the optimization of conditioning, processing, and utilization.

As materials, the most marked feature of polymers lies in their viscoelastic character. In principle, at least, polymers can be modeled by combining both viscous and elastic components. Such models serve to interpret and predict the materials properties; in turn, a correct interpretation requires a precise description of the stress and strain history. For this, a number of analytical techniques are available which can be classified as either reliability or thermostructural tests. Reliability tests are made under conditions that approximate the end use environment and serve to predict long term stability; such tests include time dependence of compliance, creep, time to rupture, brittleness, hardness, resistance to abrasion, wear, friction, and fatigue, electrical resistivity and strength. On the other hand, thermostructural tests aim at correlating structural changes that occur within the polymer with changes in a thermodynamic property; they include specific heat, thermal conductivity and expansion, degradation, static and dynamic mechanical analysis, dielectric behaviour, I.R., Raman and N.M.R. spectroscopy.

For a given application, one technique may be more adequate or practical than another but, none is universal. From this, it appears that a different analytical tool, method or technique could be of interest, especially if its field of application were broad and if it required minimum effort to implement. In this perspective, ultrasonic techniques seem to be promising. Indeed the propagation characteristics (attenuation and velocity) of ultrasonic waves are known to depend upon the elastic and viscous properties of the material in which they radiate, and therefore should also be influenced by the viscoelasticity of polymers. Also, the technique should prove to be practical since it is noninvasive and the measurement can be performed in situ, in a reactor, at the outlet of an extruder or, in a mold.

This paper is meant as an introduction to a continuing study we have undertaken, on the use of ultrasonics as a tool for the thermomechanical analysis of thermoplastic polymers. First we shall briefly describe the polymeric material, then we will review the state of the art, and finally we shall report our first results on a typical commercial polymer - polystyrene.

POLYMERS, FROM LIQUIDS TO SOLIDS

Before going any further, a brief description of the polymeric material is in order. Basically, polymers are constituted of large molecules, comprising a number (large) of repeating units which are covalently linked together to form chains. In turn, the attraction between the chains is provided by secondary bonds, which, depending on the nature of the molecules, may be Van-der-Waals forces, dipole-dipole interactions, hydrogen bonds and, if crosslinking is present, covalent bonds. At the higher temperature, typically 200°C, the configuration for the chains is random, like that of a viscous liquid. The specific volume, V, of this liquid can be described[1] as the sum of a free volume, V_f, associated with holes and packing irregularities, and that of an occupied volume, V_o corresponding to the effective volume occupied by the molecules under Brownian agitation. As temperature, T, is lowered, the thermal excitation diminishes and, this is accompanied by a decrease of the specific volume ($\alpha = d(\ln V)/dT \simeq 10^{-3}$/deg), mainly due to the decrease of free volume, V_f. This collapse of V_f limits the freedom of movement for the chains, which results in the increase of viscosity, η. Upon approaching a certain characteristic temperature[1], referred to as the "glass transition temperature", Tg, the free volume as so diminished, that the possibility of conformational rearrangements of the chains has become strongly hindered. In this region of temperature, the response to a perturbation is characterized by a finite relaxation time, τ, so that material properties such as modulus and viscosity are also time[1] and frequency[2] dependent. On going through Tg (ie. the "glass transition temperature"), the thermal expansion suddenly drops

($\alpha \simeq 10^{-4}$/deg), and the viscosity increases to such values ($\eta \simeq 10^{13}$ poises) that the structure is frozen in, and the material appears as a solid. Now, because of its high viscosity, the liquid may be undercooled to various degrees, so that the value for Tg is dependent on the cooling rate. Therefore the structure and properties of the solid are contingent not only upon the nature of the molecules, but also, to a large extent, upon the thermal history near Tg (temperature, pressure and cooling rate). According to their structure "solid" polymers are found to be either amorphous or, semi-crystalline. Amorphous polymers, such as the usual atactic polystyrene (PS), and polymethyl-metacrylate (PMMA or plexiglass), are characterized by the absence of long range ordering of the molecules. Semi-crystalline polymers, such as polypropylene (PP), and polyethylene (PE), are those where the formation of ordered regions (crystals), dispersed in the amorphous matrix, is favoured by strong intermolecular forces, and by the highly regular structure of the chains.

Such a description is sufficient to help foresee some of the problems and difficulties associated with experimentation on polymers; this is true for static and quasi static measurements[1], but even more so, for high frequency[2] experiments.

ULTRASONICS, THE STATE OF THE ART

Our objective here is to provide a rapid and yet pertinent overview of the state of the art in the use of ultrasonics as a probe of the thermomechanical properties of polymers. Even though, the literature on the subject is relatively limited, this presentation cannot be complete; therefore reference must be made to review articles and books which furnish a more thorough covering of the matter. For example Hartmann[3] has given an excellent introduction to the subject; Phillips and Pethrick[4] have discussed the physical factors which influence the propagation of sound in polymers, glasses and dielectrics; Lamb[5] described the propagation in relaxing liquids; Herzfeld and Litovitz[6] and Litovitz and Davis[7] discussed structural and shear relaxation; more recently, Christensen[8] established the sound propagation equations in terms of viscoelasticity; and finally, McSkimmin[9] gave an account of the various ultrasonic methods and techniques for measuring the mechanical properties of liquids and solids.

As could be anticipated from our brief description of the polymeric material, the first measurements with ultrasound[10,11] attested that the attenuation, A, and velocity of sound, v, in polymers behaved quite differently from that in usual materials. So as to set ideas, the characteristic features for the behaviour can be roughly sketched as follows. Upon increasing the temperature from a value where the polymer is solid; the velocity (longitudinal, $v_l \simeq 2.5$ km/sec, and shear, $v_s \simeq 1.0$ km/sec) decreases, first slowly, goes through a break, after which, it becomes frequency dependent and diminishes rapidly. At the higher temperatures, the velocity may well have fallen to less than half its value in the solid state. Correspondingly, the attenuation of longitudinal waves, A_l, goes through a strong maximum, reaching values up to 15 cm^{-1} ($\simeq$ 100dB/cm), while that of shear waves, A_s, increases up to where the signal is lost and not recovered at higher temperatures.

These observations stimulated researchers to establish methods and techniques best adapted to polymers. Experiments were made with liquid immersion and goniometer techniques[9,12-16], and with solid delay buffers[9,11,17-21]; using methods such as through transmission, impedence measurements[7,9,22] and interferometry[9,23,24] The various results were in agreement with the general description given above. However, some discrepencies were reported[25,26], which is far from surprising considering the experimental difficulties involved. Indeed, given the strong influence of both the temperature and the thermal history, a close control of these variables is a requirement. Also, for an accurate description of the velocity, the thermal expansion must be accounted for, either experimentally by fixing the sound path[9,21] using a double path technique[18] or by monitoring the thickness of the sample during the experiment[16,27], or through calculations[24].

This pioneering work, even though not systematic, was an essential step towards an understanding of basic physical mechanisms underlying the propagation of sound in polymers. In the region of temperatures below, but close to Tg, and at sufficiently high frequencies ($\omega = 2\pi f$), it was observed[19] that the dispersion for the velocity ($dv/d\omega$) was small. Therefore, it appeared reasonable to interpret the velocity in terms of elasticity [14,16,19,20,24]. Some measurements were also carried out at high pressures[24,28,29]; this allowed to establish values for an experimental equation of state, which in turn was discussed in reference to theoretical models such as the Gruneisen [3,28-30] equation of state. These experiments suggested that compression was mostly effective in directions normal to the chains and that the bulk modulus was closely related to the volume dependence of the intermolecular Van-der-Waals forces. Still in the region below Tg, it was seen that the attenuation, A, increased with temperature. This was analysed in terms of hysteresis losses, and it was shown[31] that the absorption per wavelength ($A\lambda$) decreased with increasing specific volume, V. This constituted added evidence that the propagation of sound was strongly influenced by the specific volume.

A good number of experiments were made[11-13,18,25,27] in the range covering the temperature region around the glass transition, Tg, where the break in the velocity was seen to occur. However, it appears that only few authors[12,13] have elaborated on the matter. It was noted[13] that the temperature of the break point was very close to Tg (as otherwise measured by classical static methods), and it was speculated[12] that the break could very well be associated with the sudden increase of free volume, V_f at Tg.

In the region of temperature above Tg, it was demonstrated that the velocity and the associated attenuation were frequency dependent. Such dispersion ($dv/d\omega$, $dA/d\omega$) was tentatively identified[10,11,32] as a relaxation of volume deformation of the polymer. In contrast to rheological methods which mostly involve shearing action[1], ultrasonic techniques involve both[6] shearing (shear waves) and combined shearing and compression (longitudinal waves). This raised the question[10,11] about the origin of the losses associated with longitudinal waves. In order to shed some light on the subject, further measurements were made with both[13,26,27] types of waves. The results showed that there were[18] energy or viscous losses associated with pure compression and that these structural relaxation[6] losses could in fact be comparable[25,32] or even larger[13] than pure shear losses.

Measurements at different frequencies were analysed with reasonable success[13,18,26-28] using an approach of low frequency rheology. This technique, referred to as the WLF (William, Laendel, and Ferry) time-temperature superposition scheme, serves to describe the shift of relaxation features to higher temperatures with increasing frequency. The WLF law is based on free volume concepts and illustrates the rate dependence of the glass transition phenomenon. That the ultrasonic data could be described by the WLF equation suggested that the high frequency relaxation peak also had its origin in the presence of free volume.

Sparse, but interesting attempts have been made to correlate the propagation of sound to the molecular structure of polymers. Such studies were mainly focussed on the influence of chain length (molecular weight)[22,33], chain configuration (presence of side groups)[25,34,35], crosslinking[36], plasticization[37,38] and copolymerization[33,39]. Work has also been done which pertained to chain entanglement effects. In these instances, the results were analysed in terms of Hirai-Eyring[22,25] or Rouse[40,41] molecular theories. Measurements in the temperature region far below Tg, revealed the existence of various relaxation features. These were tentatively identified either to motions of small side group units[42-44] as observed at low frequencies[2], or more generally to structural configuration defects[45]. Finally, still on the subject of molecular structure, ultrasonics was used in studies on polymer crystallization[46]. Investigations as a function of thermal history[47], drawing[48], and pressure[49], showed that the propagation of sound was sensitive to crystallinity and reflected changes in branched chain content and dimensions of the crystalline entities.

In summary it appears that the propagation of sound is sensitive to many of the important parameters which characterize a given polymer, namely the "glass transition temperature" (Tg), the specific volume (V), the free volume (V_f) and the chemical nature and morphology of the polymer chains.

EXPERIMENTAL

Sample

The polymer studied here was an atactic, amorphous polystyrene (PS). The material, manufactured by Dow Chemical of Canada is commercialized under the brand name Styron 685D and recommended where high temperature processing conditions are encountered. It is thus possible that the material might have contained additives, such as antioxidants. The glass transition temperature was not specified; instead, a Vicat softening point temperature was given: T_v = 107°C. This value, which is usually 4 to 6°C above Tg is obtained in a test (ASTM D1525-B)[50] where the sample being heated at a rate of 2°C/min. is submitted to a point load: T_v is the temperature at which a given deflection or penetration is arrived at. So as to complete the characterization of our sample we performed an analysis using Gel Permeation Chromatography (GPC)[51] to access the average molecular weight, [MW]. The number average molecular weight is defined by: [Mn] = $\sum$niMi, where ni is the mole fraction and Mi is the molecular weight of the ith fraction; the weight average molecular weight on the other hand is defined as: [Mw] = $\sum$wi Mi with wi being the weight fraction of the ith component. The values found were [Mn] = 130,000 g/mol. and [Mw] = 258,000 g/mol., indicating that our sample was a high molecular weight polystyrene. The ratio [Mw]/[Mn] is referred to as the heterogeneity index (HI) and is a measure of polydispersity. The value found here, HI = 1.98, is typical of the random distribution found in polystyrene.

Ultrasonic technique

We shall but briefly sketch out our technical approach; a full description of the apparatus is of order and will be given elsewhere. The sample in the form of thin (≃ 0.3cm) plate was confined between two aligned buffers at the opposite end of which were attached an emitting and a receiving transducer. The assembly was constructed such that the buffers could be submitted to axial clamping while allowance was maintained for their axial displacement so as to compensate for the thermal expansion of the polymer. The small clamping force which was kept constant, ensured that the transmission of sound at the buffer/sample interface was good and could easily be reproduced. The axial length of the assembly was monitored with good accuracy (∓ 1 μm); this permitted to follow changes in sample thickness, volume and specific volume (or density) with temperature. Provision to scan the temperature in the range from -150 to 300°C (≃ 150/550K) was made through heating elements and liquid nitrogen cooling circuit attached to the buffer lines. The temperature control system allowed to establish heating or cooling rates from 50°C/min. to 1 or 2°C/hr, or to stabilize the temperature to ∓0.1°C. Finally, the transducers were energized coherently by use of a gated amplifier (Matec Mod. 5000), and a super homodyne system was used for the detection, which allowed a real time measurement of both the amplitude and the phase of the signal. The entire set-up was under full computer control and the collection of all the pertinent data concerning length, temperature and acoustic signal was accomplished every 20 sec. and stored on disc. Further processing of the rough data was made to account and correct for the influence of the buffers (length, velocity, attenuation, acoustic mismatch).

RESULTS AND DISCUSSION

Here we present results that we have obtained with longitudinal waves at a frequency $f = \omega/2\pi = 2.5$ MHz, in the temperature range T = -150°C to T = 250°C. Prior to all measurements, the sample was annealed in the apparatus for ½ hour at a temperature of 190°C so as to relieve it from all of its past thermal history. Since the properties of the polymer in the solid state are strongly influenced by the thermal history in the vicinity of Tg, we needed to decide on a fixed procedure that would allow to reproduce the experiments. We found practical to proceed by first rapidly cooling (25°C/min.) the sample and then performing the measurements on heating at a constant rate. Here, we chose 2°C/min., a value which is commonly found in reference to other techniques[50].

A Classical View of Things

The results for the density ρ are shown in Fig. (1). On increasing temperature, ρ appears to diminish, first very slowly up to T ≃ -20°C; then more rapidly until a temperature Ts = (101 ∓ 2)°C is reached, where ρ suddenly falls off much faster. In the region of temperature, $T < T_s + 30°C$, the measurements can be reproduced only if the thermal history is maintained. The material is clearly out of thermodynamic equilibrium in regards to the static measuring technique. Such a behaviour is typical of the "glass transition phenomenon" and we shall therefore identify T_s as the "glass transition temperature" associated with the thermal history we have used.

The "glass transition temperature", Tg, is usually defined in terms of a kinetic process of volume contraction, and therefore dilatometry must be considered as a basic experiment for the determination of Tg. However, other more accessible techniques are often used, such as the Vicat temperature which was mentioned above: Tv = (107 ∓ 5)°C. Another commonly employed method involves calorimetry. We performed such a measurement using a Differential Thermal Analysis (DTA) technique[52] wherein the energy required to maintain the heating rate of the sample at a constant value is measured. The results for a heating rate of 2°C/min. are shown in Fig. (2). Even though the determination of Tg with this method is subject to some ambiguity[1,52], we find Tg (DTA) ≃ 102°C, in close agreement with Ts.

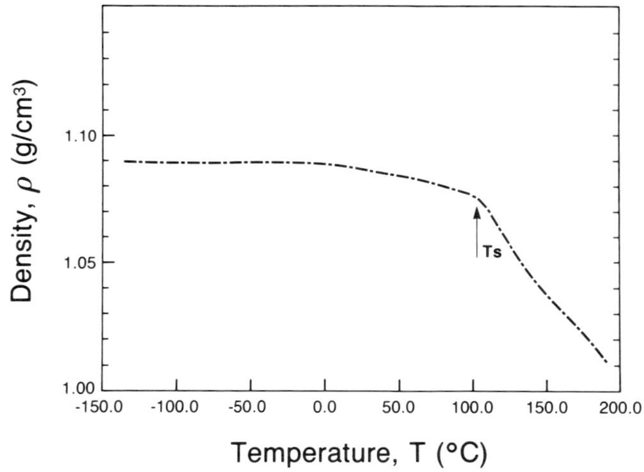

Fig. (1) Thermal variation of the density of polystyrene obtained in simultaneity to the measurement of velocity and attenuation (data of Fig. (3). The arrow (Ts) indicates the glass transition temperature on heating at 2°C/min.

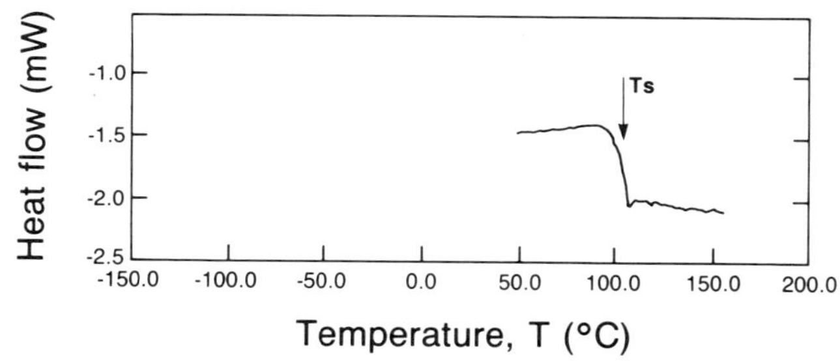

Fig. (2) Differential Thermal Analysis (DTA) signal obtained in a separate experiment, with a heating rate of 2°C/min. The arrow (Ts) points at the glass transition temperature as defined by this technique.

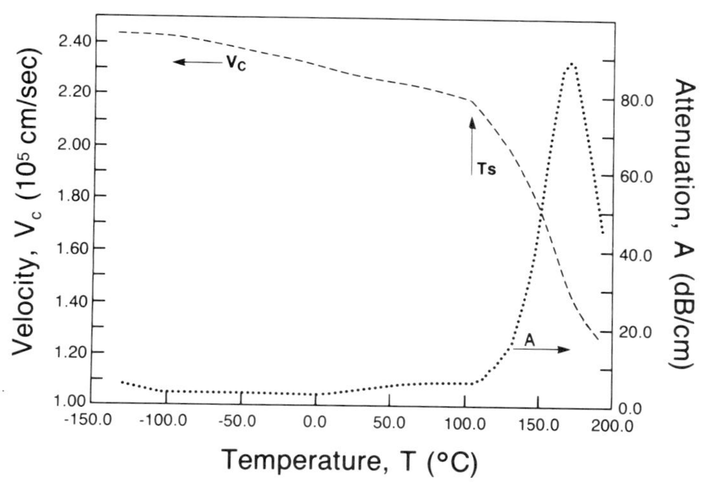

Fig. (3) Velocity and attenuation of 2.5 MHz longitudinal ultrasonic waves. The arrow points to the break in the velocity/temperature curve which is found to occur at Ts (Fig. (1) and Fig. (2)).

Going back to Fig. (1), it is seen that as the temperature is still increased, there comes a point (T ≃ 130°C) where the behaviour is no furher influenced by the thermal history. In this region the molecular structure adapts very rapidly to a change of temperature. The material appears to be in equilibrium and behaves as a "liquid". In most theories, Tg corresponds to the temperature below which the free volume, Vf, changes very slowly. In view of this, the difference between the thermal expansion, α, of the "liquid" and the "solid", i.e. $\Delta\alpha = \alpha_l - \alpha_s$, is considered as an important characteristic of the sample. The quantity $\Delta\alpha$, is often equated[1] to the thermal expansion of the free volume, α_f, which for usual polymers extends[1,53] from 1 to 12 X 10^{-4}/°C. From our data in Fig. (1), we find $\alpha_l = 5.59$, $\alpha_s = 1.63$ and $\Delta\alpha = \alpha_f = 3.9$ in units of 10^{-4}/°C, in good agreement to what is expected[1,54,55] for polystyrene.

At still higher temperature, T ≃ 174°C, there occurs another change of slope in the density curve. Such as also been observed by other authors[54,55]. However, the nature of this anomaly is still a matter of conjecture and is not our primary concern for the present.

Ultrasonic Response

The results obtained for the velocity, v (10^5 cm/sec.), and the attenuation, A (dB/cm), are shown in Fig. (3). In discussing their behaviour it is often quite instructive to consider also the corresponding modulus. For longitudinal waves, one can define a longitudinal modulus, $L = L' + iL''$, which in fact is a linear combination of the bulk, B, and shear, G, moduli, such that $L = B + 4G/3$. The real part, or storage modulus L' is given[6,7,8] by $L' \simeq \rho v^2$ and the imaginary part, or loss modulus by $L'' \simeq 2\rho A v^3/(8.68 \times 2\pi f)$. The results calculated from the data of Fig. (1) and Fig. (3) are shown in Fig. (4).

On increasing the temperature, from -150 to -100°C, one observes a small decrease in both v and A. This could well correspond to the high temperature tail of a local mode of vibration which is often observed[1,2] in rheology experiments and attributed to the rotation of side group units (the γ mode). In fact different authors[45] have pointed out that this mode could be very weak and show little effect on the ultrasonic attenuation, as is observed here.

On further increasing temperature, the velocity decreases steadily while the attenuation remains small (up to 20°C). In this region, the density, Fig. (1) decreases very little, therefore the temperature variation for the velocity can be attributed in full to a decrease of the modulus L'. The behaviour is similar to that of usual solids. As a basis for comparison we can compute (1/L') (dL'/dT): for the polymer we find -1.33×10^{-3}/°C which is of the same order of magnitude as the value, -0.8×10^{-3}/°C, found for aluminum. Still in this region of temperature, the attenuation describes a small maximum centered near 75°C. Again, we can tentatively make reference to rheology measurements where an equivalent feature is attributed to local mode torsional oscillations of the main chain[1,2] (the β mode).

As the temperature exceeds a value, T = (101 ∓ 1)°C, which appears to be "critical", the velocity suddenly decreases very rapidly. Considering the data of Fig. (1), this break point for the velocity appears to coincide with that for the density, i.e. Ts. Furthermore we found that the break point for the velocity exactly followed changes of Ts resulting from modifications to the thermal history. Under all evidence,

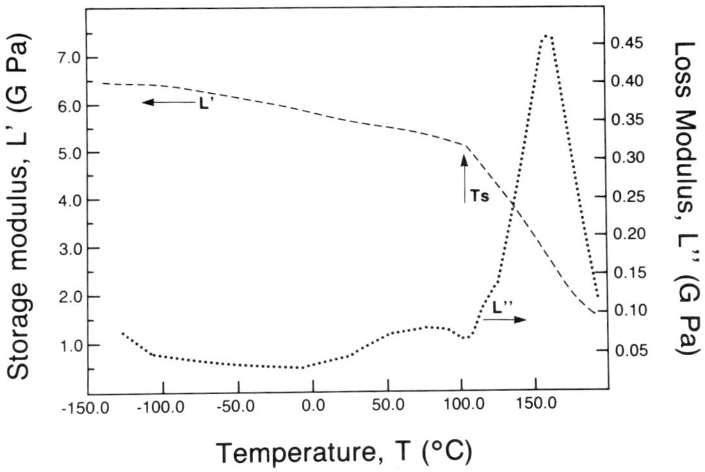

Fig. (4) Plot of the storage (L') and loss moduli (L") calculated from the data in Fig. (1) and Fig. (3).

we can identify the break point for the velocity with Ts; that is with the "glass transition temperature" (Tg) defined by basic volumetric measurements, and related to the appearance of excess free volume V_f.

The velocity in the vicinity of Ts can be described, as for the specific volume V, by the slope (1/v) (dv/dT) which on crossing Ts goes from -6.0 to -40.2 in units of 10^{-4}/°C. This effect is very important and can easily be measured. Therefore, it can be suggested that the measurement of velocity provides for a convenient means to determine Tg.

In this region of temperature (90/130°C), the relaxation time, τ, associated with structural rearrangement of the molecules is long (hours, mins., secs.) such that the properties measured on a long time scale are those of a material which is out of thermal equilibrium. However, on the time scale of the acoustic period ($1/f \simeq 0.4 \times 10^{-6}$ sec.), the material appears as if in equilibrium. The situation corresponds to the condition where $2\pi f \tau \gg 1$, and differs from that which usually prevails, $2\pi f \tau \ll 1$ for low frequency isochroneous measurements. In this range of temperature, low frequency measurements are influenced by relaxation, and are described accordingly, while ultrasonics probes the instantaneous, unrelaxed, properties of the material in their evolution with the thermal history.

On crossing Ts, the rapid change in velocity actually reflects the change in the modulus, $L' = \rho v^2$. In this region, the slope, (I/L')(dL'/dT) in units of 10^{-4}/°C goes from -13.0 below Ts to -93.0 above Ts. Here we would like to remark that the only change in going through Ts is that the slope, (I/L')(dL'/dT), decreases and becomes more typical of the liquid state. The concept of a "second order transition" (often applied to glass formation) does not hold here since L', which is a second derivative property, does not show on abrupt change in value on passing through Ts.

Finally in the high temperature region (T ≃ 130°C) the velocity, v, continues to decrease and describes a sigmoïd (inverted s) type curve. At the same time, the attenuation, A, rapidly goes through a sharp maximum (Amax = 80dB/cm, Tmax = 165°C) before decreasing. The combined behaviour for v and A is typical of what can be expected[2,5–8] for a relaxation process. In this region of temperature, the polymer is an equilibrium liquid, as evidenced by Fig. (1). As the free volume, V_f, increases, the time required for structural rearrangements, τ, becomes much shorter, and a maximum in the attenuation curve occurs when $2\pi f\tau = 1$. In the present case this corresponds to an average value of $\tau \simeq 64 \times 10^{-9}$ sec at Tmax = 165°C, characteristic of very viscous liquids[5–7]. In this region, the modulus must be analysed in terms of a relaxing modulus, which implies the use of a model for the viscoelastic properties of the polymer. For this, we can proceed by analogy with work that has been done in the field of low frequency rheology and study the distribution of relaxation times, their temperature dependence and corresponding energies of activation.

For the sake of the discussion, we have distinguished between the regime in the vicinity of Ts, and the regime in the region of the attenuation peak. In fact, both "regimes" are related to the same physical mechanism. At Ts, which we established to be the "glass transition temperature", there occurs a rapid increase of the free volue, V_f. This increase in V_f, causes the modulus, L', to collapse, and liberates frozen degrees of freedom for the molecules. The system progressively absorbs more energy as the associated effective relaxation time decreases to a value corresponding to the period of the sonic disturbance. In this sense, the "glass transition phenomenon" appears as a relaxation effect.

CONCLUSION

The primary objective here was to evaluate ultrasonic techniques as a tool for the thermomechanical evaluation of polymers. We gave a brief description of polymeric materials where we emphasized the concepts of "glass transition temperature" (Tg) and excess free volume (V_f) in their relation to the thermal history. A survey of the literature showed that even though ultrasonics had seldom been used in studies on polymers, there existed a number of convincing examples which indicated that the technique could be very sensitive to some of the distinctive features of these particular materials. This brought us to design our own experiments where we simultaneously measure the specific volume, and the velocity and attenuation of sound as a function of thermal history. We presented our first results which we obtained on a representative amorphous polymer (polystyrene). In the discussion of these results we showed that indeed both, the velocity and the attenuation of sound were governed by the thermal history of the material and the resulting specific volume. At temperatures below Tg, the propagation is quite characteristic of a solid, while above Tg, the propagation is clearly dominated by the viscoelastic nature of the material. Therefore, we are lead to conclude that ultrasonics should prove to be a useful and powerful tool for polymer studies, although a great amount of work is still required before the full potential of the technique is established.

ACKNOWLEDGMENTS

Here we need to duly acknowledge the very efficient and professionnal work of Mr. A. Hamel concerning the technical aspect of this work. Also we would like to thank Mr. C. Néron for his participation in programming. Finally, we are sincerely grateful to Dr. J.Y. Cavaillé for friendly and helpful discussions.

REFERENCES

1. J.D. Ferry, "Viscoelastic Properties of Polymers", John Wiley and Sons, New York (1970)
2. N.G. McCrum, B.E. Read, and G. Williams, "Anelastic and Dielectric Effects in Polymeric Solids", John Wiley and Sons, New York (1967)
3. B. Hartmann, "Ultrasonic Measurements" in "Methods of Experimental Physics", edited by R.A. Fava, Academic Press, New York (1980), Vol. 16-c, Chap. 12.1, pp. 59-90
4. D.W. Phillips, and R.A. Pethrick, J. Macromol. Sci. - Rev. Macromol. Chem., c 16, 1 (1977-1978)
5. J. Lamb, "Thermal Relaxation in Liquids" in "Physical Acoustics", edited by W.P. Mason, Academic Press, New York (1965), Vol. II-A, Chap. 4, pp. 203-280
6. K.F. Herzfeld, and T.A. Litovitz, "Absorption and Dispersion of Ultrasonic Waves", Academic Press, New York (1959)
7. T.A. Litovitz, and C.M. Davis, "Structural and Shear Relaxation in Liquids" in "Physical Acoustics", edited by W.P. Mason, Academic Press, New York (1965), Vol. II-A, Chap. 5, pp. 281-350
8. R.M. Christensen, "Theory of Viscoelasticity", Academic Press, New York (1971)
9. H.J. McSkimin, "Ultrasonic Methods for Measuring the Mechanical Properties of Liquids and Solids" in "Physical Acoustics", edited by W.P. Mason, Academic Press, New York (1964), Vol. I-A, Chap. 4, pp. 271-334
10. W. Mason, W. Baker, H.J. McSkimin, and J. Heiss, Phys. Rev., 73, 1074 (1948)
11. A.W. Nolle, and P.W. Sieck, J. Appl. Phys., 23, 888 (1952)
12. Y. Wada, and K. Yamamoto, J. Phys. Soc. Jpn., 11, 887 (1956)
13. R. Kono, J. Phys. Soc. Jpn., 15, 718 (1960)
14. H.A. Waterman, Kolloid-Z., 192, 1 (1963)
15. B. Hartmann, and J. Jarzinski, J. Acoust. Soc. Am., 56, 1469 (1974)
16. G.W. Paddison, Proc. IEEE Ultrasonics Symposium, 502 (1979)
17. H.J. McSkimin, J. Acoust. Soc. Am., 22, 413 (1950)
18. J.R. Cunningham, and D.G. Ivey, J. Appl. Phys., 27, 967 (1956)
19. N.D. Arnold, and A.H. Guenther, J. Appl. Polym. Sci.,10, 731 (1966)
20. J.R. Asay, and A.H. Guenther, J. Appl. Polym. Sci., 11, 1087 (1967)
21. J. Arman, Acustica, 43, 212 (1979)
22. R. Kono, and H. Yoshizaki, J. Appl. Phys., 47, 531 (1976)
23. H.J. McSkimin, and R.P. Chambers, Proc. IEEE Trans. Sonics Ultrasonics, SU-11, 74 (1964)
24. J.R. Asay, D.L. Lamberson, and A.H. Guenther, J. Appl. Phys., 40, 1768 (1969)
25. R. Kono, J. Phys. Soc. Jpn., 16, 1580 (1961)
26. Y. Wada, H. Hirose, H. Umebayashi, and M. Otomo, J. Phys. Soc. Jpn., 15, 2324 (1960)
27. E. Morita, R. Kono, and H. Yoshizaki, Jpn. J. Appl. Phys., 7, 451 (1968)
28. D.L. Lamberson, J.R. Asay, and A.H. Guenther, J. Appl. Phys., 43, 976 (1972)
29. Y. Wada, A. Itany, T. Nishi, and S. Nagai, J. Polym. Sci: A2, 7, 201 (1969)
30. B. Hartmann, Acustica, 36, 24 (1976)
31. B. Hartmann, and J. Jarzinski, J. Appl. Phys., 43, 4304 (1972)
32. R.S. Marvin, R. Aldrich, and H.S. Sack, J. Appl. Phys., 25, 1213 (1954)
33. C. Delides, and R.A. Pethrick, Eur. Polym. J., 17, 675 (1981)
34. A.M. North, R.A. Pethrick, and D.W. Phillips, Polymer, 18, 324 (1977)
35. D.W. Phillips, A.M. North, and R.A. Pethrick, J. Appl. Polym. Sci., 21, 1859 (1977)

36. B. Hartmann, J. Appl. Polym. Sci., 19, 3241 (1975)
37. A.M. North, R.A. Pethrick, and D.W. Phillips, Macromolecules, 10, 993 (1977)
38. A.S. Gilbert, R.A. Pethrick, and D.W. Phillips, J. Appl. Polym. Sci., 21, 319 (1977)
39. D.J. Hourston, and J.A. McCluskey, Polymer, 22, 405 (1981)
40. A.J. Barlow, M. Day, G. Harrison, J. Lamb, and S. Subramanian, Proc. Roy. Soc., A.309, 497 (1969)
41. W. Bell, A.M. North, R.A. Pethrick, and P.B. Teik, J.C.S. Faraday II, 75, 1115 (1979)
42. O. Yano, and Y. Wada, J. Polym. Sci: A-2, 9, 669 (1971)
43. K. Skimizu, O. Yano, Y. Wada, and Y. Kawamura, J. Polym. Sci: Polym. Phys. Ed., 11, 1641 (1973)
44. I.I. Perepechko, and V.E. Sorokin, Sov. Phys.-Acoust., 18, 485 (1973)
45. J.Y. Duquesne, and G. Bellessa, J. Physique - Lett., 40, L-193 (1979); and 42, L-491 (1981)
46. R.K. Eby, J. Acoust. Soc. Am., 36, 1485 (1964)
47. K. Adachi, G. Harrison, J. Lamb, A.M. North, and R.A. Pethrick, Polymer, 22, 1032 (1981)
48. K. Adachi, G. Harrison, J. Lamb, A.M. North and R.A. Pethrick, Polymer, 22, 1026 (1981)
49. K. Nagata, K. Tagashira, S. Taki, and T. Takemura, Jpn. J. Appl. Phys., 19, 985 (1980)
50. "Annual Book of ASTM Standards", Part 8, ASTM, Philadelphia (1986)
51. D.J. Pollock, and R.T. Kratz, "Polymer Molecular Weights" in "Methods of Experimental Physics", edited by R.A. Fava, Academic Press, New York (1980), Vol. 16A, Chap. 2, pp. 13-72
52. J. Runt, and I.R. Harrison, "Thermal Analysis of Polymers", in "Methods of Experimental Physics" edited by R.A. Fava, Academic Press, New York (1980), Vol. 16-b, Chap. 9, pp.287-338
53. D.W. Van Krevelen, "Properties of Polymers, Their Estimation and Correlation with Chemical Structure", Elsevier, New York (1976)
54. S. Matsukoa, and B. Maxwell, J. Polym. Sci., 22, 131 (1958)
55. R.F. Boyer, "Styrene Polymers, Physical Properties" in "Encyclopedia of Polymer Science and Technology", John Wiley and Sons, New York (1970), Vol. 13, pp. 251-326

POLAR CHARACTERISTICS OF THE GROUP AND PHASE VELOCITIES AS WELL AS THE FREQUENCY DEPENDENCE OF LAMB WAVES IN GRAPHITE/EPOXY COMPOSITES

Wade R. Rose[a], S. I. Rokhlin, Peter B. Nagy[b] and Laszlo Adler

Department of Welding Engineering
The Ohio State University
Columbus, Ohio 43210

The anisotropy of graphite/epoxy composite plates was studied using measurements of phase and group velocities of Lamb waves. Three types of composite structures were used: unidirectional, two-directional with orthogonal fibers, and quasi-isotropic. Two experimental methods were developed. For single mode, the phase and group velocities and the angle of deviation between them were measured using angle variable contact ultrasonic transducers which were situated on the surface of the plates. For multimodes, a new broadband single transducer immersion technique was used to examine the dispersion behavior for the phase velocity as propagation direction was varied relative to the fibers. The experimental data for an S_0 mode was compared with calculations of the angular dependence of the phase and group velocities of bulk waves in the graphite/epoxy composite material. The advantage of the Lamb wave technique over bulk waves is the ability to measure the in-plane anisotropic properties of thin composite plates such as those used in actual aircraft applications.

INTRODUCTION

The introduction of advanced composite materials into many of the new generation aircraft and spacecraft has given rise to a significant increase in all aspects of operational capability. These materials come in many forms; organic and non-organic, fiberous and particulate, and endless combinations of the above which not only give enhanced strength characteristics, but deliver them in very specific design directions. The more common uses are in both primary and secondary aircraft structures, however, considerable effort is also being put into designing composite materials for use in hostile environments such as those destined for use as jet engine turbine blades.

The obvious advantage of these materials is in their high strength and stiffness to weight ratios. The anisotropy of the material allows strength to be utilized in a given design direction without the addition of extra weight resulting from strength in unnecessary directions, as is usually the

[a]Permanent Address: Aerospace Maintenance Development Unit, Canadian Armed Forces, Trenton, Ontario, Canada.
[b]Permanent Address: Applied Biophysics Laboratory, Technical University Budapest.

case in isotropic materials. The use of these materials over their more conventional metallic counterparts can yield dramatic weight savings which can reduce fuel consumption and allow for carrying additional fuel or cargo.

The present and future importance of these materials demands that unique methods be developed for their testing and evaluation. Their inherent anisotropy and non-homogeneity create problems for standard testing procedures, however, through the use of new techniques these can be overcome.

The graphite/epoxy samples used in this work were manufactured from Hercules Corporation AS4/3501-6 Prepreg Tape. Each layer, or ply, in a given sample is made up of a thermo-set epoxy impregnated with unidirectional graphite fibers. The plys in a given laminate can all be in the same direction or the directions can be varied as desired. Due to the two components present in each ply and because of the discrete directions of the fibers, these materials are non-homogeneous and anisotropic by nature.

The first sample is an eight ply unidirectional panel with dimensions of 1.35 mm X 300 mm X 300 mm. The second is a nine ply two-directional panel with dimensions of 1.4 mm X 300 mm X 300 mm and with ply orientation of 0/90/0/90/0/90/0/90/0. The third sample is a nine ply quasi-isotropic panel with dimensions of 1.4 mm X 300 mm X 300 mm and with ply orientation of 0/+45/90/-45/0/-45/90/+45/0. The more fiber directions present in a given laminate, the more isotropic that laminate will be. In all cases the zero degree direction is considered to be the direction of the fibers in the first ply of a given sample.

THE EFFECT OF ANISOTROPY ON PHASE AND GROUP VELOCITIES

One of the interesting phenomena associated with anisotropic materials is the deviation in direction that can exist between group and phase velocities. Figure 1 demonstrates this effect with bulk waves. The transducer coupled onto the isotropic block causes a wave to propagate (group velocity) in a direction normal to the transducer face. The individual wavefronts (phase velocity) also travel in this direction. When a transducer is coupled to an anisotropic material, the resulting wave may propagate in a direction not normal to the transducer face but along some axis of symmetry in the material. However, the individual wavefronts will travel normal to the transducer face just as they would in an isotropic material. Therefore, the distances used in velocity calculations would be as shown in Figure 1, and yield different values for group and phase velocities.

Many composite applications involve the use of relatively thin panels and it is therefore advantageous to test them in their own plane as opposed

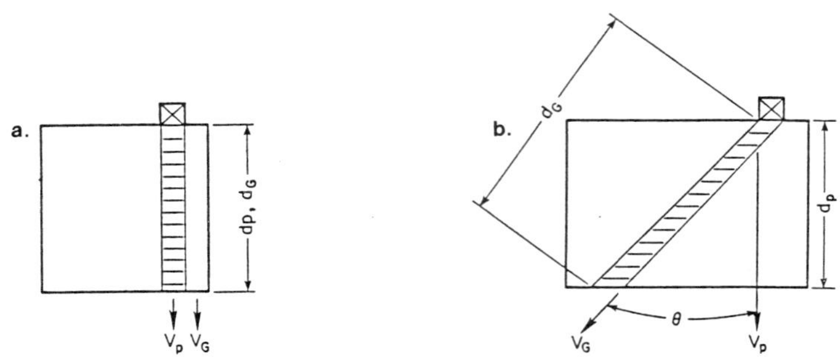

Fig. 1. (a) Wave propagation direction in an isotropic medium and (b) possible direction in an anisotropic medium.

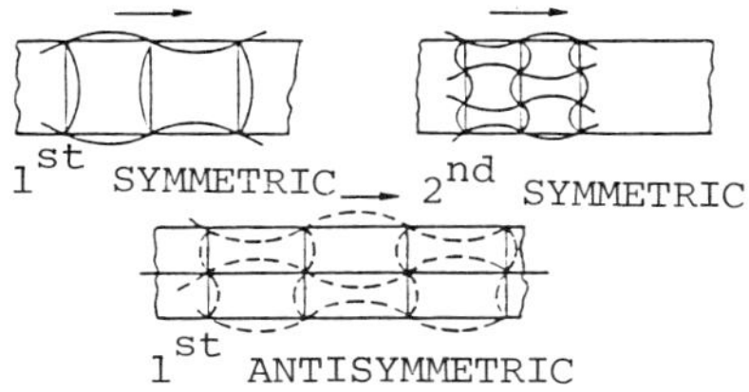

Fig. 2. Examples of Lamb mode propagation.

to through their thickness. Lamb waves travel in the planes of thin plates with particle displacement out of these planes.

Figure 2 [1] displays the motion of a few select Lamb modes in a plate. At a particular phase or group velocity, several Lamb waves will propagate in a particular plate at discrete values of the product of frequency and plate thickness.

CONTACT TRANSDUCER TECHNIQUE

One method of studying the anisotropy of a given material is to plot the variation in velocity of ultrasonic waves at various propagation directions in a sample. Using a discrete frequency toneburst at 0.6 MHz it is possible to isolate a single Lamb mode and measure its group and phase velocities. Variable angle contact transducers were used with the equipment shown in Figure 3. The Wavetek generator produced a constant frequency signal which was gated to give a toneburst and, once amplified, sent to the transmitting transducer. This generated a Lamb wave in the composite plate which was picked up by the receiving transducer, channeled through the receiver and displayed on the oscilloscope.

Figure 4(a) pictures the transmitter generating a wave in the direction of the fibers in the unidirectional plate, as well as the subsequent signal on the oscilloscope. Figure 4(b) shows the transmitter aligned at 45 de-

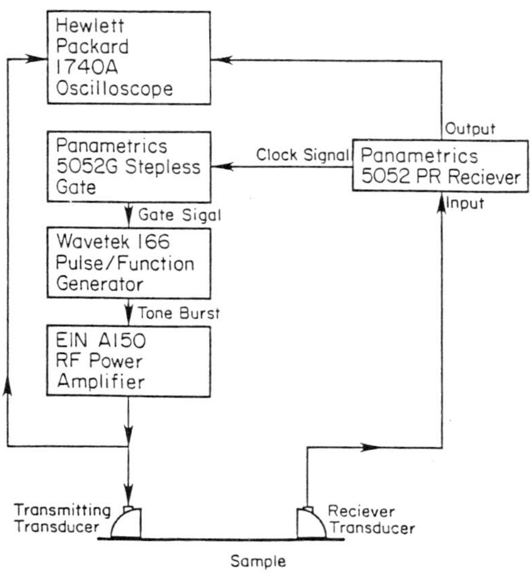

Fig. 3. Equipment setup for contact technique.

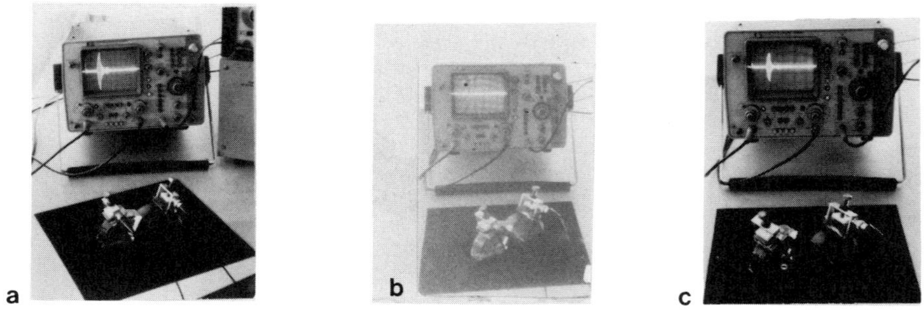

Fig. 4. (a) Propagation parallel with the fibers. (b) Transmitter at 45 degrees to the fibers. (c) Receiver is moved closer to fiber direction and signal is received.

grees to the fibers. With the receiver also aligned at 45 degrees to the fibers no signal is received. The anisotropy of the material has caused the wave to propagate closer to the fiber direction. By moving the receiver more in line with the fiber direction as shown in Figure 4(c), the signal is picked up and displayed on the oscilloscope.

Figure 5 illustrates the received signal. These constant frequency waveforms have varied amplitudes which come together to form a signal envelope with a peak. The movement of the entire signal envelope in time on the oscilloscope screen (Point 1 on Figure 5) is related to the separation of the transducers in the direction of propagation on the composite plate to calculate group velocity. The movement of a single waveform in time on the oscilloscope (Point 2 on Figure 5) is related to the separation of the transducers in a direction normal to the transmitter face, on the composite plate, to calculate phase velocity.

Figure 6 is a polar plot of the group and phase velocities for the unidirectional plate. The zero fiber direction corresponds with the horizontal axis. The small arrows indicate the velocity directions at the various angles relative to the zero fiber direction. Note that even when the transmitter is aligned at 85 degrees to the zero fiber direction, the wave propagates close to the fiber direction. Both velocities rapidly drop off as a

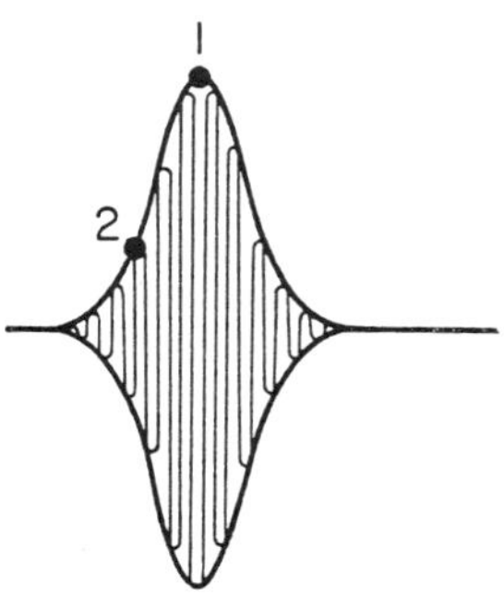

Fig. 5. Illustration of received signal.

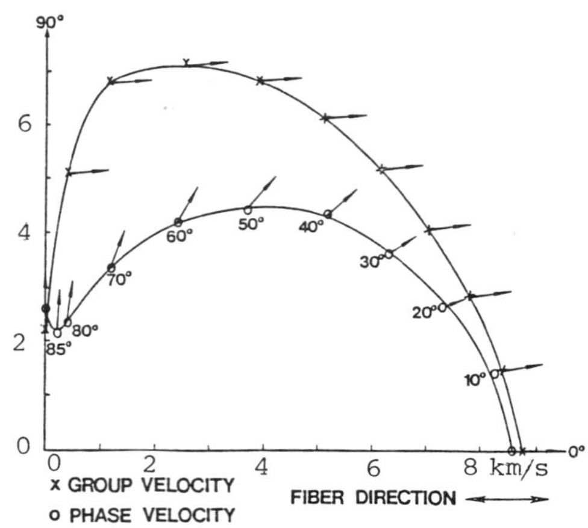

Fig. 6. Polar plot of the group and phase velocities for the unidirectional plate.

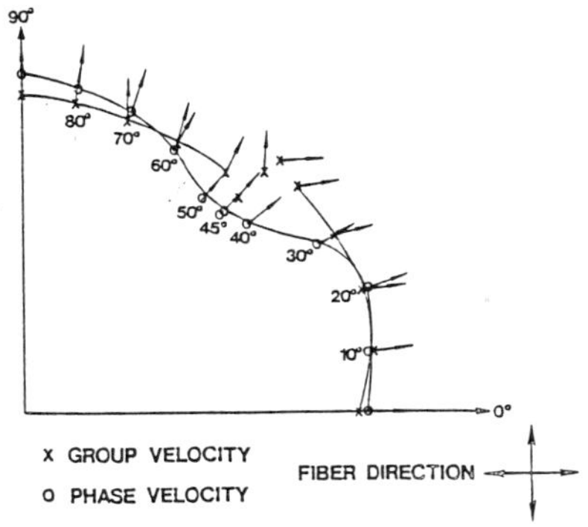

Fig. 7. Polar plot of the group and phase velocities for the two-directional plate.

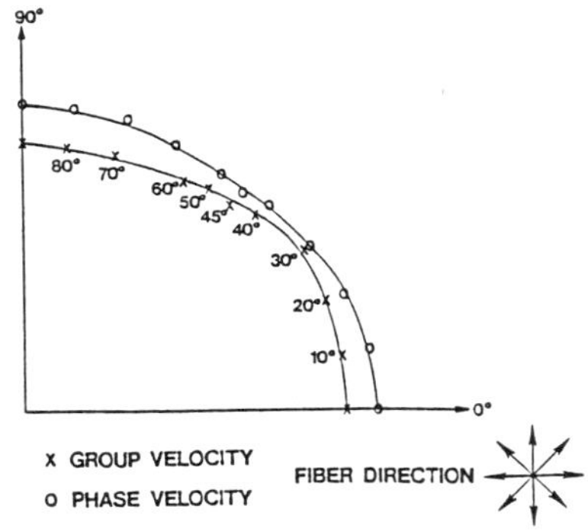

Fig. 8. Polar plot of the group and phase velocities for the quasi-isotropic plate.

direction perpendicular to the fibers is approached. Figure 7 displays the corresponding results for the plate with fibers in both the 0 and 90 degree directions. Phase velocity drops to a minimum in between the two fiber directions. At 45 degrees to the two fiber directions there are three values of group velocity. One wave turns toward the zero degree fibers, a second turns toward the 90 degree fibers, and a third propagates along the 45 degree direction between them. Because there is one extra zero degree ply, the fastest wave is in this direction with the second fastest in the 90 degree direction as would be expected. Figure 8 illustrates the results in the quasi-isotropic panel. Because this sample is more isotropic, the velocities are more constant and phase and group velocities tend to be in the same directions.

Figures 9 and 10 show the results for the unidirectional panel plotted alongside theoretically calculated group and phase velocity curves for bulk waves in a unidirectional graphite/epoxy panel. Although the waves travel at different velocities, they do share the same general shape with a significant decrease in velocity near a direction perpendicular to the fibers.

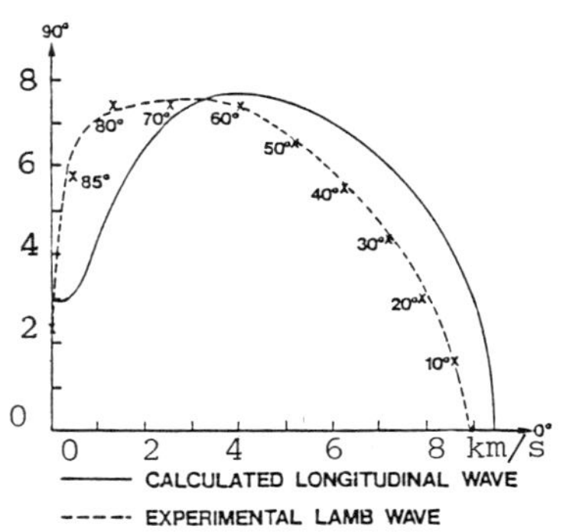

Fig. 9. Theoretical calculation for bulk waves and experimental data for Lamb waves for group velocity in the unidirectional plate.

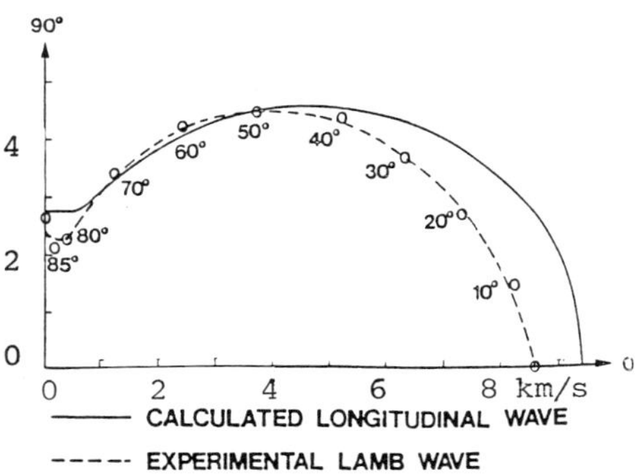

Fig. 10. Theoretical calculation for bulk waves and experimental data for Lamb waves for phase velocity in the unidirectional plate.

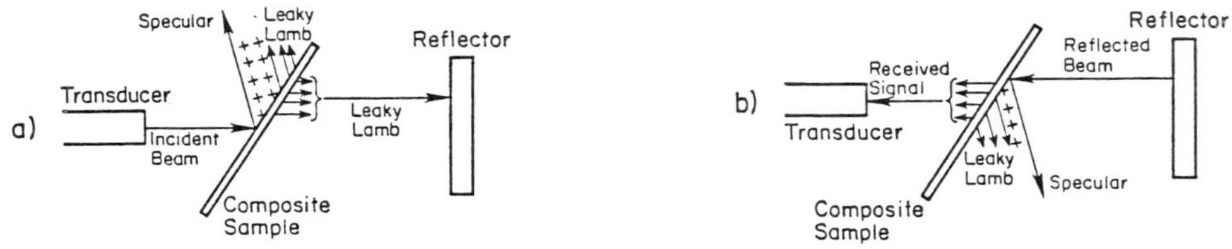

Fig. 11. Single transducer and reflector immersion technique.

SINGLE TRANSDUCER AND REFLECTOR IMMERSION TECHNIQUE

Another method of studying the anisotropy of a given material is to look at the different values of frequency times plate thickness where Lamb modes appear, at constant phase velocity, as the direction of propagation is varied in a sample. When insonifying an immersed plate at a given incident angle with a broadband signal, leaky Lamb waves within that frequency range will be excited. The leaky Lamb signal from the back face can be detected and found to contain peaks at frequencies where Lamb modes occurred in its frequency range. A single transducer and reflector method has been used to detect these back face signals and is illustrated in Figure 11.

The transducer and reflector are aligned such that their faces are parallel. The composite plate is suspended from a goniometer between them so that its face is also parallel to the reflector face. The plate can then be rotated using the goniometer to obtain any incident angle desired. In this way it is not necessary to move the transducer or reflector once they have been positioned. The plate is insonified at a given incident angle using a broadband ultrasonic signal as displayed in Figure 12(a). This causes specular and leaky Lamb signals on the front face and leaky Lamb signals on the back face. The back face leaky Lamb signals hit the reflector at normal incidence and are reflected back along their own path where they re-insonify the plate. The back face leaky signals are now on the same side of the plate as the transducer and are thus received. When this received signal is put into the frequency domain it will have peaks, such as those seen in Figure 12(b), at frequencies where Lamb waves occurred. By plotting these frequency times plate thickness values versus polar angle, the frequency dependence can be observed as propagation direction relative to the fibers is varied. Figure 13 is a plot at a 15 degree incident angle for the unidirectional plate, of the frequency times plate thickness values where Lamb modes

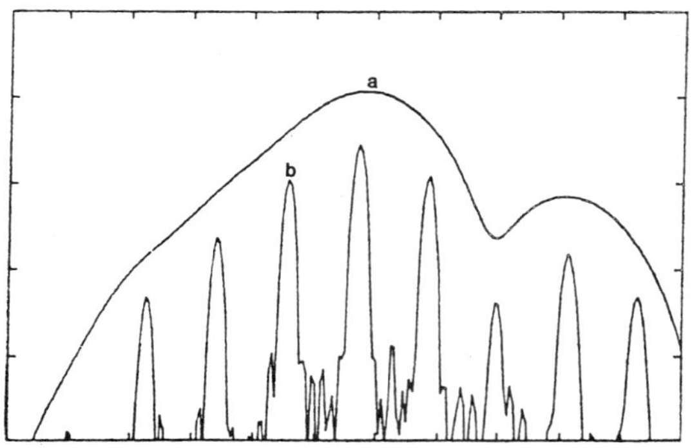

Fig. 12. Example of frequency response of (a) input signal and (b) received signal with peaks indicating Lamb modes.

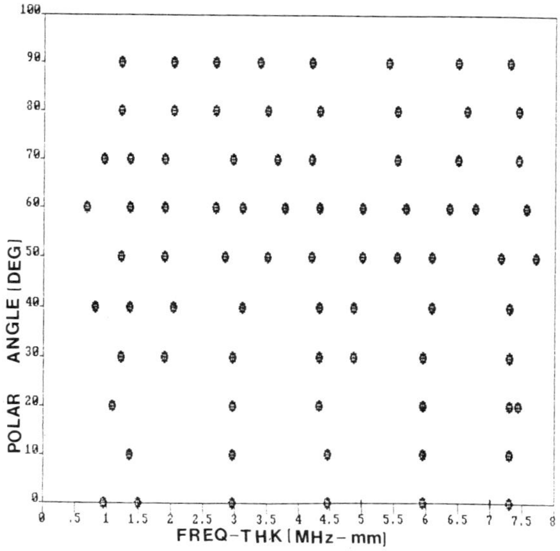

Fig. 13. Frequency dependence of Lamb modes at 15 degrees incident angle as polar angle is varied from 0 to 90 degrees.

occurred as propagation direction (polar angle) is varied from 0 to 90 degrees. The greatest number of Lamb modes seem to be excited between polar angles of 50 to 70 degrees indicating some sort of mode splitting in this region. In general, there seems to be a decrease in the frequency times plate thickness values of given Lamb modes as propagation is moved further away from the fiber direction. To illustrate this point a polar plot is shown at Figure 14 representing the frequency dependence of one mode at an incident angle of 10 degrees in the unidirectional sample. There is more than a 10% drop in the frequency at which the mode was excited as propagation is varied from parallel with, to perpendicular to, the fibers.

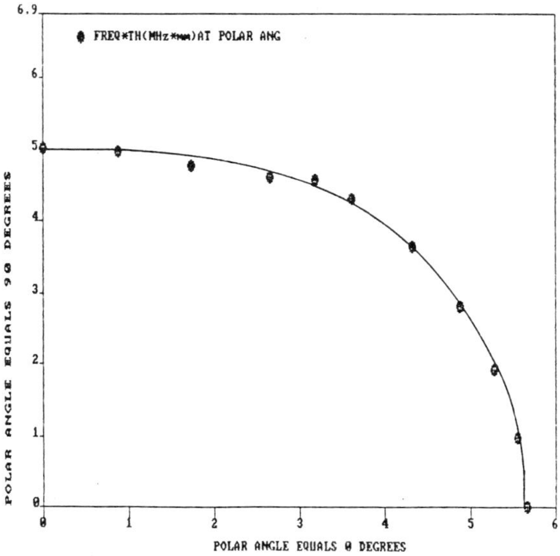

Fig. 14. Polar plot of single Lamb mode at 10 degrees incident angle as polar angle is varied from 0 to 90 degrees.

SUMMARY

This paper has examined the effect of the anisotropy of graphite/epoxy plates on the frequency dependence and velocity of Lamb waves propagating in these plates. Both contact and immersion methods have been used.

Both group and phase velocities were measured in their unique directions and found to decrease as the transmitting transducer was aligned further away from a fiber direction. The frequency times plate thickness values at which given Lamb modes occurred also decreased as the transmitter was aligned away from a fiber direction.

Although these materials present problems to conventional testing techniques, new approaches such as the use of Lamb waves may prove to be a practical alternative in the near future.

REFERENCE

1. I. A. Viktorov, "Rayleigh and Lamb Waves," Plenum Press, New York (1967).

DEVELOPMENT OF A COMPREHENSIVE APPROACH FOR ACOUSTIC EMISSION MONITORING OF CARBON-EPOXY COMPOSITES

A. Maslouhi, D. Proulx, C. Roy, and M. Tasnon

Department of Mechanical Engineering
University of Sherbrooke, Sherbrooke, Quebec, Canada

D.G. Zimcik

Communications Research Centre
Ottawa, (Ontario) Canada

ABSTRACT

The development of quantitative acoustic emission (AE) techniques has made it possible to apply signal processing analyses to recover detailed information from the detected AE signals. This paper focuses on the results of experiments following two different approaches of extracting useful information from AE signals produced in carbon-epoxy composites. Experiments to show how the two methods can be coupled to provide a comprehensive monitoring of damage accumulation in composites are described.

INTRODUCTION

The multiplicity of failure modes and the corresponding complexity of damage progression in fiber reinforced composite materials make strength prediction of these materials extremely difficult. Although a great deal of work has been performed in qualifying damage that occurs during mechanical testing, little effort has been directed towards quantitative nondestructive evaluation of composite materials concurrent with a program characterizing their mechanical properties. Such an approach has several advantages including establishing failure models that relate the measured Nondestructive Test (NDT) parameters to the instanteous mechanical properties of the composites and quantifying the damage processes responsible for the degradation of the composite material.

A wide range of NDT methods has been used to monitor internal damage progression in composites[1]. Among these, acoustic emission (AE) has been increasingly used because of relative ease of generating AE data in-situ during testing. However, careless experimentation and analysis of these data can render the data meaningless and lead to very questionable conclusions.

The work described in this paper has been concerned with extracting the AE signals produced in a carbon-epoxy composite during mechanical testing using simultaneous broad-band and resonant piezoelectric sensors. These signals were processed concurrently in real time by a microcomputer-based AE analyser and a high speed waveform digitizer in order to: a) identify failure modes (such as matrix cracking, delamination, fiber-matrix disbonding or fiber breaking) and, b) assess the contribution of individual or combined failure modes to the overall damage progression and ultimately to the actual strength.

MATERIAL AND SPECIMENS

The material used for this study consisted of a high strain graphite-epoxy (Narmco IM6 graphite fibers in a matrix of 5245C epoxy). Specimens were cut from panels of $(0°)_8$ and $(0, 90, 0, 90)_s$ laminates. The mechanical tests carried out were of two types; i) tensile to induce predominantly fiber-breaking and fiber-matrix disbonds, and ii) delamination in the Mode I interlaminar fracture (using the double cantilever beam specimen[2] to induce mainly matrix cracking, with lesser amounts of fiber-matrix disbond and fiber breakage.

Special measures were taken to minimize or discriminate extraneous A.E. noises from specimen gripping, the mechanical testing machines, etc. Particular precautions taken were:

a) A Delrin plastic coupling between the machine grips and the loading ram
b) Damping material covering of the surfaces susceptible of transmitting vibration to the specimen area.
c) Guard sensors mounted near the tab ends (see Fig. 1) to discard A.E. signals from the gripping areas.

EXPERIMENTAL APPROACH

A review of the experimental approach for a tensile test is shown in the flow chart in Fig. 1. The two analysis methods were coupled to provide a comprehensive monitoring of damage accumulation in the composite during loading.

The microcomputer-based AE system (Model 5000-B AET Corporation) counted AE events and recorded the envelope characteristics of each event namely: event counts, rise time, duration, amplitude, slope, energy, time of arrivals etc. This system could process event rates as large as 2000 events per second. The waveform digitizer (Data Precision, Model 6000) was a self-contained system that combined the high speed processing ability of a Fast-Fourier processor with the flexibility inherent in a mini-computer. Unfortunately, digitizing an AE event involved a great deal of computer time which did not permit processing all individual events which were usually generated at high rate during a normal mechanical test. However, the waveform digitizer provided time and spectral properties of AE waveforms at selected time intervals during a test. Together these two systems provided a potential means of measuring the magnitude of damage concurrently with information on the mode of damage occurring.

The characteristics of the measuring system were fixed by the frequency response of the piezoelectric sensor, the band-pass filter and the amplifier. The details of the calibration procedures for the system were described in a previous paper[3] but it can be shown that the broad-band sensor displayed a continuous dynamic response of ≈40dB over the frequency range of 0.1 to 0.8 MHz. Nevertheless, it should be emphasized that because the acoustic emissions are generally broad-band elastic transient waves at their source, the AE waveforms detected by either of the two systems were possibly highly distorted due to stress wave scattering, attenuation, internal reflections an mode conversions in the composite material. The use of resonant sensors and

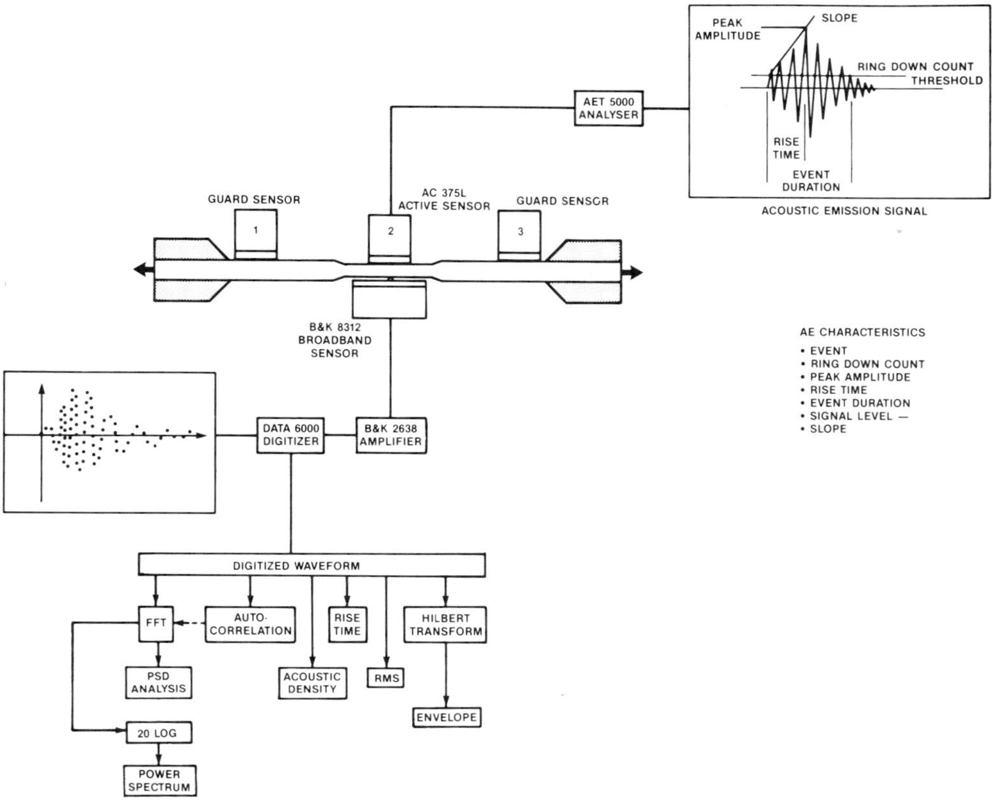

Fig. 1 Flow Chart of Experimental Program

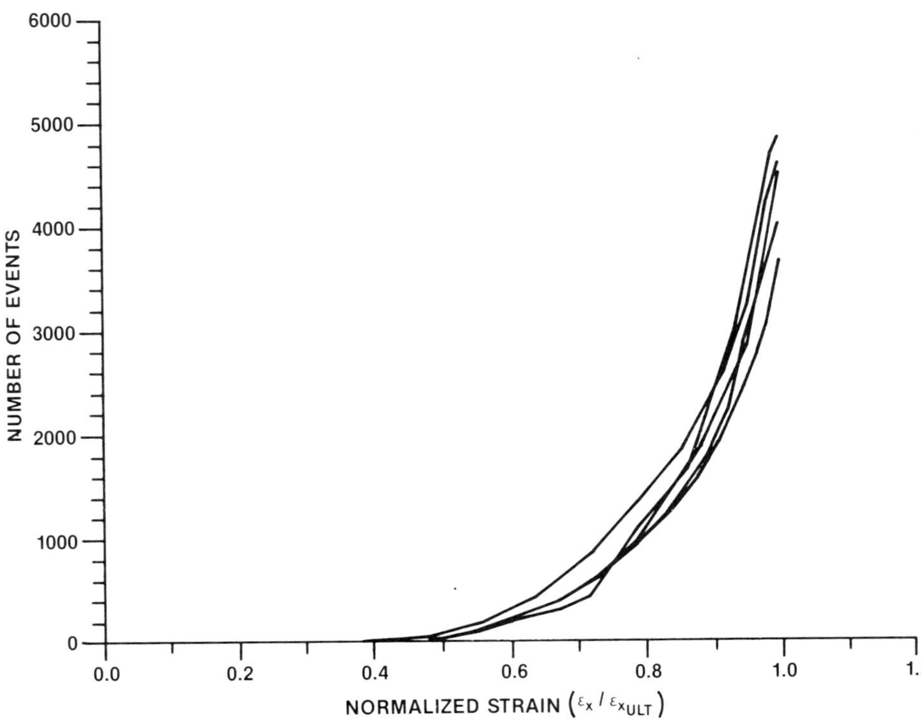

Fig. 2 Cumulative AE Count for Unidirectional Tension

analysis based on a few envelope characteristics would have the effect of distorting the detected AE signals and accordingly limit the usefulness of the information contained in the signal from any event. However, the problem was circumvented by the high speed of the AE analyser in recording each event. This second method was therefore useful in assessing the overall damage growth during a test which could be correlated to the strength degradation of the composite material. The ability to digitize wide dynamic range signals enabled the analysis of small variations within a signal and high precision measurements within the captured waveform. Therefore AE signals detected with broad-band sensors could be correlated to the computed frequency characteristics of AE sources and possibly to the various failure mechanisms in the composite materials. However, it remained to be demonstrated that for a given source process, the nature and the stacking of the composite did not modify the frequency distribution function.

RESULTS AND DISCUSSIONS

Tensile Loading of Unidirectional Material

Analysis of data from AE signals from uniaxial testing was done using several techniques. The simplest representation of data is shown in the variation of the AE accumulative event count as a function of normalized strain ($\varepsilon/\varepsilon_{ultimate}$) for five specimens illustrated in Fig. 2. In each load case the sensors to specimen coupling was precisely positioned with respect to distance between guard sensors and orientation. Accordingly, the reproducibility of the results was excellent thus suggesting that this approach could be used to predict the lifetime of a specimen under unidirectional loading. As would be expected, the first sign of damage appeared at a significantly higher strain value (≈0.4 - 0.5%) in this high strain composite system by comparison to the strain reported by Fuwa and Bunsell[4] for conventional Carbon Fiber Reinforced Plastic (CFRP) in tension (between 0.15% and 0.3%). The ultimate failure strain of the unidirectional composite occurred between 1.5 and 1.7% strain.

Damage growth during uniaxial loading is often described using a power law of the AE signal amplitude distribution first proposed by Pollock[5]. In the present work, the exponent of the power law characterizing the damage had a measured value of between 1.65 and 1.86 which was slightly lower than the value (approximately 2) reported by Valentin[6] for a conventional Carbon Fiber Reinforced Plastic.

A third AE parameter which is often used to evaluate damage growth in composite is the energy distribution up to total failure. A typical energy distribution diagram measured in test is shown in Fig. 3. Cordon and Verchery[7] reported energy diagrams showing four characteristic energy distributions which were associated respectively with matrix cracking (26-54dB), delamination (55-76dB) disbonding (77-98dB) and fiber fracture (99-111dB). However, this relationship between energy distribution and source mechanisms was not observed in the present work. In particular, although fiber fracture was an observed failure mode, the energy interval suggested above was absent. Factors such as sensor position, amplifier gain etc. rather than inherent material properties would play a major role in AE amplitude measured.

A more positive identification of failure mode is given by typical wave-forms characterizing the AE emission and their frequency content as determined from the DATA 6000 shown in Fig. 4. The evaluation of the AE spectra at discrete times during a typical tensile test is illustrated in Fig. 5. The differences between spectra were small and the majority of the observed characteristics were very reproducible. Two recurring peaks, at 240 kHz and

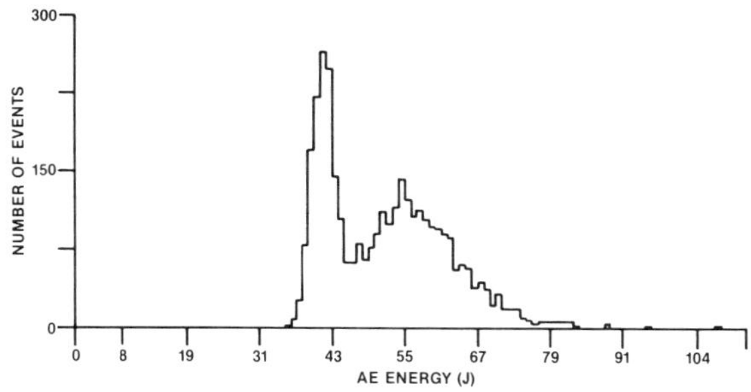

Fig. 3 Energy Distribution Histogram for Unidirectional Tension

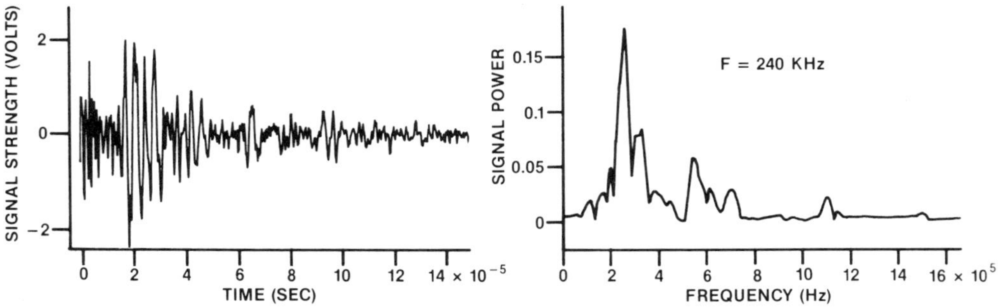

Fig. 4 Waveform and Spectral Density for Unidirectional Tension

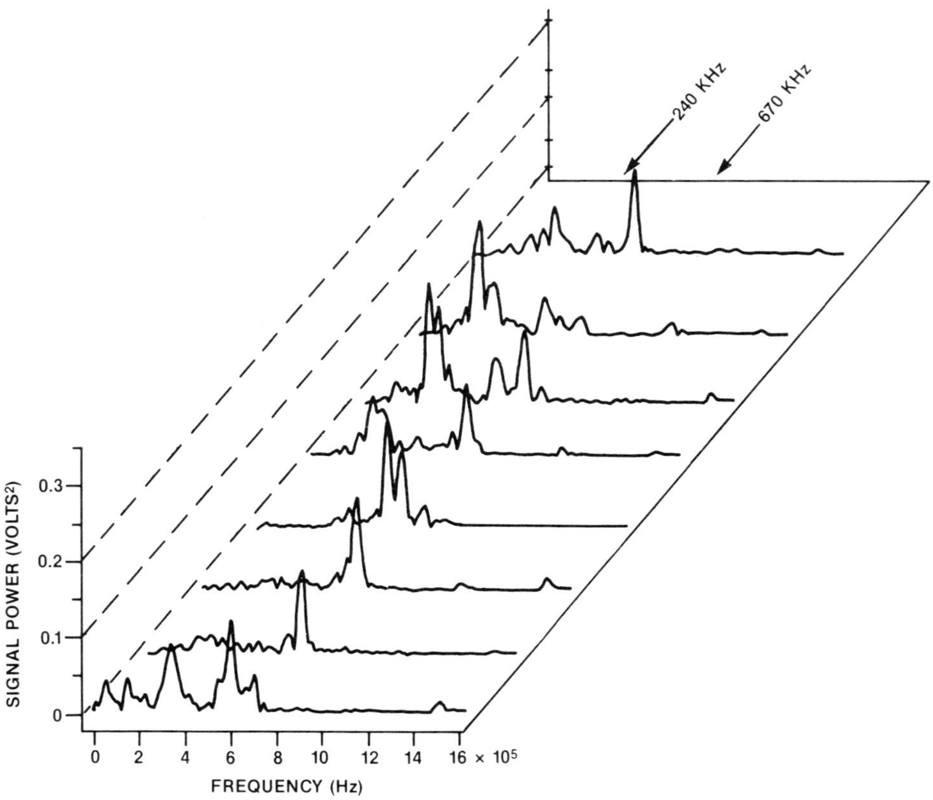

Fig. 5 Spectral Analysis of AE Signals During Unidirectional Test

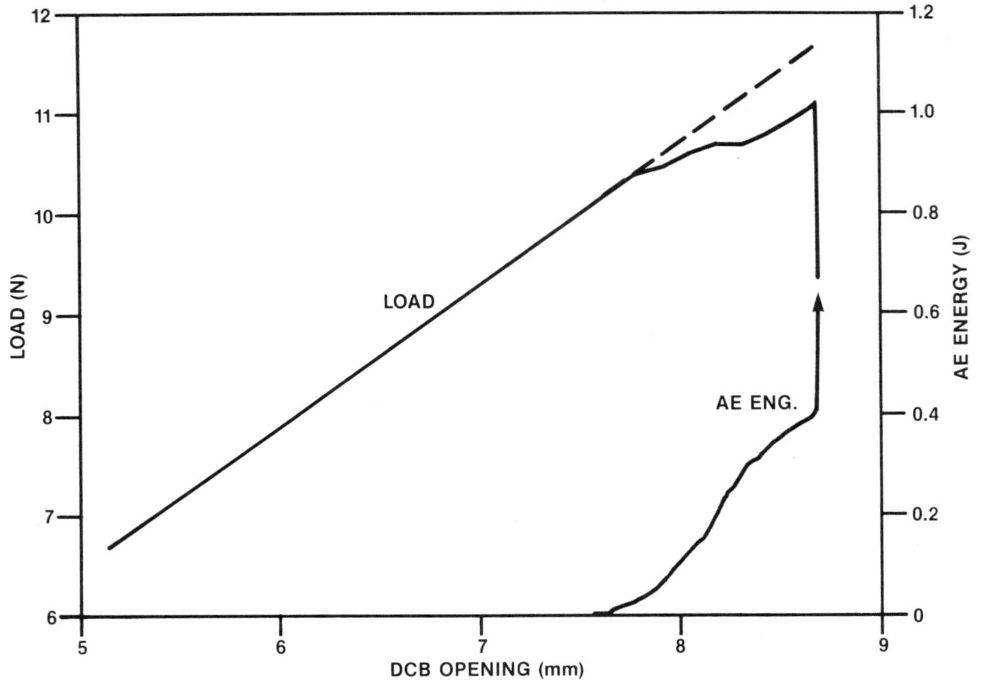

Fig. 6 Cumulative AE Energy and Load for Delamination Test

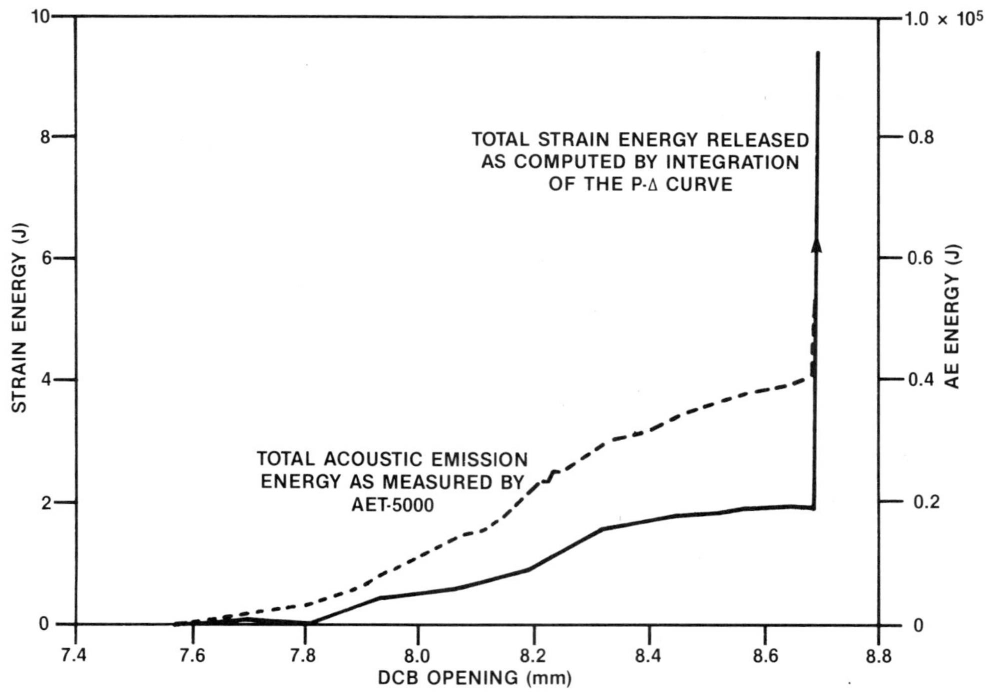

Fig. 7 Total Strain Energy and Acoustic Energy for Delamination Test

670 kHz, have in a previous study[8] been related to fiber rupture. It is however surprising that the frequency characteristic of the epoxy cracking had not appeared in the spectra which is in contradiction to previous observations of the signal analysis performed on impacted unidirectional CFRP specimens

Mode 1 Delamination Test of $(0,90°)_4$ Laminate

The initiation and growth mechanisms of interlaminar cracks in the form of Mode 1 delamination has been the object of a previous study[9]. Delamination damage first started slowly by matrix microcracking at the crack tip and subsequently lead to a rapid propagation of the main crack in the mid-plane of the specimen. Fig. 6 shows a typical measured load-deflection plot with the accompanying AE energy recorded simultaneously. It was noticed that initially, the linear loading of the specimen did not generate any AE emission but at higher load, a slight deviation of the load linearity corresponded to the generation of damage as recorded by the intensity of the AE energy emitted. The mechanical elastic energy computed (i.e. area under the load-deflection curve) and the AE energy recorded from the time delamination damage was initiated until fast crack growth occurred (at which point AE emissions saturated the AET-system) are shown in Fig. 7. The correlation between the two types of energy integration (elastic vs. acoustic) is striking and clearly suggests a total energy correlation between mechanical load and measured acoustic emission. Unfortunately, by itself two important facts are missing, viz: the scaling factor from total mechanical energy to total acoustic energy at fracture (which was also a function of the system) and the clear early indication of the onset of catastrophic failure similar to Fig. 6.

To supplement this approach, a similar analysis using the Pollock[5] power law was performed. The exponent of the power law was measured to be 3.1 to 3.4. However, as for the tensile test on the unidirectional material, information on the source mechanism that can be extracted from any of these analyses was limited.

However, more information on the source mechanism at work in the failure can be obtained from spectral analysis of the signal using the DATA 6000 digitizer. A typical A.E. waveform for the Mode I delamination failure is shown in Fig. 8. The AE waveforms were complex but each displayed similar characteristics, i.e. short duration of high frequency signal component followed by a longer lower frequency component that attenuated with time. These characteristics could be more clearly identified by application of a Hilbert Transform also shown in Fig. 8, which assisted in identifying primary AE emissions and possibly the presence of reflection in the specimen. Also, application of a Fourier Transform enabled the extraction of useable frequency characterizing the signal that might reveal the source mechanism of the AE waveforms. The evaluation of AE waveforms during a delamination test is represented in Fig. 9. Two characteristic frequencies constantly reappeared, though with variable amplitude, on the spectra; these were the 110-120 KHz and the 592-600 KHz.

The first peak corresponds to the characteristic frequency of resin cracking previously observed[1] and the second one may be associated with fiber/matrix disbond or fiber fracture (640 KHz). The repetitive nature of the AE signal characteristics verifies the discriminating power of the technique and when used in conjunction with total AE count monitoring has the potential to identify both the occurrence and mode of failure. Although not included here, the logical extension to the above is the use of multiple sensors positioned judiciously around the structure to enable triangulation of the waveforms to pinpoint the failure location.

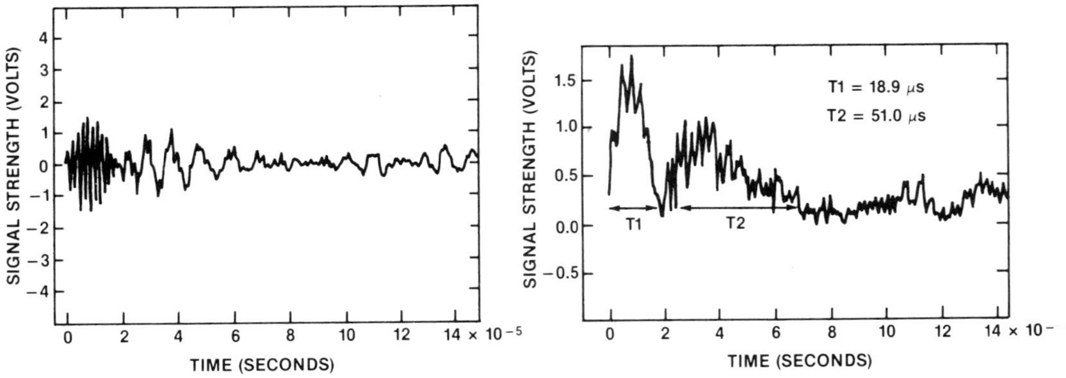

Fig. 8 Waveform and Envelope for Delamination Test

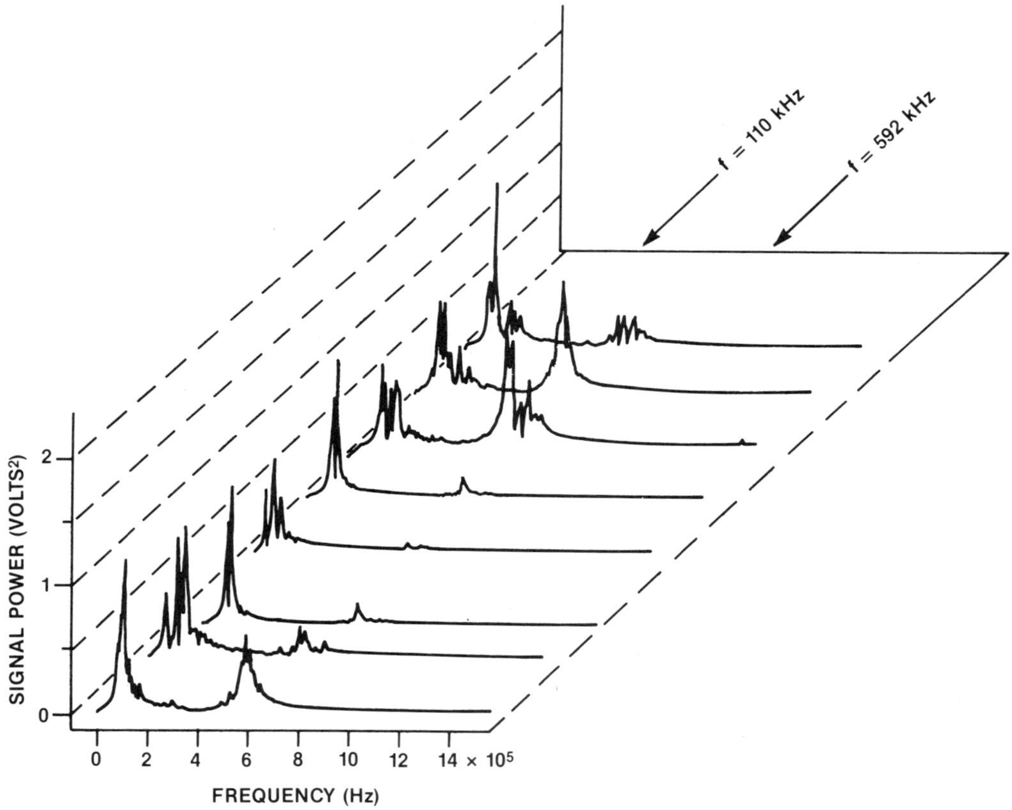

Fig. 9 Spectral Analysis of AE Signals During Delamination Test

CONCLUSION

The AE monitoring system described above which combines both quantitative and qualitative analyses of the AE emissions has the potential to provide a comprehensive damage assessment capability for carbon-epoxy systems. Measured test data for a particular high strain CFRP system showed the capability to identify the onset of failure and the mode that failure would take.

ACKNOWLEDGMENT

This work was supported in part by the Department of Communications through contract No. 37ST-36001-5-10107 and in part by the Natural Sciences and Engineering Research Council of Canada.

REFERENCES

1. Proulx D., Roy C. and Zimcik D.G., "Assessment of the State of the Art of Nondestructive Evaluation of Advanced Composite Materials", Canadian Aeronautics and Space Institute Journal, Vol. 31, No. 4, 1985 pp. 325-334.

2. Roy C., Benzeggagh M., Prel Y. and de Charentenay F.X., "The Influence of Specimen Thickness in Determining the Interlaminar Fracture Energy of Glass Fiber-Epoxy Composites", Proceedings of 39th Meeting of the Mech. Failures Prevention Group, National Bureau of Standards, Gaithersburg, May 1-3 1984, Cambridge Un. Press, pp. 205-213.

3. Maslouhi A., Roy C. and Proulx D., "Identification of Acoustic Emission Sources in Carbon-Epoxy Composite", European Materials Research Society Conference, Strasbourg, Nov. 26-28, 1985.

4. Fuwa M., Bunsell A.R. and Harris B., "Tensile Failure Mechanisms in Carbon Fiber Reinforced Plastics", J. of Materials Science, Vol. 10, 1975, pp. 2062-2070.

5. Pollock A.A.B., "Acoustic Emission Amplitude", N.D.T., Vol. 6, 1973, pp. 63-70.

6. Valentin D., "A Critical Analysis of Amplitude Histograms Obtained During Acoustic Emission Tests on Unidirectional Composites with an Epoxy and PSP Matrix," Composites, Vol. 16, No. 3., 1985, pp. 225-230.

7. Cordon A.H. and Verchery G., "Mechanical Characterization of Load Bearing Fibre Composite Laminates", Elsevier Applied Publishers, 1984, pp. 161-169.

8. Maslouhi A., Roy C. and Proulx D., "Characterization of Acoustic Emission Signals Generated in Carbon-Epoxy Composites, July 22-25, 1986, Montreal Canada.

9. de Charentenay F.X. and Benzeggagh M., "Fracture Mechanics of Mode I Delamination in Composite Materials", 4th Int. Conf. on Composite Mat., Ed. Bunsell A.D., Vol. 1, 1980, pp. 186-191.

APPLICATION OF ULTRASONICS TO THE CHARACTERIZATION OF COMPOSITES: A METHOD FOR THE DETERMINATION OF POLYETHYLENE DENSITY

Luc Piché

Industrial Materials Research Institute
National Research Council Canada
75 de Mortagne Blvd, Boucherville, Québec, J4B 6Y4

INTRODUCTION

Polyethylene (PE) is the most fundamental plastic, scientifically because of its simple structure and commercially because of its low cost and interesting balance of properties. Basically, the structure is that of long chains of $-[CH_2 - CH_2]-$ units. Actually, the details of the structure, such as the average chain length, or molecular weight and the distribution thereof, the presence of side groups (branching), etc. can be adjusted by controlling the polymerization process. This allows to fabricate different grades of polyethylene, which in turn, have specific properties. The spectrum of applications for PE is extremely broad: packaging (from garbage bags to milk pouches), household construction (from plumbing to water vapor insulation), general facilities (from gaz tubing, electrical insulation to chemical containers), automotive industry (car bumpers to gazoline tanks, wear plates, gears and bearings). In order to take advantage of the versatile character of PE, it is of primary importance for the producer and also for the processor to have means and methods of characterizing the material in a representative manner.

In this respect, it is useful to consider the intimate structure of polymers. Usually, polymers are viewed as materials marked out by the absence of long range ordering, and as such classified as amorphous. In fact, crystals may also exist in many types of polymers. Polyethylene, where the intermolecular forces are strong and the molecular structure highly regular, is a polymer where regular, organised, morphological entities (crystals) can form and coexist with a more random liquid-like (amorphous) phase.

The importance of crystallinity in determining the end-use properties of the material cannot be overemphasized. Crystallinity has a strong bearing[1] on density, elastic moduli, strength, elongation to break, impact strength, thermal conductivity, thermal expansion and hardness. On the other hand it was established that density, ρ, being closely correlated to crystallinity, was a valuable and practical criterion for the characterization of PE. This explains why the different grades of PE are most often referred to by density - very low, low, high and very high density PE (VLDPE, LDPE, HDPE and VHDPE) - even though new grades have since appeared which are better characterized by their molecular structure - linear low density PE (LLDPE) and ultra high molecular weight PE (UHMWPE), actually a medium density PE.

Typically the density for these materials ranges from 0.90 g/cm^3 to 0.97 g/cm^3, and it is usual that it be specified with an accuracy of ∓ 0.001 g/cm^3. In order to obtain such an accuracy, the most commonly used method is that of the Density Gradient Column (ASTM-1505). The basic principle of the method is sound and well established, but the technique is quite involved and tedious. For the purpose of quality or process control however, the main drawback comes from that the measurement is time consuming (anywhere from 45 mins to 2 hrs).

In a previous paper[2] we proposed that a technique based on the measurement of ultrasonic velocity would allow for an accurate, reliable, yet rapid determination of density. Here we shall describe more extended experimental results which bring us to reassert the usefulness of this new approach as a tool for quality control. Based on the measurements of both compressional and shear velocities, v_c and v_s, we shall discuss the possible origin for the behaviour of velocity, using a model for the elastic moduli that takes into account the microstructural details of the material. This will help to establish the physical foundations for the ultrasonic determination of density in polyethylene.

METHODOLOGY

It is not common practice to use ultrasonic techniques for the study of polymers[3], however, the method is practical and well documented. Polyethylene has been studied by several authors[4-8] and the various measurements tend to indicate that there exists a correlation between the velocity of sound, v, and the density, ρ. However, no systematic study has been done in the range of ultrasonic frequencies (MHz) and the nature of the correlation is neither clear nor well defined[6]. In a previous work, we established[2] that there was indeed a strong correlation between the velocity of compressional sound waves, v_c and the density, ρ, of PE. Here we shall extend our measurements and examine the behaviour of shear waves, v_s. The additional measurement is of importance since the knowledge of both the compressional velocity, v_c, and the shear velocity v_s, together with the value of density, ρ, allow to calculate the bulk modulus, B, and the shear modulus G, which for an isotropic material, are given[9] by:

$$B = \rho(v_c{}^2 - 4v_s{}^2/3) \quad (1)$$

$$G = \rho(v_s{}^2) \quad (2)$$

The actual strains involved in ultrasonic measurements are very small ($\Delta\varepsilon/\varepsilon \simeq 10^{-5}$), so the problem of strain dependence or creep, which is important for polymers, is virtually eliminated. Also, owing to various structural relaxation mechanisms in the material, the elastic properties of polymers may depend on the rate at which the strain is imposed. In polyethylene, torsion pendulum measurements[10] at 1 Hz show the existence of a relaxation process centered near 20°C. However, experience shows[7] that because of the high frequencies, ultrasonic measurements, at 20°C, are not influenced by this relaxation. The ultrasonic moduli can thus be discussed in terms of elasticity alone.

EXPERIMENTAL PROCEDURE

Samples

The measurements were made on samples which covered the whole range of commercial grades of PE. These included VLDPE, LDPE, HDPE, VHDPE, LLDPE and VHMWPE with densities ranging from 0.91 to 0.97g/cm^3. The actual samples used in the measurements are the same press-molded plaques (approximately 80 X 50 X 1.9 mm) that are normally prepared for density assessment. In most cases, the plaques were made by the resin manufacturer who also indicated the density he had obtained. The plaques were used "as were" with no further preparation; however, we repeated the measurement of density with our own density-gradient column in the standard manner. In this way, we were able to define the standard deviation for values of density obtained in different columns. We found $(\delta\rho/\rho)_{col.} = \mp 0.0007$, a value which accounts for all sources of uncertainty and gives a good idea of the reproducibility for this technique.

The ultrasonic techniques

In the present study, we used a water immersion technique because of the ease with which the measurements can be made and repeated. The samples were inserted between two highly damped transducers, one acting as an emitter, the other as a receiver, and the time of flight measured accurately. The compressional velocity, v_c, was obtained by orientating the sample with its faces perpendicular to the sound beam; the shear velocity, v_s, was found by using a different angle of incidence. The accuracy on the values of velocity was estimated to be: $v_c \mp 0.05\%$ and, $v_s \mp 1.0\%$; the overall accuracy on the moduli was in the 1% range.

Experimental Results

The results obtained in the measurement of the compressional and shear velocity, v_c and v_s, versus density, ρ, are shown in Fig. 1. The various curves correspond to theoretical modeling and will be referred to in the discussion.

The results illustrated in Fig. (1) were obtained at a frequency of 3 MHz. However, in order to investigate the possible influence of viscoelasticity, we performed measurements at other frequencies: our observations show that, indeed, within 1 or 2%, there was no dispersion in the range of 1.0 to 50.0 MHz.

Ultrasonic determination of density

In reference to Fig. (1), it can be seen that to a very good approximation, both the compressional and the shear velocities are linearly correlated to the density. This lead us to the idea[2] that the measurement of sound velocity could be used for the determination of density in PE. From a practical point of view the measurement of v_c was preferred. Using a linear model, $\rho = a\, v_c + b$, we found $a = (0.0965 \mp 0.0010) \times 10^{-5}$ g sec/cm^4 and $b = (0.7218 \mp 0.0020)$ g/cm^3. The standard error of estimate is 0.00090 g/cm^3 yielding the relative average value $(\delta\rho/\rho) = \mp 0.00095$. This value integrates errors from all sources, and yet it compares quite well with what can be expected using a density-gradient column: $(\delta\rho/\rho)_{col.} = \mp 0.0007$. Given this result, it appeared to us that the technique should effectively allow the determination of density of polyethylene with all the accuracy required for quality control purposes. Therefore, these ideas were implemented in a practical instrument, Fig. (2), which is commercially available[11].

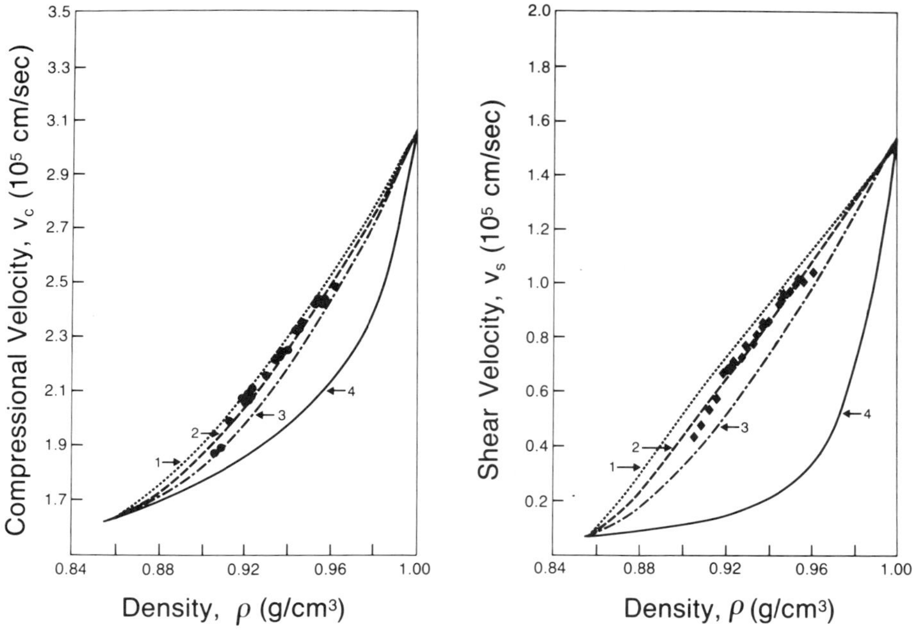

Fig. 1: Compressional (v_c) and shear (v_s) sound velocities in polyethylene versus density (ρ). The symbols represent the data. The continuous curves are from a model for the moduli (Fig. 3) and correspond to the aspect ratio (p = thickness/length) of the crystalline lamellae: 1) p = 0.020; 2) p = 0.030; 3) p = 0.050; 4) p = 1.0.

Fig. 2: Photograph of the commercial system which evaluates density from the measurement of velocity.

THE ELASTIC MODULI

Fundamentally, the velocity is determined by the modulus of the material. Now, since it has been ascertained that the velocity was not influenced by viscoelasticity, the data (v_c, v_s and ρ) can be used to compute the elastic moduli (Eq. 1 and 2). Also, PE is a semi-crystalline material, where crystalline regions coexist with a disordered liquid-like phase, and it has been established[1,12] that the value of density, ρ, is to a good approximation linearly related to the degree of crystallinity, c, (volume content for the crystalline phase) though:

$$\rho = (1-c)\rho_2 + c\rho_2 \qquad (3)$$

where ρ_1 is the density of the amorphous phase and ρ_2 that of the crystalline phase. At room temperature[1], $\rho_1 = 0.855$ g/cm^3 and $\rho_2 = 1.000$ g/cm^3. Considering PE as a composite material, for which the effective modulus depends on the relative volume contents of each phase, we present our results for the bulk modulus, B, shear modulus, G, versus crystallinity, c, in Fig. (3). In these figures, the continuous curves represent theoretical predictions which we will now discuss.

In all that follows, the properties (density, compressional and shear velocities, bulk, shear moduli) of the amorphous matrix will be written as ρ_1, v_{c1}, v_{s1}, B_1, G_1, those of the crystalline inclusion as ρ_2, v_{c2}, v_{s2}, B_2, G_2, and those of the composite as ρ, v_s, v_c, B, G; c indicates the fractional volume content of inclusions, i.e. crystallinity.

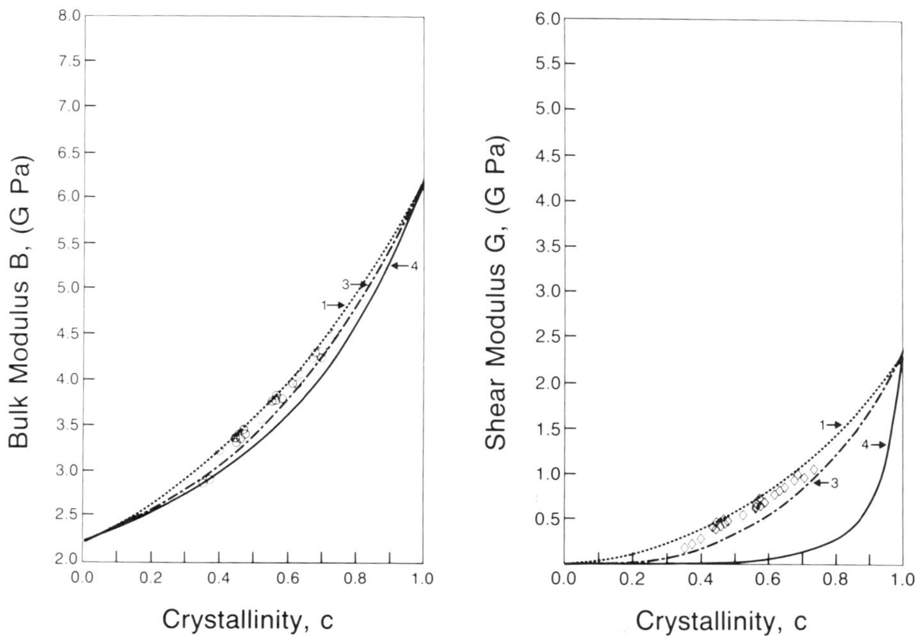

Fig. 3: Bulk (B) and shear (G) moduli versus crystallinity obtained from the data in Fig. 1. The curves are theoretical and correspond to the aspect ratio (p = thickness/length) of the crystalline lamellae: 1) p = 0.020; 3) p = 0.050; 4) p = 1.0.

The problem of calculating the effective moduli of multiphase materials, albeit not a new one, is still very much of actuality. In a companion paper[13], we reviewed some of the different models; here, for the sake of clarity, we shall recall the main key-points of the different relevant approaches. The composite material is defined as an homogeneous isotropic, infinite, elastic medium (matrix) in which are embedded a number of independent inhomogeneities which are randomly distributed and strongly bonded to the matrix. A distinction can be made between models which are based on global energy considerations and those which account for the local perturbation of the strain field in the vicinity of the individual particle. Models of the first category are the better known and have been reviewed[14]. They include the Voigt average scheme which is based on the assumption that the strain is uniformly distributed; and the Reuss average which assumes that the stress is uniform. These approaches yield extreme high (Voigt) and low (Reuss) bounds for B and G. Therefore, it was suggested that the models be combined; however, this implies the introduction of an unknown and arbitrary[14] weighing parameter. In this context, Hashin and Shtrickman[15], using variational methods were able to establish exact bounds for the moduli.

In attempt to fit our data, we used these various models. The best results were found from the Reuss (lower bound) approach, but generally speaking, the fits were not satisfactory. It was demonstrated in our companion paper[13] that such models, even though satisfactory in certain cases, are but approximations and in particular, cannot account for the microscopic details of the structure.

SPECIFICALLY POLYETHYLENE

Without going deeply in the subject, it is of interest to consider the finer details of the morphology of semi crystalline polymers. An idealized and simplified representation of the more recent concepts[1] on the subject is given in Fig. (4). The polymer crystals are constituted of long molecules which appear to fold back on themselves and align. In doing so, lamellar structures (platelets) are formed. In PE, the lamellae are of the order to 50 nm thick, 0.5 to 3 μm wide and 1 to 10 μm long, depending on the crystallization conditions. In turn the lamellae are found to give rise to a higher level of organization: spherulites. The spherulites, Fig. (4b), form by crystallization that begins at an impurity or site of local order and then proceeds radially until other growing, Fig. (4a), spherulites are met. In growing outwards from the center, each "lamella" gives rise to secondary branches with the result that, within a spherulite, the density is uniform. The space which is left out is occupied in part by bundles of molecular segments, chain folds, loops and ends that constitute the disordered, less compact, liquid-like, amorphous phase, Fig. (4c).

For the moduli, models of the second category are based on the calculation of the elastic field which surrounds each inclusion, and thus take into account the microscopic details of the material. The starting point of this complicated problem lies in the work of Eshelby[16] who produced an elegant solution for the case of an isolated particle. The inclusion, assumed to be ellipsoïdal, is characterized by its aspect ratio, p, ie. the ratio of the principal axes such that: $p > 1$ corresponds to a prolate spheroïd (fiber like), $p < 1$ to an oblate spheroïd (disk like) and $p = 1.0$ to a sphere. The solution is exact, but limited to very low concentrations. In order to treat the problem of finite volume contents, various schemes have been proposed. One of these, is the differential scheme[17] whereby the composite is progressively built up by adding single particles to the existing effective media.

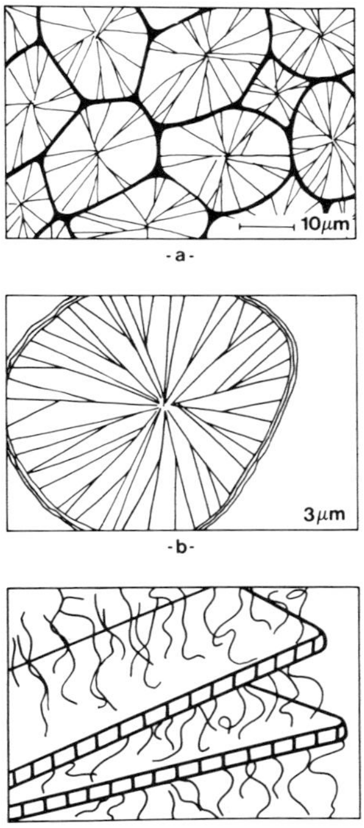

Fig. 4: Schematic representation of morphology in polyethylene. In decreasing order of hierarchy: a) the arrangement of spherulites (deformable spheres) in the bulk of the material; b) the radial and branching organization of crystalline lamellae within each spherulite; c) a 3-D view of crossing lamellae with the intersticial amorphous matter (tie loops, dangling molecules, cilias).

Since it was shown[13] that the model could be successful in describing the effects of volume content and aspect ratio, we used it as an approach to describe the moduli of polyethylene. Proceeding by trial and error we arrived at the results shown by the continuous curves in Fig. (3). The calculations using $B_1 = 2.25$, $G_1 = 0.005$ and $B_2 = 6.10$ and $G_2 = 2.30$, in units of GPa, are given for different values of p, the aspect ratio: p = 0.020 (curve label 1), p = 0.050 (curve label 3) and p = 1.0 (curve label 4).

Even though a thorough discussion requires further work, a certain number of comments can already be made. First, we note that the values found for the moduli of the crystalline phase, B_2 and G2, are consistent with the results obtained by calculating from theoretical elastic constants[18], aggregate averaged Reuss bounds. Bearing in mind that the amorphous material cannot be isolated as such, the available data suggest values for the modulus of the amorphous phase from [19, 20, 21] G_1 = 0.5 GPa down[22] to 0.003 GPa. The value we find, G_1 = 0.005 GPa, compares well to theoretical predictions[19, 23] and reflects the liquid-like aspect of the amorphous state.

Turning now to the aspect ratio p, we see that for our choice of B_1, G_1, B_2, and G_2, all the data lie within the bounds set by p = 0.050 and 0.020. These values of the aspect ratio, whereby the crystalline entities are approximated as disk-like inclusions, reflect the lamellar structure for the crystalline phase. The values found here for p are very consistant with what can be expected[22] in commercial polyethylenes.

In the whole, our approach to describe the modulus appears satisfactory. We are then in a position to calculate the compressional, v_c, and shear, v_s, velocities. The results are shown in Fig. (1). As could be expected, the experimental data is well bounded by the curves p = 0.050 (curve label 3) and p = 0.020 (curve label 1). Thus, it can be suggested that the relationship between the velocity and the density, simply reflects the interdependence that exists between the moduli and crystallinity.

CONCLUSION

We have reported measurements of both compressional and shear wave velocities in different grades of commercial polyethylene. It was established that the velocity is strongly correlated to density. In an approach to describe such a behaviour, polyethylene, was described as a composite where crystalline entities are embedded in an amorphous matrix. Existing models for the elastic moduli were referred to and evaluated. It was found that models which take into account the microstructural details of the material furnished a satisfactory description of the bulk properties. In view of this, it was proposed that the behaviour observed for the velocity has its origin in the morphology of the material and in the growing habits of its structure.

ACKNOWLEDGMENTS

The author acknowledges the close contribution of A. Hamel in performing the measurements; he also would like to express his recognition to P.Y. Kelly and D. St-Cyr of DuPont Canada who initiated this work and furnished some of the samples. The author is also grateful to C. Boisjoli and Z. Bakerdjian of Union Carbide Canada for their stimulating and friendly participation.

REFERENCES

1. B. Wunderlich, "Macromolecular Physics", Volume 3, Academic Press, New York (1980).
2. L. Piché, Polym. Eng. Sci., 24, 1354 (1984).
3. For a review: B. Hartman, "Methods of Experimental Physics", Chapter 12, Vol. 16, Part C, Academic Press, New York (1980).
4. J. Schuyer, J. Polym. Sci., 36, 375 (1959).
5. P.P. Davidse, H.I. Waterman and J.B. Westerdijk, J. Polym. Sci., 59, 389 (1962).
6. A. Levene, W.J. Pullen and J. Roberts, J. Polym. Sci., A-3, 697 (1965).
7. K. Adachie, G. Harrison, J. Lamb, A.M. North, and R.A. Pethrick, Polymer, 22, 1032 (1981).
8. B. Hartman, J. Jarzynski, J. Acoust. Soc. Am., 56, 1469 (1974).
9. E. Schreiber, O.L. Anderson, N. Soga, "Elastic Constants and Their Measurement", McGraw Hill, New York (1973).
10. K.H. Illers, H. Breuer, J. Colloid Sci., 18, 1 (1963).
11. By kind permission of Tecrad Inc., 775 av. St-Jean Baptiste, suite 105, Québec, P.Q., Canada, G2E 5G5.
12. L. Mandelkern, Polym. J., 17, 337 (1985).

13. L. Piché, A. Hamel, "Characterization of Isotropic Composites Containing Inclusions of Specified Shapes by use of Ultrasonics in the Long Wavelength Limit". Proc. This Conf.
14. T.S. Chow, J. Matl. Sci., 15, 1873 (1980).
15. Z. Hashin, S. Shtrickman, J. Franklin Inst., 271, 336 (1961).
16. J.D. Eshelby, Proc. Roy Soc., A241, 376 (1957).
17. S. Boucher, Revue M., 21, 243 (1975) and Revue M., 22, 31 (1976).
18. H. Tadokoro, Polymer, 25, 147 (1984).
19. R.H. Boyd, Polym. Eng. Sci., 19, 1010 (1979).
20. R.H. Boyd, J. Polym. Sci.: Polym. Phys. Ed., 21, 493 (1983).
21. D. Heyer, U. Buchenau, M. Stamm, J. Polym. Sci.: Polym. Phys. Ed., 22, 1515 (1984).
22. J.L. Kardos, J. Raisoni, S. Piccarolo and J.C. Halpin, Polym. Eng. Sci., 19, 1000 (1979).
23. R.J. Gaylord, Polym. Eng. Sci., 19, 955 (1979).

NONDESTRUCTIVE EVALUATION OF COMPOSITE LAMINATES BY LEAKY LAMB WAVES USING A BUBBLER DEVICE

Yoseph Bar-Cohen

Douglas Aircraft Company
McDonnell Douglas Corporation
Long Beach, California

D. E. Chimenti

Materials Laboratory
Air Force Wright Aeronautic Laboratories
Wright Patterson Air Force Base, Ohio

ABSTRACT

The application of Leaky Lamb waves (LLW) to nondestructive testing in field conditions was studied using a bubbler device. To reduce the sensitivity of LLW modes to surface curvature, the dispersion curve of LLW in graphite/epoxy laminates was used in the design of the bubbler. The results are described and discussed.

INTRODUCTION

The use of composite materials in critical aerospace structures has been increased in recent years. This trend suggests that better nondestructive evaluation (NDE) methods should be developed for detecting and characterizing defects that might be induced in these materials during production and service. A main difficulty here is the anisotropic, inhomogeneous, and layered nature of composite materials. Leaky Lamb wave phenomena have been found to provide a potential NDE method for composites.[1,2]

LLW are very sensitive to localized material variations resulting from a change in properties or the presence of a discontinuity. In this method, the amplitude of the LLW at the null zone is employed as a sensing parameter. Using these LLW phenomena in a C-scan system, the authors have been able to image defects, such as delaminations, porosity, ply gaps, and variations in resin content, throughout the entire thickness of graphite/epoxy laminates up to about 20 mm thick.

Early LLW experiments were conducted with the composite laminates immersed in water in the C-scan system with a setup of two transducers. To locate the null, the transmitted frequency was adjusted to excite LLW, and the receiver was moved back and forth horizontally to the location of minimum amplitude between the two components of the LLW phenomena. To employ the LLW phenomena in field conditions, a bubbler device was developed. In order to minimize the effect of structural angles and curvatures of tested parts on the LLW modes, the bubbler was designed to take advantage of the general characteristics of the graphite/epoxy dispersion curve. The bubbler was provided with a means of positioning the receiver transducer at the null zone by simultaneously adjusting the height of the two transducers, which in turn changes the transducers' distance from each other. Graphite/epoxy samples with controlled defects were examined to determine the bubbler's capability. The results are discussed below.

EXPERIMENTAL TECHNIQUE

The bubbler unit is designed such that the two transducers are held on a flat plate in a 15-degree angle of incidence. This angle was determined (see the dispersion curve of graphite/epoxy T300/CG914[1]) to minimize the sensitivity of the LLW modes to small changes in the angle of incidence. A schematic description of the bubbler unit is shown in Figure 1. The transducers' holding plate (top plate) is connected to a support plate (bottom plate) by a set of screws and springs. The screws are used to adjust the height of the transducers so that the receiver is positioned in the null zone. The springs are then used to maintain the desired height. A large hole at the center of the bottom plate allows the sound wave to travel directly from the transducer to the composite under test. A rubber sleeve is used to cover the opening tightly around the edges of the top and bottom plates. To prevent water leakage when the bubbler is being used, the rubber sleeve is bonded to the edges of the bottom plate. However, the top plate has been left unbonded to allow its height to be adjusted easily. A feed tube is connected to a hole in the top plate so that water can be introduced to fill the inner space of the bubbler. The transducers employed in this study were a set of broadband, 5-MHz, 12.7-mm (1/2-in) diameter devices with matched characteristics.

Tone-burst pulses were transmitted at various frequencies using a function generator. The received signals were digitized with the DATA-6000 (Analogic Corp.) at a 100-MHz digitizing rate and later were processed with the aid of the 6130 Tektronix workstation. The transducers were adjusted to optimum height at the null zone by determining the minimum amplitude in the null zone. This was accomplished by varying the transducers' height and the ultrasonic frequency until the least amplitude was measured. During this adjustment of the transducers' height (i.e., the height of the top plate), the bubbler unit was placed over a defect-free area of a $[0]_{24}$ T300/CG914 laminate. LLW spectrums were obtained by measuring the amplitude at the null zone as a function of frequency and then deconvolving the transducer response from the measured spectrum. The transducer spectral response was determined by reflecting the signal from a 1-inch-thick steel plate. The samples tested in this study were described in detail in an earlier publication by the authors.[1] These samples consisted of two laminates of $[0]_{32}$ AS4/3501-6 and $[0]_{24}$ T300/CG914 graphite/epoxy containing various controlled defects of different sizes and depths.

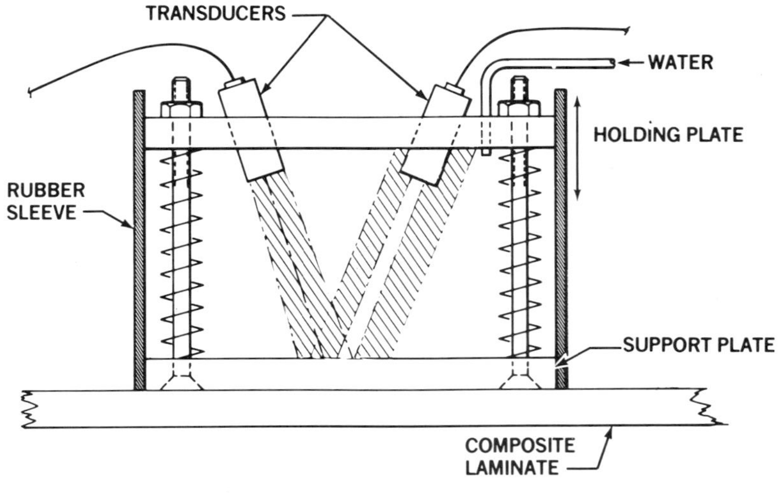

FIGURE 1. SCHEMATIC DESCRIPTION OF THE BUBBLER UNIT

RESULTS AND DISCUSSION

Responses from various laminates have been measured with the transducers oriented to transmit and receive along the fiber direction. The LLW spectral response from two AS4/3501-6 laminates are given in Figure 2. Figure 2a shows the response for a 32-layer laminate, and Figure 2b shows the response for an 8-layer laminate. From this figure, one can easily see that the frequency spacing between two consecutive minima is about four times smaller for the 32-layer laminate, as expected. Generally, results obtained with the bubbler were similar to those obtained with the sample immersed in water. The small change in modes is attributed to the difference in boundary conditions between the two coupling methods.

Using the bubbler, all the defects that were embedded in the sample were revealed through their effect on the LLW spectral response. An example of the results is shown in Figure 3, where the response from a 25-mm-diameter delamination between the sixth and seventh layers can be seen and compared to the one obtained from a defect-free area of a $[0]_{24}$ T300/CG914 laminate.

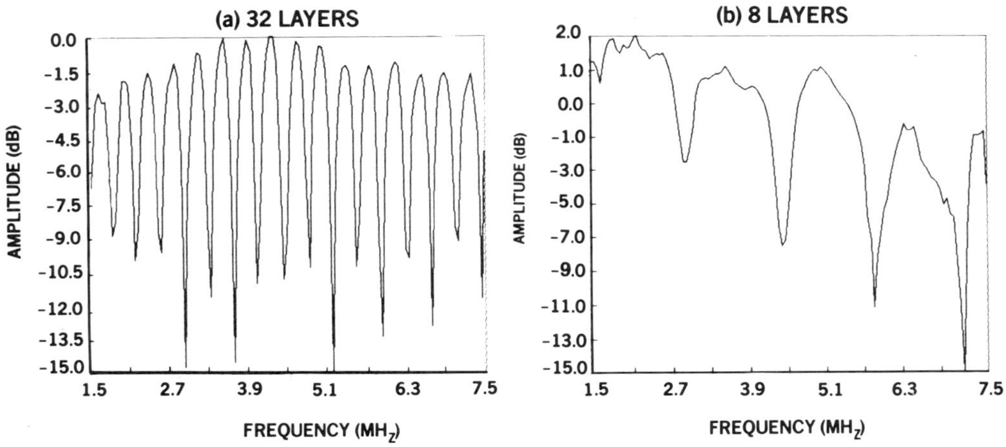

FIGURE 2. LLW SPECTRAL RESPONSES FROM DEFECT-FREE AS4/3501-6 LAMINATES

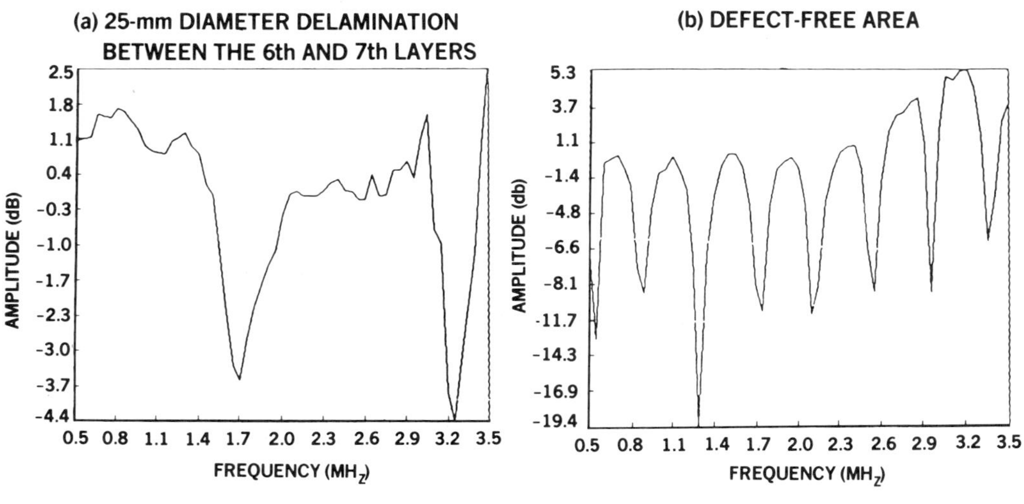

FIGURE 3. LLW RESPONSES FROM A $[0]_{24}$ T300/CG914 LAMINATE

LLW tests for defects can be performed in two ways: (a) by evaluating the amplitude changes at a given frequency of an LLW mode using tone-bursts or (b) by evaluating changes in the spectral response using a frequency analyzer and broadband pulses. Both methods were found to be very sensitive to small changes in elastic properties or plate thickness variations. Therefore, when using these methods, it is difficult to discriminate between these variations and the presence of discontinuities. To reduce the oversensitivity of the LLW phenomena to insignificant changes in plate thickness and to take advantage of the information available in a broad frequency range, a double fast Fourier transform (DFFT) process was employed.[2] Under DFFT, a periodic minimum in the frequency domain is characterized by a single value, which represents the inverse of the period of excitation, namely LLW modes. This modification provides a substantial increase in the signal-to-noise ratio (more than 20 dB) and, therefore, improved defect detectibility. The DATA-6000 signal analyzer was programmed to perform a DFFT of the received broadband pulses at the null zone. To reduce the level of noise that corrupts the signal after the DFFT process, a cross-correlation process was applied to the measured DFFT, using the defect-free signal as a reference. Examples of the response from the above sample, at a location with a 25-mm-diameter delamination between the 12th and 13th layers, as compared to a defect-free area, are shown in Figure 4. The location of the peak is related to the depth of the delamination. The closer a delamination is to the upper surface of a laminate, the smaller the value of the transformed frequency (x-axis units of the DFFT graph) of the associated peak.

Using a C-scan system that can be applied in the field, one can simplify the process of defect detection and evaluation. To demonstrate the performance of the bubbler employed on a C-scan system, the bubbler was connected to the manipulator arm of a C-scan system, and the laminate then tested. Results (see Figure 5) show that all the defects embedded in the laminate were detected, and the sensitivity to porosity was sufficiently high to allow clear imaging of its location and extent.

CONCLUSIONS

In this study, a bubbler unit was developed, taking advantage of the characteristic behavior of the dispersion curve of LLW in graphite/epoxy laminates. Defects, such as delaminations, porosity, ply gaps, and variations in resin content, were easily detected using this method. The employment of the DFFT process substantially enhances the capability to detect and characterize defects. Further, to improve the signal-to-noise levels, a cross-correlation process was applied. The response from a defect-free area served as a reference for the correlation. The study showed that, by using a field C-scan system, one can simplify the detection and evaluation process substantially.

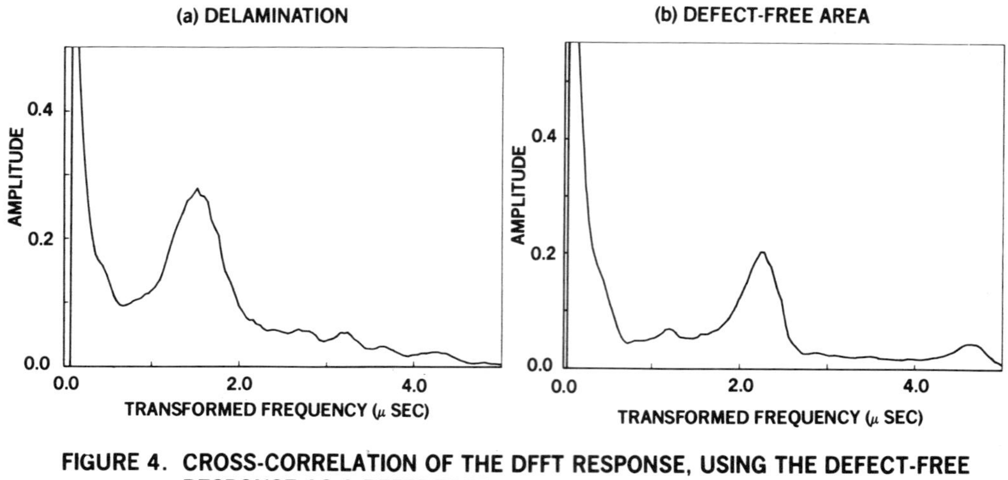

FIGURE 4. CROSS-CORRELATION OF THE DFFT RESPONSE, USING THE DEFECT-FREE RESPONSE AS A REFERENCE

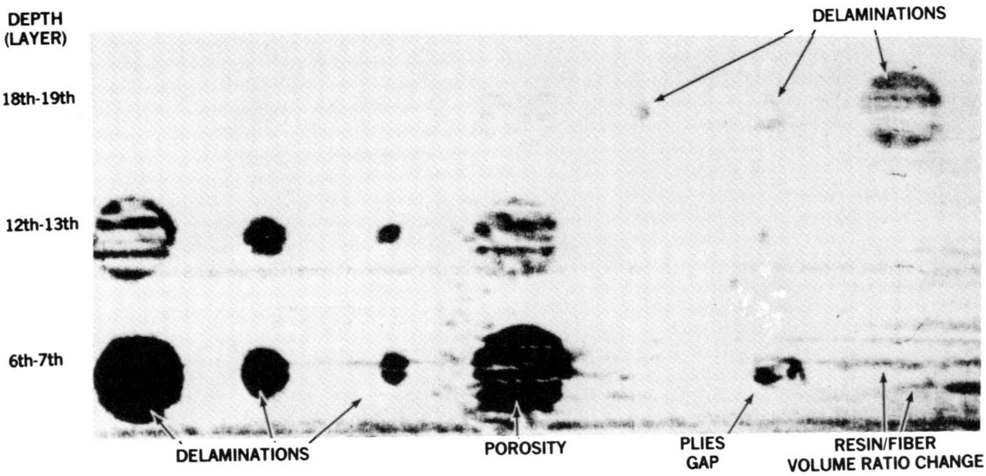

FIGURE 5. A C-SCAN OBTAINED WITH THE AID OF THE BUBBLER, TESTING THE $[0]_{24}$ T300/CG914 LAMINATE ALONG THE FIBERS DIRECTION. SCANNING SPEED IS 2.5 cm/SEC AND WITH A RESOLUTION (INDEXING) OF 0.5 mm

ACKNOWLEDGMENT

The authors would like to express their gratitude to Mr. Nishat Shah of the Douglas Aircraft Company for his support in conducting some of the measurements in this study.

REFERENCES

1. Bar-Cohen, Y., and Chimenti, D. E., Proceedings of the 11th World Conference on Nondestructive Testing, Vol. III, hosted by ASNT, Taylor Publishing Co., Dallas, Texas, (1985), pp 1661-1668.

2. Chimenti, D. E., and Bar-Cohen, Y., IEEE 1985 Ultrasonic Symposium Proceedings, B. R. McAvoy, ed., IEEE, New York, (1986).

CHARACTERIZATION OF ISOTROPIC COMPOSITES CONTAINING INCLUSIONS OF SPECIFIED SHAPES BY USE OF ULTRASONICS IN THE LONG WAVELENGTH LIMIT

Luc Piché and André Hamel

Industrial Materials Research Institute
National Research Council Canada
75 de Mortagne Blvd, Boucherville, Québec, J4B 6Y4

INTRODUCTION

The presence of an additional heterogeneous phase, dispersed in a matrix can bring about important improvements in the properties of the original material and make it appropriate for more specific purposes. Most engineering needs require the optimization of the strength to weight ratio of materials and from this point of view, polymer composites offer excellent characteristics. This factor and others such as versatility, ease of fabrication and cost effectiveness are some of the reasons why composite plastics are rapidly gaining ground over the more usual engineering materials.

The end properties of a composite will depend mainly on the properties of the matrix and of the inclusions, on the phase geometry and on the nature of the interface between the matrix and the inclusion. All of these parameters must be accounted for in the design of a composite, which corresponds to a great number of possible combinations. Because of this an empirical trial and error method of design becomes impractical. Instead it is highly desirable that the effective properties of the material can be predicted.

In this respect, being able to model the behaviour of the elastic modulus appears as a priority. Indeed, the elastic modulus is a fundamental property which basically reflects the strength of the atomic links in the solid but it is also an important and significant engineering characteristic. The elastic constants enter the description of a number of the physical properties such as stiffness, load bearing and fracture characteristics, hardness, residual and thermoelastic stress, thermal expansion, creep elongation, specific heat and thermal conductivity. For this reason, the question of the elastic moduli of composites has been given a great deal of consideration and a number of theoretical approaches have been proposed. There are however comparatively few published experimental results to which the various models can be confronted.

In this paper we shall describe and justify the use of ultrasonics for measuring the moduli, and present experimental results on model composites constituted of steel particles, of specified shapes, dispersed in a resin matrix. The results will be compared to the different theoretical predictions. The ability of these models to link the microscopic details of the composites to their macroscopic end use properties will be discussed.

MEASURING THE MODULUS OF POLYMER COMPOSITES

The usual method for measuring the elastic moduli of most engineering materials is that of stress-strain[1] experiments. Often, however, the results will depend on strain and strain-rate. For composites the strain factor is important since yielding can occur unnoticed at weak matrix particle interfaces. For a polymer, the strain rate is important because of the viscoelastic nature of the material. Both strain and strain-rate dependent effects are difficult to appreciate and their analysis can be complex[2]. In such cases, a most useful approach is provided by sound propagation techniques, where both the compressional and shear wave velocities, v_c and v_s are measured. Given the value of the density, ρ, the bulk modulus, B, the shear modulus, G, the Young modulus, E, and the Poisson ratio, ν, of an isotropic material are easily calculated[3] from the following relations:

$$B = \rho(v_c^2 - 4v_s^2/3 \qquad (1)$$
$$G = \rho\, v_s^2 \qquad (2)$$
$$E = 9BG/(3B+G) \qquad (3)$$
$$\nu = 1/2 - E/6B \qquad (4)$$

Even though it is not common practice[4] to use ultrasonics for the study of polymers, the technique is both practical, accurate and appropriate. The strains which are involved being very small ($\Delta\varepsilon/\varepsilon \simeq 10^{-5}$), strain amplitude effects are virtually eliminated. Also, the high frequencies correspond to effective measuring times which are much shorter than the dominant characteristic relaxation times of most polymers at room temperature. Thus the effect of viscoelasticity becomes negligible and the polymer matrix behaves as a linear elastic material.

In composites, the presence of inclusions causes the scattering of sound and leads to dispersion[5-7]. This can be made negligibly small by choosing a sonic frequency such that the acoustic wavelength is long compared to effective cross-section of the inclusion. When this condition is satisfied, the ultrasonic technique yields unambiguous and reproducible measurements of the unrelaxed moduli.

EXPERIMENTAL PROCEDURE

Samples

A great deal of care was exerted in the preparation of the samples. The basic constituants - a thermoplastic resin for the matrix and steel for the inclusions - were chosen because of their highly contrasted elastic properties, as this would make for a discriminating test of the various models. A compression molding technique was used and the composites were obtained in the form of plaques, 2 to 4 mm in thickness.

Samples were made with inclusions of different, but well defined geometries: spheres with an average diameter $D=(60 \mp 20)/\mu m$; disks (or flakes), equivalent to oblate spheroïds with long axis $l=(100 \mp 5)\mu m$, short axis $d = (20 \mp 5)\mu m$ and aspect ratio, $p = d/l = 0.20 \mp 0.06$; and short fibers, equivalent to prolate spheroïds with long axis, l', ranging from 500 to 2000μm, short axis, $d' = (70 \mp 20)\mu m$ and aspect ratio, $p' = l'/d' = 20.0 \mp 10.0$.

Micrographic observations showed the composites to be well compacted and free of voids and porosity; also, the inclusions were homogeneously distributed with random orientations. The micrographs revealed the presence of contiguity for volume contents exceeding, $\simeq 0.3$ in the case of spheres; and $\simeq 0.12 - 0.15$ in the case of flakes; while for fibrous composites, contiguity was present even at the lowest concentrations. Finally, samples were fractured and the fracture surfaces polished; micrographs of these areas, showed that the inclusions were left protruding while the softer matrix material had been removed during the mechanical operations. This served as an indication that the particles were well bonded to the matrix.

Measurement Techniques

The velocity measurements were made using a water immersion techniques[4]. The sample is placed between an emitting and a receiving transducer and the time of flight is measured. The compressional velocity, v_c, is obtained when the sample is oriented with its principal faces perpendicular to the direction of propagation in the water; the shear velocity, v_s, is found by orienting the sample at an angle where the compressional wave is totally reflected and only the shear wave can propagate. Digital data acquisition and computorized facilities allowed spectrum analysis of the signals to be made.

The densities ρ were determined using a picnometer technique and the results cross checked with those obtained through careful weighing and volume measurements.

The accuracy on the values of velocity was estimated to be: $v_c \mp 0.05\%$ and, $v_s \mp 1.0\%$, that on density was of the order of: $\rho \mp 1.0\%$, and the overall accuracy on the moduli in the range of 2 to 3%.

EXPERIMENTAL RESULTS

Hereafter, the properties (density; compressional and shear velocities; bulk, shear and Young moduli and Poisson ratio) of the matrix will be denoted: ρ_1, v_{c1}, v_{s1}, B_1, G_1, E_1 and ν_1; those of the inclusions: ρ_2, v_{c2}, v_{s2}, B_2, G_2, E_2 and ν_2 and those of the effective composite: ρ, v_c, v_s, B, G, E and ν. The aspect ratio will be referred to as: p, and the fractionnal volume content of inclusion material as: c.

The densities were measured, Fig. (1), and found to increase linearly with volume content, c. Thus, ρ can be described within 1% by the rule of mixtures:

$$\rho = (1-c)\rho_1 + c\rho_2 \qquad (5)$$

This behaviour suggests that the samples were near ideal compaction with low porosity; which corroborates the micrographic observations.

Preliminary velocity measurements were made, and spectrum analysis of the signals showed that the dispersive effect of sound scattering was weak at frequencies below 1.0 MHz. In view of this, the experiments were made at 0.5 MHz, which ensured that the measurements were free of dispersion. The situation corresponds to the long wavelength limit where the ultrasonic elastic constants can be identified with the more usual engineering static quantities.

The results obtained for the compressional (v_c) and shear (v_s)

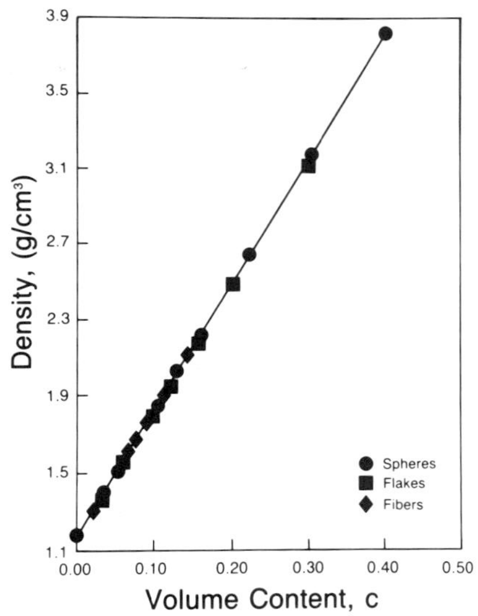

Fig. 1 Density of steel-resin composites.

velocities are shown in Fig. (2); the solid lines being guides for the eye. The data corresponds to averages obtained at different locations and orientations of the sample: the scattering of the points is within the expected accuracies indicating that the samples were homogeneous and isotropic. As can be observed, the velocity is very sensitive to the shape of the inclusions, and its behaviour cannot be easily described without the help of a model.

Using the values obtained for ρ, v_C, v_S, the bulk, B, and shear G, moduli were calculated, using Eq. (1 and 2). The results for the composites containing: spheres are shown in Fig. (3); flake-like inclusions (aspect ratio, $p = 0.20 \mp 0.06$) in Fig. (4); and fibers ($p = 20.0 \mp 10.0$) in Fig. (5). The curves correspond to theoretical predictions, which will be discussed below.

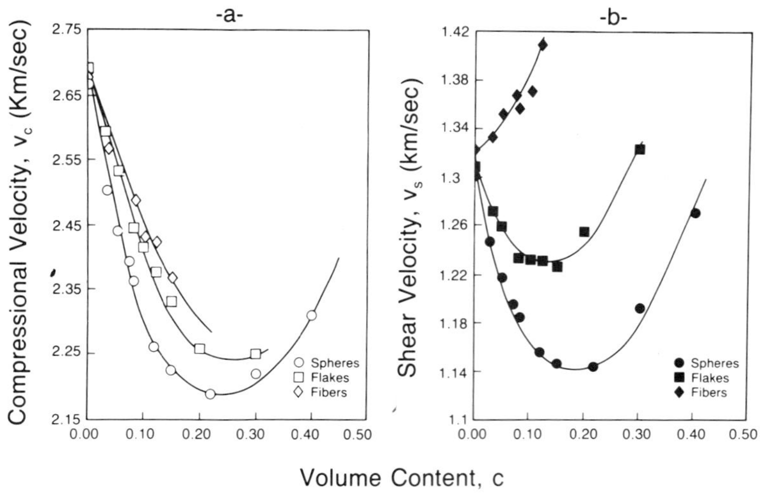

Fig. 2 Compressional and shear sound velocities versus volume content of inclusions having aspect ratios p = 1.0 (spheres), p = 0.2 (flakes) and p = 20.0 (fibers). The curves are guides for the eye.

DISCUSSION

Modelling the elastic constants

A great deal of attention has been devoted to the theoretical calculation of effective moduli of multiphase materials in general and of polymer composites in particular[8]. It is useful to distinguish models which are based on global energy considerations from those which account for the local perturbation of the strain field by the inclusion. Models of the first category include the Voigt approximation, where it is assumed that the strain is uniformity distributed (stiffness - in series model) and the Reuss average, where this time the stress is assumed to be homogeneous (stiffness-in-parallel model). The models produce extreme high (Voigt) and low (Reuss) values for the moduli, and so it has been suggested - Hill[9], Takayanagi[8] - that they be combined into a series - parallel scheme. However, this implies the introduction of an additional unknown and arbitrary parameter. On the other hand, using variational methods, Hashin and Sktrickman[10,11] arrived at exact values for bounds on the moduli.

These various schemes do not include the pertinent variables that would make them appropriate for the description of particle shape; consequently they must be adapted. In this respect, one of the better known is the Cox model [8,12], which was elaborated for paper, described as a tight web of randomly oriented semicontinuous fibers. In the absence of a matrix material as such, the load is presumed to be carried by the fibers and transferred from fiber to fiber by shear alone. Another very well known approach, is given by the Halpin-Tsai equation [8], for anisotropic short fiber composites. However, the correlation is empirical and involves parameters which need to be found through curve fitting. In another attempt, Weng and Sun[13] used classical results[14] to calculate the elastic properties of a single fiber composite.

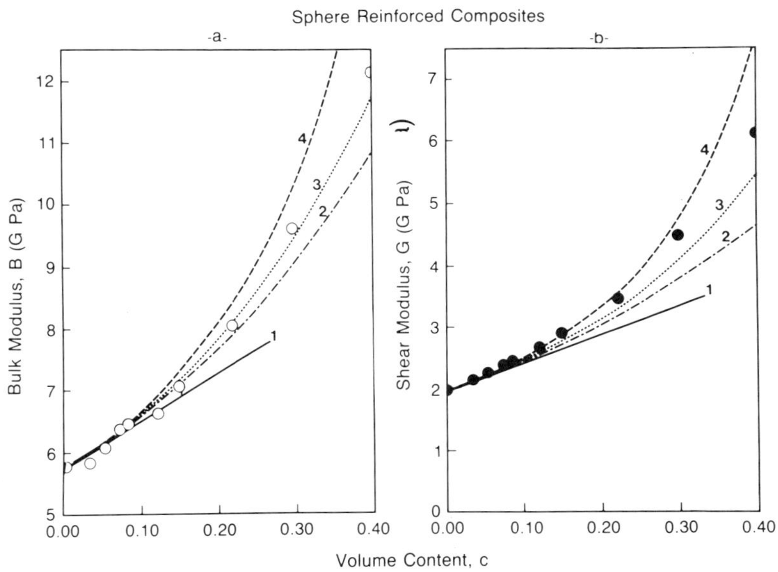

Fig. 3 Bulk and shear moduli of sphere reinforced composites. The curves correspond to different models discussed in the text.

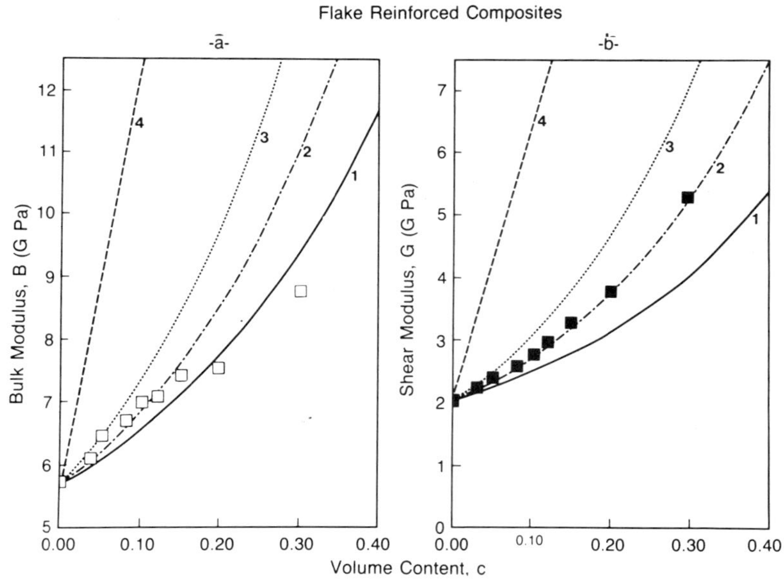

Fig. 4 Bulk and shear moduli of flake reinforces composites with aspect ratio p = 0.2. The curves are theoretical and correspond to: 1) p = 1.0; 2) p = 0.2; 3) p = 0.1 and 4) p = 10^{-4}.

These various models might, in some specific cases, provide an empirical yet, satisfactory description of experimental results. However, they cannot be translated into a unified interpretation of the

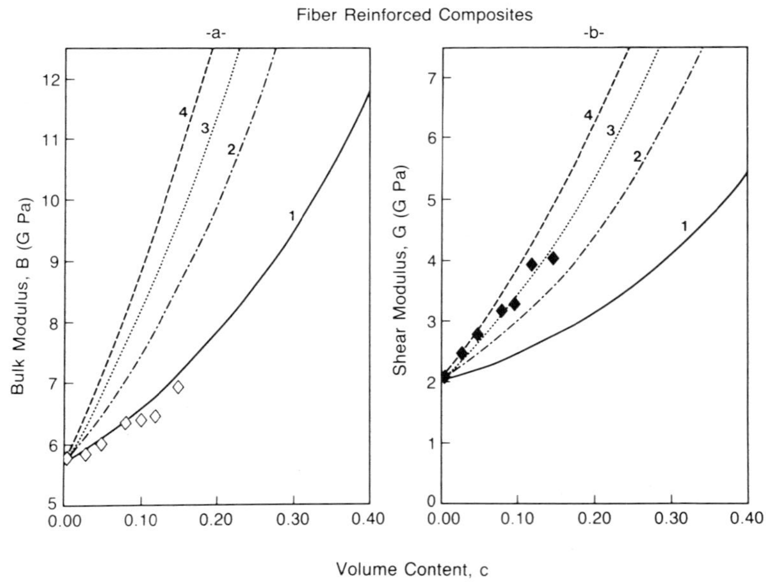

Fig. 5 Bulk and shear moduli of fiber reinforced composites with aspect ratio p = 20.0. The curves are theoretical and correspond to: 1) p = 1.0; 2) p = 10.0; 3) p = 20.0 and 4) p = 10^4.

shape factor, p. This can only be achieved through modeling the local perturbation of the elastic field in the close vicinity of the individual particle (models of the second category). The solution for the response of an isolated ellipsoïdal inclusion, of aspect ratio p, to an homogeneous static strain was given by Eshelby[15]. Other authors using different techniques, such as Greens functions to describe the problem in terms of scattering, arrive at identical results[16] in the long wavelength limit.

These approaches, however, are limited to low concentrations. A number of schemes have been proposed for the case of finite concentrations. Kuster and Toksöz[17] formulated the problem in terms of scattering and using integral equations obtained closed form expressions for the case of spherical inclusions. The same concepts of mean-fields averaging were used by Chow and Tandon and Weng[18] for unidirectionally aligned composites. Another approach is the Self-Consistant-Scheme (SCS) whereby a new particle is added to the medium which already possesses the properties of the final composite. This idea has been used by Budianski[19] and Berryman[20] for spherical particles and by Chou et al.[21] for the case of aligned short fiber composites. In a last approach, a differential scheme (DS) has been proposed[22].In this model, starting from the exact solution[15] at small concentrations, the composite is progressively built up by adding new particles. This is expressed through a system of coupled nonlinear differential equations which must be solved numerically.

We applied these microscopic models to our problem. First to the case of spherical inclusions: the results for B and G are illustrated in Fig. (3). As can be observed, Eshelby's calculations (curve 1) is indeed valid only at low volume contents. The KT model (curve 2) appears as a better approximation but at concentrations higher than 15%,it yields values which are too small. In contrast, the SCS (curve 4) gives values which are too high. In the whole the DS (curve 3) appears as the most satisfactory description of the data.

The use and application of composites containing short randomly oriented fibers or flakes, is relatively new, and both theoretical and experimental studies are scarce. In principle, the elastic properties of composites with randomly oriented inclusions can be obtained from those of aligned composites using a normalized[23] integration scheme. Actually, the calculation is simplified, if the integration over all possible orientations is performed in the zero concentration limit. Such an approach was proposed by Boucher[22] in conjunction with the differential scheme (D.S.), the theory is elegant but the detailed calculations are lengthy and will not be reproduced here.

We used this approach and the results for flake reinforced composites are shown in Fig. (4a and 4b); those for short fiber composites, in Fig. (5a and 5b). In Fig. (4) curve labels 1, 2,3 and 4 correspond to values for p of 1.0, 0.20, 0.10 and 10^{-4}; in Fig. (5) labels 1,2,3 and 4 relate to values for p of 1.0, 10.0, 20.0 and 10^4. In the case of flake-like inclusions, ($p = 0.20 \mp 0.06$), the bulk modulus, B, is rather well described by the corresponding theoretical curve for volume contents up to $c \simeq 0.12$, above which, B decreases to values expected for spherical particles. On the other hand, the shear modulus, G, is very nicely described in the whole range of concentrations. For the fiber-like inclusions, ($p = 20.0 \mp 10.0$), B is smaller than the theoretical prediction at all concentrations. In contradistinction, G is properly described by the theory, as in the case of flake-like inclusions.

The question of contiguity

Contiguity is the only factor which, while being a characteristic of some samples, is not included in the models. On the other hand, the results for the moduli seemingly show the model to diverge from the data, at values of c where contiguity was detected in the micrographs. This seams a reasonable indication that contiguity influences the elastic constants. Moreover, under the assumption that the model is otherwise satisfactory, contiguity appears to affect differently spherical particles (an increase of B and G) and flakes or fibers (a decrease of B and little influence on G).

Without going into an elaborate discussion, a number of arguments can be put forth. First, it can be easily viewed that contiguous particles when submitted to pure compression, as in the measurement of B, appear to be bonded to each other, ie. in a situation of strong interaction. Such a situation also prevails when aligned spherical particles are submitted to shear stresses polarized in the direction of alignment. On the other hand when the stresses applied to a "packet" of particles are such that the individual particles are free to slide passed each other, the interaction is weak. This prevails when flakes or fibers are submitted to shearing forces. In a situation of strong interaction, the effective aspect ratio that needs be considered is that of a "packet": p_e. For a "packet" of spheres, the broken symetry brings an increase (or decrease) of p_e away from $p_e = 1.0$, resulting in an enhancement of both B and G, which for statistical reasons is expected to be small. For a "packet" of flakes or fibers, p_e will come closer to the limiting value $p_e = 1.0$, than are the individual inclusions, resulting in a lowering of B, while G remains unaffected.

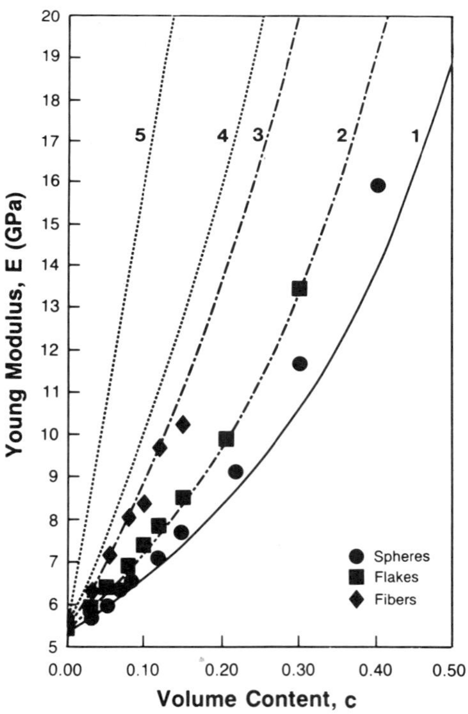

Fig. 6 Young modulus versus volume content for the composites containing spheres (p = 1.0), flakes (p = 0.20) and fibers (p = 20.0). The curves are theoretical and are given for: 1) p = 1.0, 2) p = 0.20, 3) p = 20.0, 4) $p = 10^4$ and 5) $p = 10^{-4}$.

It is instructive to represent the results in terms of the Young modulus, E, Fig. (6). As can be seen, the behaviour of E is mainly governed by that of G, such that contiguity is manifest only in the case of spherical inclusions.

CONCLUSION

Ultrasonics experiments were made on composites containing highly contrasted materials and allowed to clearly demonstrate the effects of the shape of inclusions on the affective moduli of the composites. It was established that models which account for the microscopic details of the composite provide a satisfactory description of the moduli. It can be concluded that the bulk properties of the material are governed by the actual details of its microstructure, as opposed to being governed by the mixed averaged properties of the individual constituents. The behaviour of the sound velocity is thus correlated to the morphological characteristics of the composite through careful modeling of the elastic moduli.

REFERENCES

1. See for example: "Testing of Polymers", ed. J.V. Schmitz, Interscience Publishers, John Wiley & Sons, N.Y. (1965).
2. See for example: I.M. Ward; "Mechanical Properties of Solid Polymers", John Wiley & Sons, N.Y. (1983).
3. For a review, see: E. Schreiber, O.L. Anderson, N. Soga; "Elastic Constants and their Measurements", McGraw Hill, N.Y. (1973).
4. For a review see: B. Hartman in "Methods of experimental Physics", Ch. 12, Vol. 16, part c, Academic Press, N.Y. (1980).
5. F.E. Stanke and G.S. Kino; J. Acoust. Soc. Am., 75, 3, 665 (1984).
6. V.K. Kinra, E. Kerr and S.K. Datta; Mech. Res. Commun., 9, 2, 109.
7. C.M. Sayers and R.L. Smith; J. Phys. D: Appl. Phys., 16, 1189 (1983).
8. T.S. Chow; "Review: The effect of particle shape on the mechanical properties of filled polymers", J. Matl. Sci., 15, 1873-1888 (1980).
9. For a review: O.L. Anderson "Determination and some uses of isotropic elastic constants of polycrystalline aggregates using single-crystal data" in Physical Acoustics Vol. 3-A, W.P. Mason ed., Academic Press, N.Y. (1965).
10. Z. Hashin, S. Shtrickman; J. Franklin Inst., 271 (4), 336 (1961).
11. Z. Hashin; J. Appl. Mech., 29, 143 (1962).
12. H. Cox; Brit. J. Appl. Phys., 3, 72 (1952).
13. G.J. Weng and C.T. Sun; in "Composite Materials: Testing and Design" (Figth Conference), ASTM STP 674, Ed. S.W. Tsai, American Society for Testing and Materials, pp. 149-162 (1979).
14. R. Christensen and F. Walls; J. Comp. Mater., 6, 518 (1972).
15. J.D. Eshelby; Proc. Roy. Soc. London, A241, 376 (1957).
16. A.K. Mal, L. Knopoff; J. Inst. Maths. Applics., 3, 376 (1967).
17. G. Kuster, M.N. Toksöz; Geophysics, 39, 587 (1974) and Geophysics, 39, 607 (1974).
18. G.P. Tandon and G.J. Weng; Polym. Compos., 5, 327 (1984).
19. B. Budianski; J. Mech. Phys. Solids, 13, 223 (1965).
20. J.G. Berryman; J. Acous. Soc. Am., 68, 1809 (1980).
21. T.W. Chou, S. Seicki and M. Taya; J. Compos. Mater., 14, 178 (1980).
22. S. Boucher; Revue M., 21, 243 (1975) and Revue M, 22, 31 (1976).
23. Y. Takao, T.W. Chou, M. Taya; J. Appl. Mech., 49, 536 (1982).

CHARACTERIZATION OF COMPOSITE AND ADHESIVE CURE BY ULTRASONIC WAVES

S. I. Rokhlin

Department of Welding Engineering
The Ohio State University
Columbus, Ohio 43210

INTRODUCTION

Thermosetting resins are widely used as matrices for advanced composite materials and structural adhesives. Fabrication of composite materials or adhesive joining consists of curing (polymerization), under proper thermal and pressure conditions, of resin-impregnated fiber fabrics. During the curing reaction, the thermoset resin transforms from the visco-liquid state to a gel and then vitrifies to the gelled glass. This transformation is accompanied by strong changes of the viscoelastic properties of the material which are related to the molecular network. Therefore measurements of velocity and attenuation of ultrasonic waves during cure may give important information on the extent of the thermoset curing reaction and the mechanical properties of the material.

The application of bulk longitudinal ultrasonic waves for polymer cure monitoring was first described by Sofer and Hauser[1] and then by Papadakis[2]. Lindrose[3] shows that group velocity and attenuation for longitudinal and shear waves change simultaneously during cure. Rokhlin et al.[4] studied the frequency dependence of the phase velocity and attenuation of ultrasonic waves continuously during the cure of the epoxy resin.

While bulk ultrasonic waves may be successfully used for the study of composite cure their application to the evaluation of adhesive joints has some difficulties which are related to the insensitivity of longitudinal waves to the adhesion between polymer and substrate.

To measure properties of the adhesion between the adhesive and the substrates, and to monitor the adhesive cure, one must induce shear deformation on the interface. This can be achieved by the use of interface waves. The application of interface waves for insitu study of curing of thin adhesive films was described by Rokhlin et al.[5,6].

In this paper we first describe our study of the temperature behavior of the velocity and attenuation of longitudinal waves at different stages of the cure reaction. Then we describe a simple technique which may be used for adhesive cure monitoring.

EFFECT OF TEMPERATURE ON VELOCITY OF LONGITUDINAL ULTRASONIC WAVES IN EPOXY RESIN

Experimental Procedure

A steel experimental cell was used which consisted of two delay buffer rods mounted in a cylindrical support. The gap between buffer rods forms the experimental cell which was filled with the prepared epoxy mixture after temperature stabilization of the sample. The homemade ultrasonic transmitter and receiver were coupled to the delay buffer rods. The measurements were performed with the Epon 815 epoxy resin (by Shell) which is widely used as a matrix of composite materials. The N-Aminoethylpiperazine was taken as curing agent (mixing ratio 4:1).

For measurements we used narrow band tone burst ultrasonic signals. The relative changes in the phase velocity of the longitudinal wave due to curing of the epoxy resin in the experimental cell were determined by measuring the phase shift of the rf signal filling the received ultrasonic pulse. The amplitude measurements were made by a calibrated attenuator. Due to large velocity changes during curing and therefore changes of the acoustic impedance the transmission loss factor through the epoxy was corrected by taking into account changes of the coefficient of transmission through buffer-epoxy T_{BE} and epoxy-buffer T_{EB} interfaces. The changes in density of the epoxy resin during curing are about 1-2% and their effect on the attenuation measurements is negligible. The specimen and the transducers were in a thermostatted environment (within ±1°C). The temperature is measured by means of miniature thermocouples. The thermocouple was immersed in the adhesive.

The viscoelastic properties of the resin change during the curing (cross-linking) reaction. To measure temperature behavior of the ultrasonic velocity the epoxy resin was quenched at different stages of the curing reaction (liquid nitrogen was used for cooling). The velocity and attenuation were measured as functions of temperature during heating of the sample back to the cure temperature.

Results and Discussion

The dependence of the longitudinal ultrasonic velocity on the time of cure is shown in Fig. 1 for several temperatures. The ultrasonic velocity first passes through a minimum which corresponds to a minimum of viscosity in the initial stage of curing and then increases monotonically during solidification of the resin. The corresponding attenuation data are shown in Fig. 2. Attenuation also passes through a shallow minimum in the initial stage of curing, followed by a strong maximum. The positions of the attenuation maxima are indicated by arrows on the velocity data (Fig. 1).

There is no adequate microscopic or at least phenomenological theory which can explain the dynamic chemo-rheological behavior of cured epoxy resin. The relaxation theory, which was first applied by Rokhlin et al.[5] and later by Hahn[7] to describe the behavior of the ultrasonic velocity and attenuation during cure, meets difficulties in consistently describing the temperature behavior because the spectrum of relaxation times and its dependence on time of cure are unknown.

For low frequency mechanical measurements, the data were systematized by Gilham[8] who introduced the Time-Temperature-Transformation (TTT) diagram. This diagram describes curing behavior in time of curing - temperature of curing coordinates. The mechanical frequency was not included in the model in this stage. It is not yet clear in which way the (time-superposition) WLF theory may be included in the TTT phenomenology. There are experimental and theoretical difficulties in understanding the effect of temperature on both

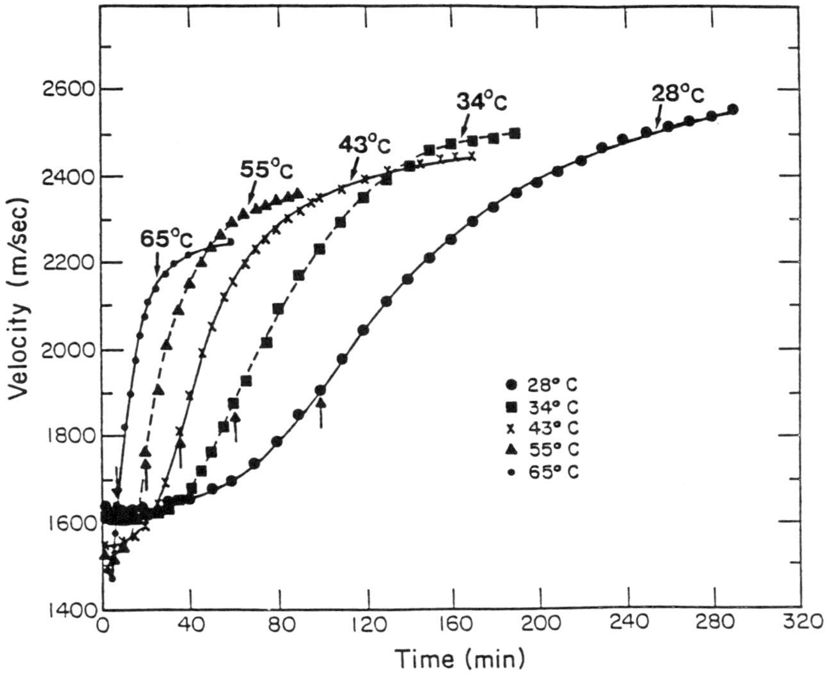

Fig. 1. The phase velocity of the ultrasonic wave versus time of curing at different temperatures.

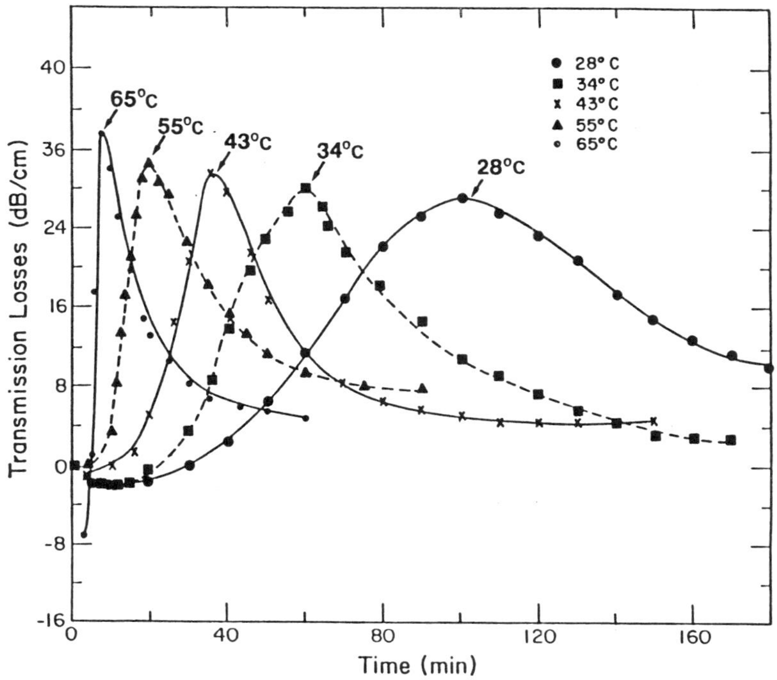

Fig. 2. The attenuation versus time of curing at different temperatures.

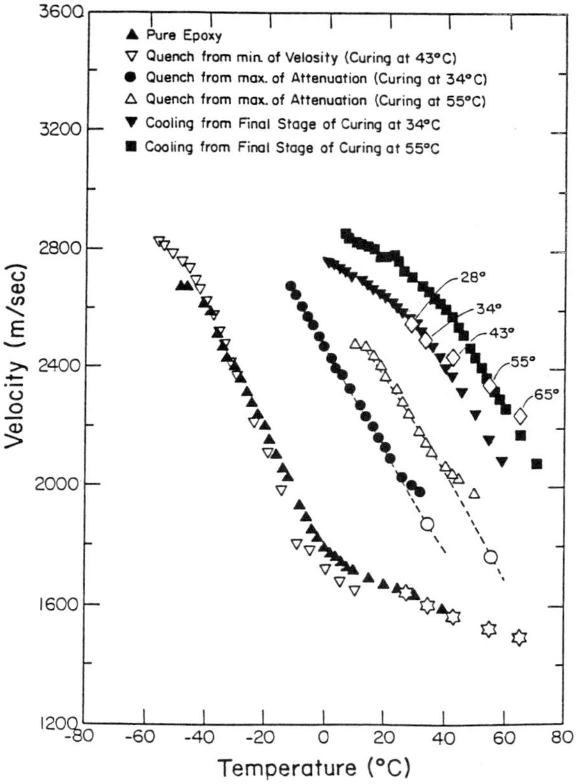

Fig. 3. The phase velocity as function of temperature at different quenched stages of cure.

the rate of the cure reaction and the viscoelastic properties of the resin at any fixed stage of the reaction.

To separate the above two effects of temperature we quench the reaction at different stages, thus fixing the chemical structure of the polymer. The velocity measurements were then done during slow heating of the sample. The experimental results are presented in Fig. 3. The left line with solid triangles corresponds to the pure epoxy (before mixing with hardening). This behavior is typical[9] for the transformation of a low viscosity liquid to the glassy state. The attenuation data (solid squares), shown together with the same velocity data (in this case shown by solid circles), are given in Fig. 4. The maximum of attenuation at around -10°C indicates a dynamic glass-transition temperature. A static glass-transition temperature is observed at about -45°C (at this point the acoustic signal disappeared due to shrinkage of the polymer and its partial separation from the buffer). This shift (35°C) in glass-transition temperature is reasonable for frequency changes of order $10^6 \div 10^7$. Above 20°C the initial epoxy resin is in a liquid state. The values of the velocity at zero time from Fig. 1 are shown by open stars in Fig. 3 and Fig. 4.

The velocity-temperature data corresponding to the quench from the velocity minimum (curing at 43°C, Fig. 1) are shown by open triangles in Fig. 3. These data are very close to the values for the initial epoxy resin. The next two lines from the left correspond to the data for the polymer state at the maxima of attenuation for curing at 34°C and 55°C (Fig. 2). Large open circles correspond to the velocity data (Fig. 1) from which the quench was

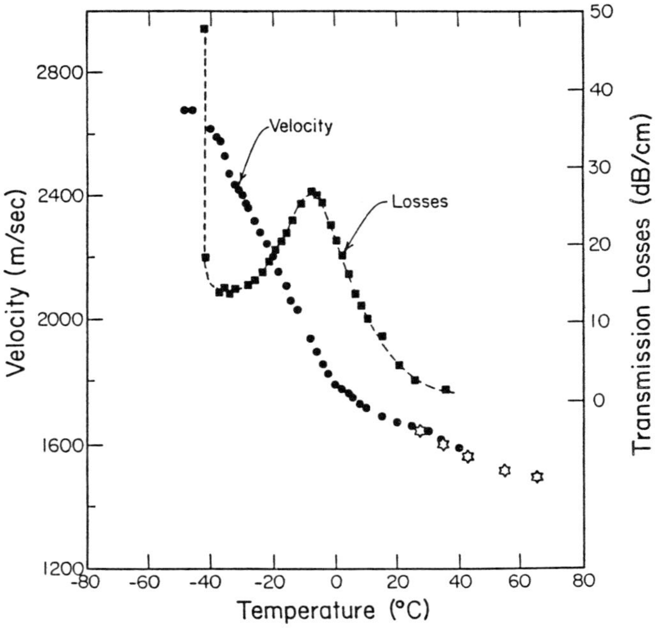

Fig. 4. The phase velocity and attenuation data as function of temperature for epoxy resin before mixing with curing agent.

done. The data in Fig. 3 may be extrapolated to these points. At the high temperature end of these curves the experimental points begin to deviate from the straight line, which is explained by the beginning of the curing reaction (recall that measurements were done from lower to higher temperature).

The two curves on the right in Fig. 3 (solid triangles and solid squares) represent the velocity-temperature data for epoxy resin with crosslinked structure corresponding to the final stages of curing at 34°C and 55°C. The open squares on these curves show velocity data in the final stages of curing from Fig. 1 (the number shows the temperature of curing). It may be useful to point out that the curing reaction stops when the thermoset achieves the crosslink structure having static glass transition temperature equal to the temperature of curing.

The velocity values for 43°C and 65°C are slightly above the curves for the resin cured at 34°C and 55°C. This means that the crosslinkage network is not fully developed at these temperatures. These last two curves also explain why in Fig. 1 the velocity changes from liquid to solid state are less at higher temperatures of curing in spite of high crosslink density at these temperatures. These velocity changes are precisely equal to the velocity difference between the points shown by open stars and open squares at corresponding temperatures in Fig. 3.

Another important observation which can be made from Fig. 3 is that the velocity-temperature curves are very nearly parallel with each other. This may help to make an evaluation of reaction kinetics from Fig. 1. As was discussed above, the temperature of curing affects the value of the velocity by changing the reaction kinetics and by affecting the viscoelastic state for a given time of cure. Therefore the velocity values cannot be compared directly from Fig. 1, since points with the same velocity on different kinetic curves will correspond to different structures. By reducing the velocity data to the same temperature using the results of Fig. 3 the data of the kinetic curves at the same velocity will correspond to the same network structure.

ADHESIVE CURE MONITORING BY LONGITUDINAL WAVE OBLIQUELY INCIDENT ON BONDED INTERFACE

Statement of the Problem

Bulk longitudinal waves are insensitive to the existence of a thin liquid layer at the interface which exhibits no shear strength resistance. To measure properties of the adhesion between the adhesive and the substrates, and to monitor the adhesive cure, one must induce shear deformations on the interface. Such shear deformations may be induced by a longitudinal wave obliquely incident on the bonded interface.

Two basic boundary conditions are possible on a solid-state interface: 1) the rigid boundary condition, which consists in continuity of tractions and displacements on the interface; and 2) the slip boundary condition, which consists in continuity only of the normal traction and the displacement, and assumes vanishing of tangential traction components. Such boundary conditions correspond to very low shear force resistance and may model absence of adhesion between adhesive and substrates. Such boundary conditions are realized when a low viscosity liquid is placed between two solids. When the liquid adhesive is applied between two solids the boundary condition will be sliplike. During curing (solidification) of the adhesive the boundary conditions change from almost slip to almost rigid. Reflectivity of the obliquely incident ultrasonic wave depends on the boundary condition and therefore the value of the reflection coefficient may be used for adhesive cure and bond evaluation.

As an example the calculated coefficient for reflection from an interface with the slip boundary condition as a function of the incident angle is shown in Fig. 5. The upper and lower substrates are steel. It is seen that for the slip boundary condition at the incident angle 68° the reflection coefficient is maximum.

The coefficient for reflection from an interface with rigid (perfect) boundary conditions is equal to zero (recall that both media are similar). This means that at this angle 68° will occur the maximum change of the reflection amplitude when the boundary conditions change from slip to rigid.

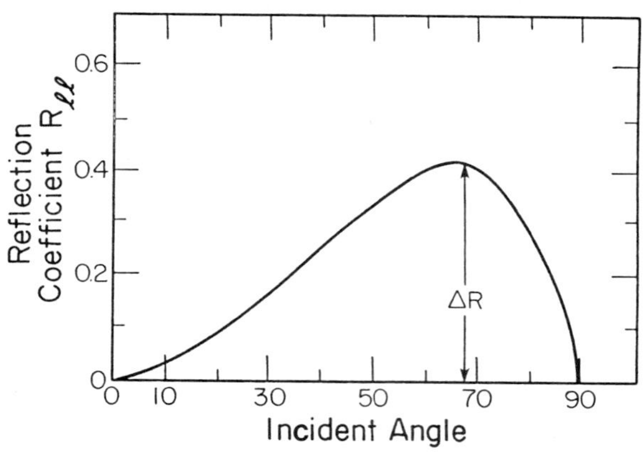

Fig. 5. Longitudinal wave reflection coefficient from slip interface between two similar solids (steel).

Such a transition of boundary conditions occurs during the polymerization of the adhesive film on the interface (conversion of the adhesive from the liquid to the solid state).

Experimental Procedure

The experimental sample was made from steel substrates bonded by a thin adhesive layer. The upper substrate was shaped in such a way that the incident ultrasonic beam makes an angle of 68° to the interface. To preset the thickness of the adhesive layer three micrometer screws were mounted in the upper substrate. We also used an upper substrate made from plexiglass. In this case the incident angle $\theta = 55°$ was selected. The sample was designed as a lap shear specimen to measure the strength of the joint after ultrasonic monitoring of the adhesive curing. The measurements were performed at the frequency 1.3 MHz at different temperatures within the 25 ÷ 55°C temperature range.

Experimental Results and Discussion

The dependence of the amplitude of the reflected signal, in dB, on the time of adhesive polymerization is shown in Fig. 6 for metal-metal bond. The thickness of the adhesive layer is taken as a parameter. The binder used was Epofix, produced by Scientific Instruments, Denmark (mixing ratio of 5:1). The temperature of measurement was held at 29°C during this experiment. It is seen that for thinner adhesive layers, the changes of amplitude of the reflected signal are larger. That means that the thinner the adhesive layer the better the conditions of slip contact (liquid adhesive) and rigid contact (solid adhesive) are satisfied. In general, the sensitivity of the method is higher for smaller ratios h/λ_t, where λ_t is the wavelength of the shear wave in the fully cured adhesive. In our case, the thickness of the adhesive h is 10μ , and h/λ_t is about 0.01.

During the curing of the adhesive both the velocity and the attenuation of the ultrasonic wave change as a function of time. The question can be raised, what kind of changes in the adhesive elastic (velocity) or inelastic (attenuation) properties correspond to changes of the coefficient of reflection from the interface? To clarify this problem we simultaneously measured longitudinal ultrasonic velocity, attenuation and changes of the reflection

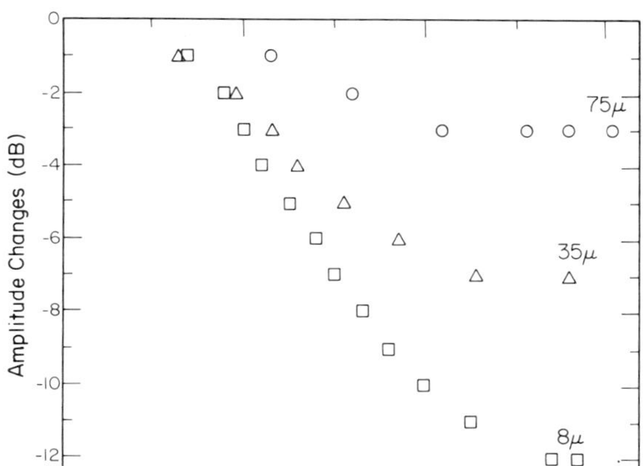

Fig. 6. The dependence of the amplitude of the reflected signal on the time of adhesive curing for different thicknesses of the adhesive.

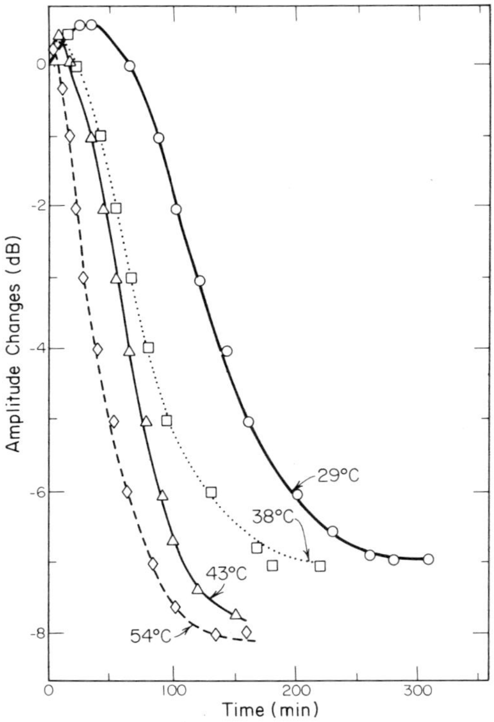

Fig. 7. Changes of the reflection amplitude from the interface during curing of the adhesive for different temperatures.

coefficient at several temperatures. The measurements were performed with Epon 815 epoxy resin (by Shell) with N-Aminoethylpiperazine curing agent (mixing ratio 4:1). This epoxy resin has wide use as a matrix for composite materials and structural adhesives.

The results for the velocity changes and the attenuation data are analogous to that described in the previous section. The changes of the amplitude of the reflection coefficient are shown in Fig. 7 (adhesive film thickness 32μ). Comparison shows that changes of the amplitude of the refelcted signal occur simultaneously with the velocity changes. The maximum in the reflection coefficient in the initial stage of curing corresponds to the velocity minimum (this minimum is accompanied by a decrease of the viscosity in this stage of the reaction).

We can conclude that the most important factor influencing the reflection coefficient is the modulus of the adhesive layer. But it is also obvious that the adhesion properties of adhesive to substrates affect the reflected signal. The loss of adhesion between the adhesive and at least one of the substrates is similar to a slip contact and therefore the reflected signal will behave as if reflected from an interface with slip contact. Another important conclusion can be drawn from these experiments: the reaction kinetics (the adhesive solidification as function of time at different temperatures) in thin adhesive layers and in bulk adhesives for the above-discussed cases are similar.

It is useful to compare the technique discussed here with the interface wave (IW) method[5,6]. The IW method has higher sensitivity to the interface properties, and due to the simple relation between the interface wave velocity and the shear modulus of the interface film[5], provides a means for shear modulus measurement. But there are some difficulties in excitation of the interface wave. The main advantage of the technique discussed in this paper is its simplicity.

REFERENCES

1. G. A. Sofer and E. A. Houser, A new tool for determination of the stage of polymerization of thermosetting polymers, J. Polym. Sci. 6:611 (1952).
2. E. P. Papadakis, Monitoring the moduli of polymers with ultrasound, J. Appl. Phys. 45:1218 (1974).
3. A. M. Lindrose, Ultrasonic wave and moduli changes in a curing epoxy resin, Exper. Mech. 18:227 (1978).
4. S. I. Rokhlin, D. K. Lewis, K. F. Graff, and L. Adler, Real-time study of frequency dependence of attenuation and velocity of ultrasonic waves during the curing reaction of epoxy resin, J. Acoust. Soc. Am. 79:1786 (1986).
5. S. I. Rokhlin, M. Hefets, and M. Rosen, An elastic interface wave guided by a thin film between two solids, J. Appl. Phys. 51:3579 (1980).
6. S. I. Rokhlin and E. Segal, Study of the cure kinetics of structural adhesives by ultrasonic interface waves, J. Mater. Sci. 20:3300 (1985).
7. H. T. Hahn, Application of ultrasonic technique to cure characterization of epoxies, in: "Nondestructive Methods for Material Property Determination," C. O. Ruud and R. E. Green, eds., Plenum, New York (1983).
8. J. K. Gilham, Torsional braid analysis of polymers, in: "Developments in Polymer Characterization," J. V. Daukins, ed., Applied Science Pub., London (1982).
9. N. S. Taskoprulu, A. J. Barlow, and J. Lamb, Ultrasonic and visco-elastic relaxation in a lubricating oil, J. Acoust. Soc. Am. 33:278 (1961).

ORIGIN OF FLAWS FROM CERAMIC PROCESSING

W. Roger Cannon
Rutgers University, Department of Ceramics
P.O. Box 909, Piscataway, N.J. 08854

Introduction

Ceramics present a particular challenge for NDE. Except for some of the most recently developed ceramic-ceramic composites, ceramics fail from pre-existing flaws. NDE is particularly difficult to use for the high technology ceramic parts such as automobile engine or other high reliability structural parts. These parts are required to withstand a high design stress which infers that the critical flaw size be very small. The flaw size is usually outside the limits of NDE detection. Since all ceramic materials of importance are essentially brittle, one can estimate the flaw size from the linear elastic fracture mechanics,

$$\sigma = K_c/\sqrt{\pi c}. \qquad (1)$$

Table I contains an estimate of the critical flaw size for sintered SiC which has a K_c value of about 4 $MN/m^{3/2}$.

Table I. Flaw sizes calculated from Eq. (1) for sintered SiC whose K_c value is about 4 $MN/m^{3/2}$

MPa	C,um
1000	5
500	20
300	55
250	80
100	500
50	2025

Typical design stresses for a turbine engine might be around 300 MPa and so a flaw size of about 55 microns must be detected. In more traditional ceramics such as white wares (porcelain sinks and toilets) and refractory bricks (used for high temperature furnace walls), stresses are much lower and flaws are on the order of 0.1 to 0.5 mm. Typically these materials are only examined for visible flaws. Such measurements as flexure strength and density are made on a regular basis to be sure that the material continues to meet specs.

The focus of this paper will be ceramic processing and the manner in which flaws enter into the ceramic part. The paper will confine itself the polycrystalline ceramics which are fabricated from the powdered metallurgy route, ie. consolidated powder compacts sintered at a high temperature (>1000°C.) Flaws may enter into the ceramic in its green (unsintered) state or after sintering. Flaws that enter in in the green state often change shape or are eliminated during sintering but most of the flaws which ultimately cause failure have their origin in the green body.

Ceramic Processing Techniques

Before discussing the origin of flaws during processing, ceramic processing techniques are briefly reviewed. Table II contains a list of widely used commercial ceramic processing techniques. In Table II is listed the moisture

Table II. Typical pressures, water and organic content for ceramic fabrication techniques

Method	Pressure	Water	Organics
Dry Pressing	35-100 MPa	0-4 vol%	1-2%
Slip Casting	capillary (.14 MPa)	30-50 vol%	1-2% dispersant
Extrusion	1-70 MPa	30-40 vol%	5-10%
Tape Casting	capillary	30-40 vol% (usually solvent)	20-30%
Injection Molding	80-120 MPa	none	30-40%
Hot Pressing	20-35 MPa	none	none
HIPing	35-200 MPa	none	none

and binder content (including plasticizers and other organics) of the powder and the pressure exerted on the powder compact to consolidate it. Included in Table II are hot pressing and hot isostatic pressing (HIPing). In these operations pressure is applied at sufficiently high temperatures to effect complete densification.

Increased applied pressure will increase the density of the compact and reduce the size of the flaws. Beyond about 100 MPa applied pressure (in dry pressing) laminations develop and so higher pressures are not usually used. Higher pressures can be used in cold isostatic pressing without producing laminations. High pressures at a high temperatures is particularly useful in eliminating defects. Recently HIPing, which usually uses Ar gas to apply the isostatic pressure, has been used successfully on pieces previously sintered to a density of closed porosity in order to reduce the critical flaw size. Hot pressing and HIPing, however, are expensive techniques and not as widely used commercially as the other techniques.

In Table II under slip casting and tape casting the applied pressure listed is the capillary pressure of water which is about equivalent to 0.14 MPa and is, therefore, lower than the others; however, interparticle forces from electrical double layers around the particles allow them to pack very well leaving probably fewer large voids than in the pressed materials. Pressure casting offers a combination of applied pressure and capillary forces and can result in more uniform densities of the material on a macroscopic level as well as a microscopic level.

Organic additives are important to the forming processes to give the green (unsintered) piece strength, to lubricate the movement of powder particles past one another, to thicken the slip to prevent settling, etc. A large number of functionally different organic additives are added commercially.[1,2] Binders are the most common. It is usually felt by manufacturers that they are burned off and so are inert to the final product; however, organic additives can also create flaws which are not eliminated when the binder burns out.

Defects may enter into the green body from a variety of sources at any of the several steps along the fabrication route. Figure 1 [3] show the stepwise schedule prior to sintering for one of the fabrication techniques listed above. The potential ware is evaluated at several steps along in the process. These, however, are usually quite simple evaluations such as powder characteristization slip rheology and green density. Improved evaluation techniques are needed in the early stages of the process to avoid more costly finished product rejection.

Typical Flaw Origins in Ceramics

The most common source of cracking in large ceramic parts such as whitewares and refractory brick originates from residual stresses produced during fabrication. These residual stresses arise from differential sintering shrinkages at neighboring locations in the body. The differential sintering shrinkages arise either because the packing density of the powder varies from one location to another, because finer powder segregates from the coarse, or temperature during sintering is not uniform throughout the body. The second of these is quite common. The fine powder shrinks more on sintering than the coarse producing residual stresses. These residual stresses may cause large cracks which are easily visible or fine cracks which can be seen only with dye penetrant. Even if cracking is not present the residual stresses may aid in growing the inherent flaws and so control of the homogeneity of the green ware is extremely important to the mechanical properties of the fired piece.

There are a variety of flaw origins which arise both during processing and handling of ceramics. Rice[4] has examined a large number of fracture surfaces to determine the fracture origin. The following sources of flaws were found: (1) pores, (2) ring cracks around powder agglomerates, (3) second phase particles (impurities), (4) large pores, (5) machining and handling flaws, and (5) oxidation pits (in Si_3N_4.

There are a variety of sources of pores. Pores from incomplete densification during sintering are most common. They are not distributed entirely uniformly and it is the regions of high pore density that are often the fracture origin. Figure 2 shows two examples of that type. Several pores may link up to form the critical flaw. The pores arise from the largest interparticle void space in the green ware. One source of porosity in ceramics comes from using spray-dried powders. Powders are often spray dried prior to dry pressing to improve the flowability of the powder. The spray-dried particles are spherical agglomerates of powder particles, maybe 100 times the diameter of the original particle and containing a binder. Porosity originates from the void between incompletely crushed spray-dried particles Also interagglomerate voids present in any one of the processes of Table I can be the source of such pores.

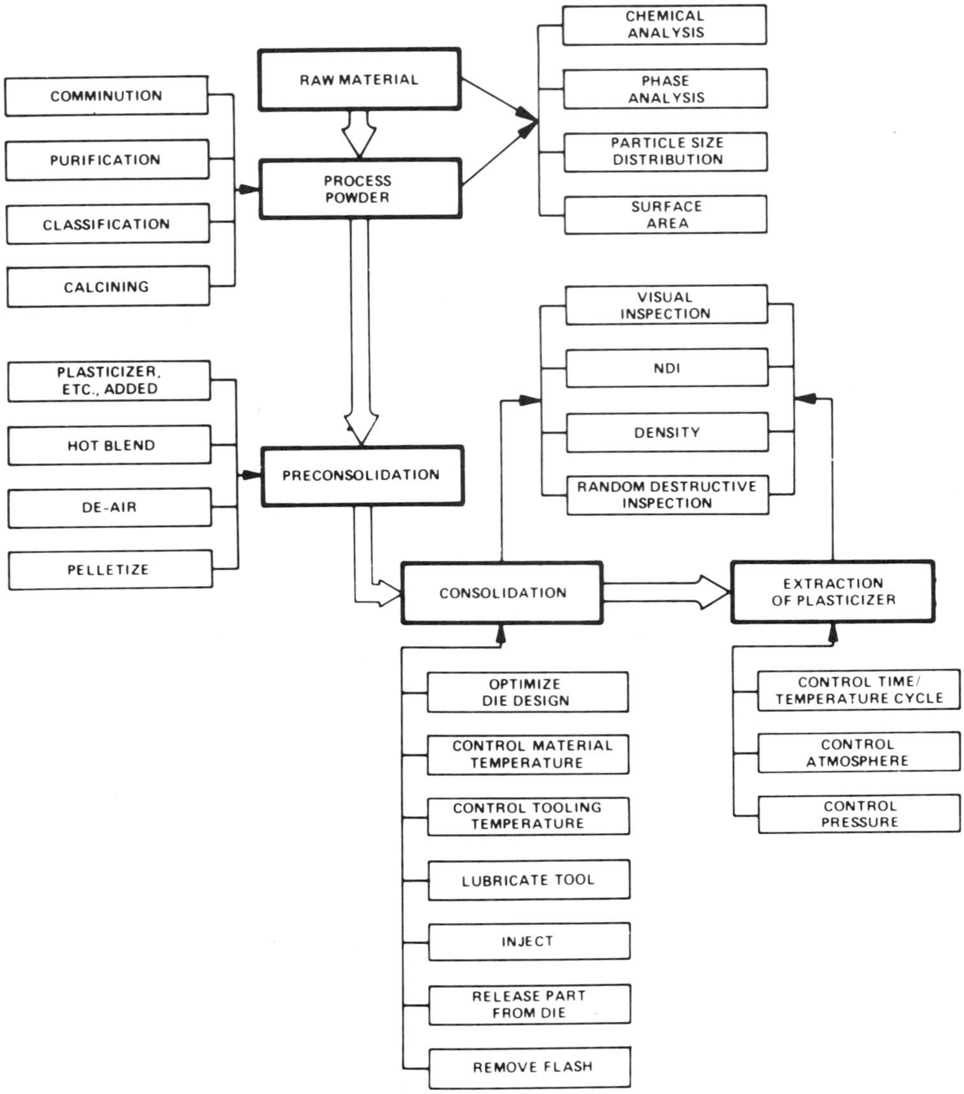

Figure 1. Injection molding processing flow sheet. Published with the permission of Deker Publishing Co. from Ref. 3.

There are a variety of flaw origins which arise both during processing and handling of ceramics. Rice[4] has examined a large number of fracture surfaces to determine the fracture origin. The following sources of flaws were found: (1) pores, (2) ring cracks around powder agglomerates, (3) second phase particles (impurities), (4) large pores, (5) machining and handling flaws, and (5) oxidation pits (in Si_3N_4.

There are a variety of sources of pores. Pores from incomplete densification during sintering are most common. They are not distributed entirely uniformly and it is the regions of high pore density that are often the fracture origin. Figure 2 shows two examples of that type. Several pores may link up to form the critical flaw. The pores arise from the largest interparticle void space in the green ware. One source of porosity in ceramics comes from using spray-dried powders. Powders are often spray dried prior to dry pressing to improve the flowability of the powder. The spray-dried particles are spherical agglomerates of powder particles, maybe 100 times the diameter of the original particle and containing a binder. Porosity originates from the void between incompletely crushed spray-dried particles Also interagglomerate voids present in any one of the processes of Table I can be the source of such pores.

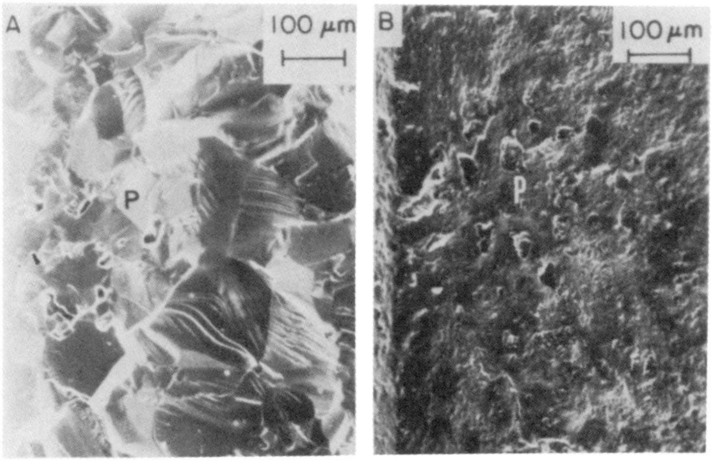

Figure 2. Fracture surfaces of flexure bars of A) dense $MgAl_2O_3$ ($\sigma_f \approx$ 210 MPa) and B) reaction bonded Si_3N_4 ($\sigma_f \approx$ 170 MPa) from regions of higher than average porosity. (Tensile surface at the left hand edges of the photos.) Published with permission of Plenum Press, Ref. 4.

Small pores are not so often the cause of failure as large pores. These pores often arise from organics which have aggregated in a local area. After burn-out a large pore remains. The pore, in the sintered ceramic may in fact be larger than the original organic piece which burned out since it will produce a large amount of gas as it burns out which expands against the constraining powder enlarging the pore. Figure 3 shows a pore many times the diameter of the average grain. Large pores may arise when the organics are not well distributed or not completely soluble in water or solvent. A second source of organics may come from the powder itself. Powder suppliers often mill their powder in polyurethane lined mills. Small pieces of the lining may end up in the powder. Thus pores arise with even the most careful processing of the powder. There are a large number of accidental sources of organics. These may include hair (human or animal), lint, dandruff, tobacco, cigarette ashes, etc.[5]

Organics are not the only source of large pores. Air bubbles can be an important source of large pores in the slip cast and tape cast ceramics. Thus the slip is often vacuum deaired prior to casting to eliminate air bubbles which are attached to particles and air which is solubilized in the water or solvent.

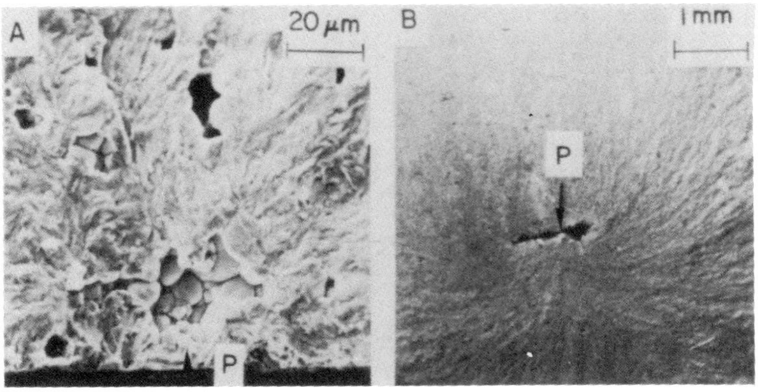

Figure 3. Failure origins for commercial PZT sonar materials from a large isolated pore (P). A) Fracture surface of a flexure bar cut from one of the highest quality commercial materials available ($\sigma_f \approx$ 115 MPa, tensile surface at photo bottom.) B) fracture surface of a poorer quality commercial sonar transducer ring failing under dynamic hoop tension loading at ≈ 17 MPa. Published with permission of Plenum Press, Ref. 4.

Ring cracks around second phase particles have recently been a subject of considerable interest because of two phase ceramic-ceramic composites. When the second phase particle (agglomerate) sinters faster that the matrix around it, a ring shaped crack results.[6] Figure 4 shows and example of such a crack. This is usually the case when a second phase powder which is composed of agglomerates of very fine powders is introduced into coarser powder matrices. An example of this is transformation toughened alumina. Also the presence of a second phase particle or fiber which sinters more slowly (or not at all) than the matrix will produce porosity in the matrix.(4)

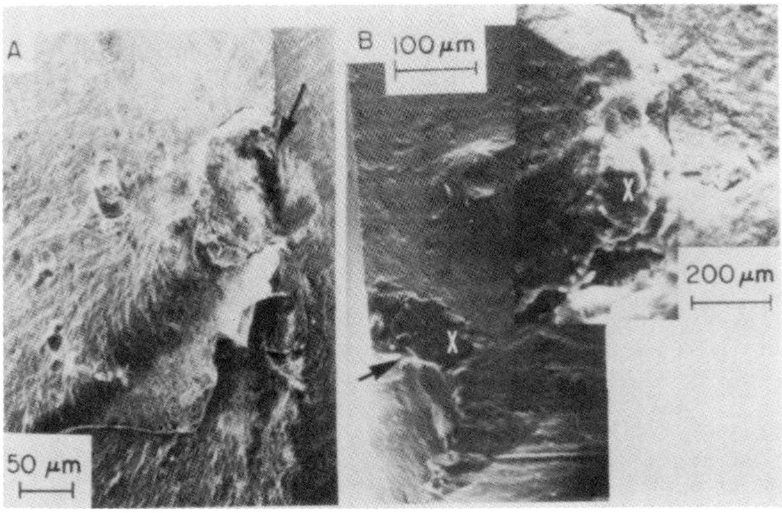

Figure 4. Failure origin of high strength partially stabilized ZrO_2 from agglomerates. A) Failure of a flexure bar from a large agglomerate (arrow) ($\sigma_f \approx$ 530 MPa, tensile surface at right edge of photo.) B) A similar sample (σ_f ≈490 MPa) failing from an agglomerate and associated spalled corner (arrow). Published with permission of Plenum Press, Ref. 4.

Impurities can often be the site of the critical flaw. Figure 5 shows an SEM micrograph of an impurity in the fracture surface of a PZT sonar device and a Al- x-ray fluorescence scan of the surface showing the particle. This particle probably entered the compact during milling. A small chip of the Al_2O_3 milling media mixed in with the powder was sintered with the compact.

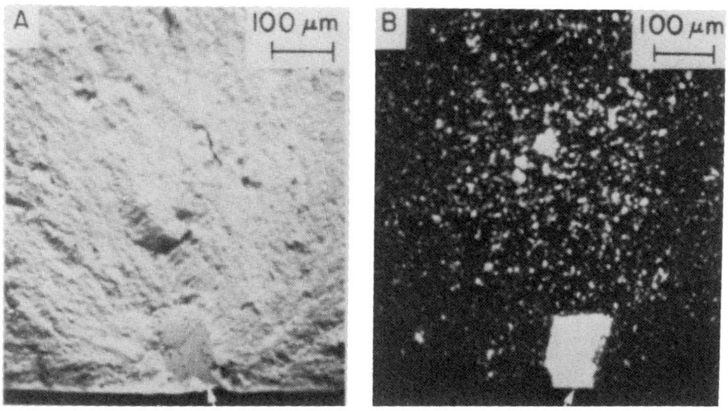

Figure 5. Failure origin of commercial PZT sonar transducer materials from a large foreign particle. A) Shows the fracture surface with fracture originating from the foreign particle (arrow) at the junction of the tensile surface and the fracture surface at the bottom of the photo. B) Al x-ray fluorescent scan of the matching fracture half showing that the particle is Al-rich. Other analysis indicates that this and the other Al-rich areas are Al_2O_3 particles which are believed to be chips from the milling media.

Figure 6 is a local area of large grains near the surface of an Al_2O_3 sintered piece. These local exaggerated grains may have resulted from a local low concentration of MgO which acts as a grain growth inhibitor. In other cases an excess of impurity may cause local exaggerated grain growth. The exaggerated grains have a low fracture energy and so may contain cracks from machining or other sources of mechanical stress.

Machining flaws are most often seen as the origin of failure in test specimens. Preparation of specimens for fracture testing nearly always involves cutting and machining the surface and so such flaws are observed at the origin of failure. Diamond machining, however, is sometimes used in preparing precision shaped or sized parts commercially and so machining flaws can be a source of failure in commercial parts also. There two types of flaws: those running transverse to the grinding direction which form periodically due to the slip-stick type of motion during grinding and the longitudinal cracks which form as a result of the high localized stress. The latter are the deeper cracks. A micrograph showing each type of crack is shown in Figure 7.

Finally handling flaws may be present in the finished ceramic due to improper care in handling the piece after manufacturing. This type of flaw may also be present in ceramics which have seen service perhaps in a wear or erosive environment.

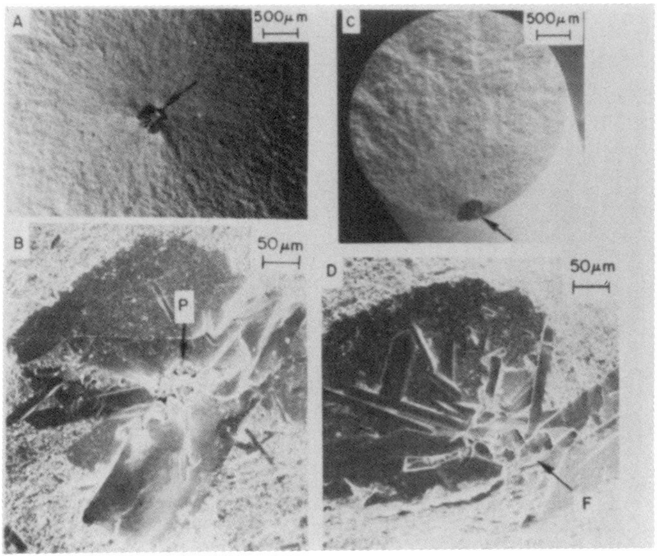

Figure 6. Failure origin of tensile tested, hot pressed Al_2O_3 from large grains. A) Tensile sample failed at $\sigma_f \approx 230$ MPa from an internal cluster of large grains (arrow), shown in higher magnification in B). Note the associated porous region (b). C) another sample failing from a cluster of large grains (arrow) at the surface of the tensile rod. D) Higher magnification of this large grain cluster; note the chipping indicative of machining flaws. Published with permission of Plenum Press, Ref. 4.

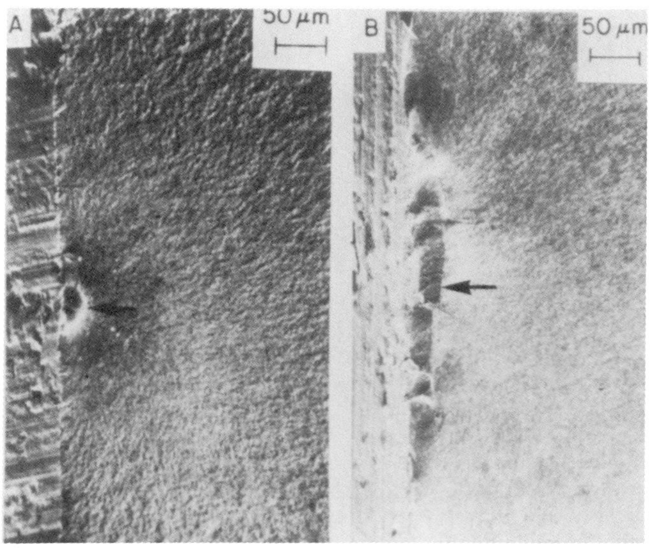

Figure 7. Characteristic machining flaws at the fracture origin: A) parallel with and B) perpendicular to the tensile axis. Two distinct types of machining flaws (arrows) extending in from the surface. Sample is MgF_2. ($\sigma_f \approx 41$ with fracture surface perpendicular to grinding direction and $\sigma_f \approx 83$ MPa parallel to the grinding direction.)
Published with permission of Plenum Press, Ref. 4.

Flaws in High Technology Structural Ceramics

An important aspect of improving the strength of ceramic is the fractography to identify flaw origins then adjusting the processing to eliminate them. Lange et al. (1986)[7] has done this during his developmental work on transformation toughened alumina. Figure 8 is interesting from the perspective of this paper since it shows the type of defects which could be eliminated to improve the strength.

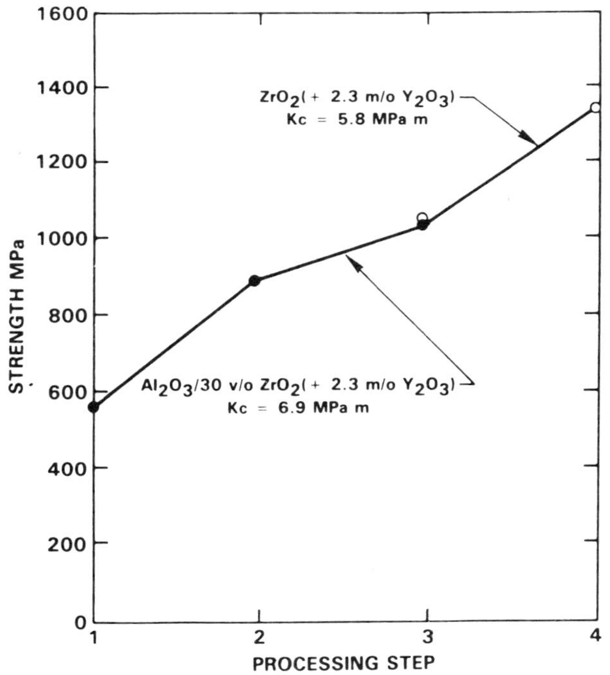

Figure 8. Flexural strength as a function of several incremental improvements in the processing. (1) Samples were dry pressed. The flaw origin was often soft agglomerates. (2) Specimens were slip cast. Flaws from soft agglomerates were eliminated but hard agglomerates persisted. (3) Powder was sedimented to eliminate hard agglomerates. (4) Organics were eliminated. Published with permission of the Am. Ceram. Soc. see Ref. 7.

In the early part of their work Lange et al.[7] wet milled the powder followed by drying, and then they dry pressed the powder. In the drying step powders soft agglomerates(those which can be broken up by milling) formed which did not break up by pressing. The type of defect shown in Figure 4 appeared at the fracture origin. This type of agglomerate disappeared if they slip cast because the interparticle forces eliminated the soft agglomerates. Elimination of hard agglomerates was achieved by sedimentation to eliminate the faster settling agglomerates from the fine powders. This resulted in a second incremental increase in strength. Next irregularly shaped voids appeared at the fracture origins which they attributed to organics. These organics were contained in the powder as it was supplied from the powder manufacturer and not introduced during their processing. Elimination of these organic particles from fine powders is difficult since heating the powders to the burnoff temperature may coarsen the powders. They chose to consolidate the powder first to the desired shape and then cool to room temperature and isostatically press the part at a high pressures to eliminate the void left by the organic particles. Finally they sinter the piece.

Another area of current interest is SiC fiber-ceramic matrix composites. The major problem in processing in these composites is to thoroughly mix the powder together with the fibers, avoiding agglomeration of either the powder or the fibers. Recent development of very tough alumina-fiber composites resulting mostly from proper mixing of the alumina with the fibers and avoiding fiber clusters.

Tape Casting of Multilayer Capacitors

During the last four years we have been carrying out research on dispersants, binders, plasticizers and other organics associated with tape casting. We have been concerned with how they effect rheology of the slip, green density and green strength of the tapes and density and strength of the sintered tapes. Tape casting is probably the most widely used technique for fabricating electronic-ceramic parts and so this has been of considerable interest to commercial manufacturers. The standard composition we used for these tapes is shown in Table III.

Table III. Composition of Slip for Tape Casting

Ingredient	Function	Weight Percentage
$BaTiO_3$	ceramic	77.2
MEK/Ethanol	solvent	10.6
phosphate ester	dispersant	0.3
Santicizer 160	plasticizer	2.2
Carbowax 400	plasticizer	2.2
cyclohexanone	homogenizer	0.4
Acryloid B-7 MEK	binder	7.1

The dielectric layers of multilayer capacitors must be very thin, on the order of 25-50 μm and so flaws must be avoided. Good dispersion of the powder is, therefore, important. In addition we found several steps to be critical in achieving a high quality tape. When using a non-aqueous system, solvent volatilization prior to casting is a problem because binders and plasticizers tend to collect around the container. It is important to eliminate these prior to casting. The slip is run through a 20 μm screen just prior to casting. Shanefield[7] has proposed a particular screen arrangement which works well. The second important step was mentioned above, that of deairing the slip prior to casting.

We found that the order of addition of the components effected the quality of the tape.[8] Three order-of-addition sequences were used: dispersant first, dispersant last, and simultaneous addition. In the dispersant first case, the solvent, dispersant, and powder were premixed, ultrasonically agitated for 2 minutes, and allowed to age for 24 hours before the remaining components were added. In the dispersant last case, all components except the dispersant were premixed, ultrasonically agitated, and aged for 24 hours before the dispersant was added. In the simultaneous addition slurry, all of the liquid components were premixed, ultrasonically agitated, aged 24 hours before adding the powder. After the final components were added, all batches were again ultrasonically agitated for 2 minutes and the slurries were placed on a slow roller mill for the duration of the study. There were no grinding media added to the slurries as the purpose of the slow rolling action was to simply keep the components from separating during the aging study. The slurries were removed from the slow roller mill only to extract samples.

Green densities shown in table IV indicate differences based on different order of addition of components. Sintered density tend to parallel the results for green tapes indicating that high porosity in the green state carries through to the sintered materials. These results tend to support the conclusion that the poor dispersion of the simultaneous addition sequence creates voids in the green state. Figures 9 and 10 show the strength of the tapes before and after firing as a function of aging time of the slips prior to sintering. Again it is apparent that the defects of the green tape are passed on to the sintered tape. The sintered surface of this specimen shown in Figure 11 clearly shows the presence of these pores. The shape of the pores may indicate that binder segregated in these specimens and bloating during sintering occurred. It is believed that the less well dispersed powder when cast contained larger voids where binder precipitated.

Table IV. Day One Density Comparisons

Addition Sequence	Green Density		Sintered Density	
	g/cm^3	% Th.	g/cm^3	% Th.
Dispersant First	2.87	52	5.13	93.3
Dispersant Last	2.76	50	5.08	92.2
Simultaneous Addition	2.69	49	4.82	87.6

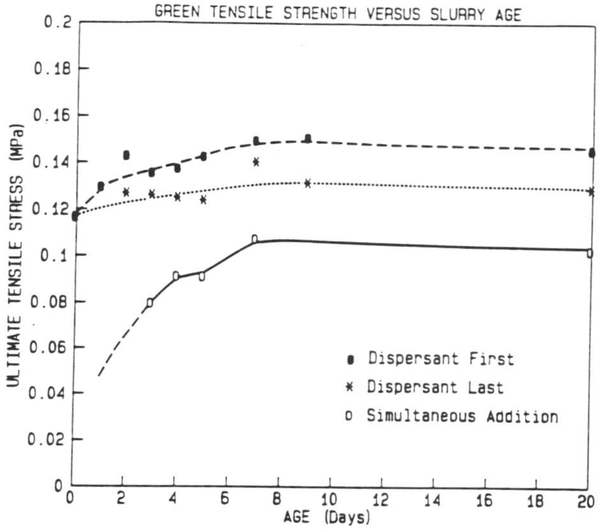

Figure 9. Ultimate tensile strength as a function of aging time of the slip prior to casting for several sequences of addition. Published with permission of the Am. Ceram. Soc. see Ref. 8.

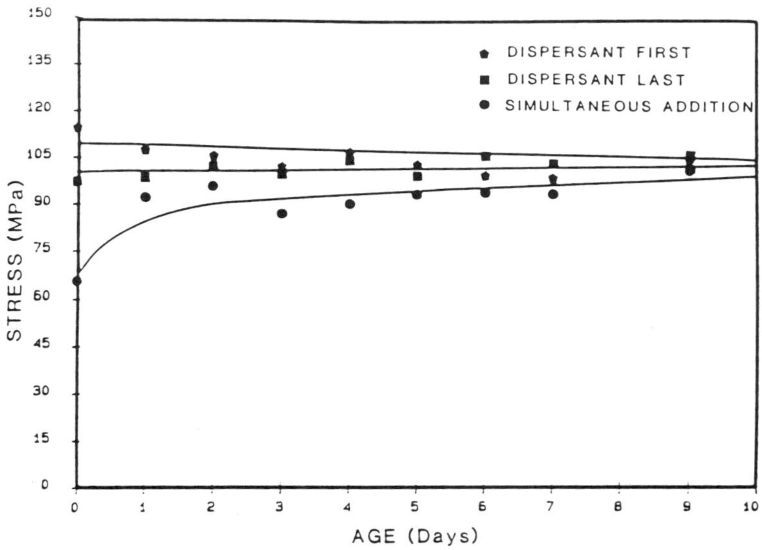

Figure 10. Strength of sintered tapes measured in biaxial flexure as a function of the addition sequence and aging time of the slips before casting. Published with permission of the Am. Ceram. Soc. see Ref. 8.

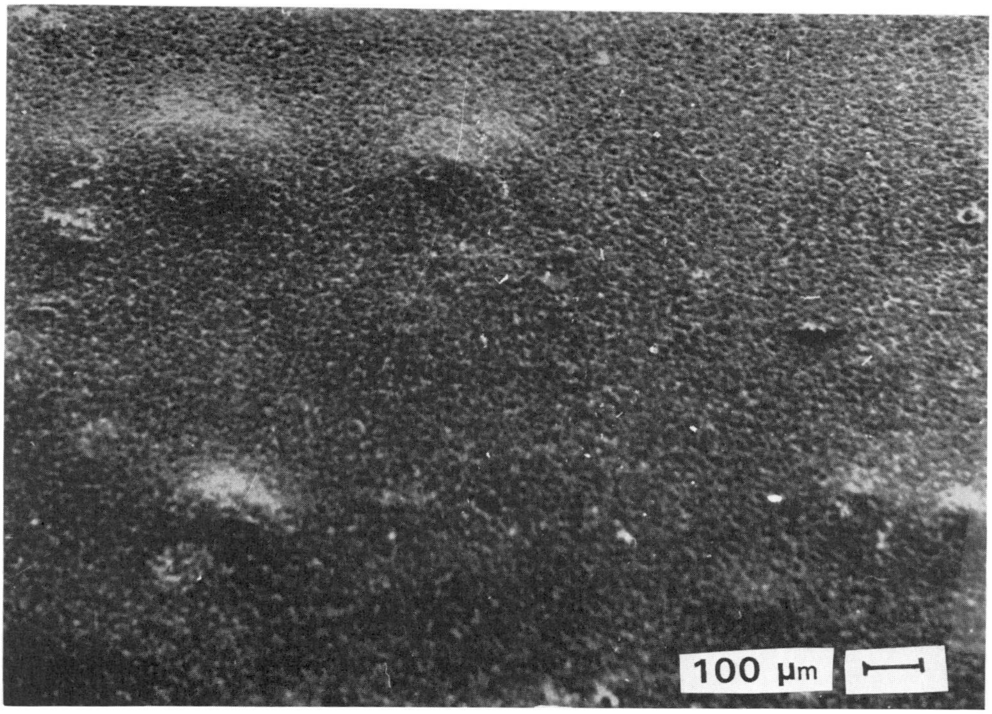

Figure 11. Micrograph showing bloating on the surface of samples prepared by the simultaneous addition sequence. Other samples did not exhibit bloating. Published with permission of the Am. Ceram. Soc. see Ref. 8.

Discussion and Conclusions

The processing specialist in ceramics is continuing to seek means processes while avoiding large flaws by better control of all processing steps. One recent trend is to use monodisperse sized powders, spherically shaped and packed in an ordered manner much as atoms in a crystalline lattice. An example is shown in Figure 12. The largest flaws are at the so-called grain boundaries. Even these would usually be eliminated during sintering. Such processing would be a quantum step improvement on the current processing techniques but these techniques are so far difficult to use in mass production.

It was shown in this paper that organics are in many cases the sources of flaws which cause failure. Care must be taken in eliminating these from the starting powders and avoiding nonuniform distribution of organics.

Finally, subsequent papers will discuss methods of NDE for the green state. Considerable savings would be realized commercially if ceramics could be rejected in their green state rather than after going through the final firing procedures.

Figure 12: Micrograph showing ordered packing of SiO_2 spherical particles. Note the grain boundary regions.

REFERENCES

1. A.G. Pincus and L.E. Shipley, 1969, Ceram. Ind., 92, 106-109.
2. S. Levine, 1960, Ceramic Age, Jan. 1960; 39-43, Feb. 1960; 25-37.
3. D. Richardson, 1982, "Modern Ceramic Engineering", Marcel Dekker, Inc., N.Y. 178-215
4. R.W. Rice, 1977, in "Processing of Crystalline Ceramics", (eds. H. Palmour, R.R. Davis and T.M. Hare) pp. 303-319, Plenum Press, N.Y.
5. W.H. Rhodes, P.L. Gernegurg, R.M. Cannon and W.C. Steele, 1973, "Microstructure Studies of Polycrystalline Refractory Oxides, "Avco Corp., Lowell, MA., Summary Report for Naval Air Systems Command, Contract N00018-72-0298.
6. F.F. Lange, 1983, J. Am. Ceram. Soc., 66: 396-398.
7. F.F. Lange, 1986, J. Am. Ceram. Soc., 69: 66-69.
8. J.M. Morris and W.R. Cannon, to be published in: "Electronic Ceramics", (eds. J.B. Blum and W.R. Cannon), Am. Ceram. Soc.
9. D.J. Shanefield and R.E. Mistler, 1976, Am. Ceram. Soc., Bul. 55, 213.

DEVELOPMENT OF NUCLEAR MAGNETIC RESONANCE IMAGING TECHNIQUES FOR CHARACTERIZING GREEN-STATE CERAMIC MATERIALS[a]

Jerome L. Ackerman,[b] William A. Ellingson,[c] Jason A. Koutcher[b,d] and Bruce R. Rosen[b]

[b]Department of Radiology, Massachusetts General Hospital, Boston, MA 02114

[c]Materials and Components Technology Division, Argonne National Laboratory, Argonne, IL 60439

Objective

Nuclear magnetic resonance (NMR) imaging holds great potential for aiding ceramics processing development. NMR is a nondestructive method of analysis which has been in routine use in chemistry, physics, and biology for over 35 years. In these areas, its main applications are in determining chemical composition, molecular or crystal structure, and molecular dynamics.[1,2] In 1973, the use of NMR in producing tomographic images of objects was reported.[3] The development of NMR imaging (NMRI or MRI) for medical diagnosis is now proceeding at a furious pace.[4] However, applications of NMR imaging in the materials sciences are only recently being explored. Our overall objective is to develop NMR methods for nondestructive evaluation in ceramics processing. The present report describes preliminary research directed toward the use of NMR for nondestructively mapping porosity and internal voids in green-state bodies. Some future extensions of this work include the imaging of pore size distribution, and the imaging of binders, plasticizers and other components of the test sample.

---

[a] Work partially supported by the U.S. Department of Energy, Office of Fossil Energy, Advanced Research and Technology Development Materials Program, under Contract W-109-Eng-38, and by the NMR Facility, Massachusetts General Hospital.

[d] Current address: Department of Medical Physics, Sloan Kettering Cancer Center, New York, NY 10021.

Background

In ceramics processing there is a need for methods of nondestructive evaluation (NDE) of parts at various stages of production. Characterization of green-state ceramic materials is cost effective because parts can be rejected before the value-adding step of final densification. Ideally, methods which produce spatial maps, rather than volume averages, of the parameter being evaluated (for example, porosity or concentration of a binder) are the most useful. Some of the properties of interest are porosity, pore size distribution, void defects originating in the mixing, fabrication and densification stages, the shapes of fabricated internal volumes, and the concentrations or distributions of ceramic powders, binders, plasticizers, sintering aids, residual solvents, and impurities.

The organic binders and plasticizers used in ceramic processing are known to affect the green-state and the final porosity. Presently, it is not possible to map the distribution of a binder or plasticizer throughout a ceramic component in a nondestructive manner. Porosity is most commonly measured with mercury porosimetry. This test method is not applicable to green-state ceramics because the test sample may not have the strength to withstand the high pressures inherent in the procedure. Furthermore, the measurement yields a porosity value averaged over the total volume of the sample; no spatial discrimination is possible.

Several imaging NDE methods are being developed in addition to the NMR work reported in this paper. One such method is X-ray computed tomographic imaging (CAT).[5] Knowledge of the stress distribution in a green-body compact is important because this affects the final sintering uniformity and impacts the overall mechanical properties. A common method of determining stress distribution within a ceramic compact is to measure the density distribution and correlate density with stress.[6] X-ray computed tomography is a method to obtain sectional images of an object, similar to NMR images. Such X-ray images are mappings of X-ray attenuation, which can be related to density. Ultrasonic through-transmission methods[7] are also being studied as means to map changes in density because of the relationship connecting acoustic attenuation and acoustic phase velocity with density. High- and low-density inclusions in green bodies, which can act as flaw sites, also may be detectable with ultrasound. Low-kV radiographic imaging with advanced image processing is also being examined as a method to detect flaws (including density gradients) in green-state ceramic bodies.

NMR analysis is nondestructive. The sample is subjected to RF and static magnetic fields, and need only be substantially nonmagnetic and possess no more than moderate (i.e., less than metallic) electrical conductivity. NMRI is inherently three-dimensional; however, because of the considerable time involved in the collection of three-dimensional data, imaging is usually performed as a series of two-dimensional views. The theory and practice of NMR spectroscopy[1,2,8] and imaging[3,4,9] have been reviewed extensively elsewhere. Briefly, nuclei with nonzero spin possess a magnetic moment, which will precess about the direction of a magnetic field. The frequency of precession is proportional to the field strength, the proportionality constant being characteristic of the particular isotope. The very small, but highly characteristic, shielding of the nucleus from the external magnetic field caused by the electron cloud of the atom or molecule in which the nucleus resides gives rise to the "chemical shift" of the precession frequency; this phenomenon accounts for the immense utility of NMR in chemical analysis. Externally applied magnetic field gradients encode spatial position in the spectrum of precession frequencies, and permit the tomographic imaging of objects containing NMR-active nuclei. Protons (hydrogen nuclei) provide the best sensitivity: the ubiquitous presence of water in biological tissues has virtually created the possibility of NMR imaging in diagnostic medicine.

In the materials sciences, NMR spectroscopy has been used extensively in the analysis of molecular structure and dynamics in polymer systems,[10,11] and to a lesser extent in the study of ceramics.[12] (NMR analysis of solid systems involves considerable instrumental and experimental complications, in contrast to its use in fluid systems.[8]) NMR imaging has been used to study the absorption of water in glass fiber/epoxy composites.[13]

It is in principle possible to have the contrast (variations in brightness) of an NMR image reflect signal intensities, relaxation times, chemical shifts or precessional phases in the different spatial regions of the sample. These detected signal parameters may in turn may be related to important material properties such as quantity or concentration, chemical identity or molecular structure, molecular dynamics (molecular motions, chemical exchange, diffusion, temperature) and bulk flow. Mathematical processing of sets of images may be used to produce spatial maps of parameters such as the concentration of a specific chemical component, or the diffusion constant of a specific chemical component.

Approach

To image the internal empty volume of a green ceramic compact, we employ a "filler" fluid as a marker giving a good NMR signal. It is the distribution of this fluid which is in fact the quantity imaged. The fluid is introduced into the test sample by vacuum impregnation. In this technique we assume adequate penetration of the filler fluid into all the internal volumes of interest, including pore spaces and internal voids.

The filler fluid approach provides a means to visualize internal volumes which are larger than the spatial resolution of the measurement, which is about 300 μm within the image plane and about 2 mm in plane thickness for the measurements presented here. Volumes smaller than the in-plane resolution (specifically, the microscopic volumes which constitute the pore structure) are not resolved. We take the fractional porosity to be the fractional image intensity normalized to the intensity of pure filler fluid.

Penetration is assisted by using a fluid of low viscosity, and of low interfacial tension with the microscopic surfaces of the sample. Desirable chemical properties of the fluid include a high concentration of the observed nuclear isotope (normally protons), and chemical compatibility with the sample (no chemical reaction, dissolution, or swelling). These properties also allow the removal of the filler fluid following NMR analysis to permit subsequent testing with other methods.

Desirable NMR properties for a filler fluid include choosing a high sensitivity isotope, protons being the best choice. The isotope with the next highest sensitivity is ^{19}F, with an intrinsic sensitivity 0.8 that of ^{1}H. The isotopic abundance of ^{19}F is 100 percent, so that all fluorine contributes to the NMR signal. Many fluorocarbons are chemically inert towards the materials found in green ceramics, and have low viscosities and interfacial and surface tensions. Despite their high cost and certain complications, they have been used effectively as NMR markers in biological systems.[14] To avoid artifacts and anomalous signal intensity variations, the filler fluid must exhibit a simple NMR spectrum, ideally containing a single sharp resonance line. Such a simple spectrum results from the absence of chemical shift differences and spin-spin (or J-) couplings within the molecule, and will be exhibited by a compound possessing a high degree of chemical symmetry (i.e., chemical and magnetic "equivalence" of nuclei). Examples of good candidates for filler fluids satisfying this criterion are water, benzene, acetone and dimethylsulfoxide for proton NMR, and hexafluorobenzene for fluorine NMR.

NMR relaxation has an important effect on the imaging process. The spin-lattice relaxation time T_1 is the characteristic time for the buildup of magnetic polarization of a nucleus, and directly affects the rate at which signal data can be accumulated from successive acquisitions of NMR signals during the production of an image. It therefore controls the total time to acquire an image and the signal-to-noise ratio achieved per unit time. By means of paramagnetic "doping" (addition of paramagnetic substances) filler fluid T_1's may be conveniently shortened from their natural values of on the order of seconds to the range of 100 msec. This offers an order of magnitude improvement in the total imaging time and signal-to-noise per unit time over undoped fluids.[15]

The spin-spin relaxation time T_2 is the characteristic time for the decay of the transverse precessing magnetic moment which is detected as the NMR signal. This relaxation time affects the achievable spatial resolution and signal-to-noise ratio. Long T_2's are desirable, and are found in compounds with minimal spin-spin coupling and which form nonviscous solutions. T_2 should be significantly greater than the echo time TE in order to avoid limiting spatial resolution and signal-to-noise ratio.

Experimental Methods

All samples were prepared at Argonne National Laboratory. Sample number 1 was a green-state MgO compact containing 309 g MgO powder and 20 percent Carbowax binder, cold-pressed at 5000 psi into a 31 mm diameter by 27 mm high cylinder. Six holes, ranging from 330 μm to 1390 μm, were drilled parallel to the cylinder axis to provide recognizable void features, as diagrammed in Figure 1. Sample number 2 was similarly prepared and drilled, using SiC powder, with a diameter of 25 mm. Sample number 3 consisted of three separate fired alumina disks with known densities of 1.648, 1.703 and 1.720 g cm^{-3}, also 25 mm diameter.

Introduction of the filler fluids and measurement of images were performed at the Massachusetts General Hospital. After some preliminary work with hexafluorobenzene, we chose to work with benzene, which was considerably less expensive, easier to dope, and sufficiently compatible with the binder. The proton molarity of benzene is 67.5 M, which provides a reasonably high proton concentration (that of water is 111 M). Paramagnetic doping was with chromium acetylacetonate, $Cr(CH_3COCH_2COCH_3)_3$, or $Cr(acac)_3$, at 10.3 $\pm$ 0.6 or 20.6 $\pm$ 1.2 mM to give proton T_1's of 237 $\pm$ 12 or 122 $\pm$ 11 msec respectively. These relaxation times were determined on a 3.52 T (150 MHz) Nicolet spectrometer, and are expected to be close to the relaxation times at the imager's field strength. Filler fluid was introduced into the test sample, contained in a 38 mm OD by 51 mm long glass tube, by vacuum impregnation.

Imaging was performed on a Technicare Corp. developmental NMR imaging instrument, with a magnetic field strength of 1.45 T (61.538 MHz proton frequency). The imaging procedure utilized a standard spin echo (SE) pulse sequence[9] with a short echo time TE of 15 msec in order to minimize signal intensity loss due to diffusion of benzene molecules through the interfacial field gradients which exist at the magnetic susceptibility discontinuities of the sample.[16] The pulse repetition time TR was around 100 to 200 msec. The digital resolution in the images was 512 x pixels by 256 y pixels, corresponding to pixel sizes 380 μm x by 350 μm y. The slice thickness (depth of the sample voxel) was about 2 mm. The slight out-of-roundness in some images is due to magnetic field inhomogeneity (spatial nonuniformity) and to magnetic field gradient nonlinearity, which are currently being improved. We determined that under our imaging conditions, the Carbowax binder gave no NMR signal. Typical imaging times were around 100 to 200 sec, corresponding to 2 or 4 signal averages per phase encoding gradient step. Occasional images were taken with up to 64 signal averages for improved signal-to-noise ratio; these required on the order of one hour. Reported intensities are region-of-interest averages calculated directly from the stored NMR data and not from video brightness or film densities.

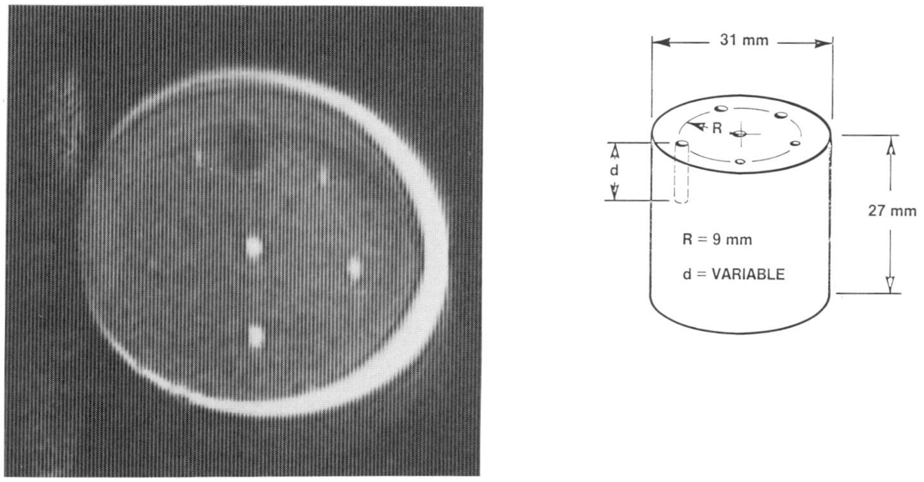

Figure 1. Transaxial spin echo NMR image of an MgO/20 percent Carbowax green ceramic compact, drilled as shown in the diagram, benzene filler fluid. The bright rim is pure benzene. The bulk of the ceramic body exhibits an NMR-derived porosity of 15.5 ± 6.2 percent, and the dark round patch a porosity of 7.4 ± 1.7 percent. The noise background value is 2.1 ± 1.1 percent.

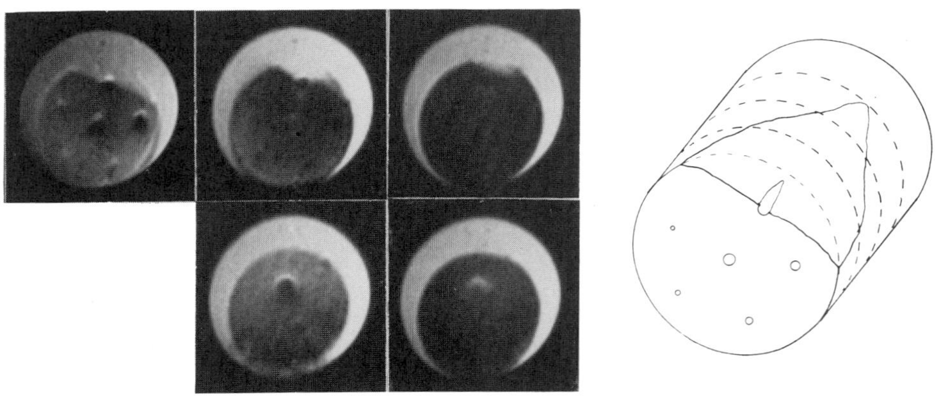

Figure 2. A sequence of five image planes of a highly fractured SiC/Carbowax sample, positioned as shown schematically in the diagram on the right. The sequence progresses inward from top left to top right, then bottom left to bottom right. The first image is made up of 32 signal averages, the others, 2 averages. The mean NMR porosity is 76 ± 11 percent. The sample is drilled similarly to that in Figure 1. The dark areas at the holes are air bubbles trapped as a result of the fracturing process.

Results

Figure 1 shows a transaxial NMR image of the drilled MgO sample, along with a diagram of the hole pattern. The plane of the image lies about one mm in from the face of the sample. The image intensity is an average over the slice thickness of about 2 mm. The smallest hole, 330 μm in diameter (on the order of the pixel size) is just barely discernible. The mean porosity of the green compact as determined from the NMR signal intensity is 15.5 ± 6.2 percent. The dark round patch is suggestive of a region of lower porosity, 7.4 ± 1.7 percent.

The image of sample number 2, silicon carbide with Carbowax binder, is shown in Figure 2. This sample was damaged by freezing of the filler prior to the NMR analysis. The body of the material is highly fractured, and exhibits a mean NMR-derived porosity of 76 ± 11 percent. The diagram shows the positions of successive image planes. The front image plane was acquired with 32 signal averages, as opposed to 2 for the others. Some bubbles formed in the drilled holes as a result of the freeze/thaw treatment; these appear as dark regions within the holes.

Sample number 3, shown in paraxial view in Figure 3, contains three fired alumina disks of varying density, with the image plane approximately containing the axes of the disks. Intensities for each disk were measured near the disk axes and near the peripheries, as shown by the two regions of interest superimposed on the same image data. Figure 4 graphs the measured disk densities vs. the NMR-derived fractional porosities. The vertical bars on the experimental points represents standard deviations within each region of interest, and the solid lines are linear least-squares fits. The lower density and increased density variance at the periphery are roughly in accord with what would be expected from the compaction process.

We have also investigated samples containing stepped binder concentrations, but have not yet seen clear correlations between binder concentration and green-state porosity.

Conclusions

We have shown it is possible to nondestructively image the internal volumes in a green-state ceramic body using filler fluids to provide NMR signal intensity. Features on the order of 300 μm can be detected in a field of view of 100 mm, and porosity may be imaged and measured as a fraction of maximum signal intensity. With improvements of the NMR apparatus specifically targeted to materials science applications, much higher resolution can be achieved: resolution approaching 10 μm has recently been reported for very small biological samples.[7] In addition to developing NMR probes optimized for ceramics processing development, we are proceeding to design experimental methods which employ cross-polarization techniques[10,11,18] to image the organic phase directly.

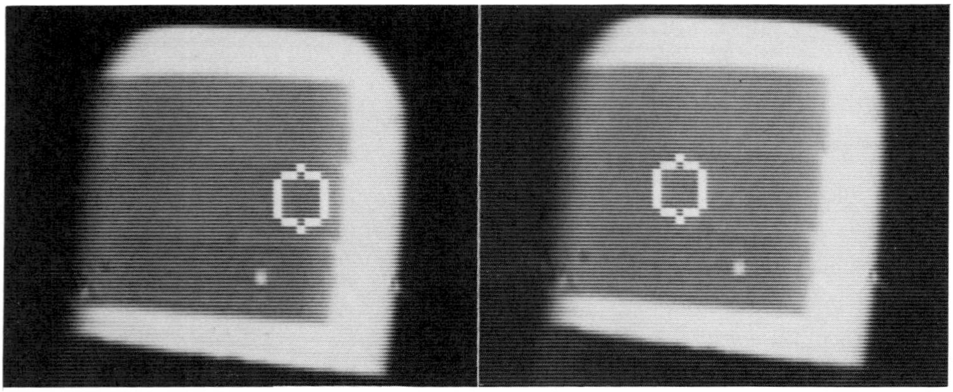

Figure 3. Paraxial view of three densified alumina disks of measured density 1.648 (top), 1.703 and 1.720 g cm^{-3}. The image plane approximately contains the axes of the disks. The two regions of interest indicate where image intensity measurements were taken.

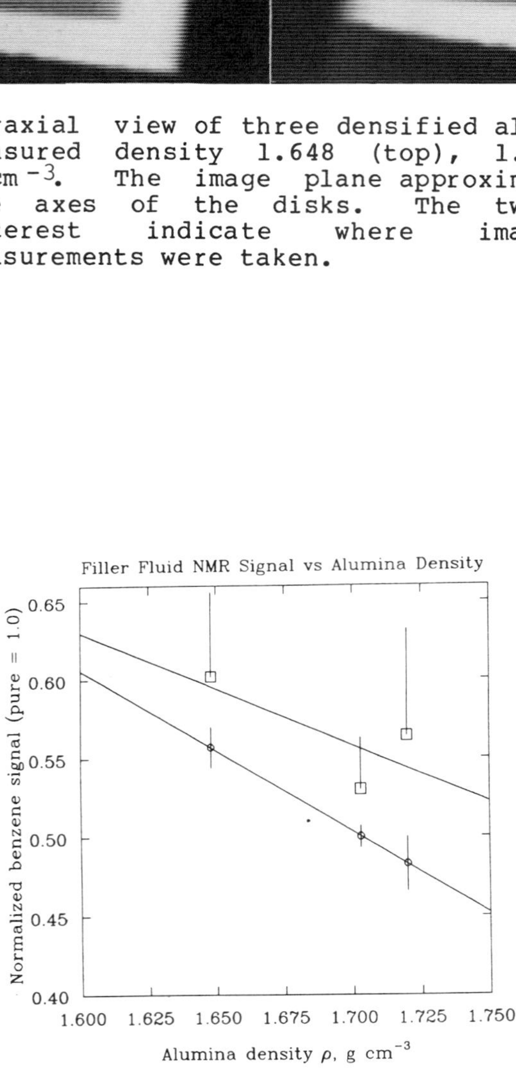

Figure 4. Graph of NMR signal intensity vs. measured density of fired alumina disks. Top: peripheral region; bottom: on-axis region. Solid lines are least-squares fits, error bars are standard deviation about the mean pixel intensity within each region of interest. (Half-error bars are used to avoid overlap with the lower graph.)

Acknowledgments

The authors wish to acknowledge the valuable assistance of J. Picciolo of ANL in the preparation of the ceramic samples, and A. Vincent of MGH for performing the T_1 measurements.

References

1. E. D. Becker, "High Resolution NMR," Academic Press, New York, 1980.
2. C. P. Slichter, "Principles of Magnetic Resonance," 2nd ed., Springer-Verlag, New York, 1978.
3. P. C. Lauterbur, Nature (London) 242, 190 (1973).
4. P. Mansfield and P. G. Morris, "NMR Imaging in Biomedicine," Academic Press, New York, 1982.
5. T. Taylor, W. A. Ellingson and W. D. Koenigsberg, "Evaluation of Engineering Ceramics by Computed Tomography," Atomic Energy of Canada, Ltd. report AECL-9005, Chalk River, Ontario, Canada (1985).
6. O. J. Whittemore, Jr., "Particle Compaction," in "Ceramic Processing Before Firing," G. Y. Onoda, Jr. and L. L. Hench, eds., John Wiley and Sons, New York, 1987, pp. 343-355.
7. R. A. Roberts, W. A. Ellingson and M. W Vannier, "A Comparison of X-Ray Computed Tomography, Through-Transmission Ultrasound, and Low-kV X-Ray Imaging for Characterizing Green-State Ceramics," in Proc. 15th Symposium on Nondestructive Evaluation, San Antonio, TX, April 23-25, 1985.
8. M. Mehring, "High Resolution NMR Spectroscopy in Solids," Springer-Verlag, Berlin, 1976.
9. P. A. Bottomley, Rev. Sci. Instrum. 53, 1319 (1982).
10. J. Schaefer, E. O. Stejskal and R. Buchdahl, Macromolecules 10, 384 (1977).
11. K. Beshah, J. E. Mark, J. L. Ackerman and A. L. Himstedt, J. Polymer Sci. B: Polymer Phys. 24, 1207 (1986).
12. G. R. Finlay, J. S. Hartman, M. F. Richardson and B. L. Williams, J. Chem. Soc. Chem. Comm. (London), 159 (1985).
13. W. P. Rothwell, D. R. Holecek and J. A. Kershaw, J. Polymer Sci. Polymer Lett. Ed. 22, 241 (1984).
14. S. R. Thomas, L. C. Clark Jr., J. L. Ackerman, R. G. Pratt, R. E. Hoffmann, L. J. Busse, R. A. Kinsey and R. C. Samaratunga, J. Comput. Assist. Tomogr. 10, 1 (1986).
15. G. C. Levy and J. D. Cargioli, in "Nuclear Magnetic Resonance Spectroscopy of Nuclei Other Than Protons," John Wiley and Sons, New York, 1974.
16. A. Villringer, B. R. Rosen, R. B. Lauffer, J. L. Ackerman, V. J. Wedeen, R. B. Buxton and T. J. Brady, 5th Ann. Meeting, Society of Magnetic Resonance in Medicine, Montreal, 1986.
17. J. B. Aguayo, S. J. Blackband, J. Schoeniger, Mark A. Mattingly and M. Hintermann, Nature 322, 190 (1986).
18. N. M. Szeverenyi and G. E. Maciel, J. Magn. Reson. 60, 460 (1984).

ULTRASONIC EVALUATION OF SPRAY-DRIED CERAMIC POWDERS DURING COMPACTION

Martin P. Jones

Materials Science Department
The Johns Hopkins University
Baltimore, MD 21218

Gerald V. Blessing

Ultrasonic Standards Group
National Bureau of Standards
Gaithersburg, MD 20899

INTRODUCTION

There is a growing need to nondestructively evaluate (NDE) compacted ceramic powder parts in the green (unfired) state. In many cases, the quality of the final sintered product is determined by the quality of the starting powder and its homogeneity after compaction. If inferior compacts could be detected, it would be possible to reject them prior to further costly processing stages [1]. Some flaws can occur during subsequent sintering [2], or machining [3], or polishing [4], etc. The quality control of ceramics can be enhanced by applying NDE methods at several stages of processing to narrow the possible origins of a particular problem.

The main goal of this work was to measure the ultrasonic wave speeds of a ceramic powder during compaction. The elastic properties of the powder were inferred from the ultrasonic measurements. The authors hypothesized that the powder's elastic properties would strongly depend upon the quality of the contact between particles. An increase in contact between particles, creating a more rigid link, should result in increased elastic values and increased ultrasonic wave speeds and amplitudes.

Previous Ultrasonic Work on Green State Ceramics

There have been many ultrasonic studies of sintered ceramics [5-8] compared to the few of green state ceramics [1,9]. This is due in large part to the difficulty in propagating ultrasonic waves in the highly attenuating powder medium. Also, special techniques must be used to couple the ultrasonic transducer to the fragile and porous green state ceramic. Liquid couplants can contaminate and/or damage the specimen, thereby affecting its subsequent sintering. Also, the absorbed liquid in the sample can unpredictably affect the wave speeds and amplitudes. One approach to dry-coupling is to apply a force large enough so as to create sufficient contact between the transducer and the specimen [9]. However, this can damage or fracture the specimen in its fragile green state. Another approach is to encapsulate the specimen in a polymeric seal and then ultrasonically scan it while both specimen and longitudinal transducer are immersed in a water tank [9]. This immersion method is cumbersome, inaccurate, and too time consuming for industrial use. Recently, the authors have avoided the above problems by using a novel elastomer, requiring minimal pressure to dry couple both longitudinal and shear wave transducers to green state alumina samples [1]. Further work has led to a new ultrasonic technique for monitoring the elastic properties of the ceramic powder as it is being formed under pressure [10] and was used to obtain the data in this paper.

Common Ceramic Processing Methods and Problems

Alumina is commonly extracted from bauxite ore by dissolving its aluminum content with caustic soda into solution. This solution is then filtered and precipitated, resulting in the formation of aluminum trihydrate particles. The trihydrate is decomposed into alumina by the addition of heat and chemicals. Smaller particle sizes, 0.5 μm, are desirable since they have more surface area to accelerate sintering. Yet, more surface area creates more friction between particles, thereby decreasing the flowability of the powder [11]. To enhance flow, the alumina powder is usually spray-dried.

Spray-drying consists of suspending the alumina powder in a water slurry containing dissolved organic binders, e.g. polyvinyl alcohol. The slurry is then atomized into hot air so as to form spherical granules, about 100 μm diameter, within which the particles cling together via the organic binder. The binder acts as a lubricant so that particles may more readily slide by each other during compaction which tends to increase the reproducibility of the quality from part to part. Also, spray-drying confines any segregation of particle sizes to the small volume of the granule, which otherwise would lead to adverse effects such as microcracking and porosity [12].

Research Goals

By monitoring the compaction, the authors hoped to characterize how the presence of binder influences particle interaction. With no binder present in a powder, particles come into direct contact with each other. In contrast, a binder will serve as an additional link for interaction between particles. Experiments were performed on powder that contained binder and on powder in which the binder was extracted by burning at 500 °C. During compaction, the applied pressure causes the particles to rearrange, increasing their packing density. In addition, as indicated earlier, the greater the applied pressure, the more rigid is the contact between particles. Therefore, not only the density, but also the extent of particle contact are factors affecting the wave speed.

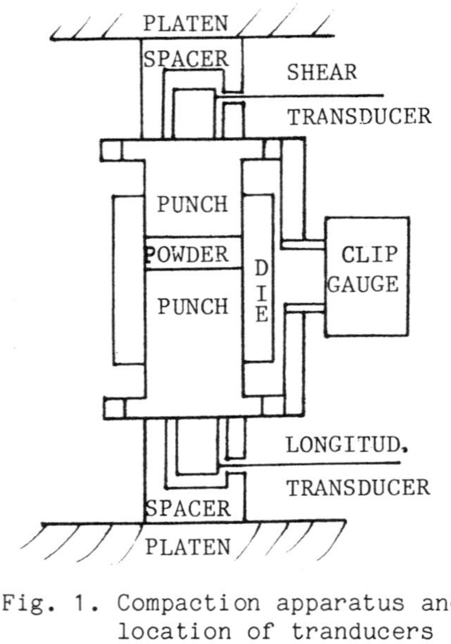

Fig. 1. Compaction apparatus and location of tranducers

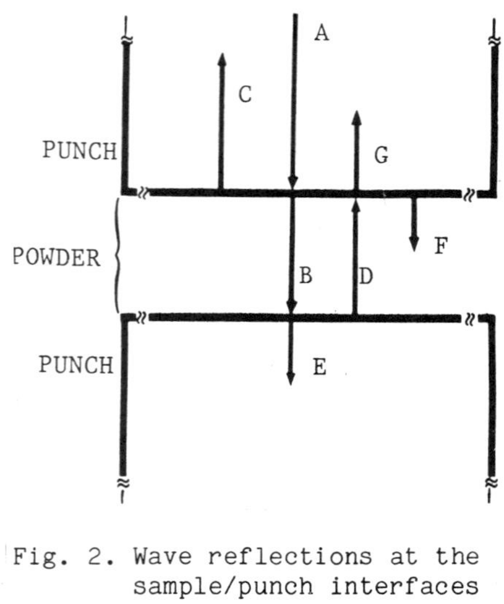

Fig. 2. Wave reflections at the sample/punch interfaces

EXPERIMENTAL PROCEDURE

Compaction Technique

Figure 1 depicts the compaction apparatus used and the attachment of the ultrasonic transducers. A commercial stainless steel mold was used consisting of a cylindrical die (29 mm inner diameter, 47 mm outer diameter and 75 mm length) and two punches having lengths of 83 mm and 19 mm. Pressures up to 100 MPa and a compaction rate of 0.2 mm/minute were used to compact 4 gram samples to an approximate thickness of 3 mm. The testing machine was equipped with load cells that allowed the pressure to be monitored accurately to within $\pm 1\%$ of the indicated value. The thickness of the sample was measured during its compaction by monitoring the relative distance between the punches with a clip gauge. Nonparallelism of the sample's faces limited the thickness measurement to $\pm$ 0.05 mm which was much greater than the error introduced by clip gauge imprecision.

Ultrasonic Measurements During Compaction

As depicted in Fig. 1, a longitudinal and a shear transducer (each 5 MHz, 12 mm diameter) were coupled to the open ends of the long punch and short punch, respectively. Transducer vibrations were channeled into the punch, generating planar elastic waves which propagated down its length. Various wave echoes from the end of the punch and sample returned to be detected by the same transducer. Shown in Fig. 2, the ultrasonic wave (A) generated by either transducer propagated down the length of the punch. Upon reaching the interface between the powder being compacted and the end of that punch, some of the ultrasonic wave was transmitted (B) into the powder and some was reflected (C) back towards the same transducer. When the transmitted wave (B) reached the opposite punch/sample interface, it was also partially reflected (D) from, and partially transmitted (E) into that punch. Subsequently, the reflected wave (D) at this interface propagated back through the sample and was similarly reflected (F) from and transmitted (G) across the first interface. Each time the wave crossed an interface or traveled through the sample its amplitude was severely reduced. Increased compaction increased the signal-to-noise ratio by improved contact between sample and punch and by decreased porosity reducing wave scattering.

Figure 3 is an oscillograph of typical longitudinal ultrasonic echoes received from various sample/punch interfaces obtained at a compaction pressure of 100 MPa. The time between echoes (C) and (G) corresponds to the time-of-flight for two traverses of the wave through the sample. In addition to this primary echo are secondary echoes (C', C'', etc.) originating from echo (C). Also present are some unwanted reflected (x) waves from the side of the punch. Although not distinguished in the oscillograph, echo (C) is phase inverted with respect to echo (G). Wave (B), traveling in the powder, is reflected (D) off the boundary of the higher acoustic impedance steel medium. No phase inversion occurs for echo (C) because the impedance conditions are opposite those of (D). A pulse-echo overlap technique was used to measure the time between echoes (C) and (G). The oscilloscope used had a pulse-echo overlap capability with a precision of $\pm$ 1 ns. Figure 4 shows the overlap condition (peaks aligned) of an electronically inverted (G) shear wave echo that had been overlapped with a (C) echo. Once the conditions were determined, the time between echoes was recorded along with the corresponding sample thickness so that the wave speed could be computed.

The amplitude of the longitudinal (G) echo was monitored continuously during compaction for possible future correlation to wave scattering due to porosity. An electronic gate was used to select an echo by delaying it relative to the pulser/receiver's synchronization. The gate width was conrolled so as to only include the echo of interest.

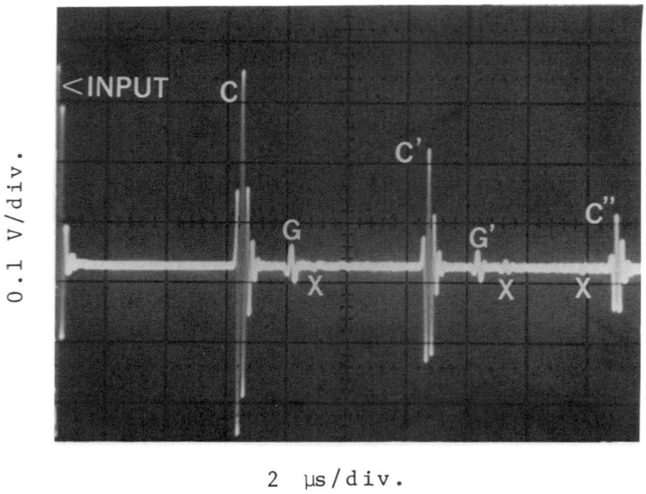

Fig. 3. Oscillograph of typical longitudinal wave echoes received from various sample-punch interfaces during compaction at 100 MPa

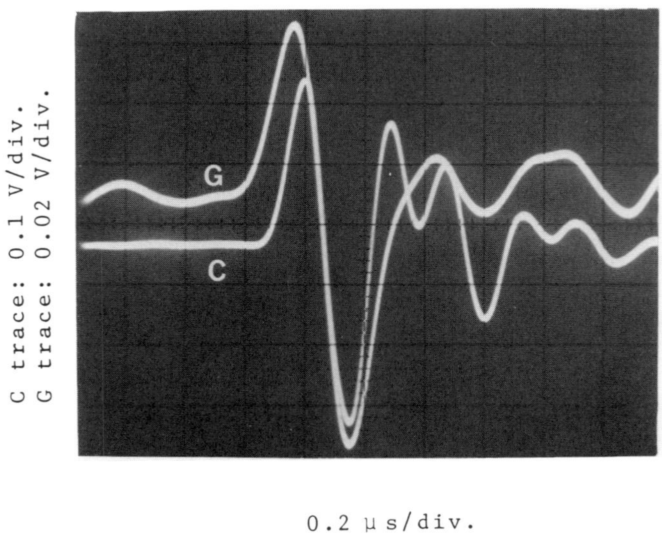

Fig. 4. Oscillograph of a typical overlap condition for shear wave echoes

RESULTS AND DISCUSSION

Wave Speeds

Compaction was stopped at intervals in order to allow for pulse-echo overlap measurements to be made. Figure 5 shows the longitudinal wave speeds of the powder containing binder during two full cycles of loading. Wave speeds remained higher for subsequent loading and unloading paths than those obtained during initial loading. This was expected because the powder had plastically deformed, achieving a higher density. However, if density was the only factor affecting the wave speeds and if all the deformation was plastic, then the wave speed attained at the highest load would remain constant during unloading. There is, of course, some elastic and anelastic expansion of the powder when pressure is decreased that would partially account for lower wave speeds. The anelastic expansion is evident by the presence of lower wave speeds for reloading than for the first unloading at equal pressures. Higher pressures were needed to overcome this expansion that had occurred with time. That the differences in wave speeds between reloading and the second unloading are less than those between the initial loading and the first unloading indicates that less plastic deformation occurred during the reloading than the first loading of the powder.

Figure 6 shows similar results for the wave speeds of powder containing no binder during two full cycles of loading. Again, the wave speeds during the first unloading remained higher than those during first loading, indicating plastic deformation and higher densities. Also, the wave speeds during reloading were lower than those at equal pressures during the first unloading, indicating that some anelastic expansion had occurred, as in Fig. 5. The differences in wave speeds between unloading and reloading were greater for the powder containing no binder than for the powder containing binder, indicating that there was more anelastic expansion during unloading of the binderless powder than for that of the powder containing binder since the cohesive property of the binder between particles resists their separation.

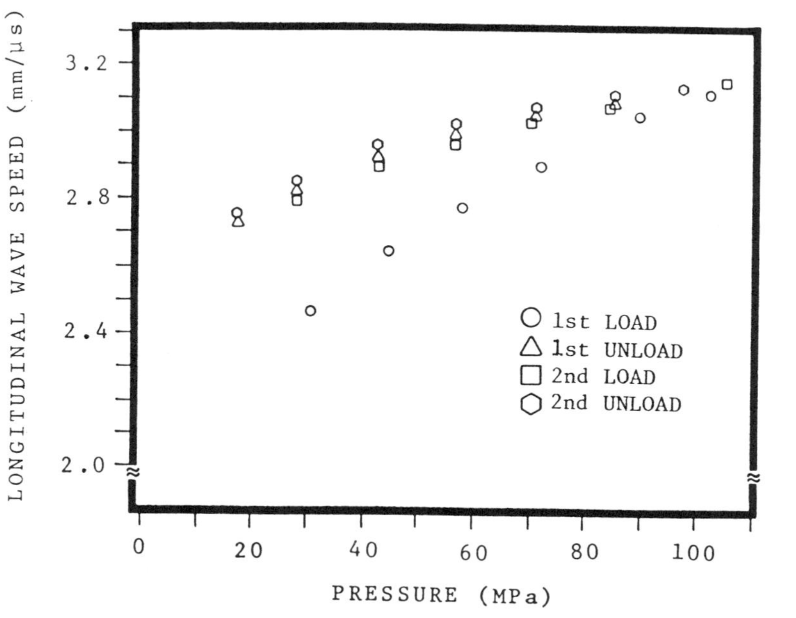

Fig. 5 Longitudinal wave speeds in alumina powder containing binder

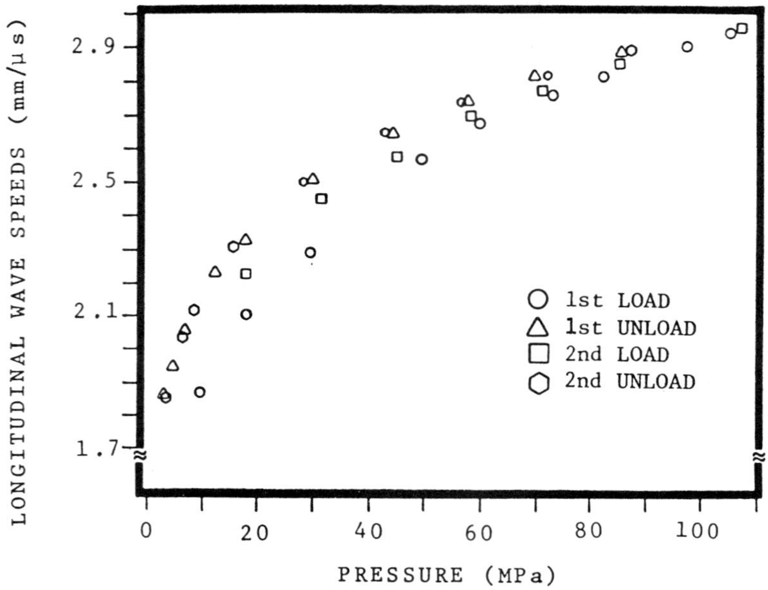

Fig. 6 Longitudinal wave speeds in alumina powder containing no binder

Note that the wave speeds in Fig. 5 remain higher for the second unloading than for the first unloading at equal pressures indicating that the particles rearranged to a higher density during the reloading path. However, as seen in Fig. 6, the wave speeds practically coincide at equal pressures for the first and second unloadings, thus indicating little rearrangement of particles during reloading for the binderless powder. Particles coated with binder are able to slide by each other during reloading to achieve higher densities. Cyclic loading of powder containing binder may be more advantageous to achieve higher densities than to resort to much higher pressures since such pressures substantially increase die wear [11] but should not be used for binderless powder since higher densities do not result.

The shear and longitudinal wave speeds, V_S and V_ℓ respectively, and densities ρ, were used to calculate the elastic moduli [13] of the powder during compaction. Figure 7 shows the calculated bulk modulus of powder containing binder during cyclic compaction. There seems to be a linear relationship between bulk modulus and pressure during loading cycles but not during unloading cycles.

Wave Amplitudes

The first measurable signal (G echo) appeared at widely different pressures (10 MPa to 30 MPa) from one sample to another of the same type of powder. One possibile cause was the variability in the contact between the powder and the punch's face. If there were little variation in the quality of contact at the face, then for larger amounts of powder, one would expect that the amplitudes of the waves would be greatly reduced. Since no clear correlation between the amount of powder and the amplitudes was observed, it was concluded that the received amplitude was dominated by variations in sample/punch contact.

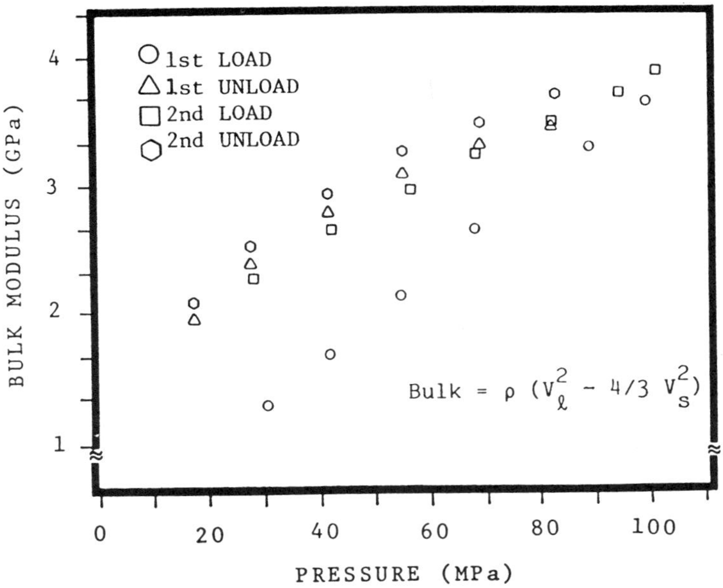

Fig. 7 Calculated bulk moduli of alumina powder containing binder

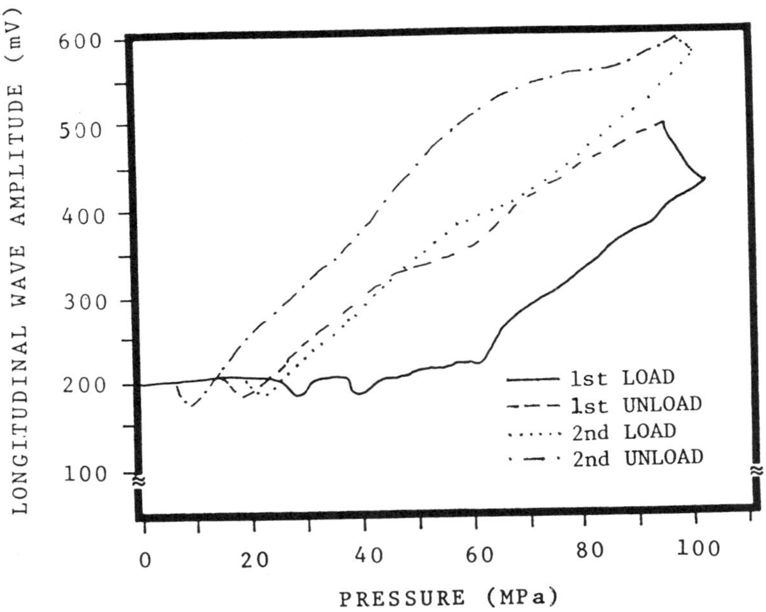

Fig. 8 Longitudinal wave amplitude of alumina powder containing binder

There were notable common features in the echo amplitude for both powders, as typified in Fig. 8, obtained from measurements on powder containing binder. Due to spurious echoes (x), the electronic peak detection system will generate an initial baseline voltage. As wave echo amplitudes within the sample increase, the baseline rises or falls since the (G) echoes and (x) echoes are comparable in amplitude and either constructively or destructively interfere with each other as the (G) echoes travel through the (x) echoes, due to increased wave speed. At higher pressures the echo amplitude dominates and the peak detected signal is proportional to it. Once the first loading is stopped, the pressure spontaneously relaxes but the amplitude continues to increase, indicating that the particles rearranged over a period of minutes into a more efficient configuration. Upon unloading, the wave amplitude decreases, generally paralleling the loading path results. The second loading path does not deviate much from the first unloading path. However, after the second loading is stopped, there is not as much relaxation of the pressure nor increase in wave amplitude as for the first loading. This indicates that most of the particle rearrangement was completed during the first loading. These amplitude trends were repeatable for every sample even though the absolute wave amplitudes were not.

CONCLUSIONS

Techniques developed in the present investigation enable both shear and longitudinal wave speeds to be measured in a ceramic powder during compaction. The measurements show how particle-particle contact and particle rearrangement influence the waves' speeds and amplitudes in the powder.

Wave speeds in a powder during compaction were reproducible under specific conditions, while the wave amplitudes were not. Wave amplitude variations were attributed to the lack of reproducibility of the contact between the sample and the punch, rather than to statistical fluctuations in particle-particle contact. Therefore, behavior of the interior particles of the specimen could not be reliably determined based on the wave amplitudes.

There were clear differences in wave speeds between the binder and binderless powders. Wave speeds were higher for the powder containing binder. This was expected since the binder, serving as a link, effectively increased the amount of contact between particles. Furthermore, powder containing binder exhibited wave speeds that decreased at a slower rate upon reduction of pressure than those of the binderless powder, indicating that the binder acted to retard particle separation when the pressure was reduced.

In conclusion, ultrasonic evaluations of ceramic powders during and after compaction can assist in understanding how particles rearrange and interact. Monitoring ultrasonic waves in a ceramic powder during compaction can be the basis for an industrial real-time sensor that helps to determine the green state quality of the part being formed.

REFERENCES

1. M.P. Jones, G.V. Blessing, and C.R. Robbins, Dry-Coupled Ultrasonic Elasticity Measurements of Sintered Ceramics and Their Green States, Mater. Eval., 44:859 (1986).
2. F.F. Lange, Structural Ceramics: A Question of Fabrication Reliability, J. Mater. for Energy Systems, 6:107 (1984).
3. K.K. Smyth and M.B. Magida, Dynamic Fatigue of a Machinable Glass Ceramic, J. Am. Ceram. Soc., 66:500 (1983).
4. J.J. Mecholsky, S.W. Freiman, and R.W. Rice, Effect of Grinding on Flaw Geometry and Fracture of Glass, J. Am. Ceram. Soc., 60:114 (1977).

5. A. Nagarajan, Ultrasonic Study of Elasticity-Porosity Relationship in Polycrystalline Alumina, J. App. Phys., 42:3693 (1971).
6. R.M. Arons and D.S. Kupperman, Use of Sound-Velocity Measurements to Evaluate the Effect of Hot Isostatic Pressing on the Porosity of Ceramic Solids, Mater. Eval., 40:1076 (1982).
7. J.J. Tien, B.T. Khuri-Yakub, G.S. Kino, D.B. Marshall, and A.G. Evans, Surface Acoustic Wave Measurements of Surface Cracks in Ceramics, J. Nondestructive Eval., 2:219 (1981).
8. W. Kreher, J. Ranachowski, and F. Rejmund, Ultrasonic Waves in Porous Ceramics with Non-Spherical Holes, Ultrasonics, 15:70 (1977).
9. D.S. Kupperman and H.B. Karplus, Ultrasonic Wave Propagation Characteristics of Green Ceramics, Ceram. Bull., 63:1505 (1984).
10. M.P. Jones and G.V. Blessing, Real-Time Ultrasonic Evaluation of Green State Ceramic Powders During Compaction, Nondestructive Testing Communications, 2(5), (1986).
11. C.A. Bruch, Problems in Die-Pressing Submicron Size Alumina Powder, Ceram. Age, 10:44 (1967).
12. F.F. Lange, Sinterability of Agglomerated Powders, J. Am. Ceram. Soc., 67:83 (1984).
13. H. Kolsky, "Stress Waves in Solids," Dover Pub., New York (1963).

ULTRASONIC CHARACTERIZATION OF CONSOLIDATED RAPIDLY SOLIDIFIED POWDERS: APPLICATION TO TYPE 304 STAINLESS STEEL

K. L. Telschow and J. E. Flinn

Idaho National Engineering Laboratory
EG&G Idaho Inc., Idaho Falls, ID 83415

INTRODUCTION

Rapid solidification processing (RSP) of materials is an active field with a potential for producing advanced materials with significantly improved properties compared to materials produced by conventional means.[1] For metal powders, RSP can promote certain beneficial features of the material which depend on the cooling rate of the powder, e.g. chemical homogeneity, very fine structures such as grain size, extension of solid solutions, and the development of metastable phases[1,2]. Consolidation of these powders into useful forms requires an in-depth understanding of the response of the microstructure to extrinsic consolidation parameters such as pressure (stress), temperature, and the time of application for these parameters. At the Idaho National Engineering Laboratory (INEL), a program is underway to consolidate and characterize the consolidation products of rapidly solidified Type 304 stainless steel (SS) powders. Powders produced by two different methods, centrifugal atomization (CA) and vacuum gas atomization (VGA), are being studied. Three approaches for consolidating the RSP Type 304 SS powders are being investigated: hot isostatic pressing (HIPping), hot extrusion, and dynamic consolidation with explosives. The first two methods are associated with fairly high temperatures and, therefore, present the possibility of altering the rapidly solidified microstructures. The dynamic consolidation approach is unique in that it is associated with relatively low bulk temperatures, even though high localized temperatures can result due to particle interaction and deformation.

Studies also are underway at the INEL to nondestructively characterize the microstructural features of these consolidated RSP 304 SS powders. The use of noninvasive ultrasonic techniques is being stressed, particularly the measurement of both attenuation and backscatter from the microstructure. The goal of these measurements is to assess the effectiveness of these ultrasonic techniques for characterization of RSP microstructures and to aid in the evaluation of the various consolidation techniques. The very small grain sizes encountered in the RSP materials present severe challenges to conventional methods of nondestructive characterization. For the case of ultrasonics, high frequencies must be used in order to obtain a measurable scattering from the microstructure. However, due to complications in fabricating high frequency transducers with sufficient sensitivity, trade-offs are required in the types of measurements that can be taken, and thus in the information gained. The attenuation measurements could be made at frequencies as high as 50 MHz, but the measurement of backscatter was limited to around 20 MHz. As much information as possible about the microstructure needs to be inferred from these long wavelength measurements.

The use of ultrasonic backscatter directly from the microstructure is a relatively new technique for characterization. It is being studied because the results depend only on scattering from the microstructure, whereas attenuation can come from absorptive processes not related to microstructural characteristics such as grain size. Also, the backscatter technique requires access to only one surface of the sample and could be adapted to non laboratory situations. One of the main aims of the characterization research is to compare the microstructural information gained from the backscatter data to that obtained from the attenuation measurements, which are better understood and correlated with known microstructural features. This paper describes the results of a particular type of backscatter measurement, called the backscatter energy, and its correlation with attenuation measurements and known microstructural features found by a variety of standard metallographic examinations. Data are presented and discussed for the hot isostatically pressed and hot extruded RSP CA 304 SS powder, but the primary emphasis is on the RSP CA and VGA Type 304 SS powders consolidated dynamically using explosives.

POWDERS

The CA powder used was obtained from a melt of Type 304 SS centrifugally atomized by a high velocity rotating cup and cooled by fast flowing helium gas. The average particle size was approximately 80 μm, based on an "accumulated weight percentage finer than" value of 50%. Additional analysis showed that the particles are primarily spherical with a dendritic microstructure. Structure determinations using x-ray diffraction showed that the particles were primarily austenitic (fcc) with a small amount of bcc phase present.

The VGA powder was also obtained from a melt of Type 304 stainless steel. Argon gas dissolved in the melt was subsequently allowed to freely expand into a vacuum resulting in atomization into the powder form. The average particle size for this powder was about 40 μm, measured as described above. The shapes of the particles were again primarily spherical but included a large number of satellites, particularly associated with the larger particles. X-ray diffraction analysis of this powder showed a predominant fcc phase with a strong component of bcc phase also present. The VGA particles appeared to have a more cellular microstructure throughout the range of particle sizes. A complete description of the two RSP Type 304 SS powders is reported elsewhere.[3]

CONSOLIDATED FORMS

Three techniques of powder consolidation were used in this study: hot isostatic pressing, hot extrusion, and dynamic consolidation using explosives. The HIPped specimens were prepared by exposing hermetically sealed CA powder filled tubes to 207 MPa for 2.5 h at 800 or 900°C. In both cases, the dendritic structure of the CA powder was preserved, i.e., there was no evidence of recrystallization. The 800°C specimen was not fully consolidated, showing about 3% porosity, whereas the 900°C specimen was fully dense.

The extrusion process consisted of preheating a degassed, hermetically sealed, can of powder to 900°C and extruding it through a die to produce an 8:1 reduction in cross-sectional area. Metallographic examination of the extruded material showed full consolidation, no remnant particle boundaries, and a grain size of 8 μm.

In the dynamic consolidation method the explosive charge was applied directly to a static flyer (cover) plate. The monoliths produced by this technique were fully dense and crack free. The microstructure was considerably different from that produced by the extrusion technique. Particle boundaries were clearly evident although severely deformed. Specimens approximately 1 X 1 X 0.4 cm were sectioned from these monoliths and subjected to one hour anneals at temperatures from 200 to 1200°C. Ultrasonic measurements from these heat treated specimens are the primary subject of this report.

ULTRASONIC MEASUREMENTS

Ultrasonic measurements have been shown to be highly sensitive to material microstructure in several ways. In particular, attenuation coefficients generally scale with grain size in metals where the grains are free of inclusions or voids and are equiaxed. The ratio of grain size to ultrasonic wavelength is the determining factor for the attenuation magnitude. Each reflected pulse was recorded separately at an appropriate gain level and Fourier transformed to the frequency domain. The spectra for the front and first two back reflections were corrected for diffraction effects in the conventional way and used to calculate the attenuation coefficient and front surface reflection coefficient. This procedure has been well outlined in the literature.[4,5] Two flat piezoelectric wideband transducers with long quartz delay lines were used for the attenuation measurements [Panametrics V3347 (100 MHz, 0.125 in., 20 μs delay) and V3348 (75 MHz, 0.25 in., 20 μs delay)]. Both transducers had very wide bandwidths of about 40-60 MHz. The lower frequency transducer was used for the highly attenuating CA material and the higher for the VGA material; cross checks were made between the two transducers to ensure accuracy. Optimizing a transducer for each material produced better, more reproducible results than using a single transducer for all the samples. This was because, for a given sample thickness, the measurement error increases both when the attenuation is very small (due to the limited digitization precision of 8 bits) and when it is large (the noise floor is reached). The data were taken with the samples in a water bath and using a scanning stage for maneuvering and alignment. This procedure allowed rapid comparisons between samples and was also useful for the backscatter measurements where averaging of many A-scans was necessary.

Backscatter measurements record the ultrasonic radiation scattered back to the transducer from the scattering centers as the wave passes through the material before it reflects from any secondary surface. The backscatter from a 20 MHz (0.25 in.) conventional wideband transducer was recorded for each sample. A trade-off was necessary in measuring backscatter since its magnitude was generally very small compared to the front and back surface reflections. Scattering usually increases dramatically with frequency, so a frequency where the wavelength is comparable to the scattering particle diameter is desired. However, for these materials this condition would dictate frequencies 50 MHz and above. Transducers at this frequency utilize delay lines as a structural part of the assembly since the piezoelectric element is very thin and fragile. These delay lines introduce extraneous ultrasonic signals from mode conversions and reflections at the boundaries. These extra signals are kept as low as possible in commercial transducers, but are still comparable or larger than the backscatter that is being measured. Therefore, backscatter measurements are usually limited to a maximum of 20-30 MHz, at present, for flat transducers.

The dynamically consolidated forms were not thick enough for extensive spectral analysis of the backscatter signal due to the limited window of data available (0.32 to 0.50 μs). However, a fixed portion of the backscattered signal could be recorded independent of the surface reflections and the total energy in this window calculated. A high gain of at least 60 dB was needed over the front surface reflection to record the backscatter. The backscatter for any one point on the sample surface resulted from the coherent addition of the waves coming from the collection of scatterers below that point. This summation produced a widely varying signal from one point on the surface to the other. Therefore, an averaging was performed by recording the power spectra from many points along the surface and averaging the results to obtain one spectrum for each sample. This procedure proved reproducible and reflected the average scattering from a given sample. Experimentally, the time signal for each point was recorded and the backscatter window was separated out synchronously with the front surface and filtered. This procedure took into account variations in the sample to transducer separation distance along with noise and DC offsets. The small sample thickness allowed for essentially uniform ultrasonic intensity throughout the sample even though it was placed in the near-field. This was verified experimentally; when the sample was displaced along the transducer axis regions could be found where little or no change occurred in the backscatter signal for translations up to 3-4 times the sample thickness. Variations in the magnitude were observed but even the shapes of the power spectra were reproducible over this range. This fact allowed the direct energy comparisons to be made between the samples which are reported here.

In general, some account must be made for the differences in experimental setup which can occur. There are two ways of normalizing the data between runs: to repeatedly measure a known standard sample with each run and to calibrate with respect to the magnitude of the front surface reflection. Both methods produce acceptable results as long as the transducer to sample distance is approximately constant. The latter method was used for this study.

RESULTS AND DISCUSSIONS

Figure 1 shows the measured backscatter energy from the "as consolidated" (NA - not annealed) forms of the RSP material. All exhibited clear evidence of remnant particle boundaries in the microstructure, except the extruded material. Figures 2, 3, 4 show the microstructure for the CA material. The HIPped form (Figure 2) exhibited moderate scattering, presumably from remnant particle boundaries. The extruded form (Figure 3) showed no evidence for any remnant particle boundaries and a very small grain size of only 8 μm; this accounts for the extremely small backscatter observed, which was only recordable at a signal to noise level of about one. The dynamically consolidated form (Figure 4) preserved the largest particle size (note scales in the figures) which was responsible for the largest amount of scattering observed for the CA material.

The VGA consolidated form exhibited a microstructure similar to that of the CA material but with smaller particle sizes, due to the smaller powder sizes used in the consolidation. Both CA and VGA powders were consolidated in the same explosive test. The CA material exhibited considerably more

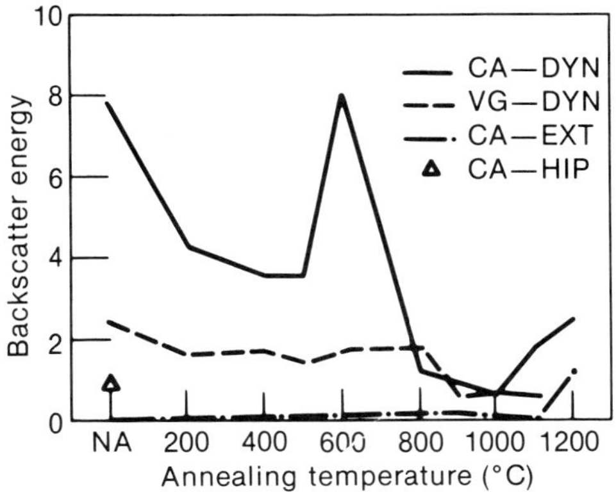

Fig. 1. Backscatter from dynamic consolidated, hot extruded, and HIPped CA and VGA powders before (NA) and after annealing at various temperatures.

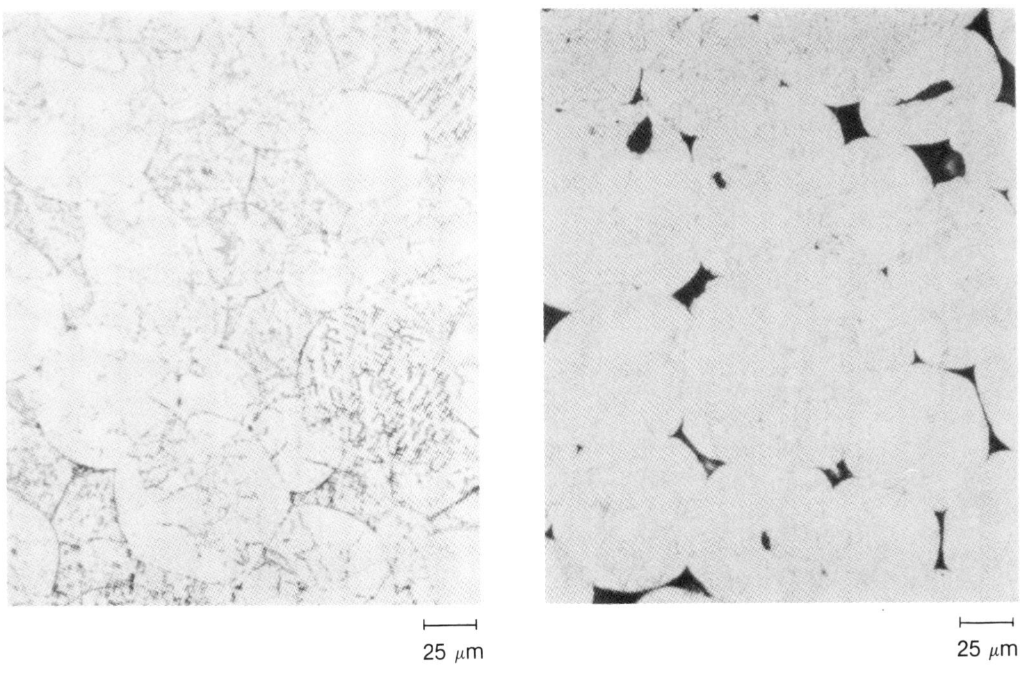

a) 900°C, fully dense

b) 800°C, 3% porosity

Fig. 2. Photographs of as-HIPped CA 304 SS specimens.

deformation, particularly in the form of particle extrusion than the VGA material. This alteration of the remnant particle shape along with the size difference accounts for the greater scattering observed in the CA material.

The measured attenuations for these materials followed essentially the same pattern as the backscatter, with that for the extruded and VGA material being so small at 25 MHz that they had to be measured at 50 MHz in order to yield acceptable error levels. On the whole it can be stated that the backscatter and attenuation reflect the remnant particle size and shape

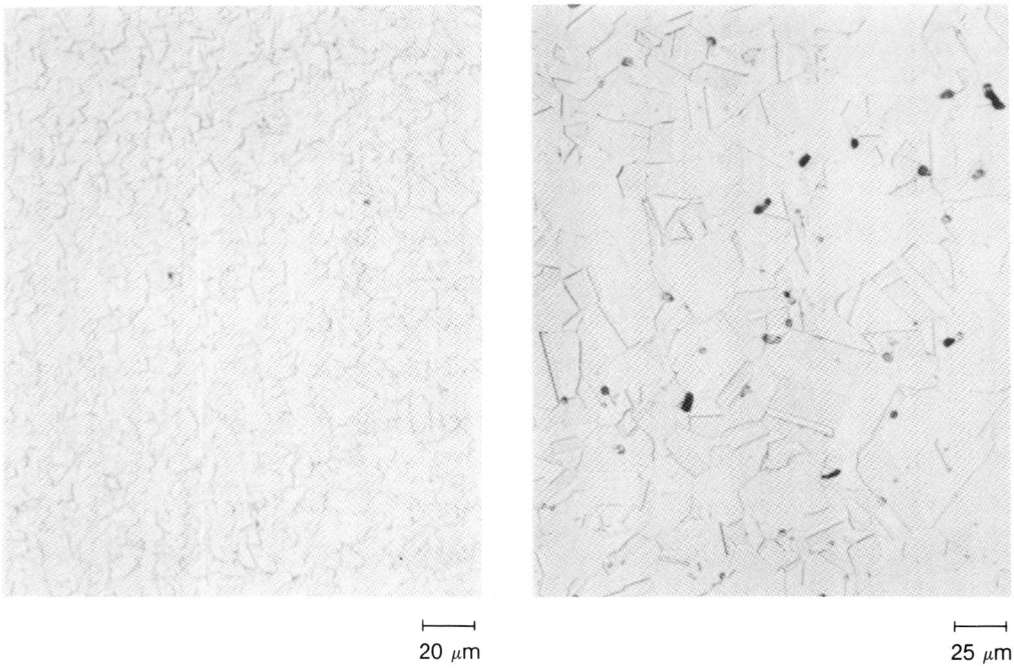

a) As consolidated　　　　b) Annealed, 1200°C/1 h　　6 4265

Fig. 3. Hot-extruded CA powder.

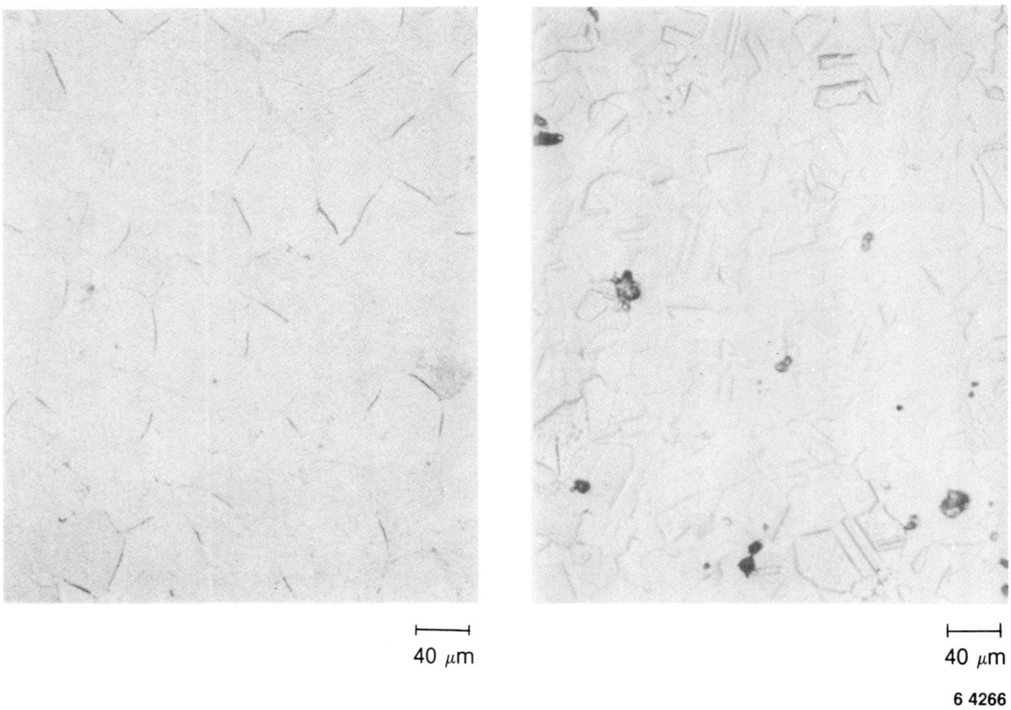

a) As consolidated　　　　b) Annealed, 1200°C/1 h

Fig. 4. Dynamically consolidated CA powder.

after consolidation. The HIPped material was an exception to this in that it exhibited a large attenuation not consistent with its relatively small backscatter. This discrepancy is not understood at present and is under further study. The fact that all these samples were fully dense was supported by density measurements and optical microscopy. There is evidence that the HIP process is less efficient at producing fully dense forms from measurements on a CA sample HIPped at 800°C which resulted in a porosity of 3%. This microstructure is shown in Figure 2 for comparison. The backscatter from this porous sample was over a factor of 10 greater than that for the explosively consolidated CA material and its attenuation was larger by a factor of 7. These results show the extreme sensitivity of the ultrasonic measurements to porosity in these materials.

The dynamically consolidated specimens were annealed at temperatures up to 1200°C for durations of one hour in order to follow changes in the microstructure. Figure 1 shows the corresponding changes in the backscatter and Figures 5 and 6 show attenuation (@ 25 MHz) along with normalized backscatter for the extrusion and dynamically consolidated CA material, respectively. Several interesting features are illustrated by these data. The CA material showed a marked evolution in its microstructure, which was also reflected in the ultrasonic measurements. Essentially three features are evident from Figure 6: large amounts of scattering from the "as consolidated" material, a pronounced effect at 600°C, and another large increase in the scattering at the highest temperatures. The scattering from the "as consolidated" material is thought to be due to the remnant particle boundaries in the microstructure, as previously discussed. No significant changes in the microstructure were seen until the 600°C annealing, where definite regions of recrystallization were observed intermixed with the "shock hardened" remnant particle matrix. This duality of microstructure results in the increased scattering. Further annealing results in marked growth of the recrystallized regions and subsequent reduction in scattering as the material is becoming more homogeneous on the whole. The microstructural evolution in the CA material is further described in other reports.[6-8] Finally, at the highest temperatures, the scattering again rises dramatically. Here, some evidence of grain growth is found in the microstructure; however, the grain size is still very small (especially compared to that which would be present in conventionally prepared material annealed in this way). Figure 4 shows the grain structure of the annealed material and many dark areas which were subsequently found to be bubbles of condensed helium gas. It is now known that substantial amounts of the helium gas used in the preparation of the CA powder remain entrained in the consolidated forms and coalesce into these bubbles at the higher annealing temperatures. Upon melting, this gas is released violently. The large scattering observed at the high annealing temperatures is thought to be due in part to this effective porosity introduced into the material by the helium bubbles.

Figures 3 and 5 show the annealing results for the extruded material. Very little evolution in the microstructure was observed until annealing at the highest temperature, where grain growth and the helium bubbles again appeared. Correspondingly, the ultrasonic results showed little change until 1200°C, where again scattering increased dramatically. It is interesting that both consolidation techniques resulted in essentially the same microstructure after annealing at 1200°C and correspondingly similar ultrasonic properties.

In contrast to the CA consolidated forms, the VGA material exhibited a very complicated microstructure. The "as consolidated" form showed remnant particle boundaries and measurable scattering. Annealing this material produced a more uniform recrystallization compared to isolated regions in the CA material. Even at 900°C, there is more evidence of remnant particle

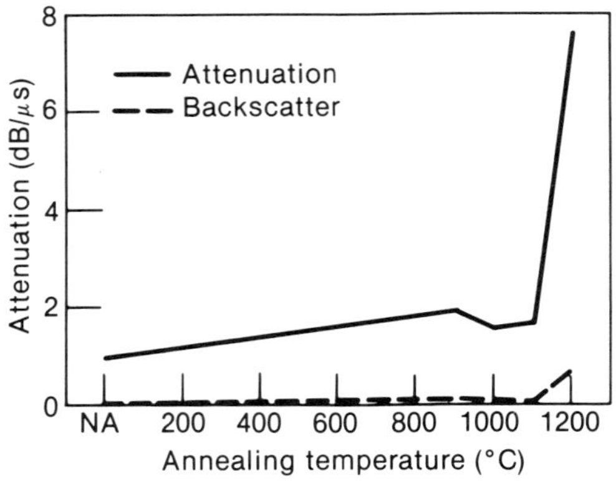

Fig. 5. Attenuation (25 MHz) and normalized backscatter energy (20 MHz) from the CA extruded material, annealed at various temperatures for 1 h.

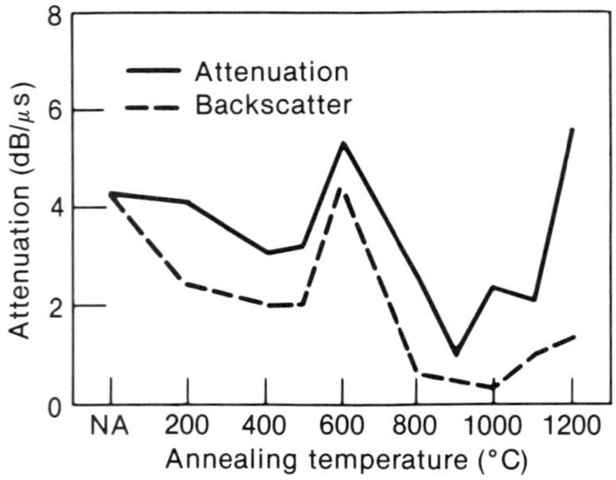

Fig. 6. Attenuation (25 MHz) and normalized backscatter energy (20 MHz) from the CA explosively consolidated material, annealed at various temperatures for 1 h.

boundaries, compared to the CA material. At 1100°C some grain growth is occurring in the VGA material and is comparable to the CA material. The ultrasonic measurements confirm this prosaic microstructural evolution as shown for the backscatter in Figure 1. The attenuation measurements correspond with the backscatter results up to the highest annealing temperatures, where the attenuation rises significantly while the backscatter remains essentially constant. This discrepancy between the two measurements is unique and not understood at present. The VGA powder was not prepared with helium gas and no significant evidence of bubbles was observed.

CONCLUSIONS

These results have shown that distinct microstructural features in the consolidated forms of these rapidly solidified powders can be observed through ultrasonic measurements. In general, the presence or absence of remnant particle boundaries could be observed with results depending on the original particle size, particle extrusion occurring during the consolidation process, and the type of consolidation process used. Annealing tests dramatically demonstrated microstructural evolution in these materials, including the onset of recrystallization zones, grain growth, helium gas condensation into bubbles, and the effects of porosity. All of these features were observable with the ultrasonic measurements in a nondestructive way. Generally, both backscatter energy and attenuation results yielded similar information, with the backscatter technique being significantly simpler to implement and particularly useful for comparative measurements where adequate normalization can be maintained.

ACKNOWLEDGMENTS

The authors acknowledge the assistance of R. A. Steele, G. L. Fletcher, Dr. G. E. Korth, Dr. R. N. Wright, and M. M. Taylor for help with the experimental aspects of this work and preparation of this manuscript. The work described in this paper was supported by the Interior Department's Bureau of Mines under Contract No. J0134035 through Department of Energy Contract No. DE-AC07-76ID01570.

REFERENCES

1. J. E. Flinn, "Rapid Solidification Technology for Reduced Consumption of Strategic Materials," Noyes Publications, Park Ridge, NJ, (1985).
2. M. Cohen, B. H. Kear and R. Mehrabian, Rapid Solidification Processing-An Outlook, in: "Rapid Solidification Processing: Principles and Technologies-II," R. Mehrabian, B. H. Kear and M. Cohen, eds., Claitor's Publ. Div., Baton Rouge, LA, (1980), pp. 1-23.
3. J. E. Flinn, et al., "Characterization of Extrusion Consolidated - Centrifugally Atomized Type 304 SS Powder," EGG-SCM-7221 (to be published.)
4. E. P. Papadakis, Scattering in Polycrystalline Media, in: "Methods of Experimental Physics," Vol.19, P. D. Edmonds ed., Academic Press, New York, (1981), pp. 237-298.
5. E. P. Papadakis, K. A. Fowler and L. C. Lynnworth, Ultrasonic Attenuation by Spectrum Analysis of Pulses in Buffer Rods: Method and Diffraction Corrections, J. Acoust. Soc. Am. 53:1336-1343 (1973).
6. K. L. Telschow and J. E. Flinn, Ultrasonic Backscatter and Attenuation in Consolidated RSP Powder, "Review of Progress in Quantitative NDE," Vol. 5B, D. O. Thompson and D. E. Chimenti eds., Plenum Pub. Co., New York, (1986), pp. 1355-1363.
7. K. L. Telschow and J. E. Flinn, "Ultrasonic Backscatter and Attenuation Measurements for Microstructure Characterization: Application to Consolidated Rapidly Solidified Type 304 Stainless Steel Powder," EGG-SCM-7198, March 1986.
8. J. E. Flinn, et al., "Explosive Consolidation and Post Annealing Response of Rapidly Solidified Type 304 SS Powders," EGG-SCM-7219 (to be published).

ULTRASONIC SCATTERING FROM VOIDS AND INCLUSIONS IN MODEL POLYCRYSTALLINE CERAMICS*

P. Mathieu, A. Stockman and P.S. Nicholson

Ceramic Engineering Research Group
Department of Materials Science and Engineering
McMaster University, Hamilton Ontario Canada

ABSTRACT

Scattering of ultrasound from voids and inclusions in glass and cystallized glass is compared. The scatterers have diameters <100 µm. Crystallized glass has small crystallites which affect the sound field around a defect as the grains of a ceramic might. The effects of these crystallites on the sound field at the defect and at the water-sample interface is discussed.

INTRODUCTION

Ceramics are ideal materials for applications where high strength, low weight, and severe environments are involved. However, they are in general brittle materials where inhomogeneities as small as 10 µm can cause catastrophic fracture.

Ultrasonic detection and characterization of inhomogeneities will play an important role in the applicability of ceramics to commercial products. Although the techniques of scanning for detection purposes are becoming more commonplace in the ceramics industry a thorough understanding of the scattering processes of ultrasound within polycrystalline materials is needed.

The work presented here is but one step in the study of scattering of ultrasound from inhomogeneities of spherical and near spherical geometry. Our objective is to be able to detect and characterize shape and composition of 10 µm flaws. From basic diffraction theory it is known that geometry information can best be obtained when the wavelength of the sound is much less than the size of the defect. This would require that frequencies in the order of gigahertz be used but to scan a component effectively a liquid couplant such as water is needed and at these frequencies the signal loss due to attenuation would be too great. Therefore, a compromise is needed wherein detection is possible and signal processing can be used to obtain geometry information. Frequencies between 10 and 100 MHz do give reasonable signal strengths while yielding wavelengths in the range of the diameter of the inhomogeneity.

* Supported by Defense Research Establishment Pacific, Victoria B.C. Canada.

Single frequency, plane wave sound scattering from spherical inhomogeneities has been examined theoretically by Ying and Truell [1] and Gaunaurd et al [2]. From these works it is clear that the amplitude of any single frequency signal scattered from a sphere will depend upon not only the host matrix but also the sound velocities and density of the inhomogeneity. But in testing components a short pulse of ultrasound is used to get information about the depth of an inhomogeneity. Furthermore, a sound focussing lens may be afixed to increase the sound intensity at the focal point of the transducer. O'Neil [3] described the sound field pattern for a single frequency focussed transducer, and it can be shown that near the focus the field can be approximated by a plane wave. Furthermore, a pulse of sound can be considered as a series of single frequency waves with different amplitudes [4]. So a mathematical model was developed to predict the backscattering of ultrasound from spherical inhomogeneities.

Since direct study of flaws in ceramics is difficult, verification of the model was initiated using microspherical bubbles in transparent glasses and comparing the optical and ultrasonic data with the model predictions [7]. At the next stage of the study spherical inclusions of zirconia and magnesia were imbedded in transparent glasses and again experimental and mathematical data compared favourably. In the present work spheres of zirconia with diameters <200 μm have been imbedded in glasses containing a crystallizing agent so that a glass ceramic can be made as a model of a polycrystalline ceramic. Unfortunately at the time of writing the data for the voids had not been analysed although that of the inclusions was completed for presentation. Information gained at this stage is to be applied to the detection and characterization of spheres in hot pressed ceramics.

THEORY

The theoretical treatment of plane wave single frequency scattering can be found in reference [5]. The scattering amplitudes are calculated for individual waves using the continuity of stresses and strains at the interface between the matrix and the sphere and are functions of the longitudinal and transverse speeds of sound and density of the matrix (C_{l1}, C_{t1}, and ϱ_1) and same parameters for the inhomgeneity (C_{l2}, C_{t2}, and ϱ_2). In the reference these scattering amplitudes are used to determine the total scattering cross-section for the wave whereas our interest lies in the scattering of the longitudinal waves which are given by the scattering amplitude A_m. These can then be used to determine the pressure at the transducer as:

$$p_r(k) = P_R\ k\ a\ (1+i)\ \exp\{-ikR\} \sum_{m=0}^{\infty} (2m+1)\ A_m \qquad \ldots 1$$

where the transducer to inclusion distance, R which is also the focal length of the transducer, is much larger than the sphere radius, a. The quantity i is $\sqrt{-1}$. The pressure, p_r, is a function of the wavenumber of the longitudinal ultrasound in the matrix, k, and P_R is the pressure at the site of the sphere. The pressure amplitude at the focus of the transducer can be determined by expanding the equations of O'Neil [3] about the focus giving:

$$p(\omega) = \varrho\ \omega^2\ S_f\ \frac{\exp[i(\omega t - k'\Delta z)]}{k'} \qquad \ldots 2$$

where:

$$k' = (k/2)\ \{1 + [1 + (D/2R)^2]^{1/2}\} \qquad \ldots 3$$

Here D is the diameter of the transducer and ω is the angular frequency of the wave such that $\omega = C_{11}k = 2\pi f$ where f is the frequency. Δz is the distance from the center of the sphere which is at the focal point of the transducer. The quantity S_f is the amplitude of particle oscillation at the site of the sphere. This value is not easily determined due to attenuation within the couplant and reflections at the sample surface. Therefore it is better to express the pressure $p(\omega)$, and hence $p_r(k)$, in terms of relative amplitudes.

Using the equations above it is possible to determine the relative amplitude of a wave of single frequency backscattered from a spherical inclusion at the focal point of a transducer. An incoming pulse of ultrasound consists of a distribution of frequencies with different amplitudes which can be estimated from the reflection off a flat surface in the focal plane of the transducer [4]. Each element of the distribution is then multiplied by the corresponding frequency in the sphere's response spectrum to obtain a spectrum of backscattered frequencies. It is easier to handle a mathematical distribution with fixed parameters for the frequency distribution of the incoming pulse. The distribution chosen to approximate that of the pulse is a Gaussian with the following form:

$$p_i(f) = p_a \exp\left\{ -4\ln(2) \left(\frac{f - f_a}{f_{FWHM}} \right)^2 \right\} \qquad \ldots 4$$

where p_i is the pressure of the incoming pulse and p_a is the pressure at the maximum amplitude which will be taken as unity. f_a is the frequency at the maximum amplitude (FMA) or center frequency of the distribution and f_{FWHM} is the frequency full width at half maximum amplitude (FWHM) of the distribution.

The total expected pressure at the transducer for a spherical inhomogeneity in the focal zone of the transducer can be written as:

$$P_T(f) = \frac{C_{11}\,a}{2\pi f} \exp\left(-i\frac{C_{11}\,R}{2\pi f} \right) \exp\left\{ -4\ln(2) \left(\frac{f - f_a}{f_{FWHM}} \right)^2 \right\} \sum_{m=0}^{\infty} (2m+1)\, A_m \qquad \ldots 5$$

where $P_T(f)$ is the relative pressure for the backscattered longitudinal wave. The summation over the scattering amplitudes will modify the Gaussian shape of the signal spectrum so the effect of density and sound velocities of both matrix and inclusion and inclusion dimension can be measured from the change in FMA and FWHM of the magnitude spectra. The differences are attributable to resonances of the sphere with the certain frequency components present in the sound pulse which physically represent creeping waves that circumnavigate the sphere [2].

EXPERIMENTAL

Samples of glasses were prepared with compositions listed in Table I. Zirconia particles of diameters 50 - 150 μm were prepared by a rolling mill process then spherical particles were selected from the batch. These were introduced to the glass melt and the melt was returned to the furnace to anneal. The glass was then poured into a mould. After cooling the sample surfaces were cut and polished. The transparent samples were examined optically and the locations of the zirconia spheres were marked by scratches on the surface near (but not above) the inclusion. The samples were scanned ultrasonically to verify the the positions of the inclusions then the radio frequency (RF) signal was recorded using a waveform digitizer.

The samples were returned to the furnace for heat treatment to grow crystallites. In some cases samples were removed from the furnace before they were fully crystallized. This was not intended but did lead to some interesting results to be described later. After cooling the surfaces were again polished and the inclusions were located by ultrasonic scanning using the scratch marks as guides to the locations. Then the ultrasonic RF signals were recorded as were the signal from the surface. In all cases the transducer height above the sample was adjusted so that the sound was focussed on the inclusion or the sample surface.

Recorded signals were filtered in the time domain to remove signals arising from other scatterers. The filtered signals were Fourier transformed and the magnitude frequency spectra were fitted by a quadratic at the peak of the curve in order to determine the FMA. On either side of the peak at points above and below the half maximum amplitude positions, a linear fit was used to determine frequency FWHM.

The parameters which are required to perform the calculations are host matrix longitudinal and transverse sound velocities and density, inclusion longitudinal and transverse sound velocities and density, incoming signal FMA and FWHM. For the host matrix the parameters were measured directly with appropriate planar transducers for sound velocities and displacement technique for density. The parameters were estimated for the zirconia inclusions using the measured porosity and linear relations for density and sound velocities as outline by Anderson [6]. These values are listed in Table II. It is known from previous work [8] that the model is very sensitve to changes in the host's sonic parameters while being relatively insensitive to small changes in the inhomogeneity's sonic parameters. Finally the frequency distribution parameters of the incoming pulse were measured form the Fourier transform of the surface signal in the same manner as described for the signals of the inclusions. Table III lists the observed values which as can be seen vary for different samples.

A computer program was used to calculate and store the scattering amplitudes for a given set of longitudinal and transverse sound velocities and density for the host matrix and inhomogeneity. Calculations were performed in terms of the the dimensionless product of wavenumber and sphere radius (ka). Then a second program used these values with the specified FMA and FWHM for the incoming pulse to calculate the expected values of these parameters for sound backscattered from spheres of diameters from 20 - 200 μm. Lastly, a third program used the scattering amplitudes to calculate the magnitude frequency spectrum and waveform for individual sphere diameters given the FMA and FWHM of the incoming signal.

RESULTS AND DISCUSSION

Using the values listed above calculations were performed to determine the theoretical FMA and FWHM curves for zirconia spheres in glass, partially crystallized glass and fully crystallized glass (glass ceramic). The curves are presented in Fig 1, 2 and 3 and in addition the experimental values have been plotted on the curves. The nonlinearity of the curves is a manifestation of the creeping wave re-inforcing the direct backscattered

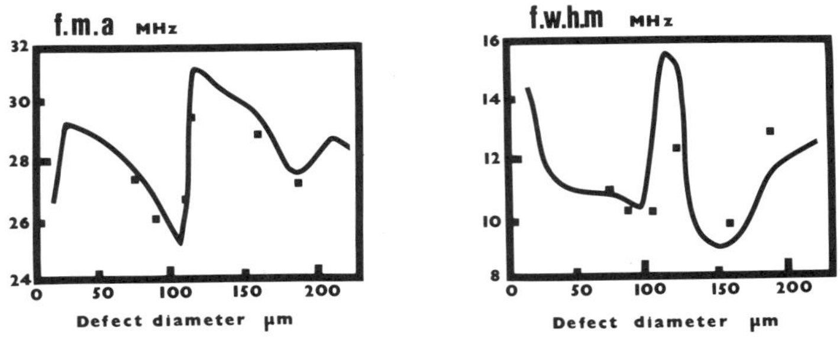

A B

Figure 1 Frequency at maximum amplitude (A) and frequency full width at half maximum amplitude (B) versus diameter for spherical zirconia inclusions in glass. The solid line represents values calculated using the model and the squares represent data measured from the magnitude frequency spectra of the backscattered signals.

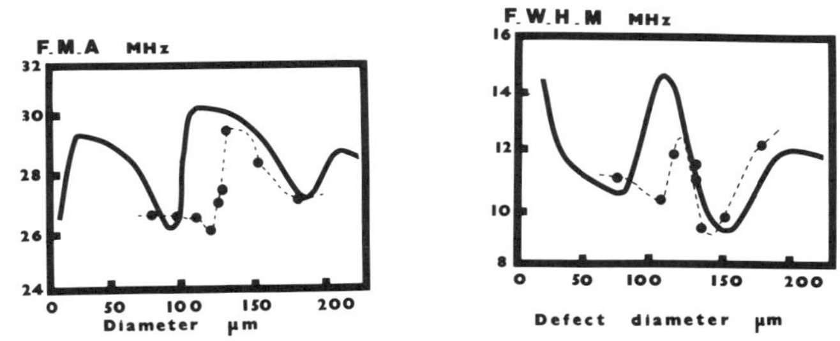

A B

Figure 2 FMA (A) and FWHM (B) versus diameter for spherical zirconia inclusions in partially crystallized glass. The solid line represents values calculated using the model and the dots represent data measured from the magnitude frequency spectra of the backscattered signals.

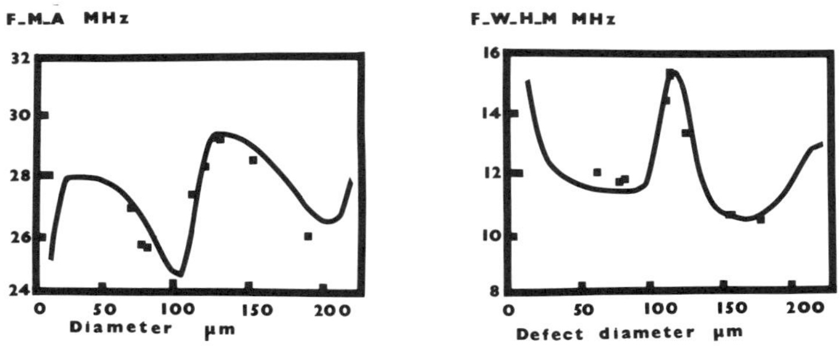

A B

Figure 3 FMA (A) and FWHM (B) versus diameter for spherical zirconia inclusions in glass-ceramic. The solid line represents values calculated using the model and the squares represent data measured from the magnitude frequency spectra of the backscattered signals.

TABLE I
GLASS COMPOSITION

SiO_2	Na_2O	Li_2O	TiO_2
55%	20%	10%	15%

TABLE II
SONIC PARAMETERS

Material	Density (Kg/m^3)	Longitudinal Velocity (m/s)	Transverse Velocity (m/s)
Glass Uncrystallized	2500	6050	3610
Partially Crystallized	2600	6350	3670
Fully Crystallized	2700	6200	3730
Zirconia Spheres	5400	5610	3620

TABLE III
PARAMETERS OF THE ULTRASONIC SIGNALS REFLECTED FROM THE SAMPLE SURFACE

Material	FMA (MHz)	FWHM (MHz)
Glass Uncrystallized	25.6	11.9
Partially Crystallized	25.6	11.9
Fully Crystallized	23.5	12.7

signal. The wavelengths which are being re-inforced, and hence are resonances, do not necessarily correspond to the central frequency of the transducer but rather to higher or lower frequencies (shorter or longer wavelengths) which distorts the shape of the magnitude frequency spectra. These resonances are also the means for determining the the character of the inhomogeneity by seeking out these resonances with a variable frequency transducer or variable bandwidth excitation [9].

Of the curve pairs presented the partially crystallized glass show the poorest fit to the data. As mentioned previously the partially crystallized samples had been removed from the furnace prematurely. When one of the samples was cut open it was discovered that there was a layer of crystallites on the surface and around the inclusions and voids but the entire sample had not crystallized. The cut samples were polished, chemically etched and electron microscopy was performed. Fig 4 is an example of crystallization around a zirconia particle. There are four distinct regions to consider in the photograph. The zirconia particle is in the center with the region surrounding it being composed of crystallites oriented radially from the inclusion. Beyond this is a zone of unoriented crystallites and finally there is the glass which had not yet crystallized. The discrepency between experimental and theoretical data can be explained by the presence of these crystallites surrounding the inclusions. Firstly, the longitudinal and transverse speeds of sound and density of the region immediately surrounding the spheres will in general be different from that measured over the entire sample. And secondly, the sound field will be strongly affected by the irregularity in structure and inhomogeneity introduced by these regions of oriented crystallites.

The theoretical curves for the FMA and FWHM versus inclusion radius fits the experimental data for the uncrystallized glass as expected from previous results [7,8]. Furthermore, the theoretical results match experiment in the case of the fully crystallized glass which suggests that the host matrix is homogeneously crystallized. To check this one sample was cut, polished, chemically etched and scanned under the electron microscope. Fig 5 is the photograph of the remains of a zirconia inclusion surrounded by the matrix of crystallized glass with some evidence of pref-

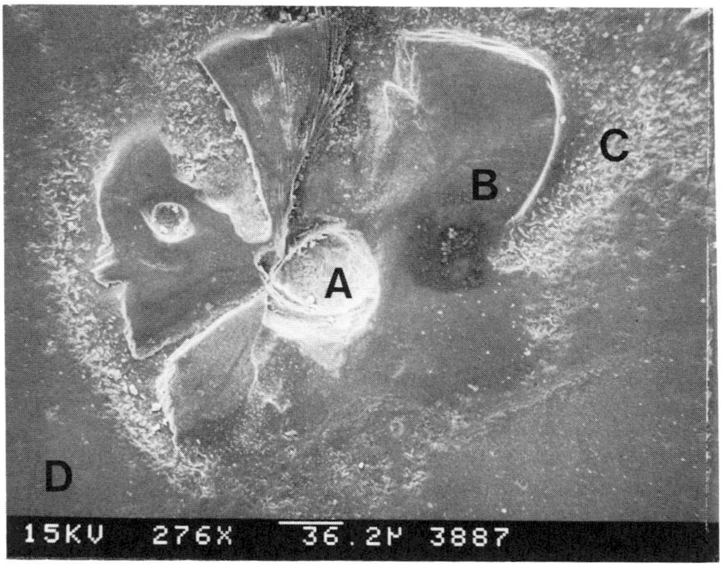

Figure 4 Electron micrograph of a zirconia particle in partially crystallized glass. Showing regions A - the zirconia particle, B - oriented crystallites, C - random crystallites, and D - uncrystallized glass.

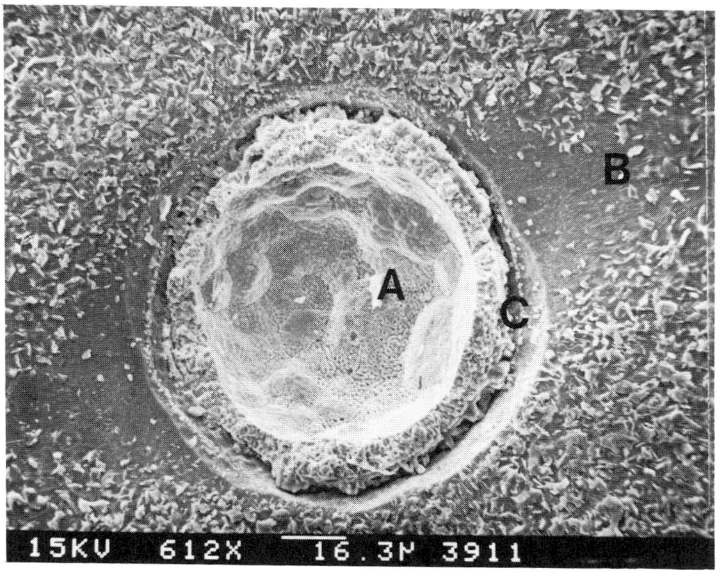

Figure 5 Electron micrograph of a zirconia particle in glass ceramic. Showing A - remains of zirconia particle, B - crystallized glass, and C - 0.5 µm shell of oriented crystallites (partially etched away) at the interface.

erential crystallite growth radially from the inclusion but for the most part the crystallites are homogeneously distributed. Around the inclusion at the interface between the matrix and the inclusion there is a zone of crystallites approximately 0.5 μm long oriented away from the sphere of zirconia. This small region does not appear to have a significant effect upon the scattering of the signal.

A further unexpected consequence of the crystallization process is the formation of a surface layer of oriented crystallites. Fig 6 is a scanning electron micrograph of the interface between the surface layer extending to the right and the fully crystallized matrix. The thickness of the surface layer is <1 mm. Originally it was expected that the frequency distribution of the incoming sound pulse should not change from sample to sample However the measured values of the FMA and FWHM for the incoming pulse can change significantly from sample to sample. The reasons for this are now a matter of study with such questions as crystallite growth and surface layer thickness being examined.

CONCLUSION

In the study of scattering of ultrasound from inhomogeneities in ceramics we have reach the stage of verifying the mathematical model with artificially introduced spherical particles of zirconia having diameters <200 μm in glass ceramics. It has been found that when the glass ceramic matrix is homogeneous so that the sound field is not greatly perturbed by differences in sound velocities and densities then the model developed to calculate scattering from spherical inhomogeneities can be used to characterize inclusions. However, when two distinct phases are present with different sound velocities and densities then the geometry of the phase encasing the inclusion will affect both the detection and characterization by modifying the radio frequency signal. Furthermore the presence of a

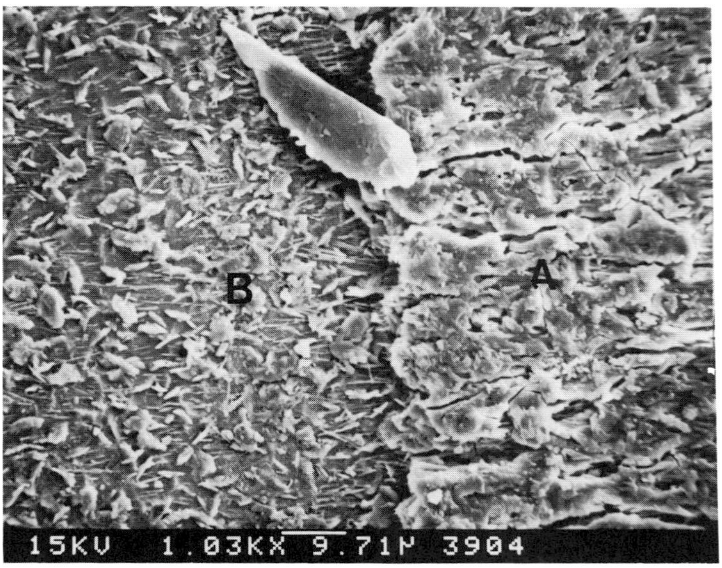

Figure 6 Electron micrograph of the boundary between the surface layer of oriented crystallites (region A) and glass ceramic (region B).

second phase at the surface affects the signal transmitted to the inhomogeneity of interest.

REFERENCES

1 Ying C.F. and Truell R., "Scattering of a Plane Longitudinal Wave by a Spherical Obstical in an Istotropically Elastic Solid", J.Appl.Phys., 27, 1086-1097, 1956.

2 Gaunard G.C., Tanglis E., Uberall H., Brill D., "Interior and Exterior Resonances in Acoustic Scattering I. - Spherical Targets", Il Nouvo Cimento, 76B, 153-175, 1983.

3 O'Neil H.T., "Theory of Focussing Radiators", J.Acoust.Soc.Am., 21, 516-526, 1949.

4 Tittmann B.R., Cohen E.R., Richardson J.M., "Scattering of Longitudinal Waves Incident on a Spherical Cavity in a Solid", J.Acoust.Soc.Am., 63, 68-74, 1978.

5 Ultrasonic Methods in Solid State Physics, R. Truell, C. Elbaum, and B.B. Chick, (Academic Press, New York N.Y., 1969), pp 161-179.

6 Anderson O.L. "Determination and Some Uses of Isotropic Elastic Constants of Polycrystalline Aggregates Using Single Crystal Data", in Physical Acoustics, Vol III - Part B, W.P. Mason (ed), (Academic Press, New Tork N.Y., 1965), pp 43-95.

7 Stockman A. and Nicholson P.S., "Ultrasonic Characterization of Model Defects in Ceramics Part I: Voids in Glass - Theory and Practice", Mat. Eval., 44, 756-761, 1986.

8 Stockman A., Mathieu P., Nicholson P.S., "Ultrasonic Detection and Characterization of Microspherical Defects in Model Ceramics" in Review of Progress in Quantitative Nondestructive Evaluation, Vol 5A, D.O. Thompson (ed), (Plenum Press, New York N.Y., 1986), pp 101-107.

9 Stockman A. and Nicholson P.S., "Ultrasonic Flaw Detection in Model Ceramic Systems" in Advances in Materials Characterization II, R.L. Snyder, R.A. Condrate Sr. and P.F. Johnson (ed), (Plenum Press, New York N.Y., 1985), pp 291-300.

BEAM-HARDENING CORRECTION METHODS FOR POLYCHROMATIC X-RAY CT SCANNERS USED TO CHARACTERIZE STRUCTURAL CERAMICS*

E. Segal** and W. A. Ellingson

Materials and Components Technology Division
Argonne National Laboratory
Argonne, Illinois 60439

INTRODUCTION

The use of medical computed tomography (CT) systems for inspecting industrial materials and components such as ceramics,[1] which have higher mass density and electron density than one encounters in medical applications, is subject to difficulties such as beam hardening and scattering. Beam hardening (BH) occurs when polychromatic radiation is preferentially attenuated by the material as a function of photon energy. The BH effect prevents accurate measurement of local material density and causes artifacts that lower the sensitivity of contrast and spatial resolution. Because BH problems cause significant complications in image interpretation, a great deal of effort has gone into addressing this problem for medical applications.[2] However, little effort has gone into developing BH corrections for industrial materials such as ceramics. The purpose of this paper is to present a BH correction method that takes into account both the photon spectrum of the X-ray head and the properties of ceramic materials. Scattering contributions to image degradation will not be considered here.

NATURE OF BEAM HARDENING IN CERAMICS

A CT scanner measures the attenuation of photons through an object along projection lines, as shown schematically in Fig. 1. The measurement of attenuation can be expressed[2,3] as

$$I(p,\theta) = I_o e^{-\int \mu(x,y) ds} , \qquad (1)$$

where

*Work supported by the U.S. Department of Energy, Office of Fossil Energy, Advanced Research & Technology Development Materials Program and the Office of Conservation and Renewable Heat Engines Program, under Contract W-109-Eng-38.

**Visiting Scientist, Technion-Israel Institute of Technology, Haifa, Israel.

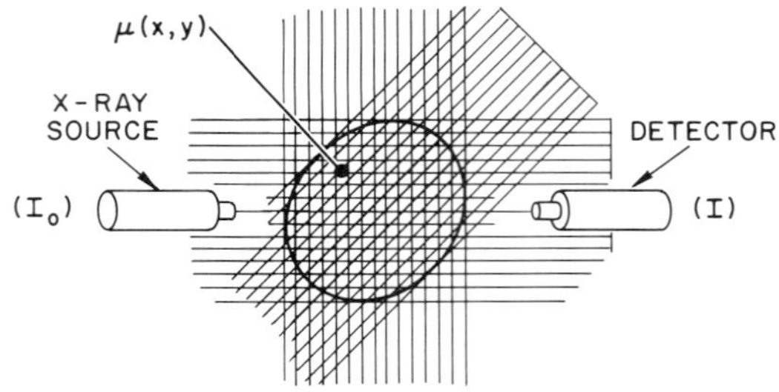

Fig. 1. Schematic Setup of a CT Scanner.

$I(p,\theta)$ = normalized photon intensity (as a function of ray projection, p, and projection angle, θ) after penetrating an object of dimension s (in cm);

I_o = normalized x-ray (photon) intensity of the source; and

$\mu(x,y)$ = total local linear attenuation coefficient (LAC) of the object in this line path, in cm^{-1}.

In the reconstruction process, the integral equations of all the ray projections are solved for the total local LAC, $\mu(x,y)$. The total LAC, μ_L, of the object (in cm^{-1}) is dependent upon the energy, E, of the X-rays; the atomic numbers, z, of the elements composing the object; and the mass density, ρ, of the object:

$$\mu_L = \mu_L(E,z,\rho) \quad . \tag{2}$$

The dependence of μ_L on ρ is linear and is given by

$$\mu_L = \rho\mu_M(E,z) \quad , \tag{3}$$

where μ_M is the mass attenuation coefficient in cm^2/g. The quantity μ_M, which is defined[2] as

$$\frac{\mu L}{\rho} = \mu_M(E,z), \tag{4}$$

is a complex function[4] of E and z. In order to calculate a total μ_M for a polychromatic photon source, the photon energy spectrum of the source must be known or assumed. For monoenergetic radiation, this is not a concern; E is known, and an effective z can be calculated according to Mayneord[4] as

$$z_{eff} = \sqrt[2.94]{az_1^{2.94} + bz_2^{2.94} + cz_3^{2.94} + \ldots} \tag{5}$$

where a, b, c, etc. are the fractional content of electrons belonging to elements z_1, z_2, z_3, etc.

Figure 2 shows the calculated μ_M as a function of energy for a few elements from which ceramic compounds are composed[5] and Fig. 3 gives the μ_M of a few common ceramic compounds. Water is included because it is a commonly used reference material for medical CT scanners. The calculated μ_M values of these compounds were bssed on the μ_M values of the constituent elements as given in Ref. 5.

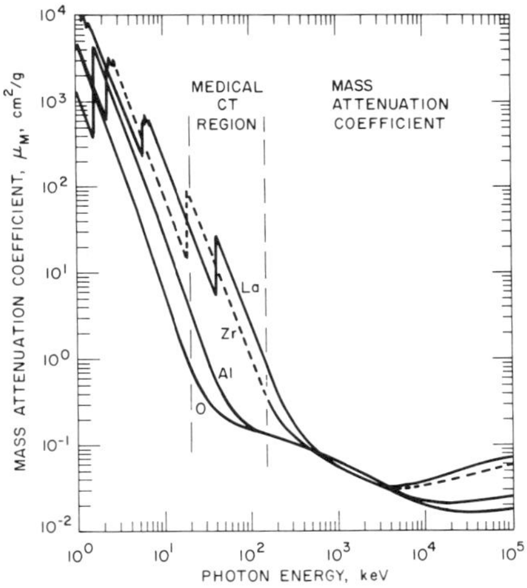

Fig. 2. Mass Attenuation Coefficient of Four Elements as a Function of Energy.

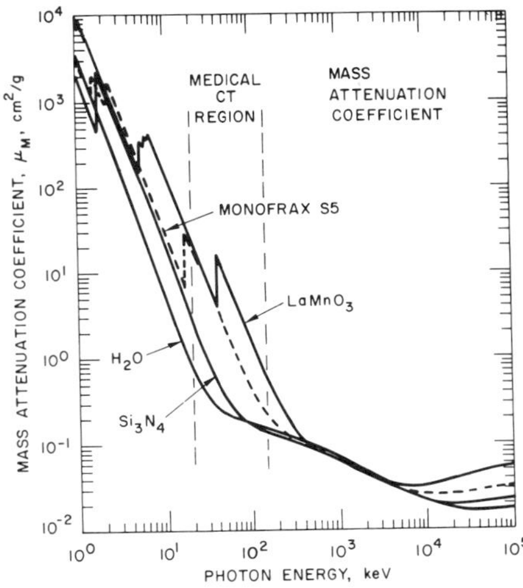

Fig. 3. Mass Attenuation Coefficient of a Few Common Ceramic Compounds.

As noted earlier, for a homogeneous material, z (or the effective z) is constant. Therefore, for a homogeneous material scanned with a mono-energetic beam, the μ_M will be constant and the calculated $\mu_L(x,y)$ will be directly proportional to the local density of the material, $\rho(x,y)$. However, for a polychromatic photon beam this is not the case and the effective total LAC, $\mu_{L_{eff}}$, for homogeneous material becomes

$$\mu_{L_{eff}}(x,y) = \rho(x,y) \int_{E_L}^{E_U} \mu_M(E)\,dE \quad , \tag{6}$$

where E_L and E_U are the lower and upper limits, respectively, of the X-ray head photon energies (see Fig. 4).

From Figs. 2 and 3, one can see that, for ceramic materials, μ_M changes by an order of magnitude or more in the energy region of a typical medical scanner. This implies that the effective LAC (see Eq. 6) is no longer proportional to the local density of the material and unless a correction is made, the measured attenuation will not be a measure of the material density.

THEORETICAL BEAM HARDENING RESULTS

In order to establish an idea of the severity of BH in ceramics imaged with a polychromatic X-ray source, theoretical calculations were made from available information about X-ray attenuation[3] and typical X-ray head spectra for medical CT scanners. The typical polychromatic X-ray head spectrum selected was that of a Siemens Model DR-H scanner, as experiments could be conducted on this machine for verification. The spectra of the X-ray source in a model DR scanner (as published by Siemens) for two different tube voltages are shown in Fig. 4.

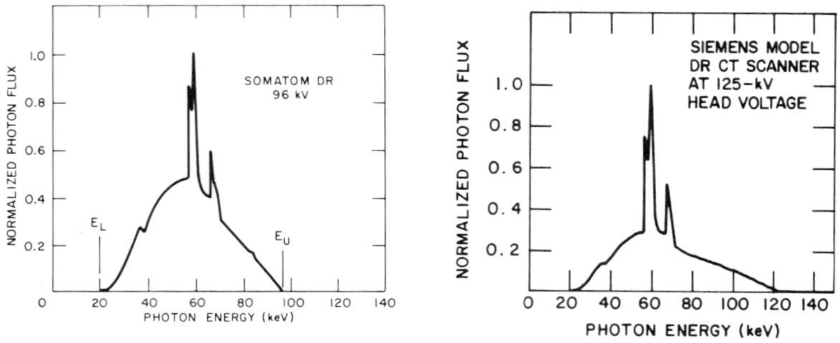

Fig. 4. Quantum Spectra for Siemens Model DR X-Ray Source at Two Different Tube Voltages.

The effect of material attenuation was calculated for several ceramics as a function of thickness. Figure 5 shows the effect of attenuation in green and dense Si_3N_4 specimens ranging from 1 to 5 cm in thickness. Green ceramics have reasonably low densities (ρ = 1.3-2 g/cm^3) relative to densified ceramics (3 to 5 g/cm^3). Of primary interest is the marked low-energy attenuation in both green and dense ceramics. This is, of course, expected from the attenuation data shown in Figs. 2 and 3. Figure 6 shows the effect of attenuation on two ceramic compounds of much higher z and mass density than the Si_3N_4 shown in Fig. 5. The thickness ranges shown in Fig. 6 are 1-10 mm for Monofrax S5 (40% ZrO_2, 48% Al_2O_3) and 50 μm to 1 mm for the electronic ceramic $LaMnO_3$. A strong effect of the La K-edge (38.9 keV) is evident in the spectra for $LaMnO_3$ (Fig. 6b); this K-edge adds a low-energy component to the total received photon intensity.

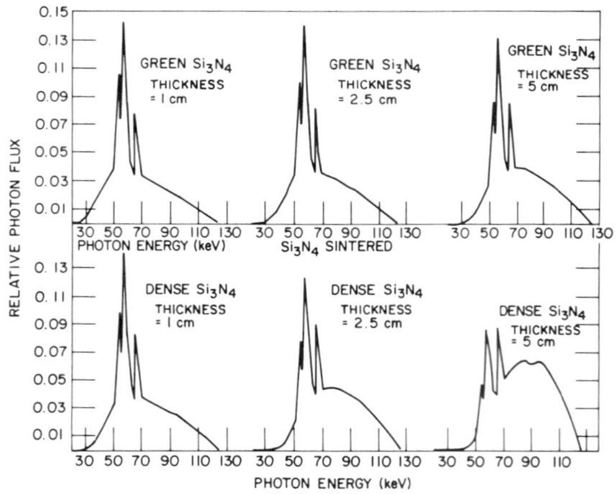

Fig. 5. Effect of Specimen Thickness (in the 1 to 5-cm Range) on X-Ray Attenuation in Si_3N_4.

EXPERIMENTAL INVESTIGATION OF BH EFFECT IN CERAMICS

To estimate the actual image-degrading effect of BH, we conducted a series of experiments with two medical CT scanners. Both rectangular- and circular-cross-section ceramic specimens were used. To reduce the process-induced density variations that normally occur in rectangular ceramic specimens, these components were cut from larger blocks at locations

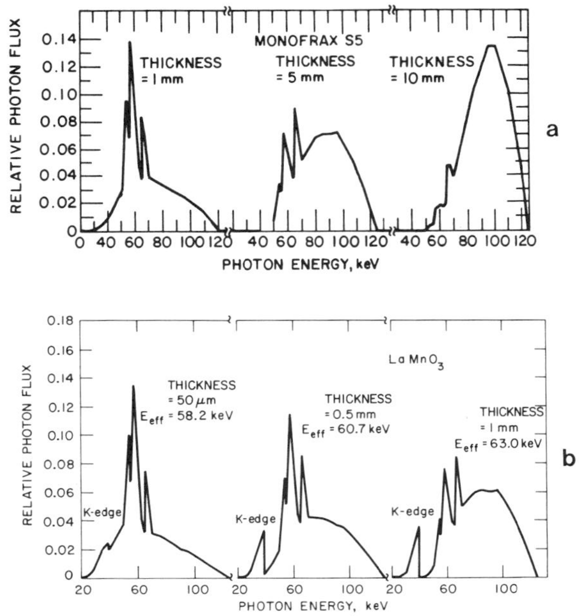

Fig. 6. Effect of Specimen Thickness on X-Ray Attenuation in Two Ceramic Compounds of Higher z and Higher Mass Density than the Si_3N_4 Specimens of Fig. 5.

where minimum density gradients would be expected. The cylindrical specimens were cold-pressed with reasonably small L/D ratios; compaction theory[7,8] predicts that density gradients will be present in such specimens.

Figure 7 shows a CT scan (1-mm slice thickness, special head kernel) of a green Si_3N_4 disk of low density (1.2 g/cm^3; z_{eff} = 12.1). The densitometer scan is nominally across the diameter of the specimen. Although there is wide variation in the density across the bulk of the body, sharp density peaks are noted near the edges. These are not real high-density regions, but primarily artifacts caused by BH. In this case, the BH effect is about 9%.* Figure 8 shows a CT scan (2-mm slice thickness, special head kernel) of a rectangular-shaped specimen of green Si_3N_4 with a slightly higher density (ρ = 1.97 g/cm^3). A BH effect of about 7.7% is observed.

As noted earlier, BH is affected not only by mass material density but also by atomic number, z. In order to experimentally assess the effect of higher z on BH, a test was conducted with liquid Freon TF. The density of liquid Freon TF is 1.565 g/cm^3 and z_{eff} is 14.4. The resulting CT image (10-mm slice thickness) is shown in Fig. 9. The liquid Freon TF was contained in a closed-cell styrofoam container which contributed little, if any, BH effect.

In order to avoid the BH problem, a correction for ceramic materials will be necessary. The approach is discussed in the next section.

---

*BH is defined here as the difference between the average low and high CT values, divided by the sum of 1000 plus the average CT value.

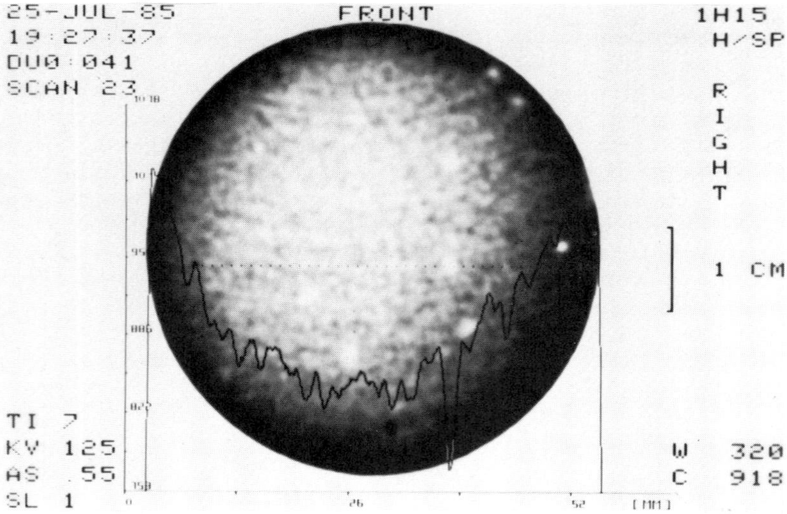

Fig. 7. CT Scan of 5-cm-diam Green Si_3N_4 Disk. The dark ring surrounding the disk is due to BH. The BH effect is about 9%.

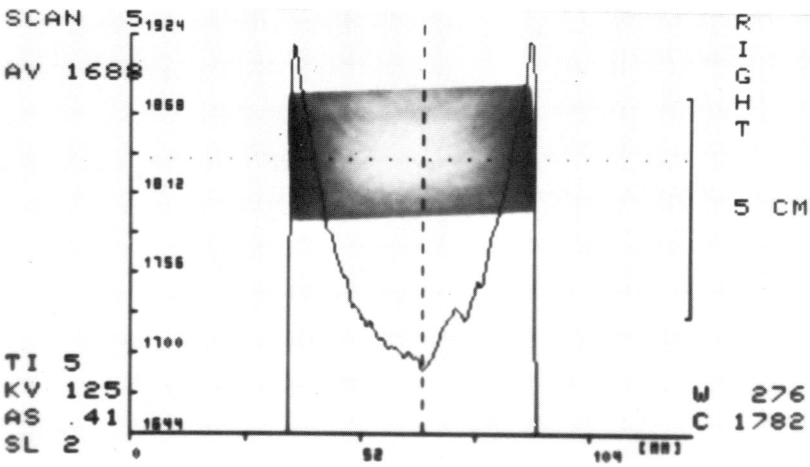

Fig. 8. CT Scan of Green Si_3N_4 Block. The BH effect is 7.7%.

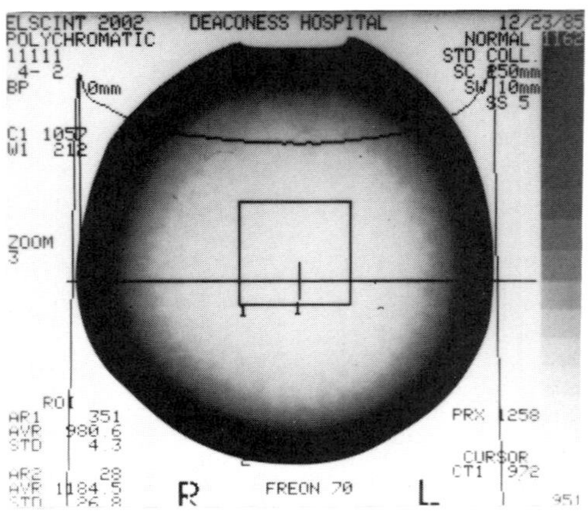

Fig. 9. CT Scan of Liquid Freon TF in a 70-mm-ID Styrofoam Container. The BH effect is 12.7%.

APPROACHES TO CORRECTION FOR THE BH EFFECT

The "Water Bag" Approach

BH is a function of the depth of penetration or the geometry of the object. In a noncylindrical object, different CT projections will undergo different BH effects because of different ray-path lengths. In the early days of medical tomographic scanning with polychromatic radiation, patients were surrounded by a water bag to avoid BH artifacts in the resulting images. A "water bag equivalent" for ceramics is a fitted symmetric structure of the same material (Fig. 10). Putting the object inside a cylinder of the same material ensures that all rays from all directions suffer the same BH effect.

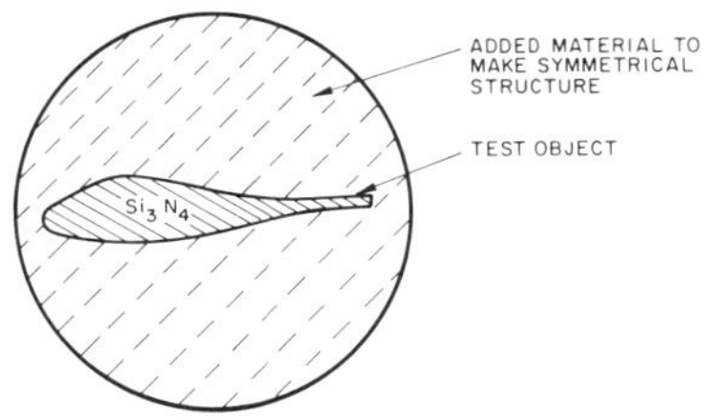

Fig. 10. "Water Bag" Method to Reduce BH Effect in Ceramics.

To estimate the effectiveness of this method, a cold-pressed MgO cylinder (ρ = 1.8 g/cm^3, z_{eff} = 10.7) was scanned with and without a ceramic "water bag equivalent" -- in this case, a Teflon ring (ρ = 2.15 g/cm^3, z_{eff} = 8.2). Figure 11 shows a CT image of the MgO ceramic scanned without a "bag"; the BH effect is about 8.7%. Figure 12 is a similar CT scan with the MgO surrounded by the Teflon ring. As the object and the ring are symmetrical, the BH is the same for all directions. The ring reduces the BH effect in the ceramic to <3%. In this case, however, since the ring reduces the number of photons reaching the object, it increases the statistical noise and reduces the contrast resolution of the reconstructed image. This approach, in principle, can only reduce the BH effect but not eliminate it.

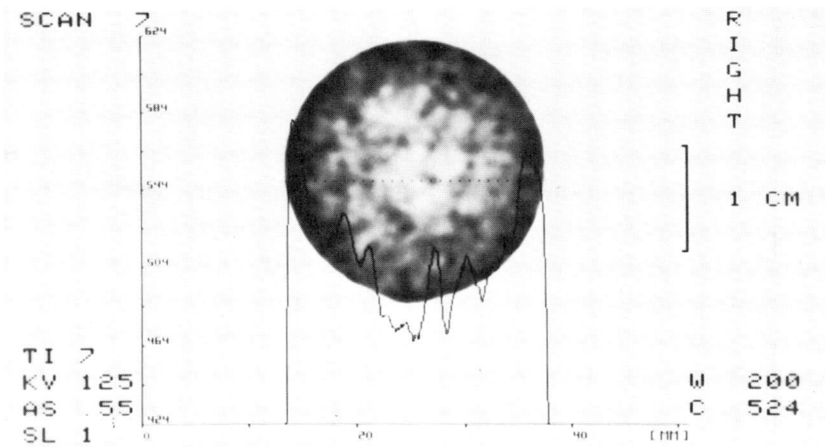

Fig. 11. CT Scan and Density Trace of MgO Specimen 1 Without Teflon Ring.

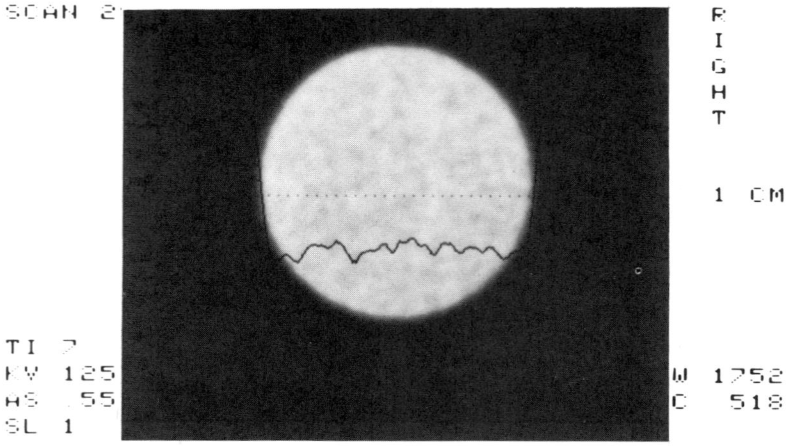

Fig. 12. CT Scan and Density Trace of MgO Specimen 1 Inside Teflon Ring. The density trace no longer shows the very high-density outer region seen within the MgO in Fig. 11.

Pre-Specimen Beam Filtering

A second approach is to make the BH correction in the machine itself prior to irradiation of the object, by use of an equivalent filter made of a standard material. The thickness and shape of the filter will vary with the material and geometry of the object, to ensure that all parts of the object have the same effective BH. For example, a cylindrical object would require the filter geometry shown in Fig. 13.

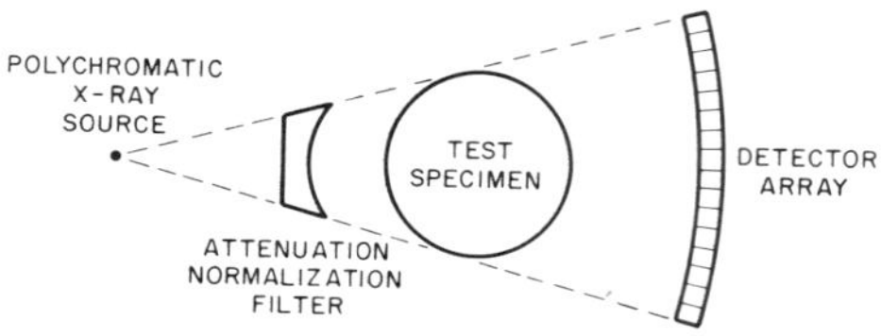

Fig. 13. Prefiltering of the X-Ray Beam to Reduce BH Effects.

The disadvantages of this BH correction approach, besides the inconvenience of designing the filter, are similar to those of the water bag approach: (1) it will reduce the BH but will not eliminate it and (2) the hardening of the beam lowers the contrast resolution and increases the noise of the picture, because the optimal energy for typical ceramic components is in the low region of the X-ray spectrum (Fig. 3).[9,10]

Proposed Linearization Correction

In this proposed method, one corrects the nonlinear preprocessed CT data by a new effective LAC, μ_L', which makes the function $\ell n\ I/I_o = -\mu_L x$ linear as a function of penetration depth x.[2] These values of μ_L' are calculated for a specific ceramic material and a specific polychromatic photon spectrum. By linearizing the function $\ell n\ I/I_o$, one obtains an effective monoenergetic beam. Images reconstructed with μ_L' will be free of BH effects. (However, this may come at the expense of increased image reconstruction time.*) Such a linearization correction requires access to the raw projection data sets in a CT scanner after normalization. (In medical CT scanners, special agreements may be required for such access.)

In order to establish the validity of this linearization approach, a test scan was made on a second-generation CT machine (Elscint Model 2002) in which a linearization correction had already been implemented. The specimen was the same green Si_3N_4 sample (ρ = 1.2 g/cm^3, z_{eff} = 12.1) shown in Fig. 7. The CT image obtained with the linearization correction is shown in Fig. 14. Unlike the image of Fig. 7, no BH is apparent at the outer edge; however, an apparent "negative BH" (i.e., higher density in the center) is present. Of course, the linearization correction used here had been optimized for tissue and water-like materials, and may not be suitable for ceramics. In practice, a correction for the particular ceramic material of interest, based on the measured attenuation coefficient of that material (see Fig. 3), would be used. Details of such a linearization correction will be reported elsewhere.[11]

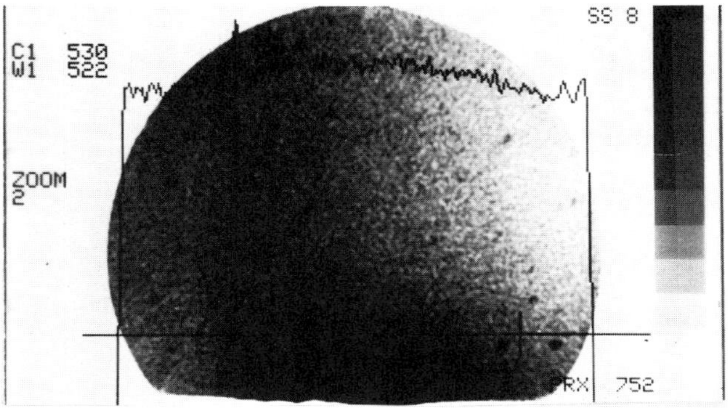

Fig. 14. CT Scan of Specimen of Fig. 7, Obtained with a Second-Generation Elscint Model 2002 Machine. No BH effect is apparent at the outer edge, but an apparent "negative BH" of about 0.5% is present.

CONCLUSIONS AND RECOMMENDATIONS

Polychromatic medical CT scanners can be used for scanning ceramic materials, but care needs to be taken to correct for BH in a material-specific manner.

ACKNOWLEDGMENTS

The work presented herein has required the cooperation of many people. We would especially like to acknowledge the efforts of Dr. M. W. Vannier, R. Knapp, and C. Offutt of the Mallinckrodt Institute of Radiology and Dr. I. Zmora of the Elscint Corporation, and to thank the Mallinckrodt Institute of Radiology and Deaconess Hospital for allowing us to use their equipment. Without this help, the work could not have been completed.

*In an experiment with a Si_3N_4 disk, no appreciable increase in imaging time was noted.

REFERENCES

1. T. Taylor, W. A. Ellingson, and W. D. Koenigsberg, "Evaluation of Engineering Ceramics by Computed Tomography," Atomic Energy of Canada Limited (AECL) Report AECL-9005, Chalk River, Ontario, Canada (1985).
2. R. A. Brooks and G. DiChiro, Beam Hardening in X-Ray Reconstructive Tomography, Phys. Med. Biol. 21(3):390-398 (1976).
3. P. Rüegsegger, Th. Hangartner, H. V. Keller, and Th. Hinderling, Standardization of Computed Tomography Images by Means of Material Selective Beam-Hardening Correction, J. Computer Assisted Tomography 2:184-188 (April 1978).
4. W. V. Mayneord, The Significance of the Röntgen, Int. Union Against Cancer, ACTA, Brussels, 2:271 (1937).
5. Y. E. Segal, Mass Attenuation Coefficients of the Elements ^{1}H to ^{100}Fm; 1 keV to 100 MeV, Technion-Israel Institute of Technology, Haifa, Israel, private communication (1985).
6. R. P. Gardner and R. L. Ely, Jr., "Radioisotope Measurement Applications in Engineering," Reinhold Publishing Co., New York (1967).
7. R. A. Thompson, Mechanics of Powder Processing: I, Model for Powder Densification, Bull. Am. Ceram. Soc. 60(2):237-243 (1981).
8. R. A. Thompson, Mechanics of Powder Processing: II, Finite-Element Analysis of End-Capping in Pressed Green Powders, Bull. Am. Ceram. Soc. 60(2):244-247 (1981).
9. Y. Segal, A. Notea, and E. Segal, A Systematic Evaluation of NDT Methods, in: "Research Techniques in NDT," Vol. III, Chapter 9, 293-321, R. S. Sharpe, ed., Academic Press, New York (1977).
10. L. Grodzings, Optimum Energies for X-Ray Transmission Tomography of Small Samples, Nucl. Instrum. Meth. 206:541-545 (1983).
11. E. Segal, W. A. Ellingson, Y. Segal, and I. Zmora, Development of Beam-Hardening Correction Method for Computerized Tomographic Imaging of Structural Ceramics, to be presented at the Review of Progress in Quantitative NDE, University of California, San Diego, La Jolla, CA, August 3-8, 1986.

MATERIAL CHARACTERIZATION FOR PROCESS CONTROL FOR ALUMINUM ALLOYS AND ADVANCED MATERIALS

P.R. Bridenbaugh, B.S. Shabel
and A.K. Govada

Alcoa Laboratories
Alcoa Center, PA 15069

INTRODUCTION

The North American aluminum industry and Alcoa are today under intense cost and quality pressures from imports and other materials as traditional customers seek to maintain competitive positions in a rapidly changing marketplace. Alcoa's response to these pressures has been the development of business strategies aimed at meeting increasingly stringent customer requirements through the discovery, design, characterization and development of cost-effective manufacturing technologies for a broad array of advanced materials systems. These systems include metals, ceramics, polymers and laminates and composites combining the inherent strengths of each. Nondestructive evaluation (NDE) and characterization procedures for these materials systems represent crucial aspects of the company's advanced manufacturing and process control technology base.

Effective process control provides improved product quality as well as greater consistency. The combination of computer-aided design (CAD) and computer-aided manufacturing (CAM), coupled with quantitative understanding of materials behavior under various processing conditions, provides the foundation for computer-integrated manufacturing (CIM). True CIM makes possible a product design process beginning with customer specifications and ending with a material whose processing history is customized to specifically satisfy structural requirements.

For many current products, experiential knowledge is sufficient to produce a satisfactory material. In these cases, quality and process control are achieved through monitoring of "external" process parameters such as product shape, temperature (material or equipment) and alloy composition. These are major applications of existing NDE technology. However, this is only part of what NDE can and will do for advanced manufacturing technology. In this paper, attention is focused on some newer areas which emphasize the growing importance of quantitative nondestructive characterization of microstructures.

NDE AND MICROSTRUCTURE/PROPERTY RELATIONSHIPS

The rapidly expanding understanding of microstructure/property relationships is the foundation of a revolution currently underway in the realm of materials science. This evolution has and will continue to lead product designers to specify new materials with properties beyond today's conventional alloys. The aluminum industry is consequently faced with the need for a new generation of aluminum-based products. For companies willing to undertake the challenges, these design criteria also offer the opportunity to create a whole new generation of highly-engineered laminate and composite structural materials. In both cases, but particularly the latter, accumulated knowledge of how changes in the external process variables of the manufacturing processes affect the material is not available. Other information, therefore, must be used to control the fabrication process.

Such information is contained in the microstructure-property relationships. These relationships are the controlling aspects of a material's behavior during manufacturing as well as in service. If one can measure and control the "right" attributes of the materials microstructure, in principle, finished products can then be produced with the required microstructure for the desired end use. The fundamental objective of process design and control thus becomes the elimination of defects and control of microstructure throughout the fabrication process to achieve the desired product performance levels. As indicated in Figure 1, this parallels the methodology of designing alloys from an end use viewpoint.

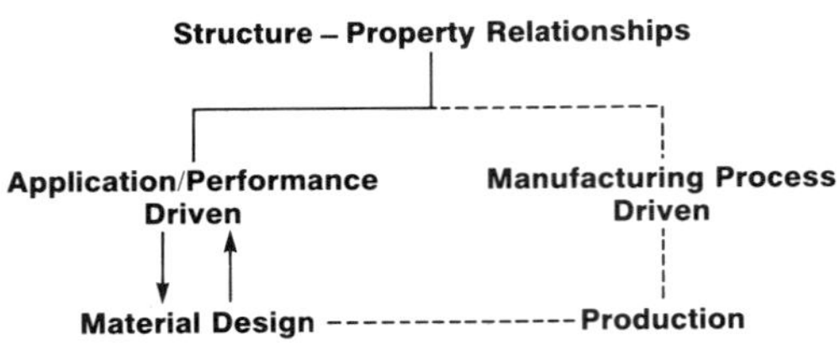

Figure 1

Relationship of Material Science to Manufacturing Process

The explosive growth in NDE and measurement technologies adds credibility to the application of structure-property understanding as an improved means of process control, improved manufacturing efficiency, flexibility and product quality.

The manufacturing system approach can also be expanded to reflect the symbiotic relationship between sensors, controls and process models (Figure 2). Microstructure-property relationships are intrinsic to the process model component of an "intelligent" system. This is a modified version of a relationship described by Mehrabian and Wadley[1] as a basis for "intelligent" or "expert" systems.

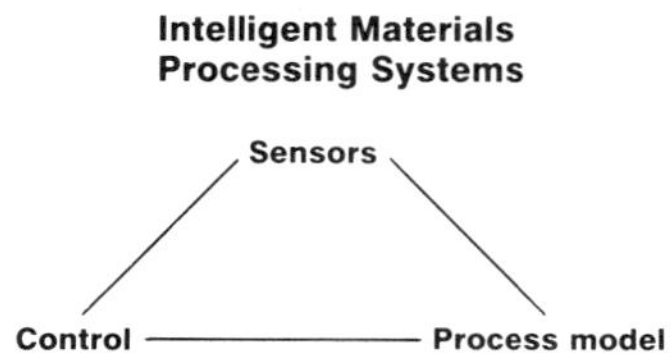

Figure 2

Components of an Automated Materials
Process Control System

INSPECTION TECHNIQUES AND SIZE/SCALE RELATIONSHIPS

The challenge for developing an effective manufacturing system is to select the most effective inspection techniques for the attribute of interest. To do this, it is also necessary to keep in mind the size of the attribute or aspect of the material's structure being measured. This can be represented in an approximate way as shown in Figure 3.

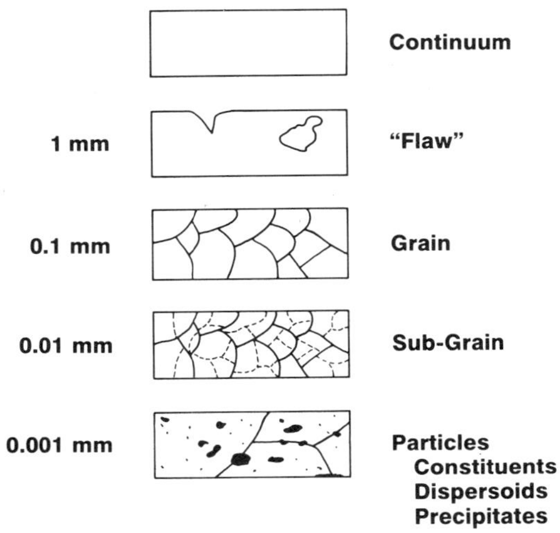

Figure 3

Structure Scale in Materials Science

The size range provides a framework for appreciating the increasing degree of resolution and sensitivity needed to quantify the fine-scale aspects of material structure. Advances in finite element methods are making it possible to use the power of continuum mechanics analysis for modeling microstructure scale behavior.

To carry out meaningful measurements of microstructural features over a range of sizes, critical evaluation of the kinds of non-destructive probes which can be used is necessary. A first approximation for what is needed to image the feature of interest is a field of either electromagnetic or mechanical waves whose length is comparable to the size of the structural feature of interest. Selected values are summarized in Figure 4.

Relationships of Feature Size to Frequency for Electromagnetic and Acoustic Probe Techniques

Feature	Nominal size mm	Frequency: Electro-magnetic	Frequency: Acoustic waves
Defect	1.0	30 GHz	6 MHz
Grain	0.1	300 GHz	60 MHz
Constituent	0.01	3 THz	600 MHz

Figure 4

One point emerging from this scale is that the use of electromagnetic waves for sub-surface feature imaging much below about 0.1 mm in "size" is not very feasible. This is due to the skin depth effects[2] which occur in alloys as electrically-conductive as aluminum. At 1 GHz, the eddy current penetration is only about 4 microns. On the other hand, the use of very high frequency ultrasonic waves becomes limited by attenuation when used for imaging applications. If extensive penetration is important, x-ray or neutron beams operating at much higher particle energies are necessary. Nevertheless, electromagnetic or acoustic waves can be used to probe samples using diffraction and quantitative absorption and scattering behavior models.

The ideal technique for process control requires speed, sampling efficiency, non-contact capabilities, precision, and the ability to function in a less-than-ideal plant environment (vibration, temperature, cleanliness, etc., shouldn't constrain the method severely).

Given these constraints, several specific areas of NDE that may be of particular value for near-term process control using microstructural correlations are: ultrasonics, x-ray, eddy current and hardness testing techniques. Some current applications, not all of which are specific to Alcoa, will be reviewed and some future prospects for these approaches examined.

FLAW DETECTION/POROSITY (100+ MICRONS)

Flaw detection is still the major basis for NDE and material inspection. There is not a trend towards "on-line" monitoring as opposed to post-manufacturing inspection. In addition, there is the drive for detection of ever-smaller defects for increasing quality, toughness and reliability in service.

Radiographic Techniques

Radiography is a major tool for "off-line" applications; e.g., inspection of castings and welds.[3] To invoke this as a process control tool would be a significant extension of conventional applications, such as thickness gauging. However, advances in x-ray technology are making it possible to image the internal structure of the material within a short enough time to permit thinking about such techniques for process control.

The usual aspect of radiography can be exemplified by its use for characterization of material homogeneity. The trend, however, is toward real-time imaging,[4-5] as indicated by this example in Figure 5.

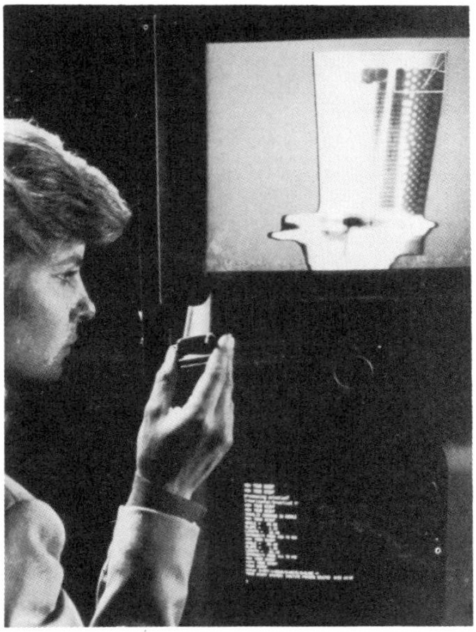

Real-time Radiography of Turbine Blades
Figure 5

Of great interest is the emerging application of tomographic techniques for inspection. These are approaching micron-scale resolution of defects in objects several feet in thickness; e.g., rocket motors.[6]

Another aspect of radiography that should be mentioned is microfocus radiography as a tool for resolving very small defects and porosity. This will be valuable for ceramic materials which, from a fracture mechanics standpoint, are even more sensitive than metals to very fine defects. An example is seen in Figure 6.[7]

Limitations of x-radiography can be overcome with neutron radiography because the depth of penetration of neutrons in aluminum is far greater than for x-rays. The sensitivity to light element segregation is also greater. This is particularly important for the new Al-Li alloys, for example. Neutron radiography, however, is not yet viable for plant applications.

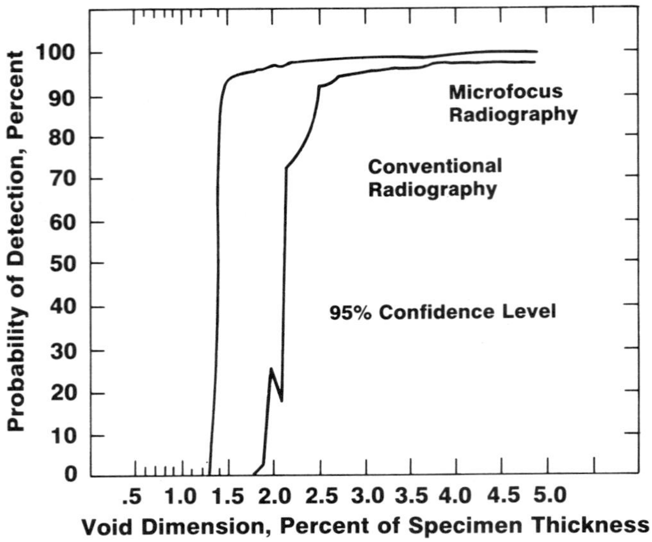

Figure 6
Example of Microfocus Radiography for Detection of Voids in Silicon Nitride

Eddy Current Techniques

For "on-line" considerations, especially for sub-surface detection, a common technique is eddy currents. The perturbation of the current by defects leads to a reduction in the material conductivity. In a more complex representation,[8] this affects the impedance of the sample. Figure 7 indicates how different effects give rise to different responses.

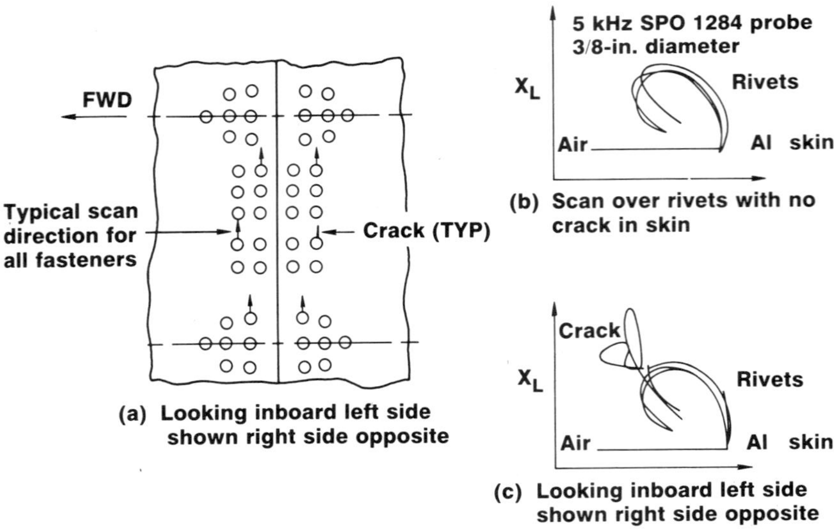

Figure 7
The Effect of a Defect on the Impedance Plane Signal

Another way of "imaging" material variations via eddy current is represented in Figure 8,[9] which indicates variations in local conductivity in heat-treated aluminum alloy plate. The maintenance of uniform quench conditions is essential in such products for minimization of residual stresses and assuring uniform as-heat-treated mechanical properties in the final product.

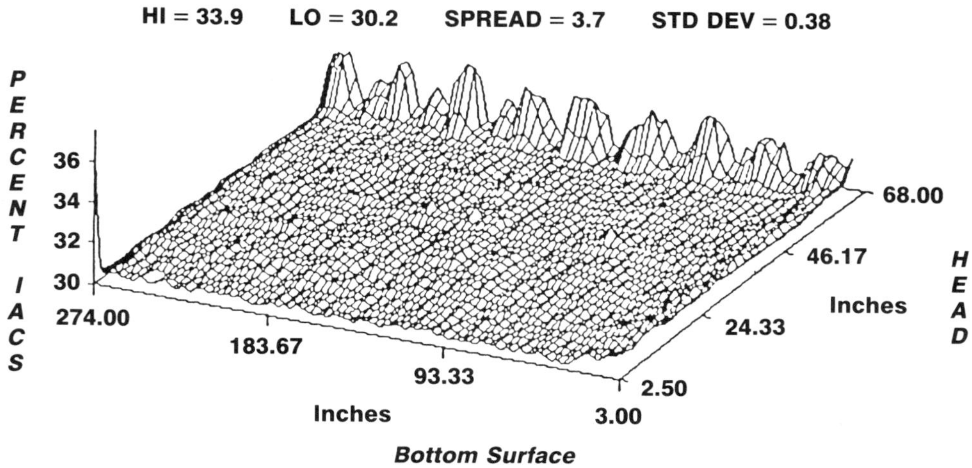

Figure 8
Variations in Conductivity in Heat
Treated Al Plate

The importance of conductivity as a probe of even finer microstructure scale uniformity will be discussed in more detail later.

Ultrasonics Techniques

The use of ultrasonic energy to probe aluminum alloys is well known for conventional defect detection and has been used for many years. Alcoa has supplied many of the aluminum standards used for calibrating flaw detection systems.

Newer uses include measuring molten metal quality.[10] By relating acoustic parameters such as attenuation or scattering amplitudes to the inclusion count of molten metal, the response to changes in filter or other molten metal conditions is observable.

Ultrasonic probes are also useful for determining the location of molten metal/solid interfaces in welds.[11] This has led to speculation that such techniques could be used for actual control of casting processes by monitoring and controlling the motion of the liquid-solid interface.[1] Given the importance of the role of interface motion and conditions on the as-cast microstructure, this is an attractive area for using fundamental material science for process control.

Acoustic techniques are also well suited for analysis of porosity in powder[12] and castings.[13] The use in powder metallurgy[12] is shown in Figure 9. Attenuation characteristics have also been related to density, as for example in SiC.[7]

The use of ultrasonic inspection for composites is well advanced. Figure 10 is a photo of an automated inspection system for ARALL, a new claminate product Alcoa has been developing for the aerospace market.

There are also useful correlations between ultrasonic characteristics and strengths of composites,[7,14] which integrate the effects of processing parameters on the composite material.

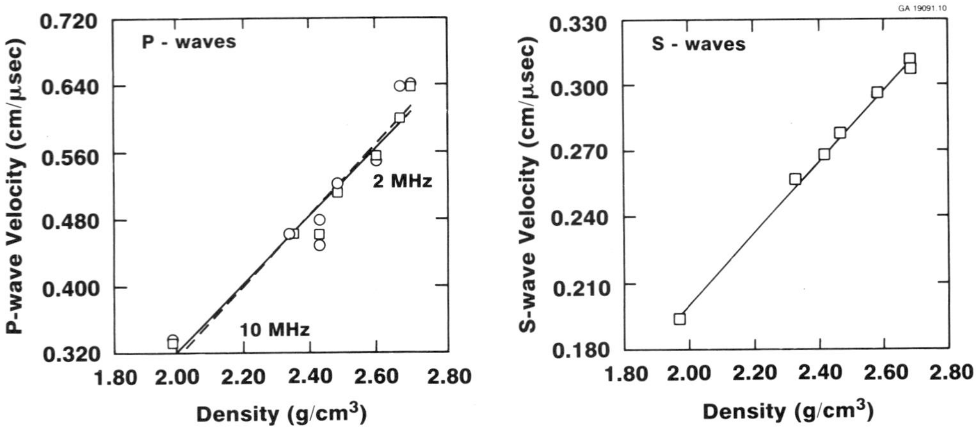

Figure 9
The Correlation of Ultrasonic Velocity with Density in Aluminum Powder Metallurgy Material

Figure 10
Multi-Channel through Transmission, Water Squirter System

In addition to these NDE systems and techniques, the aluminum industry is also working on surface flaw detection and non-contact temperature measurement systems using optical and thermographic techniques. These will have application to non-metallic as well as metallic materials. While some of these applications are "off-line", they are capable of further automation and reduced turn-around time which will facilitate their use for more "on-line" applications.

In the future, progress in the use of lasers, imaging and ultrasonics is expected. Moreover, the use of tomography-like techniques such as synthetic aperture focusing and imaging will increase the resolution and information obtained from materials inspection.

Continued improvement in defect resolution and reduction in the size of detectable defects are being driven by the need for improved reliability and fracture toughness performance specified for high-strength alloys and advanced structural materials. Another major stimulus for improved NDE techniques is the increased complexity and cost of finished parts produced by advanced manufacturing techniques. Destroying such pieces as part of a Statistical Process Control strategy is simply not feasible.

Various NDE techniques have been developed for characterization of the material on an even finer microstructural scale. As a first example, the feature size of interest is on the order of the grain size of the material.

GRAIN STRUCTURE (10-200 MICRONS)

Quantitative analysis of the attenuation of acoustic waves may soon become a practical tool for grain size measurement as a process control parameter. Excellent correlations have been shown in ferrous materials[15] and the British Non-Ferrous Metals Technology Centre has announced an on-line grain size measurement scheme for copper alloys. Alcoa Laboratories[16] has confirmed that this technique can be applied to aluminum-magnesium alloys and work is underway to apply this understanding to the behavior for other alloys. Figure 11 illustrates the effect of grain size on the ultrasonic signal.

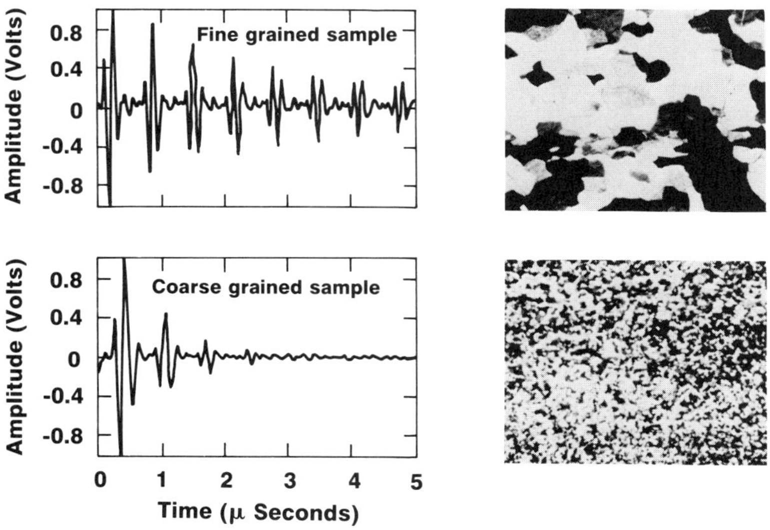

Figure 11
Microstructure and Acoustic Signals
from Al-Mg Alloys

Continued improvements in x-ray equipment for quicker determination of diffraction line profiles for residual stress determination will also be beneficial because the same equipment can be utilized to obtain diffraction profile shapes for the grain size analyses as well as the peak shift effects due to residual stresses.

Progress in resolving effects on a finer scale than grain size has also been made. In this "size" range, microstructural considerations in Al alloys are typically anisotropy, solid solution content, the condition of the matrix; i.e., degree of recrystallization or recovery, and the presence and morphology of constituent or dispersoid phases.

TEXTURE

Anisotropy of mechanical properties is of great importance even though the degree of anisotropy of aluminum is typically less than other common engineering metals; e.g., steel or copper-based alloys. The industry's premier rolled product, beverage can sheet, is produced from alloy 3004-H19 sheet that must meet extremely tight tolerances on anisotropic behavior specified by can makers. "Earing" has been related to the occurrence of specific texture components whose presence can be controlled by processing.[17] Traditionally, this has been represented by determining pole figures; i.e., maps of the relative degree of preferred orientation as seen by x-ray diffraction. Such diagrams are normally only determined under laboratory conditions. However, rapid texture determinations are very promising for improving can sheet production quality and efficiency. If the essential elements of information about the preferred orientation can be determined quickly enough, this can be used as a process control tool. To evaluate this, both x-ray and ultrasonic techniques are being explored.

The evolution of x-ray techniques has also led to a simple scheme to monitor the relative amounts of particular texture components by determining the relative intensities of specific diffraction peaks at selected points in the process. A schematic illustration of this texture device[18] is shown in this Figure 12. This ultrasonic technique takes advantage of improvements in measurement technology to determine small, but measurable, variations in acoustic wave velocities which result from variations in crystallographic orientation. These can be measured and related to the presence of specific texture components, (Figure 13).[19] Such techniques are useful for other fabricated products where properties such as formability or strength are affected by texture.

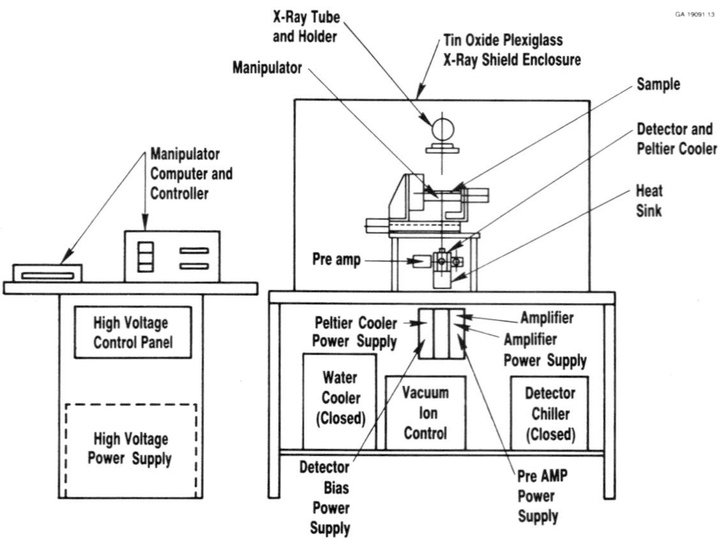

Figure 12
Schematic of New X-ray Texture Measurement System

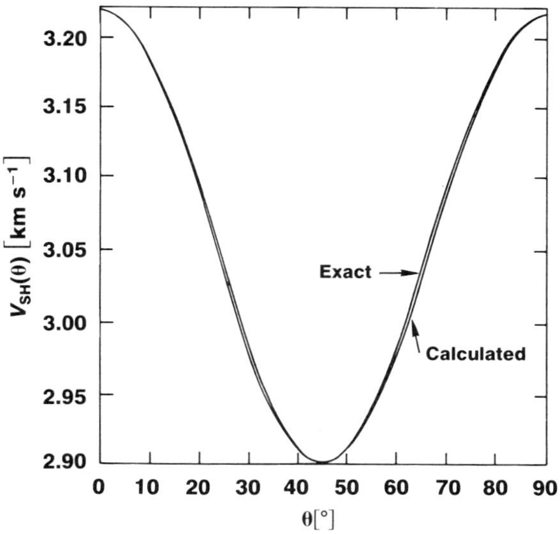

Figure 13

Comparison of Measured and Calculated Ultrasonic Velocity with Respect to Rolling Direction

Improvements in velocity measurement also open the way for this technique to separate the effects of residual stress and texture. This should also permit the improved measurement of residual stresses and would be significant as an alternative to the x-ray diffraction techniques normally relied upon for residual stress analysis.

SOLID SOLUTION CONTENT/SECOND PHASE MORPHOLOGY

Matrix composition and microstructural homogeneity are important ways of characterizing an alloy for process control. They are amenable to measurement by non-destructive techniques.

Electrical conductivity is a useful parameter for estimating matrix composition,[2] but it is not a unique measurement of composition or alloy condition. Nevertheless, it provides a good basic tool and, through eddy current techniques, is a practical means of approximate alloy identification and material uniformity on a microstructural scale. An example of its use for uniformity, with respect to quenching of heat treatable alloy plate, has been discussed.

Conductivity can be related to the extent of precipitation that is necessary for control of the aging process because it is affected not only by the solid solution content of the matrix but also by the scattering characteristics of the precipitate phase(s). In this sense, the conductivity changes reflect microstructural changes occurring on a sub-micron scale. These changes are applied during quality assurance (QA) testing.[21] (Figure 14)

Although these aspects of conductivity measurement are not a true on-line process, because they do not include feedback for process control, they nevertheless help to minimize rejects and are soundly based on known microstructure-property relationships developed for aluminum alloys.

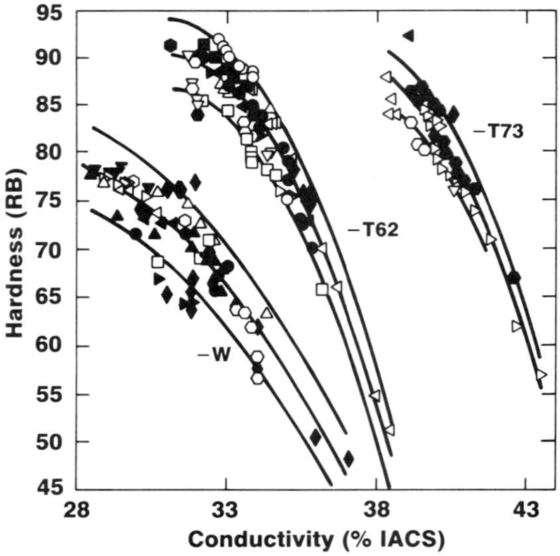

Figure 14
Examples of Conductivity vs. Hardness
for 7075 Alloy

To complement conductivity measurements, there is new interest in the use of thermoelectric power for estimation of matrix composition.[22] Conductivity testing, although a mainstay of transport/property type measurements, has drawbacks with respect to alloy identification. In particular, multiple solute effects or other interactions can occur which overlap in their effects on apparent conductivity. Thermoelectric power, which is also sensitive to matrix composition, can be an alternative technique. Here again, there is opportunity for nondestructive measurement of different characteristics to yield more information than any single technique alone.

While hardness is not a material probe in the same sense as acoustic or electromagnetic waves, it is widely recognized as an essentially nondestructive tool, especially in conjunction with other measurements. Figure 14 illustrates a problem with hardness data, variations in strength (temper) for the same hardness number. Alcoa researchers have found a way around that problem.[23] Their work improves upon the prediction of strength from hardness tests by making it possible to estimate both yield and tensile strength from a simple hardness-type test. This analysis enables the use hardness-type tests more effectively as a nondestructive tool for material characterization and process control.

In addition to the increasingly effective use of "traditional" NDE tools, progress is being made in the use of ultrasonics for microstructural characterization. An example at the "constituent particle" size level is the relationship between particle morphology and attenuation (Figure 15).[15]

Acoustic attenuation and velocity/scattering behavior also can be correlated with strength levels and microstructural changes even on the scale of precipitation hardening response, as shown in Figures 16-17 for alloys 2219 and 2024.[24-25] This sensitivity of ultrasonic velocity to subtle alterations of the structure is further supported by the correlation of second phase volume fractions to the nonlinear and higher order acoustic parameters shown in the work of Salama[26] and illustrated in

Figure 18. The influence of microstructure on attenuation of ultrasonic waves used in metallic systems can also be utilized in ceramics.[7]

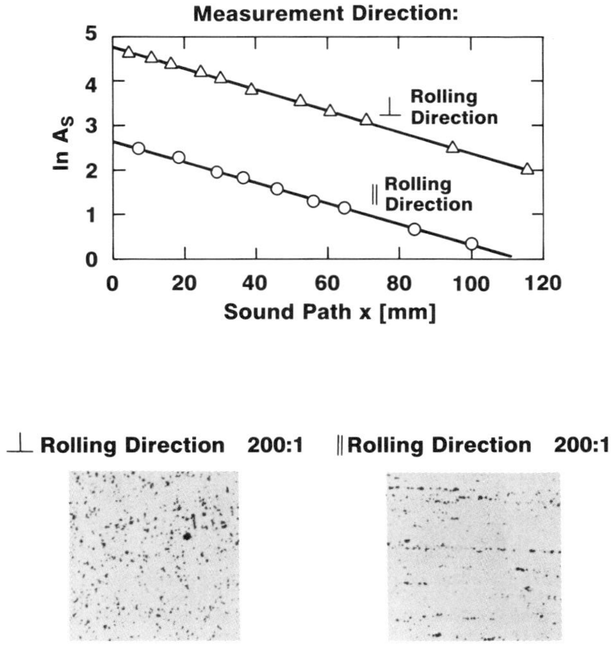

Figure 15

Relationship of Ultrasonic Attenuation to Consistent Morphology in an Al-Cu-Mg Alloy

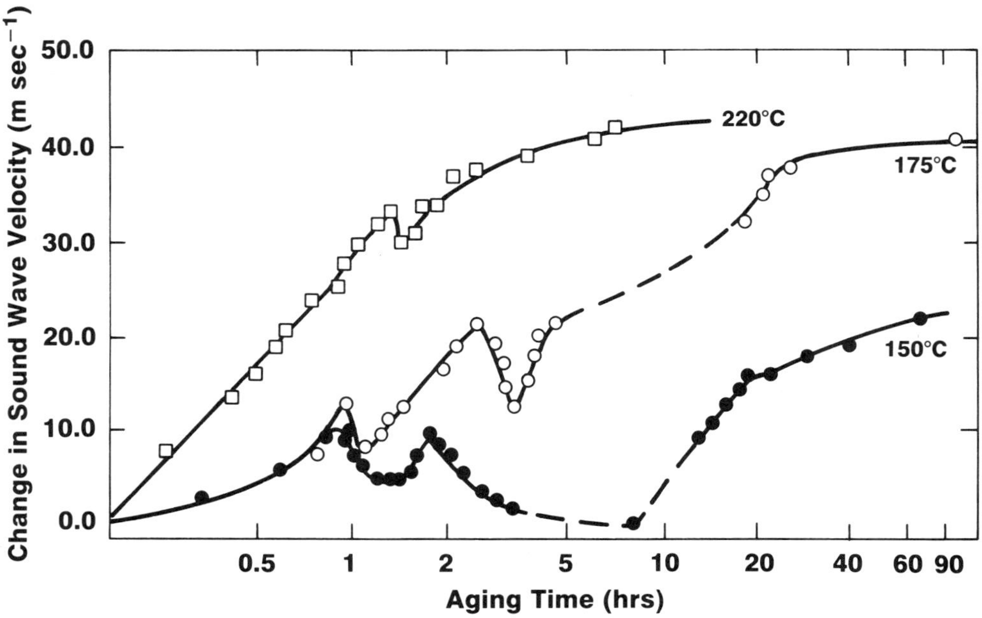

Figure 16

The Effect of Aging Time on Ultrasonic Velocity in 2219

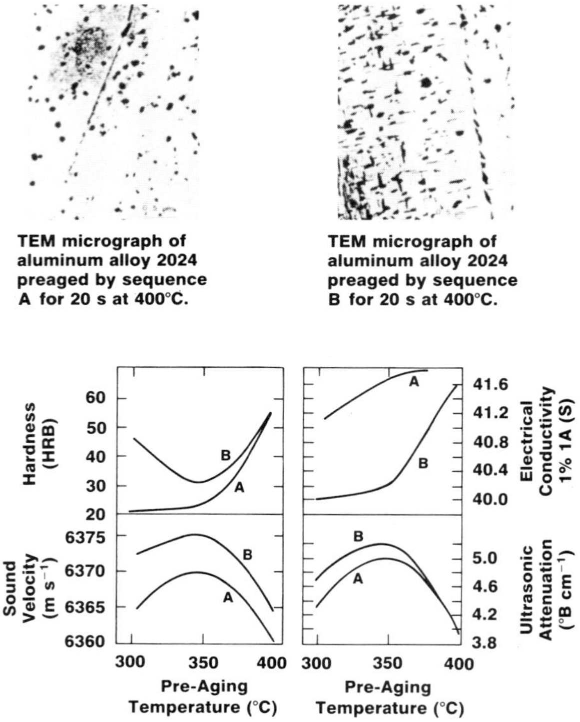

Figure 17

Relationship of NDE Measurements to Microstructure Changes in 2024

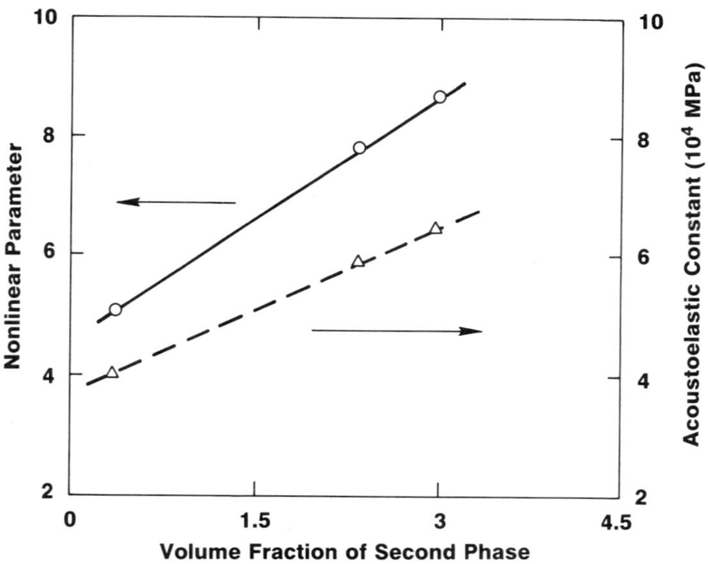

Figure 18

Relationship of Nonlinear Acoustic Parameters to Second Phase Volume Fraction in Aluminum Alloys

It appears obvious, then, that a number of aspects of microstructural uniformity can be inferred from ultrasonic NDE techniques beyond "simply" imaging. Moreover, the automation and speed of data acquisition and analysis opens the way for the use of multiple measurements to get complementary data from different techniques.

SUMMARY

There is a broad range of nondestructive techniques which are capable of interrogating materials to measure properties and attributes of microstructure. These are, or will be, useful for the manufacture of aluminum alloys and for laminate and composite systems.

Within Alcoa, a period of rapid growth in the application of emerging NDE technologies for material characterization and process control applications is underway. Automated, computer-aided nondestructive techniques for material characterization are being developed with an emphasis on their ability to probe underlying microstructure attributes. These techniques are being combined with an increasingly quantitative understanding of microstructure-property relationships to provide a fundamental basis for the development of new materials and improved manufacturing processes as the industry and the company evolve.

Alcoa, and any company hoping to compete in the materials marketplace of the 1990s and beyond, must expand their NDE techniques and capabilities to use these new and more quantitative methods. The process control needs of the aluminum industry will provide an excellent starting point for technology transfer from the NDE world to a manufacturing environment.

REFERENCES

1. R. Mehrabian, H. Wadley, Needs for Process Control in Advanced Processing of Materials, in: "Review of Progress in Quantitative NDE," vol. 48, Plenum Press, New York, c. 1984.
2. D. J. Hagemaier, Eddy Current Standard Depth of Penetration, Material Evaluation, 43:1438 (1985).
3. P. Dickerson, Quality Control in Aluminum Arc Welding, Journal of Metals, 38(5):47 (1986).
4. Metal Progress, 129(2):13 (1986).
5. R. Rudolph, E. Henry, Jr., Real Time Radiographic Imaging for Submerged-Arc Welded Pipe, in: Real Time Radiographic Imaging: Medical and Industrial Applications, ASTM STP 716, (1978).
6. Research and Development, June 1986, p. 47.
7. S. Klima, NDE of Advanced Ceramics, Materials Evaluation, 44:571 (1986).
8. D. Hagemaier, Application of Eddy Current Impedance Plane Testing, Materials Evaluation, 42:1035 (1984).
9. J. Duarte, Computer Graphics for Displaying Eddy Current Test Results From Heat Treated Al Plate, Light Metal Age, p. 6, Dec. 1983.
10. D. Mansfield, Ultrasonic Technology for Measuring Molten Aluminum Quality, in: Light Metals, (AIME), 1982.
11. L. Lott, Ultrasonic Detection of Molten/Solid Interfaces of Weld Pools, Materials Evaluation, 42:337 (1984).
12. S. Howard, J. Tani, H. Arnold, H. Schwetlick, W. Sachse, "Ultrasonic Characterization of Porosity in Powder Metals", ASM Metals Congress, 1984, paper no. 8408-025.
13. S. Wang, A. Csakany, L. Adler, Ultrasonic Determination of Porosity in Cast Aluminum, in: Review of Progress in Quantitative NDE, vol. 48, Plenum Press, New York, (1984).

14. A. Vary, Ultrasonic Measurement of Material Properties, in: Research Techniques in Nondestructive Testing, vol. IV, Academic Press, (1980).
15. K. Goebbels, Structure Analysis by Scattered Ultrasonic Radiation, in: Research Techniques in Nondestructive Testing, vol. IV, Academic Press, (1980).
16. A. Govada, B. Shabel, unpublished research (Alcoa Laboratories).
17. W. G. Fricke, Jr., H. B. McShane, Evolution of Texture During Hot Rolling of 3004 Aluminum Alloy, in: Textures in Non-Ferrous Metals and Alloys, AIME Conf. Proceedings, 1984.
18. W. Fricke, J. Ioannou, unpublished research (Alcoa Laboratories).
19. D. Allen, R. Langman, C. Sayers, Ultrasonic SH Wave Velocity in Textured Al Plates, Ultrasonics, 23:215 (1985).
20. J. E. Hatch, ed., "Aluminum:Properties and Physical Metallurgy", ASM, Metals Park (1984).
21. S. Axter, "Effect of Interrupted Quenches on the Properties of Aluminum", private communication.
22. J. M. Pelletier, R. Borrelly, Temperature and Concentration Dependences of Thermoelectric Power at High Temperature in Some Aluminum A,lloys, Materials Science and Engineering, 55:191 (1982).
23. B. Shabel, R. Young, unpublished research (Alcoa Laboratories).
24. M. Rosen, E. Horowitz, S. Frick, R. Reno, R. Mehrabian, An Investigation of the Precipitation-hardening Process in Aluminum Alloy 2219 by Means of Sound Wave Velocity and Ultrasonic Attenuation, Materials Science and Engineering, 53:163 (1983).
25. M. Rosen, L. Ives, C. Ridder, F. Biancaniello, R. Mehrabian, Correlation between Ultrasonic and Hardness Measurements in Aged Aluminum Alloy 2024, Materials Science and Engineering, 74:1, (1985).
26. K. Salama, "Nondestructive Ultrasonic Characterization of Engineering Materials", Final Report, Houston U., NASA-CR-176349, November 1984.

IMPORTANT METALLURGICAL PARAMETERS THAT MUST BE DETERMINED TO CONTROL THE PROPERTIES OF STEELS DURING PROCESSING

Hiroshi Yada* and Katsuhiro Kawashima**

*R and D Laboratories-III, Nipon Steel Corporation
1-1-1 Edamitsu, Kitakyushu City, 805 Japan
**R and D Laboratories-I, Nippon Steel Corporation, 1618 Ida
Nakahara, Kawasaki City, 211 Japan

ABSTRACT

Recently, great progress has been achieved in the field of physical metallurgy of steels. Especially, the development of controlled rolling has stimulated studies aimed at elucidating the evolution of microstructures in hot-rolled products.

Such fundamental knowledge is now being formulated in a general manner by the aid of theoretical constructions. Computer models of the evolution of microstructures and mechanical properties are being developed on the basis of such formulations. One example is outlined briefly, and important parameters that determine the properties of steels are explained.

These parameters can be classified as follows:

(i) Processing variables such as surface temperature, plate thickness, rolling speed, etc.

(ii) Variables which are processed from the above ones and which can be directly used in metallurgical formulation such as; through thickness temperature distribution, strain, strain rate and so on.

(iii) Intermediate microstructural features such as; austenite grain size, the degree of recrystallization, degree of transformation etc.

In order to utilize the above-mentioned computer modelling for the prediction and control of material properties during processing, the variables of category (ii) must be known. They can be obtained either from the processing variables of category (i) using either models, or direct measurements. At least one feature of category (iii) will probably be necessary for the construction of an on line control system for material properties, because it will greatly contribute to improving the accuracy of feed back or feed forward control.

Nondestructive measurement of the properties of steels is also discussed briefly with special reference to the on-line control processing system.

The on-line control system thus constructed will lead to important innovations in steel production.

INTRODUCTION

Conventional production facilites of steels were designed mainly so as to obtain maximum efficiency. When we needed steel with special properties we usually resorted to additional processes such as heat treatment.

Controlled rolling offered an oppotunity to change this way of thinking. A large amount of high quality steels were needed for constructing arctic pipelines, and at low cost. It was then discovered that by controlling the processing variables in hot rolling we could obtain an even better combination of properties than one possible in heat-treated steels. Many studies were made to elucidate the fundamental mechanisms involved in this technology, aud this effort brought about a great progress in physical metallurgy related to steel processing.

The oil crisis added a strong incentive to this effect. The simplification process has been pursued to save energy and cut costs. This in turn resulted in greater use of continuous casting, the development of hot charging and direct rolling technology, and increased use of non-heat treated speciality steels and so on.

In such simplified processes, each unit process has a much greater contribution to steel properties than before.

There are two seperate aspects involved in the control of properties. The first is to set certain processing variables at desired levels, as in controlled rolling. From the empirical relation shown in Fig.1[(1)] , one can choose the rolling conditions suitable for a desired combination of strength and toughness. The other aspect is regulating as is used in automatic control. For example, there usually exists a large fluctuation of properties within a hot strip coil due to the changes in rolling speed, cooling rate and so on. This fluctuation causes great losses to both the steelmakers and users. To minimize this fluctuation without a corresponding loss in production efficiency, on-line computer systems for controlling properties, as shown in Fig.2, are now being planned and developed. Modern mills are already equipped with advanced process computer systems for realizing maximum efficiency and ensuring high product shape quality. Such process computer systems offer suitable ground for realizing on-line control of properties.

Based on the advances in physical metallurgy mentioned above, computer models of metallurgical phenomena are now being actively worked out.

Computer modellings only,however, are not sufficient for developing such on-line systems. In addition feed forward and feed back control using appropriate sensors to ensure smooth operation and good accuracy are needed. What kind of metallurgical parameters then are to be monitored during processing? That is the main theme of this paper.

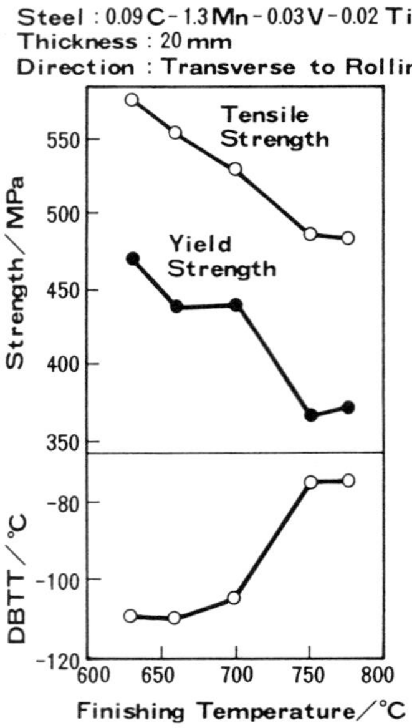

Fig.1 Effect of finishing temperature on the strength and the ductile -brittle transition temperature (DBTT) of a controlled-rolled steel.

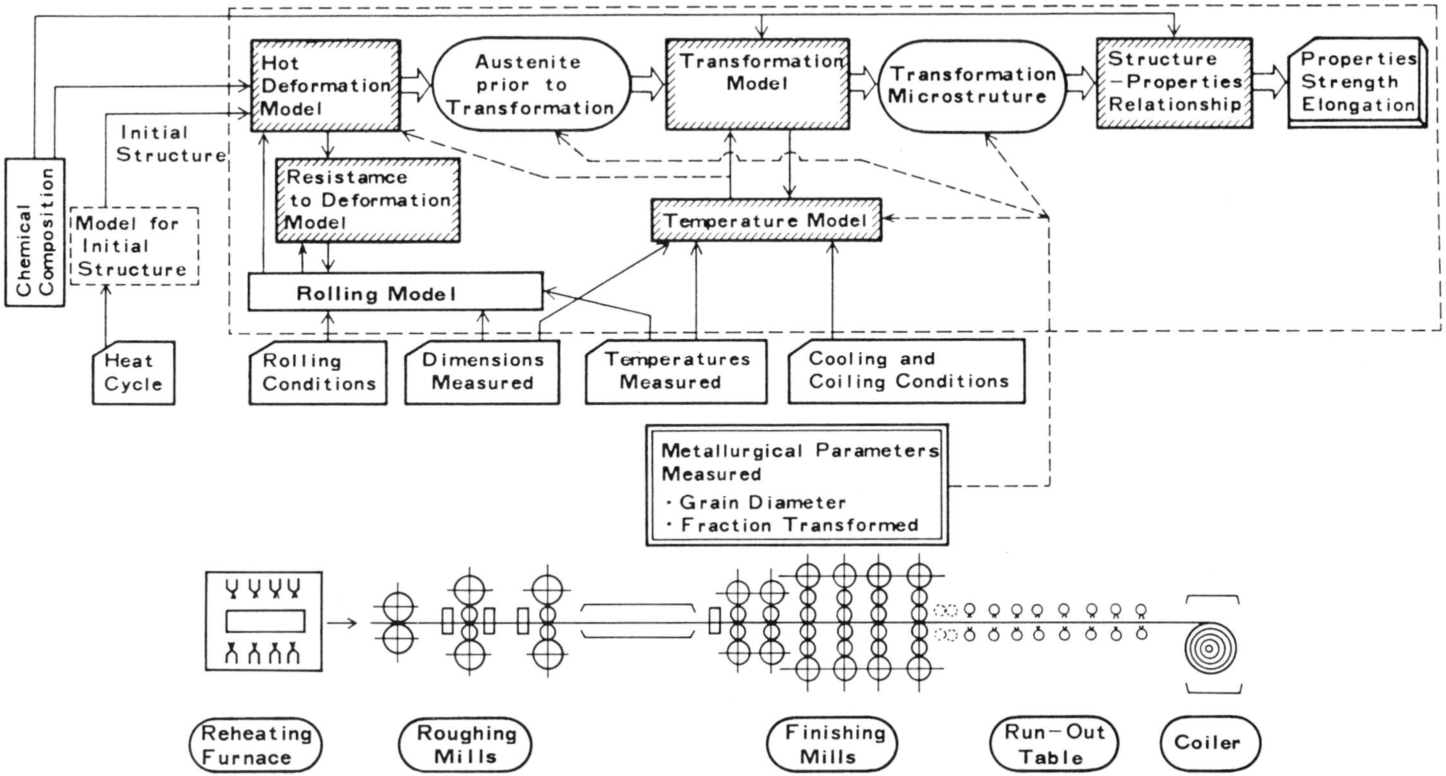

Fig.2 Concept of on-line control system of hot strip mills.

What Determines Properties Directly

Tensile strength or Hardness. Irvine and Pickering obtained the results shown in Fig.3 for the strength of low-carbon low-alloy normalized bainite steel plate(2) . In this experiment only the content of the alloying element is changed, but one can infer from this figure that the change of the transformation temperature with processing variables such as cooling rate will have a similar influence on the strenghth. A recent study on the accelerated-cooled plates indicated the truth of this assumption in mainly ferritic steels(3).

One could presume a more general relationship between strength and transformation temperature even for the individual phases in a mixed microstructure. Such a microstructure is now very common due to the faster cooling rate employed after rolling as in Fig.4.

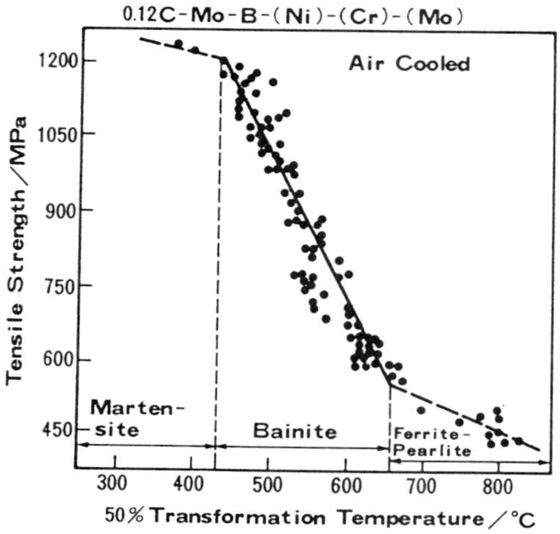

Fig.3 Effect of 50% transformation temperature on tensile strength of low carbon,low alloy steels.

Steel :0.20C-0.2Si-0.5Mn
Rolling:40 to2.4mm at 900°C (6passes)

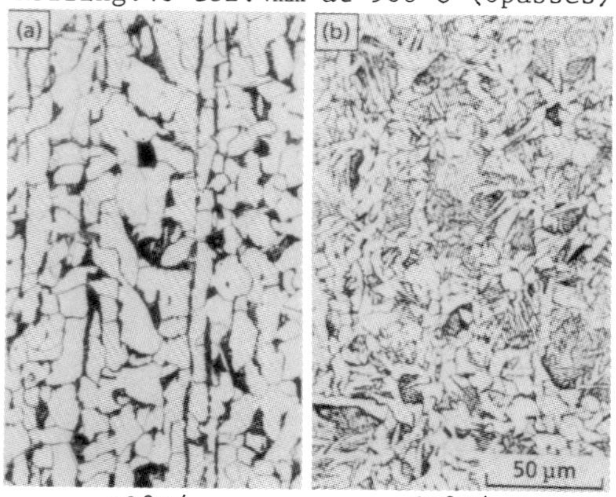

10°C/s 60°C/s

Calculation of fraction of each phase

		f_F	f_P	f_B
a (10℃/s)	Observed Value	0.73	0.27	0
	Calculated Value	0.72	0.28	0
b (60℃/s)	Observed Value	0.55	0.03	0.42
	Calculated Value	0.54	0	0.46

Fig.4 Effect of cooling rate on microstructures and comparison between actual microstructure and those predicted by the calculation model.

One of the authors has recently verified this assumption(4) . As shown in Fig.5, the microhardness of both ferrite and bainite, either with or without the presence of the other phases, has a linear relationship with their transformation temperatures which are calculated by using a model to be shown later. Solid solution hardening of silicon is added to it. Thus we can put the hardness of the steel in the usual practice, Hv, assuming the law of additivity as,

$$H_V = f_F \{-0.357\ T_F + 50(\%Si) + 0.255\ d_F^{-1/2} + 361\} + f_P \cdot H_P + f_B \{-0.588\ T_B + 50(\%Si) + 508\} \quad (1)$$

where f_F, f_P and f_B are the fraction T_F and T_B are the mean transformation temperature of each phase, F, P and B denoting ferrite, pearlite and bainite, respectively. This was proved to be true by the experiment.

We can now predict the strength of steels by using the common empirical law as :

$$T.S.\ (MPa) = 3.04\ Hv \quad (2)$$

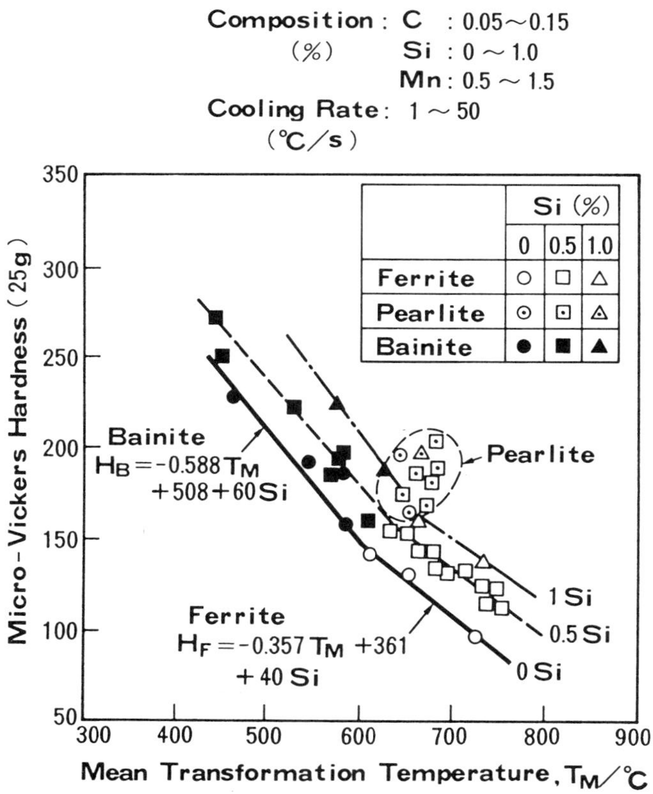

Fig.5 Effect of mean transformation temperature on each phase in mixed microstructures.

In generalizing the above equation, however, we must take the following other factors into consideration.

(a) Solid solution hardening by substitutional alloying elements (Mn, Cr,...) other than silicon must contribute to hardness, though it is not apparent in Fig.5. We can use, however, the reported values for solid solution hardening, since it is independent of microstructure.

(b) Solid solution hardening by interstitial atoms (C, N) is too delicate to be described in a simple expression. In ordinary steels, however, it does not differ greatly, because the cooling rate in the lower temperature range is not so high.

(c) Precipitation hardening must also be added if any. The general expression for it is not available now, so it must be estimated case by case.

(d) The hardness of pearlite, H, though it could not be explicity expressed from Fig.5, can be calculated by the formula derived for high carbon steels from the reported relations as :

$$H = 234 + 0.29\sqrt{723 - T_P} \qquad (3)$$

where T_P is the transformation temperature.
Actually, the hardness of pearlite is known to change with its carbon content. This carbon content can also be calculated by the transformation model.

(e) The hardness of martensite can be estimated only from the content of carbon as in fully martensitic steels. In this case also the carbon content must again be that of austenite when it is formed.

(f) The size of each phase must also be considered. The well-known grain size dependence of ferrite, as shown in Fig.6 is a typical example. In this figure accelerated-cooled steels and ultrafine-grained steels developed recently are included. In these steels great deviation from the usual relationship is observed indicating the other factors mentioned above.

(g) Dislocation density in normal hot-rolled steels is mainly introduced by transformation, and is already included in Equation 1.

In conclusion, we can write the strength of steel generally as

$$T.S. = f_F \cdot (a_1 T_F + b d_F^{-\frac{1}{2}} + c_1) + f_P \cdot \{h_P(x_P^C) + c_2\} + f_B \cdot (a_3 T_B + c_3) + f_M \cdot h_M(X_M^C) + \sum_n^{Si, Mn \cdots} A_n X_n \qquad (4)$$

,where X_P^C, X_P^C are the content of carbon in pearlite and martensite, respectively, X_n the content of substitutional alloying elements and a_1, a_3, b, c_1, c_2 and c_3 are constants.

<u>Yield strength.</u> The factors considered above also contribute to yield strength but in some different ways. For example its dependence on grain size is stronger than that of tensile strength, as shown in Fig.6. The size and distribution of the other phases can have a definite effect as the martensite in the dual-phase high-tensile steels. Also the state of interstitial solute atoms causes a marked effect on yield strength as seen in aged steels. We must beware that residual strain can have a strong influence on yield strength. Consequently at present we must resort

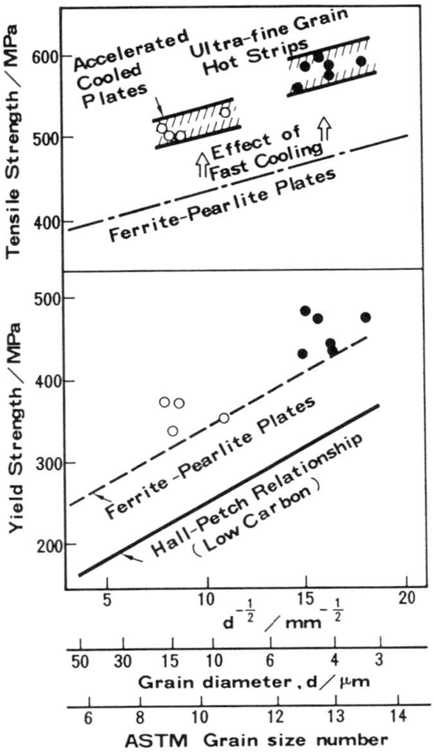

Fig.6 Grain size dependence of tensile and yield strength in low carbon Si-Mn steel.

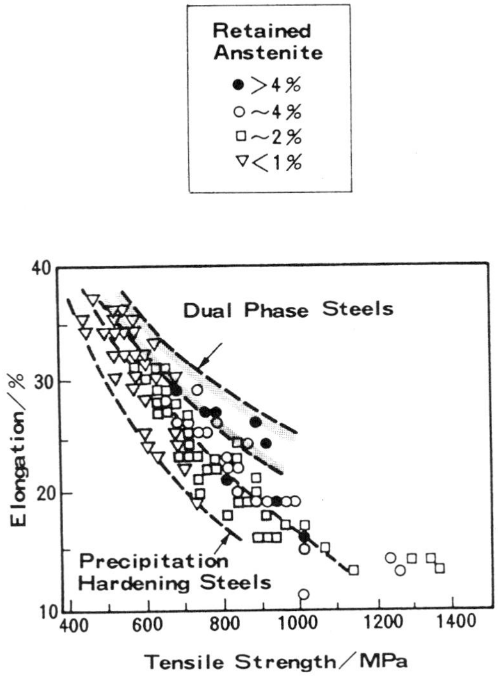

Fig.7 Strength-elongation relationship in fine grained ferritic steels and the effect of retained austenite (γ_R) on it.

to some empirical formulas. An example is the following equation(5).

$$Y.S.(MPa)=15.4[3.5+2.1(\%Mn)+5.4(\%Si)+23\sqrt{(\%N_f)}+1.13\, d^{-\frac{1}{2}}] \quad (5)$$

for low carbon, silicon-manganese air-cooled plates, where N_f is the amount of solute nitrogen. We must note this type of equation can be used in a limited range of composition and processing.

Ductility. Although metallurgical factors governing ductility are thoroughly understood, the empirical relation between strength and ductility in the groups of steels as shown in Fig. 7(8), might be a clue for future formulation. These groups seem to be distinguished by the strengthening mechanism of steels. The amount of impurity elements such as phosphorous and sulfur, and also the amount and the shape of inclusions have considerable effect on ductility. The effect of the specimen size must be considered for formulation.

Toughness. The formulation of toughness should be done separately for each type of test. If we limit ourselves to the popular 2mm V-notch Charpy test here, probably we should formulate the shelf energy and the ductile-brittle transition temperature (DBTT). The former has the same physical meaing as ductility, while the latter is determined mainly by other factors, one of which is the effective grain size, as shown in Fig.8, while another is the solid solution effect of a particular alloying element, especially nickel. The formulation of these factors will not be very difficult.

Modelling of Hot Deformation

Since diffusional transformations such as ferrite, pearlite and bainite mainly nucleate at the grain boundary and the substructure of austenite, the model of structual change of austenite during rolling is important. Although a few models have already been presented[(10)-(12)], they have not incorporated the high speed, continuous rolling practice, now prevailing in steelworks. One of the authors has developed a general model covering such commercial rolling practices by enlisting the aid of a new deformation simulator[(14)-(16)]. This model contains an explicit formulation of all the elementary metallurgical processes shown in Fig.9.

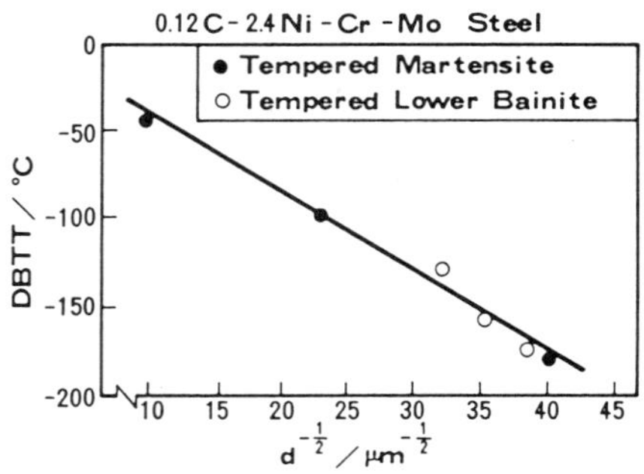

Fig.8 Relationship between ductile-brittle transition temperature and effective grain size

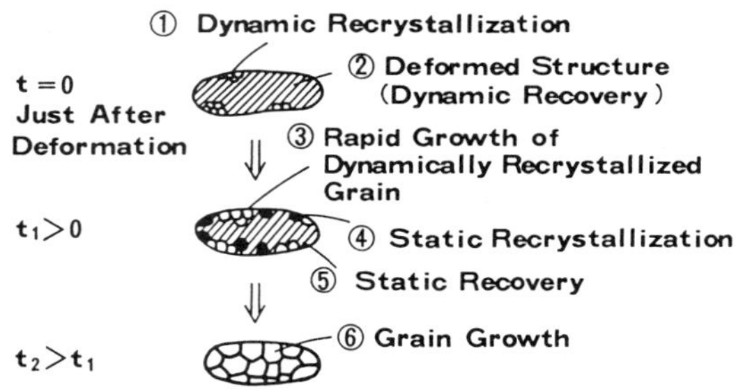

Fig.9 Schematic of elementary processes in hot-rolling

The changes in the grain size and dislocation density, which represent the substructure formed by working and which accompany these processes, are calculated and then averaged when the succeeding deformation starts and finally adopted as the initial microstructure of the succeeding deformation. An example of calculation is shown in Fig.10. A similar model is also used for predicting the resistance to deformation in the newest hot strip mills and it is now contributing to improvements in the accuracy of the thickness. Deformation stress, σ, can be easily derived from the above model because it is usually related to the dislocation density, ρ, by the following equation

$$\sigma = a\sqrt{\rho} \tag{6}$$

where a is a universal constant.

Modelling of Transformations during Cooling

Diffusional decomposition of austenite is a heterogeneous nucleation and growth process. If both the nucleation and growth rate are constant, the amount transformed, X, of such a process is generally given as[(16)]

$$X = 1 - \exp\left\{ - f(T)\frac{t^n}{d^m} \right\} \tag{7}$$

, where f (T) is a function of chemical composition and temperature, d is the grain diameter of austenite, and m,n are constants.

In ordinary steels an accurate determination of the parameters in Equation 7 is not always possible. If we can eliminate time from this

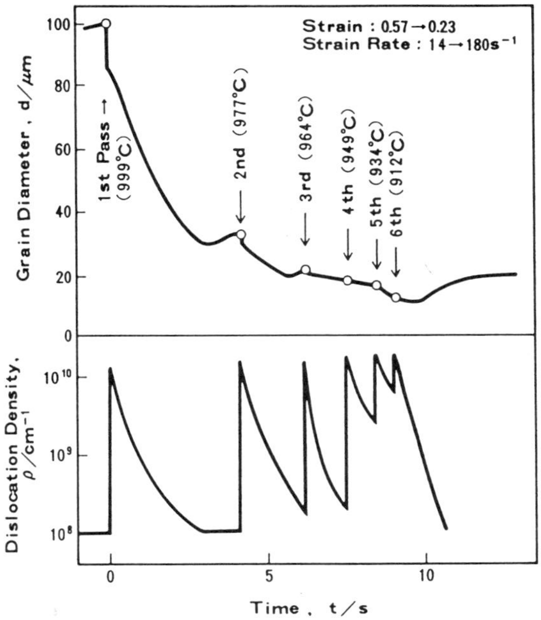

Fig.10 Change of grain diameter and dislocation density of carbon steel austenite during hot strip rolling calculated by a simulation model.

equation, we can formulate the progress of transformation from the continuous cooling experiments. The general expression would then be ;

$$\frac{dX}{dt} = g(x) \cdot h(T) \tag{8}$$

One of the authors has recently derived the following expressions from Equation 7[18], in the case of concurrent nucleation and growth

$$\frac{dX}{dt} = h_1(T) \cdot (1-x)\{\ell n \frac{1}{1-x}\}^{\frac{3}{4}} \tag{9}$$

, assuming the rule of additivity, and in the case of saturated nucleation site

$$\frac{dX}{dt} = h_2(T) \cdot (1-x) \tag{10}$$

Hurkmans et al. also derived the following expression[18],

$$\frac{dX}{dt} = h_3(T) \cdot (x+a)^p \cdot (1-x)^q \tag{11}$$

These types of equation can be used also in the case of changing nucleation and growth rate. Actually, ferrite drives out carbon atoms into untransformed austenite resulting in the gradual decrease of its growth rate, which has not been treated quantatively yet. We have been able to solve this problem and to obtain an explicit expression in h'(T) in Equation 9 as the function of temperature and steel composition. These expressions have been proved accurate in predicting the progress of transformations, as already shown in Fig.5 and the table attached to Fig.4.

A transformation model is also necessary for temperature calculation, as shown in Fig.11, because the heat of transformation greatly affects the cooling curves of steels[19].

Other Models. Besides the important models mentioned above, we also need the other models shown in Fig.2, though the importance of each model will depend on the kind of product being processed. In alloyed steels, especially in microalloyed steels now in wide use, the dissolution and precipitation model of alloying elements is vital, which is not included in Fig.2. Although these phenomena are involved in every other model, no general models are presently available for these phenomena.

Also it must be noted that the high-level rolling and temperature models are prerequisites for metallurgical models.

Total Model. Combining elementary models some attempts have already been made to construct a total model [20],[21]. One of the authors has developed a total model for determining the strength of ordinary hot-rolled steels which is also capable of predicting structural change at every moment, by combining the elementary models described above. Agreement between the calculations and the experimental results are quite good as shown in Fig.12[22].

FACTORS THAT MUST BE DETERMINED FOR CONTROLLING PROPERTIES OF STEELS

Information Necessary for Calculation of Properties and the Present State of Measurement

As described in the preceding paragraph, the total thermal history and the total work history as well as chemical composition, are necessary, as well as sufficient, for predicting properties. They are compared with the present state of measurement in Table 1. For the thermal history, which is the most important information through the whole process, only insufficient measurements are made. The measurement of temperatures within the materials, if possible, would greately improve the accuracy of control, especially since accelerated cooling is now in wide use. Elaborate models for calculating temperature are now available as shown in the example of Fig.13[23], and to a considerable extent are able to bridge the gap between the desired and available information. In order to prove the validity of such models, however, a method for measuring internal temperatures is earnestly desired.

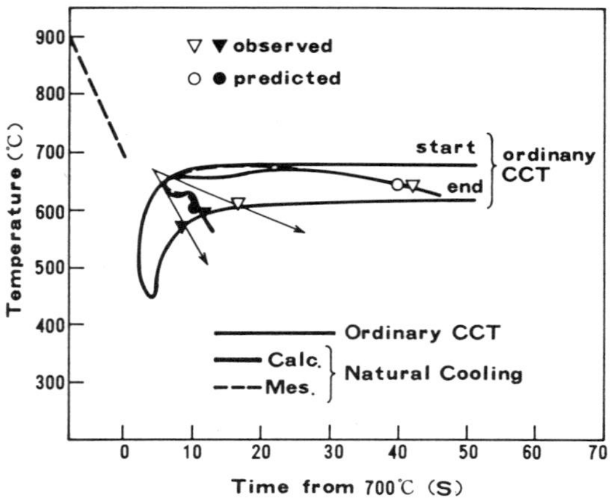

Fig.11 Comparison between observed and calculated cooling curves and the end temperature of transformation in 0.57C steel.

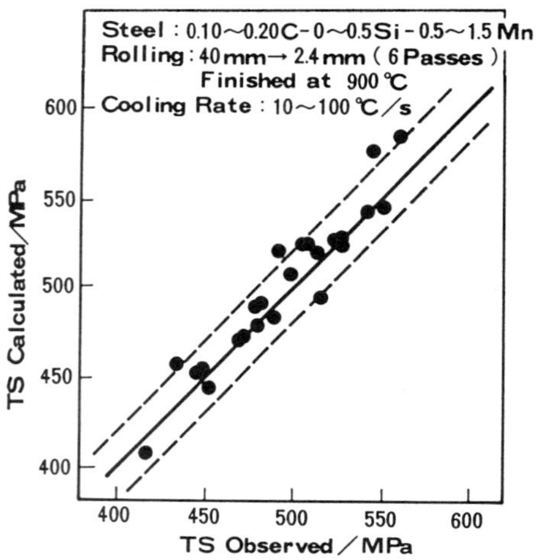

Fig.12 Comparison of observed tensile strength and that predicted from the total calculation model [22]

Table. Informations necessary for calculation of properties and present state of measurement.

Information Necessary for Calculation	Desired Items for Measurement	Calculation Model Necessary for Processing Indirect Information	Present State of Measurement
Total Thermal History	Through Thickness Temperature Distribution [or Average or Internal Temperature]	Temperature Model [including deformation and transformation calculations]	Surface Temperature at particular points [where no disturbance by water and so on exist]
Total work History [Effective Strain Effective Strain Rate]	Effective Strain Distribution during Rolling (or Total Strain)	Deformation Model [Resistance to Deformation Model]	Material Thickness Plate Profile Rolling Speed Material Speed Rolling Load

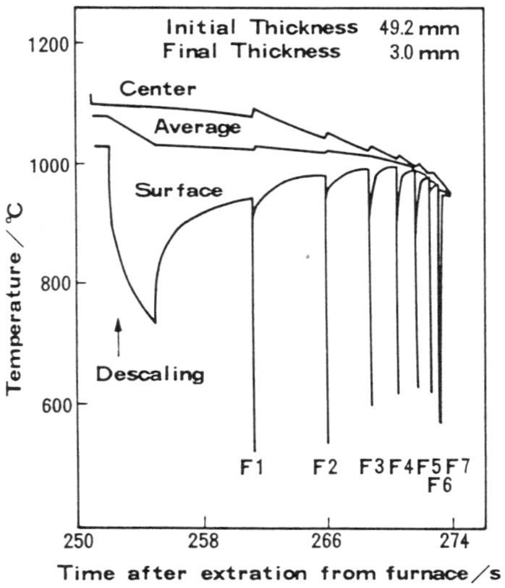

Fig.13 Calculation of temperature change during hot strip rolling.

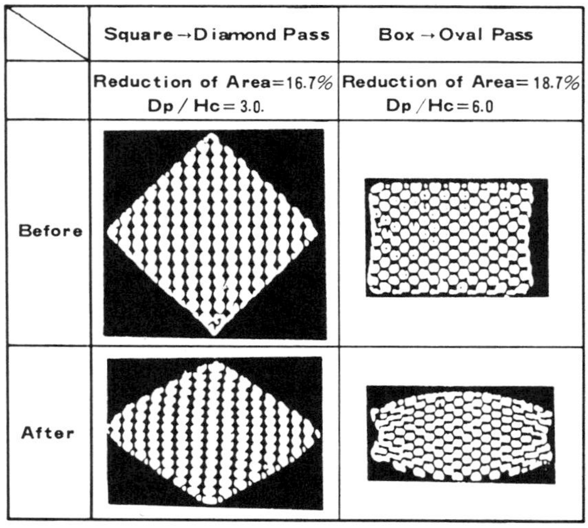

Fig.14 Examples of material flow in rod rolling by plasticine experiment.

In sheets and strips effective strain is not so different from the nominal strain derived from measurements of thickness. In thicker steels, especially in shapes and rods, however, the former is much greater than the latter, and exhibits complicated cross sectional distribution. As in Fig.14[24], the changes in shapes during caliber rolling, other than the nominal strain calculated by the reduction of area, also contribute to the change in microstructure, because the compression and shear strain have the same effect on the movement of dislocations.

In-line Monitoring of Microstructures during Hot Processing

There are many causes of errors in calculating properties by using the models described above. Even if we could succeed in describing metallurgical factors completely, we must usually simplify such sophisticated models to some extent in order to shorten calculation time for using it in the on-line control system. As described above, informations now available

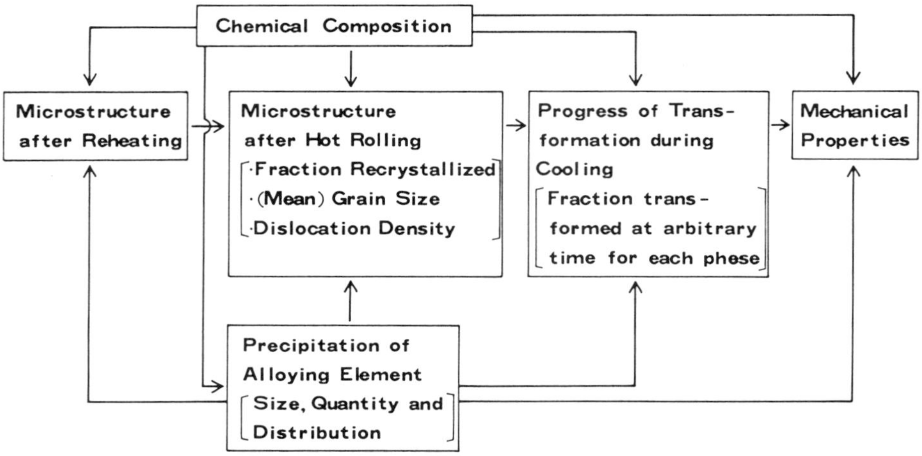

Fig. 15 Metallurgical factors that must be determined at each stage of calculation

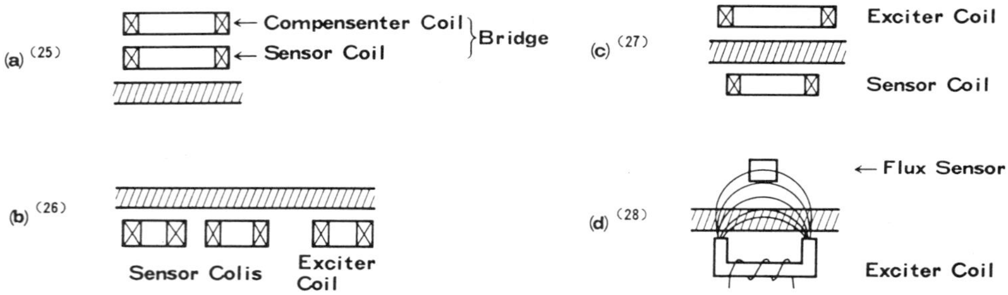

Fig.16 Magnetic detectors of $\gamma \rightarrow \alpha$ transformation

are far from satisfactory for accurate calculation. Errors from such inevitable causes in each model may accumulate in the final results. Thus some information of microstructure at intermediate stages would greatly improve the accuracy of prediction and enhance the reliability of the total system. Actually they might be essential for constructing the system.

The microstructural parameters that prescribe the structural changes in the succeeding stages are summarized in Fig.15. Since the transformation during cooling is usually the most dominating factor over properties, the microstructure before transformations is the most important information for feed forward control, though effective methods for its measurement have not been proposed yet.

The fraction transformed at some particular points is the most direct information for feed-back and, perhaps, also for feed-forward control. Fortunately, the transformation from austenite to ferritic phases below the Curie temperature accompanies drastic changes in magnetic properties, so magnetic sensors are very effective. The examples are shown in Fig.16 [25], [28]. Some of them have already received practical application.

The precipitation behavior, which is vital to alloyed steels, appears especially difficult to be measured, since the quantity of precipitates is usually small.

In-Line Monitoring of Properties and Microstructures around Room Temperature

Nondestructive measurement of properties and microstructures, after steels are cooled, also can be used as the means of in-line monitoring. Since in most mills steels undergo a series of finishing processes such as straightening, cutting and surface conditioning after rolling, to reject the disqualified products before finishing contributes to cutting costs. If we can know mechanical properties accurately in a short time by non-destructive methods, we can expect a much greater effect by eliminating mechanical testings as well as by realizing the guarantee of the steel properties at the desired portion.

Aiming at this target many studies are being done, but they have hardly received practical applications yet. Although some relationships between mechanical properties and physical properties such as magnetic coercive force have been found(30), the lack of reasonable explanation of them from the physico-metallurgical point of view seems to limit their practical use.

In the example of Fig.17, a clear relationship between the ultrasonic attenuation and the effective grain diameter, shown in Fig.8, was presented(30). This result seems reasonable since the effective grain boundary corresponds the large angle change of the orientation between neighbouring grains and also makes it possible to extend the idea of the effective grain size to ferrite, and possibly, to pearlite.

Accumulation of such fundamental studies will promote the understanding of the relation between the mechanical and physical properties leading the way to the practical use of nondestructive testings of steel properties.

CONCLUDING REMARKS

Let us imagine a rolling mill around the year 2000. The orders from customers are input to a supervising computer which carries out calculation to obtain the best combination of processing variables under the cost-minimum condition. The results of the calculation are automatically input

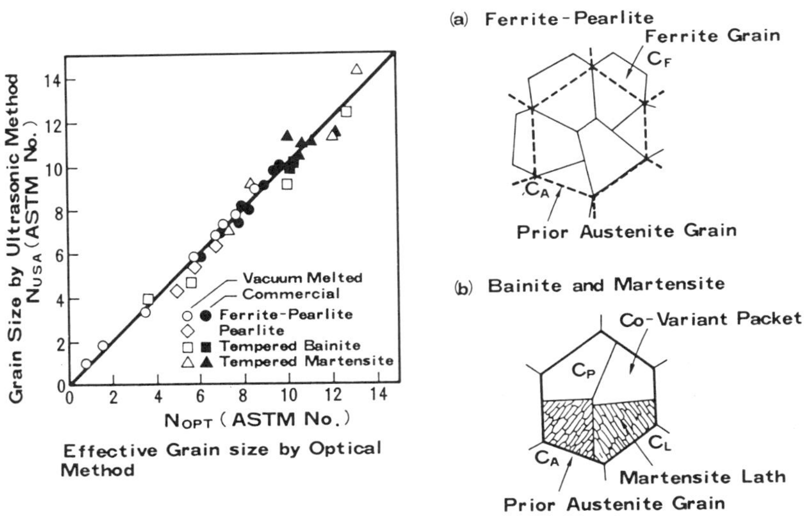

Fig. 17 Relationship between grain diameter measured by ultrasonic method and that measured optically as shown in (a) and (b)

to the process computers of the mill. Fluctuation of processing variables as well as chemical compositions are monitored in line by sensors and properties are automatically adjusted to the desired levels by the feed-back and feed-forward control. Important properties are inspected by means of nondestructive testings with good accuracy.

As has been described in this paper, such a mill image is no more a fantastic story but a concrete target that we are aiming at. It is also an answer to the demands for diversified products at low cost. Thus this metallurgical control system will play the roll similar to FMS (Flexible Manufacturing System) now being introduced to machinery factories.

As already pointed out, there remains much work to be done to realize the target mentioned above. In the field of nondestructive testing the need for new sensing systems of metallurgical information is very strong. In developing such systems close co-operation between physicists and metallurgists is necessary.

ACKNOWLEDGEMENT

The authors are much indebted to our coworkers, Dr.T. Senuma, Mrs.M.Suehiro, Y.Matumura, Y.Tsukano, K.Sato and M.Hatta for many helpful discussions.

REFERENCES

1. H.Matsuda, Y.Onoe, K.Moriyama and H.Sekine, Nippon Steel Tech. Rep., No.21, 217 (1983).
2. K.J. Irvine and F.B. Pickering, JISI, 187, 292 (1957).
3. H.Morikawa and T.Hasegawa, Int. Conf. on Accelerated Cooling of Steel, 19-21 Aug. 1985, Pittsburgh, Pa.
4. Y.Tsukano, M.Suehiro, K.Sato and H.Yada, Tetsu-to-Hagané, 72, S536 (1986).
5. F.B. Pickering, "Physical Metallurgy and the Design of Steels", Applied Science Publishers (1978).
6. T.Gladman, I.D. McIvor and F.B. Pickering, JISI, 210, 916 (1972).
7. C Zener, Trans. AIME, 167, 550 (1946).
8. Y.Matsumura and H.Yada, Tetsu-to-Hagané, 72, S542 (1986).
9. S.Matsuda, T.Inoue, H.Mimura and Y.Okamura, Trans ISIJ, 12, 325 (1972).
10. C.M. Sellars and J.A. Whiteman, Met. Sci., 13, 187 (1979).
11. W.Roberts, Int. Conf. on the Strength of Metals and alloys, 12-16 Aug. 1985, Montreal.
12. P.Choquet, A.LeBon and C.Perdrix, ibid.
13. H.Yada, N.Matsuzu, K.Nakajima, K.Watanabe and H.Tokita, Trans ISIJ, 23, 100 (1983).
14. T.Senuma, H.Yada, Y.Matumura and C.Futamura, Tetsu-to-Hagané, 70, 2112 (1984).
15. H.Yada and T.Senuma, Journal of JSTP, 27(300), 34.
16. J.W.Cahn, Acta Met., 4, 449 (1956).
17. M.Suehiro, H.Yada, Y.Matusmura and T.Ariyoshi, Tetsu-to-Hagané, 71, S1492 (1985).
18. A.Hurkman, G.A. Duit, T.M. Hoogendoorn and F.Hollander, ibid to ref.3.
19. H.Yada, J.Tominaga, K,Wakimoto and N.Matuszu, ibid.
20. Y.Saito, M.Tanaka, T.Sekine and H.Nishizaki, Int. conf. HSLA steels, Aug. 1984, Wollongong, Australia.
21. T.Takahasi, J.Wakita, O.Kohno and K.Esaka, Tetsu-to-Hagané, 72 S538 (1986).
22. M.Suehiro, K.Sato, Y.Tsukano, H.Yada, Y.Matumura and T.Semma, ibid 72, S537 (1986).

23. S.Hamauzu, T.Kikuma, K.Nakajima and N.Hosomi, Proceedings of 1980 Spring Joint Meeting of Plastic Working, The Japan Soc. for Tech. of Plasticity, Tokyo, 53 (1981).
24. H.Suzuki, T.Ashiura, K.Aoyagi, H.Fujii and K.Tanabe, Tetsu-to-Hagané, 72, 587 (1986).
25. M.Lacroix, J.Abrigo and L.Arnanlt, Proceedings of Int. Conf. on Steel Rolling, 29 sep.- 4 Oct. 1980, Tokyo, Vol.2, 1286.
26. M.Morita and O.Hashimoto, ibid. to ref. 3.
27. K.Kawashima et al, to be published in Tetsu-to-Hagané.
28. A.Ogasawara, K.Soejima, Y.Ohno and K.Uchino, Japn Pat. Appl. No.54-160610 (1979).
29. K.Miyagawa, Y.Sasaki and N.Matusda, Japan J. of Nondestructive Testing, 31, 11 (1982).
30. H.Takafuji, S.Sekiguchi, T.Iuchi and S.Matsuda, Proceedings of Ultrasonic Int. conf. '85, London.

NONDESTRUCTIVE ANALYSIS OF STRUCTURE AND STRESSES BY ULTRASONIC AND MICROMAGNETIC METHODS

P. Höller

Institut für zerstörungsfreie Prüfverfahren

Saarbrücken, West Germany

INTRODUCTION

The objective of materials testing is to assess **fitness for purpose.** This holds for any type of materials testing: the classical testing for strength, toughness, and other properties by applying load, the structure analysis by using different types of microscopes or diffractometers and also for nondestructive testing which traditionally is devoted to detection and characterization of defects, especially cracklike ones. Assessment also is necessary for safety. Layout, design and fabrication of machinery and components primarily consider purpose and function but as a conditio sine qua non safety must be guaranteed. This means that the envelope of requests for fitness, purpose, function and safety is the basis for layout, design and fabrication.

During operation of machinery or use of components, strength and toughness of materials are slowly degraded by load induced fatigue, creep or corrosion. Here, temperature is a very important parameter. Any degradation starts with changes of the microstructure of the materials and only grows slowly to cracklike defects, significant from the fracture mechanic point of view; but already changes of microstructure have influence on strength and toughness, especially of high strength steels. Therefore materials testing for quality assurance is needed not only during fabrication but also during operation.

Testing for properties during fabrication can be performed destructively, but in safety related cases more and more the quality of the real components (not only small specimens) has to be tested nondestructively.

Any testing for quality assurance during operation must be performed nondestructively. Nondestructive testing for mechanical and most other properties in a direct way is impossible; but nondestructive structure analysis - including the microstructure - is possible and falls under the heading of this conference.

In addition, it is necessary to determine both after fabrication and during operation whether stresses in components are below levels bearable by the materials under operational conditions. There is a strong demand for nondestructive measurement of residual and load induced stresses.

The **Fraunhofer-Institute** for **nondestructive Testing** (IzfP) in Saarbrücken, West Germany, since its foundation in 1972 has considered **defects, structure** of materials, residual and load induced **stresses** as **equivalent objectives of NDT.** This contribution will give a survey of these activities during the last few years. Several specific papers presenting more details given by Theiner, Conrad and Willems, will be included in the printed version of the conference proceedings.

Different NDT tools in the above mentioned sense for structure and stress analysis of materials are well-known:

- optical and electron microscopy of replica taken from the surface of components for structure analysis
- x-ray diffraction for stress measurement
- ultrasonic testing at high frequencies for structure characterization
- microradiography and -tomography for structure analysis and detection of very small defects
- multiparameter micromagnetic microstructure and stress analysis (**3MA**)
- ultrasonic measurement of macrostresses by polarized shear waves.

In Fig.1 the very wide range of linear dimensions of structure parameters involved in the context of this conference is presented in the lower part, the examination methods are shown in the upper part. Precipitates and dislocations with linear dimensions between nanometer and micrometer are the most important microstructure parameters for strength and toughness of steels. Magnetic Bloch-walls have comparable linear dimensions; they are not important for strength and toughness, but if they oscillate for testing purposes they very strongly interact with dislocations and precipitates. The micromagnetic interaction can be measured nondestructively and in the last years this has given a great impact to nd-microstructure analysis. Methods to measure ultrasonic scattering realistically cannot contribute to microstructure analysis in this range, while ultrasonic absorption and also velocity have a high potential for this purpose.

Above one micron, inclusions, grain sizes, texture, anisotropy, and stress fields cover a wide range of more than four decades of linear dimensions. From the upper part of Fig.1 it can be recognized that ultrasonic resolution has come down to microns by gigahertz ultrasonic microscopy. The same holds for photoacoustic microscopy. Ultrasonic scattering only starts slightly below 10 microns in materials with low absorption like high density ceramics.

STRUCTURE AND STRESS ANALYSIS BY ULTRASONICS

Interaction between ultrasonic waves and structure has to be considered in terms of the diameter d of the scattering particle related to the wavelength λ. Whereas defect detection and analysis mainly are done on the resonant area ($d \approx \lambda$), the structure analysis because of many scatterers interacting at the same time can be performed even in the very low Rayleigh region ($d \ll \lambda$). For precipitates or inclusions the interaction is very similar to pores in the Rayleigh region but extremely different in the resonance domain[1].

It was shown many years ago that grain size can be determined in polycrystalline materials by measuring backscattering in the Rayleigh region[2].

An instrument, using a double frequency method to extract from the backscattering measurement the scattering coefficient and correlate that to grain size is realized[2,3]. An advanced approach by Arnold[4] uses a laser source for a wideband pulse excitation which leads to several backwall echoes even in very thin specimens. The backwall-echoes are picked up by a piezoelectric probe or a laser interferometer. The spectra of the backwall-echoes are evaluated and lead to quantitative data for grain sizes.

Creep damage starts with small micropores, Fig.2. It is important to detect these micropores by nd means in an early stage of damage before cracking occurs. One method well approved and applied is the microscopic investigation of replica taken from components[6], but also nd ultrasonic testing has a potential. A calculation of Hirsekorn shows the change of velocity due to pore concentration on the ordinate and in addition the size dependency of this effect along the abscissa, Fig.3. Fig.4 gives an experimental result showing the influence of creep damage on the ultrasonic velocity in a ferritic steel. If one relates the measured velocity changes with measured density changes (due to porosity) the results are in good agreement with the theoretical calculations[7].

There is still further potential of ultrasonic testing for microstructure analysis including porosity. An overview is given in Fig.5: on the left the scanning laser acoustic microscope (SLAM), developed and produced in the USA by Sonoscan, has been used successfully for several years. In the center part the scanning acoustic microscope (SAM) also developed in the USA by Quate but now produced as well in our country as in England and in Japan, produces a C-scan with very high resolution. In the right part the photoacoustic and photothermal device (PAM) is explained schematically. Fig.6 shows the detection and imaging of fatigue damage by PAM.

Elasticity in materials is not exactly linear. The deviation from linearity behaviour is unimportant for mechanical engineering but it is the basis for stress measurement by ultrasonics. If we superimpose the very weak stress field of an ultrasonic wave on the load induced or residual stress field of comparatively great amplitudes we observe weak but measurable dependencies of the ultrasonic velocity on the macroscopic stresses. Basic work on this phenomenon has been performed especially by Hughes and Kelly[8] and by Egle and Bray[9].

This is the basis for the birefringence stress measuring approach[10]. In ultrasonics it is very easy to measure precisely time-of-flight but it is very difficult to measure velocity because path length normally is not precisely known and also very difficult to measure with an accuracy of 10^{-4}. Therefore the most practical approach is to measure time-of-flight for two wave modes which have the same path. Then the relative difference of both is independent of the path. Fig.7 shows residual stresses aside a weldment. Fig.8 shows a special probe design for bolt stress measurements. The inner pancake coil of the probe emits a shear wave and the piezoelectric ring a longitudinal wave. In Fig.9 ultrasonic measurements of strains observed with this device in the bolt are plotted against strain gage measurements[11].

The most important subject still under research and development is the separation of stress and structure induced velocity changes.

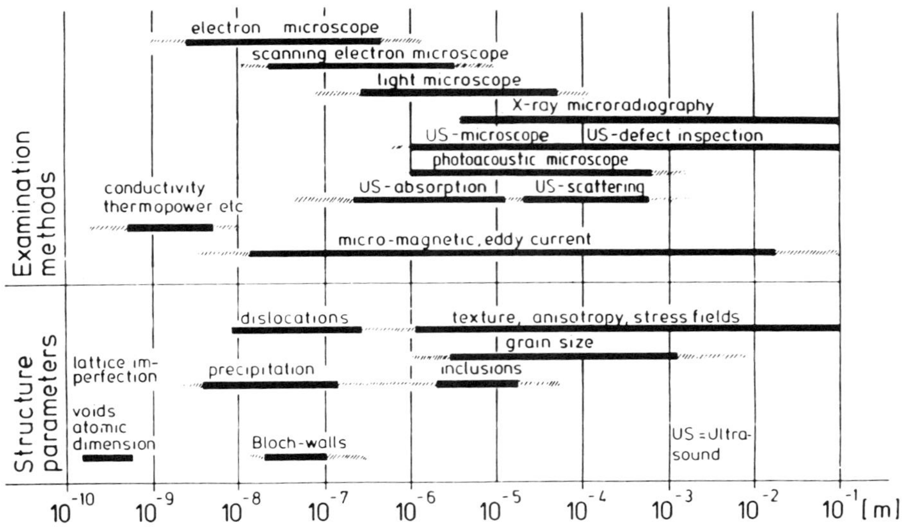

Fig. 1: Linear dimensions of structure parameters and examination methods.

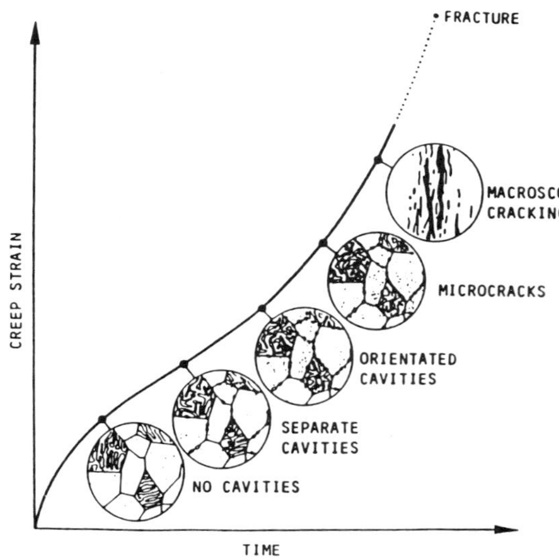

Fig. 2: Damage formation during creep (schematically) after[5].

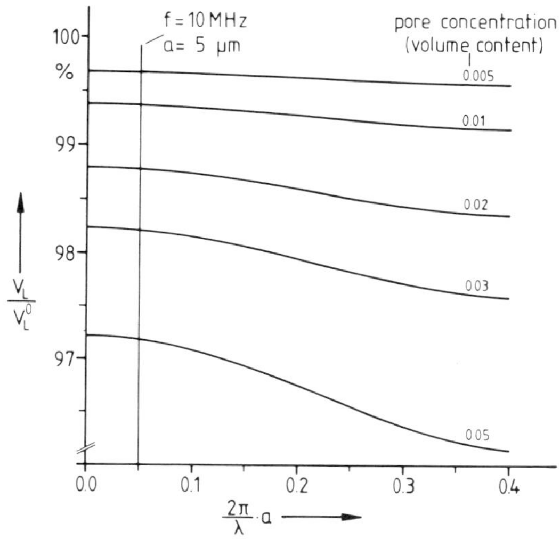

Fig. 3: Relative change of ultrasonic velocity V_L / V_L^o in iron as a function of pore concentration, pore radius a and ultrasonic wavelength λ (after Hirsekorn).

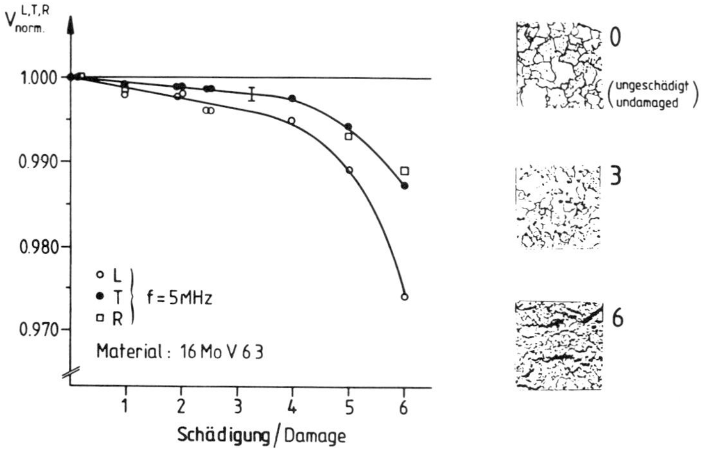

Fig. 4: Influence of creep damage on ultrasonic velocity v. The damage grade is defined in terms of micropore formation and microcrack formation, respectively. Examples are shown in the micrographs on the right side. For damage grade 4 the density change was 0.8 %.

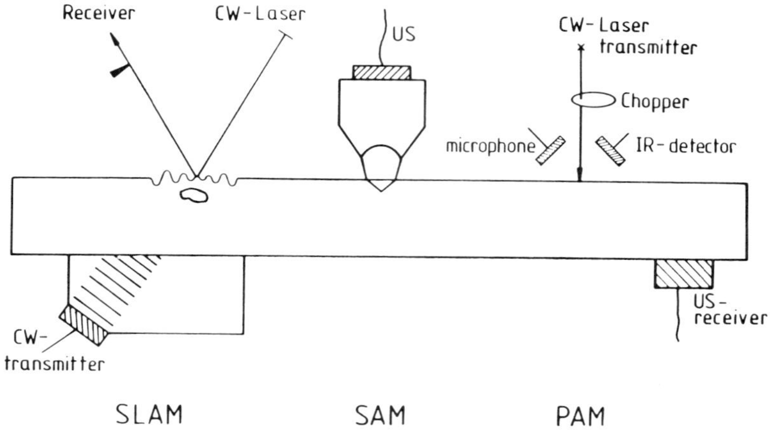

Fig. 5:

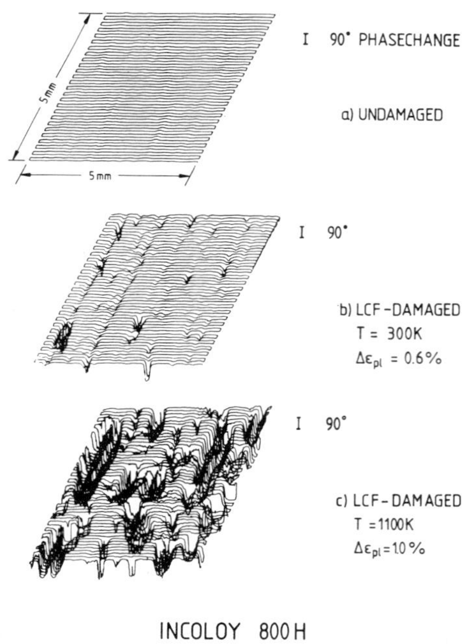

Fig. 6: PAM images of fatigue damaged Alloy 800H.

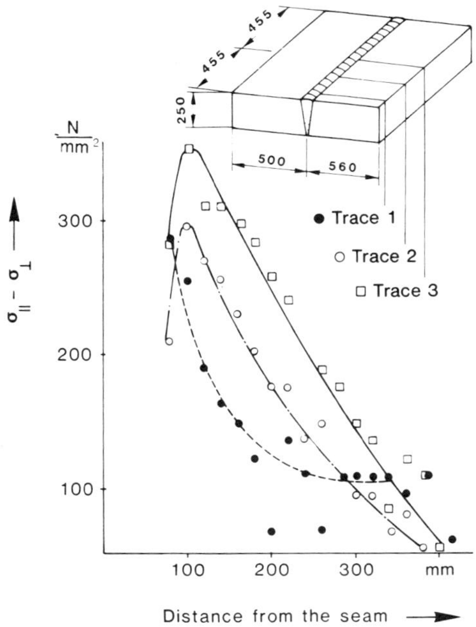

Fig. 7: Residual stress difference $\sigma_{\parallel} - \sigma_{\perp}$ versus distance from a ferritic weld seam.

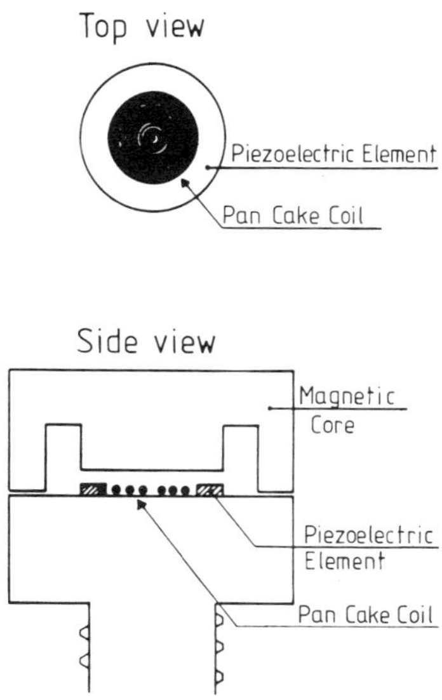

Fig. 8: Probe design for ultrasonic stress measurements in bolts.

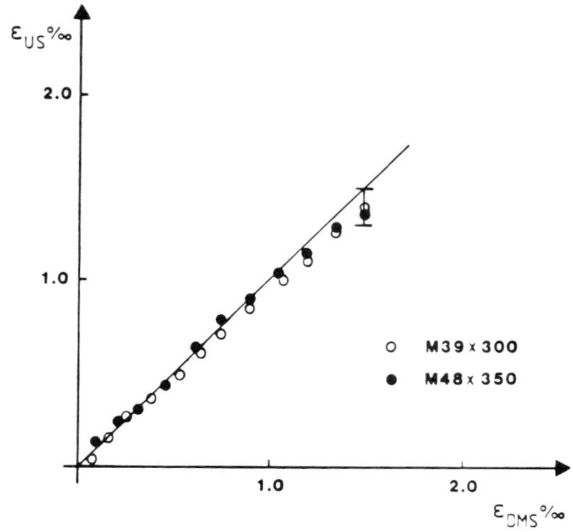

Fig. 9: Comparison between strain measurements in bolts by ultrasonics (ε_{US}) and by strain gage (ε_{DMS}).

MICROMAGNETIC MULTIPARAMETER MICROSTRUCTURE AND STRESS ANALYSIS (3MA)

As mentioned at the beginning, dynamic interactions between Bloch-walls and microstructure parameters can be measured.

Barkhausen-noise for instance has been investigated for a long time. Basic work came from Matzkanin[12] and Otala, Tiitto[13,14]. The basic set up for this type of measurement is shown in Fig.10. Under the magnetic yoke, which is rather small and fits in a human hand, several sensors can be placed. The philosophy in our institute from the very beginning was that we cannot rely on one measurement of a single micromagnetic measurement quantity. Each magnetic measurement quantity depends at least on several structure and/or stress influences.

To find a solution one easily can use several micromagnetic techniques under the same magnetization. But the informations obtained from the different techniques must be independent from each other. We prefer the following techniques:

- magnetic Barkhausen-noise

- acoustic Barkhausen-noise

- incremental permeability

- dynamic magnetostriction

and other testing quantities derived from these, for instance the coercivity.

Fig.11 shows qualitatively the influence of stress- and microstructure-changes on the hysteresis loop; Fig.12 shows a magnetization loop and the measured rectified and low-pass-filtered magnetic Barkhausen-noise as a function of the magnetizing field. A sharp peak at the coercivity is observed which makes it possible to measure the coercivity nondestructively. Fig.13 shows the measurement of the incremental permeability by superimposing on the hysteresis loop an alternating magnetic field with a small amplitude but a high frequency. This superposition is performed using eddy-current probes. Since a peak in this curve also occurs at the coercivity, there are two methods to measure the coercivity, both Barkhausen-noise and incremental permeability may be used. Fig.14 shows the instrument and a probe with integrated yoke and sensor.

For analysing hardness profiles normal to the surface different micromagnetic quantities can be used. E.g. if the spectra of Barkhausen noise signals are analyzed or frequencies of eddy-current probes are changed it is possible to determine the thickness of near surface layers. The skin effect of eddy-currents is well-known and need not be explained. Magnetic Barkhausen signals in the material excite eddy-currents depending on their spectrum and thereby they are attenuated on the signal path between the source and the sensor. Schematically this is explained in Fig.15. Typical Barkhausen signals for a two layer specimen are presented by Fig.16 for different analysing frequencies f_A. The near surface layer is laser hardened and therefore has a much greater coercivity (large peak-distance) than the soft base material (smaller peak-distance). From a wideband- or single-frequency-measurement one cannot derive the thickness of the hardened layer; but frequency analysis of the Barkhausen-noise shows, that the ratio of signals from the soft material to those of the hardened surface layer increases strongly when the analyzing frequency is decreased from 100 kHz to 2 kHz[16,17].

An example for determining the lateral extension of a flame hardenened zone and residual stress on a turbine blade using Barkhausen-noise combined with coercivity in a 3MA-algorithm was presented by Theiner in Hershey[10,15].

A more complicated case is the thickness evaluation of a thin new-hardened layer which can occur during grinding. Conrad in his contribution will refer the details[18] applying incremental permeability-measurements, Fig.17. In Fig.17 three different hardness profiles normal to the surface which can be obtained by grinding procedures are given. Section A shows e.g. an extended new hardening zone (NHZ) joined by annealed microstructure states. The nondestructive results of these states (A, B, C) using the coercivity $H_{C\mu}$, deduced from the incremental permeability versus the magnetic penetration depth $D_{MAG} \sim (\sigma_{el}\, \mu_\Delta\, f_\Delta)^{-1/2}$, compare well with these hardness profiles.

In addition to the description in this overview, acoustic Barkhausen-noise and dynamic magnetostriction are under strong investigation and development for materials characterization. The technology for measuring acoustic Barkhausen-noise with the acoustic emission technique is available. The same holds for the dynamic magnetostriction by exciting an ultrasonic surface wave magnetostrictively and measuring its amplitude by piezoelectric or electromagnetic UT; but at this time the instrument still is bulky and needs further development to fit into a 3MA device. Whether and in which cases these additional parameters give significant contributions or at least redundant information we have yet to learn.

SUMMARY

- There is a great industrial demand for nondestructive materials characterization as a tool for quality assurance during production and operation of machinery and components.

- It is not possible to determine directly properties of materials, especially mechanical properties by nondestructive means.

- There are many possibilities to measure and analyse structure and stresses of materials with resolutions down to nanometers by nondestructive testing procedures consisting of a handy probe and a transportable and affordable package of instrumentation and software for control, pick up the data and evaluate them.

- Industry strongly supports the development of new technology for materials characterization and their transfer into practical use.

- Unfortunately the capacities for R a. D and the funds for basic work in this area are far below levels for nondestructive testing for defects.

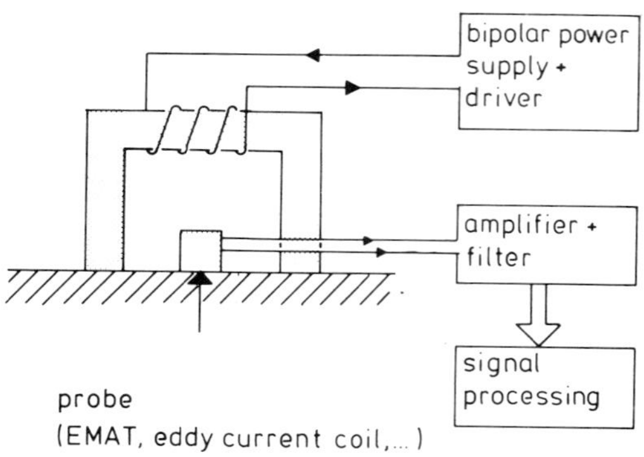

Fig. 10: Transmitter and receiver system schematically.

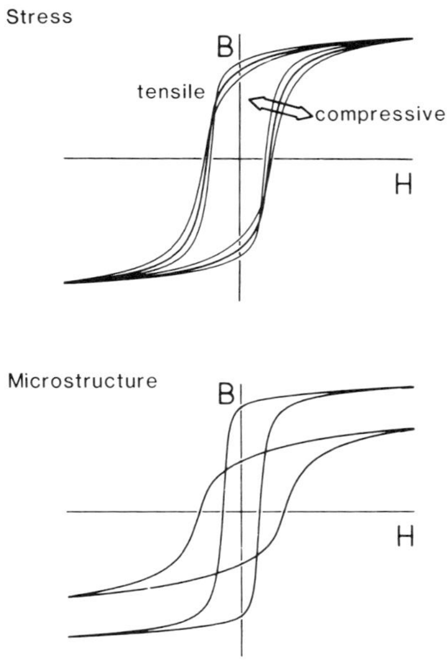

Fig. 11: Change of hysteresys as function of different stress- and microstructure states.

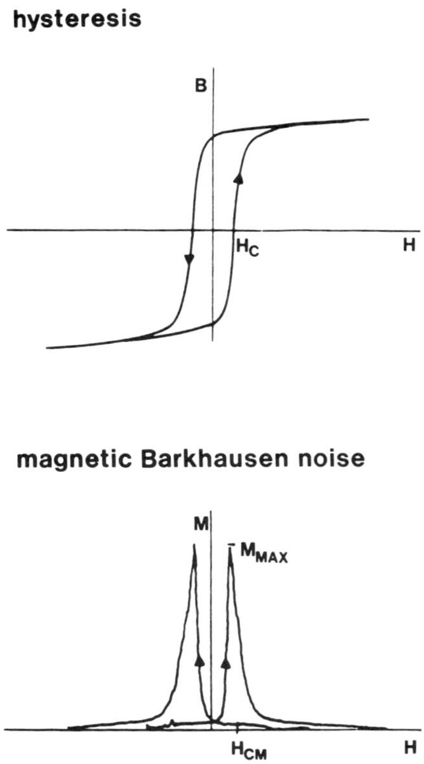

Fig. 12: Magnetization loop and the rectified magnetic Barkhausen-noise signal as function of the magnetic field strength.

incremental permeability (μ_Δ)

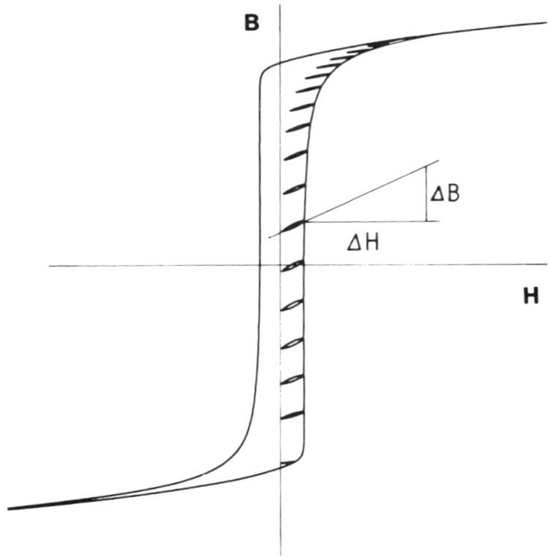

$$\mu_\Delta = \frac{1}{\mu_0} \quad \frac{\Delta B}{\Delta H}$$

Fig. 13: Incremental permeability.

Fig. 14: 3MA-Prototype device which uses the magnetic Barkhausen-noise and incremental permeability.

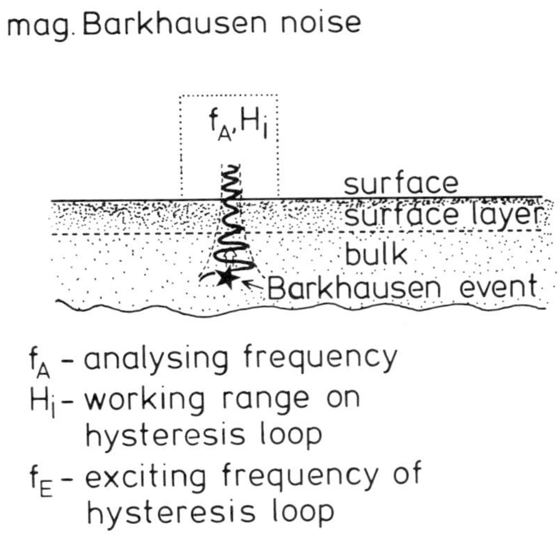

Fig. 15: Determination of surface near microstructure states.

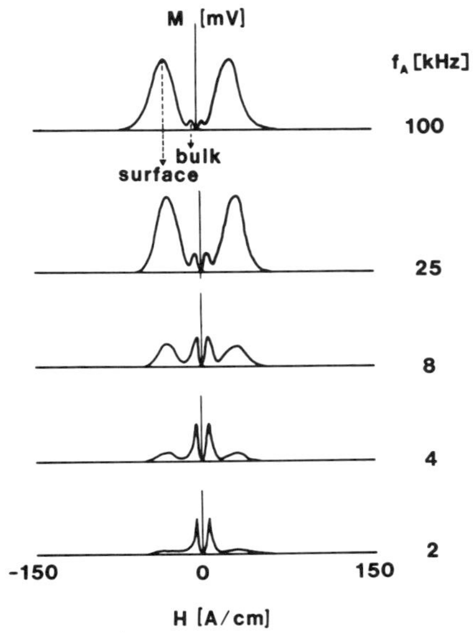

Fig. 16: Barkhausen signals of a laser hardened specimen at different analysing frequencies f_A.

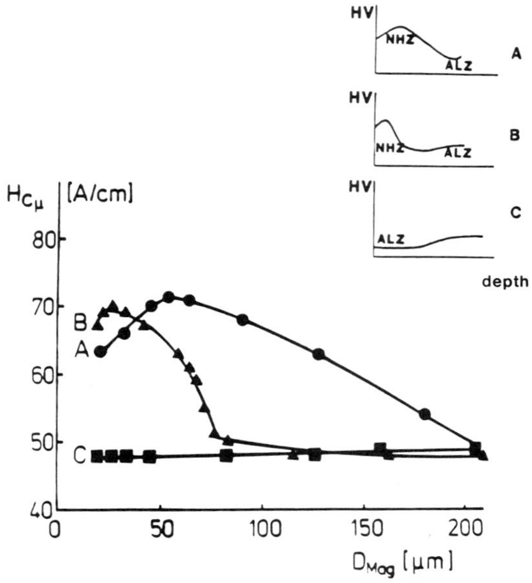

Fig. 17: Magnetic hardness profile of three different grinding states.

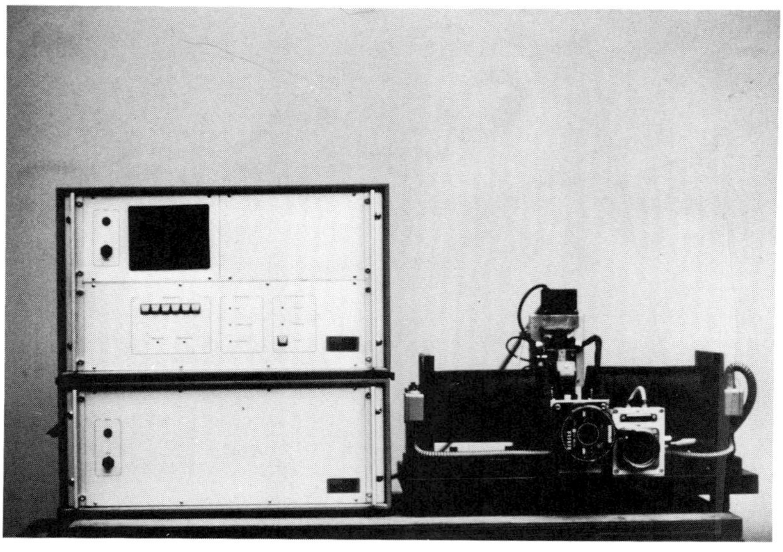

Fig. 18: 3MA device for detecting grinding defects.

REFERENCES

1. Hirsekorn, S.; Streuung von ebenen Ultraschallwellen an kugelförmigen, isotropen Einschlüssen in einem isotropen Medium unter Berücksichtigung von Mehrfachstreuung. IzfP-Bericht Nr. 790218-TW, Saarbrücken (1979)
2. Goebbels, K.; Research Techniques in Nondestructive Testing, Vol.IV. R.S. Sharpe (Ed.), Academic Press, London 1980
3. Theiner, W.A./ Willems, H.; Determination of microstructural parameters by magnetic and ultrasonic quantitative NDE. in: Nondestructive methods for material property determination. (C.O.Ruud and R.E.Green Eds.) Plenum Press, New York 1984, 249-258
4. Arnold, W./ Betz, B.; Deutsches Patent 3412615
5. Auerkari, P.; Eurotest Technical Bulletin E51, Dec. 1983
6. Bendick, W./ Bruening, B./ Weber, H.; Sonderbände der Praktischen Metallografie Nr. 16, 439-450, Dr. Riederer Verlag Stuttgart 1955
7. Willems, H./ Bendick, W./ Weber, H.; Nondestructive evaluation of creep damage in service exposed 14MoV 6 3 steel. paper submitted to: Symposium on the Nondestructive Characterization of Materials, Montreal, July 21-23, 1986
8. Hughes, D.S./ Kelly, J.L.; Phys. Rev. 92 (1953) 5, 1145-1149
9. Egle, D.M./ Bray, D.E.; Measurement of acoustoelastic and third-order elastic constants for rail steel. J.Acoust.Soc.Am. 60 (1976) 741-744
10. Schneider, E./ Altpeter, I./ Theiner, W.; Nondestructive determination of residual and applied stress by micro-magnetic and ultrasonic methods. aus: Nondestructive methods for material property determination, Plenum Press, New York and London (1983) 115-122
11. Altpeter, I./ Goebbels, K./ Pongratz, H.J./ Theiner, W.A.; Weiterentwicklung von Ultraschall- und ferromagnetisch-magnetoelastischen Verfahren zur Bestimmung der Einhärtetiefe und der Eigenspannungen an Werkstückoberflächen sowie der Lastspannungen in Schrauben. IzfP-Bericht Nr. 860209-TW, Saarbrücken (1985)
12. Matzkanin, G.A./ Beissner, R.E./ Teller, C.M.; The Barkhausen effect and its applications to nondestructive evaluation. Southwest Research Institute,San Antonio, Texas, NTIAC-79-2, Okt.1979
13. Otala, M./ Säynäjäkangas, S.; A New Electronic Grain-Size Analyzer for Technical Steel. J.Phys.E.Sci.Instrum. 5 (1972) 669-672
14. Tiitto, K.; Solving internal stress measurement problems by a new magnetoelastic method. aus: Nondestructive methods for material property determination. Plenum Press, New York and London (1983) 105-114
15. Theiner, W.A./ Altpeter, I.; Determination of residual stresses using micromagnetic parameters. in: New Procedures in NDT. P. Höller (Ed.), Springer Verlag, Berlin 1983
16. Theiner, W.A./ Kern, R./ Conrad, R.; Determination of surface integrities by ferromagnetic quantities. 1st Int. Conference on Surface Engineering, Brighton, England, June 26-28, 1985, Paper 25.
17. Theiner, W.A./ Altpeter, I./ Kern, R.; Determination of subsurface microstructure states by micromagnetic ND-techniques. paper submitted to: Symposium on the Nondestructive Characterization of Materials, Montreal, July 21-23, 1986
18. Conrad, R./ Reimringer, B./ Theiner, W.A.; ND Determination of grinding defects using the 3MA prototype device. paper submitted to: Symposium on the Nondestructive Characterization of Materials, Montreal, July 21-23, 1986

A VIEW OF THE NEED AND TECHNOLOGIES FOR IN-PROCESS MATERIALS PROPERTY EVALUATION

Clay Olaf Rudd and Chandra Vikram

The Material Research Laboratory
The Pennsylvania State University
University Park, PA 16802

ABSTRACT

The productivity of industry from basic production of raw metallic shapes through fabrication of components and manufacture of multi-component structures would be greatly enhanced by constant and thorough interrogation of the relevant properties of the material. However, even though great strides have been made in interrogation technologies, for example, nondestructive detection of cracks and voids in components, sparse practical progress has been made in nondestructive interrogation for material property measurement. Advancements in this field will be of great benefit, but the lack of broadly multi-disciplinary and multi-technique research has been lacking. Further, the need and possibilities of success of broader approaches seem ripe for exploitation. This paper is concerned with the need and application of techniques for nondestructive, in-process measurement of metallic properties. The investigation has reviewed a broad spectrum of interrogating energies, means to apply them, the effect of the energy on the propagating media (metallic component), and the effect of various types of metallic media on the propagating energy. Several systems consisting of a combination of technologies have been identified as promising and one concept of teamed techniques will be described.

INTRODUCTION

U.S. industry uses millions of pounds of wrought metallic alloys in the fabrication of everything from electronic components to amored personnel carriers. However, the important mechanical and physical properties upon which fabrication characteristics depend, and upon which materials are selected, are vague in that the properties can only be described within a broad band on uncertainty. For example, the Metals Handbook [1] shows that 6061-T6 aluminum varies over a percentage range of 22, 26, and 54 percent in tensile strength, yield strength, and ductility, respectively. Hundreds of other similar examples can be cited for steel, copper, nickel, titanium alloys, etc. The result of the property uncertainty is that design must be to the poorest property, and poor tolerances and excessive rework in subsequent fabrication must be accepted; hence one experiences lowered overall quality, substantial waste, and degraded reliability and life expectancy of the fabricated components.

There is a great deal of interest in automated processing of materials which would require, for instance, adjustment of forming dies during processing of the metal, and this so called "intelligent automated processing" may be automatically controlled by computers. Such automated processing certainly represents a path to improved productivity, higher quality materials and components, and reduced costs. However one of the necessary conditions for this automated processing is precise information on the forming characteristics of the feed material; in addition there must be continuous monitoring of the physical and mechanical properties of the products emerging from the process. Thus, in-process measurement of physical and mechanical properties and fabrication characteristics of wrought metallic mill and forged products is a necessary precursor for process control for the successful application of "intelligent processing of materials."

BACKGROUND

The quality and reliability of metallics from basic production of raw shapes through fabrication of components and manufacture of multicomponent structures would be greatly enhanced by constant and thorough in-process interrogation of their relevant characteristics. However, such comprehensive nonintrusive evaluation of material is not yet possible due to a vast range of problems, not the least of which is that the most versatile methods, e.g., ultrasonic testing, cannot independently measure those characteristics of a material that affect a particular property. Papadakis [2] discussed the interdependent effects of material characteristics on the propagation of ultrasound in structural materials and noted the complexity of isolating the change of a single characteristic, e.g., grain size, through measuring ultrasonic velocity and attenuation. An acoustic planar wave traveling at a speed c through a homogeneous velocity medium is described by

$$A = A_o e^{-\alpha x} e^{i\omega(t-x/c)},$$

where ω is the frequency, t the time, x the position coordinate, and α the attenuation coefficient. However, a metallic alloy is not homogeneous in any of the characteristics that affect the propagation of the acoustic wave, and thus α and c are, to an extent, different at every position coordinate, where that coordinate is measured on the order of micrometers to tenths of millimeters.

Mignogna et al. [3] showed that common structural materials such as aluminum plates, which for most engineering purposes are normally idealized as isotropic, must in fact be considered anisotropic when considering the propagation of ultrasound. The degree of anisotropy was noted to affect a number of elastic moduli. For instance, rolled aluminum plates of the same alloy and temper were found to have different moduli for different amounts of rolling, i.e., different thicknesses; the variety of texture condition was demonstrated by X-ray diffraction (XRD) measurements. Ledbetter et al. [4] demonstrated the vast range in acoustic velocity found in components made from a single alloy of stainless steel and were unable to isolate one or even a few specific characteristics of the material causing that scatter.

The problem of isolating property controlling characteristics of materials, e.g., metals and ceramics, is a problem of major concern to scientists and engineers interested in nondestructive characterization. However, in spite of great strides made in the evolution of techniques for the detection of cracks and voids, little progress has been made in nondestructive materials characterization. This concern is the subject of a series of symposia, the first of which has had the proceedings published [5]. It has been noted that not only acoustic (ultrasonic) methods are plagued with the problem of interdependent effects of the characteristics on the interrogation signal, but other methods, such as eddy current and magnetic (e.g., Barkhausen), also suffer from the same unknowns.

SIGNATURES

The isolation and independent measurement of all of the characteristics of the metal that affect the forming processes are not yet attainable [6]. Accomplishing these for even one alloy system requires a considerable effort in fundamental research and empirical technological development. However, a detailed understanding of the quantitative effects of a metal's characteristics on the interrogating energy may not be necessary. McCauley [7] has introduced a signature describing a specific mechanical property for a ceramic as a complicated function of chemistry, microstructure and processing defects. For wrought metallic alloys the signature for a particular property would more appropriately be written as a complicated function of elemental content, phase content, dislocation density, crystallographic texture, grain size and shape, dislocation density and distribution, stacking fault density, internal elastic strain, etc. Further, the function defining this signature might be developed through an interactive process of theoretical and empirical relationships using the results of destructive and nondestructive testing. Thus, if a component, or team, signature from several nondestructive interrogation techniques, which are sensitive to the relevant characteristics of the material in a complimentary fashion, could be obtained then adjustment and control of the forming processes could be accomplished a priori.

It is not implied that the isolation and measurement of a specific characteristic, e.g., dislocation density, could be accomplished through use of a team signature. For example, if the acoustic velocity of a metal feed stock, e.g., sheet for stampings, were continuously monitored in several locations across its width, it must be recognized that changes in that parameter might be affected by variations in alloy elements in the matrix, cold work, grain size, residual stress, texture, etc. and not just dislocation density. No attempt would be made to isolate the characteristic affecting the change, but other complimentary techniques which are sensitive to the several characteristics but to different degrees would be used in concert, e.g., XRD. Thus, what is implied is that a signature representing a property of the material and dependent upon several nondestructively determined parameters, would be developed which would indicate changes in that property as the material was scanned. Therefore, through empirically establishing composite (or team) signatures for feed stock property variation using teamed instrumentation, automatic adjustment of forming dies might be accomplished.

SPECIFIC OPPORTUNITY

The semi-empirical nature of the proceeding approach must focus upon a specific type of metal alloy system and class of forming processes so as to restrict the investigation to a managable number of variables. Thus, for example, a copper alloy or stainless steel strip, such as used in the fabrication of small components, might be selected as the material; and the stamping of these small parts selected as the forming process. However, the technology, algorithms, and database developed for these alloys could be generally applicable to a wide range of polycrystalline metal alloys.

Total automation of stamping machines, which produce hundreds of components per minute, is frustrated by the variability of the mechanical properties of the feed stock. Presently, electronic switch and contact manufacturers must adjust stamping and forming dies for each reel of feed stock to produce components that are within the required shape and physical (e.g., contact pressure) tolerances. Also, fine adjustments are often required due to property changes from one end to the other of a single reel. At present, there is no way to judge a priori what die settings each reel of feed stock might require. Thus, the reel must be started through the stamping machines, samples of the component inspected, and then the dies changed to accommodate the out of tolerance condition; this procedure is repeated until components within tolerance are produced, an exercise that often produces hundreds of parts that are subsequently scrapped, is labor intensive, and thus wastes time as well as material.

There is an urgency to develop improved (intelligent) processing for many products such as electronic sheet parts because the demand for smaller and smaller components with increasingly more precise tolerances is accelerating, and this demand will provide more impetus for total automation of stamping and forming processes. However, in order to accomplish the degree of automation that will be required, automatic adjustment of the dies for variation in feed stock properties must be provided or else the feed stock must be pre-selected (classified) to insure only a very narrow band of uniform properties. To accomplish either of these alternatives, continuous and thorough in-process interrogation of the relevant material properties is required.

INVESTIGATIVE APPROACH

The preceeding has offered a view of the need to develop in-process interrogation of wrought alloy products so as to continuously provide a characterization signature to determine the instantaneous properties and forming characteristics of the metal. An investigative approach for copper alloys or austenitic stainless steels might use a combination of three concerted nonintrusive techniques, i.e., eddy current, EMAT ultrasound, and X-ray diffraction. The first two individually have seen limited applications while the latter has never been applied in the real-time frames necessary. Significant reasons for the suggested combination of these three techniques are that they are non-contacting, very rapid, and each is especially sensitive to different characteristics in complimentary ways. One example of their complimentary nature is that eddy current is quite sensitive to elements in solid solution in the matrix, x-ray is especially sensitive to microstrain caused by coherent precipitation from the matrix, and acoustic attenuation is sensitive to incoherent precipitation. Thus the team signature could provide a sensitive indicator of thermo-mechanical effects in the microstructure.

REFERENCES

1. AMS, Metals Handbook, Vol. 2, 9th Ed., p. 258, 1979.
2. Papadakis, E.P., "The Inverse Problem in Materials Characterization through Ultrasonic Attenuation and Velocity Measurements," Nondestructive Methods for Material Property Determination, C.O. Ruud and R.E. Green, Eds., Plenum Press, New York, 1984.
3. Mignogna, R.B., A.V. Clark, B.R. Rath, and C.L. Vold, "Effects of Rolled Plate Thickness on Anisotropy with Application to Acoustic

Stress Measurement," Nondestructive Methods for Material Property Determination, C.O. Ruud and R.E. Green, Eds., Plenum Press, New York, 1984.
4. Ledbetter, H.M., N.J. Frederick, and M.W. Austin, "Elastic Constant Variability in Stainless-Steel 304," J. Appl. Phy. 51, 305-309, 1980.
5. Ruud, C.O. and R.E. Green, Eds., Nondestructive Methods for Material Property Determination, Plenum Press, New York, 1984.
6. Ruud, C.O. and C. Vikram, "Morphological Investigation of Fundamental Methods for In-Process Materials Evaluation for Productivity Improvement," Proceeding of the 12th Conf. on Prod. R. and D., Nat. Science Foundation, Pub. by Soc. of Manf. Engr., 1 SME Dr., Dearborn, MI, pp. 345-352, May 1985.
7. McCauley, J.M., "The Role of Characterization in Emerging High Performance Ceramic Materials," AMMRL TR 84-49, Army Mat. and Mech. Res. Cntr., Watertown, MA, Dec. 1984.

DETERMINATION OF SUB-SURFACE MICROSTRUCTURE STATES BY MICROMAGNETIC NDT

W.A. Theiner, I. Altpeter, and R. Kern

Institut für zerstörungsfreie Prüfverfahren

Saarbrücken, West Germany

INTRODUCTION

The functional behaviour of many components is determined by the intrinsic properties - residual stresses and microstructure - of sub-surface near zones. Therefore, in order to improve the static and dynamic mechanical properties of these surface near states, mechanical and/or thermal treatments such as laser hardening, inductive hardening, case hardening, nitriding, grinding and shot peening are applied. Depending upon the workpiece/component and the employed technique and machining parameters, the residual stress- and microstructure states will be changed between a few micrometers and several millimeters. To control the quality of these sub-surface near states, metallographic inspections, x-ray methods and hardness measurements have been applied. Because all these conventional testing methods are expensive, time consuming and not suitable for real time control,[1-4] there is a considerable need of nondestructive testing (ndt) methods. So far ndt procedures are mainly restricted to eddy current methods,[5,6] ultrasonic techniques[7,8] and - in the case of ferromagnetic materials - micromagnetic techniques[1,9-11].

RESULTS

As outlined in Hershey[12,13] micromagnetic measuring quantities are able to indicate in a direct manner changes of residual stresses and microstructure states. It was shown there that superimposing microstructure states or superimposing microstructure and stress states can be characterized by the use of independent micromagnetic quantities. These quantities are deduced from the magnetic- and acoustic Barkhausen noise, the dynamic magnetostriction and from the incremental permeability. The method used is called 3MA (micromagnetic multiparameter microstructure and residual stress analysis) and will be explained on different layered microstructure states as they are produced by laser hardening- or inductive hardening processes.

The principal demand for a successful application of 3MA is that every sub-surface layer should have different magnetic properties in comparison to the bulk material state.

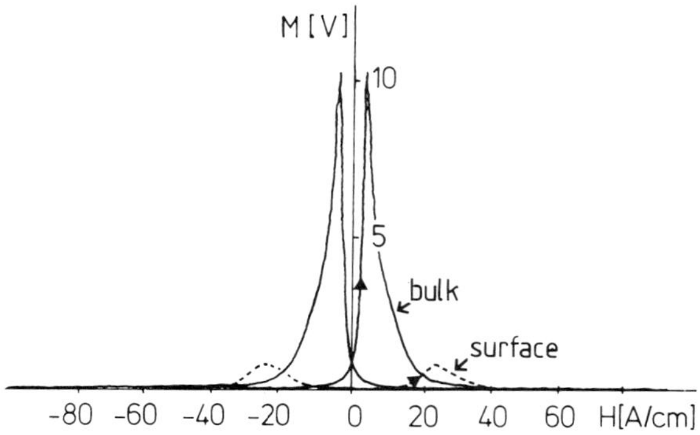

Fig.1. Magnetic Barkhausen noise M(H) curves of a soft bulk material state (e.g. annealed martensite) and of a hard surface state (e.g. martensite).

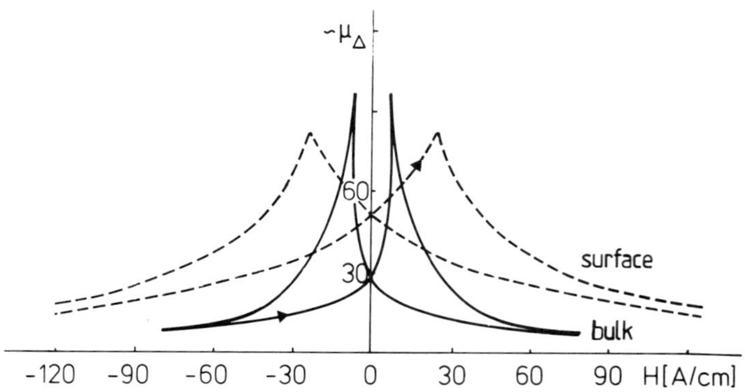

Fig.2. Incremental permeability curves μ_Δ (H) for the same microstructure states as in Fig.1.

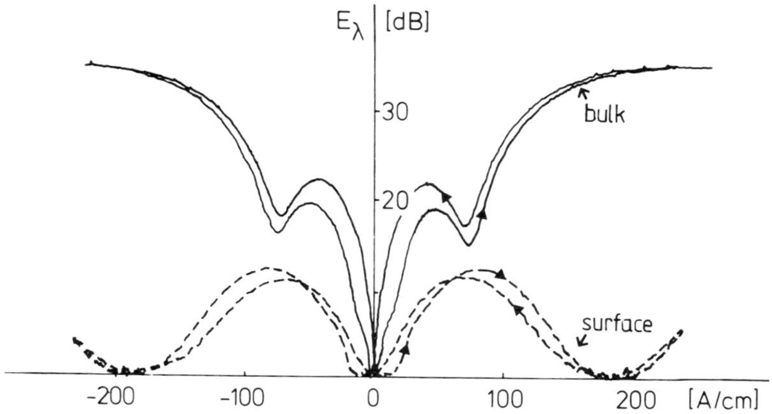

Fig.3. Dynamic magnetostriction curves E_λ (H) for the same microstructure states as in Fig.1.

From Fig.1., the rectified magnetic Barkhausen noise signal, one can see that the experimental results of the soft bulk material state is quite different from the hard surface state.

In this case one can use for 3MA the maximum of the rectified noise signal M_{MAX} and the coercivity H_{CM}, deduced from the M_{MAX}-position. The results given in Fig.1 are measured on homogeneous states, which means that within the interaction volume determined by the probe-geometry and the analysing frequency f_A, there are no gradients in microstructure or residual stress states.

If some layered structures or gradients normal to the surface have to be examined, one can change the analysing depth by the analysing frequency f_A range of the Barkhausen noise signals which are detected by an e.g. air coil or tape recorder head at the surface. Whatever the chosen interaction volume, the harder surface states or the softer bulk material states contribute mainly or in a superimposing way to the M(H)-envelope. Using this method the 3MA quantities/variables are:

$$M_{MAX}, \quad H_{CM}, \quad f_A, \quad \frac{dM_{MAX}}{dH}, \quad H$$

The same microstructure states shown in Fig.1 have been examined in Fig.2 by the incremental permeability $\mu_\Delta(H)$ and in Fig.3 by the dynamic magnetostriction $E_\lambda(H)$. From the $\mu_\Delta(H)$-curve one can see that both microstructure states are showing quite similar maxima values $\mu_{\Delta MAX}$, whereas the $\mu_{\Delta MAX}$-value of the hard surface state can lay below or above the bulk material value. This behaviour is the reason why eddy current techniques are very often not successful in detecting microstructure- or hardness-gradients. Plotting the incremental permeability μ_Δ versus the magnetic field strength (Fig.2) several quantities such as the coercivity $H_{C\mu}$, deduced from the peak position, $\mu_{\Delta MAX}$, the broadening of the μ_Δ (H)-curve $\Delta\mu_\Delta$ and the slope $d\mu_\Delta/dH$ can be used. The main variables are the exciting frequency f_Δ and the magnetic field strength H. Employing this method, the 3MA main quantities/variables are:

$$\mu_{\Delta\,MAX}, \quad H_{C\mu}, \quad \Delta\mu_{\Delta}, \quad \frac{d\mu_{\Delta}}{dH}, \quad f_{\Delta}, \quad H$$

In the case of the dynamic magnetostriction E_λ (Fig.3), one can use the fact that all martensitic- or solution treated states show only a very small magnetostriction and therefore also small E_λ -amplitudes. The penetration depth can be changed by the exciting frequency f_λ of the electromagnetic ultrasonic-transducer (EMUS) . In this case the 3MA quantities are:

$$E_{\lambda(Hi)}, \quad \frac{dE_\lambda}{dH}, \quad f_\lambda, \quad H$$

If the carbon content is changed by case-hardening or decarbonization processes, different microstructure parameters can be changed. If the cementite concentration is changed continuously, magnetic Barkhausen noise quantities such as the coercivity H_{CM} or the rectified noise amplitude at $H=H_C$ $M(H_C)$ are very sensitive, Fig.4.

In Fig.5-7 laser hardened CK45 specimens have been characterized by magnetic Barkhausen measurements. As explained above (Fig.1), the slope of the M(H) curves can be explained by the different penetration depth, determined by the analysing frequency f_A, and the characteristic irreversible magnetic response of bulk and surface states.

At f_A=100 kHz the magnetic penetration depth δ will be about 100 μm so that the hard surface state gives the main contribution to the M(H)-signal. If one lowers the analysing frequency f_A, the interaction volume will be enhanced and at f_A=8 kHz ($\delta \sim 350$ μm) the contribution of the bulk material states becomes dominant. This means that soft ferromagnetic states (Fig.1) are detected mainly in the low frequency range. Soft states are characterized by great single Barkhausen events (duration and amplitude), and by the fact that the frequency content of their spectrum is shifted to lower frequencies in contrast to magnetic hard materials. Additionally, Barkhausen events will lower their frequency content by eddy current damping during propagation through the hard surface states.

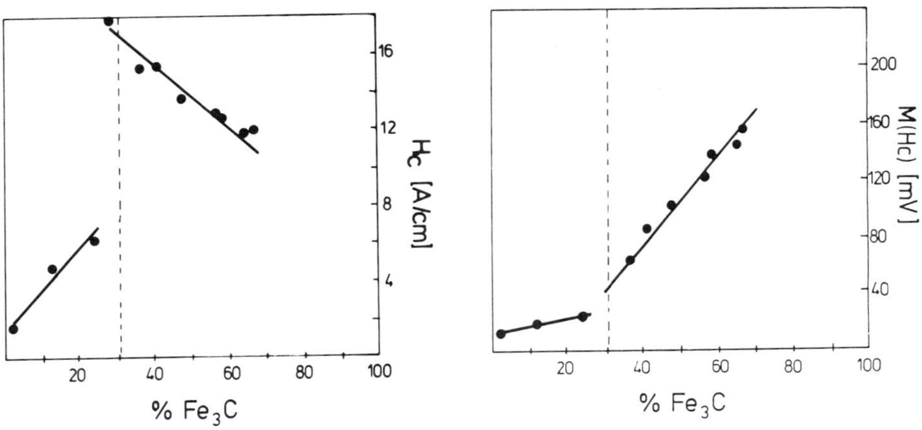

Fig.4. Barkhausen noise quantities H_{CM} (coercivity) and $M(H_C)$ (amplitude at H_C) for different cementite concentrations. The dotted line marks the phase transition steel ↔ cast iron. 14

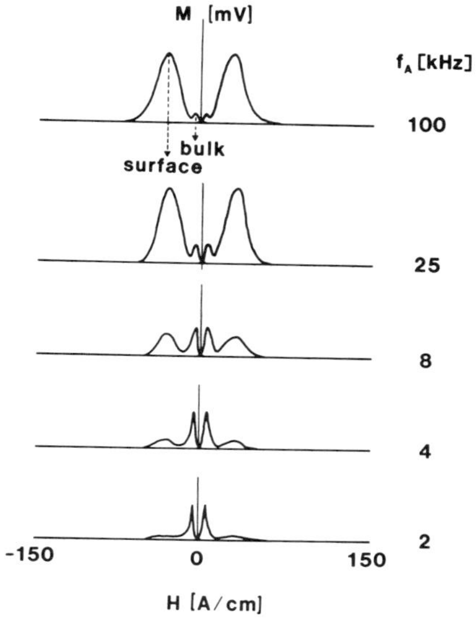

Fig.5. Rectified magnetic Barkhausen noise signals for different analysing frequencies. Material: CK45. Laser hardening depth: 0,70 mm.

If one determines the magnetic penetration depth $D_{nd}=(\sigma \mu_{\Delta} f_A)^{-\frac{1}{2}}$ (σ =5 m/Ω mm^2, μ_{Δ} =50), where the amplitude of the bulk- and surface states becomes equal ($M_{bulk}=M_{surface}$) one gets the result shown in Fig.6.

This linear response can be used for quantitative measurements.

If there are some expected values which should be maintained during one manufacturing step, another procedure can be chosen. As demonstrated in Fig.7, the analysing frequency f_A is constant. If the expected laser hardening depth value will be about 0.6 mm, one has to chose $f_A \sim 12.5$ kHz and one can decide immediately whether the hard surface layer becomes thinner

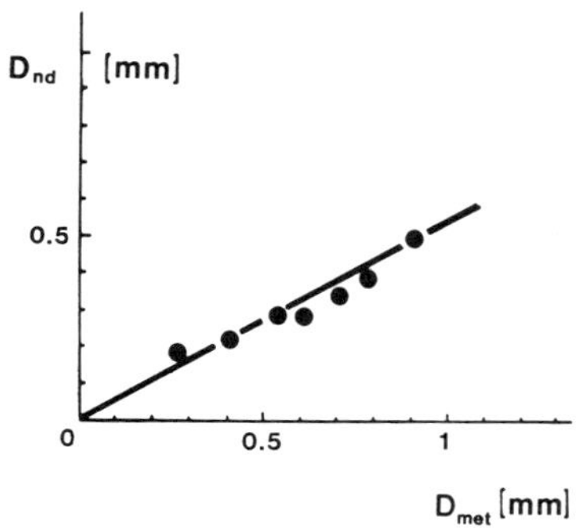

Fig.6. Magnetic determined hardening depht D_{nd} over metallographically determined values D_{Met}.

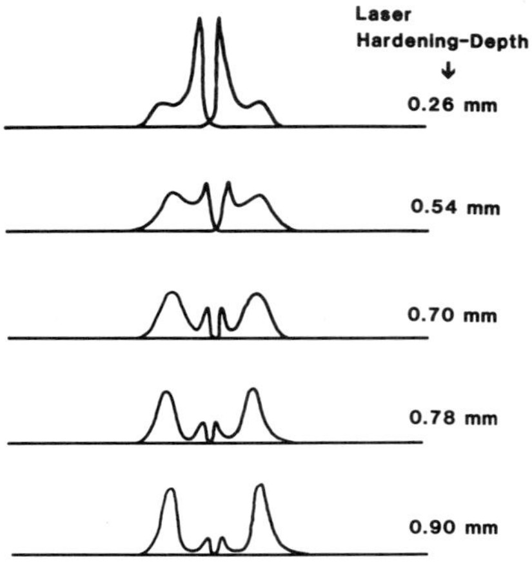

Fig.7. Magnetic Barkhausen noise curves for a fixed analysing frequency f_A=12.5 kHz and different laser hardening depths.

or thicker by evaluating the quotient of the Barkhausen noise amplitude of the bulk and surface states $M_{bulk} : M_{surface}$ 1.

Using the incremental permeability μ_Δ , mainly reversible magnetization processes are detected. One application of inductive hardened specimens is given in Fig.8. The results show the superposition of the two states introduced in Fig.2 at a constant interaction volume (f_Δ =1 kHz, constant geometry). As expected, with increasing hardening depth the contribution at $H = H_{C\ surface}$ becomes more and more dominant.

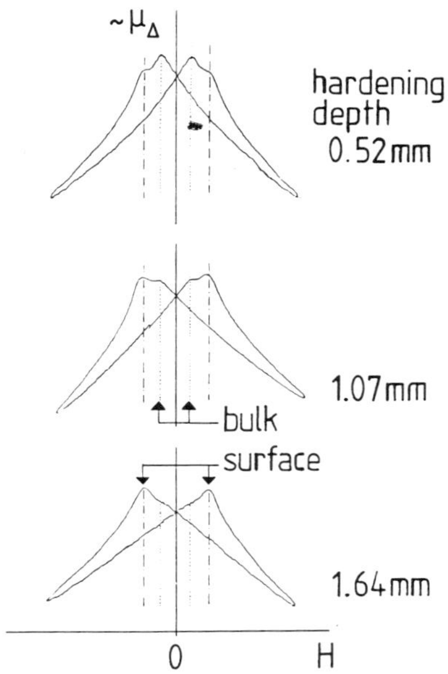

Fig.8. Incremental permeability measurements on inductive hardened 34Cr4 specimens with different hardening depths. Testing frequency: f_Δ =1 kHz.

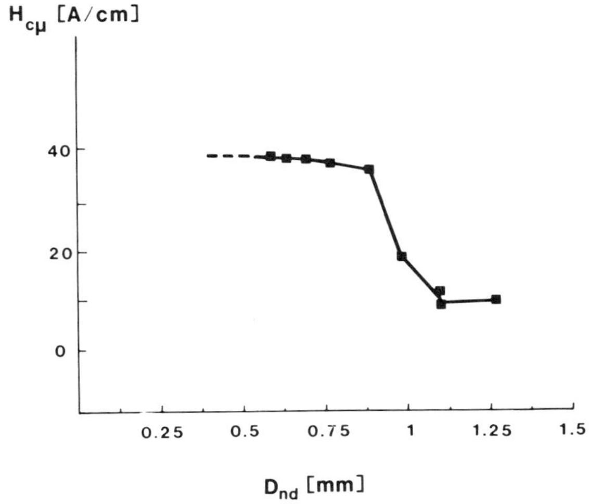

Fig.9. Determination of hardness gradients by incremental permeability measurements. $H_{C\mu}$-coercivity. $D_{nd}=(\sigma\mu_{\Delta} f_{\Delta})^{-\frac{1}{2}}$. Material: cast iron, Process: laser hardening.

Hardness profiles can be measured by tuning the f_{Δ}-frequency and recording the μ_{Δ} (H) curves. One possibility is to measure the coercivity $H_{C\mu}$ which is given by the peak position of the greatest μ_{Δ}-value. Doing this, one gets $H_{C\mu}(D_{nd})$-hardness profile curves as shown in Fig.9.

This compares well with the conventional destructive measurement of hardening depth.

CONCLUSIONS

It has been shown that micromagnetic nondestructive methods can be used for the characterization of the sub-surface ferromagnetic state. Different methods with different penetration depth can be used: the magnetic Barkhausen noise and the incremental permeability in the range of $0<\delta\,[mm]<3$; the dynamic magnetostriction in the range of $0<\delta\,[mm]<0.50$; the acoustic Barkhausen noise in the range of $0<\delta\,[mm]<10$.

The first applications measuring hardness profiles and other micromagnetic features have been successful using the 3MA equipment. (See e.g. paper R. Conrad et al.)

ACKNOWLEDGEMENT

This contribution is based on work performed with the support of the German Ministry for Research and Technology.

REFERENCES

1. Brinksmeier, E.; Schneider, E.; Theiner, W.A.; Tönshoff, H.K. Nondestructive Testing for Evaluating Surface Integrities. Annals of the CIRP 33 (1984)2, 489-509
2. Field, M.; Kahles, J.F. Review of Surface Integrity of Machined Components. Annals of the CIRP 20 (1971) 2, 153-163
3. Field, M.; Kahles, J.F.; Cammett, J.T. A Review of Measuring Methods for Surface Integrity. Annals of the CIRP 21 (1972) 2, 219-238
4. Hauk, V.; Macherauch, E. Eigenspannungen und Lastspannungen. HTM-Beiheft, Eds., Carl Hanser Verlag, München 1982
5. Larsson, H. Testing of Hardening Depth by an advanced eddy current method. 3rd European Conference on Nondestructive Testing, 16-18 Oct. 1984, Firence
6. Becker, R.; Rodner, Ch. Eddy current testing. IzfP-Bericht 820123-E, 1982
7. Good, M.S.; Rose, J.L. Measurement of thin case depth in hardened steel by ultrasonic pulse-echo angulation techniques. from: Nondestructive methods for material property determination. Plenum Press, New York and London 1983, 192-203
8. Mesure des gradients de dureté. Prospectus: CETIM 60304 Senlis Cedex, France, 1985
9. Rulka, R.; Pawlowski, Z. Evaluation of the Physical State of Surface Layers in Steel using Magnetic Noise Measurement. 9th World Conference on Nondestructive Testing, Paper 4AB, Melbourne 1979
10. Persch, H.; Theiner, W.A. Vorrichtung zum Messen von Werkstoffeigenschaften. European Patent 831067103
11. Theiner, W.A.; Kern, R.; Conrad, R. Determination of surface Integrities by ferromagnetic quantities. First International Conference Surface engineering, Brighton 25-28 June 1985, Paper 25, 25-1 - 25-11
12. Theiner, W.A.; Willems, H.H. Determination of microstructural parameters by magnetic and ultrasonic quantitative NDE. from: Nondestructive methods for material property determination. Plenum Press, New York and London, 1983, 249-258
13. Schneider, E.; Altpeter, I.; Theiner, W.A. Nondestructive determination of residual and applied stress by micro-magnetic and ultrasonic methods. from: Nondestructive methods for material property determination. Plenum Press, New York and London, 1983, 115-122
14. Altpeter, I.; Kern, R. Einfluß ferromagnetisch 2ter Phasen auf die Eigenspannungsmessung und die Gefügecharakterisierung bei Nutzung magnetischer und magnetoelastischer Meßgrössen. IzfP-Bericht 840133-TW, 1984

DETECTION OF COARSE GRAINS BY NONDESTRUCTIVE MAGNETIC METHODS

S. Segalini C. Dunand, M. Putignani*, and S. Genet**

*IRSID, France 78105 Saint-German-en laye
**USINOR DUNKERQUE, Service métallurgie qualité
B.P. 2508 59381 Dunkerque France

INTRODUCTION

Superficial coarse grains (100 µm - 300 µm) (figure 1) appear with a significant frequency on hot rolled low carbon aluminum killed steel coiled at high temperature[1]. On forming, these coarse grains will give rise to the so called defect "orange peel defect".

The suppression of this defect by a regulation of the industrial process is difficult at the present time and a detection by destructive methods (micrography, forming test...) is heavy due to the difficulty to take samples. So a nondestructive on-line detection system would ensure quality and would help for the monitoring of the line.

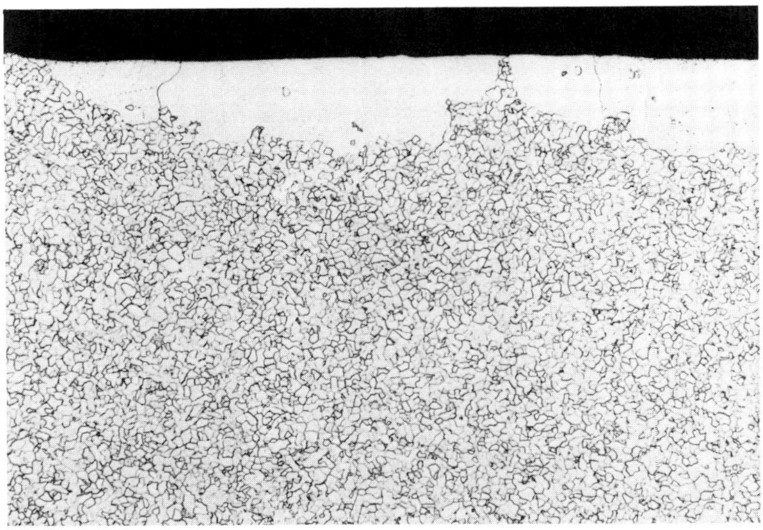

Figure 1 - Example of coarse grain layer (x 100).

It is well known that magnetic properties depend on microstructure and in particular grain size[2]. As, generally, superficial coarse grains appear in layers and pollute an important part of the coil, the detection by electromagnetic methods, in spite of the small area of the head, is conceivable. So we have tested, in the laboratory, the ability of Barkhausen noise and incremental permeability to solve the problem. These tests show that Barkhausen noise gives better results than incremental permeability.

In this paper, after a short description of the Barkhausen noise method we present the results obtained on 43 sheets both by destructive and non destructive testing.

BARKHAUSEN NOISE SET UP

The principle of non destructive testing by Barkhausen noise[3] is well known and the set up was described elsewhere[4]. So, in this section, we only give a short recall on the method and the specifications utilized in this study.

Our equipment permits the recording, with the help of a mini computer, of the magnetic noise enveloppe vs magnetic field. The apparatus and the measuring head ensure three functions (figure 2):

- Generation of magnetic field: the magnetization loop is created by an electromagnet driven by a power amplifier. The magnetization frequency is 1 Hz.

- Measurement of magnetic field ensured by a Hall sensor and a teslameter.

- Barkhausen noise analysis: a reception coil picks up the noise which is analysed by a specific device. In the case of coarse grain detection, the gain is 100 dB and the analysis bandwith is 25 kHz-35 kHz. The pick up coil diameter is 20 mm. During reported experiments, the measuring head was placed on the sample without lift-off.

SAMPLE CHARACTERISTICS

Samples

The study concerns 43 sheets including:

- 13 sheets sampled every 50 meters in a first coil (coil n° 1, boron grade, format: 1000 x 3,6 mm2) rejected for coarse grains occurence.

- 6 sheets sampled in a second coil (coil n° 2, no boron, format: 2000 x 2,4 mm2) rejected for coarse grains occurence.

- 24 sheets issued from different unaffected coils.

All steel coils were coiled at high temperature (600-700°C).

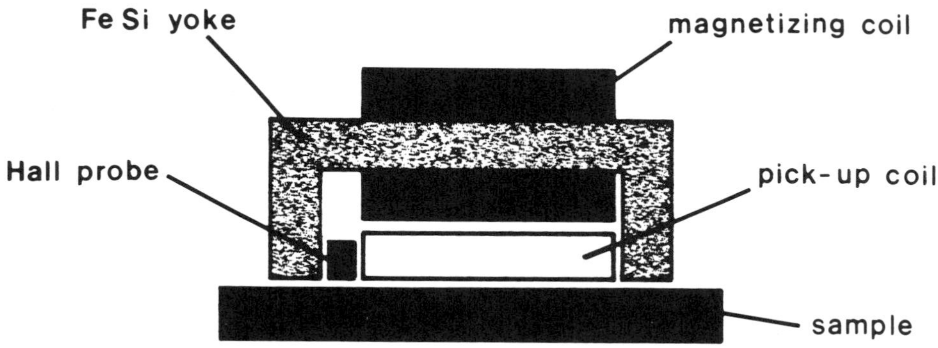

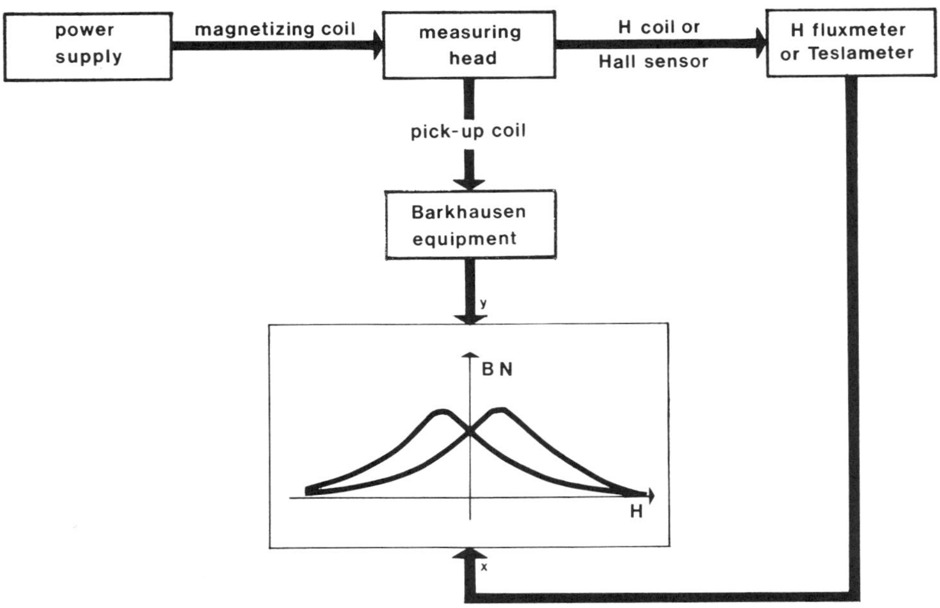

Figure 2 - Experimental set up and measuring head.

Figure 3 - Orange peel defect revealed by tensile testing.

Destructive testing

Destructive testing is performed by tensile test which reveals the "orange peel defect" (figure 3). A ranking in five classes is then possible

A : unaffected
B : a few coarse grains
C : numerous coarse grains
D : very numerous coarse grains
E : continuous layers of coarse grains.

After the test, all the 28 sheets sampled in unaffected coils were classified A. As for coil n° 1 and n° 2 (figures 4 et 5), a cartography shows that edges are always unaffected while large areas of coil centers are contaminated. Intermediate classes (C, D and B) dot not appear in a large number.

BARKHAUSEN NOISE DETECTION

Magnetic anisotropy

Magnetic anisotropy is an important phenomenon which disturbs detection. In fact, depending on the electromagnet direction, Barkhausen noise response curves may be different, and, moreover, the degree of anisotropy depends on which side of the sheet the test is performed.

The sheets are sampled at the end of the pickling line, but they keep a residual curvature from the coil. On the convex side, anisotropy is very strong (figure 6) and the sharpest curve is obtained in the transverse direction. Anisotropy is weaker on the opposite side (figure 7). Sharper curves are often obtained in the longitudinal direction but exceptions are numerous.

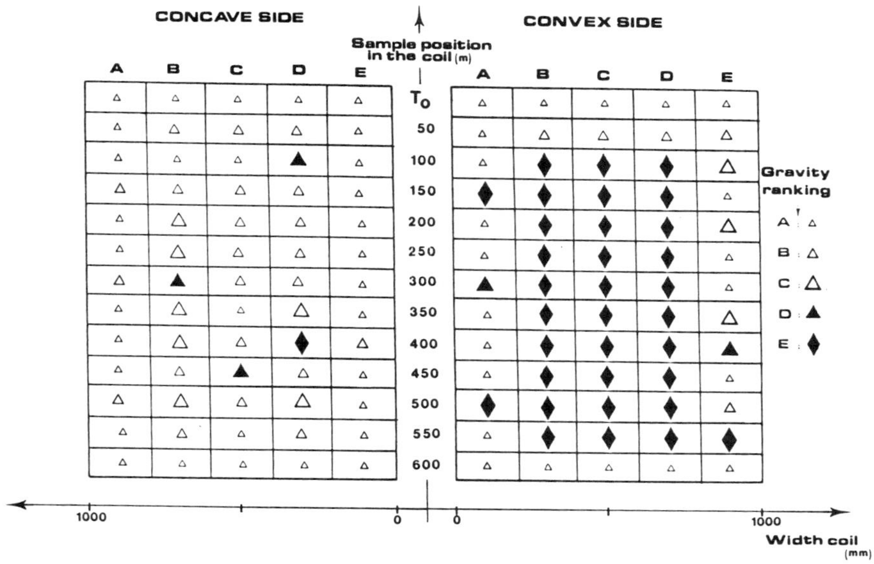

Figure 4 - Cartography of coil n°1 obtained by tensile testing.

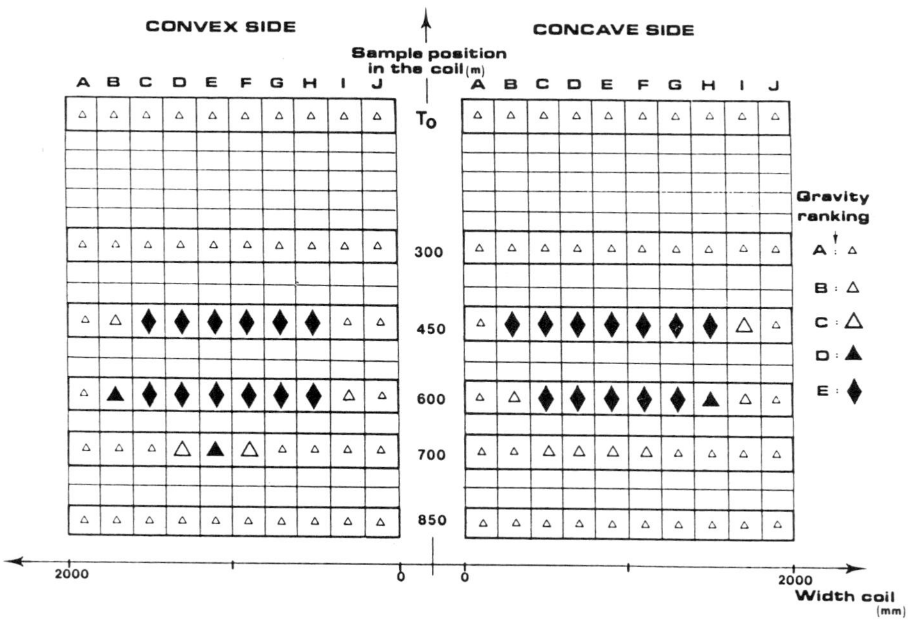

Figure 5 - Cartography of coil n° 2 obtained by tensile testing.

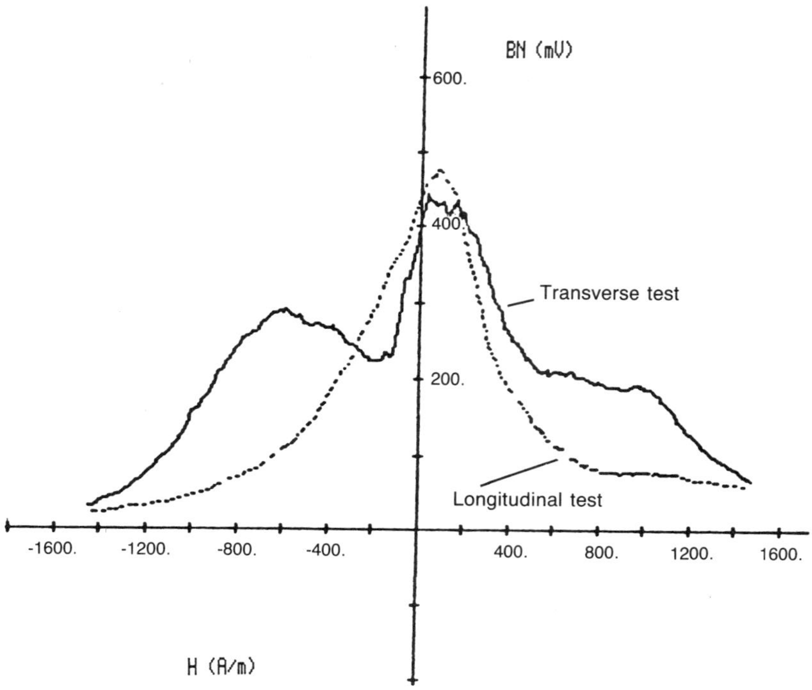

Figure 6 - Magnetic anisotropy observed on convex side.

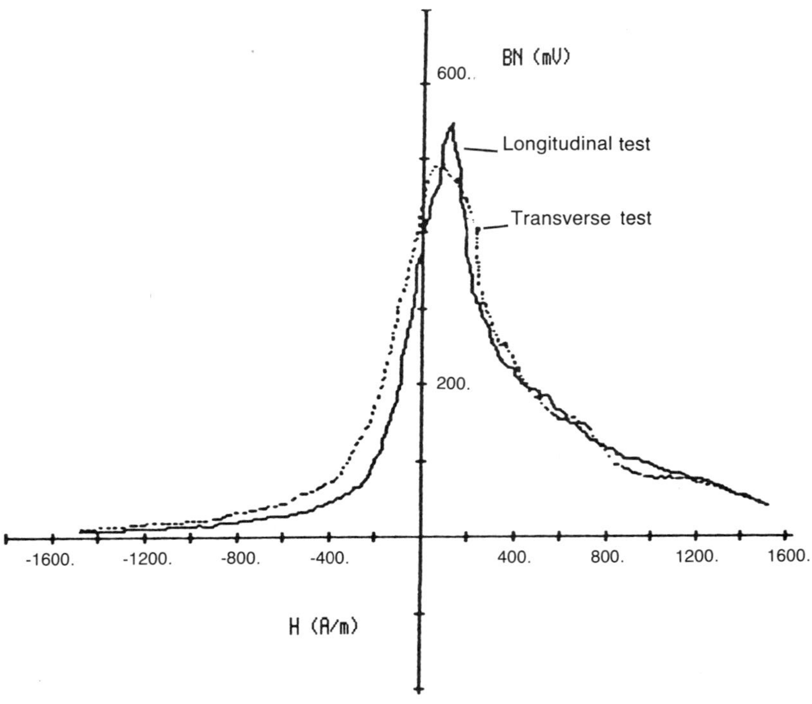

Figure 7 - Magnetic anisotropy observed on concave side.

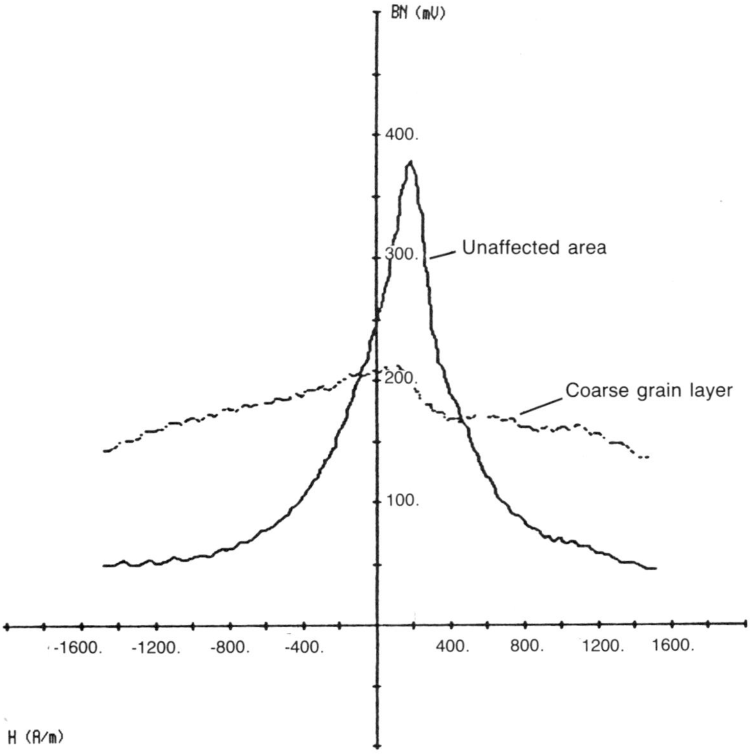

Figure 8 - Barkhausen modifications due to coarse grains.

To obtain good reliability in detection, we must eliminate this spurious phenomenon. In this purpose, we have decided to always take into account the sharpest curve. So on the convex side, the testing direction is perpendicular to the rolling direction. Yet the opposite side must be tested in two orthogonal directions and the suitable curve must be chosen lest false alarms would occur with a significant frequency.

RESULTS

Results concern 422 measurement points:

- coil n° 1 and n° 2: 1 measurement point every twenty centimeters on each side
- other sheets : 3 or 6 measurements on each side depending on the format.

After analysis, it appears that an important presence of coarse grains (class D and E) changes the sharpness of Barkhausen curves: smaller maximum and bigger width (figure 8).

A histogram of maximum values (figure 9) shows a reliable detection (5 undetected affected areas) and no false alarms. A histogram of widths (figure 10) shows a reliable detection too, but with 5 false alarms.

Total	7	6	11	9	9	4	12	21	28	35	57	51	53	33	36	16	10	13	6	5
Ranks D and E	7	6	11	9	9		1			2	1			1						
Ranks A, B and C						4	11	21	28	33	56	51	53	32	36	16	10	13	6	5

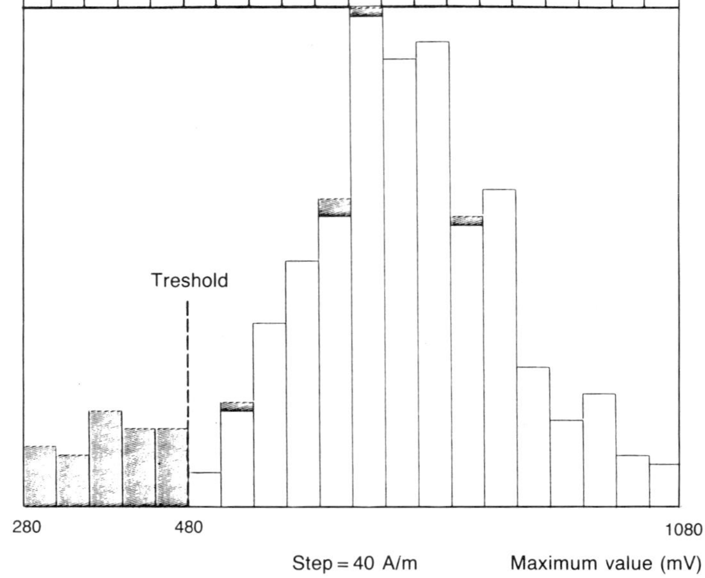

Figure 9 - Histogram of maximum values.

Total	63	163	105	33	10	7	13	9	12				1		1	1	1	1		2
Ranks D and E	1	2		2		6	11	8	10				1		1	1	1	1		2
Ranks A, B and C	62	161	105	31	10	1	2	1	2											

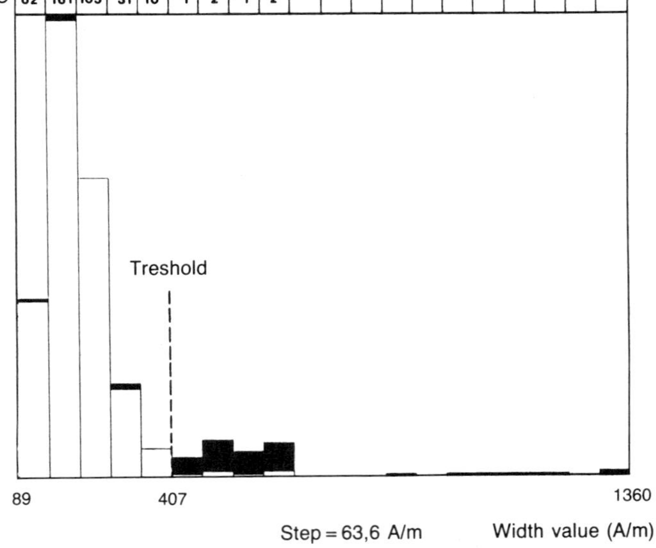

Figure 10 - Histogram of width values.

All the nondetection events come from a single specimen and so do all the false alarms. However, the measurement dispersion in each big class defined by nondestructive testing is important, which has two consequences:

- there is no gap between the two classes
- finer discrimination (e.g. between A, B and C) is impossible.

CONCLUSION

This laboratory study shows that Barkhausen noise permits the detection of coarse grains with a good reliability. However, magnetic anisotropy is an important spurious phenomenon which imposes measurements in two orthogonal directions. Moreover, as the origin of Barkhausen noise modification is not known, a risk subsists that other parameters, unknown at present time may disturb future measurements. Yet, laboratory results are sufficiently extensive (different grades, thicknesses, positions in the coil) and conclusive to think of on line testing.

REFERENCES

1. Y. Tokunaga et al, Occurence of coarse grains in hot rolled low carbon aluminum killed steel coiled at high temperature, Testu to Hagane, 1984, 70, 15, pp. 2136-2143.
2. S. Tiitto, On the influence of microstructure on magnetization transitions in steel, Acta Polytechnica Scandinavica, Applied Physics series 119 - Helsinki, 1977.
3. R. L. Pasley, Barkhausen effect, an indication of stress, Materials Evaluation (July 1970) pp. 157-161.
4. S. Segalini, M. Mayos, M. Putignani, Utilisation des méthodes électromagnétiques pour le contrôle de la microstructure des aciers, Revue Métallurgie, 10 (Octobre 1985), pp. 569-575.

ULTRASONIC DETERMINATION OF FRACTURE TOUGHNESS

A.N. Sinclair and H. Eng

Department of Mechanical Engineering
University of Toronto, Canada, M5S 1A4

BACKGROUND

The characteristics of sound propagation through any material are dependent on virtually all physical aspects of the medium: atomic constituents, grain size, shape, and structure, phase boundaries, elastic constants, presence of flaws or dislocations, etc. At least in theory, a detailed knowledge of a material's structure should allow the determination of its acoustic transfer function or impulse response; this would enable the calculation of the acoustic signal arriving at any point on the structure given an arbitrary system input. Unfortunately, the transfer function corresponding to most material characteristics is extremely complicated. Once it has been calculated, an even more difficult problem is determining an approximate inverse transfer function, assuming that one exists.

Nondestructive evaluation of a material's plane strain fracture toughness K_{1C} is a difficult task because there is no direct relationship between this property and readily-measured acoustic parameters, such as velocity or attenuation[1,2], i.e., the exact dependence of the acoustic transfer function (and its inverse) on K_{1C} is unknown. The task is further complicated by the basic lack of understanding of the fracture process itself[3–5]. Under such circumstances, purely empirical techniques are unreliable, due to uncertainty as to their applicable range, or sensitivity to perturbations in material preparation and environment. Instead, an attempt is made to determine critical microstructural parameters, and from these evaluate K_{1C}.

Based on these observations and preliminary experiments, ultrasonic determination of K_{1C} can be divided into three broad steps, none of which is trivial, but all of which have seen considerable progress over the last few years:

1. Measurement of an ultrasonic propagation parameter. For the case of a specimen with no gross flaws, the two parameters of major interest are the frequency-dependent velocity and attenuation. These measurements must typically be made to an accuracy of a fraction of 1% over a wide frequency range that strains the capabilities of conventional ultrasonic equipment. The dependence of both K_{1C} and attenuation on grain morphology and second phase particles indicates that attenuation should be the ultrasonic parameter of interest in fracture toughness studies[6–9].

2. Evaluation of Material Structure. The functional dependence of absorption and scattering coefficients, α_a and α_s respectively, on frequency and particle size, is well established[10–11]. Although absorption is predominant in the biomedical

field, the metallurgist commonly finds that scattering from grains or other inhomogeneities is the major contributor to sound attenuation. The inverse problem - determining the specimen morphology from the frequency dependence of attenuation - can be solved in theory, provided data is available over a sufficiently wide frequency band.

3. Evaluation of Fracture Toughness from Microstructure. This task is the most challenging of the three, due to the lack of understanding on the fracture process itself. The interrelationship between many material parameters, e.g., heat treatment, yield strength σ_y, grain size, etc., make it unclear as to whether the attenuation spectrum can uniquely indicate K_{1C} for an arbitrary specimen. However, tests have indicated that an evaluation of attenuation could indicate the perturbation in toughness from that of a reference sample[12-13].

The current status of these three components to evaluation of K_{1C} shall be briefly reviewed. Experimental results for Aluminum alloy A357 will be presented to illustrate the salient points.

ULTRASONIC ATTENUATION MEASUREMENT

The problems in making precise measurements of signal amplitude are rooted in its dependence on several parameters which are difficult to precisely evaluate, or maintain constant. However, due to the high accuracy required in the attenuation measurement, these elusive factors must be accurately accounted for in some way. The dominant parameters in question, each dependent on frequency, are:

1. Beam profile (geometric divergence and diffraction due to the finite size of the probe).
2. Pulser strength and equipment amplification factors.
3. Reflection and transmission coefficients at material interfaces.

Evaluation of the transfer function of the system electronics and transducer is normally avoided by using a comparison of two signals to determine the attenuation. Corrections for beam divergence have been extensively investigated in the bio-medical profession, where depth profiling of sound scatterers is essential[14-15]. A complicating factor is that it may not be possible to use a single probe or experimental configuration over the entire range of frequencies to be interrogated. In addition, it may be necessary to go to thinner samples as the frequency is increased, to avoid loss of accuracy when attenuation becomes very large.

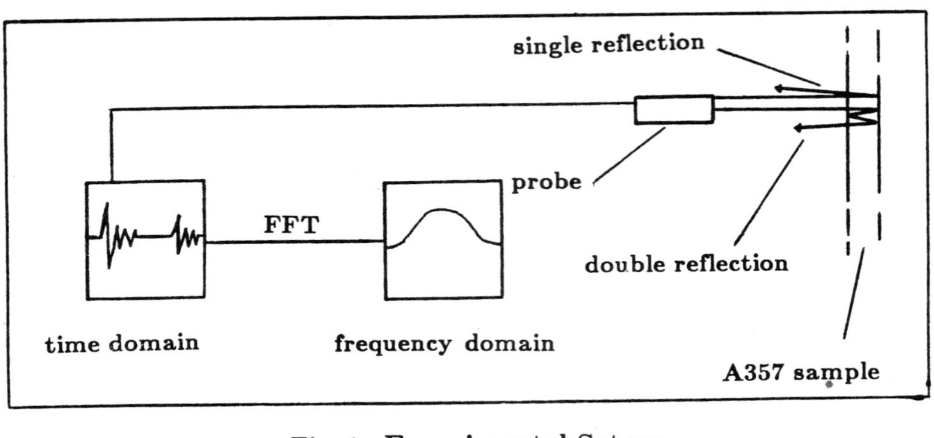

Fig. 1. Experimental Set-up

Figure 1 illustrates an experimental setup that will yield the required attenuation data. Surface finishes should be very smooth (roughness $<<$ wavelength λ), such that reflection and transmission coefficients R and T approach their theoretical limits[16–17]. The echo corresponding to a multiply reflected beam is deconvolved with that of a single reflection. With appropriate corrections for beam divergence, the attenuation coefficient is obtained as a function of frequency f.

DETERMINATION OF STRUCTURAL MORPHOLOGY

The functional dependence of attenuation on the characteristic dimension of grains or other inhomogeneity is seeing practical field application. Adopting the notation of Reynolds and Smith[18],

for Rayleigh scattering ($\lambda \gg \overline{D}$),

$$\alpha_s = C_1 \frac{<D^6>}{<D^3>} f^4$$

for the stochastic regime ($\lambda \approx \overline{D}$),

$$\alpha_s = C_2 \, \overline{D} \, f^2 \tag{1}$$

and for diffusion scattering ($\lambda < \overline{D}$),

$$\alpha_s = C_3 / \overline{D}$$

where the symbol D represents the dimension of an individual scatterer (e.g., mean chord length), $\overline{D}$ is the specimen averaged value for D, and $<D^i>$ refers to the mean of the ith power of D. The constants C_1, C_2, and C_3 can be determined analytically only for simple shapes such as spheres; in practice these constants are found experimentally.

For our case, measurements are to be performed primarily in the Rayleigh regime, for which Eq. 1 clearly demonstrates the dominance of the largest scatterers on the value of α_s. It is evident that analysis of the data is considerably simplified by the assumption of a narrow distribution in the size of sound scatterers.

FRACTURE TOUGHNESS EVALUATION

Early attempts to correlate fracture toughness with ultrasonic attenuation were largely empirical in nature. Typically, a number of specimens would be manufactured under some carefully controlled prescription, by which only a single parameter was to be varied, e.g., alloy content, heat treatment, or environmental conditions. However, the interdependence of many material properties made it difficult to evaluate the individual influences of several variables on the values of attenuation and K_{1C}.

To address this problem, a model is required to describe the dependence of fracture toughness on a material's microstructure. By necessity, a simplistic approach must be taken, and several factors must be ignored, or assumed to be invariant. A model developed by Vary[1,8,12], and later expanded by Fu[2,19], is based on the interaction of stress waves with potential crack nucleation sites in the material. Stress waves are known to be generated by the fracture or sudden deformation of a grain, second phase particle, or other discontinuity under an applied stress. These waves radiate through the material much as would externally stimulated ultrasonic waves, and follow the same rules for diffraction and attenuation. According to the model, stress waves emitted by crack nucleation at one site can prompt crack nucleation at a nearby site. If a chain reaction or avalanche effect results, then unstable crack propagation (fracture) occurs.

Numerical implementation of the model requires several approximations. The mathematics are simplified by assuming the stress waves to be originally broadband in nature, and from Eq. 1 are attenuated according to a simple power law :

$$\alpha_s = cf^m = c\left(\frac{v_l}{\lambda}\right)^m , \tag{2}$$

where and c and m are material constants measured at a "critical" wavelength λ_c, and v_l is the longitudinal wave velocity. The spectrum of stress wave energy, $e_s(\lambda)$, arriving at a potential nucleation site can then be calculated, and integrated:

$$E_s = \int e_s(\lambda)\, d\lambda = A_1 \frac{\sigma_\omega^2}{E} \left(\frac{v_l \beta}{m}\right) , \tag{3}$$

where $\beta = \frac{\partial \alpha}{\partial f}$ is evaluated at λ_c. The incremental stress σ_ω required to initiate fracture at a statically loaded site is assumed to be a fixed fraction of yield stress σ_y, i.e.,

$$\sigma_y = A_2\, \sigma_y \tag{4}$$

The integration of Eq. 2 is performed only over the long wavelength (Rayleigh) regime; the short wavelength components are assumed to be absorbed or stochastically scattered close to their point of origin, and are therefore unable to prompt crack initiation at a neighboring site.

According to linear elastic fracture mechanics (LEFM), the strain energy density in the area ahead of a crack is K_C^2/σ_y (or K_{1C}^2/σ_y at the onset of plane strain fracture). Vary assumed that a small portion of this energy, A_3, is made available upon fracture to induce crack nucleation at a neighboring site, i.e.:

$$E_s = A_3 \frac{K_{1C}^2}{\sigma_y} \tag{5}$$

Combining the above expressions, and using LEFM to express Young's modulus E in terms of K_{1C} and σ_y yields:

$$\left(\frac{K_{1C}}{\sigma_y}\right)^2 = \psi \left(\frac{v_l \beta}{m}\right)^{0.5} \tag{6}$$

A crucial assumption is that the parameter ψ is invariant among the expected range of samples to be inspected; this in turn is dependent on consistency of parameters A_1 through A_3.

The relationship described by Eq. 6 is by no means unique. Fu used a mode of stress wave interaction between layers of material[19] to predict an exponent of 1 instead of 0.5 on the right side of Eq. 5. A good fit with experimental data ha been obtained with an exponent somewhere between these two values[1,8].

The choice of wavelength λ_c at which to evaluate β and m is also not clear. Vary and Fu suggest that λ_c ought to correspond to a microstructural characteristic length, such as the diameter D of a potential nucleation site or grain, and suggest that its value can be determined ultrasonically. However, good experimental results have been achieved for a wide range of frequencies. Equipment limitation figure prominently in this issue, as one is confined to deep within the Rayleigh regime when testing fine-grained metals with conventional ultrasonic testing apparatus. Experimental data for the parameter m, ranging from $m = 1.1$ to ove 3, indicate that the scattering regime is not well-defined.

A good linear fit (correlation coefficient > 0.9) was obtained[8] by plotting lc $v_l\beta/m$ versus $K_{1C}^2/\sigma y$ for several samples of maraging steel and Ti-8Mo-8V-2Fe-3A manufactured using a range of aging temperatures from 700 to 867 K. Nadeau e al[20] conducted fracture toughness measurements on samples of 403 stainless steel at temperatures both above and below the nil ductility transition temperature. They found that Vary's selection of parametric constants for Eq. 6 gave very po correlation with their own data.

Table 1. A357 Cast Aluminum Alloy: Properties and Chemical Composition[a]

Constituents (wt%)	Si: 6.5-7.5 Mg: 0.4-0.7 Fe: 0.2 Cu: 0.2 Ti: 0.1-0.2 Mn: 0.1 Zn: 0.1 Be: 0.04-1.0
Tensile Strength	317 *MPa*
Compressive Yield Strength	241 *MPa*
Fracture Toughness[b] K_{1C}	22-29 $MPa\ m^{1/2}$

a) from Metals Handbook, ASTM (1983), and Cercast Test Data, Montreal (1985).
b) Specimen width was insuffucient for strict conformance to ASTM standard E399.83. Quoted values are for apparent plane strain fracture toughness.

It must be noted that determination of $(K_{1C}/\sigma_y)^2$ was not the ultimate goal of this work. Ultrasonic evaluation of K_{1C} requires a semi-empirical correlation to estimate σ_y from the sound velocity. This area is also still under development, so that dependable nondestructive estimates of K_{1C} are far from being assured at the present time.

EXPERIMENT

A first set of experiments has been performed on A-357 cast material, whose properties are listed in Table 1. There are several reasons why nondestructive measurements of K_{1C} for this material are particularly desirable:

(a) A-357 material is extensively used in the aerospace industry, where quality control standards may demand 100% testing.
(b) For destructive tests, determination of fracture toughness of a compact tension specimen according to *ASTM* standard procedures requires a certain minimum thickness. Such specimens may be impossible to manufacture from thin-walled components.
(c) The A357 microstructure can lead to crack arrest or crack tip bowing during the fatigue pre-cracking phase of destructive measurements of K_{1C}. This complicates efforts to comply with ASTM testing standards.

Eight compact tension specimens were divided into two groups *A* and *B*, and subjected to a heat treatment of 815 *K* for 24 hours, quenched in a glycol solution at room temperature, followed by aging at 450 *K* for 5 hours. Group *B* was subsequently reheated to 815 *K* for 8 hours, and then slow-cooled to room temperature and aged. following the heat treatment (as is done for the commercial product), while group *B* was slow-cooled at room temperature.

Samples from each group were polished and etched with a 0.5% HF solution. Micrographs are shown in Figures 2a to 2d. Figure 2b illustrates the enlargement of the dendrite arms, and the generally coarser structure of group *B*, as compared to group *A* (Figure 2a). In the enlargements of Figures 2c and 2d, the average dendrite arm spacing (*DAS*) can be estimated: approximately 40 μm for group *A*, and 50 μm for group *B*. Research has shown that the *DAS* is greatly influenced

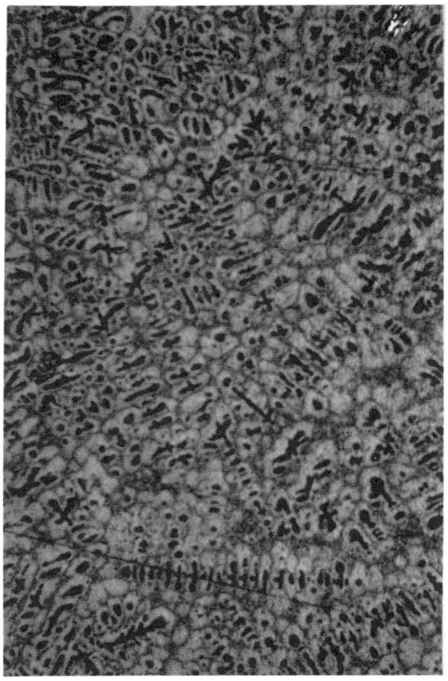

Fig. 2a 0 250 μm

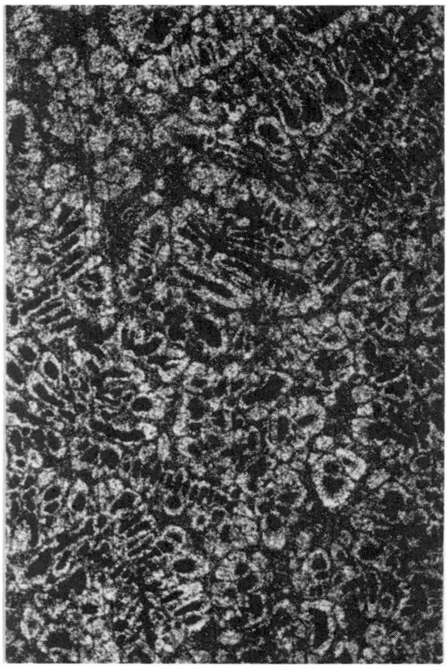

Fig. 2b 0 250 μm

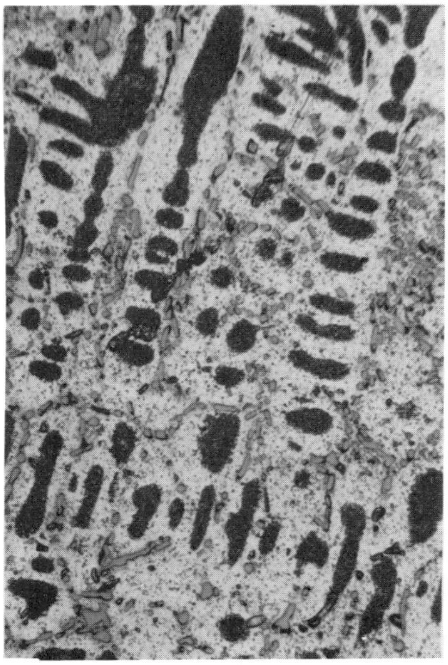

Fig. 2c 0 100 μm

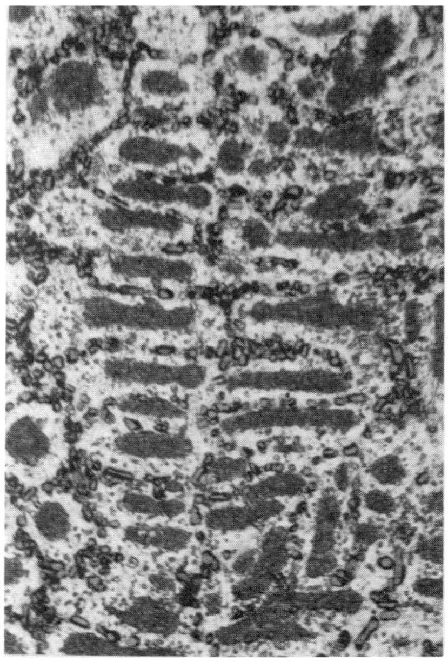

Fig. 2d 0 100 μm

Fig. 2: **Micrographs of typical specimens from Groups *A* and *B*. Figure 2a illustrates the finer structure of the dendrite arms (dark patches) and grain structure of group *A*, relative to group *B* (Figure 2b). The enlargements of Figure 2c (Group *A*) and 2d (Group *B*) allow an evaluation of the average dendrite arm spacing (DAS).**

Table 2. Experimental Results

Parameter	Group A	Group B	Source
v_l	$6.727 \times 10^3\ m/s$	$6.670 \times 10^3\ m/s$	measurement
λ_c	$4.0\times10^{-5}\ m$	$5.0\times10^{-5}\ m$	Figure 2
m	1.73	1.98	Figure 2
c	1.25×10^{-11}	2.10×10^{-11}	Figure 2
$\beta\|_{\lambda_c}$	$2.36\times10^{-5}\ Np.s/m$	$3.81\times10^{-5}\ Np.s/m$	$\beta = mc(v_l/\lambda_c)^{m-1}$
K_{1C}	$2.51\times10^7\ Pa\ m^{1/2}$	$2.67\times10^7\ Pa\ m^{1/2}$	Destructive Test[a]
$\sigma_y(0.2\%)$	$2.92\times10^8\ Pa$	$1.90\times10^8\ Pa$	Destructive test
ψ	$2.44\times10^{-2}\ m$	$5.88\times10^{-2}\ m$	Eq. 6

a) Specimen width was insufficient for strict conformance to ASTM standard E399.83. Quoted values are for apparent plane strain fracture toughness.

by the quenching rate, and can be used as an indicator of material properties such as fatigue strength[21]. The similarity of the *DAS* between the two groups indicates that the second heat treatment for group *B* had little effect on this parameter.

Both the longitudinal wave velocity, and attenuation spectra from 5-32 *MHz* were measured for both sets of specimens. Finely polished surfaces were used on all specimens to ensure that reflection coefficients were not dependent on frequency. The relatively low frequency cut-off of 32 *MHz* was selected due to the large size of the grains and other heterogeneities in the cast specimens.

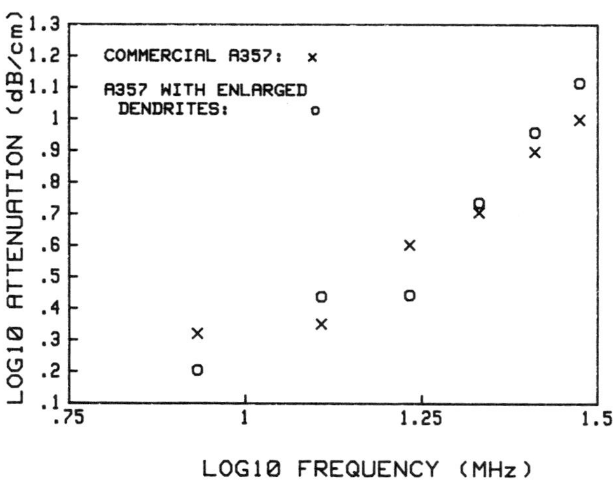

Fig. 3. Results of Attenuation Measurements on Groups *A* and *B*

The data were processed by a standard 2048 FFT package on a HP 9920A mini-computer to compute the frequency spectra. Values for the parameters c and m were determined by a least squares fit for both sets of specimens. The attenuation data are plotted in Figure 3, and are comparable to the results of Lewis et al to within experimental error and expected specimen-to-specimen variation[22].

Values for σ_y and K_{1C} were destructively measured for both groups A and B, and the results presented in Table 2. The value of the parameter ψ, of Eq. 6, was then calculated for the two specimen groups; as shown in Table 2, there is a large discrepancy, of more than a factor of two, between the two values for ψ.

Based on the work performed to date, it is not possible to numerically evaluate the factors that led to such a wide discrepancy in the two values of ψ. However, the scatter of data in Figure 3 indicates a large uncertainty in the calculated values of c and m, and hence β. In addition, there were inadequate samples available to properly evaluate the variability in destructive measurements of K_{1C} for the group B specimens.

A more basic problem is the strong indication that the model of Eqs. 2-6 is not adequate to describe the large differences in material properties between groups A and B. This is similar to the conclusion reached by Nadeau et al[20], who found that the parameters of Eq. (6) corresponding to one temperature were invalid at a substantially different temperature.

DISCUSSION AND CONCLUSIONS

The problems encountered in determining K_{1C} from measurements of ultrasonic attenuation can be divided into two main categories. The first category, labeled "engineering problems", represents challenges to the suppliers of ultrasonic testing equipment. These problems can now be overcome (or at least compensated for) in the laboratory; they include achievement of a consistent level of acoustic coupling, standardized transducer design and performance, and broadband equipment capable of rapid frequency domain calculations.

The "conceptual problems" are more fundamental in nature, and pose a more long-term uncertainty to nondestructive measurements of K_{1C}. These include:

1. Dependence of K_{1C} on material structure and environment. A more accurate model than that of Eqs. 2-6 is required to accurately describe the influence of grain size, second phase particles, temperature, heat treatment, etc., on fracture toughness. Alternatively, acceptable results might be achievable if the parameter ψ is generated from test specimens that bracket the range of material properties and environment to be encountered in an inspection program.
2. Selection of an appropriate frequency range. This will depend on the results of item #1, in particular, the size of the material characteristic(s) that determine fracture toughness. This could lead to a signal processing problem, as it may be impossible to isolate the parameters of interest from those that have little effect on K_{1C}. The values of m shown in Table 2 indicate that the frequency range used for this study was too high for pure Rayleigh scattering.
3. Reflection Coefficient R at Back Surface. There may be no method to accurately measure R as a function of frequency for items where the back surface is inaccessible. If this is the case, it may be necessary to consider attenuation calculations by back-scatter measurements.
4. Evaluation of σ_y. Nondestructive yield strength measurements have been attempted using correlations with ultrasonic velocity. However, these would require an accurate measurement of the component thickness - an impractical proposition if there is no access to the back surface.

REFERENCES

1. A. Vary and D.R. Hull, Interrelation of Material Microstructure, Ultrasonic Factors, and Fracture Toughness of a Two-Phase Titanium Alloy, Mat. Eval. 41:309 (1974).
2. L.S. Fu, "On the Feasibility of Quantitative Ultrasonic Determination of Fracture Toughness - A Literature Review", NASA Report # 3356 (1980).
3. F.H. Froes, J.C. Chesnutt, C.G. Rhodes, and J.C. Williams, Relationship of Fracture Toughness and Ductility to Microstructure and Fractographic Features in Advanced Deep Hardenable Titanium Alloys, in: "Toughness and Fracture Behaviour of Titanium", ASTM STP 651 (1978).
4. R.W. Hertzberg, "Deformation and Fracture Mechanics of Engineering Materials", J. Wiley & Sons, New York (1983).
5. P. Beaudet, "Study of the Effect of Dwell Time on the Fatigue Crack Propagation Rate in Ti-6Al-4V Alloy", MASc. thesis, University of Toronto (1986).
6. A. Vary, Correlations Between Ultrasonic and Fracture Toughness Factors in Metallic Materials, in: "Fracture Mechanics", ASTM STP 677 (1979).
7. R.L. Smith, K.L. Rusbridge, W.N. Reynolds, and B. Hudson, Ultrasonic Attenuation, Microstructure, and Ductile to Brittle Transition Temperature in Fe-C Alloys, Mat. Eval. 41(2):219 (1983).
8. A. Vary, Correlations among Ultrasonic Propagation Factors and Fracture Toughness Properties of Metallic Materials, Mat. Eval. 26(7):55 (1978).
9. A. Vary and D.R. Hull, "Ultrasonic Ranking of Toughness of Tungsten Carbide", NASA Technical Memorandum #83358, Cleveland (1983).
10. L.S. Fu, Mechanics Aspects of NDE by Sound and Ultrasound, Appl. Mec. Rev. 35(8):1047 (1982).
11. S.W. Flax, N.G. Pelc, G.H. Glover, F.D. Gutman, and M. McLachlan, Spectral Characterization and Attenuation Measurements in Ultrasound, Ultrasonic Imag. 5:95 (1983).
12. A. Vary, "Ultrasonic Nondestructive Evaluation, Microstructure, and Mechanical Property Interrelations", NASA Technical Memorandum #86876, Cleveland (1984).
13. E.R. Generazio, Ultrasonic Verification of Microstructural Changes Due to Heat Treatment, in: "Analytical Ultrasonics in Materials Research and Testing", NASA Report #CP-2383 (1984).
14. G.B. Devey and P.N.T. Wells, Ultrasound in Medical Diagnosis, Sci. Am. 238:98 (May 1978).
15. S. Leeman, L. Ferrari, J.P. Jones, and M. Fink, Perspectives on Attenuation Estimation from Pulse-Echo Signals, IEEE Trans. Sonics and Ultrasonics, SU-31(4):352 (1984).
16. E.R. Generazio, "The Role of the Reflection Coefficient in Precision Measurement of Ultrasonic Attenuation", NASA #83788, Cleveland (1984).
17. N.F. Haines, "The Theory of Sound Transmission and Reflection at Contacting Surfaces", CEGB Report #RD/B/N4744, Berkeley Laboratories (1980).
18. W.N. Reynolds and R.L. Smith, Ultrasonic Wave Attenuation Spectra in Steels, J. Phys. D: Appl. Phys., 17:109 (1984).
19. L.S. Fu, On Ultrasonic Factors and Fracture Toughness, Eng. Frac. Mec., 18(1):59 (1983).
20. F. Nadeau, J.F. Bussiere, and G. Van Drunen, On the Relation Between Ultrasonic Attenuation and Fracture Toughness in Type 403 Stainless Steel, Mat. Eval. 43(1):101 (1984).
21. A. Wickberg, G. Gusyafsson, and L.E. Larson, "Microstructural Effects on the Fatigue Properties of a Cast Al7SiMg Alloy (A356)", SAE Paper #840121, Warrendale (1984).
22. K. Lewis, S.W. Wang, and L. Adler, Ultrasonic Characterization of Aluminum Cast Materials, in: "Review of Progress in Quantitative NDE", Vol. 3B, D.O. Thompson and D.E. Chimenti, eds., Plenum Press, New York (1984).

CORRELATION BETWEEN ULTRASONIC ATTENUATION AND

FRACTURE TOUGHNESS OF STEELS*

G. Canella and M. Taddei

Centro Sperimentale Metallurgico S.P.A.
Via di Castel Romano - 00129 Roma (I)

SUMMARY

An ultrasonic index has been found for direct NDT estimation of the fracture toughness of three different steels; one forged bainitic and two rolled: ferritic-pearlitic and acicular ferritic. Direct correlations have been found between ultrasonic attenuation and Critical Stress Intensity Factor in plain strain (K_{IC}), Ductile Fracture Appearance Transition Temperature of Charpy-V curve (FATT) and Shelf Energy of the Charpy-V curve itself (KV).

A direct correlation has also been found between fracture toughness indices and grain size. To obtain these results a new method has been developed for measuring ultrasonic attenuation at room temperature on thin specimens even at high absorption.

INTRODUCTION

The characterization of steels in terms of fracture toughness is a matter of fundamental importance, but it is a long, difficult operation, calling for the sacrifice of large quantities of material in destructive tests. Various attempts have been made to surmount this difficulty by correlating the K_{IC} measurement with other mechanical tests, such as the Charpy-V for instance[1-3].

In actual fact, many mechanical properties of steel, including fracture toughness indices, Charpy-V curve and K_{IC} are dependent on the grain size, type of structure, concentration of second phases, and type and quantity of nonmetallic inclusions.

The same factors that influence the mechanical properties of steel also influence ultrasonic attenuation within the material. In recent years various experimental studies have been made to correlate mechanical properties and ultrasonic attenuation in steels or in other polycrystalline metals[4-8].

(*) The work was supported for under the terms of art. 55 of the ECSC convention.

However, because these studies cover only part of the problem, it was decided not to consider the results as something to be verified, but to use them to develop an autonomous approach for checking on the existence of direct correlations between fracture toughness indices and ultrasonic attenuation.

RESEARCH APPROACH

Appropriate samples of various types of steel were selected and subjected to mechanical, quantitative metallographic and ultrasonic characterizations.

Mechanical characterization was by means of tensile tests, K_{IC} at room temperature and Charpy-V curve as a function of temperature. When direct measurements of K_{IC} were not valid, the value was obtained by measuring J_{IC}, which is the value of the J-integral at the start of ductile fracture.

Metallographic characterization involved measurement of the mean size of the grain whose edges, under wide angle conditions, scatter ultrasound. Consequently, in the case of steels with ferritic or ferritic-pearlitic structure, the mean size of the ferrite grain was measured, while in the case of steels with a bainitic structure, the original austenite grain was measured. Second phase concentration and the inclusion content were also measured.

Ultrasonic characterization consisted in measuring the attenuation coefficient α at room temperature in the 5-25 MHz range.

SAMPLES

Three steels, their chemical compositions are in Table 1, were used.

Steel A: This was an API X60 normalized, tempered steel with ferrite-acicular ferrite microstructure, rolled in three thicknesses: 30 mm (Sample A30), 45 mm (Sample A45) and 60 mm (Sample A60). Five of the 60 mm samples, were renormalized at 930°C for one hour and air cooled (Sample AN), while another five of 60 mm samples were hardened and tempered at 920°C for one hour, water quenched and tempered at 620°C for eight hours followed by air cooling (Sample AB) thus obtaining a mixed martensite-bainite structure.

Steel B: This was an ASA F52 normalized, tempered steel with ferrite-pearlite microstructure, hot rolled to 80 mm thickness (Sample BN). Five samples were hardened at 940°C for one hour, water quenched and tempered at 620°C for eight hours followed by air cooling (Sample BB), thus obtaining a martensite plus tempered bainite structure.

Table 1: Chemical analysis of steels.

STEELS	C%	Si%	S%	P%	Mn%	Cr%	Ni%	Mo%	V%	Nb%	Cu%	Al%
A	.12	.36	.008	.011	1.67	.06	.44	.09	.068	.032	.03	.024
B	.18	.23	.009	.007	.59	.15		.32	.005	.006		
C	.21	.04	.009	.009	.48	.44	3.50	.53	.10			

Steel C: This was a forged, hardened and tempered Ni-Mo-V rotor steel with bainite microstructure. Five 150 x 150 x 60 mm samples (large face at right angles to rotor axis) were prepared of this steel. Heat treatments were performed on groups of five samples, obtaining bainite structures with different austenite grain size.

Sample C1: Material "as-is"; bainite structure; heat treatment not known but original austenite grain very enlarged ($\bar{D} \approx 400$ µm).

Sample C2: 1005°C for one hour, furnace cooled, tempered at 620°C for twenty-four hours, air cooled; bainite structure with trace of martensite; austenite grain $\bar{D} \approx 220$ µm.

Sample C3: 905°C for one hour, cooled and tempered as per C2; bainite structure with austenite grain $\bar{D} \approx 30$ µm.

Sample C4: 810°C for one hour, cooled and tempered as per C2; bainite structure with austenite grain $\bar{D} \approx 21.5$ µm.

Sample C5: 1105°C for one hour, cooled and tempered as per C2; bainite structure with austenite grain $\bar{D} \approx 355$ µm.

METALLOGRAPHIC CHARACTERISTICS

To complete the microstructural characterization, quantitative measurements were made under the optical microscope as per the ASTM E 112-81 method along the longitudinal, transverse and surface directions. The dimensions of the original austenite grain were determined for Group C steels and for Structures AB and BB, while the size of the ferrite grain was determined for Structures A30, A45, A60, AN and BN.

Quantitative inclusion analysis was performed on the three basic structures. The values obtained were very close, so it can be assumed that the three steels are of the same cleanness.

MECHANICAL CHARACTERIZATION

Mechanical characterization was performed by means of tensile tests at room temperature on 7 mm diameter cylindrical specimens taken in the transverse direction for the plates (Groups A and B) and in the tangential direction for the forged (Group C), as well as by Charpy-V value curves and measurement of K_{IC} at room temperature. The K_{IC} measurement tests were run on precracked CT type specimens modified for J-integral tests as per ASTM E 813-81, with the specimen taken in the TL direction for Groups A and B and the CR direction for Group C, as for Charpy-V value specimens.

Test results are given in Table 3. For Group A structures the K_{IC} (J_{IC}) value is the same in all three rolling conditions: 30, 45 and 60 mm. For the remaining structures, instead, there is a significant difference among values as a result of the heat treatments given. Taken overall, the K_{IC} ranges between 50 MPa $\sqrt{m}$ and a maximum of 330 MPa $\sqrt{m}$.

MEASUREMENT OF ULTRASONIC ATTENUATION COEFFICIENT

The techniques normally used for measuring ultrasonic attenuation, i.e. direct contact or immersion pulse reflection[9], do not permit measurements exclusively in the far field (over 3N). The diffraction terms thus have to be corrected or else the measurements have to be made under high attenuation conditions where the former are negligible. This approach has been adopted by various workers, especially Papadakis[10,11]. In the first case it ensued that the corrections were too large compared with the values to be measured (up to ten times), which meant that there were serious doubt as to the validity of the results.

In the second case measurements were needed at higher ultrasonic frequencies (30-70 MHz) than those generated in first harmonic by the piezoelectric or ceramic probes now on the market (1-25 MHz). Both these difficulties were overcome by Papadakis by using probes with delay line that permit the ultrasound that reaches the sample to be examined under far field conditions[10]. An improvement on the Papadakis method has been adopted in the work reported here[12]. This consists in sending the ultrasonic pulses through a delay line whose length is such that they reach the sample under far field conditions. The backwall echoes reflected inside the sample are received by a second probe placed in direct contact, at the other side, as schematized in Fig. 1.

The literature reports various functional relations linking ultrasonic attenuation and frequency at different $\bar{D}/\lambda$ values:

$$\alpha = A \cdot f + B \cdot f^4 \tag{1}$$

to be applied when $\bar{D} \ll \lambda$, [13];

$$\alpha = k \cdot f^2 \tag{2}$$

to be applied when $\bar{D} \approx \lambda$, [14];

Table 2: Attenuation versus frequency curve coefficients according to the three models for each sample.

STRUCTURE	MODEL 1 $\alpha = c \cdot f^n$ ($\ln\alpha = \ln c + n \cdot \ln f$)			MODEL 2 $\alpha = K \cdot f^2$		MODEL 3 $\alpha = A \cdot f + B \cdot f^4$		
	c	n	r	K (dB/m. .MHz^2)	R_2	A (dB/m. .MHz)	B (dB/m. MHz^4) (.10^{-5})	R_3
A30	.110**	2.2**	.97	.19**	.99	2.7**	10	.99
A45	.055**	2.5**	.97	.24**	.98	2.3**	26**	.98
A60	.035**	2.7**	.99	.29**	.99	3.5**	23	.98
AN	.225**	2.0**	.99	.20**	.99	2.9**	12**	.98
AB	.350**	1.7**	.97	.15**	.99	2.4**	8*	.99
BN	2.85**	1.8**	.98	1.7**	.99	16**	190**	.99
BB	2.15**	1.8**	.93	1.2**	.94	20**	0	.96
C1	3.80**	2.0**	.99	3.6**	.99	51**	230**	.99
C2	3.45**	2.1**	.99	3.8**	.99	40**	450**	.99
C3	.55**	2.3**	.98	1.4**	.99	11**	190**	.99
C4	2.05**	1.8**	.99	1.3**	.99	14**	150**	.99
C5	6.80**	1.8**	.98	3.8**	.99	41**	410**	.99

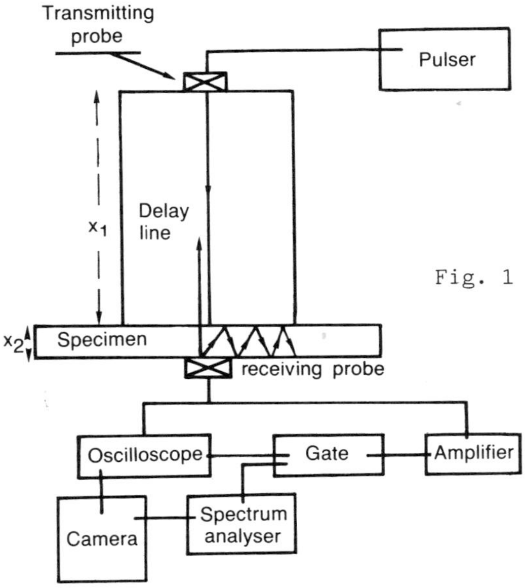

Fig. 1 - CSM transmission-reflection method for u.s. attenuation measurements.

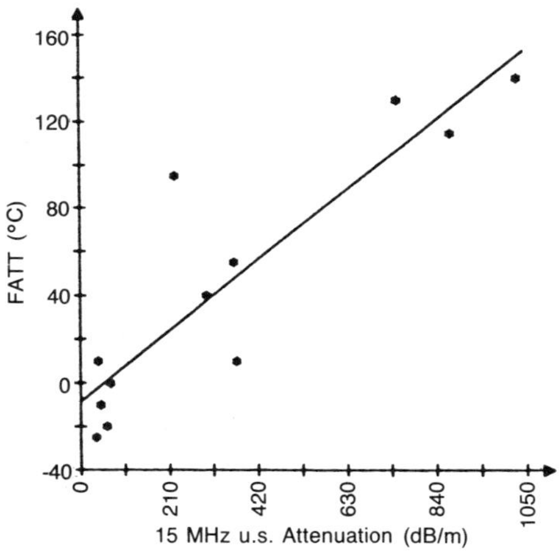

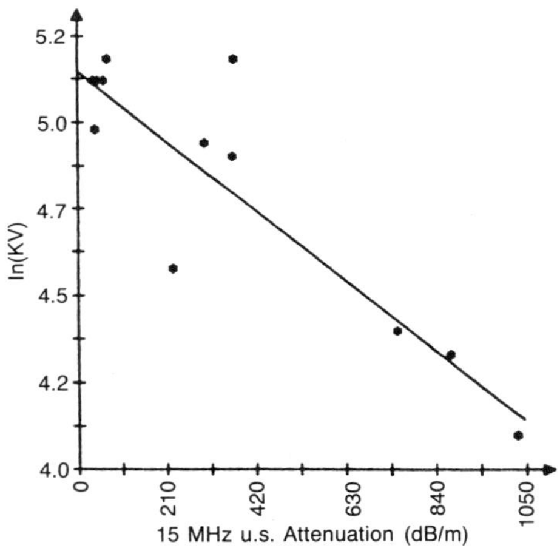

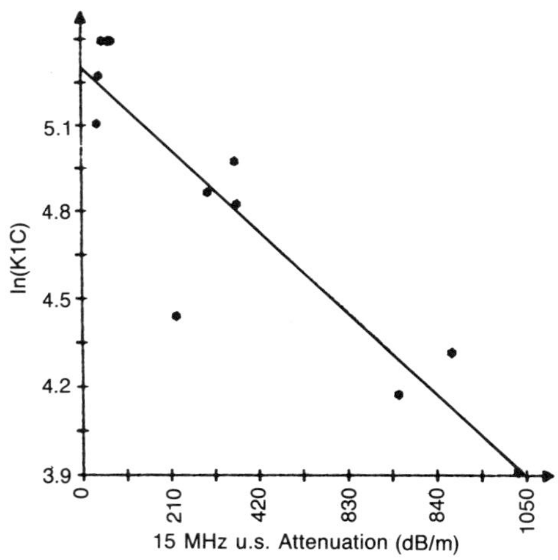

Fig. 2

- Correlations between u.s. attenuation at 15 MHz and the FATT, ln KV and ln K_{IC} values of all samples.

Vary[4], however, proposes an empirical formula for application for any D/λ value:

$$\alpha = C \cdot f^n \tag{3}$$

The data obtained in the research reported here (Table 2) have been processed by the least squares method, separately for each structure examined. In the case of Model 1 of Table 3 (which uses equation 3), the calculation was made by the logarithmic linear regression method, according to the formula:

$$\ln \alpha = \ln (C) + n \cdot \ln(f) \tag{3a}$$

The three models used (equations 1-3) provide for $\alpha = 0$ when $f \rightarrow 0$, which is certainly in line with the experimental values.

The results of the processing are given in Table 2 in which the coefficients with statistical significance level (better than 99% and 95%) are marked by two asterisks and one asterisk, respectively. It will be observed that Models (2) and (3) always have two asterisks, while in the case of model (1) this occurs only with Group C and BN steel, while in the remaining cases, one of the two coefficients either has only one asterisk or none at all (significance level less than 95%).

In short, there is difficulty in applying (1) in the case of its theoretical validity, namely for $\bar{D} \ll \lambda$. This may depend on the low values of α for these structures with an increase in the relative percentage error. A further comparison can be made among the three models, by the correlation coefficients R of eq. (1) and eq. (2) and the linear correlation coefficient r of eq. (3a).

The differences are very small, however, and it is impossible to select among the three models on this basis. Nevertheless, it is felt that (2) is to be preferred, not least because the exponent n of eq. (3) is always ≈ 2, thus corroborating eq. (2) itself, but also for its simplicity.

As eq. (2) has only one parameter, it ensues that the whole attenuation-frequency curve provides a characterization of the sample under examination equal to that obtained by measuring attenuation at

Table 3: Summarized data stored in computer on statistic file.

CODE	FATT (°C)	KV (J)	K_{IC} (Mpa m)	D (μm)	ALFA15 (dB/m)
A30	-10	160	220	10	48
A45	-20	160	220	11	62
A60	0	170	220	12	71
AN	10	140	195	13	42
AB	-25	160	165	20	37
BN	55	130	145	34	365
BB	10	170	125	42	370
C1	140	60	50	400	1030
C2	115	75	75	220	875
C3	95	95	85	30	225
C4	40	135	130	21	300
C5	130	80	65	355	750

only one frequency, provided that in this last case the measurement error is small compared with the measured value. In the case in point, a frequency of 15 MHz was chosen. The average values measured at this ultrasonic frequency, for every structure, are given in Table 3 under the item ALFA15, expressed in dB/m.

CORRELATION BETWEEN FRACTURE TOUGHNESS INDICES AND ULTRASONIC ATTENUATION

A parameter characteristic of the attenuation-frequency curve can be chosen on the basis of the theoretical model utilized. The data for correlation were loaded into a statistical file in a computer and are reported in Table 3 under the headings:

CODE	Material code
FATT	Charpy-V Fracture Appearance Transition Temp. in °C
KV	Shelf Energy, in J
K_{IC}	Fracture toughness index K_{IC}, in MPa $\sqrt{m}$
D	Size of ferrite grain or original austenite grain, in μm
ALFA15	Attenuation coefficient α at 15 MHz, in dB/m

The correlations found between mechanical parameters and ultrasonic attenuation at 15 MHz are as follows:

$$FATT = .154 \cdot ALFA15 - 8.5 \quad \text{with } r = .89^{**} \tag{4}$$

$$KV = 167 \cdot EXP\,(-9.2 \cdot 10^{-4} \cdot ALFA15) \tag{5}$$

$$(\ln KV = -9.2 \cdot 10^{-4} \cdot ALFA15 + 5.1) \qquad (\text{with } r = -.89^{**})$$

$$K_{IC} = 200 \cdot EXP\,(-1.34 \cdot 10^{-3} \cdot ALFA15) \tag{6}$$

$$(\ln (K_{IC}) = -1.34 \cdot 10^{-3} \cdot ALFA15 + 5.3) \qquad (\text{with } r = -.91^{**})$$

Fig. 2 illustrates the variables in relation to one another and the relevant regression lines.

Equations (4), (5) and (6) can be used to calculate the corresponding mechanical parameters, starting from the respective ultrasonic attenuation values.

CORRELATIONS WITH GRAIN SIZE

The link between mechanical properties and ultrasonic attenuation is affected mainly by the grain size, as per the following relation:

$$ALFA15 = 252 \cdot \ln \bar{D} - 570 \tag{7}$$

(where $\bar{D}$ = mean grain size) with r = .97**

The plot of ALFA15 - ln $\bar{D}$ is shown in Fig. 3a together with the regression line. Eq. (7) provides a causal link to justify the correlations between fracture toughness indices and ultrasonic attenuation, namely a means of estimating the value of $\bar{D}$ starting from ultrasonic attenuation measurements.

The direct correlations between toughness indices and average grain size are

$$FATT = 40 \cdot \ln \bar{D} - 100 \quad (\text{with } r = -.90^{**}) \tag{8}$$

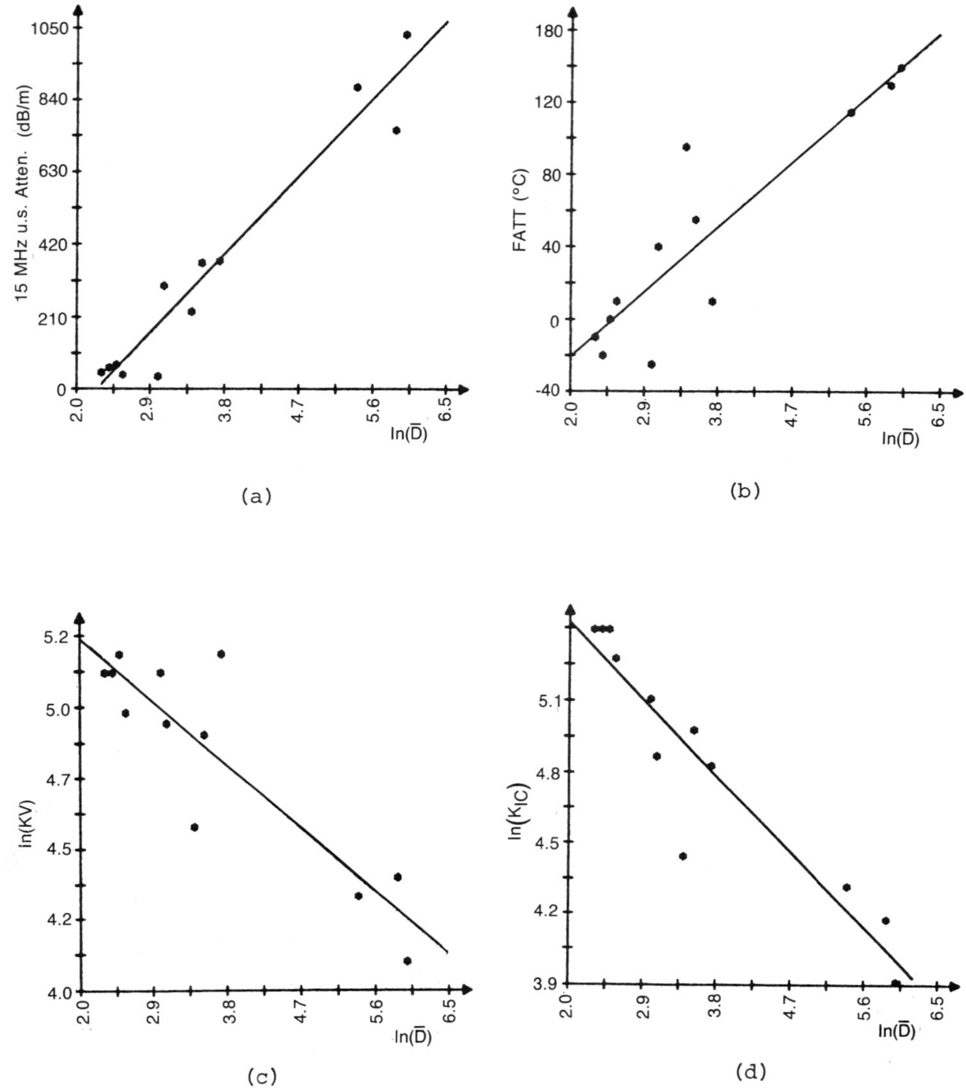

Fig. 3 - Correlations between logarithm of mean grain diameter (ln D) and the ultrasonic attenuation at 15 MHz, FATT, ln KV and ln K_{IC} values of all samples.

$$KV = 290/\bar{D}^{.24} \quad (9)$$

$$(\ln KV = -.24 \cdot \ln \bar{D} + 5.67 \qquad \text{with } r = -.89^{**})$$

$$K_{IC} = 465/\bar{D}^{.357} \quad (10)$$

$$(\ln K_{IC} = -.357 \cdot \ln \bar{D} + 6.14 \qquad \text{with } r = -.94^{**})$$

The FATT, lnKV and $\ln K_{IC}$ values in terms of $\ln\bar{D}$ are plotted in Figs. 3b, 3c and 3d along with the relevant regression lines.

CONCLUSIONS

It has been demonstrated that for the three steels examined there exist direct correlations between fracture toughness indices, such as K_{IC}, FATT and Shelf Energy and ultrasonic attenuation.

The average grain size is moreover the common factor between toughness indices and ultrasonic attenuation.

REFERENCES

1. S.T. Rolfe and S.R. Novak, ASTM STP 436, 1970, pp. 124-147.
2. J.M. Barson and S.T. Rolfe, ASTM STP 466, 1970, pp. 281-302.
3. D.M. Norris et al.,ASTM STP 743, 1981, pp. 207-217.
4. A. Vary, Materials Evaluation, June 1978, pp. 55-64.
5. R. Klinman et al., Materials Evaluation, October 1980, pp.26-32.
6. R.L. Smith and W.N. Reynolds, Journal of Materials Science, 17, 1982, pp. 1420-1426.
7. A. Vary and D.R. Hull, Materials Evaluation, 41, March 1983, pp. 309-314.
8. F. Nadeau et al., Materials Evaluation, 43, January 1985, pp. 101-107.
9. R.L. Roderick and T. Truell, Journal Applied Physics 23, pp. 267-279, 1952.
10. E.P. Papadakis, The Journal of the Acoustical Society of America, 44 n. 3, 1968.
11. E.P. Papadakis et al., The Journal of the Acoustical Society of America, 53 n. 5, 1973.
12. G. Canella and M. Taddei, Ultrasonic attenuation in direct contact in the far field and at high frequencies in this samples. 3rd European Conference on Nondestructive Testing, Florence, 15-18 October 1984, Vol. 5, p. 234.
13. W.P. Mason and H. J. McSkimin, Journal of Applied Physics, 19, Oct. 1948.
14. L.D. Landau and E.M. Lifshitz, Fluid Mechanics, Pergamon Press, 1959.

A METHOD TO CALCULATE THE RESISTIVITY OF STRIP CAST 3000 SERIES AL ALLOYS BASED ON THEIR NOMINAL SOLUTE CONTENT

Omar S. Es-Said* and J. G. Morris**

*Department of Mechanical Engineering, Loyola Marymount University, Los Angeles, CA 90045
**Department of Metallurgical Engineering and Materials Science, University of Kentucky, Lexington, KY 40506

ABSTRACT

The initial supersaturation level of 13 different 3000 series Al alloys was characterized by electrical resistivity measurements using a four-point probe method. The prediction of these electrical resistivity values as a function of electron scattering from a particular solute in solid solution was accomplished by fitting a least-squares linear regression equation to the data. The agreement between the experimental and theoretical results is within a ± 5% relative error.

BACKGROUND

The principal microstructural feature of strip cast 3000 series Al alloys is the high degree of solute supersaturation. The supersaturation level of the 3000 series alloys has a 3-fold effect. The first effect is that it determines the size, shape and distribution of the large intermetallic particles present in the cast alloy matrix. The latter is known to affect the subsequent nucleation (and shape) of recrystallized grains and to affect the processing of strip cast 3000 series Al alloys in can stock application(1,2) or in stretch formability(3). The second effect of the initial supersaturation level is the determination of the subsequent recrystallization kinetics in terms of the retarding force of the segregating foreign atoms(4,5). The third effect of the supersaturation level is that it leads subsequently through appropriate heat treatments to the development of several particle states (unimodal, bimodal and trimodal size distributions) in the homogenized and non-homogenized conditions of these alloys(6).

Much research has been carried out on understanding the resistivity changes of Al alloys due to microscopic factors such as phonons, the density of dislocation lines, vacancies(7-9), fine precipitate particles, large intermetallic phases(10), mechanical deformation(11-14) or Guinier-Preston zones(15-17). The electrical resistivity of an Al alloy can be expressed as:

$$\rho_T = \rho_{(\tau)} + \rho_{(i)} + \rho_{(d)} \qquad (1)$$

where ρ_T is the observed resistivity, $\rho_{(\tau)}$ is the resistivity due to lattice vibrations, $\rho_{(i)}$ is the residual resistivity due to solute atoms, $\rho_{(d)}$ is the resistivity due to electron scattering from imperfections such as dislocations, vacancies, grain boundaries, etc. Equation 1 is a simple form of the Matthiessen-Vogt rule(18).

The residual resistivity component ρ_i (the difference between the resistivity of the alloy and that of the pure matrix) has received the attention of many physicists and metallurgists since the early 1930's(19,20) whence the scattering of conduction electrons by solute atoms is thought to be due to the distortion of the lattice around the foreign atoms and to the departure of the lattice field from periodicity in the neighborhood of each solute atom. When impurities with increasing valencies are substituted in the Al lattice their residual resistivity increases roughly as the square of the difference of the valency between Al and the solute elements. Fig. (1) illustrates this rule, Norbury's rule(21). However, the residual resistivity does not increase monotonically but rather periodically, illustrating that the residual resistivity of an impurity depends mainly on its valency and to a lesser degree on the row it belongs to in the periodic table. Willey(22) has confirmed the results of Fig. (1), Table (1). In the present paper a simple analytical procedure was followed to predict the electrical resistivity values as a function of electron scattering from a particular solute in solid solution using current data available. This procedure serves as a rapid analytical evaluation of the supersaturation level of strip cast 3000 series Al alloys. It also forms a basis for further theoretical work relating electrical resistivity values to other parameters like the solid solubility limit of different solutes in aluminum.

EXPERIMENTAL

Thirteen different 3000 series Al alloys were used for this study. The compositions of the alloys are given Table 2. All of the these alloys were cast by a continuous strip-casting method. The electrical resistivity measurements were carried out using a four-point probe method(23), by measuring the voltage drop over a constant gauge length with a constant current. The voltage drop was read with a highly sensitive digital nanovoltmeter.

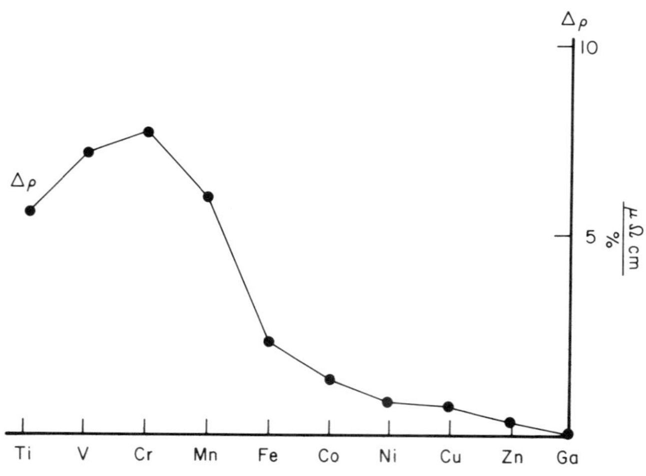

Fig. 1. Residual Resistivity Δρ (in μ Ω cm per wt) % for Various Substitional Impurities in Aluminum (from Reference 21)

The Calculation

Nes and Embury(24) calculated the electrical resistivity of an Al-Mn alloy (0.9% Mn, 0.35% Fe) in the as-cast state using the following equation:

$$\rho_{Al,Mn} = KC_{Mn} + \rho_{Al} \quad (2)$$

where $\rho_{Al,Mn}$ is the resistivity of the alloy in the as-cast state, ρ_{Al} is the resistivity of pure Al at 20°C (2.64 μΩcm) and C_{Mn} is the amount of Mn in solid solution (which was assumed to be the nominal wt% of Mn). The coefficient K(μΩcm/wt%) in the above linear equation has been given slightly different values by different investigators. According to Altenpohl(25) it has a value of 3.6, while a value of 3.065 was given by Kutner and Lang(26), a value of 2.94 was used by Willey(22) and a suggested value of 3.2 by Andersson(27).

In the present work the experimental electrical resistivities of the 13 alloys were numerically fitted to a modified form of equation (2). With the aid of equation (3), next section, the resistivity of any 3000 series Al alloy can be predicted.

Assumptions

1. This calculation is valid only for an as-cast structure (continuous cooling with a high cooling rate) at room temperature.
2. The elements individually go into solid solution and their effects on resistivity are additive; no clustering or interaction between solute atoms is assumed(22).
3. Any solute of wt.% < 0.1% is neglectd.
4. Intermetallics are insoluble.
5. The contribution to the resistivity from vacancies, dislocation structures, stacking faults, grain boundaries is negligible(7).
6. (a) Mg, Cr and Cu are in S.S. (solid solution).
   (b) 0.025 wt% Si is in S.S., the rest is out of solution or is segregated.
   (c) 0.07 wt% Fe is in S.S., (high Fe alloys, all alloys in Table 2, except alloy #11).
   (d) 0.05 wt% Fe is in S.S., (low Fe alloys, #11 in Table 2).
7. Mn goes into solid solution according to an equation of the form:

   $$C_s = (C - a)/b \quad (3)$$

   where C is the weight percent Mn and C_s is the weight percent actually in solid solution, while a and b are constants.
8. All experimental measurements in this study were carried out using the same equipment. Hence, any error differences from one experimental procedure to another is eliminated.

Method

Equation (2) suggested by Nes and Embury (24) is modified to the following form:

$$\rho_{Al-Mn} = KC_s + \rho' \quad (2')$$

where K = 3.6 μΩcm/wt %, (25), C_s = amount of Mn in S.S., and, (28),

$$\rho' = \rho_{Al} + \Sigma C_j^o \rho_j^{in} + \Sigma(C_j - C_j^o)\rho^{out}, \; C_j \geq C_j, \text{ or}$$

$$\rho' = \rho_{Al} + \Sigma C_j^o \rho_j^{in}, \; C_j \leq C_j^o. \quad (4)$$

where $\rho_{Al(25°C)}$ = 2.71 μΩcm, (22). C_j^o = the maximum (or actual) solubility in wt % of the jth solute (not according to table I, (22), but according to the observed microprobe analysis of Furrer and Hausch (29).

$(C_j - C_j^o)$ = The amount of solute precipitated.

ρ_j^{in} = The average increase in resistivity (incremental) due to solute in S.S.

ρ_j^{out} = Average increase in resistivity (incremental) due to solute out of S.S.

Example

By applying equations (2'), (4) and assumptions 1-8, to alloy #1 in table 2 (#638), also from table 1:

$$\rho' = 2.71 + 0.26 \times 4.0 + 0.025 \times 1.02$$

$$+ (0.23 - 0.025) \times 0.088 + 0.07 \times 2.56$$

$$+ (1.13 - 0.07) \times 0.058 + 1.36 \times (0.54)$$

$$= 4.76862 \ \mu\Omega cm.$$

From table 3 the measured resistivity = 6.1318 μΩmcm.

Substituting in equation (2'), ∴ C_s = 0.378661.

Note how equation (2') is satisfied,

6.1318 = 0.378661 x 3.6 + 4.76862.

Similarly for the rest of the alloys:

	Measured Resistivity	K x C_s	ρ'
X	6.4170	1.97762	4.43938
IX	7.0587	2.3523599	4.70634
VI	6.3943	2.1748399	4.21946
II	6.8255	2.7272999	4.0982
VII-CA	7.0860	2.7539204	4.33208
VI-C	7.3762	3.1367002	4.2395
Coil 28	7.1010	3.0794198	4.02158
VIII	7.3883	2.8449799	4.54332
I	7.5141	3.4825799	4.03152
624	6.795	3.3022598	3.49274
VAW	6.52	3.0795199	3.44048
S	6.72	3.7034201	3.01658

From the above table both sides of equation (2') are equal. The important calculation is to predict theoretically from the composition of the alloy its resistivity in the as-cast state at room temperature. From table 3 by simple linear* regression between columns 5 and 7 we

*A quadratic, cubic, quartic and quintic regression did not increase R^2 (correlation coefficient) value significantly.

obtain an equation of the form:

$$C_s = (C - 0.1031694)/1.0335085 \qquad (4)$$

with a correlation of 0.966859, refer to assumption 7.

From C, the nominal composition, if we insert its values, column 5, table 3, in equation (4), we get the predicted values of C_s (column 8, table 3). Then C_s is plugged into equation (2'). Hence, we obtain the calculated resistivity values in column 10. In column 11 the relative error,

$$\frac{\rho_{measured} - \rho_{calculated}}{\rho_{calculated}}$$ is included, with the worst error $\leqq \pm 5\%$.

Table 1. Effect of Elements In and Out of Solid Solution on the Resistivity of Aluminum (from reference 22)

Element	Max. Solubility in Al,%	Average Increase in Resistivity: In Solution	Average Increase in Resistivity: Out of Solution
Cr	0.77	4.00	0.18
Cu	5.65	0.344	0.030
Fe	0.052	2.56	0.058
Mg	14.9	0.54	0.22
Cu	5.65	0.344	0.030
Mn	1.82	3.6*	0.34
Si	1.65	1.02	0.088

*is from reference (25) but all the rest are from reference (22)

Table 3. The Measured and Predicted Resistivities of Strip Cast 3000 Series Al Alloys in μΩcm

		1	2	3	4	5	6	7	8	9	10	11
	Alloy	Cr	Si	Fe	Mg	Mn	Cu	Calculated C_s	Predicted C_s	Measured Resistivity	Calculated Resistivity	Relative Error $\rho_m - \rho_c / \rho_c$
1	638	0.26	0.23	1.13	1.36	0.48		0.378661	0.3646129	6.1318	6.0812266	0.83%
2	X	0.22	0.10	0.83	1.10	0.67		0.5493389	0.5484528	6.4170	6.4138099	0.05%
3	IX	0.27	0.30	0.75	1.20	0.71		0.6534333	0.5871559	7.0587	6.8201011	3.5%
4	VI	0.17	0.15	0.69	1.07	0.80		0.6041222	0.6742379	6.3943	6.6467164	-3.7%
5	II	0.15	0.17	0.60	1.0	0.88		0.7575833	0.7516441	6.8255	6.8041188	0.31%
6	VII-CA	0.21	0.21	0.62	0.98	0.89		0.7649778	0.7613199	7.0860	7.0728316	0.19%
7	VI-C	0.18	0.16	0.61	1.04	0.93		0.8713056	0.8000230	7.3762	7.1195828	3.6%
8	Coil 28	0.12	0.17	0.51	1.09	0.96		0.8553944	0.8290503	7.1010	7.0061612	1.35%
9	VIII	0.23	0.22	0.54	1.23	1.02		0.7902722	0.8871050	7.3883	7.7368981	-4.5%
10	I	0.12	0.18	0.48	1.11	1.07		0.9673833	0.9354839	7.5141	7.3992621	1.6%
11	624	-	0.32	0.21	1.10	1.08		0.9172944	0.9451597	6.795	6.8953149	-1.46%
12	VAW	-	0.22	0.46	0.90	0.98		0.8554222	0.8484019	6.52	6.4947269	0.4%
13	S	-	0.24	0.67	-	1.20	0.14	1.0287278	1.0612690	6.72	6.8371486	-1.7%

Table 2. Continuously Cast Materials

	Alloy	Cr	Si	Fe	Mg	Mn	Cu
1	638	0.26	0.23	1.13	1.36	0.48	
2	X	0.22	0.10	0.83	1.10	0.67	
3	IX	0.27	0.30	0.75	1.20	0.71	
4	VI	0.17	0.15	0.69	1.07	0.80	
5	II	0.15	0.17	0.60	1.0	0.88	
6	VII-CA	0.21	0.21	0.62	0.98	0.89	
7	VI-C	0.18	0.16	0.61	1.04	0.93	
8	Coil 28	0.12	0.17	0.51	1.09	0.96	
9	VIII	0.23	0.22	0.54	1.23	1.02	
10	I	0.12	0.18	0.48	1.11	1.07	
11	624	-	0.32	0.21	1.10	1.08	
12	VAW*	-	0.22	0.46	0.90	0.98	
13	S*	-	0.24	0.67	-	1.20	0.14

*Alloy 12 is from references (11) and (12) while alloy 13 is from reference (4).

CONCLUDING REMARKS

In the present work the authors have applied a simple analytical procedure to predict the electrical resistivity of strip cast 3000 series Al alloys. This procedure serves as a simple and rapid estimation of the degree of solute supersaturation which in turn is a vital controlling factor in the subsequent processing of strip cast 3000 series Al alloys. The agreement between the experimental values and the predicted values encourages further work to develop a theoretical procedure which can relate the electrical resistivity values of aluminum alloys to other parameters like the solid solubility limit of different solutes in Al and the change in lattice parameter of aluminum when alloyed with different solutes.

REFERENCES

1. O. S. Es-Said and J. G. Morris, The Interaction Effect of Second Phase Particles and Recrystallization Behavior and the Consequent Effect on Processing of Aluminum Strip Cast Material for Can Stock Application, in: "V Yugoslav International Symposium on Aluminum", Andrej Paulin, ed., Tiskarna DDU Univerzum, (1986).
2. H. D. Merchant and J. G. Morris, Production of Aluminum Alloy, U.K. Patent GB2 123319A, (1983).

3. Y. Kwag and J. G. Morris, The Effect of Structure on the Mechanical Behavior and Stretch Formability of Constitutionally Dynamic 3000 Series Aluminum Alloys, Mat. Sci. and Eng., 77:59 (1986).
4. O. S. Es-Said and J. G. Morris, The Effect of Second Phase Particles on the Recrystallization Behavior, Texture and Earing Behavior of Strip Cast 3004 Al Alloy, in: "Aluminum Alloys - Their Physical and Mechanical Properties", E. A. Starke, Jr. and T. H. Sanders, Jr., ed., EMAS, West Midlands, U.K. (1986).
5. K. Holm and E. Hornbogen, Annealing of Supersaturated and Deformed Al-0.042 wt% Fe Solid Solutions", J. Mat. Sci., 5:655 (1970).
6. L. Chen and J. G. Morris, The Precipitation Behavior of Strip Cast 3004 Al Alloy, Scr. Metall. 18:1365 (1984).
7. F. R. Fickett, A Review of Resistive Mechanisms in Aluminum, Cryogenics, 11:349 (1971).
8. R. A. L. Drew, W. B. Muir, and W. M. Williams, Differential Resistivity Measurement for Monitoring Annealing, Met. Trans. A, 14A:175 (1983).
9. C. Panseri, F. Gatto and T. Federighi, Interaction Between Solute Magnesium Atoms and Vacancies in Aluminum, Acta. Met., 6:198 (1958).
10. E. Louis and C. G. Cordovilla, Resistivity of Aluminum Binary Alloys in Annealed Condition, Met. Sci., Dec:597 (1980).
11. R. V. Tilak and J. G. Morris, Studies of the Effect of Thermomechanical Treatments on the Supersaturation Content of Strip-cast Aluminum Alloy 3004, Mat. Sci. and Eng., 73:139 (1985).
12. J. G. Morris and R. V. Tilak, The Criticality of Retained Supersaturation Content on Superstrengthening Processes in Aluminum Alloys, Scr. Met., 19:587 (1985).
13. G. A. Hassan and F. H. Hammad, A Resistivity Decrement in Deformed Aluminum, Phys. Stat. Sol., 37:K209 (1976).
14. J. Schrank, M. Zehetbauer, W. Pfeiler and L. Trieb, Effect of High Deformation on Electrical Resistivity in Pure Aluminum, Scr. Met., 14:1125 (1980).
15. E. Holmes and B. Noble, Resistivity Examination of Artificial Aging in an Al-Cu-Cd Alloy, J. Inst. Met., 95:106 (1967).
16. J. Lendual, T. Ungar, I. Kovacs and G. Broma, Correlation Between Resistivity Increment and Volume Fraction of G. P. Zones in an Al-3.2 wt% Zn-2.2 wt% Mg Alloy, Phil. Mag., 33:209 (1976).
17. J. Lendual, T. Ungar and I. Kovacs, Electrical Resistivity of η' Precipitates in an Al-Zn-Mg Alloy, Phil. Mag., 35:1119 (1977).
18. K. Schroder, Experimental Aspects of Electrical and Thermal Properties of Metals, in: "High Conductivity Copper and Aluminum Alloys", E. Ling and P. W. Taubenblat, ed., Met. Soc. of AIME, Warrendale (1984).
19. N. F. Mott, The Electrical Resistance of Dilute Solid Solutions, Proc. Phys. Camb. Phil. Soc., 32:281 (1936).
20. E. Kovacs, F. J. Kedues and L. Gergely, Temperature Dependence of Impurity Resistivity in Dilute Al-Mn Alloys Containing Precipitates, Phys. Stat. Sol.(a), 15:57 (1973).
21. J. Friedel, On Some Electrical and Magnetic Properties of Metallic Solid Solutions, Cand. J. Phys., 34:1190 (1956).
22. L. A. Willey: quoted by W. A. Dean, Effects of Alloying Elements, in: "Aluminum" Vol. 1, K. R. Van Horn, ed., Metals Park, Ohio, ASM (1967).
23. G. T. Meaden, "Electrical Resistance of Metals", Plenum Press, New York (1965).

24. E. Nes and J. D. Embury, The Influence of a Fine Particle Dispersion on the Recrystallization Behavior of a Two Phase Aluminum Alloy, Z. Metallk., 66:589 (1975).
25. D. Altenphol, "Aluminum and Aluminum-legierungen", Springer-Verlag, Berlin (1965).
26. F. Kutner and G. Lang, Einflub von zusatzelementen und Warmebehandlung auf den spezifischen elektnischen Widerstand von Aluminum, Aluminum, 52:322 (1976).
27. B. Andersson, Modelling of Properties of Strain Hardened Qualities of Strip Cast Al(Mn) FeSi, in: "Aluminum Alloys - Their Physical and Mechanical Properties", E. A. Starke, Jr. and T. H. Sanders, Jr., ed., EMAS, West Midlands, U.K. (1986).
28. Prof. Dr. P. Gillis, University of Kentucky in Lexington, Personal Communication, 1984.
29. P. Furrer and G. Hausch, Recrystallization Behavior of Commercial Al-1% Mn Alloy, Met. Sci., 13:155 (1979).

ULTRASONIC CHARACTERIZATION OF CANDIDATE NUCLEAR USED FUEL CONTAINER WELD MATERIALS

M.D.C. Moles[1], J. Imada[2] and M.P. Dolbey[1]

[1]Nondestructive & Fracture Evaluation Section, Ontario Hydro Research Division, 800 Kipling Avenue, Toronto, Ontario
[2]Now at Babcock & Wilcox, Cambridge, Ontario

SUMMARY

Ultrasonic inspection of weld cylinders of candidate nuclear used fuel container materials (copper, titanium grades 2 and 12, stainless steel 316, Inconel 625, Hastelloy C-276) using immersion longitudinal waves showed that titanium grades 2 and 12 exhibited essentially isotropic behaviour (little beam skewing). The other materials all showed significant anisotropy, which will probably present inspection problems for final closure welds. Metallography showed that the stainless steel, Inconel and Hastelloy all had long columnar grains, but that the copper showed columnar grains round each weld pass with equiaxed grains in the centres. Altering weld interpass temperatures produced no consistent trend in ultrasonic behaviour.

INTRODUCTION

Nuclear Used Fuel Disposal Program

The current Canadian concept for disposal of used fuel fron CANDU nuclear reactors is to isolate the fuel in corrosion-resistant containers and to place the containers at a depth of 500 to 1000 m in plutonic rock. As part of its Technical Assistance Program within the Nuclear Fuel Waste Management Program of Atomic Energy of Canada Limited, Ontario Hydro is performing containment and immobilization studies. Part of this program involves developing remote nondestructive evaluation (NDE) techniques for the final closure weld of the used-fuel immobilization container. Ultrasonic inspection, which can be performed remotely, is potentially the most useful NDE technique for weld inspection, possibly supplemented by other techniques. The objective of the NDE inspection is to ensure the mechanical integrity of the weld, and that a necessary minimum corrosion barrier exists[1].

Material Characterization

Although several designs are under consideration, most containers consist of a metal cylindrical body and a lid. A weld between these two parts is required in order to seal the container, and this must be performed remotely due to high radiation fields. Inspection is then necessary, and ultrasonics has been established as the most viable method of volumetric inspection. However, difficulties in inspecting welds ultrasonically have been encountered with some of the candidate container materials due to substantial beam scattering and attenuation (see below). This behaviour has been related to the formation of large grains in welds with a preferred orientation, producing an acoustically anisotropic microstructure.

Several materials have been selected as candidates for used fuel containers, primarily on the basis of their corrosion resistance[2] (see Table 1). Welds of these materials, plus a mild steel weld and rod, were manufactured and examined ultrasonically to determine anisotropy. Since smaller grain sizes are generally considered more beneficial than large sizes, two welds of each material (except 316 stainless and titanium grade 2) were manufactured, one at a high interpass temperature, the other at a lower interpass temperature to inhibit grain growth.

In contrast to other workers who have been primarily interested in either characterizing or improving inspectability of materials, the objective of this work was to rank the various materials in terms of their ultrasonic inspectability. Consequently, numerical/statistical analyses have been developed and performed on the data generated.

PREVIOUS WORK

A considerable amount of work has been performed by other workers on characterizing the ultrasonic behaviour of welds in austenitic-type material. Most of the work has been performed on 300 series stainless steels, with the balance on Inconel 82. The work has ranged from theoretical and experimental analysis to modelling beam skewing to developing techniques to improve inspectability.

Several workers[3-5] have shown that austenitic welds typically consist of large grains growing with their < 001 > direction in the z (out of weld) direction. In contrast to ferritic steels, no phase transformations occur in austenitic steels on cooling, so no recrystallization/randomization of the microstructure occurs. This preferred orientation in austenitics can have three effects[6]: 1) ultrasonic attenuation increases due to grain boundary scattering, though this would not be a major problem for thin-walled containers; 2) mode conversions can occur, and 3) most important, beam skewing/steering can occur because ultrasonic beams are refracted by changes in velocity between grains of different orientations.

Kupperman et al[3] have shown that beam skewing is a function of wave type. Longitudinal waves are less affected than SV waves (shear waves polarized in a vertical plane), which are usually used for inspections. Silk[7] has developed a finite-element beam skewing model. Other workers have attempted to improve inspectability by signal processing or other means, eg, adaptive learning, signal averaging, special transducers or procedural improvements.

EXPERIMENTAL PROCEDURE

Cylinder Preparation

Thick section GTA welds were made following standard welding procedures, except for interpass temperatures. All welds had butt joints, single V-grooves and were 25.4 mm thick. The parent plate and filler metal were the same material in all cases. Weld interpass temperature was varied (see Table 1 for details). Cylinders of 76 mm length and 19.1 mm diameter were machined from each weld. One end of each cylinder was polished, etched and metallographically examined.

Table 1. Weld Preparation

Material	Interpass Temperature	Number of Passes
Mild Steel	99-260°C	17
316L Stainless Steel	29-149°C	21
Titanium Grade 2	66-93°C	56
Titanium Grade 12 LT[1]	21°C	56
Titanium Grade 12 HT[2]	66-93°C	56
Copper C11000 LT	149°C	17
Copper C11000 HT	260°C	16
Inconel 625 LT	135-154°C	53
Inconel 625 HT	260-371°C	54
Hastelloy C-276 LT	21°C	53
Hastelloy C-276 HT	148°C	49

1. LT - Low Interpass Temperature
2. HT - High Interpass Temperature

Scanning Procedures

The test procedure developed by Tomlinson et al [19] was used. The equipment set-up is shown in Figure 1. The transmitting probe was focussed on the test cylinder, and rotated with it through 180 degrees. Ultrasound transmitted through the cylinder was detected by the stationary receiver probe. The transmitter and receiver probes were in line after 90 degrees of rotation. After each scan, the cylinder was rotated 10°, and the scan repeated. Thirty-six scans in total were taken from each cylinder, and the results plotted as signal amplitude isometrics. The equipment consisted of an Apple II Plus computer, stepping motor driver, turntable, water-tank, and a Krautkramer USIP11 ultrasonic flaw detector. The transmitter probe was a 25 MHz Aerotech Alpha, 6.3 mm active element, focussed at 50 mm in water, and the receiver probe a 10 MHz Aerotech Alpha, also 6.3 mm diameter and 50 mm focus. The USIP11 receiver was set at 2-8 MHz frequency.

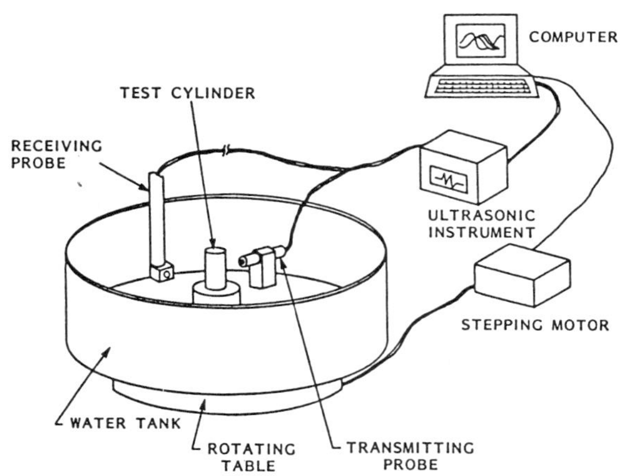

Fig.1. Schematic of equipment setup.

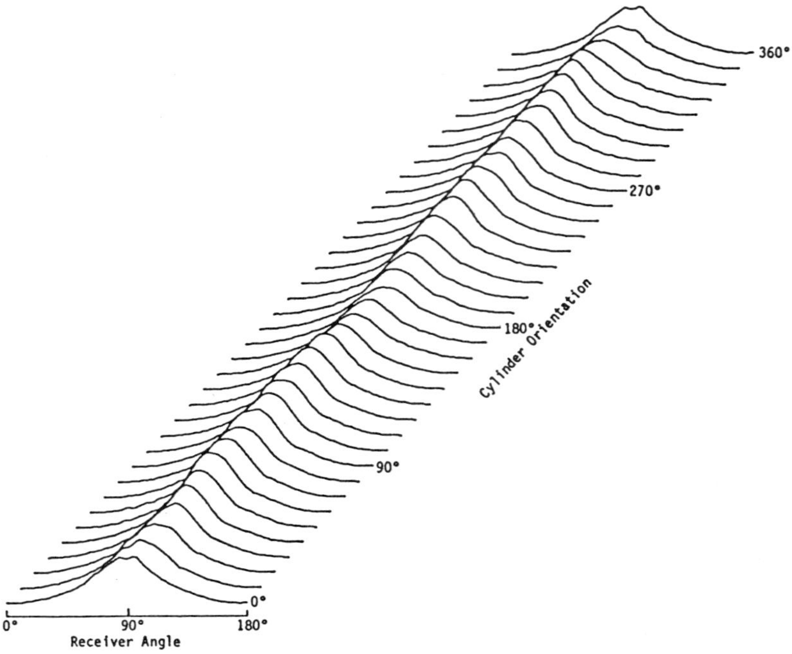

Fig. 2. Isometric plot from mild steel weld, showing little amplitude variation with orientation.

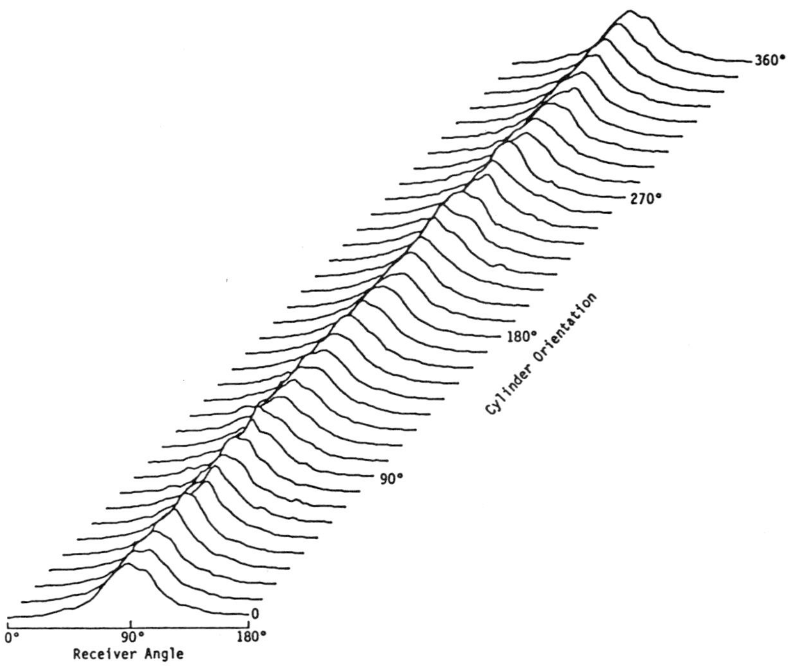

Fig. 3. Isometric plot from titanium Gr.12 Lt, showing relatively little amplitude variation with orientation.

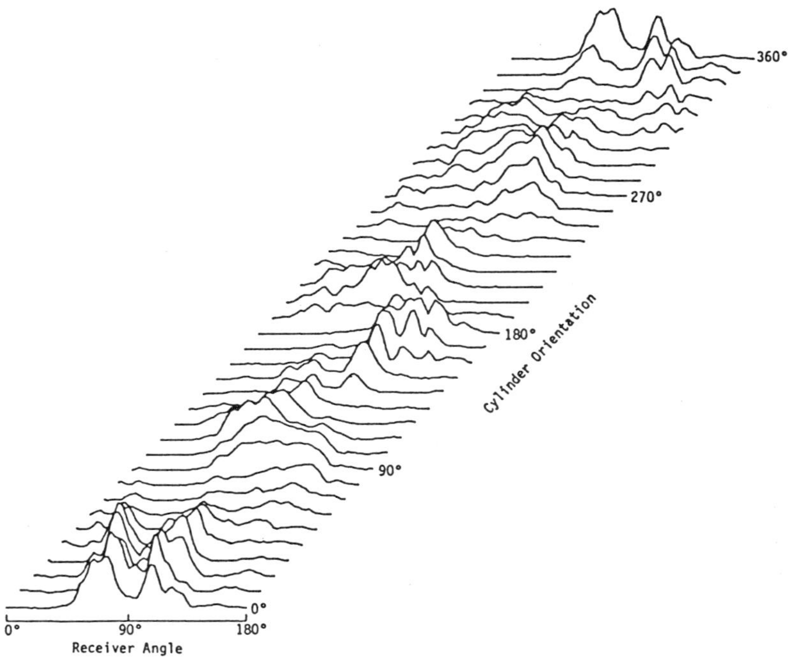

Fig.4. Isometric plot of stainless steel, showing considerable amplitude variation with both receiver angle and cylinder orientation.

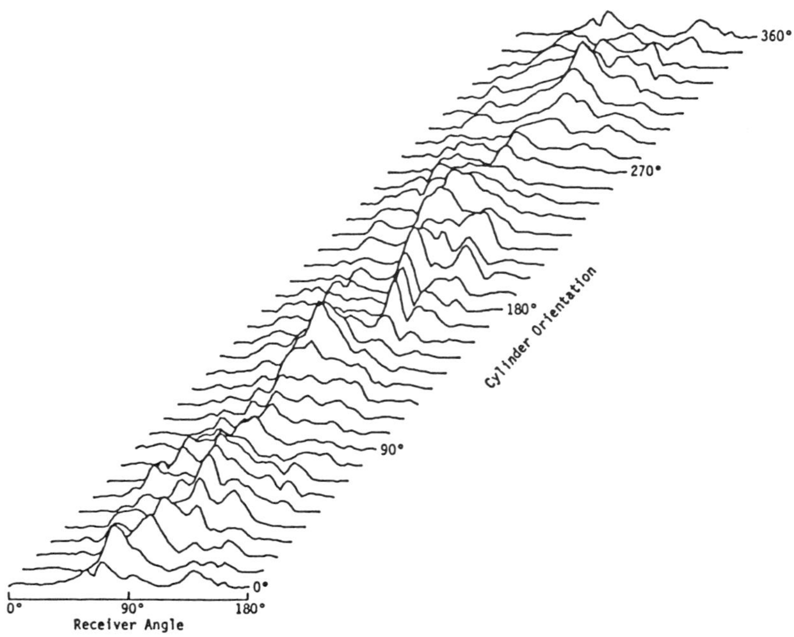

Fig.5. Isometric plot of inconel 625 Lt, also showing considerable amplitude variation with receiver angle and cylinder orientation.

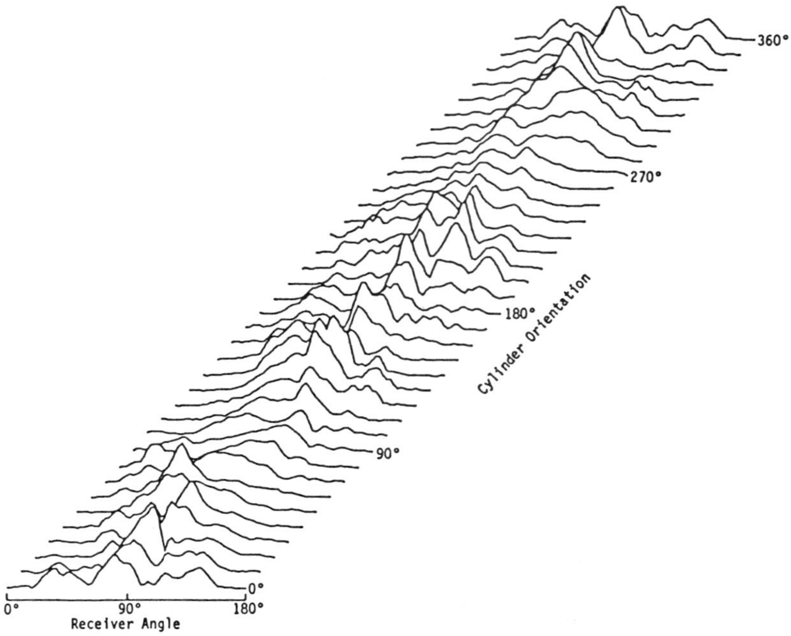

Fig.6. Isometric plot of hastelloy C-276 HT, showing extensive anisotropy.

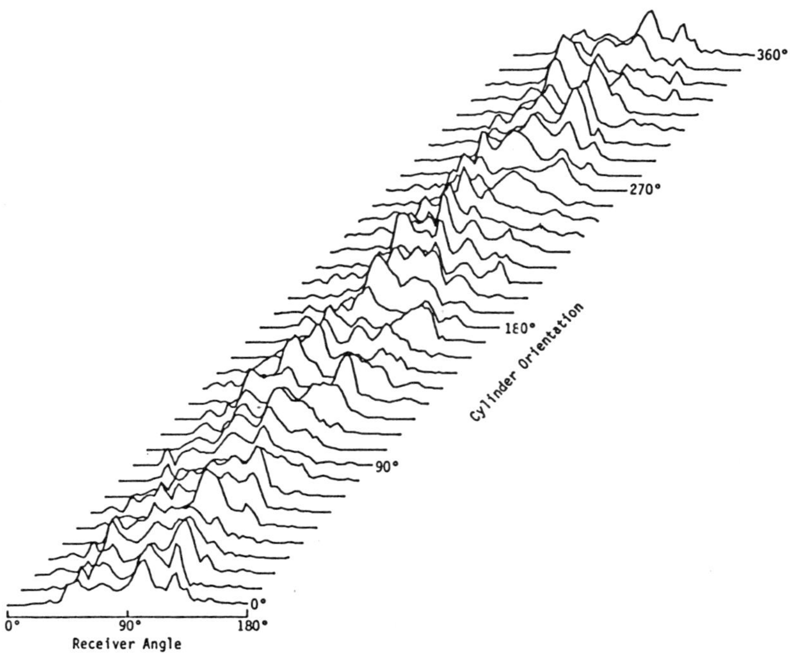

Fig.7. Isometric plot of copper LT, showing beam skewing.

Analysis Procedures

Three parameters were measured on each scan: plot width, skew angle, and number of peaks. Plot width was defined as the signal width 6 dB below the maximum signal amplitude, measured in degrees of rotation. The skew angle was measured as the number of degrees the maximum signal amplitude occurred away from the in-line position. Mean and standard deviations were calculated for both plot width and skew angle. The number of peaks is fairly self-explanatory, though the criteria used were essentially arbitrary. For example, any peak had to be not more than 6 dB down from the highest signal.

The results of the scans and experience with in-service inspections indicated that beam skewing was the most important criterion. Since the mean value of skew angle could be close to zero, standard deviations was considered a more meaningful parameter.

Table 2. Summary of Measurements for Each Cylinder

Cylinder	Plot Width (°)		Skew Angle (°)		Weighted Avg. No. of Peaks	Beam Skew Ranking	Overall Ranking
	Avg.	Std. Dev.	Avg.	Std. Dev.			
Mild Steel Rod	50.81	0.82	1.23	1.23	1.0	1	1
Mild Steel Weld	55.14	5.22	2.33	2.24	1.0	3	4
Stainless Steel	74.93	36.53	8.15	27.53	2.5	12	12
Titanium Gr. 2	50.57	1.13	1.39	1.40	1.0	2	2
Titanium Gr. 12 LT	49.64	3.55	2.11	3.47	1.0	5	5
Titanium Gr. 12 HT	47.80	2.26	2.19	3.42	1.0	9	3
Copper LT	69.66	17.77	3.70	18.05	2.6	11	9
Copper HT	56.94	15.69	4.85	11.40	1.9	7	6
Inconel 625 LT	71.51	25.20	0.68	10.97	2.1	6	7
Inconel 625 HT	76.94	24.14	0.00	15.23	2.6	9	10
Hastelloy C-276 LT	76.56	27.28	3.57	17.71	2.5	10	10
Hastelloy C-276 HT	72.17	24.64	3.49	14.02	2.0	8	7

RESULTS

Ultrasonic Scans

Figure 2 shows the isometric plot from the mild steel weld, essentially the reference material. Titanium grade 2 was similar to the mild steel weld. Figure 3 shows the isometric from the titanium grade 12 LT weld, also showing essentially isotropic behaviour. In contrast, the stainless steel weld in Figure 4, the Inconel 625 LT in Figure 5 and the Hastelloy C-276-HT in Figure 6 show extensive anisotropy. The two copper welds also showed severe beam skewing (see Figure 7).

The analysis of the results are shown in Table 2, with both the beam skew ranking based on standard deviation, and the overall ranking based on all three criteria.

Metallographic Results

Predictably, the mild steel and titanium welds showed fine, equiaxed, recrystallized grain structures. The stainless, Inconel and Hastelloy welds all showed elongated grains in the z-axis direction. However, the copper welds showed relatively large equiaxed grains in the centre of each weld pass, with columnar grains oriented towards the grain centres. This microstructure is typical of a cast metal, and is consistent with the very high heat transfer rates of copper.

DISCUSSION

The mild steel and both grades of titanium alloys appear readily inspectable. All show little anisotropy with insignificant beam skewing or spread and only a single peak. In contrast, the stainless steel 316, Inconel 625 and Hastelloy C-276 all show severe anisotropy, with the stainless steel being the least inspectable. This is not unexpected from the microstructures due to the presence of elongated grains in the z-axis direction. Kupperman et al[3] have shown the stainless steel behaves rather less anisotropically for longitudinal waves than for shear waves. This suggests it would be better to inspect austenitics with longitudinal rather than shear waves, which may present practical problems.

The copper welds are somewhat different from the stainless, Inconel and Hastelloy welds. Though considerable anisotropy is apparent (Table 2) it appears to come from a different mechanism. Instead of elongated, columnar grains in the z-axis direction, both copper microstructures consist predominately of fairly large, equiaxed grains surrounded by columnar grains from each weld pass. Copper is known to be a fairly "noisy" material to inspect due to its fairly large grains, which agrees with these results.

There was no consistent trend with changing weld interpass temperature in this study, based on the standard deviation of skew angle. While the predicted trend in interpass temperatures may exist, it may be eclipsed by other factors. Put another way, "every weld is different".

From an inspectability viewpoint, the tests should be repeated using shear waves since most inspections are performed using these. For the titanium welds, few inspection problems are anticipated except for possible oxygen/hydrogen pick-up. However, the stainless, copper, Inconel and Hastelloy could present even more severe beam skewing problems. Techniques such as spatial signal averaging, multiple beam techniques or adaptive learning might improve inspectability, as might grain refinement. At this concept assessment phase, further development could wait until a container design and material have been selected.

CONCLUSIONS

1. Ultrasonic inspections of cylinders of weld metal from candidate used fuel container materials using longitudinal waves showed the following:

    (a) titanium grade 2 and 12 showed essentially isotropic behaviour;

    (b) commercially-pure copper, 316 stainless steel, Inconel 625 and Hastelloy C-276 showed significant anisotropy (ie) beam skewing and spread;

(c) altering interpass temperatures had no consistent effect.

2. Numerical/statistical procedures were used to rank these materials. Using the most important criterion (beam skew) the order was titanium grade 2 (best)>mild steel>titanium Gr 12>HT titanium Gr 12 LT>>Inconel LT Copper HT>Hastelloy HT>Inconel LT>Hastelloy LT>Copper LT>stainless steel (worst).

3. Metallography showed that the stainless steel, Inconel and Hastelloy welds all had large columnar grains in the out-of-weld plane. In contrast, the copper showed columnar grains round each weld pass and equiaxed grains in the weld pass centres.

4. Titanium (either grade 2 or 12) fusion welds should present few material inspection problems, but copper, stainless, Inconel or Hastelloy will. Further development work can be anticipated for the latter materials.

ACKNOWLEDGEMENT

This project is part of Ontario Hydro's Technical Assistance Program with Atomic Energy of Canada Limited on the Canadian Nuclear Fuel Waste Management Program.

REFERENCES

1. K. Nuttall, J.L. Crosthwaite, P.J. McKay, P.M. Mathew, B. Teper, P.Y.Y. Maak and M.D.C. Moles, The Canadian Container Development Program for Fuel Isolation, Materials Research Society 6th International Symposium on the Scientific Basis for Nuclear Waste Management, Boston, MA, November 1-4, 1982.

2. D.J. Cameron, J.L. Crosthwaite and K. Nuttall, The Development of Durable Man-Made Containment Systems for Fuel Isolation, Can. Met. Quar., 22 (1):87 (1983).

3. D.S. Kupperman, K.J. Reimann and D.I. Kim, Ultrasonic Characterization and Microstructure of Stainless Steel Weld Metal, Proc. of Symp. on "Nondestructive Evaluation: Microstructural Characterization and Reliability Strategies", TMS-AIME and ASM, Pittsburgh, PENN, October 5-9, 1980, Ed. O. Buck and S.M. Wolf, p. 199.

4. J. Tomlinson, A. Wagg and M. Whittle, Ultrasonic Inspection of Austenitic Welds, Proc. of Int. Conf. on NDE in the Nuclear Industry, Salt Lake City, Utah, February 13-15, 1978. p. 64.

5. L. Adler, K.V. Cook and D.W. Fitting, Ultrasonic Characterization of Austenitic Welds, First Int. Symp. on Ultrasonic Materials Characterization, Gaithersburg, MD, June 7-9, 1978, p. 533.

6. D.S. Kupperman and K.J. Reimann, Ultrasonic Wave Propagation and Anisotropy in Austenitic Stainless Steel Weld Metal, IEEE Trans. Sonics and Ultrasonics, SU-27(1):7 (1980).

7. M.G. Silk, Computer Model for Ultrasonic Propagation in Complex Orthotropic Structures, Ultrasonics, 19(5):208 (1981).

MAGNETOELASTIC CONTRIBUTION TO ULTRASONIC ATTENUATION IN STRUCTURAL STEELS

Pierre Langlois and Jean F. Bussière

Industrial Materials Research Institute
National Research Council of Canada
75 de Mortagne Blvd., Boucherville, Québec J4B 6Y4

INTRODUCTION

During recent years, there has been sustained interest in the use of ultrasonic attenuation to characterize the microstructure and mechanical properties of metals[1-4]. However, most of these studies have focussed on attenuation associated with scattering and have either neglected or treated in an empirical and simplified manner other contributions to attenuation. In the present study, we show that absorption of magnetoelastic origin can give rise to substantial attenuation in common structural steels and investigate its dependence on ultrasonic frequency and magnetic permeability.

Although magnetoelastic relaxation is described in some textbooks[5,6] and early works on Nickel have been published more than thirty years ago[7-9], few experimental results are available for carbon steels at high frequencies. Bratina[10] reports some results on a few carbon steels, at fixed frequencies, as a function of applied stress without reference to the relative importance of scattering. More recently, Deka and Eberhardt[11] have published absorption measurements on Fe-based binary alloys in the 0.1-1.0 MHz range using a resonance technique. They found that the predominant part of absorption in all their alloys was magnetoelastic in nature. However, the frequencies that offer an interest for microstructure characterization are often above 1 MHz. Magnetoelastic absorption measurements, in the frequency range 0.5-20 MHz, have been reported recently by Monchalin and Bussière[12] for A36 steel, using shear waves. Like Deka and Eberhardt, they found that magnetic absorption was more important than nonmagnetic absorption, which was determined using a new technique based on the infrared detection of the heat produced by intensity modulated ultrasound. Moreover, Monchalin and Bussière found that the attenuation due to magnetoelastic damping was more important than scattering from grain boundaries below 7 MHz.

In their work on the correlation of ultrasonic attenuation, microstructure and ductile to brittle transition temperature in very low carbon steels, Smith and Reynolds[2] found that the attenuation was anomalously high when compared to normal low carbon steels and could not be interpreted in terms of simple scattering models. They assumed that this was probably due partly to magnetoelastic interaction and partly to grain size distributions. In fact, they pointed out the need for a better knowledge of the relative importance of scattering and magnetoelastic contributions to the attenuation, as well as a knowledge of their frequency dependence.

In the present work we have measured the magnetoelastic absorption in the 4-12 MHz range and the total attenuation (including scattering) in the 10-25 MHz range for various structural steels with different carbon contents. A technique based on ultrasonic spectroscopy with wide band longitudinal transducers[13] was used. The average grain size, d, the initial magnetic premeability, μ_o, and the electric resistivity, ρ, were measured for each sample.

THEORY

Current theory explains magnetoelastic absorption by the vibration of magnetic domain walls and the induction of micro-eddy currents[7,8,10]. This theory predicts a relaxation type loss producing an attenuation

$$a_m = C \frac{\mu_o}{I_s^2} f \frac{f/f_o}{1 + f^2/f_o^2} \tag{1}$$

where

$$f_o = \frac{\pi \rho}{24 \mu_o D^2} \tag{2}$$

with

f = frequency
ρ = resistivity (in CGS emu units)
μ_o = initial magnetic permeability
I_s = saturation magnetization
D = magnetic domain thickness (cm)

C is a constant depending on the elastic and magnetostrictive properties of the materials and the mode of excitation.

With reasonable values of D, f_o is found to be around 1 MHz for our polycrystalline steel samples, but if one takes into account domain rotation which is associated with a much smaller initial permeability, higher values for f_o are found[8].

When the material is in a saturating field, the magnetic domains are lined up and more micro eddy-currents can no longer be induced by the ultrasound wave. The magnetoelastic absorption thus vanishes in a saturating magnetic field. Then, by taking the difference between the attenuation in zero field and the attenuation in a saturating field, one can obtain the magnetic absorption contribution.

EXPERIMENTAL

Our experiments were performed on four carbon steel samples from hot rolled bars of SAE 1005, 1025, 1045, 1090 and one High Strength Low Alloy (HSLA) steel of the Lloyd's LT60 grade (Ladle analysis: 0.12% C, 0.28% Si, 1.39% Mn, 0.017% P, 0.01% S, 0.17% Cr, 0.15% Cu, 0.15% Ni, 0.027% Aℓ, 0.04% Nb, 0.05% V). The sample's dimensions were 3 cm x 7 cm and the thickness varied between 7.6 mm and 20 mm.

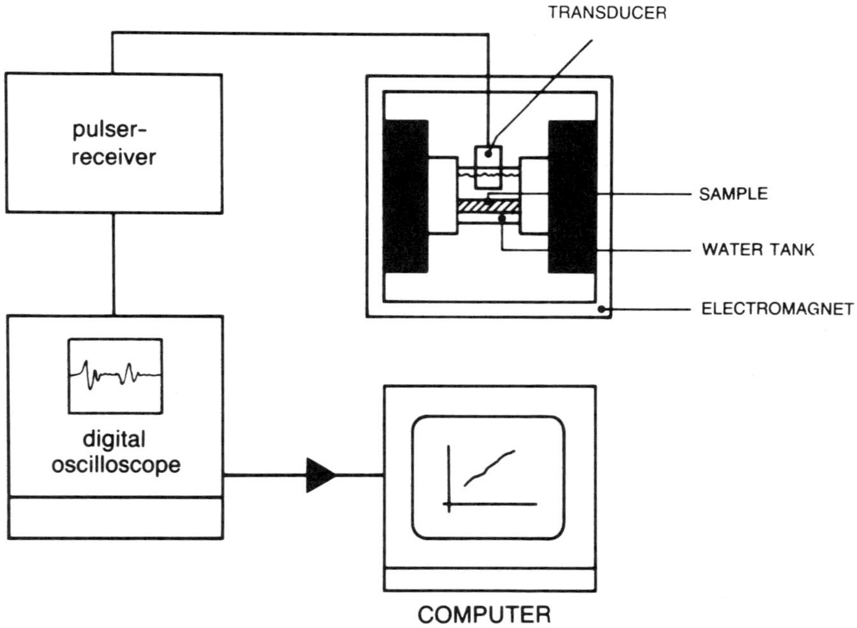

Figure 1: Experimental arrangement.

As illustrated in Figure 1, our samples were placed in a water tank between the poles of an electromagnet. A few wideband longitudinal wave transducers with different central frequencies were used in a pulse-echo mode. The successive echoes, coming from the reflections between the two faces of a sample, were digitized by a Tektronix 7854 programmable oscilloscope and processed in a VAX computer.

The attenuation measurements (see Figure 2) were carried out with the three echoes technique using wideband longitudinal transducers and a buffer[13]. The sensitivity of this method is limited mainly by the diffraction corrections, for which we took the Rogers and Van Buren[14] expression. Indeed when the diffraction corrections become larger than the attenuation to be measured, the results are less reliable due to some possible departure from the ideal piston source, which would involve some variations[15] on the Lommel diffraction correction integral. For this reason, our total attenuation curves in Figure 2 are limited to values over 0.5 dB/cm, which corresponds to the frequency range 10-25 MHz. Each segment of the attenuation curve for the same sample corresponds to a different transducer.

As for the magnetoelastic absorption, it is possible to get more sensitivity than for the total attenuation since there is no need for diffraction corrections. Indeed, as we have seen, the magnetoelastic absorption is the difference between the attenuation measured in a zero field (H=0) and the attenuation measured in a saturating field (H=2500 Oe for our case) for a sample initially demagnetized. The geometrical position of the transducer being the same for these two acquisitions, diffraction corrections are cancelled out. The estimated lower value attainable due to our instrument limitations is about 0.005 dB/cm and the overall experimental precision is around 15%. So the ripples on the lower magnetic absorption curves do not have any physical significance. The magnetoelastic absorption in SAE 1090 steel was too low to give significant spectral curves and its level is below 0.01 dB/cm.

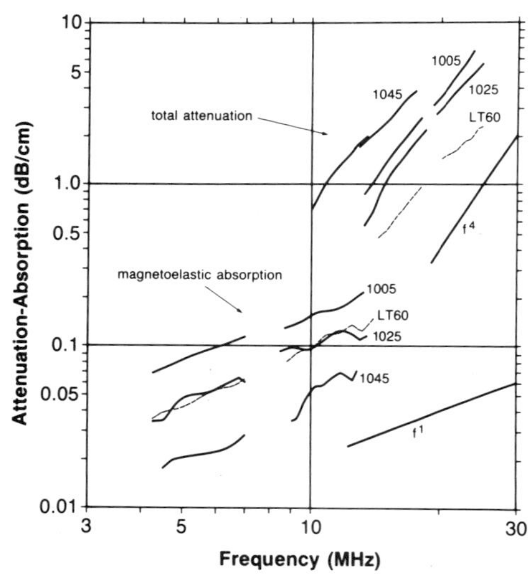

Figure 2: Total attenuation and magnetoelastic absorption measurements as a function of frequency.

For each sample we measured the average grain size, d, according to the procedure described in ASTM specification E112-63, "Standard Methods for Estimating the Average Grain Size of Metals". The initial magnetic permeability, μ_o, and the electrical resistivity, ρ, were also measured and the results are gathered in Table 1.

TABLE 1: Sample specifications

SAMPLE	$\bar{d}$ (μm)	μ_o	ρ ($\mu\Omega$-cm)	a_{m5} (dB/cm)
1005	23	447	13.1	0.082
1025	11	174	18.2	0.044
1045	42	95	18.5	0.02
1090	35	58	19.1	$\lesssim$ 0.01
LT60	9	156	23.8	0.041

d = average grain size,
μ_o = initial permeability,
ρ = electrical resistivity,
a_{m5} = magnetoelastic absorption at 5 MHz.

Figure 3 shows a graph of the magnetoelastic absorption at 5 MHz, a_{m5} as a function of the initial permeability μ_o.

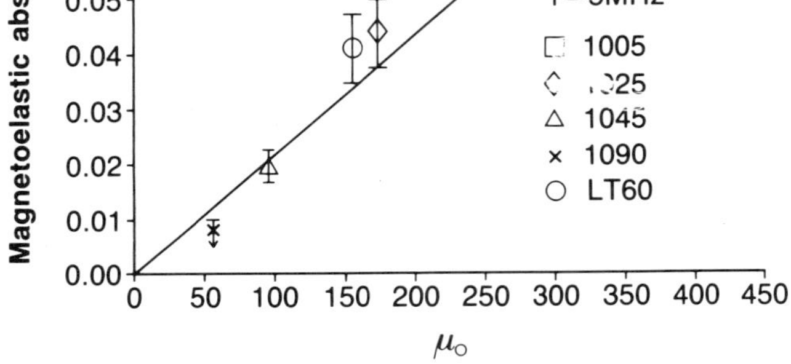

Figure 3: Magnetoelastic absorption at 5 MHz, a_{m5}, as a function of the initial magnetic permeability.

DISCUSSION

The total attenuation measurements (see Figure 2) present a f^4 dependence characteristic of Rayleigh scattering with a higher level for a higher grain size, d, (see Table 1). This was to be expected since the smallest wavelength (at 25 MHz) is about 250 μm and the average grain sizes vary between 9-42 μm.

The frequency dependence of the magnetoelastic absorption goes as f^1 which, according to equation (1) would imply that we are near the maximum of the function.

$$F = \frac{f / f_o}{1 + f^2/f_o{}^2} \qquad (3)$$

This makes sense since there is in fact a domain size distribution which considerably broadens[8] the bell shaped function F.

Another interesting point is the fact that the maximum value of the function F is given by

$$F_{max} = 1/2 \qquad (4)$$

That is, F_{max} is independent of f_o and thus, of the domain size D and the resistivity ρ. Consequently, the last two parameters, D and ρ, should not have a significant importance for the frequency range that we have investigated.

The saturation magnetization I_s that appears in the expression for the magnetoelastic absorption a_m given by equation (1), presents only a small variation among our samples. Indeed we have verified that with a saturating field of 200 Oe, the magnetic induction of the five samples is equal to 19000 Gauss ± 3%. As for the modulus of elasticity of these samples, they are all equal to 205 GPa to a fraction of one percent.

Thus, by referring to equation 1, and in the light of the above discussion, it appears that, for the frequency range 4-12 MHz, the only parameter of importance to characterize the magnetoelastic absorption of our sample is the initial magnetic permeability μ_o (since we were demagnetizing the samples before the absorption measurements). This fact is rendered explicit in Figure 3, where we have plotted the magnetoelastic absorption at 5 MHz, a_{m5} as a function of the initial permeability μ_o. As can be seen, the linear relationship is verified within our experimental accuracy indicated by the error bars, except perhaps for the SAE 1090 steel for which we would need more precision.

Finally, by referring to Figure 2, one can conclude by extrapolating the attenuation curves, that the magnetoelastic absorption becomes more important than the scattering at frequencies below 5 to 9 MHz depending on the grain size of the sample (involved in the scattering) and the magnetic permeability (involved in the magnetoelastic absorption).

So when correlations are attempted between attenuation measurements and steel microstructures at frequencies below 10 MHz, one should be very cautious, especially for low carbon steels having high permeability.

CONCLUSION

From our experimental results on the various structural steels analyzed, we can say that:

1. The contribution of magnetoelastic absorption to the total attenuation is more important than scattering for frequencies below 5 to 9 MHz depending on the grain size and the magnetic permeability of the sample. Thus one should give the proper importance to this magnetic contribution when searching for correlations between attenuation measurements and microstructure below 10 MHz.

2. The magnetoelastic absorption frequency dependence goes as f^1 in the 4-12 MHz range investigated which agrees with the results of Monchalin and Bussière[12] for A36 steel.

3. For frequencies between 4 and 12 MHz, the only significant parameter for the magnetoelastic absorption is the magnetic permeability of the steel to which the magnetoelastic absorption is proportional within our experimental accuracy.

ACKNOWLEDGEMENTS

The authors are grateful to Denis Hardy for preparation of the software.

REFERENCES

1. R. Klinman, G.R. Webster, F.J. Marsh and E.T. Stephenson, "Ultrasonic prediction of grain size, strength and toughness in plain carbon steel", Mat. Eval., Oct. 1980, p. 26.
2. R. L. Smith and W.N. Reynolds, "The correlation of ultrasonic attenuation, microstructure and ductile to brittle transition temperature in very low carbon steels", J. Mat. Sc., 17, p. 1420 (1982).
3. F. Nadeau, J.F. Bussière and G. Van Drunen, "On the relation between ultrasonic attenuation and fracture toughness in type 403 stainless steel", Mat. Eval., Jan. 1985, p. 101.
4. E.R. Generazio, "Scaling attenuation data characterizes changes in material microstructure", Mat. Eval., Feb. 1986, p. 198.
5. R. Truell, C. Elbaum and B. B. Chick, "Ultrasonic methods in solid state physics", Academic Press, New York (1969).
6. A. S. Nowick and B.S. Berry, "Anelastic relaxation in crystalline solids", Academic Press, New York (1972).
7. W. P. Mason, "Domain wall relaxation in nickel", Phys. Rev., 83, p. 683 (1951)
8. W. P. Mason, "Rotational relaxation in nickel at high frequencies", Rev. Mod. Phys., 25, p. 136 (1953).
9. S. Levy and R. Truell, "Ultrasonic attenuation in magnetic single crystals", Rev. Mod. Phys., 25, p. 140 (1953).
10. W. J. Bratina, "Internal friction and basic fatigue mechanisms in body-centered cubic metals, mainly iron and carbon steels" in "Physical Acoustics" Vol. III part A, p. 223, Academic Press, New York (1966).
11. M. Deka and N. Eberhardt, "Internal friction of Fe-based binary alloys at high frequency". Proc. of a Symposium on Nondestructive Methods for Material Property Determination, Hershey, Pennsylvania, April 6-8, 1983, ed. C.O. Ruud and R.E. Green, Jr., Plenum Press, New York, p. 135 (1984).

12. J. P. Monchalin and J. F. Bussière, "Infrared detection of ultrasonic absorption and application to the determination of absorption in steel", Proc. Review of Progress in Quantitative NDE, San Diego, CA., July 8-13, 1984, ed. D.O. Thompson, Plenum Press, New York, p. 965 (1985).
13. E. P. Papadakis, "Absolute measurements of ultrasonic attenuation using damped nondestructive testing transducers", J. Testing and Evaluation, JTEVA, 12, p. 273 (1984).
14. P. H. Rogers and A. L. Van Buren, "An exact expression for the Lommel diffraction correction integral", J. Acoust. Soc. Am., 55, p. 724 (1974).
15. E. P. Papadakis, "Effects of input amplitude profile upon diffraction loss and phase change in a pulse-echo system", J. Acoust. Soc. Am., 49, p. 166 (1971).

CYCLIC PLASTICITY OF PURE ALUMINUM STUDIES BY CONTINUOUS ACOUSTIC EMISSION MEASUREMENT

A. Slimani, P. Fleischmann and R. Fougères

Groupe d'Études de Métallurgie Physique et de Physique des Matériaux (U.A. CNRS341), INSA DE LYON, Bât. 502
69621 Villeurbanne Cedex, France

INTRODUCTION

Continuous acoustic emission (AE) observed during plastic deformation of pure metals such as Aℓ or Cu is currently used for experimental verification of theoretical models which link the acoustic waves to microscopic mechanisms acting as emission sources. Indeed, in such materials, these mechanisms were well studied and only attributed to the dislocation movement. In a pure polycrystalline material, at the beginning of plastic deformation, Frank's dislocation sources generate a great number of dislocations which can move over large distances in relation to the grain size. However, as the plastic deformation increases, these distances diminish as a result of the mutual interaction between dislocations.

The different AE models which were proposed are very similar in that they connect the AE signal to the goemetrical and dynamical properties of dislocation motion[1-4]. These models lead to a relation between the displacement of the ultrasonic wave and the characteristics of the emission source.

$$u_{L\delta} \propto \ell\, v \qquad (1)$$

where ℓ is the dislocation length, v the dislocation velocity and $u_{L\delta}$ the ultrasonic displacement. By integration on a large number, $\dot{N}$, of sources per unit time acting simultaneously without spatial or temporal relation together, one can show that, in the low frequency range, AE can be described by the following relation:

$$P \propto \dot{N} A^2 \qquad (2)$$

where P is the power of the AE signal, $\dot{N}$ is the number of active sources per unit time and A is the mean area swept by an elementary source. The number of active sources $\dot{N}$ can be deduced from Orowan's law:

$$\dot{N} = \dot{\varepsilon}_{pmac} V/Ab \qquad (3)$$

Therefore the relation (1) reduces to:

$$P \propto \dot{\varepsilon}_{pmac} VA \qquad (4)$$

where V is the plasticized volume, $\dot{\varepsilon}_{pmac}$ the macroscopic strain rate and b the Burger's vector. Most frequently, the AE is measured by rms voltage which is the square root of the signal power. Relation (4) leads to a credible explanation of all the results available in the literature on AE studies of materials such as Aℓ or Cu strained under monotonic conditions.

- the proportionality of P to $\dot{\varepsilon}_{pmac}$ has been verified by Hatano[5] and Hamstad[6] and to the plastic volume V, by James and Carpenter[7].

- the effects of grain size and material purity have been explained from the evolution of the factor A [2,8,9].

Sankar[10] and Raouadj[11] have studied the AE during unloading of pure strained copper. Kishi[12] has studied the AE during the cyclic deformation of fcc metals. These last results cannot however be explained with the above mentioned models. For this reason, a systematic study of the AE during cyclic deformation of Aℓ 99,999% pure has been carried out and the purpose of this paper is to present these results.

EXPERIMENTAL

The deformation is obtained with a hydraulic tension/compression machine controlled by a PDP11 computer. Polycrystals of aluminum having a purity of 99.999% and a grain size between 1 and 2 mm, (obtained by annealing for two hours at 450°C) were used in this study. The AE transducer is fixed on one head of the sample. The transducer is operated in a frequency range of 0.3 to 1.5 MHz. The amplification value is 60 dB and the rms value of the signal is measured with a METRIX voltmeter. Experiments were carried out at room temperature on annealed or saturated states. The plastic strain amplitude, ε_p is between 1.5×10^{-4} and 2×10^{-3}.

RESULTS

Acoustic emission during one fatigue cycle

At the second fatigue cycle the recorded rms voltage is given in Fig. 1. The observed behaviour is similar to the one described by Kishi[12]. AB and DE peaks, respectively in tension and compression, appear at the corresponding macroscopic yield strength. Moreover, in the fully plastic range, AE voltage exhibits a stationary level.

Effect of the cycle number on AE

According to the number of cycles applied, both the peak amplitude and the stationary level, previously described, exhibit three stages of variation:

- for about the first four cycles, stage 1 is characterized by a fast increase of both the peak amplitude and the stationary level.
- stage 2 is characterized by a slow decrease of these two parameters as the cycle number is increased until the maximal cyclic stress reaches a constant value.
- stage 3 is called "saturated stage" because all AE parameters remain unchanged as the cycle number is increased.

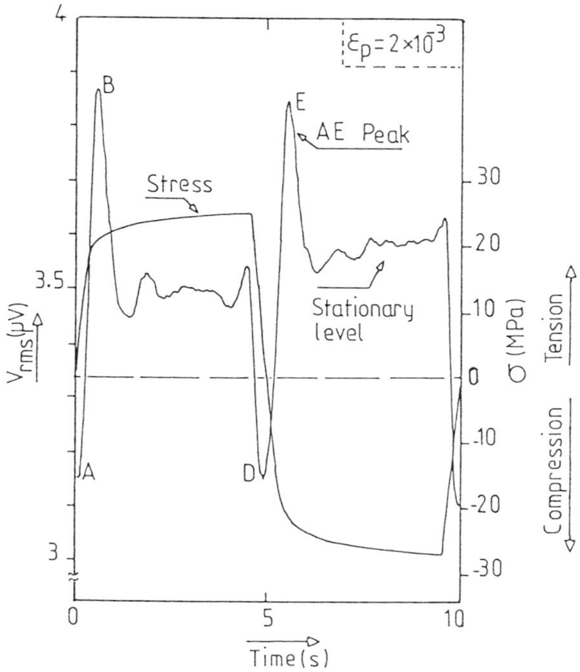

Fig. 1. AE rms voltage and fatigue stress versus time. Cyclic deformation conditions : plastic strain ε_p = $2x10^{-3}$; period test = 10s ; cycle number = 60.

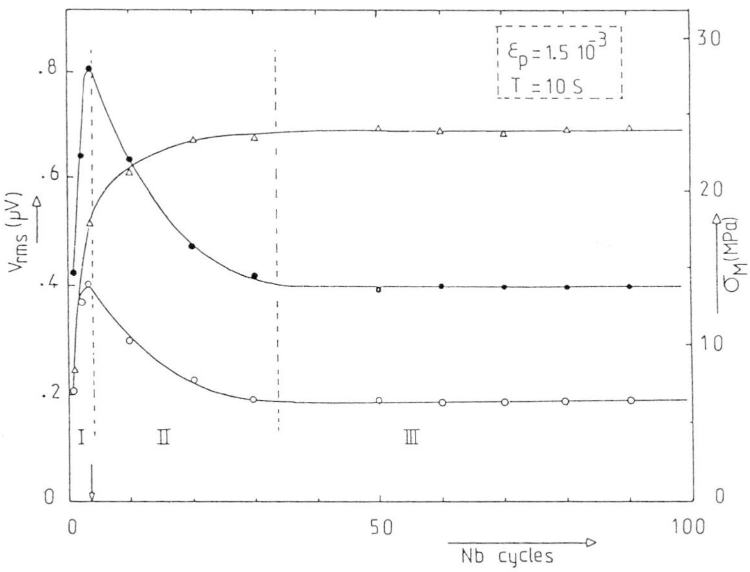

Fig. 2. Fatigue maximum stress, AE peak amplitude and AE stationnary level versus cycle number. Annealed sample ; ε_p = 1,5 x 10^{-3} ; period test = 10 s

Effect of strain rate on AE

The strain rate effect is measured in the saturated stage where all cycles are stable. Thus the influence of the strain rate can be determined precisely. The effect of strain rate is determined by the modification of the plastic strain period. Fig. 3 shows the results we obtained. It is observed that for plastic strain amplitudes smaller than 10^{-3}, the law $V_{rms} \alpha \sqrt{\dot{\varepsilon}_{pmac}}$ is well verified, whereas for plastic strain amplitudes above 10^{-3} the rms voltage increases more rapidly than predicted by this law and $V_{rms} \propto \dot{\varepsilon}_{pmac}^{0.7}$.

DISCUSSION

Immediately after the first cycle, a cellular structure of dislocations has been observed by T.E.M. (Fig. 4). The microstructure reveals high dislocation density in cell walls and fairly few dislocations in the interior of the cellular structure. This microstructure is slowly modified with the advancement of the cycle number. Cell walls are easier to observe as the cycle number is increased. It is interesting to note that the plastic strain amplitude has little effect on the microstructure[13]. The deformation is caused by dislocations which move from one side to the opposite side of the cell. It is clear from T.E.M. studies that walls of the cellular structure act as dislocation sources.

In the tension or compression range, the shape of the AE rms voltage-time curve (Fig. 1), is very similar to the one obtained in monotonic deformation. However, the stationary level is lower in monotonic deformation than in a cyclic one. Experimental AE results are frequently explained in terms of competition between two factors acting

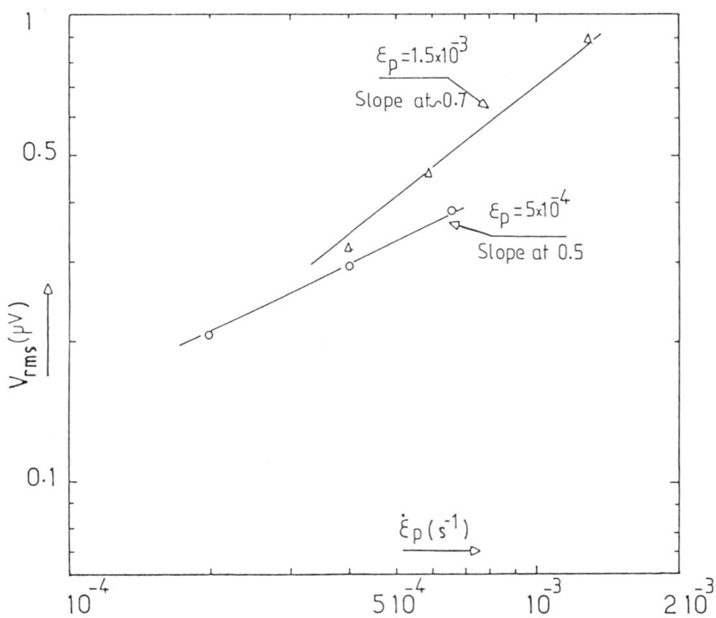

Fig. 3: AE voltage peak magnitude versus plastic strain rate. Results were obtained for plastic strain amplitudes equal to 5×10^{-4} and 1.5×10^{-3}.

in an opposite way: (i) a very rapid increase of $\dot{\varepsilon}_p$ (or $\dot{N}$) at the beginning of the plastic range which gives the fast increase of the AE signal, (ii) a decrease of the swept area A which leads to the decrease of AE (see relation 4).

In the case of plastic deformation of pure aluminum, it appears that this explanation is not suitable. Effectively, in the saturated stage, where the cellular structure remains stable with increases of the plastic strain amplitude and of the cycle number, the observed AE decreases after the AE peak cannot be understood by these assumptions. Moreover, ultrasonic attenuation measurements[14] show also that both the mobile dislocation density and the dislocation length remain at a constant value for conditions we have considered here: $\varepsilon_p = 1.5\ 10^{-3}$. So, the swept area does not vary significantly as the plastic strain increases. Because $\dot{\varepsilon}_p$ is maintained at a constant value in the whole plastic range, and the relation (4) cannot explain the AE decrease.

In monotonic deformation, the proportionality between the power of the AE signal and the strain rate is always explained by relation (4). This relation results from athermal behaviour of the mechanisms which control the plastic deformation. So, the duration of these generated mechanisms are directly proportional to the plastic strain rate. In cyclic deformation, this assumption can be retained so the law $V_{rms} \propto \sqrt{\dot{\varepsilon}_{pmac}}$ is satisfied ($\varepsilon_p < 10^{-3}$). For plastic strain amplitudes higher than 10^{-3}, the $V_{rms} \propto \dot{\varepsilon}_{pmac}^{0.7}$ law is observed: the AE signal increases more rapidly than the value predicted by relation (4) and consequently a particular behaviour exists in cyclic deformation.

To explain results obtained in cyclic deformation experiments, the following assumptions are proposed. The plastic strain ε_{pmac} is divided into two components: (i) ε_{p1} due to dislocation movement which leads to AE in the measured frequency range, (ii) another term ε_{p2} due to very slow dislocation motion which leads to negligible AE. Hence:

$$\varepsilon_{pmac} = \varepsilon_{p1} + \varepsilon_{p2}$$

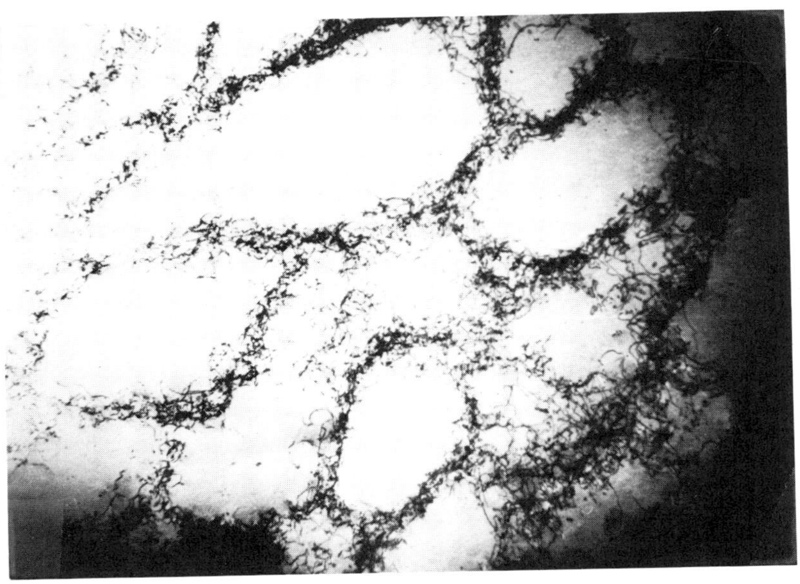

Fig. 4: TEM observation of the cellular microstructure. Plastic strain amplitude $\varepsilon_p = 1.5 \times 10^{-3}$; number of cycles = 200 cycles.

Using this last relation the relation (4) becomes:

$$V_{rms} \; \alpha \; [\dot{\varepsilon}_{pmac} (1 - \dot{\varepsilon}_{p2})/ \dot{\varepsilon}_{pmac}]^{\frac{1}{2}} \qquad (5)$$

Due to the application of small stresses, a dislocation segment begins to bend until an unstable position (Fig. 5) and a dislocation loop is created. In the unstable situation an area A is very quickly swept by the dislocation and consequently measurable AE occurs. These dislocations are called type 1 dislocations. During the unstable motion, a plastic strain equal to ε_{p1} is created. When the applied stress increases, the previously created loop can move in the cell (Fig. 5), with a low speed which does not give a measurable AE in the frequency range studied. These last dislocations are called type 2 dislocations. This motion leads to a plastic deformation equal to ε_{p2}.

At the beginning of the plastic deformation, few dislocations are inside the cells. So dislocations in unstable positions, which are first emitted from cell walls, are responsible for the major part of the plastic deformation: in this case $\dot{\varepsilon}_{p1}$ is near $\dot{\varepsilon}_{pmac}$. Then, as the plastic deformation becomes more important, $\dot{\varepsilon}_{p2}$ component increases because dislocations are more numerous inside cells. They move slowly and therefore give a negligible AE signal. So the total AE decreases according to relation (5), and then leads to a stationary level. In this stage, the fraction of dislocations of type 1 and type 2 remains constant and the $\sqrt{\dot{\varepsilon}_{pmac}^{0.7}}$ law is also satisfied.

On the contrary, at high strain amplitude, the number of moving dislocations per unit time increases. Simultaneously the distance between two glide planes decreases as a result of the increase of the moving dislocation number. So, the annihilation probability of dislocations increases, especially for type 2 dislocations. The previous equilibrium is broken and the AE is increased as a result of the increase of the number of type 1 dislocations, necessary to assume the plastic deformation, and the law $\dot{\varepsilon}_{pmac}$ is then observed.

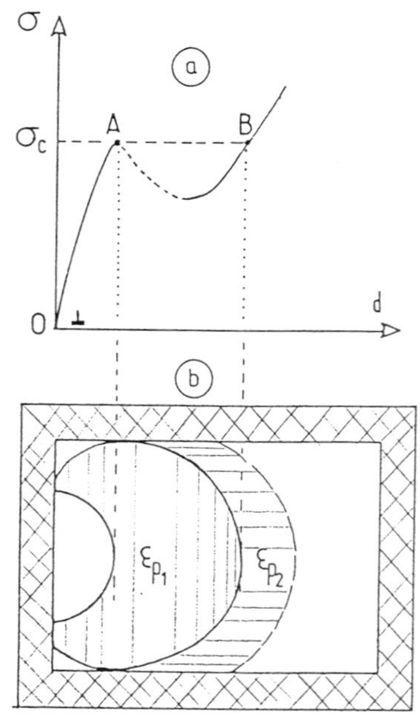

Fig. 5: Schematic description of unstable movement of dislocation loops. (a) applied stress on the dislocation versus dislocation displacement (b) movement of dislocations inside the cell.

This assertion can be verified in the case of a very special mechanical loading. Effectively, it has been shown by Hamel[15], using very fine stiffness measurements, that the dislocation annihilation occurs at the end of unloading, near the zero stress state. Fig. 6 shows the AE results obtained during such experiments.

During the fatigue test, the cyclic tests are always stopped at the B1 point. The B1 point position belongs to the AE stationnary stage where type 1 and type 2 dislocations are present. Just after the stop, two situations are imaginable:

i) the fatigue test can be restarted from point B1. During the stop under fatigue stress (point B1) no dislocation annihilation occurs and AE again reaches the previous maximum level obtained before the interruption.

ii) from point B1, the specimen is unloaded to zero stess, then reloaded to point B1 and cycled so that plastic deformation can be produced when unloading (near zero stress), and part of the type 2 dislocations are annihilated. The reloading then leads to a maximum AE peak larger than the one obtained during type i) interruption, due to the fact that plastic deformation must first be assumed by type 1 dislocations.

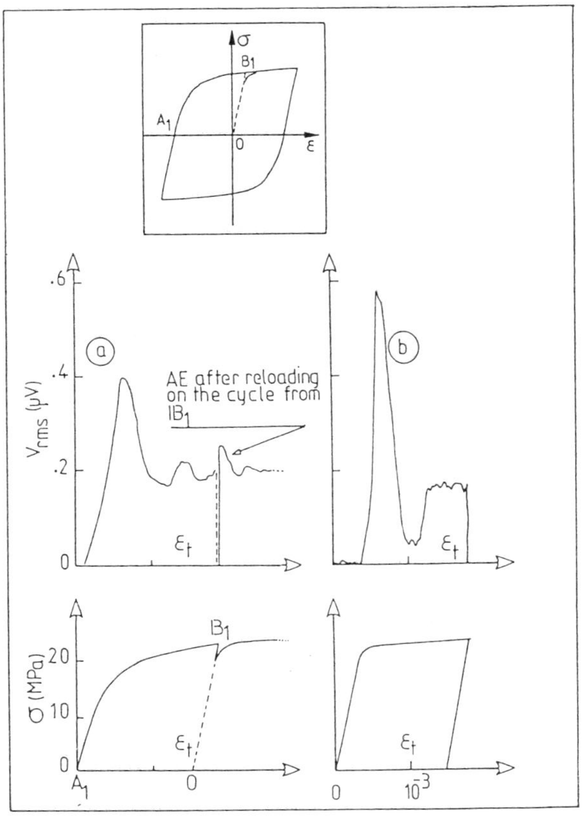

Fig. 6: AE rms voltage and fatigue stress versus strain. In case (a) the experiment is stopped at point B1 (see upper hysteresis loop), it is then restarted from the same point without unloading. In case (b) after stopping at point B1, the stress (and strain) are reduced to zero from point B1, prior to reloading.

The AE observed during cyclic hardening (Fig. 2) is a very complex phenomenon. Slimani[16] has shown in his thesis that the AE development can be understood in terms of opposite effects of two mechanisms: for the first few cycles, an increase of the number of dislocation sources is observed and a decrease of the mean swept area A as the cycle number increased.

CONCLUSION

In order to have an experimental verification of theoretical models, the study of AE during cyclic deformation of an fcc metal exhibits a great advantage as compared to monotonic deformation: Effectively, in the saturated stage, AE signals are reproducible from one given cycle to another. Experimental fatigue parameters can be modified on the same sample. On the other hand, the microstructure in the saturated state is clearly defined. Under these conditions, a good investigation of the AE behaviour during cyclic deformation of pure aluminum can be carried out. In particular, in the frequency range studied, it has been suggested that only unstable dislocations are AE sources. Dislocations which move slowly do not lead to measurable AE. In fact, these results do not contradict previous models. Effectively, at low dislocation speed, all models provide a weak AE signal in the low frequency range.

Our results however put into question the classical explanations for AE observed in monotonic deformation and demonstrate the presence of AE active and pasive dislocations within the cell walls.

REFERENCES

1. C.B. Scruby, H.N.G. Wadley and J.J. Hill, "Dynamic elastic displacement at the surface of an elastic half-space due to defect sources", J. Phys. D: Appl. Phys., 16, 1983, 1069-1083.
2. N. Kiesewetter, P. Schiller, "Acoustic emission from moving dislocation in Aluminum", Phys., stat. sol. (a) 38, 569-576 (1976).
3. N.N. Hsu, Y.A. Simmons, S.C. Hardy, "An approach to acoustic emission signal analysis theory and experiment", Materials Evaluation, October 1977, 100-106.
4. D. Rouby, P. Fleischmann, C. Duvergier, "Un modèle de sources d'émission acoustique pour l'analyse de l'émission continue et de l'émission par salves", Phil. Mag. A, 1983, 47, 671-687 et 689-705.
5. H. Hatano, H. Tanaka, R. Horiuchi and N. Niwa, "Stress wave emission during plastic deformation in pure aluminum", J. Jap. Inst. Met., 1975, 39, 7, 675-679.
6. M.A. Hamstad and A.K. Mukherjee, "The dependence of acoustic emission on strain and strain rate for a dispersion strengthened aluminum alloy", Proceeding 4th Int. Conf. on the Strength of Metals and Alloys, Nancy, 1976, 574-578.
7. D.R. James and S.H. Carpenter, "Relation between acoustic emission and dislocation kinetics in crystalline solids", J. of Phys., 42, 12, 1971, 4685-4697.
8. H.N.G. Wadley and R. Mehrabian, "Acoustic emission for materials processing: a review", Mater. Sc., and Eng., 65, 1984, 245-263.
9. P. Fleischmann, D. Rouby, F. Lakestani and J.C. Baboux, "Spectral and energetical analysis of a moving ultrasonic source. Application to acoustic emission during plastic deformation of aluminum", Mat. Sc. and Eng., 29, 1977, 205-210.
10. N.G. Sankar, Y.R. Frederick and D.K. Felbek, "Acoustic emission from metals during unloading and its relation to the Bauschinger effect", Met. Trans. 1 (1970) 2979-2985.

11. R. Rahouadj "Instabilité de l'état écroui et émission acoustique du cuivre pur mono et polycristallin" Thèse de Doctorat, 1985, Université de Technologie de Compiègne, France.
12. T. Kishi, Y. Obata, H. Tanaka and Y. Sakakibara, "Acoustic emission peak under cyclic deformation", Journal of Japan Institute of Metals, 5, 40, 1976, 492-498.
13. G. Guichon, J. Chicois, C. Esnouf and R. Fougères, "Study of dislocation structure in a polycrystalline pure aluminum strained under fatigue conditions", Fatigue 84, Chameleon Press, Vol. 1, 1984, 31-37.
14. A. Vincent, A. Hamel, J. Chicois and R. Fougères "Dislocations mobility in 5N aluminum during push-pull cycling studied by ultrasonic attenuation measurements. J. de Phys. Colloque C10, supplément au No. 12, Tome 46, (1985), C10-321.
15. A. Hamel, A. Vincent, J. Chicois, C. Mai and R. Fougères, "On the fatigue behaviour of pure polycrystalline aluminum studied by unloading stiffness evolution", Fatigue 84, Chameleon Press, vol. 1, 1984, 72-82.
16. A. Slimani, "Étude par émission acoustique de la déformation cyclique de polycristaux d'aluminum 5N sollicités en traction compression" Thèse de Doctorat, 1986, Institut National des Sciences Appliquées de Lyon, France.

APPLICATION OF THE DTA TO THE ANALYSIS OF THE RESIDUAL AUSTENITE TRANSFORMATION DURING TEMPERING

Ignacy Wierszyllowski, and Jerzy Rys

Institute of Metallurgy Poznan Technical University
Pl. Sklodowskiej-Curie 5, Poznan Poland

INTRODUCTION

In quenched steel, the structure usually consists of two major components: hard, brittle and ferromagnetic martensite, and soft, ductile and paramagnetic residual austenite (RA). A proper ratio of these constituants gives required properties for machinery parts manufactured from the steel. This is a reason why the amount of residual austenite (RA) has to be known. There are several methods to determine the amount of RA, such as X-rays, quantitative metallography, magnetic analysis, dilatometry, and differential thermal analysis (DTA). The most popular are X-ray and magnetic methods, the DTA is seldom used, despite its certain advantages. Some expected advantages of DTA are: the possibility of a complex analysis of the tempering process simultaneously with a determination of the amount of RA. Moreover, it is also possible to evaluate the three following values simultaneously using the DTA curve.

- the amount of RA
- the heat effects produced during the RA transformation
- the activation energy of the transformation

The goal of this study was to carry out experimental verifications of the three last possibilities.

EXPERIMENTAL PROCEDURE

The samples for the DTA analysis were taken from 100Cr6 steel spheroidize-annealed. The samples were machined, and after austenitizing for 40 min in a furnace with a protective atmosphere, were quenched from 800, 850, 900, 950, 1000 and 1050°C in oil. The DTA analysis was performed with use of two standard apparatuses a Q DERIVATOGRAPH and SETARAM-GDTD 16. Both were calibrated according to the method described elsewhere[1,2]. As an example, the Q DERIVATOGRAPH calibration curve is shown in fig. 1. Determinations of the activation energy of RA transformation were performed by means of the Piloyan[1,2] and Kissinger[1,2] methods. According to the Piloyan method, the activation energy (E_{ap}) of a transformation is expressed by:

$$\frac{\ln \Delta T - C}{1/T} = - \frac{E_{ap}}{R} \qquad (1)$$

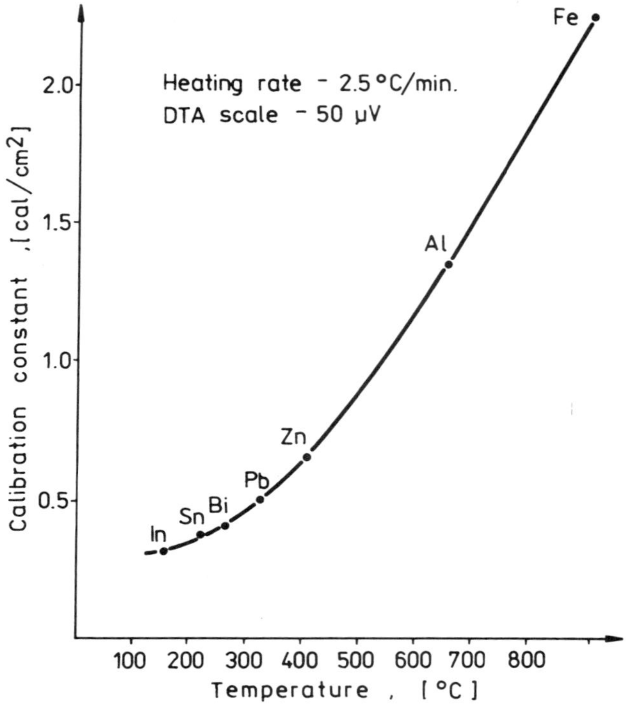

Fig. 1 The Q DERIVATOGRAPH calibration curve.

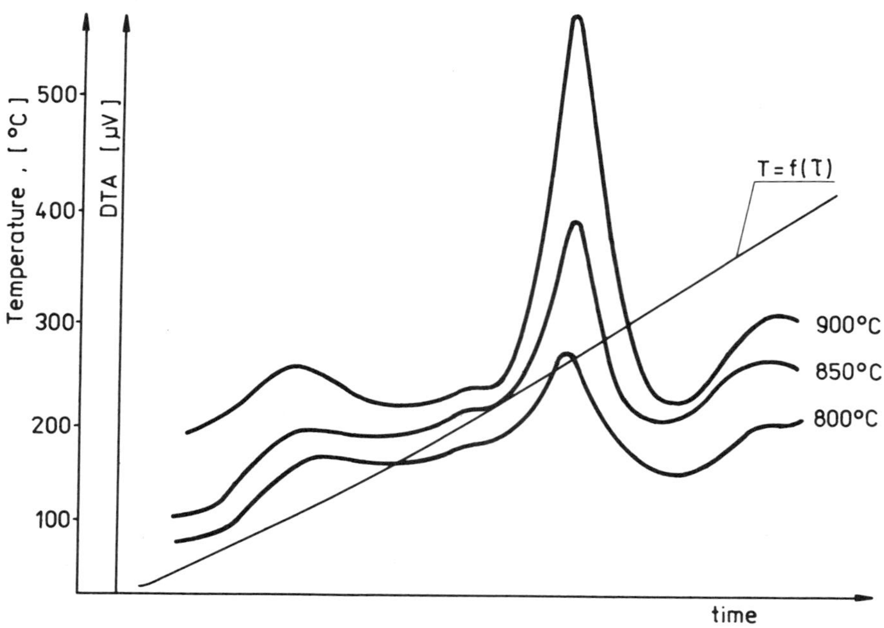

Fig. 2: Examples of DTA diagrams for isochronal tempering of 100 Cr6 steel (from Q DERIVATOGRAPH).

where ΔT is the difference between the base line and the DTA line for a range of 5-80% of the transformation, T, is temperature, R and C are gas constant and constant, respectively. According to the Kissinger method[1,2] the activation energy (E_{ak}) of a transformation is expressed by:

$$\frac{d \ln \frac{\emptyset}{Tm^2}}{d(1/Tm)} = - \frac{E_{ak}}{R} \qquad (2)$$

where $\emptyset$ is a constant heating rate, Tm, the temperature of the DTA peak (the temperature of the maximum heat effect), R, gas constant.

The heating rate applied to experiments in the case of the Kissinger method varied was from 6°C/min to 19°C/min, in case of the Piloyan method the heating rate held constant at 4°C/min. After an analysis of each equation one can see that the activation energy can be calculated from one DTA diagram according to the Piloyan method and from at least three DTA diagrams, obtained for different heating rates, according to the Kissinger method.

The heat produced (ΔH) by the RA transformation can be calculated using a calibration curve[1,2].

$$\Delta H = \frac{K \cdot F}{m} \qquad (3)$$

$$\Delta H = c \cdot X \qquad (4)$$

Where: K - calibration constant for the temperature of the beginning of the heat effect
F - area under the DTA line and base line
c - specific heat of the transformation
X - amount of the transformed RA
m - sample mass

From equations (3) and (4) X can be determined.

$$X = \frac{K \cdot F}{m \cdot c} \qquad (5)$$

Because reliable literature data of the specific heat of the RA transformation were not known by the authors, it appeared necessary to calculate it. The specific heat of the RA transformation was calculated as a medium value from six DTA specimens quenched from different temperatures. The amount of RA in each specimen was determined by means of the X-ray method[3]. The determined medium value of the RA transformation specific heat is 143.0 $\pm$ 18.2 kJ/g and is about two times bigger than that for the high temperature austenite transformation[4].

RESULTS AND DISCUSSION

The examples of the DTA diagrams for isochronal tempering of 100Cr6 steel are shown in fig. 2 for different quenching temperatures. The results of the RA amounts and values of activation energy are collected in table 1. From fig. 2 one can learn that the RA peak is increasing with an increase of the quenching temperature and respectively the RA

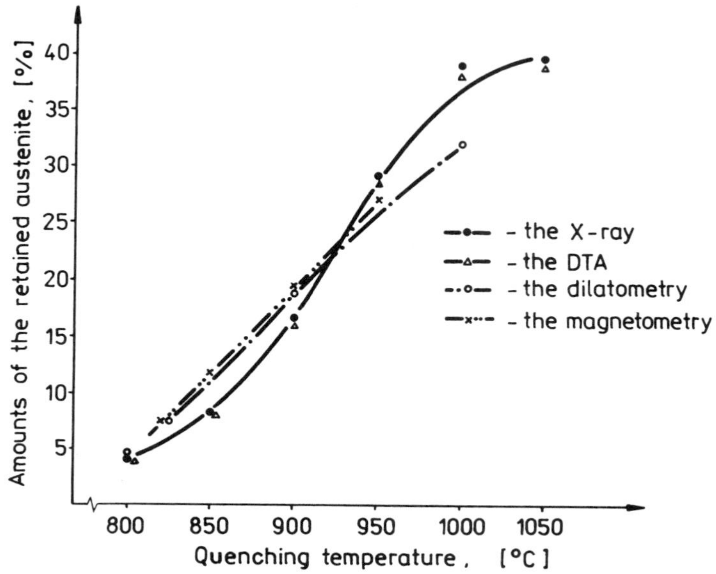

Fig. 3: Comparison between the amounts of RA determined by means of DTA and other methods for the same steel (6).

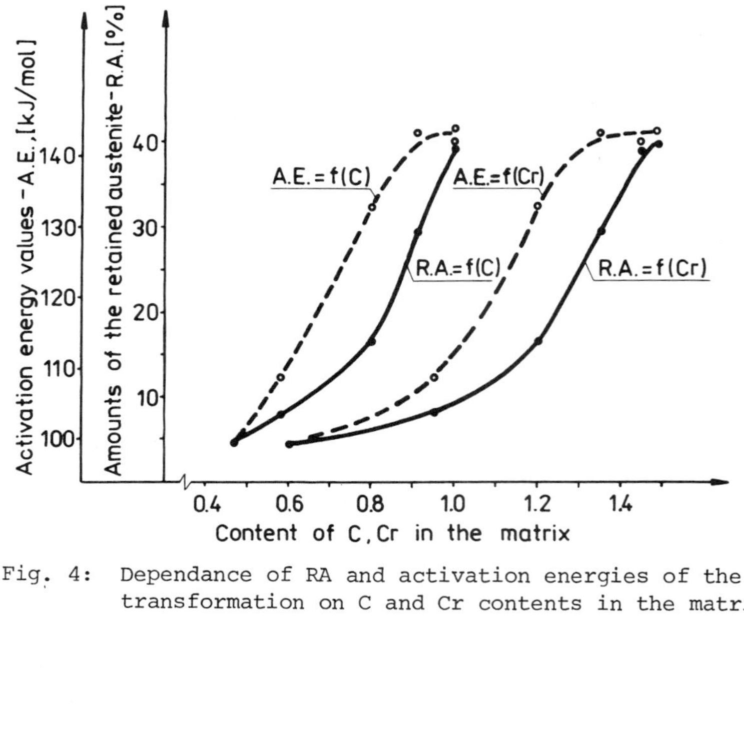

Fig. 4: Dependance of RA and activation energies of the RA transformation on C and Cr contents in the matrix.

TABLE 1: Results of RA amounts determination and values of activation energies

Austenitizing temperature °C	Amount of C and Cr in the matrix[4] (%)		Mass of the specimen m (g)	Calibration constant K J/cm^2	Area under austenite peak F (cm^2)	Specific heat of the RA transformation (J^c/g)	Amount of RA (%)	Activation energy of the RA transformation (kJ/mol) according to	
	C	Cr						Piloyan	Kissinger
800	0.47	0.6	3.233	1.634	12.06	143.03	4.2	99.4	91.4
850	0.58	0.95	3.126	1.634	21.90	143.03	8.0	108.7	98.9
900	0.8	1.20	3.088	1.655	44.10	143.03	16.5	133.3	120.3
950	0.91	1.35	3.069	1.655	77.83	143.03	29.3	143.8	129.4
1000	1.01	1.45	3.054	1.655	102.27	143.03	38.7	141.9	129.2
1050	1.01	1.49	3.092	1.655	104.72	143.03	39.2	143.5	132.0

amount has to increase. The same can be noticed from table 1. The RA amount increase with temperature is steep in the quenching temperature range up to 950°C and much less steep in the range higher than 950°C. The reason is the C and Cr content in the matrix of the quenched steel (column 2 and 3 of table 1), which also increases more steeply in the temperature region up to 950°C and very slowly in the higher temperatures. A similar phenomenon was found in another paper[6]. The comparison of the RA amounts determined by means of the DTA and other methods for the same steel[6] is shown in fig. 3. The results of the DTA, X-ray, magnetometry and dilatometry are almost the same (see fig. 3) which supports our conviction that the results of the RA amount determination by means of DTA are true and reliable. Activation energies of the RA transformation obtained by the Kissinger method (see table 1) are about 10% lower than the ones obtained by the Piloyan method. The discrepancy might be explained in two ways. First, the RA transformation is not, or very little, dependent on the heating rate[4], second, the specimens used for the Kissinger method were too big for the heating rates applied to experiments. This could have caused that the real difference between the temperature of the maximum transformation rate, Tm, and the temperature of the DTA peak increased with the increase of the heating rate, because of heat resistance of the big specimen. In such circumstances, the Kissinger method could not be applied. The results of the Piloyan method appear to be correct (see table 1) because the RA transformation activation energies are similar to those obtained by other methods[8,9] and a majority of the results indicate that the transformation is controlled by diffusion of carbon in austenite[9]. Lower activation energies obtained for specimens quenched from 800 and 850°C indicate a diffusion of carbon in ferrite[9] as a reason of the transformation. This could be the case as evidenced by the chemical composition of the matrix after quenching (see table 1). The amounts of C and Cr are relatively low and RA can transform into martensite[7] which is immediately tempered because of a transformation temperature. Activation energies of such a process[8] are of the order of 100 kJ/mol and equal to the activation energy for diffusion of C into alpha iron[9]. The results of the RA amounts calculation and the RA transformation activation energies measurements according to Piloyan method are summarized in fig. 4. Both results depend on C and Cr contents in the matrix after quenching. The RA amounts depend to a much larger extent on even a small difference in a C and Cr content in the matrix than activation energies (taking into consideration the activation energy (AE) scale).

CONCLUSION

The DTA method can be successfully applied for the RA amount determination and a simultaneous calculation of the activation energy of the RA transformation according to the Piloyan method. Heat effect produced during the transformation can be also determined as well as previous values. Activation energy values and RA amounts are dependent on a C and Cr content in the matrix after quenching.

REFERENCES

1. W. Wendlandt "Thermal Methods of Analysis" A. Wiley, Interscience Publication, John Wiley and Sons, New York. London-Sydney. Toronto. 1974.
2. D. Schultze "Differential thermoanalyse" VEB Deutscher Verlag der Wissenschaften. Berlin 1971.
3. J. Karp and I. Pofelska-Filip "Rentgenowska ilosciowa analiza fazowa RIAF austenitu w stalach" Hutnik Nr 6, 1979.

4 B. G. Lifszyc "Fiziczeskije svojstva metallov i splavov" Maszgiz. Moskva 1956.
5. Z. Glowacki, Cz. Niescierowicz and I. Wierszyllowski "Wplyw temperatury i czasu austenityzowania na strukture i twardosc stali LH15SG" Zeszyty Naukowe Politechniki Poznan, skiej Nr 13/87 - Mechanika 1971.
6. I. Wierszyllowski, J. Jakubowski, J. Rys, P. Malecki, Z. Szczepanski, D. Miklaszewska and B. Pietraszewska "Zastosowanie wybranych method fizycznych do badania procesu odpuszczania stali lozyskowych" Prace z Inzynierii Materialowej, 1980. Poznan, 1982.
7. P.G. Shewmon "Transformations in Metals" McGraw-Hill Book Company, New York, St. Louis, San Francisco, London, Sydney, Toronto, Mexico, Panama, 1969.
8. J.W. Christian "The Theory of Transformations in Metals and Alloys" Pergamon Press, Oxford, London, Edinburgh, New York, Paris, Frankfurt, 1965.
9. M.A. Krishtal "Diffusion Processes in Iron Alloys", Israel Program for Scientific Translation, Jerusalem, 1966.

THE DILATOMETRIC AND THERMOMAGNETIC ANALYSIS OF TEMPERING PROCESS OF QUENCHED Fe-N ALLOYS

Ignacy Wierszyllowski, Leszek Maldzinski, and Marek Hrebeniak

Institute of Metallurgy Poznan Technical University
Pl. Sklodowskiej-Curie 5, Poznan Poland

INTRODUCTION

The majority of published studies concerning tempering of quenched Fe-N alloys were performed by means of the TEM and X-ray methods[1-6], In addition, microhardness[5-7] and resistivity[5] were also studied. Incomplete thermomagnetic investigations were presented in two papers[8] and one can find fragmentary dilatometric investigations in a few other papers[7,8]. Considerations concerning thermal effects accompanying the tempering of quenched Fe-N alloys were presented in one paper[7]. On the basis of previous studies, the tempering process of quenched Fe-N alloys can be described in the following way:

- In the first stage of tempering till about 200°C the tetragonal ordered α" ($Fe_{16}N_2$) phase precipitates[2,5,6]. Precipitation of α" can appear as early as after one week tempering at room temperature[7,9]. Simultaneously, the content of nitrogen in the matrix decreases. Most authors call the tempered matrix ferrite[2,5,6]. The precipitation is accompanied by a significant thermal effect[7].

- In the next stage of tempering, which begins at temperatures slightly lower than 200°C, α" progressively disappears and is replaced progressively by γ' (Fe_4N)[2,5,6]. During this stage of tempering the hardness suddenly decreases[7]. In dilatometric diagrams significant contraction is visible[7,8], thermal effects are small[7], and magnetic effects are difficult to interpret unambiguously[8]. In the case of the presence of residual austenite (RA) after quenching, its transformation begins at a tempering temperature of the order of 230°C [2,8]. The RA transformation is connected with an increase of magnetization[8] and with liberation of heat[7]. The RA transformation may also cause a certain increase of hardness[7]. It is believed that the nitrogen content in ferrite attains the equilibrium level after the γ' (Fe_4N) precipitation.

From this short review of literature data, one can learn that the magnetic effects are not recognized enough, as far as the dilatometric effects are concerned, the lack of any effect in the temperature range of the first transformation seems to be rather strange. The aim of this study is to answer the questions mentioned above.

EXPERIMENTAL

Specimens were obtained by gas nitriding in the austenite region in a mixture of ammonia and ammonia dissociated before introducing it into the furnace. After nitriding specimens were immediately quenched into brine and afterwards cooled in liquid nitrogen (LN). Specimens were kept in LN until the moment of the magnetic or dilatometric study. Dilatometric studies were performed with a LK02 ADAMEL-LHOMARGY dilatometer on thin tube - like specimens. The thickness of the wall varied from 0.2 to 0.25 mm. The applied heating rate was 0.03°C/sec, the magnification of the elongation axis was 4000 X.

Thermomagnetic studies were performed by means of the Faraday balance method(10), using the SETARAM GDTD 16 thermobalance and the RADIOPAN electromagnet. Small specimens of mass m_o were introduced into the furnace located between the electromagnet pole pieces. Saturation magnetization of induction of 1.2 T was used, the applied heating rate was 0.01°C/sec. The magnetic force, F, and dF/dt, were registered as a function of time, t, and temperature, T. The quantitative study was carried out on the basis of the balance of magnetic forces per unit mass, M_T.

$$M_T = F / m_o \qquad (1)$$

The activation energies, AE, of the first and second transformations were calculated from dilatometric and thermomagnetic diagrams of isothermal tempering according to the method described elsewhere[11]. The kinetics of each isothermal diagram can be expressed by the Avrami equation and AE was calculated as a function of the fraction transformed in the temperature region of the first and second transformation.

RESULTS

Examples of dilatometric and thermomagnetic diagrams are shown in fig. 1 - 3. Two contractions are visible in the dilatometric diagram (see Fig. 1). A small contraction appeared in the temperature range 50-100°C, and a significant contraction in the temperature range 210-270°C. A small and significant decrease of the magnetic force appears in the temperature region of 90-140°C, a significant decrease of the magnetic force in the temperature region of 190-280°C. The Curie temperature, Θ, of about 480°C appeared in all diagrams, and in some of them, from specimens heated to the temperature higher than 600°C an A_1 temperature appears (the magnetic force decreased to the 0 level). A quantitative analysis of the dilatometric results was performed according to the method presented by Roberts et al.[12] employing the following equations:

$$\frac{\Delta l}{l_0} = \frac{1 \, \Delta V}{3V_0} \qquad (2)$$

$$v_i = \frac{w_i}{1.6602 \, z_i/55.8 + 14.007/n_i} \qquad (3)$$

Δl - elongation change
l_0 - length of the specimen at the beginning of the experiment
ΔV - volume change
V_0 - volume at the beginning of the experiment
v_i - specific volume of the i-phase
w_i - volume of the unit cell
z_i - number of Fe atoms in the unit cell
n_i - number of Fe atoms on one N atom in the unit cell

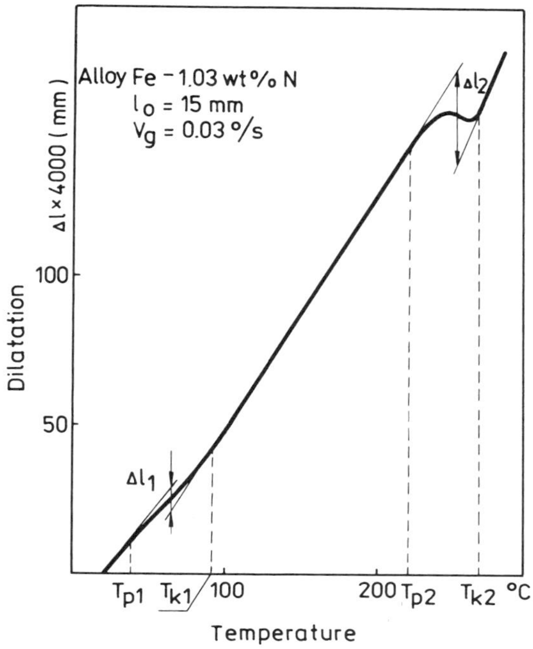

Fig. 1: The dilatometric diagram of isochronal tempering of Fe-1.03 wt. % N alloy.

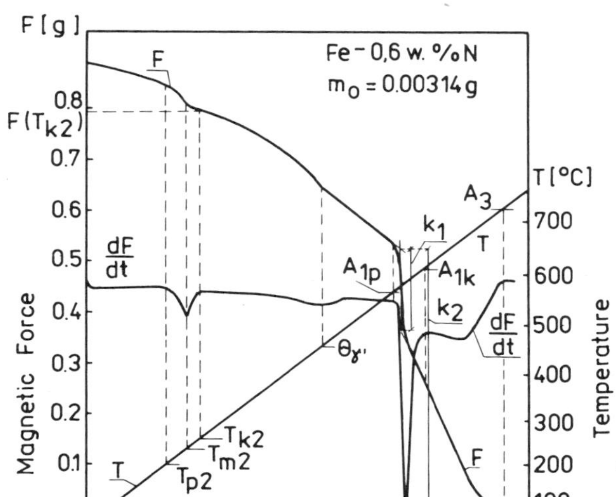

Fig. 2: The thermomagnetic diagram of isochronal tempering of Fe-0.6 wt. % N alloy.

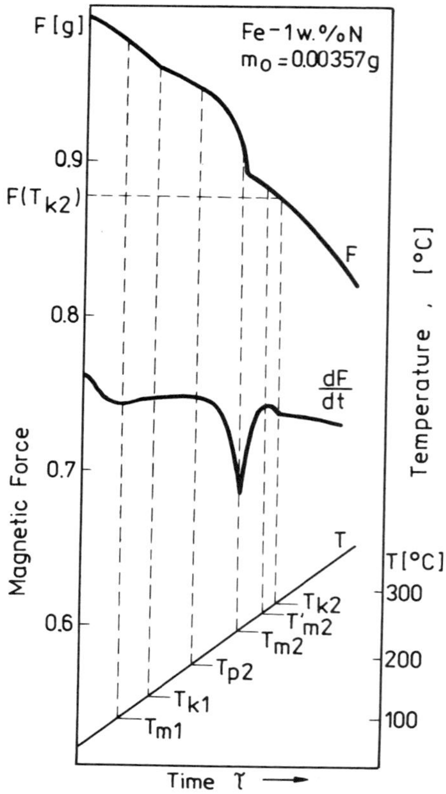

Fig. 3: The thermomagnetic diagram of isochronal tempering of Fe-1.0 wt. % N alloy

Details concerning the calculation are presented in another work[10]. Results of the quantitative dilatometric analysis are shown in table 1. In order to execute quantitative magnetic analysis two standards were produced, one of γ' (Fe_4N), the second of ferrite with nitrogen. On the basis of a thermomagnetic analysis it was easy to determine M_T/γ and $M_{T/F}$ for γ' and ferrite respectively. The results of the quantitative magnetic analysis are shown in table 2. The AE changes during progress of the transformation are shown in fig. 4. The AE of the first stage of tempering is lower than that of the second stage and both increase with the progress of the transformation.

DISCUSSION OF RESULTS

The obtained results could be achieved only because the specimens were kept in liquid nitrogen (LN) until the beginning of the experiment (dilatometric or thermomagnetic). This is why small dilatometric and thermomagnetic effects accompanying the α" ($Fe_{16}N_2$) precipitation appears at relatively low temperatures of tempering (40-140°C), at which it was difficult to identify this by means of diffraction methods[5-7]. The temperatures of the α" ($Fe_{16}N_2$) precipitations (100-140°C) are closer to the previous diffraction results in case of these magnetic investigations than the dilatometric ones, however, the sensitivity of the magnetic methods is lower than of dilatometry considering the α" ($F_{16}N_2$) precipitation. Temperature regions determined by means of both methods are similar when the γ' phase (Fe_4N) precipitates. The region is from 190°C to 270°C and is in agreement with other studies[2,5,6,7]. Proper results of the dilatometric quantitative analysis have been

TABLE 1 RESULTS OF THE QUANTITATIVE DILATOMETRIC ANALYSIS

Specimen	wt % N in the specimen	v (cm^3/g)					l_0 (mm)	l_1	l_2	$x\alpha_2$	$x\gamma'$	$x\gamma$	$x\alpha_1$	$x\alpha''$	wt. % N in the solid solution	
		α	α''	γ'	α'	γ		X 4000 (mm)							α_1	α_2
1	0.6	0.127	0.135	0.139	0.128	-	16.0	-7	-31	0.9	0.1	-	0.821	0.179	0.08	0.01
2	0.6	0.127	0.135	0.139	0.128	-	16.0	-7	-28	0.895	0.102	-	0.821	0.179	0.08	0.01
3	1.03	0.127	0.135	0.139	0.129	0.125	15.0	-8	-30	0.827	0.173	0.028	0.672	0.300	0.15	0.01
4	1.44	0.127	0.135	0.139	0.129	0.125	13.0	-9	-40	0.757	0.243	0.039	0.540	0.421	0.22	0.01
5	1.58	0.127	0.135	0.139	0.131	0.126	12.8	-9	-49	0.733	0.267	0.043	0.500	0.457	0.28	0.01

α - ferrite, α'' - $Fe_{16}N_2$, γ' - Fe_4N, α' - martensite, γ - austenite

α_2 - ferrite after the second stage of tempering,

α_1 - regular martensite after the first stage of tempering,

x_i - fraction of the i-phase

TABLE 2 RESULTS OF THE QUANTITATIVE MAGNETIC ANALYSIS

Specimen No	wt.% N in the specimen	m_o mg	T_{K2} °C	$M\gamma$ g/g	$M\alpha$ g/g	χ_γ	χ_{α_2}	Calculated wt. % N in the specimen
1	0.6	3.14	265	190.4	258.6	0.102	0.898	0.61
2	0.6	3.21	270	188.7	257.4	0.100	0.900	0.60
3	0.6	3.97	276	187.1	256.3	0.095	0.905	0.57
4	0.6	4.18	270	188.7	257.4	0.095	0.905	0.57
5	1.0	3.06	270	188.7	257.4	0.170	0.830	1.01
6	1.0	3.96	274	187.1	257.0	0.175	0.825	1.04
7	1.0	3.57	282	184.4	255.7	0.166	0.834	0.99
8	1.0	3.45	282	184.4	257.7	0.171	0.829	1.01

m_o - mass of the specimen
T_{K2} - temperature of end of the second transformation
$M\gamma$; $M\alpha$ - specific magnetic forces γ' and ferrite respectively
x_i - fraction of the i-phase

obtained only assuming that after α'' ($Fe_{16}N_2$) precipitation the unit cell of martensite is regular and that after the γ'' (Fe_4N) precipitation the nitrogen content in ferrite achieved the equilibrium level (0.01 wt. %N). It is noticed that, according to calculations, the N content in the matrix after the α'' ($Fe_{16}N_2$) precipitation increases with an increase of a total amount of nitrogen in the specimen (see table 1).

Similar results, as in table 1, for the specimen containing 0.6 wt. % N/0.08 wt. % N have been obtained in another study[13], where the amount of nitrogen in the matrix after the α'' precipitation was of the order of 0.09 wt. % N. When the amount of nitrogen in the specimens is increasing, the fraction of the precipitated α'' and γ' also increases. Fractions of the precipitated γ' (Fe_4N) determined according to both methods are almost the same (see table 1 and 2). The nitrogen contents in the specimens calculated according to the results of the dilatometric and magnetic analyses are very similar to the ones obtained by means of the other methods (microanalysis, weight). The retained austenite (RA) amounts obtained on the basis of the dilatometric and thermomagnetic analysis[10] are a little bit lower than those obtained by Bell and Brough[5], but taking into account the presence of RA is necessary in order to obtain proper results during the quantitative analysis. The AE values of the first stage of the transformation during tempering are similar to those of diffusion of nitrogen in ferrite[14]. The AE values of the second stage of transformation during tempering are too small for diffusion of Fe in ferrite and too large for diffusion of N in ferrite[14]. These values can indicate coupled diffusion of Fe and N in alpha iron[15]. A similar phenomenon was found during high temperature tempering of Fe-C martensite[16].

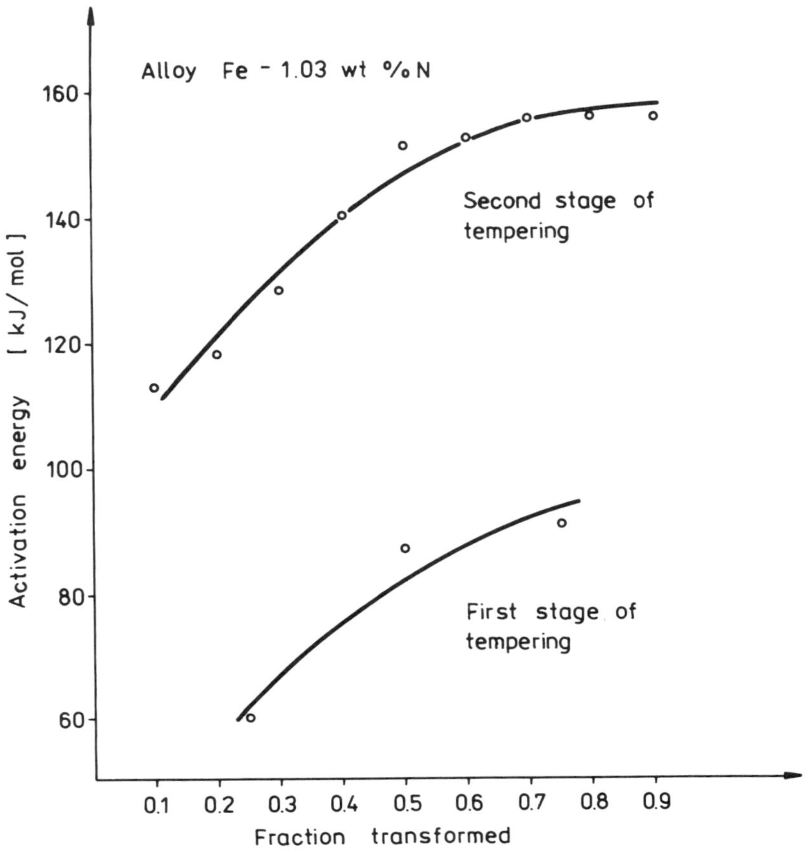

Fig. 4: Changes of activation energy during the first and second stages of tempering of Fe-1.03 wt. % N alloy.

The results of the dilatometric and thermomagnetic analyses are interesting because of the possibility to quantitatively determine the amounts of precipitated phases, the confirmation of contraction during the γ' precipitation, observation of a decrease of the magnetic force during the γ' precipitation and a slight decrease of magnetic force, and a slight contraction during the α'' precipitation.

REFERENCES

1. K.H. Jack "The iron-nitrogen system: the preparation and the crystal structures of nitrogen-austenite and nitrogen-martensite" Proc. Roy. Soc. A/208, 1951.
2. K.H. Jack "The occurance and the crystal structure of "-iron nitride; a new type of interstitial alloy formed during the tempering of nitrogen martensite", Proc. Roy. Soc. A/208 1951, p. 216-224.
3. M.G. A. Biswas and I. Codd "An electron transmission study of iron-nitrogen martensite" Journal of The Iron and Steel Institute, 1968, p. 494-497.
4. T.Bell and W.S. Owen "Martensite in iron-nitrogen alloys" Journal of The Iron and Steel Institute, 1967, p. 428-434.
5. T.Bell and D. Brough "The Tempering of Iron-Nitrogen Massive Martensite" Metal Science Journal, 1970, Vol. 4, p. 171-177.

6. R.D. Garwood and G. Thomas "The Tempering of Martensite in an Fe-1.5 pct. N. Alloy" Metallurgical Transactions.
7. E.I. Mittemeijer, M. Van Rooyen, I. Wierszyllowski, H.C.F. Rozendaal and P.F. Colijn "Tempering of Iron-Nitrogen Martensite" Zeitschrift für Metallkunde Bd. 74, 1983, H7, p. 473-483.
8. M.W. Belous, W.T. Czerepin and M.A. Wasiliew "Prevraszczenia pri otpuskie stali" Metallurgija 1973, Moskva, p. 91-100.
9. L. Maldzinski, I. Wierszyllowski and J. Putniewicz "Production of homogenous martensitic Fe-N alloys" (in Polish) Inzynieria Materialowa, No. 3 (26), 1985, p. 82-86.
10. I. Wierszyllowski, L. Maldzinski and M. Hrebeniak "Efekty dylatometryczne i magnetyczne podczas odpuszczania stopow Fe-N". Materialy konferencyjne "Wegliki, Azotki, Borki" Poznan-Kolobrzeg 1984, p. 68-79.
11. I. Wierszyllowski and J. Jakubowski "The Influence of Transformation Progress on Activation Energy Changes During Low-Temperature Tempering of Quenched Steel" Scripta Metallurgica 20, 1986, p. 49-54.
12. C.S. Roberts, B.L. Averbach and M. Cohen "The Mechanism and Kinetics of the First Stage of Tempering" Transactions American Society for Metals, Vol. 45, 1953, p. 576.
13. B.A. Fuller and R.D. Garwood "Initial Tempering of Iron-Manganese-Nitrogen Mattensite" Metal Science 1975, Vol. 9, p. 213-216.
14. M.A. Krishtal "Diffusion Processes in Iron Alloys" (in Russian), Metallurgizdat. 1963, Moskva.
15. H. Schmalzried "Solid State Reactions" Verlag Chemie, GmbH, Weinheim/Berystr. 1974, p. 162.
16. K.M. Vedula and R.W. Heckel "Spheroidization of Binaary Fe-C Alloys over a Range of Temperatures" Metallurgical Transactions 1970, p. 9-18.

ANALYSIS OF ELEVATED TEMPERATURE PROPERTY OF HEAT RESISTANT MATERIALS BY INTERNAL FRICTION METHOD

Jumpei Shioiri*, Osamu Furuta* and Katsuhiko Satoh**

* College of Engineering
Hosei University
Koganei-shi, Tokyo, Japan

** Faculty of Engineering
University of Tokyo
Bunkyo-ku, Tokyo, Japan

INTRODUCTION

Designs of advanced heat resistant materials are highly complicated and at high temperatures various strengthening and weakening mechanisms are operative simultaneously. However, usual characterization methods such as creep tests are not necessarily suitable for analyzing the contributions of the individual mechanisms. In this paper, as a nondestructive way of the above type of analysis, the internal friction method is proposed, and details of the technique devised for measurements at very high temperatures are presented together with results of applications to ODS (oxide dispersion-strengthened) NiCr alloy (MA-754) and sintered silicon nitrides.

The temperature-, frequency- and strain amplitude-dependencies of the internal friction can provide important information on the individual deformation mechanisms. For example, as was shown by Kê in his pioneering work,[1] viscous slip along grain boundaries, which is one of the dominant weakening mechanisms in polycrystalline materials at high temperatures, appears as a typical relaxation peak in the internal friction spectrum. Simultaneously, the inelastic behavior of the crystal grains appears as a background component of the internal friction, which is also often accompanied by relaxation peaks if the deformation mechanism of the grains has relaxation mechanisms. Further, the frequency dependency of the temperature at which the internal friction peak appears gives the activation energy of the rate process associated with the relaxation mechanism, which is often a powerful information for identifying the relaxation mechanism.

EXPERIMENTAL TECHNIQUE

Since, at high temperatures, the internal friction often becomes very high, the longitudinal and bending oscillation methods and the ultrasonic methods are not suitable because of the difficulty in exciting the specimen up to a necessary amplitude. In this work, therefore, the torsion pendulum method was used.

Figure 1 shows a standard inverted torsion pendulum type apparatus devised for measurements at high temperatures. By changing the moment of inertia of the pendulum but without changing the specimen, measurements over a frequency range about 0.5 to 40 Hz are possible. At present, the measure-

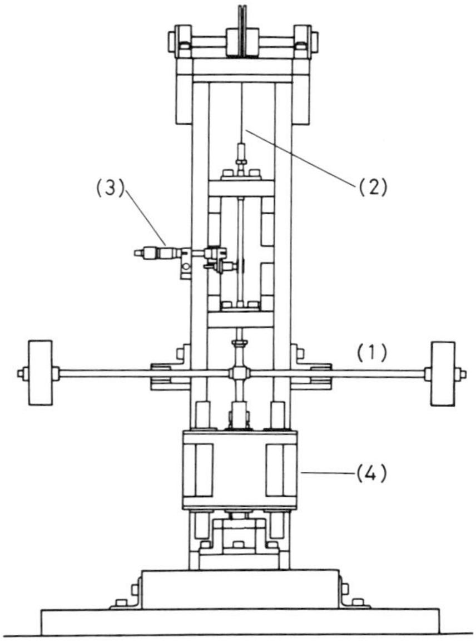

Fig. 1. Inverted torsion pendulum apparatus: (1) pendulum; (2) tension wire; (3) angular displacement sensor; (4) infrared furnace.

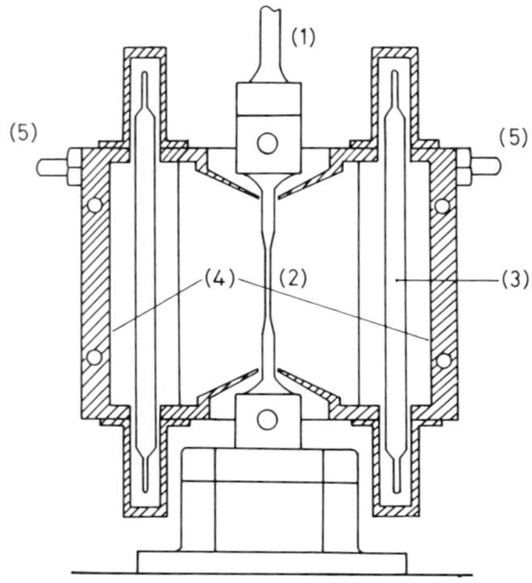

Fig. 2. Infrared furnace: (1) pendulum axis; (2) specimen; (3) infrared tube; (4)elliptic mirror; (5) cooling water pipe.

ments are made in air. For the purpose of mechanical stability, the axis of the torsion pendulum is supported by two miniature bearings. The frictions of the air and the bearings cause additional decrements. Those additional decrements are, however, of the order of 10^{-5} or less and usually very small compared with the decrement due to the internal friction of the specimen, especially at high temperatures.

For heating the specimen, the pendulum apparatus is equipped with an infrared furnace, which is composed of four infrared tubes and four elliptic mirrors. The tubes and mirrors are so arranged that the infrared rays from the tubes are focused along the specimen. The wall of the tube is made of quartz glass and the filament is of tungsten. The maximum electric power input of the furnace is 3.2 kW. The elliptic mirrors and endplates are gold plated in order to obtain good reflectivity, but the water cooling of the furnace wall is necessary. The inside of the furnace is shown in Fig. 2. For temperature control a digital controller is used, which is so designed to fit the quick response characteristic of the infrared furnace. The temperature fluctuation of the specimen is kept below ±1° K. The temperature sensor is a fine (0.2 mm in diameter) thermocouple attached to the surface of the specimen. The highest attainable temperature on the specimen is 1873° K.

As seen in Fig. 2, the grips of the specimen are located outside of the furnace. By this method, the difficulty in gripping at high temperatures can be avoided, but the temperature gradient along the specimen becomes a problem. The test section with a smaller cross-sectional area and of appropriate length is put at the central part of the specimen where the temperature gradient is small. The torsional deformation and accordingly the energ

upper: prototype specimen
lower: standard specimen for ceramics

standard specimen for metallic materials

Fig. 3. Specimens

dissipation due to the internal friction occur, practically speaking, only in this part. The length of the test section is determined taking into account the thermal conductivity of the specimen materials. The shapes of the specimens for metallic materials and ceramics are shown in Fig. 3.

The measurements of the decrement and elastic rigidity are made by the free oscillation method. The amplifier of the angular displacement signal of the pendulum has a 10-stepped attenuator. Attenuation per step is -6 dB. This attenuator is automatically switched by the control signal from the computer in response to the level of the amplitude signal. Thus, through a peak-to-peak holding circuit, the amplitude of the free damped oscillation is continuously digitized with a high accuracy over a wide range of the amplitude and stored in the memory. The dynamic range of this system is 6 (dB/step)×10 (step)=60 dB. The decrement is automatically calculated using the stored data as a function of the strain amplitude.

APPLICATION TO ODS ALLOY

The ODS (oxide dispersion-strengthened) superalloys are regarded as promising candidates for the next-generation heat resistant alloys for gas turbine engines, since the dispersed oxide particles are stable up to temperatures near the melting points of the matrix alloys.

The internal friction analysis was applied to MA-754 alloy which is one of the simplest ODS superalloys.[2] The chemical composition of this alloy is essentially Ni-20%Cr-0.6%Y_2O_3. In order to clarify the effect of the oxide dispersion, measurements were made also for Nimonic-75 alloy which has a very similar chemical composition to the matrix of MA-754 alloy but has no dispersed particles. Those two alloys tested were commercial products supplied by Inco Ltd.

In both alloys, the spectrum of the internal friction against the temperature was composed of a peak component and a background component which increases monotonically with increasing temperature. In Fig. 4 (a), as an example, the temperature dependencies of the internal friction and elastic rigidity of MA-754 alloy are shown. In Fig. 4 (b), the internal friction spectrum of Fig. 4 (a) is separated into the peak component and the background component. From the close correlation with Kê's results for polycrystalline aluminum,[1] the peak component and the background component may be related to the stress relaxation due to the grain boundary slip and the anelasticity of the crystal grains, respectively.

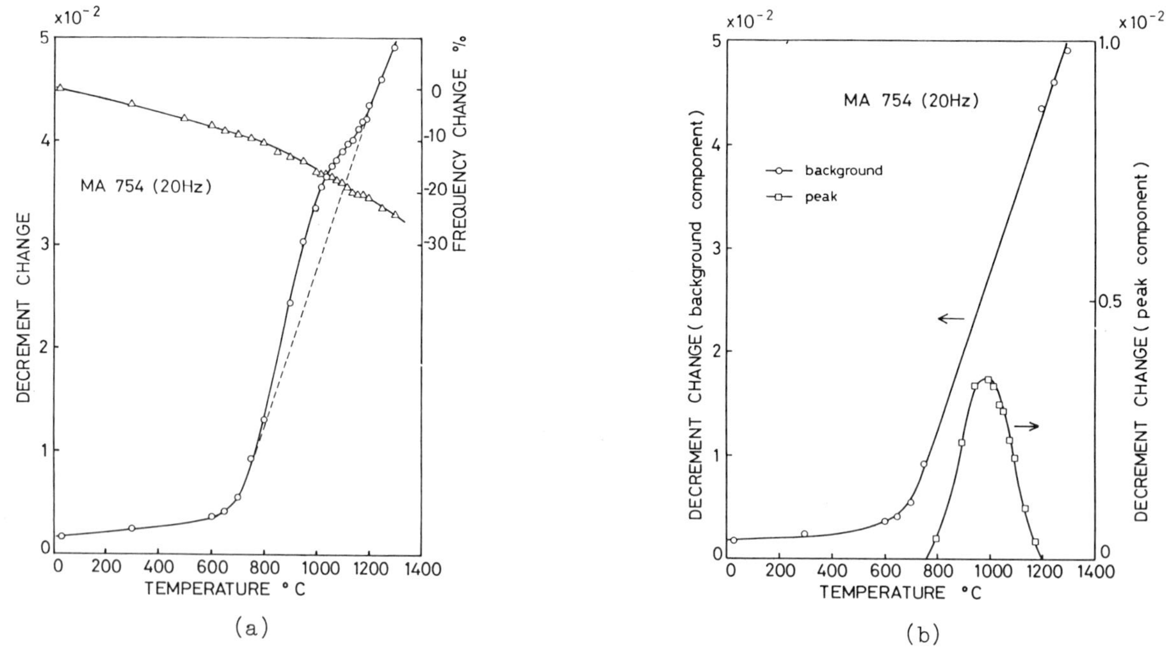

Fig. 4. Variations of internal friction and elastic rigidity with temperature and separation of internal friction spectrum into peak and background components. Internal friction and elastic rigidity are shown in terms of logarithmic decrement and pendulum frequency. MA-754. Amplitude of maximum shear strain is 3×10^{-5}.
(a) Measured internal friction and elastic rigidity. (b) Internal friction spectrum separated into peak and background components.

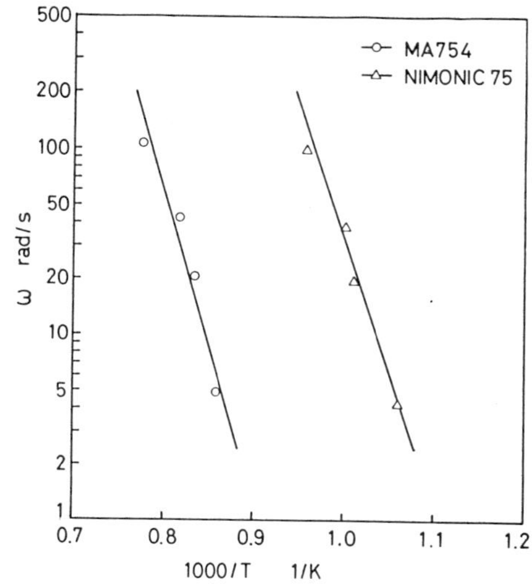

Fig. 5. Arrhenius plots of temperature at which the internal friction peak appears against pendulum frequency for MA-754 and Nimonic-75. Amplitude of maximum shear strain is 3×10^{-5}.

Figure 5 shows Arrhenius plots of the temperature at which the internal friction peak appeared against the pendulum frequency for both alloys. The activation energies calculated from these plots, which should be regarded as the activation energies of the grain boundary viscosity, were 75 and 66 kcal/mol for MA-754 alloy and Nimonic-75 alloy, respectively. Although the value for MA-754 is a little higher than the value for Nimonic-75, presumably this difference is due to the difference in impurity control and the dispersed oxide particles seem to have very little effect upon the grain boundary viscosity.

On the other hand, a large difference between these two alloys was observed in the background component to be related to the anelasticity of the grains. Figure 6 shows that the background component of MA-754 increases much more gradually with increasing temperature and strain-amplitude than that of Nimonic-75 especially at very high temperatures. This implies that in MA-754 the motion of dislocations is strongly suppressed by the dispersed oxide particles up to very high temperatures and strain-amplitudes.

The above results show that the high temperature strength of crystal grains is considerably improved by dispersed oxide particles while that of grain boundaries is not. This implies that the directional structure composed of parallel high aspect ratio grains, which is formed by directional recrystallization,[3,4] is indispensable for ODS alloys which are expected to be used at very high temperatures. In the above directional structure, the weakness of grain boundaries is covered at least for the tensile stress in the longitudinal direction of the elongated grains.

It should be also noted concerning the measurements in the present work that, although the MA-754 specimen had the directional structure in the longitudinal direction, grain boundaries were taking an active direction for torsional loading and accordingly the effect of the viscous slip along the grain boundaries could be detected.

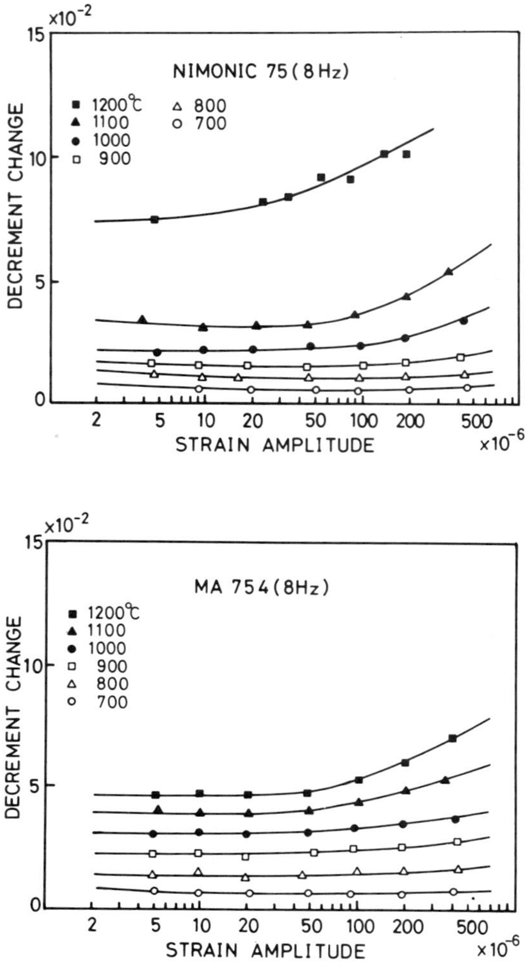

Fig. 6. Variation of the background component of internal friction with temperature and strain amplitude. Comparison between MA-754 and Nimonic-75.

APPLICATION TO SILICON NITRIDE

The covalent crystal ceramics such as silicon nitride and silicon carbide have excellent physical and chemical stabilities at elevated temperatures. However, the above stabilities make it difficult to sinter them in a pure state and usually some additives are used to promote sintering. In the case of silicon nitride, metallic oxides are used as additives and the sintered products have a composite structure composed of silicon nitride grains and a grain boundary phase which is the residue of the reaction between the surface of the silicon nitride grains and additives. This phase is usually glassy and at high temperatures the viscous behavior of this phase exerts a strong effect upon the mechanical properties of the sintered products. For the purpose of obtaining experimental information on the viscous properties of this phase at high temperatures, Shioiri et al tried the internal friction analysis.[5] In the above work, the internal friction measurements were made at a low frequency around 1 Hz by the torsion pendulum method and at a high frequency around 10 MHz by the ultrasonic method. However, the determination of the activation energy of the viscosity of the grain boundary phase, which is the most important factor characterizing the temperature dependency, was made not necessarily in a satisfactory manner. This was due to the following reasons: (i) the frequency range and the strain amplitude level were too different between these two method to measure the internal friction of the same origin; (ii) furthermore, in the ultrasonic measurements the relaxation peak due to the viscous flow in the grain boundary phase was not detected owing to the high attenuation of the ultrasonic pulse at high temperatures and the begining part of the attenuation rise was used in determining the activation energy.

In the present work, the torsion pendulum measurements were made over an appropriately wide range of the pendulum frequency and the determination of the activation energy was made. Figure 7 shows an example of the temperature dependency of the internal friction and elastic rigidity. The internal friction spectrum is composed of a clear relaxation peak and a back-

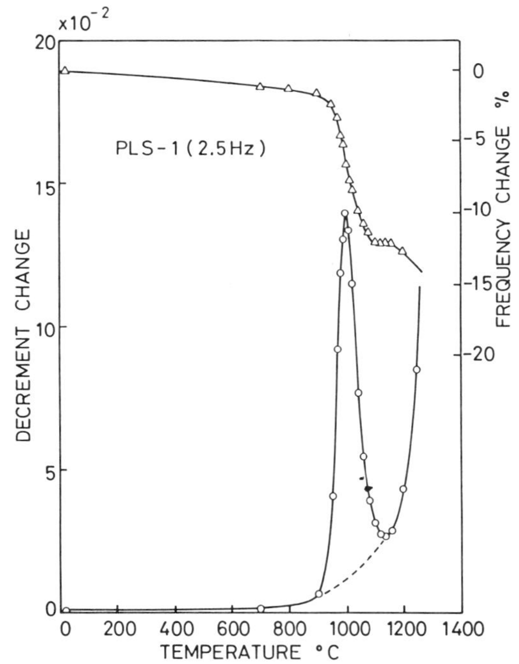

Fig. 7. Variations of internal friction and elastic rigidity with temperature. Internal friction and elastic rigidity are shown in terms of logarithmic decrement and pendulum frequency, respectively. Silicon nitride pressureless-sintered with Al_2O_3-Y_2O_3 as additives.

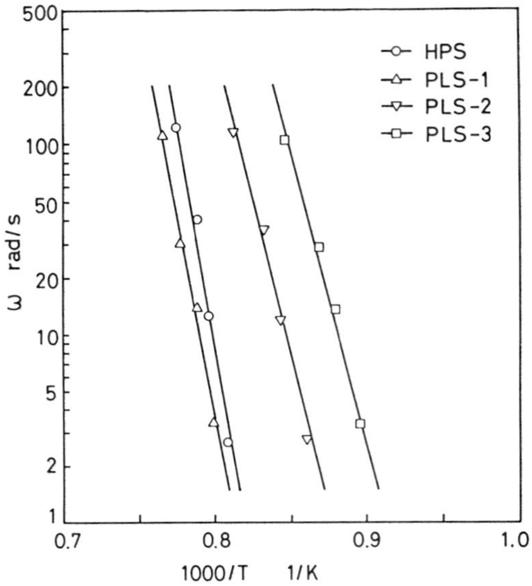

Fig. 8. Arrhenius plots of temperature at which the internal friction peak appears against pendulum frequency for sintered silicon nitrides.

ground component, and the elastic rigidity varies also in accordance with the internal friction spectrum. The Arrhenius plots of the temperature at which the internal friction peak appears against the pendulum frequency are shown in Fig. 8. From these plots the activation energy of the viscous flow of the grain boundary phase can be determined. Measurements were made for four sintered products which differ in the sintering method or in the additives composition. The sintering method, additive composition and the obtained activation energy of these four sintered products are as follows:

Abbreviation	Sintering	Additives	Activation Energy
HPS	Hot press	Al_2O_3-Y_2O_3	207 kcal/mol
PLS-1	Pressureless	Al_2O_3-Y_2O_3	200 kcal/mol
PLS-2	Pressureless	Al_2O_3-MgO	147 kcal/mol
PLS-3	Pressureless	SrO-MgO-CeO_2	134 kcal/mol

As seen above, HPS and PLS-1 which have Al_2O_3-Y_2O_3 type additive composition have high activation energy compared with the other two. The high temperature strength of the sintered products with Al_2O_3-Y_2O_3 type additives is excellent,[6] and it may be concluded that there is a close correlation between the high temperature strength and the activation energy of the viscous flow of the grain boundary phase.

CONCLUSION

A technique for measuring the low frequency internal friction at very high temperatures over wide ranges of frequency and strain amplitude was developed, and measurements were made for ODS alloy (MA-754) and sintered silicon nitrides. In the case of MA-754 alloy, the effects of the dispersed oxide particles upon the grain boundary viscosity and the anelasticity of the crystal grains were evaluated separately; the dispersed particles improve the high temperature strength of crystal grains greatly but have practically no effect upon the strength of grain boundaries. For sintered silicon nitrides the activation energy of the viscosity of the grain boundary phase was evaluated and a good correlation was obtained with the high temperature strength. Summarizing the above, it can be concluded that the internal friction method is a promising means for nondestructive characterization of the high temperature properties of heat resistant materials. This method is especially suitable for analyzing the effects of the individual strengthening and weakening mechanisms which are operative simultaneously in the advanced heat resistant materials with complicated design.

REFERENCES

1. T. S. Kê, Experimental Evidence of the Viscous Behavior of Grain Boundaries in Metals, Phys. Rev., LXXI: 533 (1947).
2. J. Shioiri and K. Satoh, Elevated Temperature Internal Friction in an Oxide Dispersion-Strengthened Nickel-Chromium Alloy, ASTM STP 864: 648 (1985).
3. J. S. Benjamin and J. M. Larson, Powder Metallurgy Techniques Applied to Superalloys, J. Aircraft, 7: 613 (1977).
4. T. K. Glasgow, An Oxide Dispersion-Strengthened Alloy for Gas Turbine Blades, NASA TM-79088 (1979).
5. J. Shioiri, K. Satoh and Y. Fujisawa, Elevated Temperature Internal Friction Due to Composite Structure in Sintered Silicon Nitrides, Proc. IVth Int. Conf. Composite Materials, Japan Society for Composite Materials, 1289 (1982).
6. K. Matusue, Y. Fujisawa and K. Takahara, Tensile Strength of Pressureless-Sintered Silicon Nitride at Elevated Temperature, TR-753, National Aerospace Laboratory, Tokyo (1983), in Japanese.

A NEW PROCEDURE FOR THE RAPID DETERMINATION OF YIELD AND TENSILE STRENGTH FROM HARDNESS TESTS

Barrie S. Shabel and Robert F. Young

Alloy Technology Division
Alcoa Laboratories
Alcoa Center, PA 15069

ABSTRACT

A new procedure for estimating the yield and tensile strengths of metallic alloys from hardness type measurements is described. The method has been successfully applied to alloys with different crystal structures and a wide range of strengths. The procedure is based on the calculation of modified Meyer hardness numbers from load and ball indentor penetration depth data. The computer-aided data acquisition and analysis includes corrections for machine deflection and elastic specimen recovery effects, providing a unifying framework for tests using a variety of loads and indentors. The method is also suitable for continuous load-depth data acquired in a single hardness test.

INTRODUCTION

Hardness tests are commonly employed in metallurgical development and quality assurance testing to provide an indication of material strength. The convenience of using small scale, essentially non-destructive, tests in place of more expensive machined or otherwise specially prepared tensile test specimens is the major driving force for the use of such tests. This, in turn, has led to the development of a wide variety of hardness tests and hardness scales for correlation with strength.[1-3]

The Rockwell-type is one of the most common tests in use, being especially favored for its simplicity and minimal surface preparation requirements. In addition, by being defined in terms of a depth of penetration, Rockwell hardnesses had the historic advantage of being obtained directly from dial gauge readings without the measurement of impression diameters in a microscope required by the conventional Brinell or Vickers procedures.

Unfortunately, for our purposes, there are several disadvantages to the Rockwell test. One is that no single scale adequately spans the whole range of interest for aluminum alloys. Each scale has an optimal application in terms of material strength (temper) and minimum thickness needed to avoid the so-called "anvil effect," i.e., the effect of the substrate or support on the observed hardness. In addition, plots of both yield strength and tensile strength vs. hardness show asymptotic behavior at either high hardness or low strength, which can be approximated by the expression for yield strength or tensile strength of the form:

$$\text{strength} = A - B \ln[1 - (HR/C)] \qquad [1]$$

where HR is the particular Rockwell scale hardness and A, B, and C are regression parameters for yield strength or tensile strength and hardness values. The non-linearity of this relationship makes it difficult to develop wholly satisfactory working relationships for alloy evaluation.

A more significant criticism of such equations is the fact that a test on a single Rockwell scale does not provide an unambiguous prediction of strength! This criticism also applies to the Brinell and Vickers tests as well and is most serious with respect to predictions of yield strength. This ambiguity arises largely because the phenomenon of work-hardening leads to an increasing resistance to penetration during the course of the hardness test. Following previous workers,[1,4] this can be shown in the form of ratios of yield and tensile strength to Vickers hardness:

$$\text{yield strength/hardness} = 0.33(40)^{-n} \qquad [2]$$

$$\text{tensile strength/hardness} = 0.333(12.5n)^{n}\exp(-n) \qquad [3]$$

These relationships assumed a stress-strain relationship of the Ludwik-Hollomon type,[5] i.e., $\sigma = K \varepsilon^{n}$, where n is the strain-hardening coefficient. They also show that the yield strength-to-hardness ratio is much more sensitive to the degree of strain-hardening (n value variations) than the tensile strength-to-hardness ratio. Equation [3] is similar to the results derived by Tabor[1] for Brinell hardnesses.

This inherent imprecision in the estimation of strength from hardness has historically precluded extensive hardness testing where reliable strength estimates were needed for aluminum alloys. However, advances in computerization and measurement technology have prompted a reconsideration of this problem. A reliable strength estimate from a hardness test would be highly cost-effective where the more expensive tensile testing procedures could be replaced.

An alternative approach to the use of equations [1]-[3] is based on the insight developed by Meyer.[6] He noted that the relationship between the applied load, L, and the impression diameter, d, for a spherical indentor, could be expressed in the simple form:

$$L = B \cdot d^{s} \qquad [4]$$

where s = Meyer index. For tests with different ball diameters, it was noted that the constant, B, depended on the ball diameter, D, so that, following Tabor (1), the results could be expressed as:

$$L/d^{2} = B\ (d/D)^{s-2} \qquad [5]$$

The similarity of this to the Ludwik-Hollomon[5] form of the true stress-true strain relationship is the basis for estimating a material's plastic flow curve from hardness impression diameter measurements. Here a significant contribution was made by Tabor in assuming that the effective strain beneath the indentor was proportional to the quantity (d/D). This has also been used in the modeling of the hardness test by Lee et al.[7] and Richmond et al.[8] as well as the experimental studies of Francis[9] and Au et al.[10]

Equation [5] was also used by George et al.[11] to estimate the yield strength of various steel alloys using a nomograph constructed from Rockwell hardnesses. They computed d from an estimate of the depth of the Rockwell impression and subsequently estimated the yield strength from the regression equation:

$$\text{yield strength} = 325B \qquad [6]$$

The above approaches did not explicitly address the question of estimating both yield and tensile strengths from the hardness test. In addition, the nomographic procedure described by George et al. appears somewhat open to question because it was limited to steels, and they did not account for the changes in impression contour due to elastic recovery effects. This could be a problem in estimating the impression diameter from a Rockwell hardness number alone, given the range of major and minor loads associated with the different Rockwell scales and the corresponding variations in elastic recovery under different test conditions.

The hypothesis behind our approach was that a modified version of equation [5], with precise measurements of the impression depth and computer-aided analysis of the springback or elastic recovery effects, would enable us to estimate the yield and tensile strengths of metallic materials with acceptable accuracy. The results of our investigation essentially confirmed that hypothesis and provide a basis for more effective use of hardness testing.

EXPERIMENTAL PROCEDURE

Our procedure is based on the availability of hardness testing equipment which measures the depth of the hardness impression during loading. For this work, we used the Verimatic 8800 tester built by K. J. Law, Inc., which we modified to provide either "static" Rockwell test conditions (specified minor and major loads, penetration depths at major and minor load) or continuous load vs. penetration depth data.

Given the spherical geometry, a simple relationship exists between the depth of the impression, h, the impression diameter, d, and the ball (penetrator) diameter, D, of the form:

$$d = 2D\,[(h/D)\,(1 - (h/D))]^{1/2} \qquad [7]$$

An initial comparison of measured and calculated impression diameters suggested a machine compliance effect. Diameters estimated from total penetrator motion measurements were systematically greater than the measured diameters. A calibration procedure was devised using a high strength steel sample and carbide indentor to minimize the sample's plastic deformation as a contribution to the total deflection. Penetrator motion was measured under loading conditions which produced no net plastic deformation of the specimen. We estimated the specimen's elastic deflection using both Hertz[1] and Boussinesq[12] equations. These yielded essentially equivalent results, so the former was used to calculate specimen deflections, which were subtracted from the total motion to provide an estimate of the machine deflection.

The results were further refined by estimating the sample's elastic springback or recovery, based on Rickerby's analysis.[13] The springback is manifest as a reduction in the depth of the impression and a corresponding increase in its radius of curvature. This was done to express the results in terms of <u>plastic</u> strain. Our critical assumption is that the net plastic strain is proportional to the quantity (d/D^*), where D^* is twice the "relaxed" radius of curvature of the impression.

This followed from the assumption of (d/D) as proportional to the total strain, i.e., the elastic plus plastic strain, and the use of the Rickerby model to estimate the change in the radius of curvature of the impression as the load is removed. This model seems physically plausible, since D* is usually at least slightly greater than D for our test conditions. It follows that (d/D*) will be < (d/D). The relevant equations are summarized in the Appendix. Since the impression diameter is essentially unchanged when the load is removed, the change in radius can be used to estimate the change in depth of the impression as well. The changes in depth due to recovery were used to calculate final "relaxed" depths, which should be the depths corresponding to Rockwell hardness numbers (ignoring, for simplicity, the depth due to the minor load in the usual Rockwell test). Rockwell hardness numbers calculated this way were in excellent agreement with experimental data. In addition, calculated radii of curvature were compared with the measured radii of curvature using the test data for various materials cited by Tabor.[1] The agreement is quite good, as shown in Figure 1, with the exception of the two points for the hardened .5% C steel.

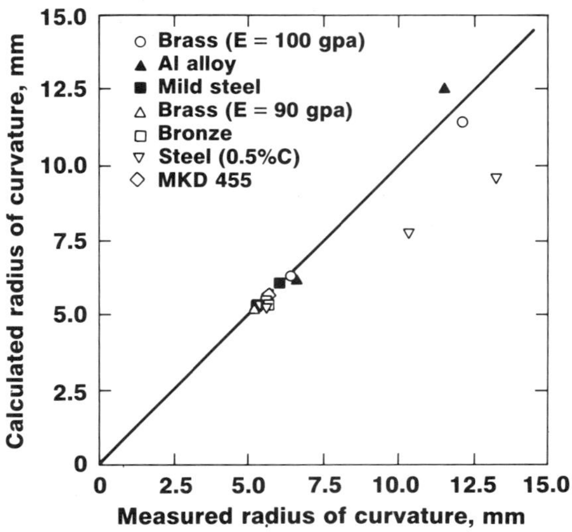

Fig. 1: Comparison of Measured vs. Calculated Radius of Curvature for Spherical Hardness Indentations

The corrections for machine deflection and specimen springback were incorporated in computer programs which were used to analyze the raw load vs. penetrator motion data on a point-by-point basis. The resulting mean pressure and effective plastic strain values were then analyzed via a "modified" Meyer equation:

$$L/d^2 = A\ (d/D^*)^m \qquad [8]$$

where A is now the value of mean pressure at (d/D*) = 1.0. The springback calculations revealed that (d/D*) approached (d/D) when (d/D) > 0.3. The range of greatest relative springback occurred at (d/D) < 0.2. It is interesting to note that this is near the lower limit of usefulness of the original Meyer model. The assumption of the proportionality between plastic strain level and the quantity (d/D*) also led to the idea that the yield strength should be correlated with the value of mean pressure at a "low" strain level. In this work, this "low" strain value was selected as (d/D*) = 0.1 and the corresponding mean pressure was designated as A′.

Tensile tests were carried out on a variety of aluminum alloys while the hardness tests were conducted near the grip ends of the specimens prior to testing. The hardness tests were conducted both as static Rockwell tests and as "continuous" compression tests in which we monitored the load and penetration depth throughout the test. The resultant A and A′ values derived from equation [8] were then correlated with tensile and yield strengths. The working relationships so derived were subsequently validated by tests carried out on a number of additional materials, e.g., C260-0 brass, CW67-T6, a high strength aluminum P/M alloy for aerospace applications, several Mg and Mg-Li alloys (to compare results on metals of different crystal structures) and cold-worked 2024 and 7075 aluminum alloys (to confirm the ability of our procedure to follow changes in yield strength even where only small or negligible changes in tensile strength occurred).

RESULTS AND DISCUSSION

Examples of the logarithmic pressure vs. effective strain (d/D*) curves derived using equation [8] are shown in Figure 2. These are essentially plastic flow curves for each material. The data were self-consistent in that results from static and continuous tests agreed and, in the continuous test, pressures based on load and impression diameter data from different indentor diameters agreed with each other at the same values of (d/D*).

The resulting strength correlations are shown in Figure 3. The equations for the yield strength (ys) and tensile strength (ts), in ksi are (σ = standard deviation, R^2 = correlation coefficient):

$$ys = 0.689A' - 4.75 \quad (\sigma = 5.6,\ R^2 = 0.939) \qquad [9]$$

$$ts = 0.436A + 3.48 \quad (\sigma = 4.7,\ R^2 = 0.950) \qquad [10]$$

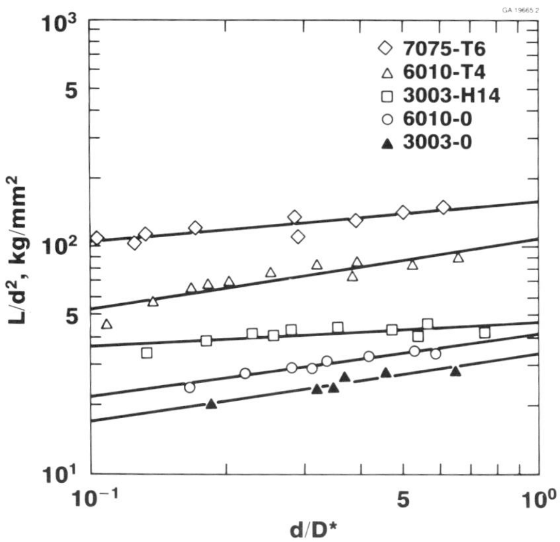

Fig. 2: Application of the Modified Meyer Equation of Various Aluminum Alloys

The data in Figs. 4-5 show the subsequent application of our method to other alloys, including a high-strength Al powder metallurgy (P/M) type, as well as to non-fcc Mg-based alloys (hcp and bcc crystal structure types). The tests included samples of different orientations, so some of the scatter in these plots is undoubtedly due to this normal variation of strength with direction. The accuracy of the predictions for the Mg alloys was especially gratifying, indicating that the analysis and equations may provide a general tool for relating hardness and strength for different materials.

The data for the P/M alloy also illustrate the ability to follow the decline in both yield and tensile strengths associated with slow quenching from the solution heat-treated temperature. This procedure, in taking account, to some extent, of the effects of changing work-hardening characteristics, provides a clearer picture of the changing material behavior than we have found with the usual "single-scale" hardness tests. This has particular advantages in dealing with processing variations, such as cold working or age-hardening, which typically affect yield strength to a greater extent than tensile strength. The effect of cold work on the 2024 and 7075 alloys is also shown in Figs. 4-5.

Further confirmation of this analysis was established by using the model to generate Rockwell hardness conversion relationships for both normal and superficial sales for a variety of alloys.[14] The calculated values were in reasonable agreement (within about 5-7 Rockwell numbers) with available data.[15] This has provided a simple but effective tool for selecting Rockwell tests from prior information for more efficient quality assurance testing.

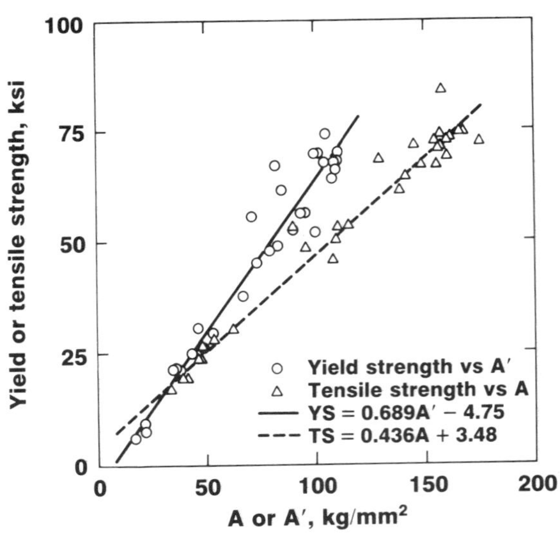

Fig. 3: Yield and Tensile Strength of Various Aluminum Alloys vs. the A and A′ Parameters Derived from Hardness Testing

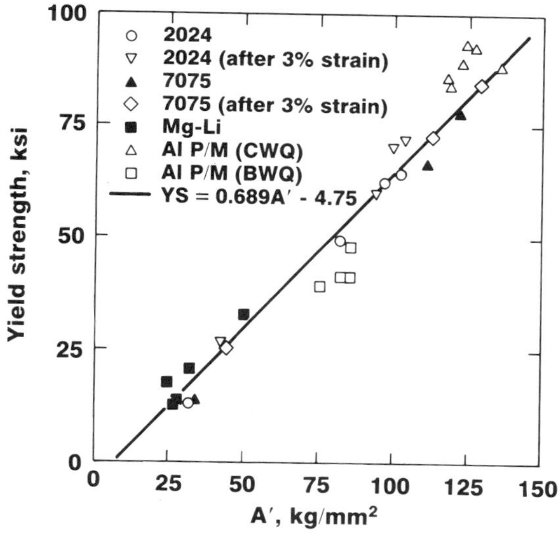

Fig. 4: Verification of the Dependence of Yield Strength on the A Parameter Using Equation [9]

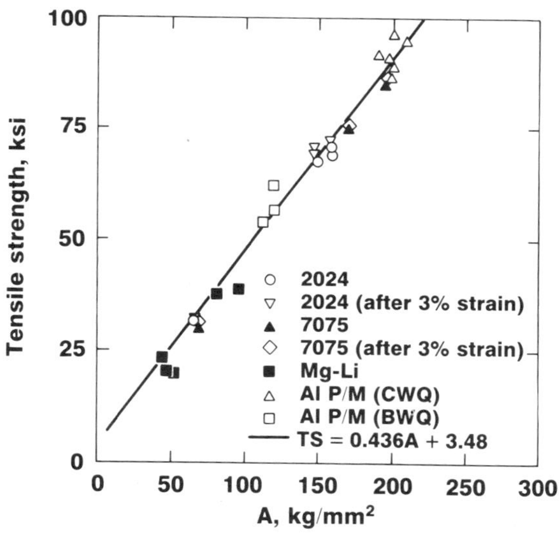

Fig. 5: Verification of the Dependence of Tensile Strength on the A Parameter Using Equation [10]

Thus, our results show that the underlying "model" is reasonably accurate. Moreover, a "multiple hardness impression" version of this technique has been patented.[16] In addition, finite element simulations at our Laboratory[17] have confirmed the accuracy of the Rockwell calculations based on this springback analysis. This FEM simulation also corroborated the hypothesis that the A and A′ parameters were proportional to the tensile and yield strengths. The simulations are continuing and will be used as a tool to further refine our data analysis procedures. This should also enable us to examine the potential effects of differences in behavior manifest as "sinking in" or "piling up" of material around the perimeter of the hardness impression on our results. We will then be able to make more explicit comparisons of our work with that of Richmond et al.,[8] whose analysis included estimation of the magnitude of these effects.

CONCLUSIONS

The results of this program show that ball hardness-type testing can be used to generate useful estimates of yield strength and tensile strength levels of metallic alloys. By incorporating a relatively simple springback model, we can use the framework of the classical Meyer analysis to generate a plastic flow curve from which parameters for estimating the yield and tensile strengths are obtained. The application of these results to a variety of alloys yielded reasonable agreement with observed strengths over approximately a 90 ksi strength range. The ease of analysis suggests that this procedure could be incorporated readily into hardness testing schemes for alloy development and quality assurance testing.

REFERENCES

1. D. Tabor, "The Hardness of Metals," Oxford (Clarendon Press) (1951).
2. J. H. Westbrook and H. Conrad, "The Science of Hardness Testing and its Research Applications," ASM (1973).
3. E. R. Petty, "Hardness Testing," in: "Techniques of Metals Research: Measurement of Mechanical Properties," Vol. 5, Part 2, R. F. Bunshah, ed., Interscience Publishers, NY (1971).
4. J. R. Cahoon, W. R. Broughton, and A. R. Kutzak, The Determination of Yield Strength from Hardness Measurements, Met. Trans. A 2:1979 (1971).
5. J. H. Hollomon, Tensile Deformation, Trans. AIME, 162:268 (1945).
6. E. Meyer, Zeits. d. Verienes Deutsch. Ingenieure, 52: 645, 740, 835 (1908).
7. C. H. Lee, S. Masaki, and S. Kobayashi, Analysis of Ball Indentation, Int. Jnl. of Mechanical Science, 14:417 (1972).
8. O. Richmond, H. L. Morrison, and M. L. Devenpeck, Sphere Indentation with Application to the Brinell Hardness Test, Int. Jnl. of Mechanical Science, 16:75 (1974).
9. H. A. Francis, Phenomenological Analysis of Plastic Spherical Indentation, Trans. ASME, 272 (1976).
10. P. Au, G. E. Lucas, J. W. Sheckherd, and G. R. Odette, Flow Property Measurements from Instrumented Hardness Tests, in: "Non-Destructive Evaluation in the Nuclear Industry," ASM (1980) 10.
11. R. A. George, S. Dinda, and A. S. Kasper, Estimating Yield Strength from Hardness Data, Metal Progress, May 1976, p. 30.
12. E. H. Yoffe, "Elastic Fields Under a Spherical Indentor," Research Report, AD R&D 4024-R-MS, Cambridge U. (1983).
13. D. G. Rickerby, Elastic Recovery in Spherical Indentations, Materials Science & Engineering, 56:195 (1982).
14. B. S. Shabel, Alcoa Laboratories, unpublished research.
15. ASTM E-140-79, Standard Hardness Conversion Tables for Metals, ASTM Standards, Part 10 (1981).
16. B. S. Shabel, "Rapid Determination of Metal Strength from Hardness Tests," U.S. Patent 4,530,235 (85/07/23).

17. J. R. Brockenbrough and C. M. Claypool, Alcoa Laboratories, unpublished research.

APPENDIX: ESTIMATION OF THE SPRINGBACK EFFECT

The springback correction is based on Rickerby's modification of the Hertzian analysis. The latter starts out with the elastic relationship based on a spherical penetrator of diameter D:

$$d^3 = 6Lr1r2/(r2-r1)[E1^{-1}(1-\nu 1^2) + E2^{-1}(1-\nu 2^2)] \quad [A\text{-}1]$$

where L = load, d = impression diameter, r1 = radius of curvature of the impression under load (= D/2), r2 = radius of curvature after the load is removed, and ν_1, E1, ν_2 and E2 are the Poisson ratio and Young's Modulus of the ball and sample, respectively. This equation can be rearranged to yield:

$$r2 = r1/(1-\alpha LD/d^3) \quad [A\text{-}2]$$

where α contains the elastic constants and conversion factors to express r1 and r2 in mm. If d is unchanged on load removal, then Rickerby's correction term can be written as:

$$\Delta r' = 2L(1 + \nu_2)r2/\pi E2d^2 \quad [A\text{-}3]$$

and the final radius of curvature, r_f (=D*/2), in mm, is:

$$r_f = r_1 + (1-d/D)\ (r2-r1) - (d/D)\Delta r' \quad [A\text{-}4]$$

The resultant "relaxed" depth, h^* (mm), can be computed from equation [7] by substituting D* for D for the given impression diameter (d). This depth can then be used to calculate a Rockwell hardness, assuming L and D correspond to standard Rockwell scale conditions:

$$HR\ (\text{normal scale}) = 130 - 500h^* \quad [A\text{-}5]$$

or, for superficial Rockwell scale conditions:

$$HR\ (\text{superficial}) = 100 - 1000h^* \quad [A\text{-}6]$$

MECHANICAL PROPERTIES OF COMPOSITE MATERIALS

STUDIED BY INTERNAL FRICTON

R. Schaller, J.J. Ammann, and P. Millet

Institut de génie atomique, Swiss Federal Institute of Technology, PHB-Ecublens, CH-1015 Lausanne, Switzerland

INTRODUCTION

Mechanical properties of materials depend strongly on their structural defects (point defects, dislocations, grain boundaries, precipitates). Structural defects can be studied by classical mechanical tests such as tensile, compressive, hardness, creep or rupture mechanics tests. These techniques provide direct information on the mechanical properties of the material, like yield stress, tensile strength, toughness, ductility, but they are destructive by nature and the microstructure is strongly altered by the test.

The anelastic behaviour[1] of crystalline solids is caused by the motion of structural defects around their equilibrium position. To induce such a motion small applied stresses are only needed which has the consequence that the microstructure is not modified. Anelasticity is an interesting property which is at the basis of nondestructive internal friction techniques for studying the structural defects of crystalline materials[2].

Internal friction is a technique very sensitive to the mobility of the structural defects, such as dislocations, in the range of very low strain amplitudes (10^{-8} - 10^{-4}). It is then well suited for studing the mechanisms of microplastic deformation which determine the mechanical properties of hard and brittle materials having limited plasticity. In this paper, we report the study of two types of composite materials: grey cast iron and WC-Co hardmetals.

Grey cast iron is often used in engineering because of its high damping capacity[3]. But, generally this advantage is associated with poor mechanical properties. Grey cast iron is usually specified by its tensile strength, but the tensile strength does not account for the damping capacity, an important property of this material[4]. Damping capacity is due to internal friction mechanisms. The purpose of this work was to study the internal friction mechanisms in grey cast iron, in order to develop new types of grey cast iron which present simultaneously a high damping capacity and a good mechanical resistance. Such developments are obviously possible only if the internal friction mechanisms are independent of the hardening phenomena.

The WC-Co composite alloys are cemented carbides with outstanding properties[5] (high values of hardness and elastic modulus, good wear resistance) and are consequently well suited for cutting tools. During machining, the cutting tools are effectively subjected to strong mechanical and chemical solicitations, which limit their "life time". From tests on such materials, it appears that toughness plays an important role in the "life time" of the tools. Internal friction has been then used to study the microscopic mechanisms which are responsible for the mechanical properties (toughness) of WC-Co hardmetals. The nondestructive character of the technique has revealed itself very advantageous for measuring brittle specimens in a reproducible manner under various experimental conditions.

ANELASTICITY AND INTERNAL FRICTION

Anelasticity of a solid can be described by the following experiment (fig. 1): A mechanical stress σ is applied abruptly at time $t = 0$ and held constant while the strain ε is recorded as a function of the time. One observes:

- the instataneous elastic strain: $\varepsilon_e = J_u\ \sigma$, where J_u is the unrelaxed compliance;

- the anelastic strain ε_a, which increases with time from zero to an equilibrium value ε_a^∞. When equilibrium is reached, it is possible to write:

$$\varepsilon = \varepsilon_e + \varepsilon_a^\infty = J_R\ \sigma \quad (J_R = \text{relaxed compliance}).$$

This evolution of the solid from one equilibrium state to a new one, under a mechanical applied stress, is called anelastic relaxation, and can be defined by two parameters:

- the <u>intensity of relaxation</u> $\Delta = \varepsilon_a^\infty / \varepsilon_e = (J_R - J_u)/J_u$

- the <u>relaxation time</u> τ appearing in the phenomenological expression for ε:

$$\varepsilon = J_u\ \sigma + (J_R - J_u)\ \sigma\ (1 - \exp(-t/\tau)) \qquad (1)$$

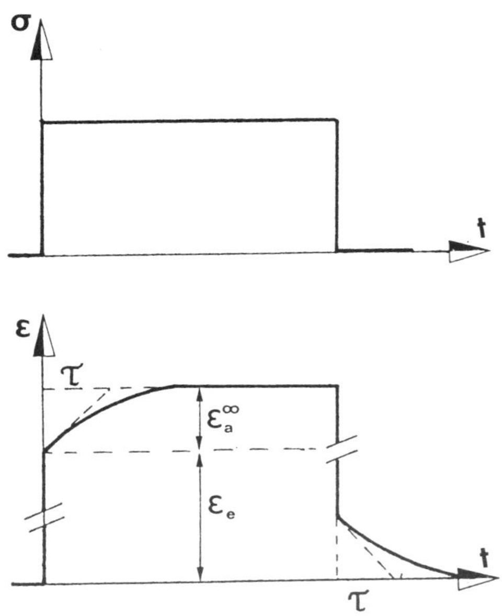

Fig. 1 Anelastic relaxation in a solid submitted to a constant stress σ.

If after a certain time delay t the stress is removed, we observe the instantaneous restauration of the elastic strain followed by the restauration of the anelastic strain with the same relaxation time τ. In such an experiment, the initial strain can be completely recovered.

From eq. 1, we can derive the equation of the Standard Anelastic Solid:

$$\varepsilon + \tau \dot{\varepsilon} = J_R \sigma + \tau J_u \dot{\sigma} \qquad (2)$$

Measurements such as described by figure 1 present the difficulty to require to resolve very small anelastic strains (10^{-9} to 10^{-8} assuming elastic strain of the order of 10^{-6}). For that reason, it is more convenient in practice to use dynamic methods for measuring relaxation parameters. In this case, an alternative stress is applied to the system: $\sigma = \sigma_0 \exp(i\omega t)$. The linearity of the stress-strain relations ensures that the strain is periodic with the same frequency: $\varepsilon = \varepsilon_0 \exp(i\omega t - i\delta)$, where δ is the phase lag of the strain with respect to the stress, caused by anelasticity. If these expressions of σ and ε are introduced into eq. 2, the following expression is obtained:

$$\varepsilon = J^*(\omega)\,\sigma = (J_1(\omega) - i\,J_2(\omega))\,\sigma \qquad (3)$$

where J_1 and J_2 are the real and the imaginary part of the complex compliance J^*, respectively.

$$J_1 = J_u + \frac{J_R - J_u}{1 + \omega^2\tau^2} \qquad (4)$$

$$J_2 = (J_R - J_u)\frac{\omega\tau}{1 + \omega^2\tau^2}$$

It is possible to show that the internal friction Q^{-1} of the material is related in the linear cases to the loss angle δ by the following relation:

$$Q^{-1} = \frac{1}{2\pi}\,\frac{\Delta w}{w} = \operatorname{tg}\,\delta = \frac{J_2}{J_1} \qquad (5)$$

where Δw is the energy dissipated during one cycle of vibration and w the maximum stored energy. Then, taking into account (4) and (5), it follows:

$$Q^{-1} = \Delta\,\frac{\omega\tau}{1 + \omega^2\tau^2}$$

$$(6)$$

and $$\frac{\delta J}{J} = \frac{J_1(\omega) - J_u}{J_u} = \Delta\,\frac{1}{1 + \omega^2\tau^2}$$

$\delta J/J$ is the variation of the dynamical modulus due to the anelastic relaxation.

The expressions 6 show that internal friction Q^{-1} presents a relaxation peak (Debye type) centered at $\omega\tau=1$. Using these expressions, we can deduce:

- the intensity of the relaxation Δ from the height of the peak;
- the relaxation time τ by the position of the peak on the $\omega\tau$ axis.

When the mechanism is thermally activated, $\tau=\tau_o \exp(E/kT)$, it is possible to measure the internal friction as a function of temperature while keeping ω constant. The peak appears at a temperature Tp defined by $\omega\tau_o \exp(E/kTp) = 1$, and expression 6 for Q^{-1} can be transformed into:

$$Q^{-1} = \Delta \left(2\mathrm{ch} \frac{E}{K} \left(\frac{1}{Tp} - \frac{1}{T}\right)\right)^{-1} \qquad (7)$$

where E is the activation energy.

Examples of relaxation mechanisms (see Nowick and Berry[1])

a) Point-defect relaxations: point-defects can give rise to anelastic relaxation only when they create elastic dipoles in the lattice. The relaxation is then due to the reorientation of such dipoles under the effect of the applied stress. Snoek relaxation refers to the point-defect relaxation produced by interstitial atoms like C, O, N, H which are in dilute solution in bcc metals. Zener relexation is produced by pairs of substitutional solute atoms.

In general, the relaxation intensity Δ is proportional to the number No of elastic dipoles per volume unit and to the specific variation of strain $\Delta\lambda$ due to the reorientation of the dipoles. The relaxation time depends on the diffusion energy E of the point defects. We have the following relations:

$$\Delta = \frac{\Delta\lambda^2 \cdot No}{JukT} \quad \text{and} \quad \tau = \tau_o \exp(E/kT) \qquad (8)$$

b) dislocations relaxations: Anelastic deformation occurs when the applied stress produces a bowing of the dislocation segments between the pinning points. The relaxation parameters can be written:

$$\Delta = (\Lambda b^2)/(KJu) \quad \text{and} \quad \tau = B/K \qquad (9)$$

where Λ is the dislocations density, b the Burgers vector, B and K are coefficients due to the dragging and restoring forces respectively. B depends on the interactions of the dislocation with phonons or with obstacles such as Peierls valleys[6], immobile or mobile point defects[7].

From the above examples it is possible to conclude that the relaxation parameters (Δ, τ) give two kinds of informations: Δ gives some information on the density and arrangement of structural defects and τ on their mobility.

When several independent relaxation mechanisms are activated, the superposition principle allows one to write

$$Q^{-1} = \sum_i \Delta_i \frac{\omega\tau_i}{1+\omega^2\tau_i^2} \qquad (10)$$

where i refers to the mechanism (Δ_i, τ_i). The material then exhibits an internal friction spectrum composed of internal friction peaks due to the various microscopic mechanisms.

INTERNAL FRICTION OF A COMPOSITE MATERIAL

To simplify, we consider only a two-phase material. The relaxation mechanism is either localized in any one of the phases or localized at the interfaces.

In the first case, an internal friction technique can provide some information on the intrinsic properties of a second phase dispersed in a matrix. For instance, an internal friction peak has been observed in Al-Ag alloys, the origin of which is the relaxation of elastic dipoles in the γ precipitates (Ag_2Al-hcp)[8]. The height of the peak $\Delta/2$ is proportional to the volumic fraction of the precipitates, and the relaxation time τ depends on the diffusion energy in the γ phase. In general, when the stresses are well transmitted through the interface, it is possible to write:

$$Q^{-1} = \frac{1}{2\pi} \frac{V_1 \Delta w_1 + V_2 \Delta w_2}{V_1 w_1 + V_2 w_2} \tag{11}$$

where Δw_i and w_i are the specific energies respectively dissipated and stored in the phase i of volume V_i.

We assume now that the dissipation of energy in the matrix Δw_2 is negligible in comparison to Δw_1. The internal friction level of the composite is found to be between the following values:

$$Q_\sigma^{-1} = Q_p^{-1} \left(1 + \frac{E_p}{E_p} \frac{V_m}{V_p}\right)^{-1} \tag{12}$$

and

$$Q_\varepsilon^{-1} = Q_p^{-1} \left(1 + \frac{E_p}{E_p} \frac{V_m}{V_p}\right)^{-1} \tag{13}$$

Q_σ^{-1} refers to a series configuration of the two phases (stress is the same in the two phases), and Q_ε^{-1} refers to a parallel configuration (strain is the same in the two phases). E_p, V_p and E_m, V_m are the modulus and the volumic fraction of the precipitates and the matrix, respectively.

EXPERIMENTAL PROCEDURE

The internal friction of grey cast iron and WC-Co has been measured by means of resonant systems vibrating at a natural frequency. From the free decay of the vibrations, it is possible to deduce the internal friction Q^{-1} by:

$$Q^{-1} = \frac{1}{n\pi} \ln \frac{A_i}{A_{i+n}} \tag{14}$$

where A_i and A_{i+n} are respectively the ith and the $(i+n)^{th}$ amplitudes of the free decay. In the low frequency range (0.2 - 4 Hz), the measurements can be performed by an inverted torsion pendulum.

The grey cast iron specimens we have used are machined wires of 2 mm in diameter and 100 mm in length, cut from industrial blocks supplied by Von Roll SA, Switzerland. The WC-Co specimens are ribbons 0.2 x 2.5 x 100 mm in size, cut by electroerosion from bars sintered by Stellram SA, Switzerland.

The internal friction and dynamic elastic modulus of each specimen were measured as a function of temperature. The internal friction spectra were then compared with the spectra of the components, in order to identify the internal friction mechanisms.

RESULTS FOR GREY CAST IRON

Grey cast iron can be considered as a two-phase material composed of graphite precipitates in an iron matrix. In order to reveal the contribution to the damping capacity and to the mechanical properties of each phase, internal friction measurements were performed on many specimens of grey cast iron differing from one another by the morphology of the graphite precipitates[9,10].

The results show that grey cast iron exhibits a typical internal friction spectrum (fig. 2, curve a), the main characteristic of which is an abrupt increase of the internal friction at about 200 K associated with a change of slope in the decrease of the vibrational frequency.

These characteristics are also found in the spectrum of pure graphite (fig. 2, curve b). On the other hand, the Q^{-1} spectrum of white iron is completely different: internal friction is weak and constant over all the temperature range investigated (fig. 2, curve c). It can then be concluded that the damping capacity of grey cast iron is caused by internal friction mechanisms taking place in the graphite precipitates.

An interesting feature which appears from the curves of figure 2 is the level of the internal friction in grey cast iron, compared to that of pure graphite. It is possible to understand this phenomenon by taking into account the above relations (12) and (13), in which the matrix modulus $E_m = 2 \cdot 10^{11}$ N/m^2 and the graphite modulus $E_p = 10^{10}$ N/m^2 are introduced. For a fraction of graphite of 10% per volume, we obtain $Q_\sigma^{-1} = 0.66$ Q^{-1} of pure graphite and $Q_\varepsilon^{-1} = 0.05$ Q^{-1} of pure graphite.

Fig 2 shows that the internal friction level of grey cast iron is about half the level of pure graphite. This fact tends to confirm that the stresses are well transmitted through the interfaces ($Q^{-1} = Q_\varepsilon^{-1}$) and then can activate mechanisms of energy dissipation in the graphite phase.

Measurements made at different frequencies have shown that the internal friction increase at 200 K is not due to an anelastic relaxation, but rather to a modification of the graphite mechanical properties. Transmission electron microscopy has revealed that this critical temperature of ≈ 200 K corresponds to an increase of the dislocation mobility in graphite. A model has been developed which shows that the internal friction is mostly caused by interaction of the dislocations

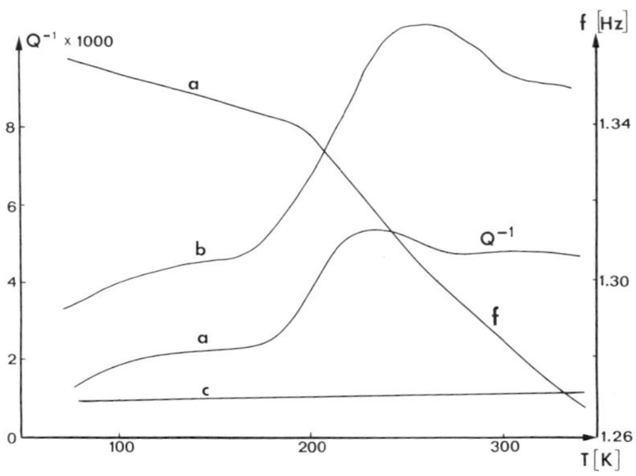

Fig. 2: Internal friction spectrum of:
a) grey cast iron; b) graphite; c)white iron.

with intercalated impurities[11]. The increase of the internal friction at ≈ 200 K is due to the breakaway of the dislocations from the pinning sites produced by the intercalated impurities. At room temperature, the internal friction is

$$Q^{-1} = \frac{\Lambda J u}{36 b^2} l^4 (H + \omega B) \qquad (15)$$

where Λ is the dislocation density, l is the dislocation loops length, H and B are respectively the coefficients of solid and viscous friction between dislocations and intercalated impurities.

High damping

Eq. 15 shows that internal friction increases rapidly with the length l of the dislocation loops. As a consequence, a high damping capacity will be favoured by large lamellar graphite precipitates[12]. This idea is supported by the results of fig. 3. The results obtained with the industrial globular GGG^{40} (curves 1) and lamellar GG^{25} (curves 2) grey cast irons correspond to what is generally expected: high damping in GG^{25} is associated with a low level of the tensile strength. On the other hand, the properties of the grey cast iron composed of long lamellae of graphite (curves 3) are quite remarkable: The mechanical resistance is comparable to that of GGG^{40} and the room temperature internal friction is three times higher than in GG^{25}.

RESULTS FOR WC-Co HARDMETALS

The characteristic internal friction spectrum[13] of WC-Co composite alloys is presented in figure 4. This spectrum is mainly composed of a peak superimposed to an exponential increase of the internal friction with the temperature. This peak is identified as a relaxation peak, because it shifts in temperature when the frequency is changed. An activation energy of 2.7 eV can be deduced, which is very close to the self diffusion energy in Co (2.8 eV). The internal friction, peak and exponential background, increases with the Co concentration. In addition, the relaxation peak shifts towards lower temperatures. These

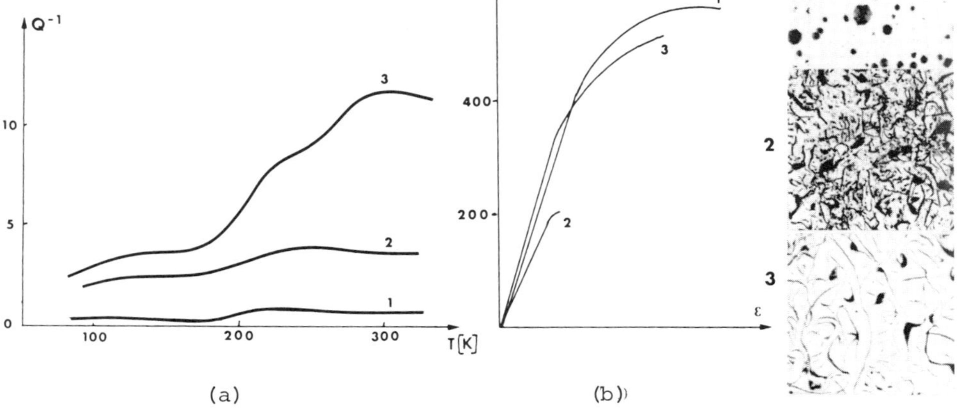

Fig. 3: Internal friction (a) and tensile tests curves (b) of three types of grey cast iron: 1) globular GGG^{40}; 2) lamellar GG^{25} and 3) lamellar directional solidification.

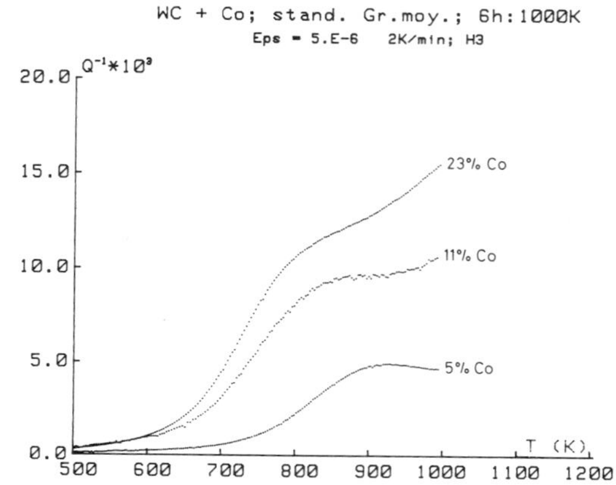

Fig. 4: Internal friction of WC-Co 5, 11 and 23 Wt %.

results tend to show that internal friction is associated with mechanisms which take place in the Co phase. Like in grey cast iron, internal friction measurements have also been performed on the two components: Co and WC. In figure 5 these results are compared to the spectrum of WC-11 % Co.

It can be readily seen that the relaxation peak is not caused by a microscopic mechanism in the WC particles or by the allotropic transition of cobalt[14] which give rise to the sharp peak observed at ≈ 700 K. The peak is then certainly associated with Co, but Co is playing the special role of the binder phase.

Mechanical properties

The internal friction peak observed at low frequency in WC-Co is located in the same temperature range where an increase of the critical stress intensity factor has been observed and attributed to a brittle/ductile transition of the material[15,16]. The peak is then certainly associated with such a transition and corresponds to the increase of toughness caused by a mechanism controlled by the Co binder phase. Further investigations would be needed to precise this mechanism of prime importance for the mechanical behaviour of WC-Co cemented carbides.

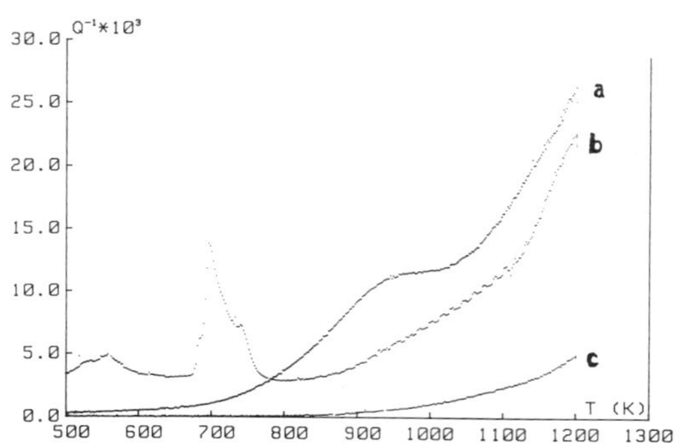

Fig. 5: Internal friction spectrum of a) WC-11 wt % Co; b) sintered Co; c) sintered WC.

CONCLUSIONS

Internal friction is a useful technique for the nondestructive evaluation of brittle composite materials. It permits to study the microscopic mechanisms, which determine the mechanical properties, and to identify their origins. These mechanisms can be traced to the interfaces between phases or, as we have seen for the case of grey cast iron, to the second phase precipitates. For WC-Co cemented carbides, the results have shown that such mechanisms which are likely to affect the toughness are activated in the Co binder phase.

ACKNOWLEDGMENTS

The supports of the Swiss National Science Foundation and the Swiss "Commission pour l'encouragement de la recherche scientifique" are acknowledged.

REFERENCES:

1. A.S. Nowick and B.S. Berry, "Anelastic Relaxation in Crystalline Solids, Academic Press, New York (1972).
2. W. Benoit, G. Gremaud and R. Schaller, Anelasticity and dislocation Damping in: "Plastic Deformation of Amorphous and Semi-Crystalline Materials", B. Escaig and G. G'sell, ed., Les Éditions de Physique, Paris (1982).
3. E. Plénard, Intérêt pratique de la grande capacité d'amortissement des fontes, Fonderie 177: 419 (1960).
4. M.A.O. Fox and R.D. Adams, Correlation of the Damping Capacity of Cast Iron with its mechanical Properties and Microstructure, J. Mech. Eng. Sci., 15: 81 (1973).
5. C. Bonjour, Nouveaux développements dans les outils de coupe en carbure fritté, Wear, 62: 83 (1980).
6. G. Fantozzi and I.G. Ritchie, Internal Friction caused by the Intrinsic Properties of Dislocations, J. Physique 42: C5-3 (1981).
7. K. Lücke and A.V. Granato, The Rigid Rod Model of dislocation Resonance Including Applications to Point Defect Drag, J. Physique 42: C5-327 (1981).
8. R. Schaller and W. Benoit, Internal Friction Associated with Precipitation in Al-Ag Alloys, in: "Internal Friction and Ultrasonic Attenuation in solids", C.C. Smith, ed., Pergamon Press, Oxford and New York (1980).
9. P. Millet, R. Schaller, W. Benoit, Characteristic Internal Friction Spectrum of Grey Cast Iron, J. Physique 42: C5-929 (1981).
10. P. Millet, R. Schaller, W. Benoit, Study of the internal Friction Spectrum of Grey Cast Iron, J. Physique 44: C9-511 (1983).
11. P. Millet, R. Schaller, W. Benoit, High Damping in Grey Cast Iron, J. Physique, 46: C10-405 (1985).
12. R. Schaller, W. Benoit, Développement d'alliages biphasés à fort amortissement, un exemple: les fontes grises, Material und Technik 13: 63 (1985).
13. R. Schaller, J.J. Ammann, A. Kulik, et al., Internal Friction Spectrum of WC-Co Composite Alloys, J. Physique 46: C10-387 (1985).
14. J.E. Bidaux, R. Schaller, W. Benoit, Internal Friction Associated with the Allotropic Transformation of Cobalt, J. Physique, 46: C10-601 (1985).
15. G. Fantozzi, H. Si Mohand and G. Orange, High Temperature Mechanical Behaviour of WC-6 wt % Co cemented Carbide, Inst. Phys. Conf. Ser. 75: 699 (1986).
16. B. Johannesson and R. Warren, Fracture of Hardmetals up to 1000°C, Inst. Phys. Conf. Ser. 75: 713 (1986).

Characterization of Thin Films and Layered Structures by Pulsed Photothermal Radiometry

A.C.Tam and H.Sontag

IBM Almaden Research Center
Dept. K06-803(E)
650 Harry Road
San Jose, CA 95120-6099

I. Introduction

Pulsed photothermal radiometry (PPTR) relies on the use of a short optical pulse to quickly heat up a sample, and the detection of the transient thermal radiation from the sample surface. Similar to pulsed photoacoustic sensing techniques[1], PPTR has applications in spectroscopy, coating thickness measurment and powder aggregation detection[2], thin-film thickness or thermal diffusivity measurements[3], pigment characterization[4] and fiber composite strength characterization[5]. This paper examines how the PPTR signal shape can be analysed to determine quantitatively the thickness of subsurface airgaps in layered structures. Furthermore, we demonstrate the effect of moisture uptake in polymer films on the PPTR signal.

The shape and amplitude of the PPTR signal is a function of the optical absorption lengths at the excitation and observation wavelengths, excitation pulse duration, thermal diffusivity, heat capacity and IR emissivity of the surface layer, and subsurface structure. The PPTR signal shape has been calculated analytically for the case of a semi-infinite uniform solid or a uniform film[6]. For the case of negligible heat diffusion during the excitation pulse, as is true for our experimental conditions, the amplitude of the PPTR signal is fully determined by the heat capacity, the optical absorption lengths and the emissivity of the top layer.

II. Measurement of subsurface airgap thickness

In the following, we examine the PPTR signal for a layered structure of an opaque film separated from a thick substrate by an airgap of thickness δ. The PPTR signal shape at certain delay times from the excitation pulse is found to be very sensitive to the value of δ up to a few hundred micron in thickness. We show how the signal shape can be deconvoluted to give δ, thus providing a quantitative non-destructive tool for delamination mapping. Also, since a small airgap behaves in essence like a thermal resistance R, the present technique can also be used to quantify the value of R between a coating and a substrate; if R is correlated with the adhesion strength, as we would expect (with R decreasing for better adhesion), the measurement of R also provides a new nondestrucive detection method for adhesion strengths.

The experimental arrangement is shown in fig. 1. The sample is an opaque polycarbonate film of 20 μm thickness containing 27% carbon black particles. At the back, the airgap between the film and a polished Germanium substrate is controlled to 1 μm accuracy by a piezoelectric translator. The front side is irradiated by an unfocused 8 ns pulse from a nitrogen laser (pulse energy 1 mJ). The center of the uniformly heated region is imaged onto a HgCdTe IR detector (sensitive from 7-12 μm, 1 MHz bandwidth), using an off-axis parabola mirror and a ZnSe lens, with a Ge plate in the collimated part of the beam to supress nitrogen laser straylight. The transient PPTR signal is recorded on a transient waveform recorder (Analogic DATA PRECISION 6000) and subsequently processed on an IBM PC.

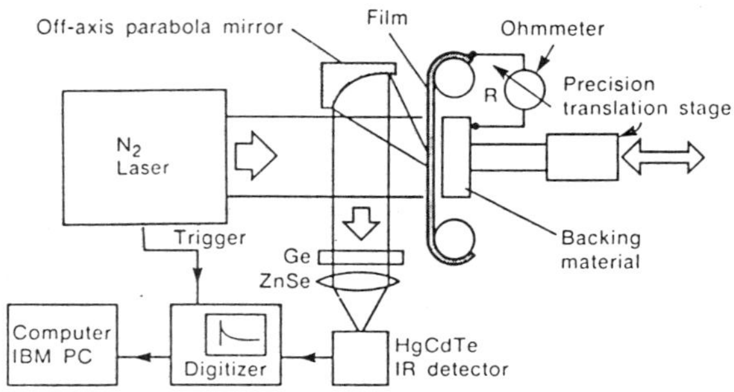

Figure 1 : Experimental arrangement to show PPTR measurement of airgap thickness betwen a thin opaque film and a thick backing material (substrate). The position of contact between the backing and the film is determined by measuring the electrical resistance between the conducting film and the backing.

Maximum surface temperatures reached directly after absorption of the nitrogen laser pulse are <5 K above T_0. After 2 ms however, the temperature is almost constant across the film and less than 0.4 K above T_0. Under these conditions convection is not expected to affect our measurements, which are taken on a time scale of 100 ms. On this time scale, lateral heat diffusion within the film can be neglected, since the laser spot is large (2 cm^2 in area) at the sample. However, any nonuniformity in the laser intensity profile can contribute to observable lateral diffusion effects.

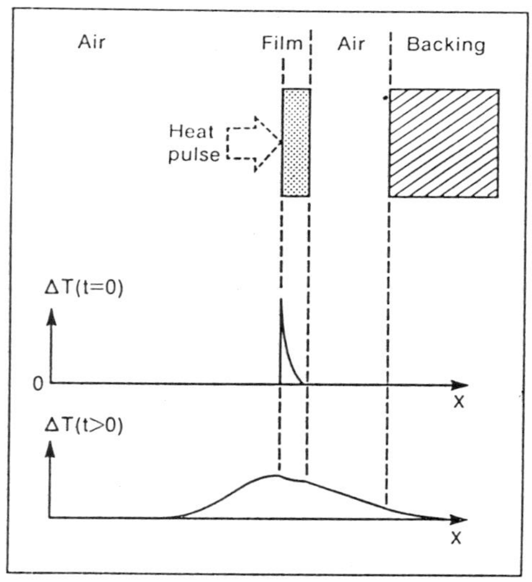

Figure 2 : Temperature distributions in the vicinity of the opaque film at time t=0 (i.e. immediately after the short excitation laser pulse) and at a later time.

To calculate the radiometry signal for our experimental set-up, we consider first the following analytical appproximation. Fig.2 gives the model with the temperature profiles at initial (t=0) and later (t>0) times. If the thermal resistance of the film is much smaller than that of the airgap, the temperature drop within will be negligible after a short time, and we can therefore use an average film temperature T_f. The semi-infinite backing material has a high thermal diffusivity, such that the surface temperature is not raised significantly above the ambient temperature T_0. Heat loss of the film is then governed by the heat flow through the airgap, which is given by the temperature gradient T_f-T_0. Taking into account radiation losses and neglecting additional cooling due to heat conduction on the other film surface, the energy loss per unit area is given by

$$C\,h\,\dot{T}_f = (\frac{\lambda}{\delta} + 8\sigma\varepsilon T_o^3)\,(T_f - T_o) \qquad [1]$$

where C is the film heat capacity, h is its thickness, λ the conductivity of air, δ the width of the airgap, σ is Stefan-Boltzmann's constant and ε is the film emissivity. Eq.(1) can be solved to yield

$$T_f = T_0 + T_e \exp(-\frac{\frac{\lambda}{\delta} + 8\sigma\varepsilon T_0^3}{C\,h}\,t) \qquad [2]$$

where T_e is the initial averaged film temperature rise. From eq.(2), it is obvious that for a known film thickness h the gap can be determined quantitatively by analyzing the slope of the radiometry signal on a logarithmic plot. It also turns out that for air as a gap medium, radiation loss is smaller than heat conduction loss for an airgap thickness less than 1 mm. Since most of our investigations have been performed on airgaps less than 200 μm in thickness, radiation losses can be neglected here.

Deviations from the above approximation become apparent under three circumstances: (i) the thermal diffusivity of the backing material is small, causing appreciable surface heating of the backing material and a smaller heat flux compared to eq. (1); as a result, the ribbon stays hotter at later times; (ii) for small airgaps, the temperature variations in the film and in the the backing cannot be neglected any more; and (iii) for large airgaps cooling due to heat conduction on both film surfaces has to be considered. To account for these cases we have chosen a numerical procedure employing the explicit finite-difference formula technique[7].

We define $T_{i,j} = T(x_i, t_j)$ as the temperature at position $x_i = x_0 + \sum_{\nu=1}^{i} \delta_\nu$ and time $t_j = j\,\tau$. Here, δ_ν is the separation between points x_{i-1} and x_i and is a constant within a homogeneous medium, while τ is the time interval in the calculation process. Within a homogeneous medium, characterized by a thermal diffusivity D, the temperature at position x_i and time t_{j+1} is iteratively calculated as

$$T_{i,j+1} = T_{i,j} + M\,(T_{i-1,j} - 2T_{i,j} + T_{i+1,j}) \qquad [3]$$

where $M = \frac{D\tau}{\delta^2}$. At an interface, the heat flux must be conserved. For the case of a continuous temperature profile, this yields the interface temperature as

$$T_{s,j+1} = \alpha T_{s-1,j} - (\alpha + \beta - 1)T_{s,j} + \beta T_{s+1,j} \qquad [4]$$

where $$\alpha = 2\tau \frac{\frac{\lambda_1}{\delta_1}}{\frac{\lambda_1\delta_1}{D_1} + \frac{\lambda_2\delta_2}{D_2}}$$

and $$\beta = 2\tau \frac{\frac{\lambda_2}{\delta_2}}{\frac{\lambda_1\delta_1}{D_1} + \frac{\lambda_2\delta_2}{D_2}}$$

$\lambda_{1,2}$ and $D_{1,2}$ are the thermal conductivity and diffusivity of the media 1 and 2, respectively.

At an interface with a thermal resistance R, a temperature drop occurs proportional to the heat flow. This yields the temperatures $T_{r,j+1}$ and $\overline{T}_{r,j+1}$ at the interface in media 1 and 2, respectively, as

$$T_{r,j+1} = 2\,M_1\,(T_{r-1,j} + \frac{\delta_1 f}{\lambda_1}) - (2\,M_1 - 1)\,T_{r,j} \qquad [5]$$

$$\overline{T}_{r,j+1} = 2\,M_2\,(\overline{T}_{r+1,j} - \frac{\delta_2 f}{\lambda_2}) - (2\,M_2 - 1)\,\overline{T}_{r,j} \qquad [6]$$

where $f = (\overline{T}_{r,j} - T_{r,j})/R$ is the heat flux across the temperature discontinuity at the interface. It should be noted, that to ensure mathematical stability of this method, the relation $M<0.5$ should hold for each layer[7]. Fig.3 gives the corresponding front surface temperatures as a function of airgap width. Eq.3-6 can be used to calculate transient temperature profiles in multi-layered structures, including possible thermal resistances at interfaces.

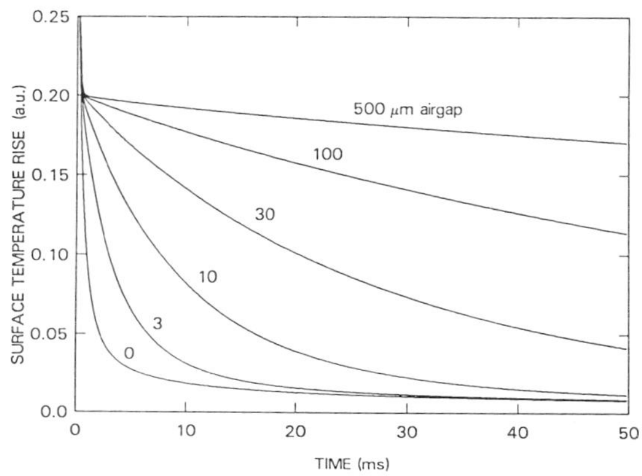

Figure 3 : Computed film surface temperature variations for a 20 μm film separated from a Ge substrate by various airgap widths, as indicated. The surface temperature at t=0 (after absorption of the heating pulse) is normalized to unity.

Experimentally, we clearly observe the effect of the backing from the long-time decay of the transient radiomatry signal. As shown in fig. 4 the early decay is independent of the airgap width, and is also independent of the film thickness, as the optical penetration depth is much smaller. We can easily distinguish different airgap widths up to 500 μm, with 1 μm accuracy.

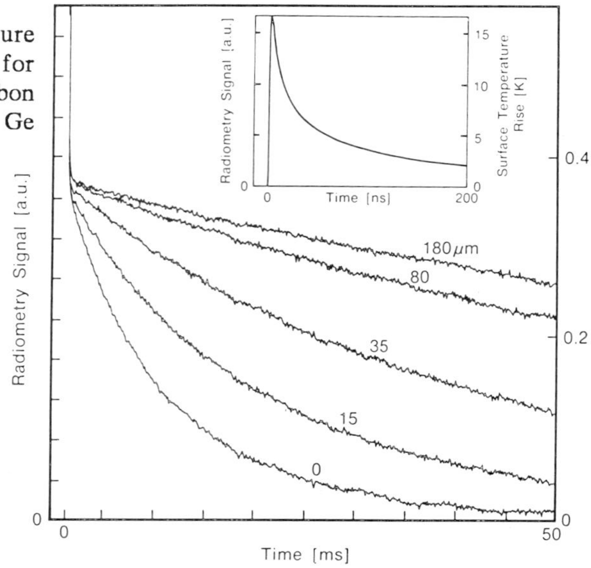

Figure 4 : Observed surface temperature (proportional to the PPTR signal I_R) for a polycarbonate film (with 27% carbon loading) at various separations from a Ge substrate, as indicated.

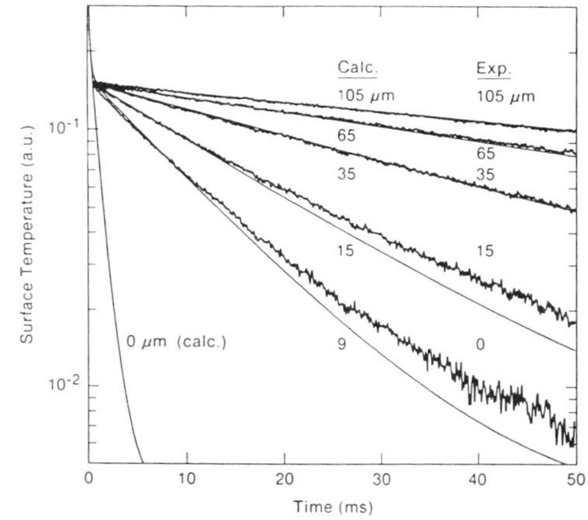

Figure 5 : Comparison between the calculated and experimental PPTR signal I_R for several airgap thicknesses, as indicated.

Quantitative agreement with the model outlined above, however, is only good for airgaps larger than 30 μm in width. Fig. 5 gives a comparison between calculated surface temperatures and observed PPTR signals which are proportional for small surface temperature rise. It should be noted, that the theoretical shape is fully determined by the material properties, and only the amplitude was scaled to compare the two sets of data. While the agreement is very good for large airgaps, it becomes increasingly worse for small airgaps below 30 μm. Several effects, which were not accounted for in the simple model, cause these deviations: (i) when the airgap width becomes comparable to the mean free path in air, thermal transport cannot be described by a continuous theory anymore, but is mainly determined by desorption rates from the surfaces involved. (ii) The surface roughness of both film and substrate do not allow a thermal contact at all points; hence, even at 0 μm gap width, there is a residual thermal resistance. Our data indicate that in our system, with an estimated surface roughness of 1 μm, this thermal resistance corresponds to a 10 μm air layer. Figure 6 gives a plot of the inverse slope of log I_R at t=5 ms, which according to eq.(2) should be proportional to both the airgap width and the ribbon thickness. We observe good agreement with the calculated slope (which is a zero parameter fit according to eq.(2)) over a wide range, while deviations again become apparent at low separations.

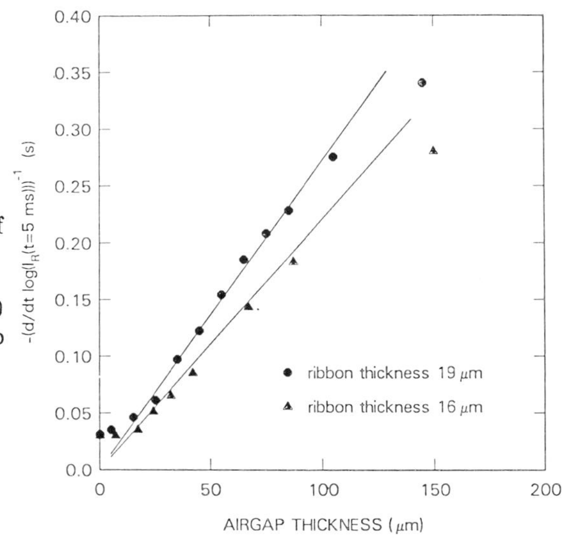

Figure 6 : Dependence of $y = -[\frac{d}{dt} \log I_R(t = 5ms)]^{-1}$ as a function of airgap thickness δ. Eq.(2) indicates that y should be proportional to δ.

III. Monitoring of moisture uptake by polymer films

Polymer films have numerous applications for insulation or protection or as sensors for various physical properties, e.g. pressure or humidity. Moisture uptake can have a profound effect on the actual performance, as it may affect the corrosion protection or physical properties of these films. For monitoring moisture uptake and the changes of physical properties associated with it, mainly contact techniques such as weight measurements or detection of electrical or mechanical properties have been applied. Noncontact measurements have been performed measuring the change in refractive index associated with moisture uptake or by direct detection of diffusion through a membrane. In the following we demonstrate the use of PPTR to detect humidity effects on thermal properties of polymer films. Moisture uptake may yield changes in the heat capacity or optical absorption lengths. Under transient ambient humidity conditions, this enables us to obtain information on sorption or desorption of water from the polymer films.

Figure 7 : Two PPTR signals as obtained from carbon loaded polycarbonate film (thickness 20 μm) for ambient relative humidities of 0% and 95%. The two signals (normalized to the same heights) are almost indistinguishable from each other.

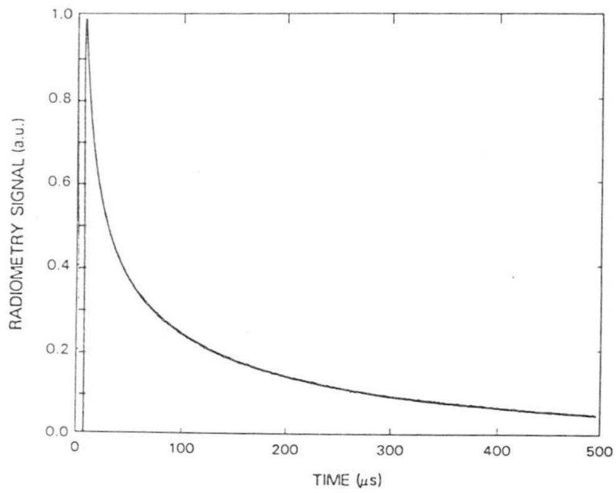

Various thin film materials, such as carbon loaded polycarbonate or polyvinyl acetate films, were used because of their strong absorption at the nitrogen laser wavelength (337.1nm). The film is mounted flat in a humidity controlled chamber. Excitation and detection of the PPTR signal occur as outlined above.

Figure 8 : PPTR signal magnitude of polycarbonate (sample 1) and polyvinyl acetate films as a function of external relative humidity (T=22°C).

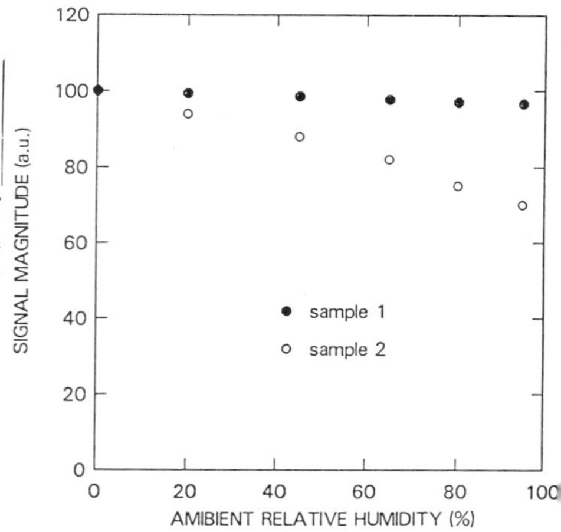

As shown in fig.7, the PPTR signal shape is hardly affected for ambient humidities ranging from 0% to 95 % at a temperature of T=22°C. The thermal diffusivity is obviously almost independent of water content. However, the signal magnitude, as expected, shows a strong dependence on the external humidity conditions for those materials, that are known to absorb moisture. In addition to measurements under equilibrium conditions, as shown in fig.8, we are able to determine the speed of water uptake and release, which for the films of thickness 20-200 μm occur on a time scale of several minutes, typically. Fig.9 shows a typical measurement, where the signal amplitude has been monitored after abrupt changes in the ambient humidity. Typical diffusion constants for water vapor in polymers[8] range from 10^{-9}-10^{-7} cm^2s^{-1}. This is in agreement with the time constants observed in our transient radiometry measurements, as outlined above. For an unambiguous quantitative correlation between water uptake and these radiometry measurements, additional work is needed to characterize water diffusion, *i.e.* by weight or volume change measurements. In contrast to these techniques, however, PPTR monitors only the properties of the surface skin layer and can therefore yield additional spatial information.

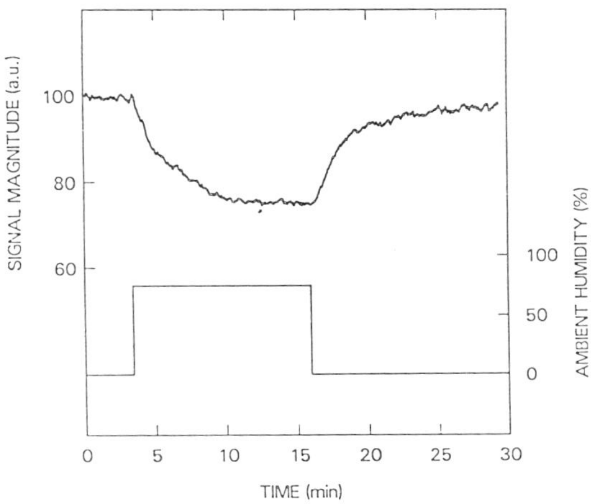

Figure 9 : Transient PPTR signal magnitude as obtained from a 120 μm polyvinyl acetate film after abrupt change in ambient relative humidity from 0% to 77% (at t=3.5 min) and back to 0% (at t=15.5 min). Moisture absorption and release occur on a time scale of several minutes.

IV. Summary

We have demonstrated two novel applications of PPTR. Shape analysis of the PPTR signal allows a characterization of opaque, multi-layered structures. This has been demonstrated by determining the thickness of subsurface airgaps between an opaque film and a substrate. Absorption and release of water in polymer films can be monitored by analyzing the magnitude and shape of the PPTR signal. The thermal diffusivity in the films investigated is found to be hardly changed by the water content.

This work is supported in part by the Office of Naval Research.

References

1. See, for example, A.C.Tam, *Rev.Mod.Phys.* **58**, 381 (1986)
2. A.C.Tam and B.Sullivan, *Appl.Phys.Lett.* **43**, 333 (1983)
3. W.P.Leung and A.C.Tam, *Opt.Lett.* **9**, 93 (1984)
4. R.E.Imhof, D.J.S.Birch, F.R.Thornley, J.R.Gilchrist, and T.A.Strivens, *J.Phys. E* **17**, 521 (1984)
5. P.Cielo, *J Appl.Phys.* **56**, 230 (1984)
6. W.P.Leung and A.C.Tam, *J.Appl.Phys* **56**, 153 (1984)
7. J.Crank, "The Mathematics of Diffusion", Clarendon Press, 2nd ed. (1975), p.137
8. A.W.Myers, J.A.Meyer, C.E.Rogers, V.Stannett, and M.Szwarc, *TAPPI* **44**, 58 (1961)

EDDY CURRENT SIZING OF CASE DEPTH IN BEARING COMPONENTS

R. Palanisamy and K. E. Jackson

Nondestructive Evaluation and Sensor Technology Group
The Timken Company
Canton, Ohio

INTRODUCTION

Encircling eddy current coil techniques are being used for the nondestructive evaluation of case depth in bearing components (Figure 1). These techniques, because of their ability to measure only bulk properties in test specimens, lack the necessary spatial resolution to detect local variations in case depth leading to eddy current imaging of the case depth. Further development in this area requires the knowledge of eddy current properties as a function of depth from the raceway (i.e. rolling contact surface). This paper describes the experimental determination of magnetic permeability and electrical conductivity of induction hardened case and unhardened core of a tapered roller bearing cup.

To obtain information regarding the variation of eddy current properties as a function of depth, response of a 2MHz absolute pencil probe was recorded for various case depths obtained through progressive

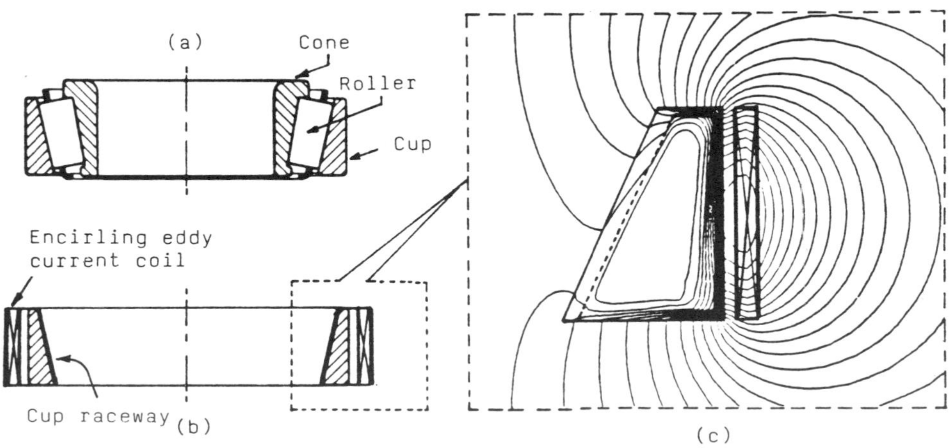

Fig. 1. Encircling eddy current coil technique for the inspection of bearing components: a) components of a tapered roller bearing, b) encircling coil, and c) flux plot.

grinding of a) an induction hardened cup (AISI 1080M steel) and b) a carburized and hardened cone (AISI 8219 steel). In the case of the cone, response of a U-core probe was recorded at 100 Hz and 1 KHz.

Finally, finite element predicted magnetic field penetration and eddy current distribution in the induction hardened specimen are presented for different frequencies, and for both U-core and pencil probes.

EXPERIMENTAL DETERMINATION OF MAGNETIC CHARACTERISTICS

Encircling coil techniques are based on empirical results and they lack the necessary spatial resolution to measure case depth accurately. A better approach would be to scan a rotating test specimen with an eddy current probe, achieving the desired spatial resolution and 100% inspection. Prerequisite to the development of such an advanced case depth sizing system is the knowledge of magnetic permeability and electrical conductivity as a function of depth from the raceway. Experimental determination of B-H characteristics of induction hardened case and unhardened core of a tapered roller bearing cup is described in this section.

Sample Preparation

Two thin rings were cut from a tapered roller bearing double-cup that had been induction heated to form the case region as shown in Figure 2. The cross-section of the hard ring was chosen close to the raceway and the soft ring well within the unhardened core, avoiding the case-to-core

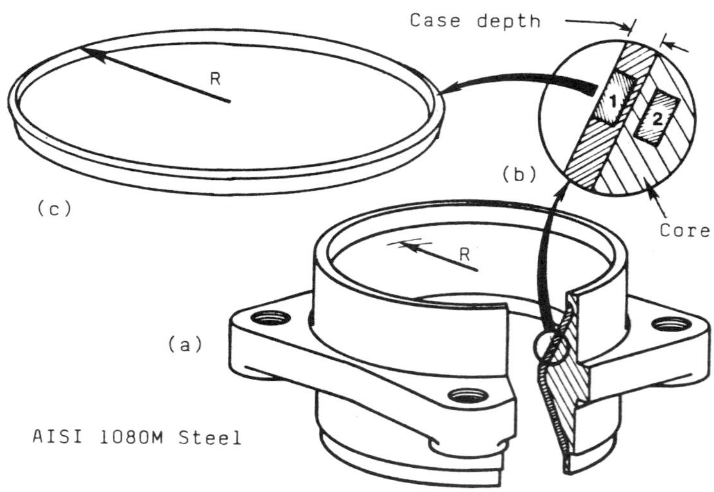

Fig. 2. a) Cross-section of a flanged double-cup of a tapered roller bearing.

b) Induction hardened case depth and unhardened core. The rectangles 1 and 2 are the cross-section of circular rings ("hard ring" and "soft ring") cut from the case and core respectively (samples for the experimental determination of magnetic characteristics and electrical conductivity of the case and core separately).

c) Ring (No. 1) cut from the hardened case ("hard ring").

interface. Because of the induction hardening process, eddy current properties could be expected to be constant within the cross-section of the hard ring (results of the experiments described in the next section confirmed this expectation).

	Hard Ring	Soft Ring
Mean diameter (cm)	9.746	10.478
Cross-section (cm x cm)	0.131	0.143
Primary turns	319 (20 AWG)	343 (20 AWG)
Secondary turns	1176 (32 AWG)	995 (30 AWG)

B-H characteristics of the rings were determined experimentally at the Technology Center, The Timken Company, using Model MH-20 Hysteresisgraph System (Walker Scientific, Inc.). The results are presented in Figures 3, 4 and 5. The data obtained from these graphs are given below:

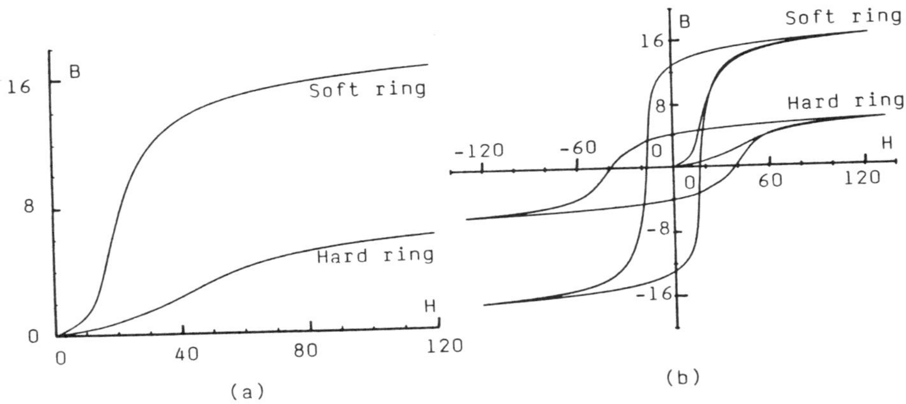

Fig. 3. a) Initial-magnetization curves, and b) major hysteresis loops for the hard and soft rings (H = magnetizing force in Oersteds, B = flux density in Kilogausses).

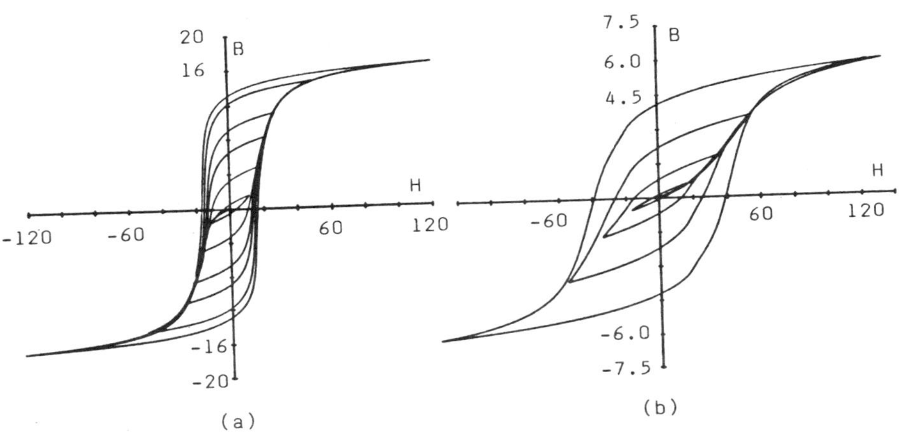

Fig. 4. Normal hysteresis loops for a) soft ring and b) hard ring.

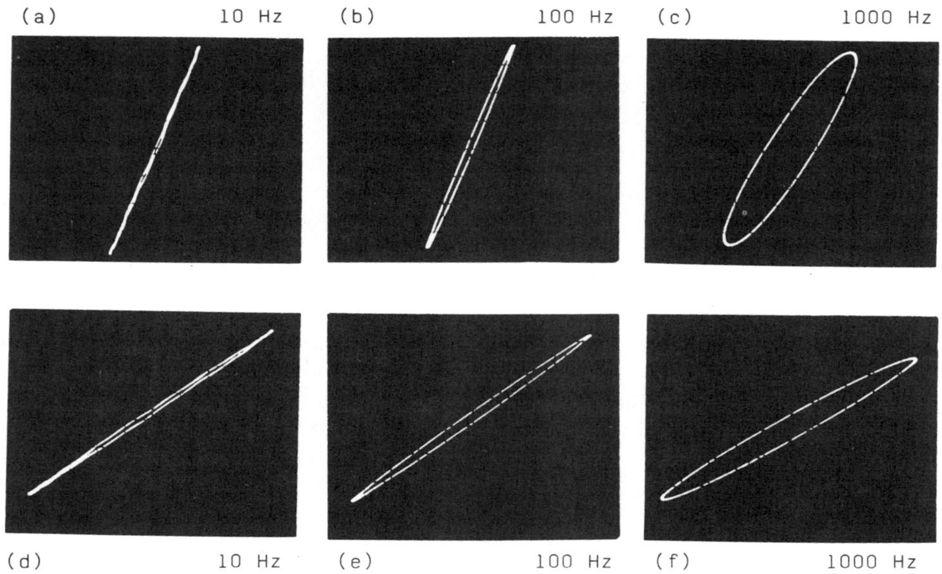

Fig. 5. B-H characteristics of soft (a, b and c) and hard (d, e and f) rings at 10, 100 and 1000 Hz. The peak-to-peak B value is 200 gausses in these traces.

	Hard Ring	Soft Ring
Initial relative permeability	26	78
Maximum relative permeability	68	400
Saturation flux density (Kilogausses)	6.0	16.4
Coercivity (Oersteds)	39	16.8
Retentivity (Kilogausses)	3.9	12.8

EXPERIMENTAL DETERMINATION OF ELECTRICAL CONDUCTIVITY

After completing the above magnetic measurements, windings were stripped from each ring. Rings were then cut into three equal sectors 120 degrees apart. Electrical conductivity of these six samples were experimentally determined at the Thermophysical Properties Research Laboratory (TPRL), School of Mechanical Engineering, Purdue University, using the four probe technique. The results are:

Electrical conductivity of hard ring: 3127000 mhos/m
Electrical conductivity of soft ring: 4662000 mhos/m

PROFILING EDDY CURRENT PROPERTIES OF THE CASE MICROSTRUCTURE AS A FUNCTION OF DEPTH

Magnetic permeability and electrical conductivity are the only two physical properties of the test specimen that control the eddy current response. Any deviation in the desired condition of the specimen (such as hardness, stress, alloy composition, dimension, etc.) affects these properties directly, and the eddy current response indirectly. (Other factors such as probe design, lift-off, and probe/specimen test configuration also affect the eddy current response. However, these parameters do not provide any additional information about the test specimen).

The eddy current properties as a function of depth can be profiled indirectly by recording the response of an eddy current probe for decreasing case depth obtained through progressive grinding until the core is reached. Although the influence of permeability and conductivity on the eddy current signal amplitude (vector sum) are unseparable, their combined effect on the eddy current response could provide information as to their combined variation as a function of depth.

Cross-section of the samples and probes are shown in Figures 2 and 6. Results of case depth measurements are plotted in Figures 7 and 8. The variation in case depth was obtained by carefully grinding the samples (avoiding heating). Using masters the stability of the eddy current instruments was checked prior to each measurement.

FINITE ELEMENT MODEL ANALYSIS

Magnetic field penetration and eddy current density in induction hardened case and unhardened core of AISI 1080M steel were analyzed using the experimentally obtained permeability and conductivity values as input to the finite element numerical model. The geometry of the pencil probe (Figure 6b) was analyzed using an axisymmetric finite element code and the geometry of the U-core probe (Figure 6c) using a 2-D finite element code. The partial meshes of these geometries are shown in Figure 9.

Plots of Figure 10 show clearly that the design of the pencil probe (coil-ferrite arrangement) does not produce adequate flux penetration at the case depths of interest in this experiment as expected.

Magnetic field penetration in the case of the U-core probe is given in Figure 11 for four frequencies. Contours of constant induced current density (magnitude) perpendicular to the plane of analysis are plotted in Figure 12. Higher magnetic permeability and electrical conductivity of the unhardened core are the cause for the increased eddy current density in the core near the case-to-core interface as shown in Figure 12b.

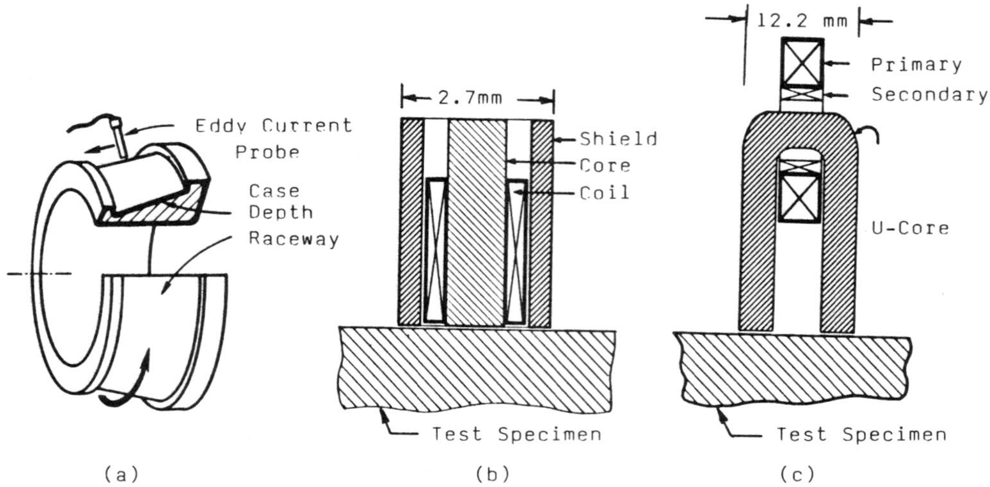

Fig. 6. a) Pencil probe over a cone raceway (carburized and hardened).
b) Cross-section of the pencil probe used in the experiment.
c) Cross-section of the U-core probe used in the experiment.

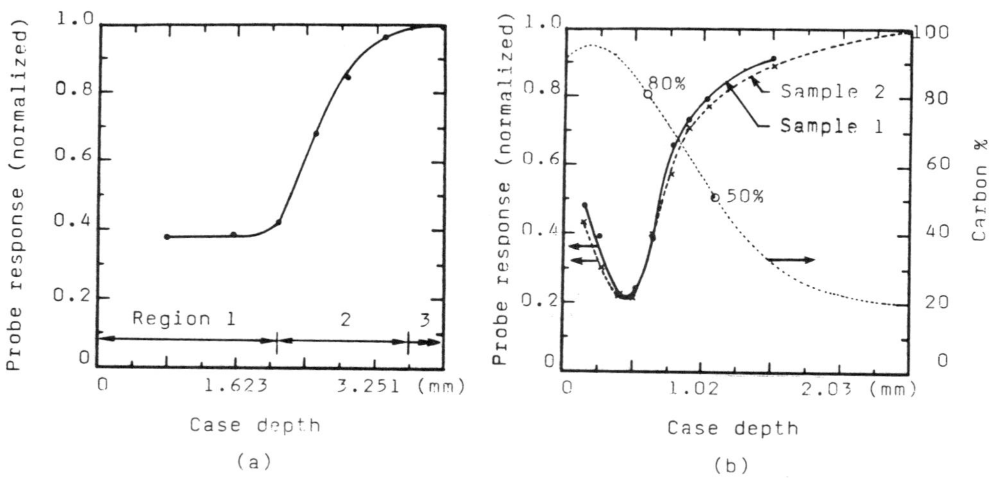

Fig. 7. a) Plots of experimental data showing the 2MHz absolute probe response (vector sum) for decreasing values of case depth in the flanged double-cup of a tapered roller bearing. The decreasing case depth was obtained by progressively grinding the raceway of a cup having nominal case depth (induction hardened AISI 1080M steel).
(Region 1 = hardened layer, 2 = transition region, 3 = Core)

b) Plots of experimental data showing the 2MHz absolute probe response (vector sum) for decreasing values of case depth in a tapered roller bearing cone. The decreasing case depth was obtained by progressively grinding the raceway of a cone with nominal case depth (carburized and hardened AISI 8219 steel). The percentage carbon vs case depth is a standard curve while the 80% and 50% depths shown were experimentally determined.

Plots of eddy current density as a function of depth for the U-core probe are given in Figure 12b.

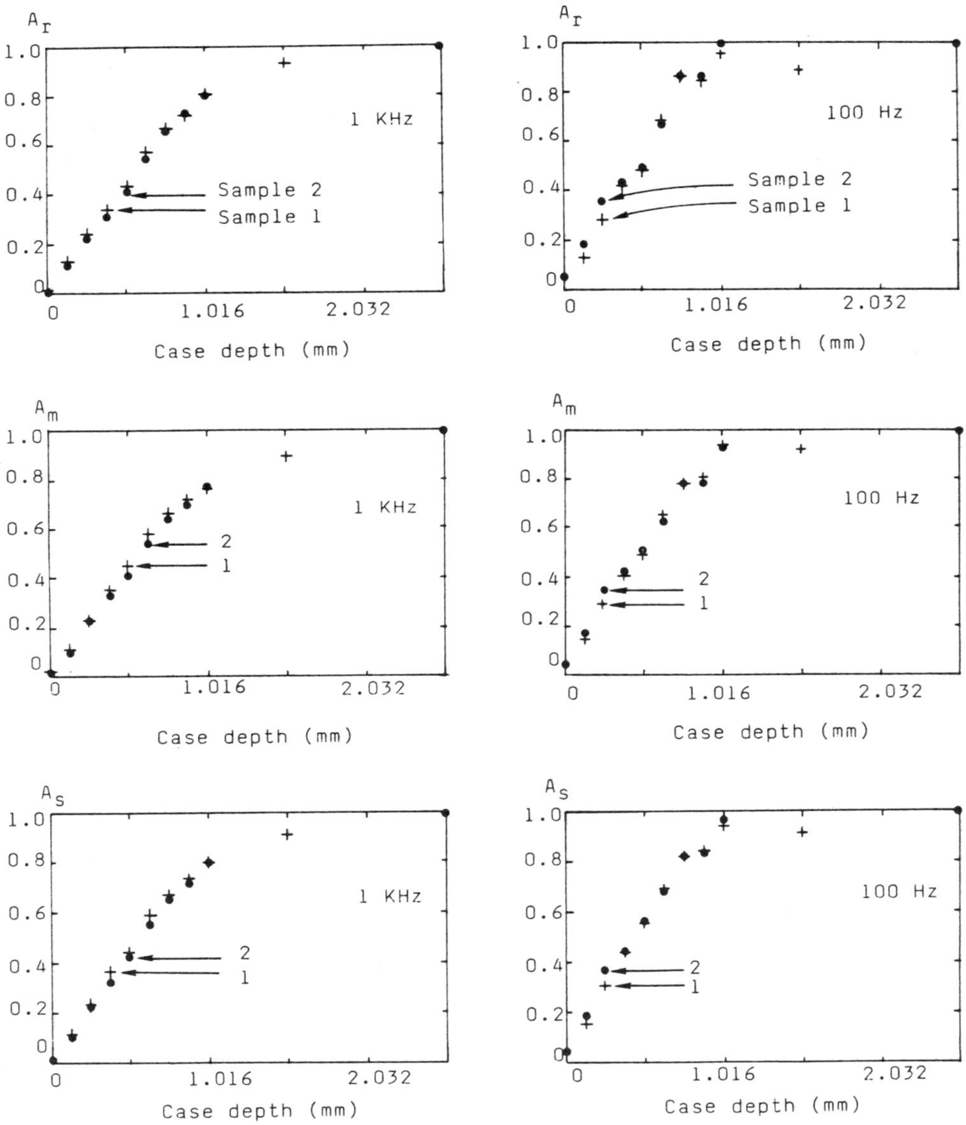

Fig. 8. U-core eddy current probe response (carburized and hardened AISI 8219 steel cone). A_r and A_m are the normalized resistive and reactive components of the signal. A_s is the normalized vector sum of A_r and A_m.

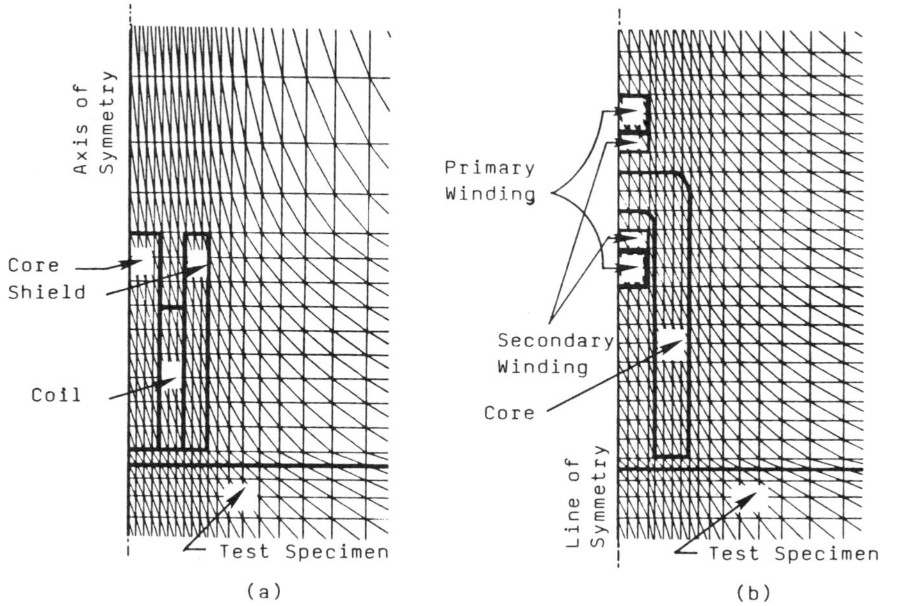

Fig. 9. a) Finite element discretization (partial mesh) for the axisymmetric analysis of the pencil probe. Because of the circular symmetry only half the test geometry needs to be analyzed.

b) Finite element discretization (partial mesh) for the two-dimensional analysis of the U-core probe. Because of the symmetry about the vertical central line only half the test geometry needs to be analyzed.

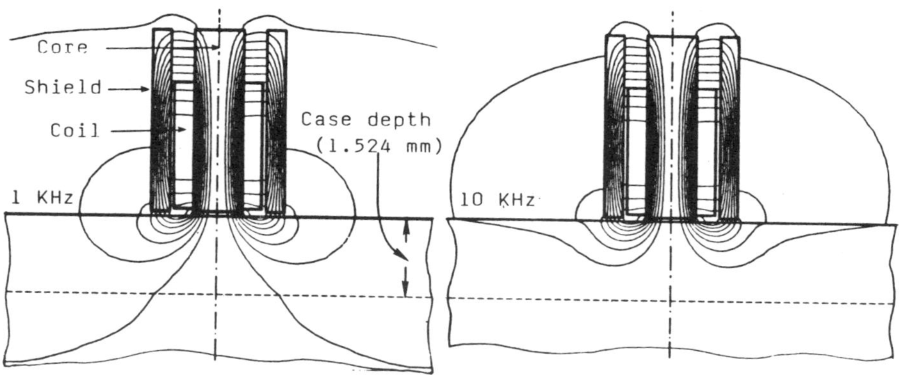

Fig. 10. Magnetic field plots for the pencil probe over AISI 1080M steel with 1.524 mm (0.06 inches) induction hardened case depth.

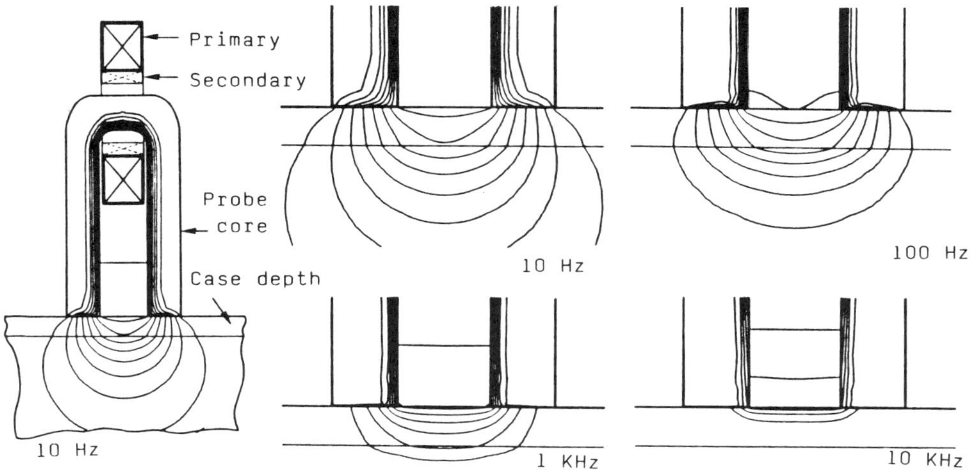

Fig. 11. Magnetic field plots for the U-core probe over AISI 1080M steel with 2.032 mm (0.08 inches) induction hardened case depth.

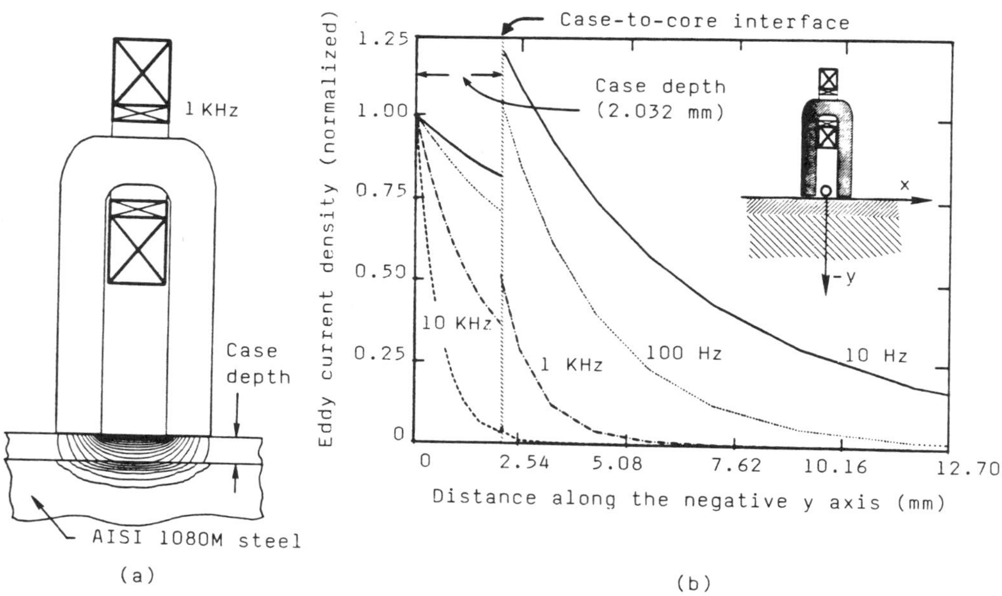

Fig. 12. a) Contours of constant eddy current density (magnitude) in AISI 1080M steel and 2.032 mm (0.08 inches) induction hardened case (excitation frequency of the U-core probe is 1 KHz).

b) Eddy current density (magnitude) as a function of depth for the above test geometry.

ACKNOWLEDGMENT

The authors wish to thank M. N. Curran and R. J. Vuksta of the NDE and Sensor Technology Group, D. A. Rohrer of Research Support Services, The Timken Company, for their assistance in preparing the samples and performing the experiments, and C. J. Morris, Manager-NDE and Sensor Technology Group for his valuable suggestions and general assistance throughout this investigation.

CONCLUSION

Magnetic permeability and electrical conductivity of induction hardened case and unhardened core of a tapered roller bearing double-cup have been determined experimentally. Variation of eddy current properties along the radius of an induction hardened cup, and, carburized and hardened cones, have been profiled using two types of eddy current probes. These experimental results and the finite element predictions show the feasibility of developing a multifrequency/multiparameter eddy current system for the real-time and 100% inspection of bearing components for case depth.

NONDESTRUCTIVE DETERMINATION OF HARDENING DEPTHS

WITH ULTRASONIC SURFACE WAVES

J. Rivenez, A. Lambert, and C. Flambard

Centre Technique des Industries Mécaniques
60300 Senlis, France

INTRODUCTION

Surface treatments are standard practice in mechanical engineering for producing metal parts subjected to operating stresses (such as fatigue, shocks, wear, abrasion ...). Their penetration inside the materials varies from several hundreths of millimeters to several millimeters, depending on process implementation, but also on specific process-related features, such as:

- the type of heating means used in surface hardening processes, i.e. induction, torch, electron bombardment, laser, etc.

- the type of element that is diffused in case hardening to modify chemical composition of surface layers and provide them with superior resistance through thermo-chemical treatment, i.e. carbon, nitrogen, sulfur, boron, etc.

Independently of the method used, producing a treated layer with precisely the expected properties requires first of all to determine the treatment operating procedure by experiment and then to comply with it very strictly. Precise control of the operation does not eliminate inspection of the actual finished treatment. The extent of this inspection depends on the application, in some cases hardness may be tested simply by file scratching, whereas in the case of structural components, a most thorough quantitative analysis is needed. Most current methods to gain quantitative data about the entire treated layer are based on destructive principles: as a result they can be applied only to a limited quantity of sampled workpieces, which has the drawback to require long and meticulous preparation.

A nondestructive testing method has been investigated at CETIM for several years now (1,2), to assess the quality and, which is the most important, the thickness of treated layers in the range from several tenths of a millimeter to about two millimeters. This range corresponds to the thicknesses generally encountered for case-hardened carbonitride-treated and laser hardened steel. The method which has been developed is based on the accurate measurement of the change of the velocity of Rayleigh ultrasonic waves propagating along the treated surface. The equipment we have developed is compact and easy to use, owing to the latest micro-computer and computing developments.

This paper includes first an outline of the underlying theory, which is followed by a description of the implementation of the method and several results.

RAYLEIGH WAVE INTERACTION WITH SURFACE LAYERS

Surface Waves (background)

Several types of ultrasonic waves can propagate in a solid body:

- longitudinal waves in which the particle displacement is parallel to the direction of propagation, and

- transverse waves in which the particle displacement is perpendicular to the direction of propagation.

There are also other types of wave propagation, depending on sample geometry. In particular, at the surface of a body, a wave can propagate without penetrating the body as a whole: it is called surface wave. The most common surface wave is the Rayleigh wave, characterized by elliptical particle motion, including therefore both longitudinal and transverse vibrations. The effective penetration depth of Rayleigh waves below the surface of the material is about one wavelength. Its velocity of propagation is dependent upon the elastic constants (Young's modulus, E, Poisson's ratio, ν) of the medium.

Basic Principle of Developed Method

In a homogeneous medium, the velocity of Rayleigh waves is independent of frequency. This is not however the case when mechanical properties change between the surface and the core, as in surface-treated materials. In these treated layers, hardness decreases with depth and can be correlated to a variation of elastic properties, such as the modulus of elasticity.

The principle of the method consists in measuring the propagation velocity at various wave frequencies, for a given hardened layer (see figure 1). The depth of interaction within the material depends upon the frequency or the wavelength, as shown schematically in fig. 2. By measuring the propagation velocity at each frequency a diagram of the velocity versus depth is obtained which can can be correlated to the hardness gradient.

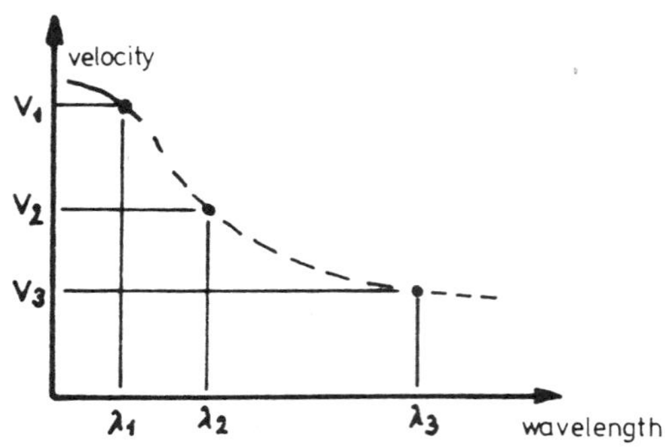

Figure 1: Velocity versus wavelength.

Figure 2: Depth of penetration of surface wave versus wavelength.

Theory

The surface-treated body is modelled by superposing an appropriate number "n" of homogeneous layers (1). Each layer is given the features of the actual material at the same depth (in particular Young's modulus "E" was assumed to be the most significant parameter and to vary like hardness when the types of steel and treatment are strictly defined). The wave is assumed to propagate in a single direction, parallel to the (plane) interface between layers, the wavefront being perpendicular to this direction. Harmonic motion is also assumed.

General laws of mechanics are applied to all layers, and the boundary conditions at the various interfaces are taken into acount (continuity of displacement and stress, in particular, the free surface is not stressed). For "n" layers, this leads to a system of 2n - 2 equations with 2n - 2 unknowns which is solved by making the main determinant equal to zero. In practice, the solution can only be obtained by numerical computation and a FORTRAN software has been developed for this purpose.

Determination of the values which should be attributed to the various coefficients of each layer (E, ν) is the main difficulty since those can only be obtained by experimentation. Extreme values, for treated and untreated material, are first determined from velocity measurement of longitudinal and transverse waves. Then various distributions, i.e. various gradients within the layer, are defined. They can be determined either experimentally or, in a general manner, by taking a schematic model as basis (figure 3).

A characteristic depth of treatment X_H is defined as the depth which corresponds some preselected hardness between minimum and maximum values. The preselected hardness is taken below as the mean between minimum and maximum values. The gradient is defined by the parameter G equal to the h/X_H ratio, where h is the distance between the depth where hardness starts to decrease and the characteristic depth.

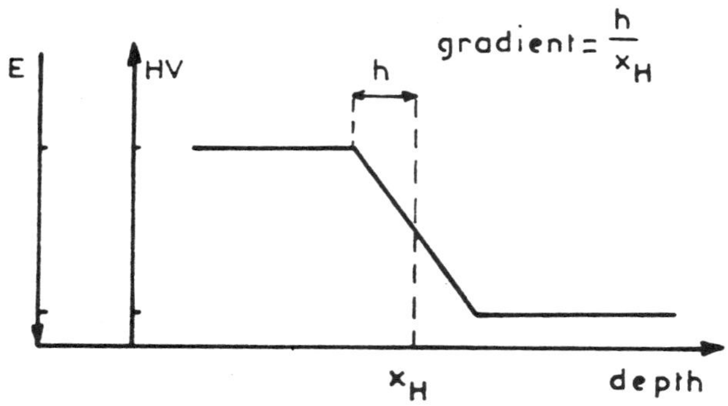

Figure 3: Schematic model of gradient

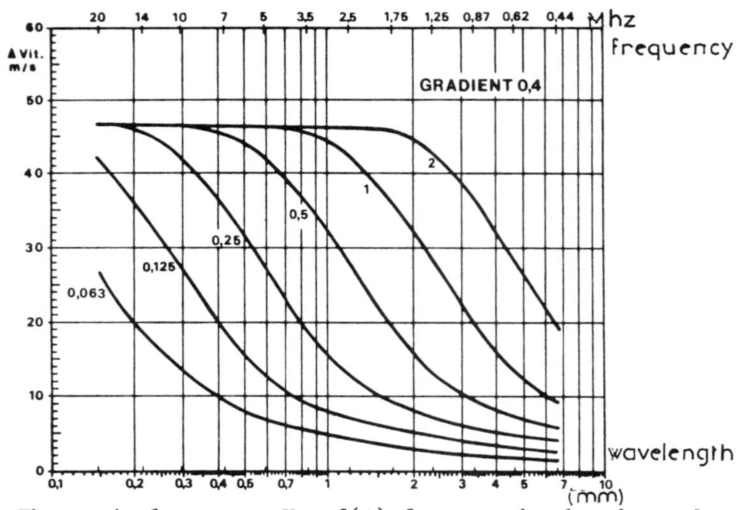

Figure 4: Theoretical curves $\Delta V = f(\lambda)$ for case-hardened steel XC10. The results are shown for 6 mean hardening depths X_H ranging from 2 to 0.063 mm.

Results from a numerical calculation are presented in Fig. 4 for a case-hardened XC10 steel with a hardness gradient of .4. The results depend but little on the value of G, as long as G remains lower than about .6. The results are shown in a semi-logarithmic plot with the logarithm of wavelength as abscissa and the velocity difference ΔV between the measured velocity and the velocity of the untreated material as ordinate. Several curves are shown for several mean treatment depths X_H

METHOD IMPLEMENTATION

Principle of Measurement

Our application, like others, especially the nondestructive determination of residual stresses (3, 4, 5), require very accurate measurements of ultrasonic velocity. Several methods are known to exist. We have chosen a solution which consists in analyzing the propagation of an ultrasonic pulse of very short duration and broad frequency-band spectrum (approximatively 1-10 MHz), thus making unnecessary any frequency scanning.

Experimental Method

The method consists in measuring the difference in transit time over the same path on a treated and untreated workpiece surface for each frequency component of a very short pulse and in deriving then the difference in velocity. This difference is obtained by computing the phase of the cross energy spectrum of two signals obtained on the treated and untreated materials.

Experimental set-up

The experimental set-up includes an ultrasonic unit (transducer), an acquisition unit for signal digitalization and a data processing unit. The transducer is of the single-piece type, and includes two piezoelectric components set with respect to the surface at the angle corresponding to Rayleigh wave generation. Various probe shapes are used, according to the geometry of the workpiece.

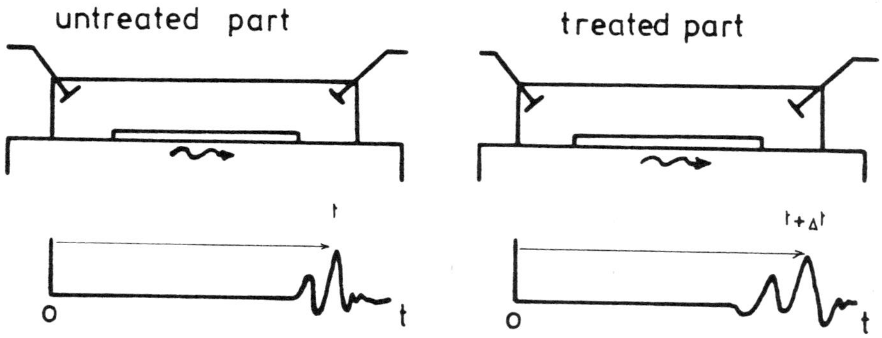

Figure 5: Principle of measurement

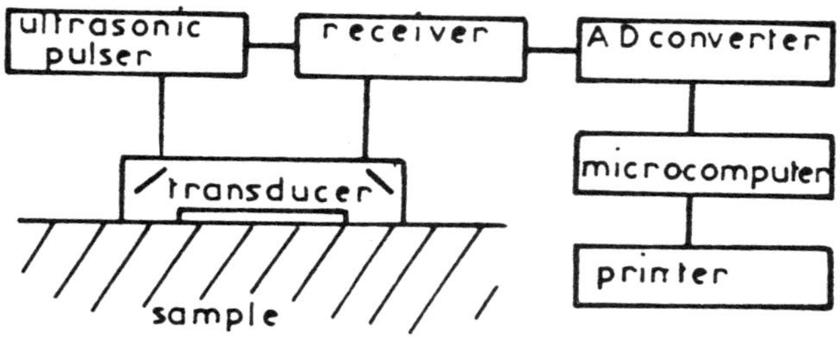

Figure 6: Block diagram of experimental set-up

Data interpretation procedure

First, time delay values are computed as a function of signal spectrum frequencies and are printed out. For easy interpretation, the corresponding velocity changes are then plotted on a graph where the theoretical curves for several mean depths have been plotted for a treatment and a steel grade as close as possible as those of the specimen. From the position of the experimental curve, the mean depth of treatment (horizontal position) can be read, as well as the hardness achieved (vertical position).

RESULTS

The validity of the method we have described for characterizing case-hardened layers has been confirmed by tests on more than 300 specimens of various steel grades in laboratory conditions. Its implementation in industry has also solved industrial problems. We present below as examples the results of several tests on:

- case-hardened specimens,
- carbonitride-treated specimens,
- laser-hardened specimens.

In figure 7 a-d, two curves are shown for each example: the left curve shows the microhardness relationship as determined by destructive testing; the right diagram shows the experimental curve deduced from ultrasonic measurements.

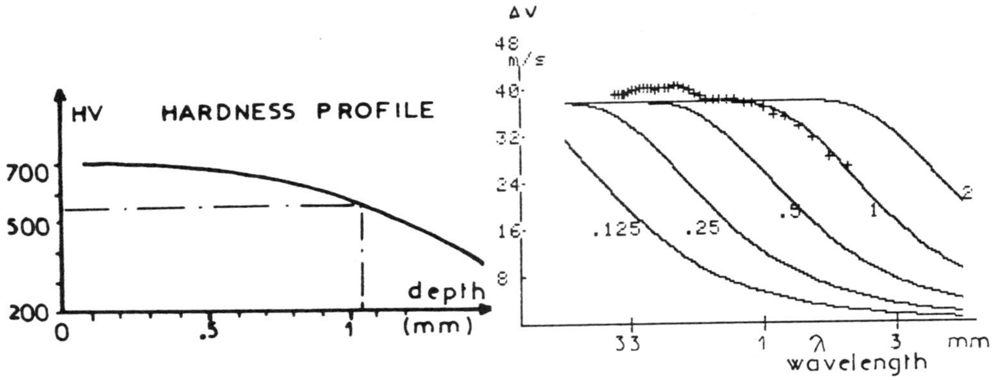

Fig. 7a : Case Hardening of 16NC6 steel specimen

. Carbon content of treated layer : .71 %
. Hardening
. Annealing at 150° C for 1 h

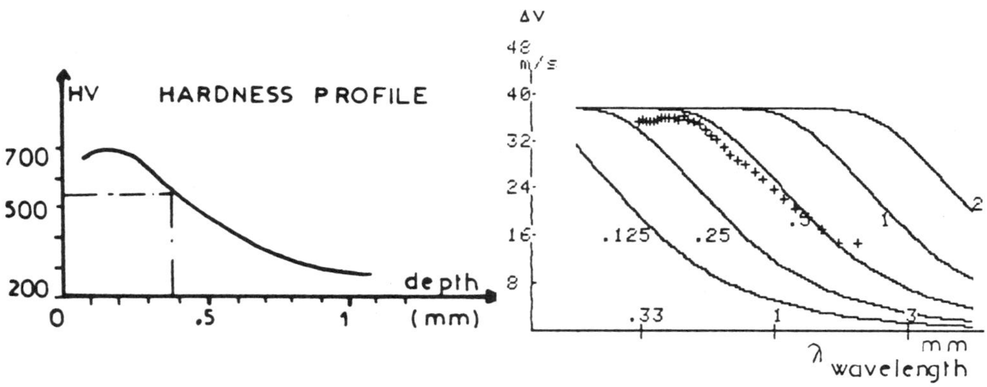

Fig. 7b : Case hardening of 16NC6 steel

. Carbon content : .65 %
. Hardening
. Annealing at 180° C for 1 h

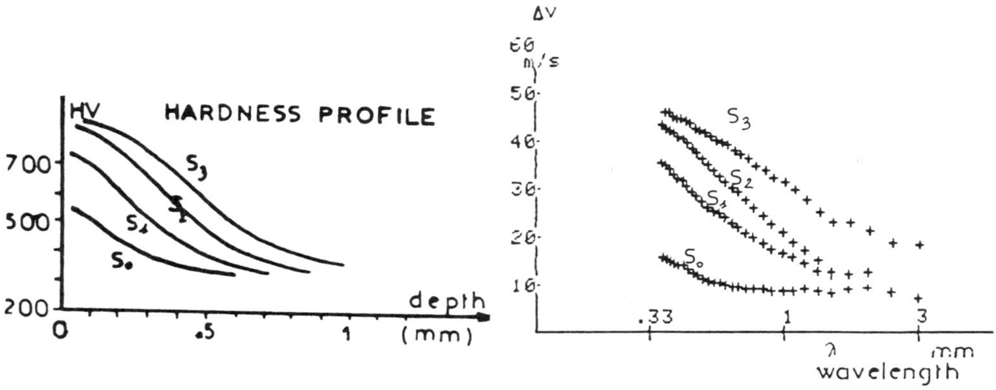

Fig. 7c : Carbonitriding of 17CD4 steel

S_0 = diffusion 5 mn - 850°

S_1 = diffusion 30 mn - 850°

S_2 = diffusion 50 mn - 850°

S_3 = diffusion 100 mn - 850°

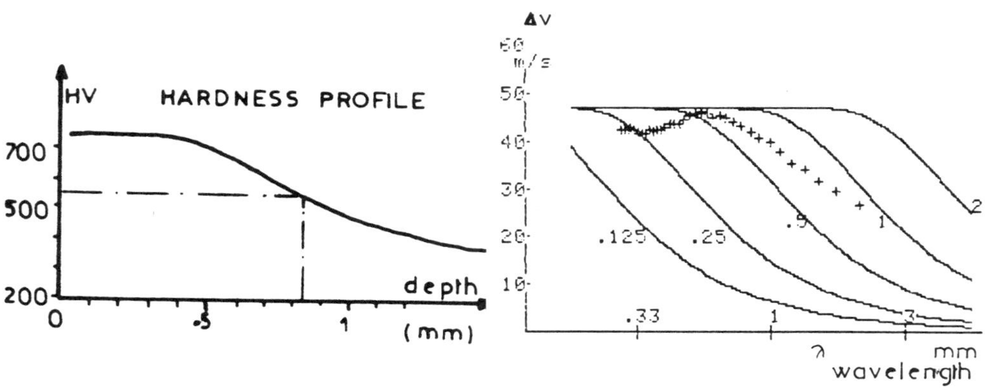

Fig. 7d : Surface hardening by laser treatment XC48 steel
Laser operated at .3 cm/s ; power 1.5 kW/cm²

CONCLUSION

We have shown that surface treatments can be inspected nondestructively with a method based on the dispersion of Rayleigh waves and we have constructed a measuring unit well adapted to this purpose. Signal analysis is based on a cross correlation method to determine the velocity change for each frequency and on a theoretical model which predicts velocity dispersion for a given hardness distribution.

This method applies to thermochemically hardened surfaces, but also to numerous other surface treatments as well, such as laser hardening, mechanical treatments, etc. It permits not only to characterize treated parts as accurately as conventional hardness tests, but is also nondestructive.

REFERENCES

1. A. Lambert "Recherche théorique et expérimentale en vue du développement d'une méthode non-destructive de mesure de la variation de dureté dans les couches superficielles des métaux". Rapport DGRST 72-7-0666 (1975).
2. A. Lambert, G. Bourse, M. Frémiot et al "Contribution des mesures de vitesse des ondes de Rayleigh par traitement numérique du signal à l'étude de l'état superficiel des métaux" 3rd European Conference on NDT (1984).
3. P.J. Noronha, J.R. Chapman and J.J. Wert "Residual Stress Measurement and analysis using ultrasonic testing" Journal of Testing and Evaluation, pp. 209-214.
4. D.R. Allen, W.H.B. Cooper "A Fourier transform technique that measures phase delays between ultrasonic impulses with sufficient accuracy to determine residual stresses in metals" NDT International, Vol. 16 No. 4, (1983).
5. D. Husson, S.D. Bennett and G.S. Kino "Measurements of stress with surface waves" Materials Evaluation, 43, pp. 92-100 (1985).

THE CHARACTERIZATION OF DEPTH PROFILES OF THIN FILMS AND INTERFACES BY NUCLEAR SCATTERING TECHNIQUES

C. Chauvin, J.F. Currie, S. Poulin-Dandurand, E. Sacher and A. Yelon

Groupe des Couches Minces and Département de Génie Physique
Ecole Polytechnique de Montréal
C.P. 6079, succursale "A"
Montréal, Québec H3C 3A7
Canada

P. Aubry, L. Lemay, S. Gujrathi and J.-P. Martin

Groupe des Couches Minces and Laboratoire de Physique Nucléaire
Université de Montréal
C.P. 6128, succursale "A"
Montréal, Québec H3C 3J7
Canada

INTRODUCTION

In an effort to reduce both space and cost, modern packaging technology has turned to miniaturization. This necessitates the development and use of techniques capable of measurements on both thin films and their interfaces. This is important for spun-on films such as polyamic acids, where surface tension effects during cure [1,2] lead to changes in both physical and chemical properties [1,3]. It is important, as well, for plasma-deposited films such as hydrogenated amorphous silicon (a-Si:H), where both physical and chemical properties of deposited films depend on the deposition parameters [4-6].

The type of chemical information sought is the identification of structure as a function of depth, to study the surface, volume and underlying interface. The reasons for this are varied: to study surface contamination, the chemistry of plasma deposition, interfacial reactions, etc. Unfortunately, no such all-encompassing single technique exists, to our knowledge. One is forced to use several complementary techniques. One technique available to us, which comes closests to the requirement of structure as a function of depth, is nuclear scattering. Using the techniques of elastic recoil detection (ERD) and Rutherford backscattering (RBS), we are able to probe to depths of up to 2 μm, obtaining atomic profiles having resolutions of less than 10 nm and elemental sensitivities of about 0.1 atomic percent. A distinct advantage of nuclear scattering is its ability to give hydrogen profiles, an absolute necessity in materials such as hydrogenated amorphous silicon [4,6].

In the next section, the ERD and RBS techniques will be discussed in detail. Following that, examples will be given of their use by our Groupe des Couches Minces (Thin Film Group) to obtain depth profiles on materials of current interest.

EXPERIMENTAL

Our nuclear scattering studies are carried out at the Laboratoire de Physique Nucléaire of the Université de Montréal. The ERD (forward) and RBS (backward) scattering techniques have been the subjects of several papers [4,6-8] and a book [9]. While the reader is referred to them for specifics, the generalities of the techniques will be described with reference to Figure 1.

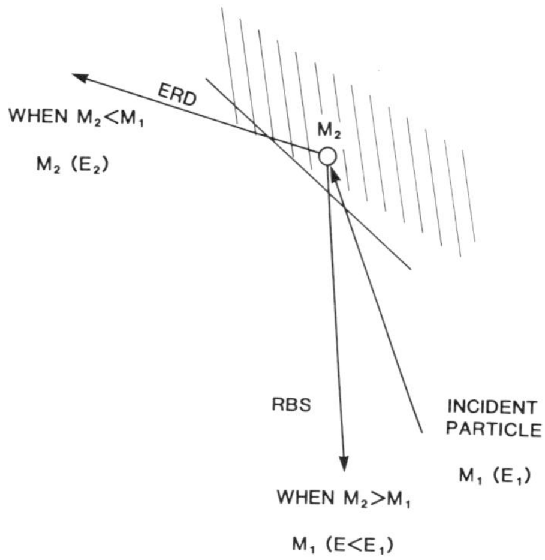

Fig. 1. A schematic of the scattering process. Illustrated are elastic recoil detection (forward scattering) and Rutherford (back) scattering.

The dual tandem Van de Graaf generator provides a monoatomic, monoenergetic beam of ^{35}Cl at 30MeV, which is collimated before impinging on a target. Most of the beam passes through the target. However, when a ^{35}Cl atom strikes a target atom, a nuclear scattering event occurs which depends on the mass of the target atom. When the mass of the target atom is greater than that of the probe atom, the latter is backscattered and the mass and depth of the target atom are obtained from the standard RBS deconvolution [9]. On the other hand, when the mass of the target atom is less than that of the probe atom, the target atom is itself forward scattered; it passes through a microchannel plate TOF setup, from which the mass is obtained, before striking an energy-sensitive beam stop [6,7]. Assuming elastic collisions and knowing the stopping power of the ion [7], the energy of the forward scattered atom becomes a measure of its depth.

The depth, resolution and sensitivity are functions of the mass and energy of the probe ion, as well as the angles used for measuring the nuclear scattering events. While any one may be increased, the others suffer. We have optimized the system to our own needs, with the following resultant values:

depth probed: material-dependent, up to 2 μm
depth resolution: mass-dependent, between 5-8 nm

atomic sensitivity: ~0.1 atomic percent
mass resolution: < 1 dalton (ERD)

EXAMPLES

Polymer-Metal Interface

Multilayer devices consist of alternating layers of metal and polymer (polyimide). As previously noted [7], the deposition of polymer onto metal differs fundamentally from that of metal onto polymer. This is because the polymer is applied as its precursor polyamic acid solution onto solid metal, while the metal is evaporated or sputtered onto cured polyimide. While the bonding of the polyamic acid to the underlying metal may be aided through the use of adhesion promotors, there are no adhesion promotors commercially available for the bonding of metal to underlying polyimide. One must rely on the heat released during metal condensation, of which there is certainly enough [7], to cause chemical reaction at the interface. One must be assured that the bonds thus formed hold during the hostile treatments of the manufacturing process, as well as during service. One method of studying the interface, both before and after hostile exposure, is with nuclear scattering techniques [7,10].

This is done by depositing layers of metal onto polyimide and profiling atoms characteristic of both the metal (e.q., Au, Cu, etc., by RBS) and the polyimide (e.g., C, O, by ERD). Using this technique, we have found [7] that, at the interface, each component diffuses into the other, forming an interphase. The thickness of this interphase depends on the metal but not on the method of deposition. That is, whether the evaporated metal releases ~450 kJ/mole of condensation energy or the sputtered metal releases ~3500 kJ/mole of impact energy, the interfacial interaction appears to be the same. More recently [10], we have explored the effect of sample conditioning, both before and after metal deposition. It is well known [11] that exposure of polyimide to humidity hydrolyzes the surface, giving a high concentration of polyamic acid, while heating the polyimide converts any surface polyamic acid to polyimide.

As seen in Figure 2, for Ni on polyimide, there is little interfacial change on conditioning, begging the question of how the extent of interphase formation influences metal adhesion. This was answered by the standard Scotch tape adhesion test, in which the tape is applied and, after a few seconds, removed. Metal removal by the tape indicates poor metal-polyimide adhesion. The results are found in Table I. Being valid for all the metals used (Al, Au, Ag, Cu, Ni), each having interphases of different thicknesses, one must conclude that no evidence exists that the dimensions of the interphase have any influence on adhesion. Rather, the controlling parameter seems to be the chemical structure in the interphase. This was verified by cycling back and forth between the latter two treatments in Table I: humidification always produced poor adhesion, which was always improved on heating.

Table I. Metal Adhesion to Polyimide

Treatment	Adhesion
A. Humidification before metallization	poor
B. Heat treatment before metallization	good
C. Humidification after metallization	poor
D. Heat treatment after metallization	good

Dopant Distribution in a-Si:H

MIS devices based on a-Si:H have been produced in our Laboratory [6] having among the highest efficiencies of any such devices (~5.5%). Nonetheless, they have only about 40% of the efficiencies of graded-gap, tandem and heterojunction cells [12] currently being produced in the laboratory. Such structures require doping, normally done through the addition of dopant gases to the feed mixture. This is self-limiting, given the fact that there is only a limited number of such gases (i.e., diborane, arsine and stibine). Our approach [6] has been to introduce metal vapors into the plasma, where the metal is then ionized through impact and incorporated into the plasma-deposited a-Si:H. The ionization has been verified for thallium, having an ionization energy of 589 kJ/mole, by the presence of a characteristic green glow. Doping has been verified [5,6] for thallium, indium and antimony by electrical measurements.

The problem arises of ensuring a uniform distribution of dopant throughout the volume (0.5-1 μm). This may be done with the aid of nuclear scattering, as shown in Figure 3. While it indicates a nonuniform distribution of dopant, the figure also suggests an inverse relationship between hydrogen and dopant, suggesting that the dopant uses hydrogen sites.

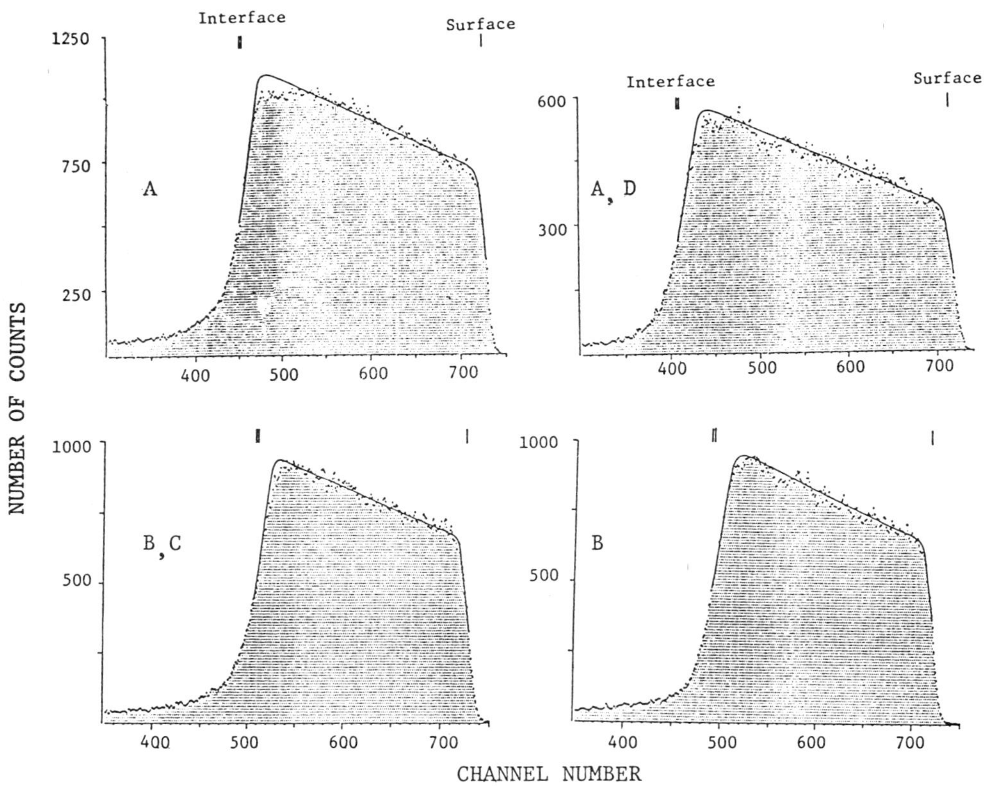

Fig. 2. Nuclear scattering profiles for nickel on polyimide subject to treatments A,B,C and D, as identified in Table I.

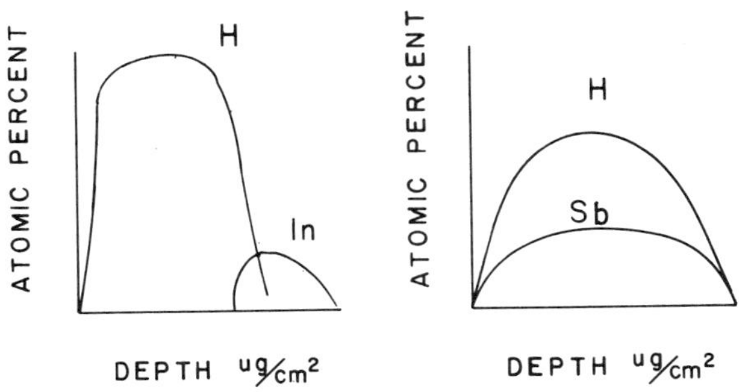

Fig. 3. Nuclear scattering profiles of a-Si:H doped with a) indium and b) antimony.

Contamination of Semiconductor Surfaces

It is well known that transition metals may absorb unsaturated or aromatic organic molecules at low temperatures through the use of their d orbitals in establishing $d\pi$-$p\pi$ bonds [13-15]. On warming to room temperature, bonds are broken and remade. Water contact angles of upwards of 100° suggests that these latter films are reasonably thick and paraffin-like.

The $d\pi$-$p\pi$ bonds are made through the overlap of a filled ligand p orbital and an empty metal orbital to form a σ bond, and the overlap of a filled metal d orbital and an empty ligand orbital to form a π bond; silicon has neither of the orbitals necessary, yet both crystalline silicon [16] and a-Si:H [17] have surface contaminants. ERD has been used to show [17] that two types of carbon-containing contaminants exist on a-Si:H: a surface layer, about 4 nm thick, lying on the surface, contains only carbon and hydrogen while carbon, associated with oxygen, exists as a contaminant within the bulk of the a-Si:H. There is little doubt that the surface layer is a hydrocarbon since a photoacoustic IR study, presently in progress [18], using a technique which can distinguish between surface and subsurface sources, shows a surface peak near 1880 cm^{-1}, due to vinyl groups [19].

CONCLUSIONS

Nuclear scattering techniques (RBS and ERD) are useful in providing depth profiles of atomic species. We have successfully applied these techniques to surface, bulk and interfacial studies, both organic and inorganic. These techniques may be used alone but are especially useful in conjunction with other techniques which provide structural detail.

ACKNOWLEDGMENTS

The authors wish to thank the Natural Sciences and Engineering Research Council of Canada and the Fonds FCAR of Québec for funding this work.

REFERENCES

1. E. Sacher and J.R. Susko, J. Appl. Polym. Sci., 23, 2355 (1979).
2. E. Sacher, J. Polym. Sci.: Polym. Lett. Ed., 21, 111 (1983).
3. C. Chauvin, E. Sacher and A. Yelon, J. Appl. Polym. Sci., 31, 583 (1986).
4. J.F. Currie, R. Depelsenaire, J.-P. Huot, L. Paquin, M.R. Wertheimer, A. Yelon C. Brassard, J. L'Ecuyer, R. Groleau and J.-P. Martin, Can. J. Phys., 61, 582 (1983).
5. B. Ranchoux and J.F. Currie, Can. J. Phys., 63, 54 (1985).
6. J.L. Brebner, R.W. Cochrane, R. Groleau, S. Gujrathi, D. Kéroack, Y. Lépine, J.-P. Martin, M. Vanacek, C. Aktik, M. Aktik, A. Azelmad, J.F. Currie, S. Poulin-Dandurand, B. Ranchoux, E. Sacher, C. Tannous, M.R. Wertheimer and A. Yelon, Can. J. Chem., 63, 786 (1985).
7. J.F. Currie, P. Depelsenaire, R. Groleau and E. Sacher, J. Coll. Interface Sci., 97, 410 (1984).
8. J. L'Ecuyer, C. Brassard, C. Cardinal and B. Terrault, Nucl. Instrum. Meth., 149, 271 (1978).
9. W.-K. Chu, J.W. Mayer and M.-A. Nicolet, "Backscattering Spectrometry", Academic Press, New York, 1978.
10. C. Chauvin, E. Sacher, A. Yelon, R. Groleau and S. Gujrathi, in "Surface and Colloid Science in Computer Technology", edited by K.M. Mittal, Plenum, in course of publication.
11. H.J. Leary, Jr. and D.S. Campbell, Surf. Interface Sci., 1, 75 (1979).
12. See, e.q., Proc. 10th Internat. Conf. Amorph. Liquid Semiconductors, edited by K. Tanaka and T. Shimizu, Tokyo, 1983; J. Non-Cryst. Solids, 60, (1983).
13. F.A. Cotton and G. Wilkinson, "Advanced Inorganic Chemistry", Wiley, New York, Fourth Edition, 1980, chapters 3 and 25.
14. J.E. Huheey, "Inorganic Chemistry", Harper & Row, New York, Second Edition, 1978, chapter 13.
15. N. Sheppard, J. Electron. Spectr. Relat. Phenom., 38, 175 (1986).
16. J.F. Cain and E. Sacher, J. Coll. Interface Sci., 67, 538 (1978).
17. E. Sacher, J. Klemberg-Sapieha, M.R. Wertheimer, H.P. Schreiber and R. Groleau, Phil. Mag. B, 49, L47 (1984).
18. P. Lours, L. Bertrand and E. Sacher, to appear.
19. L.J. Bellamy, "The Infrared Spectra of Complex Molecules", Wiley, New York, 1959, p. 34.

NONDESTRUCTIVE HIGH-RESOLUTION MEASUREMENT OF CHARGE, POLARIZATION AND PIEZOELECTRICITY DISTRIBUTIONS IN THIN DIELECTRIC FILMS

G. M. Sessler, R. Gerhard-Multhaupt,[+]
J. E. West,[++] and A. Berraissoul

Technical University of Darmstadt, Darmstadt, F. R. G.

[+]Present address:
Heinrich-Hertz-Institut fuer Nachrichtentechnik,
Berlin, F. R. G.

[++]Permanent address:
AT & T Bell Laboratories, Murray Hill, NJ

ABSTRACT

Ultrasonic pulses, generated by thermoelastic effects and ablation in an opaque coupling layer deposited on the sample surface, are utilized to nondestructively determine charge, polarization and piezoelectricity profiles in thin dielectric films with a resolution of 1 to 2 μm. The method utilizes a 70 ps laser light pulse to launch a less than 500 ps long pressure pulse in the sample. Propagation of the laser-induced pressure pulse (LIPP) through the film causes electrode currents which directly yield the distributions. Results for charge profiles in Teflon FEP and piezoelectricity profiles in PVDF are described. The LIPP method has also been used to measure ultrasonic velocity and attenuation up to 1 GHz in various biased or surface charged polymer films.

INTRODUCTION

Ultrasonic methods are widely used for nondestructive materials characterization.[1] The advantage of these methods lies in the fact that solids are transparent for ultrasonic waves in certain frequency ranges and that their structure and faults can therefore be made visible with appropriate ultrasonic techniques. Most of the existing ultrasonic methods are, however, devoted to the detection of mechanical and structural parameters, while the investigation of electrical properties has been somewhat neglected.

In the present paper, a recently developed ultrasonic method for the nondestructive evaluation of the distribution of space charge, polarization, and piezoelectricity in the thickness direction of dielectric films with thicknesses in the μm-range is described and results achieved with this method are discussed. The method is based on the generation of an ultrasonic pressure pulse by interaction of a laser pulse with a metal absorber on the surface of the sample to be investigated.[2-6] Apart from its use for distribution measurements, the laser-induced pressure pulse (LIPP) can also be employed to determine ultrasonic attenuation and velocity in solids.

METHOD OF MEASUREMENT

Experiments were performed with polymer films metalized on one or both surfaces. The measuring setup, as used for one-side metalized samples, is schematically shown in Fig. 1. Light pulses of 30 to 70 ps duration and 1 to 10 mJ energy are generated with a Nd:YAG laser and directed to an absorbing layer on the front electrode of the sample. This layer consists either of an approximately 2 μm thick painted-on graphite coating or of a 100 to 500 nm thick evaporated Zn, Cd, Pb, Bi, or other metal layer.

The surface region of the metal layer is heated by the absorbed laser light. This causes stress effects and eventually ablation of target material. The stress effects and the ablation-generated recoil launch a pressure pulse of <500 ps duration[6] which propagates through the sample (direction of propagation x) with the sound velocity c.

The pressure pulse generates a current I(t) in the external circuit if the sample is charged or polarized (see below). The current is amplified and displayed on a Tektronix 7104 oscilloscope (1 GHz cutoff). Results are obtained by evaluating photographs of the oscilloscope display.

It is assumed that the sample has a piezoelectric constant $e(x) = dP(x)/dS$ in its thickness direction and a local charge density

$$\rho(x) = \rho_r(x) - dP(x)/dx, \tag{1}$$

where $\rho_r(x)$ is the real-charge density (permanent and electrode charges), P(x) the polarization (permanent and instantaneous), and S the strain. Then the pressure pulse generates, under short-circuit conditions, the current signal[4,5]

$$I(t) = \frac{Ap\tau}{\rho_0(s+\varepsilon g)}\left[\left[1 + \frac{(\varepsilon_\infty+2)(\varepsilon_\infty-1)}{3\varepsilon}\right]\rho(x) - \frac{de(x)}{dx}\right]_{x = ct}, \tag{2}$$

where A is the sample area, p the pressure amplitude of the pulse, τ the duration of the pressure pulse, ρ_0 the density of the sample material, s the sample thickness, g the air-gap thickness, and ε and ε_∞ the static and the high-frequency dielectric permittivity, respectively. The method is similar to techniques recently developed in other laboratories.[7-11]

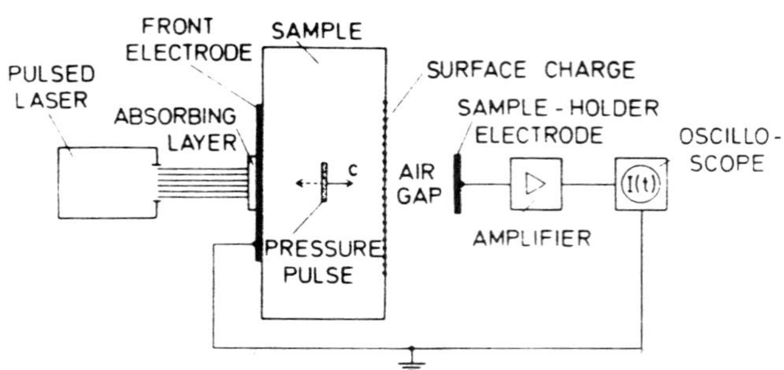

Fig. 1. Experimental setup for the laser-induced pressure-pulse (LIPP) method for one-side metalized samples. In the arrangement for two-side metalized samples, the free surface is also electroded.

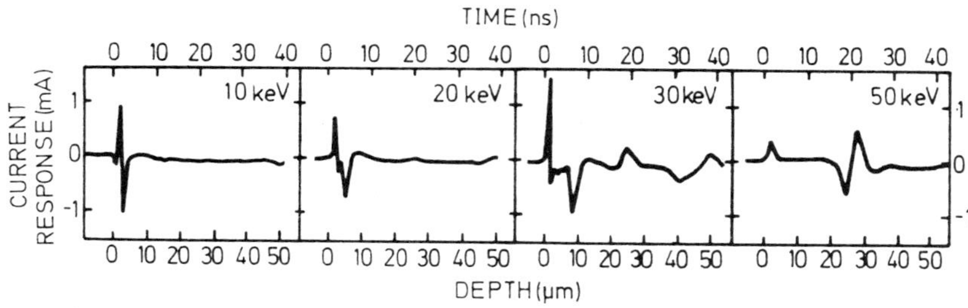

Fig. 2. LIPP responses, corresponding to charge distributions, for 25 μm Teflon FEP charged with electron beams of energies indicated. The positive spikes at 0 and 20 ns are due to induction charges on front and rear electrodes, respectively, while the negative spike represents the trapped space charge deposited by the electron beam.

SPACE CHARGE DISTRIBUTION IN TEFLON FEP

FEP does not assume a permanent polarization and is not piezoelectric. Thus, the LIPP method generates, according to Eqs. (1) and (2), a current signal proportional to $\rho_r(x)$. LIPP measurements were performed on a number of FEP samples charged with electron beams, corona discharges or liquid contacts. Results for samples charged with 10 to 50 keV electron beams through the electrically floating front electrode, depicted in Fig. 2, show relatively narrow charge layers close to the respective CSDA ranges of the electrons.[3] The small width of the layers is due to the field direction during charging, which tends to pull the injected electrons deeper into the sample. The charge density integrated over the thickness of these samples is approximately 10^{-8} C/cm^2.

Negatively corona- and liquid-contact charged FEP films exhibit surface-charge layers which become mobile at temperatures above 120 °C. At higher temperatures, their schubweg is a sizeable fraction of the sample thickness.[12] Thus, only little trapping within the sample volume occurs, as has already been shown by other LIPP experiments.[13]

PIEZOELECTRICITY PROFILES IN PVDF

According to Eq. (2), the LIPP method yields a current signal proportional to $(1+\gamma)\rho(x)-de/dx$ with $\gamma = -(1/\varepsilon)d\varepsilon/dS \approx (\varepsilon_\infty+2)(\varepsilon_\infty-1)/(3\varepsilon)$. Since PVDF has a conductivity on the order of 1 pS/m, its Maxwell relaxation time is about 100 s. Thus, in a short-circuited sample any electric field E within the dielectric will cause conduction currents which will compensate the field over periods of the order of minutes. Since E is given by Poisson's equation as

$$\varepsilon\varepsilon_0 \frac{dE}{dx} = \rho(x) \qquad (3)$$

the disappearance of E means $\rho(x) = 0$. Thus, according to Eq. (1), in a short-circuited, relaxed sample, the permanent polarization is screened by real and instantaneous polarization charges.[14] This holds also for the case where a small external field (not causing space-charge effects) is applied to a homogeneous sample since this field does not generate a field gradient dE/dx within the dielectric. Thus, any current signal observed in LIPP experiments on PVDF films having reached an equilibrium condition is due to the term de/dx in Eq. (2).

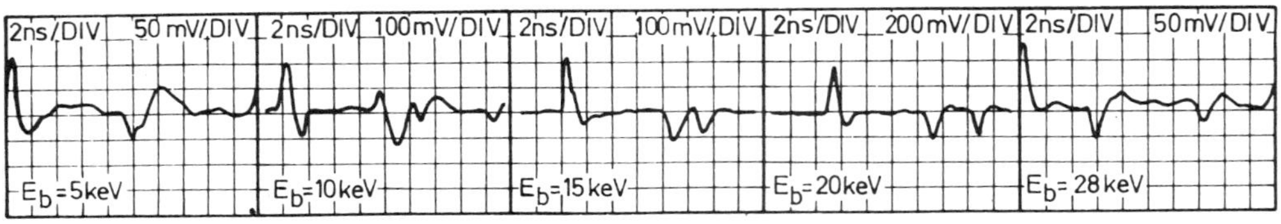

Fig. 3. LIPP responses, corresponding to charge distributions, for 22 μm PVDF poled with electron beams of different energies E_b. The first positive spike is due to rise of piezoelectricity at electrode-polymer interface while the two negative spikes following at 10 -Δt and 10 +Δt ns are due to drop of piezoelectricity.

The LIPP method has recently been applied to PVDF samples poled with a new method[15] utilizing electron-beam charging of one-side metalized samples through their nonmetalized surface. Since the electrons are stored in the polymer, the surface potential reaches eventually a value equal to the potential of the electron beam. Thus, a field is generated between the plane of deposition of the electrons and the rear electrode. This field polarizes the sample. The thickness of the polarized zone thus depends on the energy of the electron beam. In the polarized zone, the sample is also piezoelectric.[15]

Results obtained with the LIPP method on 25 μm PVDF samples poled with electron beams of different energies E_b are depicted in Fig. 3. The left-hand (positive) signal in the oscillograms is due to the penetration of the LIPP through the interface between the metal electrode and the polymer, where de/dx is finite, while the two (negative) signals following later are due to the penetration of the direct and the reflected LIPP, respectively, through the end of the polarized zone. On a length scale, the distance of these two signals thus equals twice the penetration depth r of the electron beam. In the region between the electrode and the electron range, de/dx equals almost zero, i. e. the piezoelectricity has a constant value. From the area of the spikes, e may be calculated if the amplitude of the pressure pulse is known.

The spacing r corresponding to the thickness of the nonpiezoelectric zone is plotted in Fig. 4 as a function of electron beam energy E_b. The straight-line dependence on the double-logarithmic plot indicates a power-law relationship between r and E with the exponent equal to 1.6, closely corresponding to a published value of 1.62.[16] However, the actual value of the penetration depth is only about half of that found before. The present results are in good agreement with penetration data in Mylar, which has approximately the same density.[17, 18]

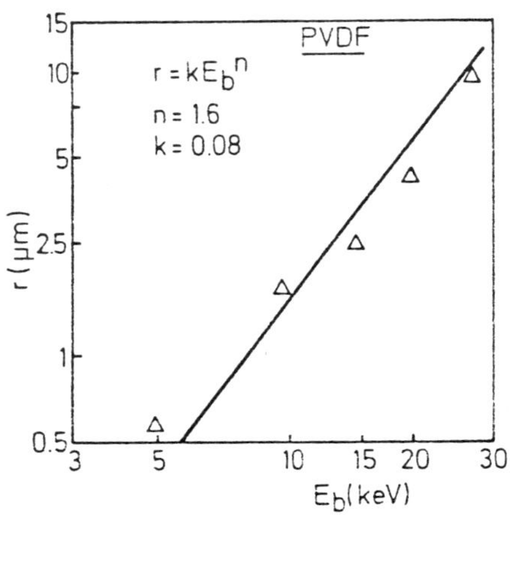

Fig. 4.
Thickness r of non piezoelectric zone in 22 μm PVDF poled with electron beams of energy E_b. The data is obtained from the LIPP responses in Fig. 3 and similar data.

Fig. 5.
Piezoelectric e-constant (relative units) of PVDF charged with the electron-beam method as function of annealing time at three annealing temperatures.

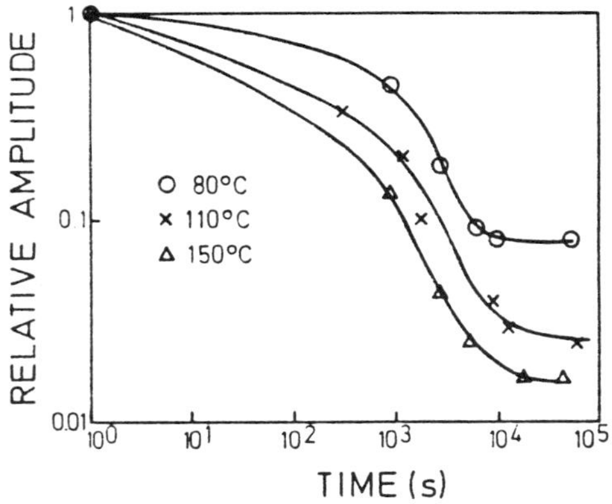

The piezoelectric e-constant decays with time at elevated temperatures. LIPP measurements of the decay of e have been performed on samples charged with the electron-beam method and then annealed at various temperatures. The results, depicted in Fig. 5, show a nonexponential decay, as was also found by other authors.[19, 20] A study of the shape of the current spikes indicates that the dropoff of the polarized zone retains its steepness during annealing.

ULTRASONIC ATTENUATION

For LIPP-measurements of ultrasonic velocity and attenuation in the polymer materials of interest, the samples are charged on their nonmetalized surfaces with the liquid-contact or the corona method. Upon every reflection of the pressure pulse from the acoustically soft rear and front surfaces of the sample, a current spike is generated. While the ultrasonic velocity can be simply calculated by measuring the delay of successive reflections, an accurate evaluation of the ultrasonic attenuation requires either the use of multiple pulses[21, 22] or the application of spectral-analytic methods.[14, 22] Only the second approach will be discussed here.

To determine the absorption constant with this method, the spectrum of a pulse after several transits is compared with the spectrum of the original pulse. This makes it possible to evaluate the absorption constant over a wide frequency range from a single measurement.

Fig. 6 shows the spectra of a single pulse obtained after one and two transits through a PETP sample as evaluated with an FFT computer routine.

Fig. 6.
Spectra of the current signals generated upon the first and second reflection at the rear surface of PETP sample.

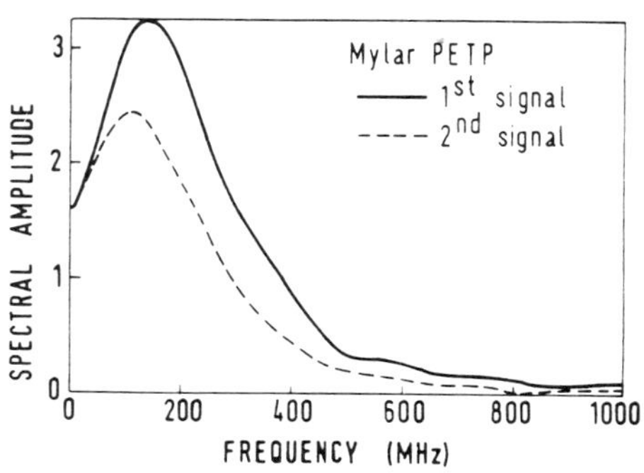

Fig. 7.
Transmission function for the spectra shown in Fig. 6.

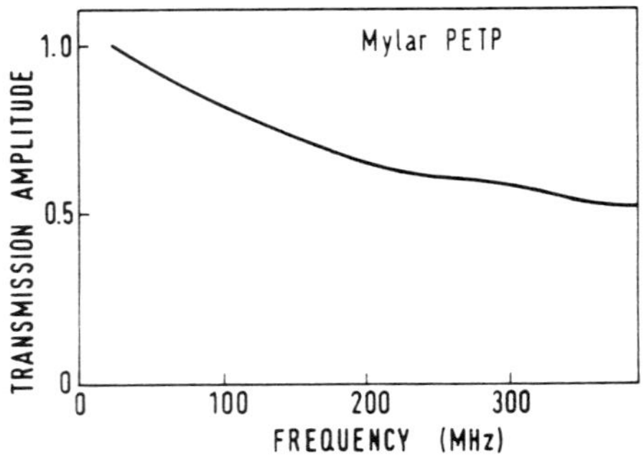

The transmission function H(ω) of this sample, defined as ratio of the spectral amplitudes of n-th and (n-1)-th signals at frequency ω, is plotted in Fig. 7. From this plot, the absorption constant α(ω) follows by means of $\alpha(\omega) = 1/(2s)\cdot\ln[H(\omega)]$, where s is the sample thickness.

Results obtained with this method for several polymer materials are shown in Fig. 8. The data reveals an almost linear rise with frequency, indicating hysteresis-like dissipation.[22, 23] A similar behavior has been found for PVDF and PE by other laboratories while no previous data are available for the polymers investigated here.

Knowledge of the absorption constant is important for determining the charge or polarization distributions accurately. For example, a 0.5 ns pressure pulse with its spectral center of gravity at about 600 MHz, suffers an average attenuation by a factor of 1.6 in a 25 μm PETP sample. In additio spectral distortion is expected. The effects are even more severe in FEP. These facts have to be considered when evaluating charge, polarization, or piezoelectricity distributions.

CONCLUSIONS

Laser-induced pressure pulses allow one to nondestructively determine charge, polarization, or piezoelectricity profiles in dielectrics with a resolution of 1 to 2 μm and measure the ultrasonic attenuation in these materials. The advantages of this method are the ease in generation and detection of the signals: The pressure pulses are generated optoacoustically and detected electrically, in both cases without the use of ultrasonic transducers. The examples of measurements of charge and piezoelectricity distributions and of ultrasonic attenuation in polymers show the practicability of the method.

Fig. 8.
Frequency dependence of the absorption coefficients in FEP, PETP and PI.

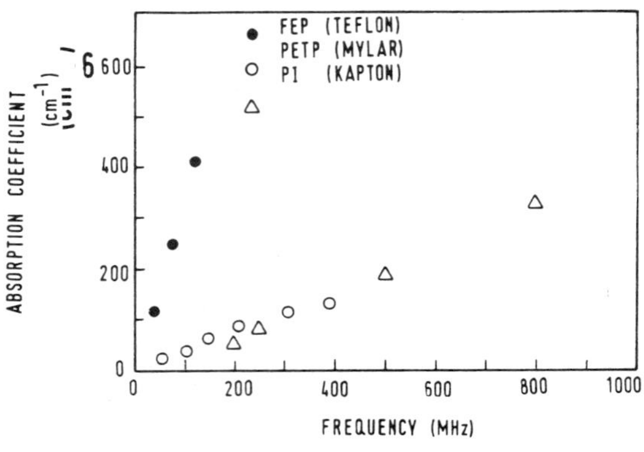

REFERENCES

1. D. O. Thompson, D. E. Chimenti (Eds.): Review of Progress in Quantitative Nondestructive Evaluation, Vol. 1 - 3, Plenum Press, New York, 1982/83/84.
2. G. M. Sessler, J. E. West, R. Gerhard, Phys. Rev. Lett. **48**, 563 (1982).
3. G. M. Sessler, J. E. West, R. Gerhard-Multhaupt, H. von Seggern, IEEE Trans. Nucl. Sci. **NS-29**, 1644 (1982).
4. R. Gerhard-Multhaupt, Phys. Rev. B **27**, 2494 (1983).
5. R. Gerhard-Multhaupt, G. M. Sessler, J. E. West, K. Holdik, M. Haardt, W. Eisenmenger, J. Appl. Phys. **55**, 2769 (1984).
6. G. M. Sessler, R. Gerhard-Multhaupt, J. E. West, H. von Seggern, J. Appl. Phys. **58**, 119 (1985).
7. A. G. Rozno, V. V. Gromov, Sov. Tech. Phys. Lett. **5**, 266 (1979).
8. C. Alquié, G. Dreyfus, J. Lewiner, Phys. Rev. Lett **47**, 1483 (1981).
9. W. Eisenmenger, M. Haardt, Solid State Commun. **41**, 917 (1982).
10. A. Migliori, T. Hofler, Rev. Sci. Instrum. **53**, 662 (1982).
11. R. A. Anderson, S. R. Kurtz, J. Appl. Phys. **56**, 2856 (1984).
12. H. von Seggern, J. Appl. Phys. **50**, 7039 (1979).
13. C. Alquié, J. Chapeau, J. Lewiner, 1984 Annual Report CEIDP, pp. 488 - 494.
14. G. M. Sessler, R. Gerhard-Multhaupt, H. von Seggern, J. E. West, IEEE Trans. Electric. Insul. **EI-21**, 411 (1986).
15. R. Gerhard-Multhaupt, B. Gross, A. Berraissoul, to be publ.
16. B. Gross, H. von Seggern, R. Gerhard-Multhaupt, J. Phys. D: Appl. Phys. **18**, 2497 (1985).
17. G. M. Sessler, J. E. West, H. von Seggern, J. Appl. Phys. **53**, 4320 (1982).
18. B. Gross, R. Gerhard-Multhaupt, K. Labonte, A. Berraissoul, Colloid & Polym. Sci. **262**, 93 (1984).
19. L. L. Blyler, G. E. Johnson, N. M. Hylton, Ferroelectr. **28**, 303 (1980).
20. A. G. Kolbeck, J. Polym. Sci. Polym. Phys. **20**, 1987 (1982).
21. G. M. Sessler, R. Gerhard-Multhaupt, J. E. West, Proceed. 11th International Congress on Acoustics, Paris, 1983, Vol. 2, p. 195.
22. R. Gerhard-Multhaupt, G. M. Sessler, J. E. West, Conf. Proceed. "Ultrasonics International 85" (Guilford: Butterworth Scientific 1985), pp. 317 - 322.
23. G. M. Sessler, R. Gerhard-Multhaupt, J. E. West, Proceed. 12th International Congress on Acoustics (Toronto 1986).

NONDESTRUCTIVE CHARACTERIZATION OF THIN ADHESIVE BONDS

M. Lethiecq, J.C. Baboux, and M. Perdrix

Laboratoire de Traitement du Signal et Ultrasons, INSA
Bât. 502, 69621 Villeurbanne cedex, France

INTRODUCTION

Modern cyanoacrylate adhesives are very efficient and are used more and more often to bond materials like aluminium, steel or ceramics. One of the factors that hampers their use is the problem of reproductibility of the bonds : one can never guarantee that a given bond is good, i.e. its mechanical resistance is greater that a fixed value.
There is a great need for nondestructive test methods that are able to predict the quality of a bond. Many techniques have given good results in controlling classical (epoxy) bonds [1-5], but up to now none has proved efficient to test very thin joints as those we have with cyanoacrylate adhesives.
The aim of our study is to find out if the combination of ultrasonic and electrical measurements can characterize such a bond.
We have made a relatively great number of aluminium ceramic bonds, all with the same brand of cyanoacrylate adhesive. Each sample was submitted to ultrasonic Longitudinal (L) and Shear (S) wave tests as well as electrical measurements. Finally a destructive test gave us the shear mechanical resistance of each bond. In this paper we present the experimental methods which give us the primary signals, the processing of these signals to obtain the different parameters we think can best characterize the bonds and the results of an automatic data analysis algorithm : the principal component analysis.

EXPERIMENTAL DEVICE

Bond Description

We have made 60 aluminium-ceramic bonds. They consist in an aluminium cylinder (diameter 20 mm, height 10 mm) and a ceramic cylinder (diameter 15 mm, height 3 mm) ; the two flat parts being bonded together.
Before making the bonds we have submitted the elements to many measurements in order to determine their acoustical impedance, their surface roughness and their electrical characteristics. For each bond we note the surface treatment and the pressure applied during the adhesive cure.

Ultrasonic test

We use a device composed of an ultrasonic wide band generator-receiver amplifier associated either to an L wave broadband transducer (center frequency 15 MHz) or to an S wave broadband transducer (center frequency 5 MHz). Signals are observed and stored on a digital scope, (60 MHz band-pass), the whole system being controlled by a micro computer.

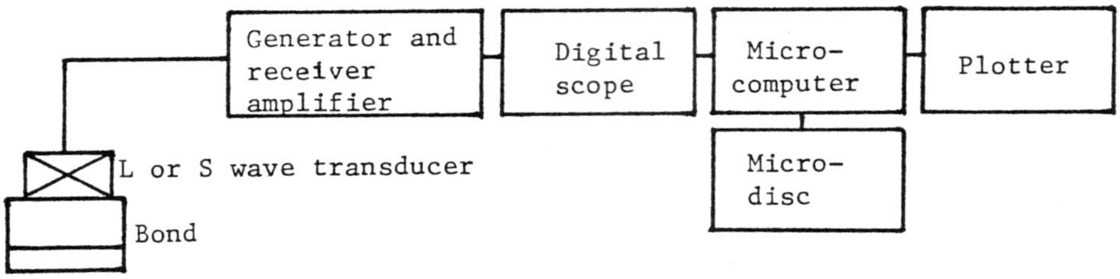

Figure 1:Ultrasonic device

We use a pulse echo method and the echoes are those shown on figures 2 and 3. The e1 echo corresponds to a reflection on the glue line and e2 has been twice transmitted through it. Let us note that the adhesive layer is so thin that to have separated echos corresponding to the "top" and "bottom" of the glue line, it would be necessary to work at frequencies of several hundreds of MHz.

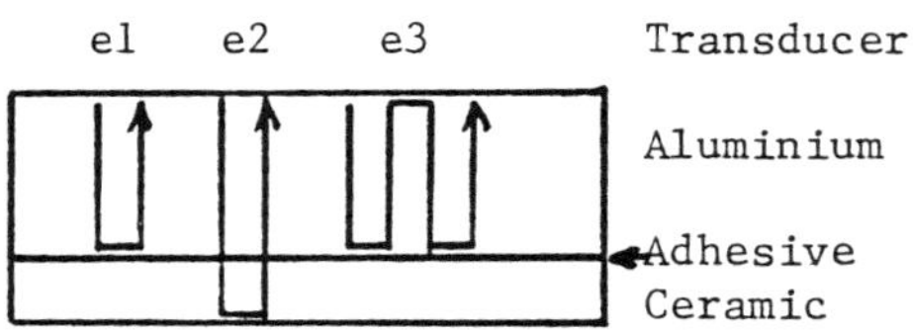

Figure 2:Echoes in the bond

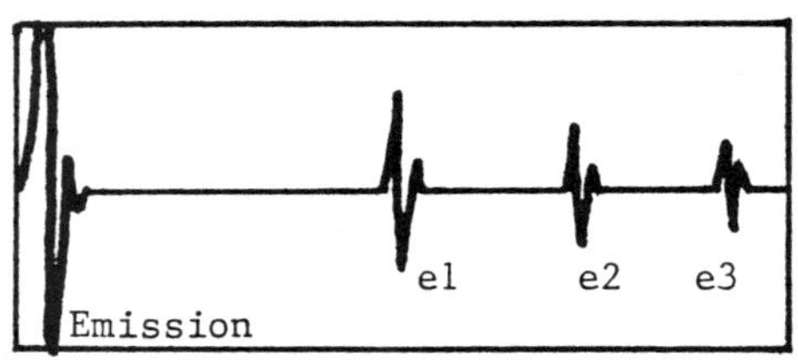

Figure 3:Time domain curve

Electrical measurements

The aluminium-ceramic bond can be considered as a capacitor : the aluminium has a very small electrical impedance so the bond can be reduced to two serial capacitors : the ceramic and the glue line. Before making the bond we have measured the ceramic's capacity Cc, and later the total capacity of the bond Cb.

Destructive test

We have decided to characterize the mechanical resistance of a bond by the force applied in order to obtain a shear rupture. This force is measured by an INSTRON traction machine which pulls in the plane of the glue layer.

PROCESSING OF THE MEASUREMENTS

Ultrasonic signals

The echoes given by the ultrasonic device depend on the generator, the emission transfer function of the transducer, the transducer-aluminium coupling, the acoustical impedance and attenuation of the aluminium and of the ceramic, the receiver transfer function of the transducer and of the adhesive layer itself. The processing of the signals must give us a signal that eliminates all information except the one concerning the adhesive layer.

Echos e1 and e2 are the result of the convolution of many time functions :

$$e1(t) = fe * fR * fr$$

$$e2(t) = fe * fTR1 * fc * fTR2 * fr$$

or in frequency domain, the convolution becoming a multiplication

$$E1(\nu) = Fe \times FR \times Fr$$

$$E2(\nu) = Fe \times FTR1 \times Fc \times FTR2 \times Fr$$

- Fe is the transfer function of the whole emitting system : generator, cables, transducer, coupling and travel through the aluminium cylinder.
- FR is the transfer function corresponding to a reflection on the adhesive.
- Fr is the transfer function of the whole receiving system : travel through the aluminium, coupling, transducer, cables and receiver amplifier.
- Fc corresponds to a return trip through the ceramic.
- FTR1 and FTR2 correspond to a transmission by the adhesive layer (aluminium to ceramic and vice-versa).

A division of the frequency curves gives us :

$$F(\nu) = \frac{E1(\nu)}{E2(\nu)} = \frac{FR}{FTR1 \times FTR2 \times Fc}$$

The $F(\nu)$ curve has only kept the useful information, except for the term Fc which corresponds to the return trip through the ceramic (which is done by echo e2 but not by echo e1).

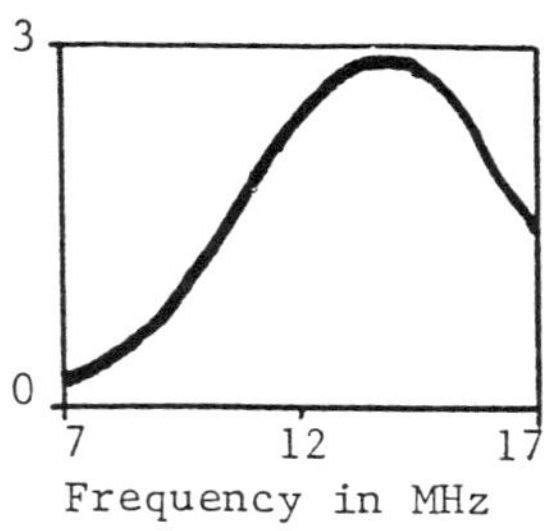

Figure 4: Fourier transform of echo e1, Lwaves

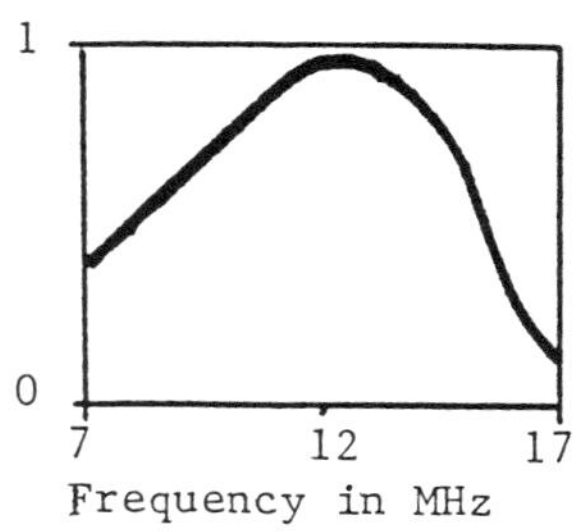

Figure 5: Fourier transform of echo e2, L waves

A simulation of the adhesive joint by a parallel faced layer as studied in a previous paper [6] shows that the influence of the ceramic can be reduced to a multiplicative factor depending on the acoustical impedance of the ceramic. The F(ν) curves are corrected by this factor.
Note : If one of the echos e1 or e2 is smaller than the noise, it is replaced by a simulated echo, whose amplitude is near the noise amplitude and whose shape is similar to the other echoes.

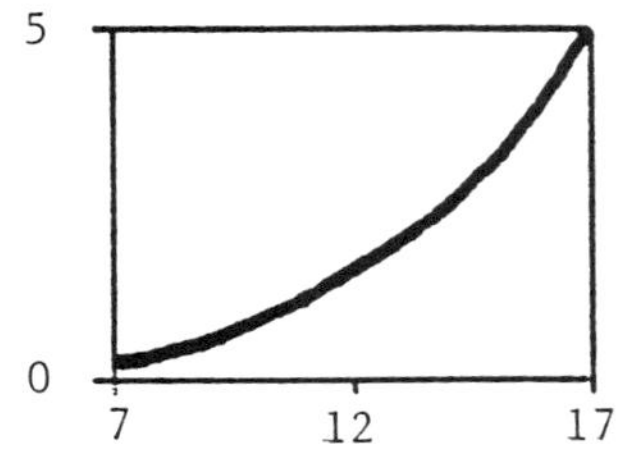

Figure 6: F(ν) curve,L waves

Electrical measurements

We have determined the ceramic's capacity Cc and the bond's capacity Cb. We can deduce the adhesive layer capacity Ca :

$$\frac{1}{Cb} = \frac{1}{Cc} + \frac{1}{Ca}$$

Ca is a characteristic of the adhesive joint itself.

CHOSEN PARAMETERS

We have decided to take into account seven parameters deduced from our measurements :

Chosen parameter	Name
Peak to peak amplitude ratio of echo e1 to echo e2 determined on the time domain curve. for L waves for S waves	 RAML RAMT
Value of the F(ν) curve at 12MHz for L waves at 4MHz for S waves	 FLMO FTMO
$\frac{F(\nu max) - F(\nu min)}{F(\nu mid)}$ for L waves νmax=17MHz,νmin=7MHz,νmid=12MHz for S waves νmax= 6MHz,νmin=2MHz,νmid= 4MHz	 PENL PENT
Capacity of the adhesive joint (Ca)	CAPA

These seven parameters are known for all 60 bonds and will be taken into account in the data analysis.

THE PRINCIPAL COMPONENT ANALYSIS (PCA)

The algorithm

The principal component analysis is an algorithm that allows us to study complicated data by using simple representations.
In this case, we have 60 individuals (bonds) each characterized by 7 parameters. We can represent the data by 60 points in a 7 dimensional space. Such a space is too difficult to handle, so the PCA algorithm allows to find a two dimensional space on which it projects the 60 individuals. The problem is to find the plane in which the individuals have relative positions that are closest to the ones they have in the 7 dimensional space. This plane is defined by the two principal axes and is called S1. A second part of the algorithm does the same thing with the 7 parameters in a 60 dimensional space : we then have a plane in which the 7 parameters are represented ; it is called S2. It can be shown[7] that the position of a parameter P in the plane S2 can give us the correlation between P and the other parameters, and that it also gives us an idea of the importance P has taken in the choice of the best plane S1.

The PCA method also allows to represent other parameters (called illustrative) that are not used to determine the best plane but that can be represented.

We have first defined an illustrative parameter called FRUP which is the shear rupture force of the bond.

We have also defined some illustrative nominal parameters :

Parameter	Modalities	Meaning
Mechanical surface treatment	RECT	Both faces (aluminium and ceramic) are rectified
	PONS	Both faces are sanded
Surface cleaning	GRGR	Both faces are greased
	LOLO	Both faces are degreased with a specific cleaner
	ACAC	Both faces are degreased with acetone
	FRFR	Both faces are rubbed with alumina
	FRAC	The aluminium is rubbed with alumina and the ceramic is degreased with acetone
Pressure during the adhesive cure	PERS	High pressure
	MAIN	Low pressure (by hand)
Quality of the bond	BONC	Good bond (FRUP $>$ 150 KgF)
	MAUV	Bad bond (FRUP $<$ 150 KgF)

RESULTS

Active parameters

Figure 7 gives a representation for the active parameters and the shear rupture force FRUP.

Interpretation : each parameter defines a straight line between it's position and the center of the plane (point (0,0)). The cosine of the angle between two lines gives the correlation between the two corresponding variables and the distance between a variable and the center of the plane gives the significance of the variable.

The four variables RAML, FLMO, RAMT and FTMO are very correlated to one another : choosing only one or two of them would probably not change the representation : the information is redundant.
The adhesive capacity CAPA also gives the same type of information since it is opposite to these 4 variables :for most bonds when FTMO has a big value, CAPA has a small value and vice-versa.
Here again there is redundant information. PENL and PENT are also quite far from the center of the plane : they also had an important role in the choice of the best plane. However the most important parameter is FRUP, for this is the value we would like to predict by the non destructive measurements. FRUP is quite correlated with CAPA and the four parameters RAML, FLMO, RAMT and FTMO.
It is slightly correlated to PENT, but not at all correlated to PENL. This last parameter would be useless to predict the mechanical resistance of a bond.

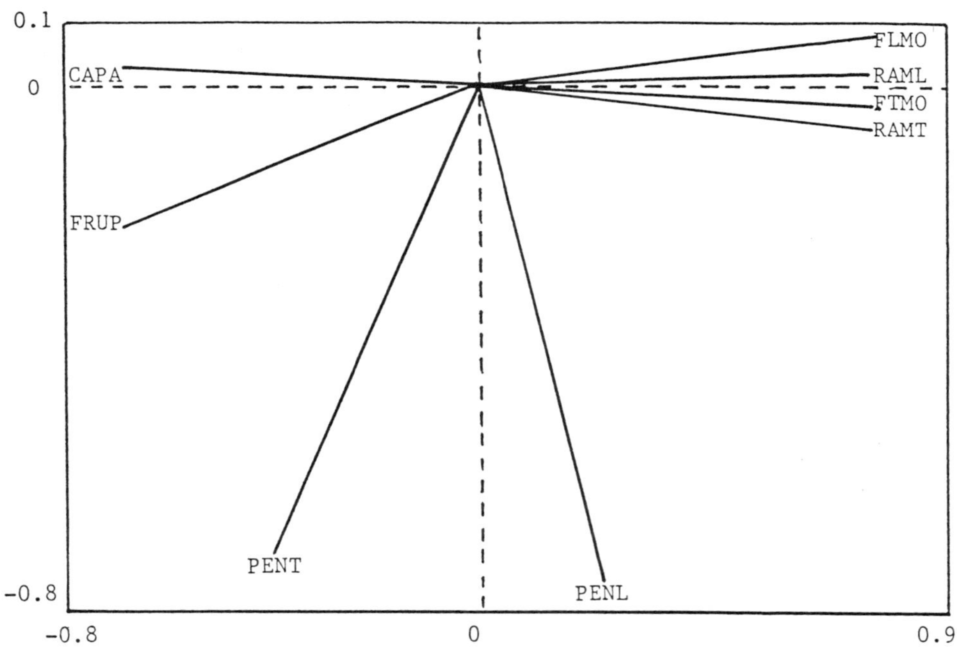

Figure 7: Active parameters and FRUP

Illustrative parameters

Figure 8 gives a representation of the illustrative nominal parameters.

We can see for example that the 4 parameters BONC, RECT, FRFR and MAIN are all correlated : this means that good bonds are usually those made with rectified surfaces, rubbed with alumina and cured under low pressure.
This representation of illustrative parameters is not specifically useful for the elaboration of a NDT method, but it can help people who want to make better bonds.

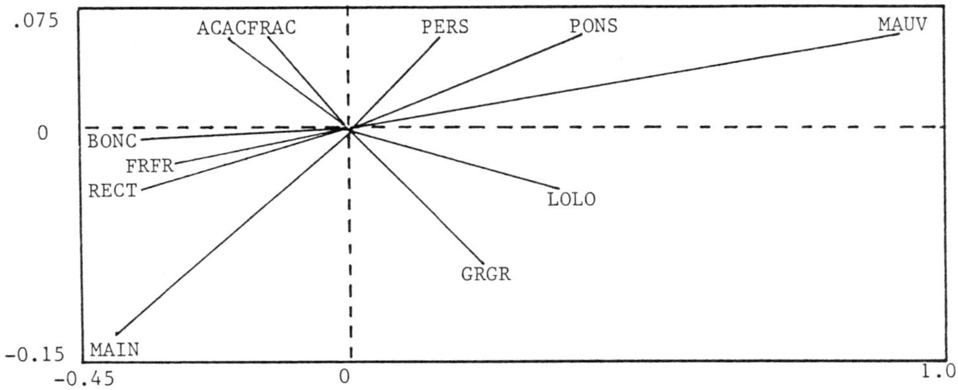

Figure 8: Illustrative nominal parameters

Representation of the individuals

On figure 9 we can see the different individuals in the best plane of the 7 dimensional variable space : O : good bonds, X : bad bonds. Their is a relatively good separation of the two groups.

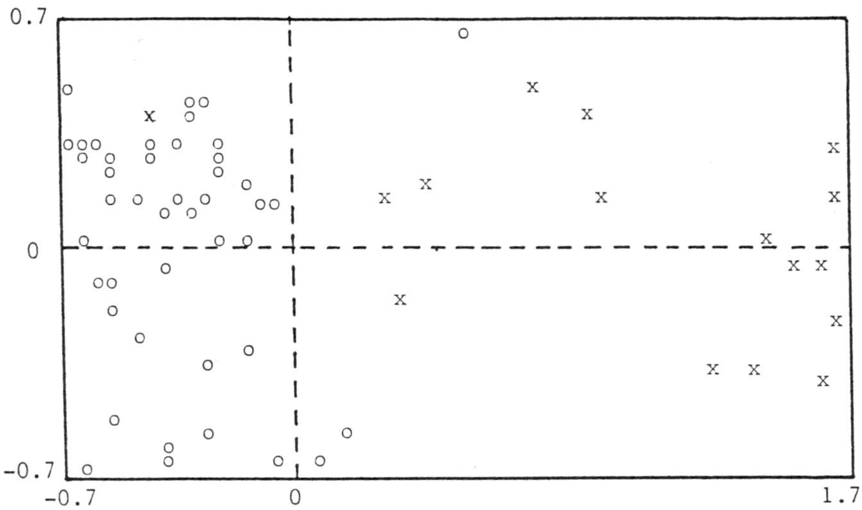

Figure 9: Individuals (o=good bonds , x=bad bonds)

CONCLUSIONS

We have studied the correlation between different non destructive measurements and the mechanical resistance of a bond.
We have shown that certain parameters are quite correlated to the shear rupture force of a bond, wheras others are not. We have also shown that there is a good separation between good and bad bonds, so we can hope that an automatic discrimination test will be able to predict the quality of a bond with a small percentage of error.
Our future work will consist in studying other parameters deduced from our measurements to see if they can bring more information about the bond, and to choose the best ones out of them in order to elaborate an automatic test.

ACKNOWLEGMENTS

This study was supported by the NDT Laboratory of the Joint Research Center (Commission of the European Communities), Ispra, Italy (Contract N° 2426 84 07 ED ISPF).

REFERENCES

1 . R.J. Schliekelmann, Non destructive testing of adhesive bonded joints, in "Adhesion and Adhesive Joints course", C.E.I. Europe, Finspang, Sweeden (2-6 september 1985).
2 . G.J. Curtis, Non destructive testing of adhesively bonded structures with acoustic methods, in : "Ultrasonic Testing", Ed. Szilard, (1982).
3 . E. Segal, J.L. Rose, Non destructive testing techniques for adhesive bond joints, in : "Research Techniques in NDT", Ed. R.S. Sharpe (1982).
4 . J.L. Rose, G.H. Thomas, The fisher linear discriminant function for adhesive bond strength prediction, in : British Journal of NDT, vol 21, n°3, (May 1979).
5 . G.A. Alers, R.K. Elsley, Application of ultrasonic signal analysis to adhesive bond strength prediction, Structural Adhesives and Bonding, El Segunds, California, U.S.A. (1979).
6 . M. Lethiecq, J.C. Baboux, Y. Jayet and M. Perdrix, Influence of very thin layers on transmission and reflection of ultrasonic longitudinal and shear plane waves, Ultrasonics International 85, London G.B. (July 1985).
7 . A. Morineau, Panorama des méthodes d'analyse des données, in : "Cours des Journées Modulad" ENST et INRIA Paris, France (13-16 may 1986).

INTERFEROMETRIC OPTICAL FIBER MEASUREMENTS OF INTERNAL STRAIN IN ADHESIVE JOINTS

R. O. Claus, K. D. Bennett, and K. T. Srinivas

Fiber and Electro-Optics Research Center
Department of Electrical Engineering
Virginia Polytechnic Institute and State University
Blacksburg, VA 24061

ABSTRACT

Strain inside adhesive lap joints has been measured using imbedded optical fiber sensors. Cylindrical glass-on-glass optical fibers with 125-micron outer diameters and an 0.85-micron single mode cutoff wavelength have been imbedded within expoxy lap joints between metal substrates. Strain integrated along the length of one imbedded fiber has been determined by interferometrically measuring the strain-induced phase modulation of multiple optical modes supported within the optical guide for 0.633-micron transmission. Differential measurements of the strains integrated along two fibers imbedded in different orientations within the lap joints have also been obtained using differential interferometric techniques modified due to the multimode nature of the waveguides. Results are compared with analytical predictions. Limitations of the imbedded fiber technique are considered. Extensions of the technique to local measurements of strain along the length of the imbedded fiber and measurements of the internal local strain tensor are discussed.

INTRODUCTION

Optical fiber waveguides are thin cylindrical strands of glass or plastic which are designed to guide light efficiently from one end to the other. To perform this type of guiding, optical fibers have cylindrical inner core regions and concentric outer cladding regions with slightly different indices of refraction. A minimally higher index for the material in the core of the fiber allows much of the light, properly injected into the fiber, to propagate without significant attenuation via total internal reflection.

The combined core and cladding sections of an optical fiber typically have an outer diameter on the order of 100 microns. This core and cladding optical waveguide structure is coated with a protective jacketing of material, usually a UV-cureable polymer, with a thickness also on the order of 100 microns. The purpose of this jacketing material is to protect the waveguiding region from chemical impurities, which lead to optical attenuation and stress corrosion cracking, and from strain, which causes optical signal loss as well as the possible mechanical failure of the fiber.

Optical fibers have been applied to both long-haul and local area network telecommunication systems for about fifteen years and applied to sensing for about ten years. Whereas optical fiber communication systems are designed to reduce interference due to external perturbations such as temperature and strain, optical fiber sensing systems are designed to detect such perturbations. Fiber sensing systems are categorized according to which property of the transmitted optical field is detected to determine the perturbation. Specifically, optical fiber amplitude (intensity), phase (interferometric), polarimetric, wavelength, time delay, and modal domain sensors have been developed [1,2] for the sensing of pressure, temperature, electric and magnetic fields, rotation rate, chemical concentrations and other environmental effects.

Fiber sensors have specifically been applied to the nondestructive characterization of materials for about five years. Due to the inherent similarity of unjacketed glass-on-glass optical fibers and individual graphite fibers in graphite/epoxy composites in particular, a number of investigators have considered the use of optical fibers as sensors which may be imbedded directly within composite laminae. The present status of this research is that the effects of temperature and strain integrated along the length of the sensor fiber in a composite specimen can be determined using a variety of simple methods. Spatial resolution of such quantities along the imbedded fiber in length may be obtained using several more complicated distributed fiber sensing techniques. Strain tensor quantities may be determined by both presuming accurate models of the applied stress and knowing the photoelastic and mechanical properties of the imbedded fiber [3].

Some of the results of work applied to the characterization of composites may be directly utilized for similar measurements inside adhesives. This includes studies of chemical pretreatments of the fibers which maximize the adhesion between the fiber and epoxy matrix materials, methods for curing specimens containing fiber sensors without damaging the sensors, and methods of mounting and subsequently loading fiber-instrumented specimens.

This paper discusses the results of initial measurements of strain in adhesive lap joints obtained using imbedded single and multimode optical fibers and several interferometric detection techniques. The following section of the paper models the detection process for both single-input and differential interferometric measurements. The next section describes the specimens, measurements, and results. The paper concludes with a brief discussion of potential extensions of these results.

INTERFEROMETRIC FIBER SENSING

Interferometric optical fiber sensor systems measure changes in the phase of an optical field which propagates in a fiber. This phase is a function of the length of the fiber, the index of refraction of the core of the fiber, and the cross-sectional dimensions of the fiber. Each of these factors is in turn dependent upon the mechanical interactions between applied temperature, pressure, or strain perturbations and the fiber. Length variations, for example, are caused either directly by thermal expansion or by the application of longitudinal strain, and indirectly by the application of hydrostatic pressure resulting in longitudinal expansion by the Poisson effect. Index variations are caused either by changes in temperature or by changes in strain via the photoelastic effect. Fiber dimension variations are caused by changes in radial strain produced by a pressure field, longitudinal strain via Poisson's ratio, or by thermal expansion.

Since this paper concerns measurements of strain, let us assume that temperature variations may be neglected. The total phase change produced by stress alone may then be written in general as

$$\Delta\phi = \frac{2\pi L}{\lambda} \left\{ \epsilon_1 - \frac{n^2}{2} (P_{11}+P_{12})\epsilon_r + P_{12}\epsilon_1 \right\}, \tag{1}$$

where L is the length of the sensing section of the fiber, λ is the optical guide wavelength, ϵ_1 and ϵ_r are the longitudinal and radial strain, respectively, n is the core index, and P_{11} and P_{12} are the photoelastic constants of the silica fiber [4]. This relation indicates that longitudinal and radial strain contributions cannot be resolved from a single phase measurement. Moreover, if the strain varies along the length of the fiber, the total phase change must be calculated from (1) by integrating individual contributions along this length.

Phase measurements may be interpreted in terms of strain components if the geometry of the strain field is known. For example, in the case of pure axial strain the radial strain is

$$\epsilon_r = -\epsilon_1 \upsilon, \tag{2}$$

where υ is Poisson's ratio, and (1) simplifies to

$$\frac{\Delta\phi}{\epsilon L} = \frac{2\pi}{\lambda} \left[1 + P_{11} \frac{n^2}{2} \upsilon + P_{12}\left(\frac{n^2}{2}\upsilon + 1\right)\right]. \tag{3}$$

Substituting values for pure silica ($n = 1.458$, $P_{11} = 0.126$, $P_{12} = 0.274$, and $\upsilon = 0.17$) and the helium-neon laser light free space wavelength 633nm, the phase change per meter of strained fiber is 2.2×10^7 radians, or 0.22 radians per centimeter of fiber per microstrain. The corresponding theoretical minimum detectable strain may be determined by expressing the noise produced by an optical signal current within a photodetector circuit as

$$i_{noise} = (2eBI)^{\frac{1}{2}}, \tag{4}$$

where e is the charge of an electron, B is the electronic bandwidth of a heterodyne receiver circuit, and I is the photocurrent, given by αP_o, where α is the detector sensitivity in A/W and P_o is the optical power. For small strain, the corresponding signal current is

$$i_{signal} = \alpha P_{signal} \tag{5}$$

$$= \alpha \frac{\Delta\phi}{\pi} P_o .$$

The minimum detectable phase change may be found by equating (4) and (5). For $P_o = 0.1$ mW, $\alpha = 0.4$ A/W, and $B = 1$ Hz, this yields

$$\Delta\phi_{minimum} = 2.8 \times 10^{-7} \text{ radians}, \tag{6}$$

or the signal produced in approximately one millimeter of fiber by a strain of 10^{-12}.

There are clearly many other situations which may be considered, including combinations of radial and axial strain as in (1), strain fields which produce optical birefringence in an initially isotropic fiber core, and optical fibers which have initial nonuniform index distributions or which are multimode. In each of these cases a strain-induced optical phase change may in principle similarly be determined.

The discussion thus far has assumed a fiber sensing system similar to that shown in Figure 1. Here, the sensor fiber is imbedded inside the specimen to be studied and a second reference fiber shunts the specimen. By interferometrically observing the difference between the stationary phase of the signal in the reference fiber and the varying phase of the signal in the sensing fiber, strain may be determined. As in most interferometric systems, the sensitivity of such an arrangement is very high and thus subject to random effects of environmental conditions which independently perturb the fibers. If, instead, the two fibers are both imbedded in the specimen, as shown in Figure 2, the difference in the phases between the fields in the two may be measured and such common mode noise reduced. Calibration of this differential interferometric arrangement requires knowledge of the strain field geometry in the vicinities of both fibers but allows an improvement in absolute sensitivity of as much as 3dB over the single fiber system [5].

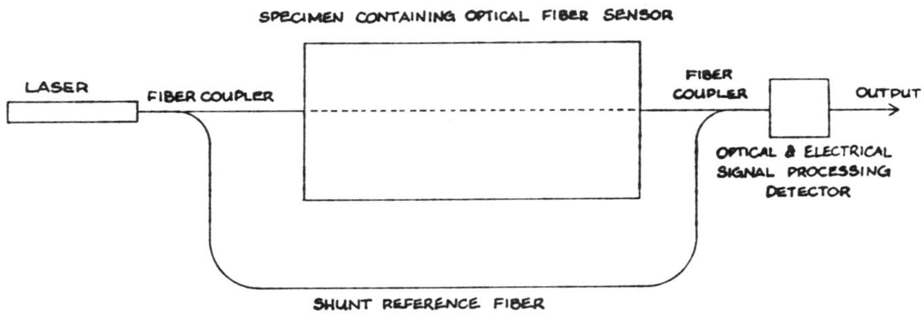

Figure 1. Simple interferometric fiber sensing system.

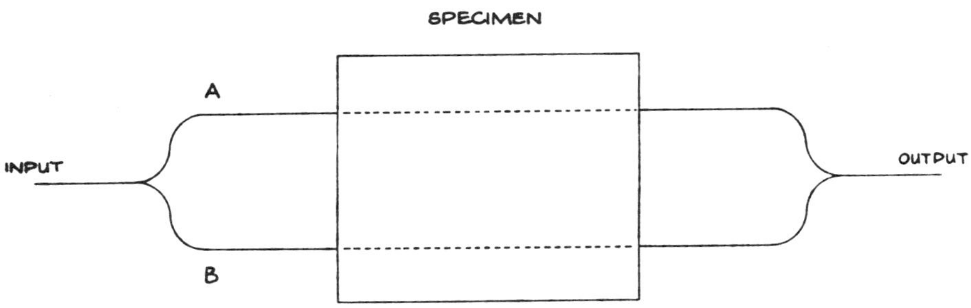

Figure 2. Differential fiber sensor arrangement which measures difference of perturbations in fibers A and B.

EXPERIMENT

A simple adhesive lap joint containing a section of optical fiber was constructed and tested to study the performance of the system model. The joint of two-part five-minute epoxy, mixed in a one-to-one ratio and cured at room temperature, measured 3.2 cm by 2.6 cm by 285 microns thick as shown in Figure 3. The jacketing of the sensor fiber was removed over the imbedded section positioned approximately equidistant between the aluminum plates and parallel to the line connecting the centers of the load frame bolt holes as shown. The imbedded fiber had an outer diameter of 125 microns and cutoff wavelength of 850 nm.

By mechanically loading the specimen, approximately axial strain was applied to the section of imbedded fiber. Resulting optical phase change, calibrated using a similar cantilever beam geometry similar to that described in [4], is plotted in Figure 4. The maximum measured strain in the specimen was 6×10^{-7}.

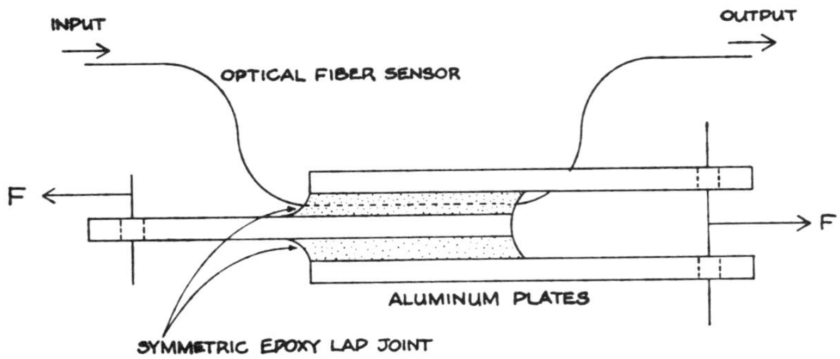

Figure 3. Geometry of fiber sensor inside lap joint.

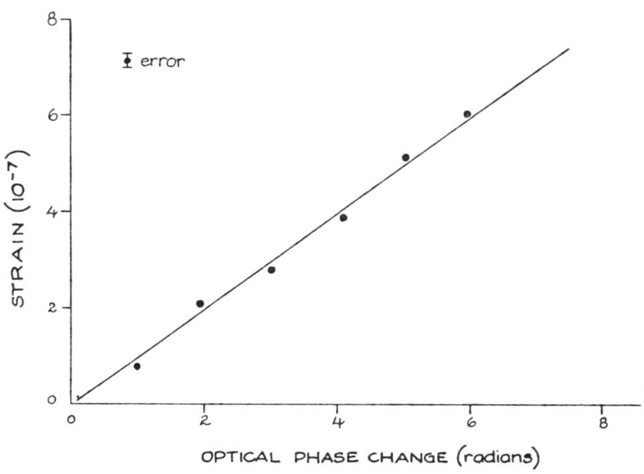

Figure 4. Measured strain versus optical phase change.

RESULTS AND CONCLUSIONS

Optical fibers may be used to measure strain in adhesives using interferometric techniques. If the core material of the sensor fiber is isotropic, as considered here, effects due to axial and radial strain cannot be distinguished. Thus, knowledge of the geometry of the strain field is necessary to determine its magnitude. The potential advantage of this method is its excellent spatial and strain amplitude resolution.

The particular measurements described here also demonstrate that multimode optical waveguides supporting a few modes may be used instead of single mode fiber for the interferometric sensing of slowly-varying strain. The principle of such fiber sensing may be interpreted as "self-referencing," "differential interferometric" or "modal domain," and is implicit in several similar methods [6-9]. These methods hold promise for the characterization of arbitrary strain tensor fields.

ACKNOWLEDGEMENTS

This work has been supported by the NASA Langley Research Center and Simmonds Precision. The authors appreciate the encouragement and suggestions of H. F. Brinson and W. T. Freeman.

REFERENCES

[1] B. Culshaw, Optical Fiber Sensing and Signal Processing (Peter Peregrinus, Ltd., 1984).
[2] T. G. Giallorenzi et.al., IEEE Trans. Microwave Theory and Techniques MTT-30, 472 (1982).
[3] R. O. Claus, B. S. Jackson, and K. D. Bennett, Proc. SPIE 566, 60 1985.
[4] C. D. Butter and G. B. Hocker, Appl. Opt. 17, 2867 (1978).
[5] C. H. Palmer, J. Acoust. Soc. Am. 53, 948 (1973).
[6] B. W. Brennan, W. B. Spillman, and J. R. Lord, Proc. 3rd SEM Conf. on Hostile Env. and High Temp. Measurements (Cincinnati, OH), March 1986.
[7] G. Meltz and J. R. Dunphy, Proc. SPIE 566, 159 (1985).
[8] S. C. Rashleigh, J. Lightwave Technology LT-1, 312 (1983).
[9] K. D. Bennett, R. O. Claus, and M. J. Pindera, Proc. Review of Progress in Quantitative NDE (San Diego, CA), August 1986.

AUTOMATED WELDING PROCESS SENSING AND CONTROL

J. A. Johnson, N. M. Carlson, J. O. Bolstad, H. B. Smartt, M. B. Ward, R. T. Allemeier, L. A. Lott, and D. C. Kunerth

Idaho National Engineering Laboratory
P. O. Box 1625
Idaho Falls, Idaho 83415

A welding machine requires three additional basic components for truly automated operation:

Sensors which detect physical properties of the weld such as reinforcement area or fracture toughness, or which detect parameters that can be related to those properties such as depth of penetration or cooling rate.

A model of the welding process which relates the controllable welding parameters such as current, voltage, welding speed, etc., to the physical properties of the weld.

A control system which takes the signals from the sensors and converts them into a form which can be used for feedback control of the weld machine.

A research program at the Idaho National Engineering Laboratory (INEL) is developing electro-optic and ultrasonic sensors to detect the physical properties of the weld and a model of the welding process to relate these properties to parameters which can be controlled. This paper discusses first the model of the gas metal arc (GMA) welding process and then the sensors used to detect the physical properties of the weld.

GMA WELDING MODEL

The gas metal arc (GMA) welding process employs a consumable wire electrode passing through a copper alloy contact tip. Electrical current, imposed on the wire by a voltage drop between the contact tip and the metal to be welded (base metal), supports an arc between the wire end and the base metal. The electrode is melted by internal resistive power and heat transferred from the arc. Molten metal droplets are detached and transferred from the wire to the weld pool. Heat is transferred to the base metal directly from the arc and also by the molten metal droplets. The electrode wire, molten droplets, weld pool, and solidified weld bead behind the weld pool are protected from oxidation by a shielding gas.

A model of the heat and mass transfer to the base metal was developed[1] for welding thick section carbon or alloy steel. The model consists of six equations. Two relate the heat input and the reinforcement area to various welding parameters including the arc voltage and current, the welding speed, and the wire size and speed. The others describe the weld machine power supply and the physical properties of the wire and arc. These equations may be solved to give the current, welding speed, and wire speed in terms of the heat input and reinforcement area and the other welding parameters. The model does not contain information regarding the formation, detachment, or transfer of the droplets from the electrode wire to the weld pool. Implementation of the model is discussed in detail in Reference 1. A block diagram of an automated welding control scheme based on the model is given in Figure 1.

Control of Heat Input and Reinforcement Area

The first feedback control scheme was developed to control heat input and reinforcement area by measuring the current. Referring to Figure 1, the desired reinforcement area and heat input are input, along with the other welding parameters. The actual welding current is measured and compared to the value predicted by the model, and the difference is used to change the model. The welding speed is simultaneously adjusted to maintain the correct reinforcement area. In the block diagram (Figure 1), only the top feedback loop is presently implemented.

Welding trials in this work using the model to control reinforcement area and heat input were made using 0.89 mm diameter type E70S-6 wire, with 98% Ar-2% O_2 shielding gas, on 12.7 mm thick type A-36 steel plate in a bead-on-plate configuration. The contact tip to work piece distance was maintained constant at 15.9 mm and the power supply open circuit voltage was 32 V. All welding involved spray transfer of the molten electrode wire material to the weld pool.

The accuracy of the system has been demonstrated by measuring reinforcement areas and calculated heat inputs obtained during welding for various machine settings. The reinforcement areas are within ±5% of the desired values, the heat inputs are well within ±1% of the desired values. The reinforcement areas were measured from macrophotographs of transverse weld sections using an image digitizing program. The heat inputs were calculated from current, voltage, and travel speed values by the control computer during welding, and assume a 75% heat transfer efficiency.

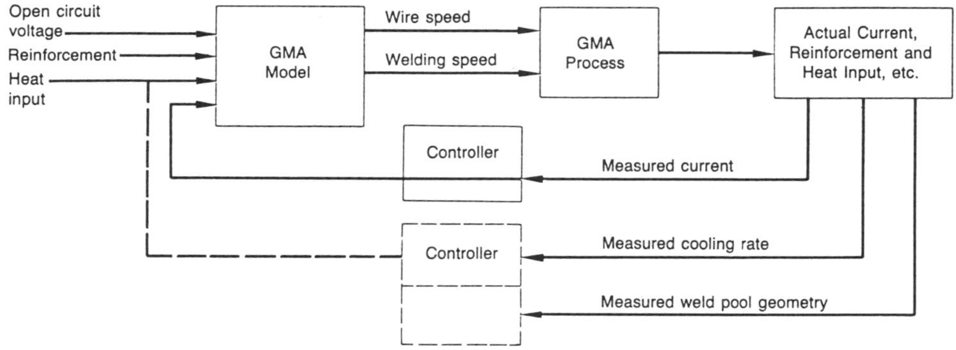

Fig. 1. Automated Welding Control Block Diagram.

Weld Bead Cooling Rate and Fracture Toughness

Fracture toughness depends on weld bead cooling rate.[2] The feasibility of controlling weld bead cooling rate in real time, independent of the reinforcement area, was examined using a set of welds made at heat inputs from 1050 to 1500 J/mm using reinforcement areas in the range of 30 to 40 mm^2.[1] The weld bead cooling rates were measured by plunging thermocouples into the weld pool at the top center of the weld bead. Although the range of values included in this study is limited, it is clear that the weld bead cooling rate is related to the heat input. Future work on this problem will involve real time measurement of the weld bead cooling rate with an infrared camera in a manner similar to Lukens and Morris[2], and direct closed-loop control of the cooling rate with the error in cooling rate being used to signify the required change in weld bead heat input. Implementation of this scheme is shown as the second feedback line of Figure 1.

MEASUREMENT AND CONTROL OF WELD POOL PENETRATION

In a similar manner the system will be used to control the penetration (or side wall fusion) of welds using appropriate sensing. The penetration may be sensed indirectly by measuring the maximum width of the weld pool optically, and then calculating the weld pool volume and depth in a manner similar to that done previously for gas tungsten arc welding at INEL.[3] The weld heat input may then be adjusted, thus increasing or decreasing the weld pool volume as needed. Alternatively, the sensing may be done directly by measuring the weld pool subsurface geometry using an ultrasonic transducer. Both approaches are shown schematically as the last feedback loop of Figure 1.

Weld Vision System

Electro-optical techniques are used for the indirect measurement of penetration. These techniques are attractive because contact with the workpiece can be avoided and problems with physical wear and heat transfer to the sensor are thereby minimized. Such systems can also accommodate a variety of weld joints and bead geometries by use of adaptive image processing, etc. On the negative side, one can expect optical systems to be relatively intolerant of smoke or aerosol generation at the weld site and to liquid metal spatter. However, a variety of schemes are available to deal with these problems, including the use of a purge gas in the optical path, the use of mechanical shields, and, of course, positioning optical surfaces at an appropriate distance to achieve significant cooling of incident spatter.

In 1982 the INEL began exploratory work on the use of machine vision for electric arc welding. A review of earlier work revealed that the welding arc light is a very severe impediment to formation of good imagery and must be greatly suppressed and/or replaced by illumination from an external light source. Figure 2 is a simple schematic of the experimental arrangement. The goal is to obtain enough peak optical power from the xenon flash lamp to overwhelm the welding arc emission during the brief 2-3 μs interval of the flash. A video camera system equipped with an image intensifier tube is used in a time-gated mode as a very high speed electro-optical shutter. The shutter is synchronized with the flash lamp and therefore acts to accept most of the flash energy reflected from the weld site, but at the same time accepts only a very small fraction of the continuous emission from the welding arc. The synchronized flash and shutter are driven at 30 pulses per second to yield a single flash exposure per video frame. On this basis then, the percentage duty time of the shutter is simply the ratio of shutter duration (3.0 μs) to video frame

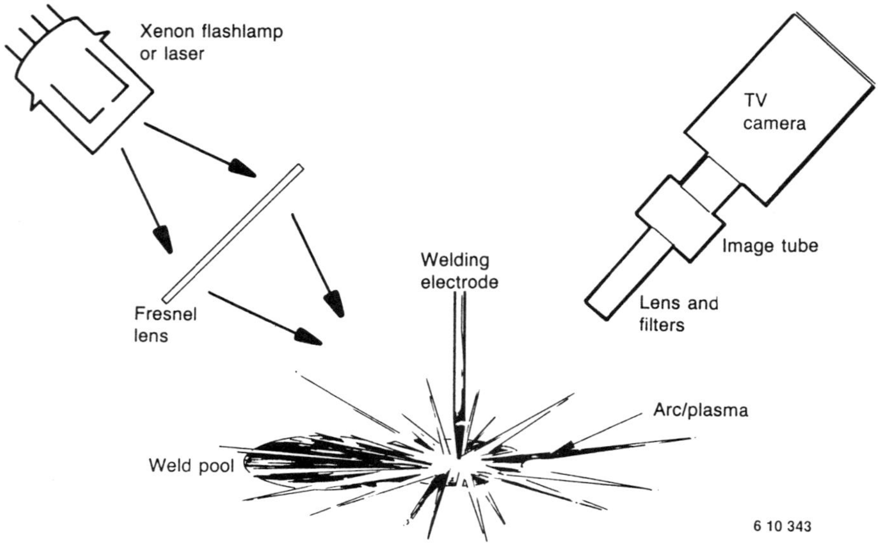

Fig. 2. Schematic of Weld Vision System.

period (33.3 ms), or roughly 0.01 percent. It follows that the degree of welding light suppression is equivalent to this factor, i.e., one part in 10,000, when compared to an ordinary video camera system which collects incoming light continuously.

These early experiments were done with a small gas tungsten arc (GTA) welder operating at 60 to 70 A arc current on stainless steel. Figure 3 shows both the conventional nonshuttered mode and the same welding run using strobe illumination and shuttering with photos taken from the video recording. This is a simple bead-on-plate weld without the use of filler wire and with the workpiece traveling to the right in the photographs. The welding pool is quite flat and the camera view angle is about 45° from the vertical, looking in beneath the rim of the gas cup of the welding torch. These photographs cannot depict the dynamics of the actual video recording, but it is evident that the welding arc light is almost totally eliminated. Visibility through the arc is greatly enhanced, with the weld pool in black contrasting sharply with the solid material in grey. The pool appears black because the liquid metal surface is a very good specular reflector and the xenon flash energy, which is incident from the right, is reflected directionally to the left and not into the video camera field-of-view. (Note that the specific viewing geometry shown in Figure 2 would be rather undesirable in this respect.) With the welding pool in black, one can also readily observe solid contaminants on the pool surface (to the rear of the welding arc in Figure 3b) and can see the reflected image of the tungsten electrode and the rim of the gas cup (in the foreground of the pool). However, the important result is a quality, high-contrast, video image that can be interfaced with a digital image processing system to automatically characterize the welding pool geometry and geometrical relationship between pool, electrode, and seam.

The INEL welding vision work was expanded in 1985 to improve system performance and to engineer techniques more suitable to industrial use. A variety of pulsed laser systems were studied as candidates to replace xenon flash illumination, and the nitrogen laser was selected for its relatively high reliability and reasonable price. A pulsed laser is attractive for this application because of the very intense peak optical power levels achievable, the single-wavelength emission which allows the use of narrow

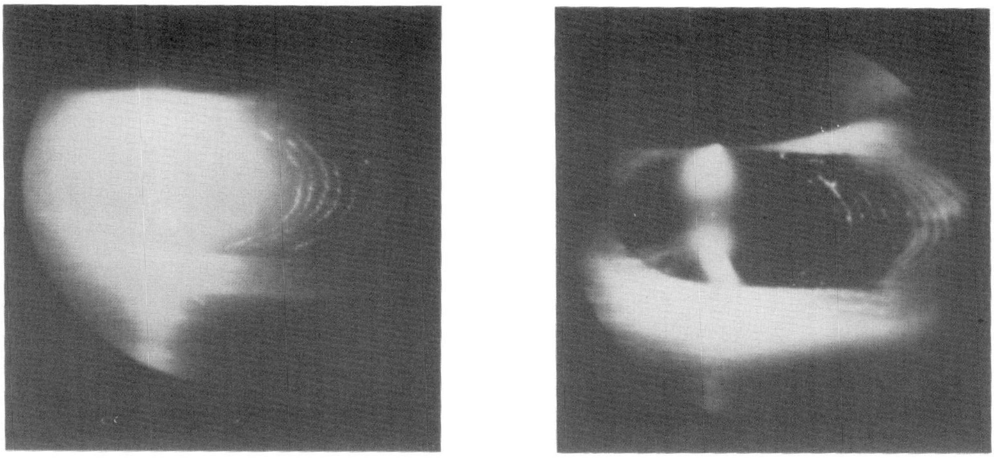

a) Ordinary video camera.

b) Flash lamp illumination and shuttering

Fig. 3. GTA Welding of Stainless Steel.

band spectral filtering, and the very good focusing characteristics of the beam, which allows transfer of the energy to the torch via optical fiber.

The nitrogen laser radiates in the near ultraviolet region at 337 nm. In general, one would like to choose a laser wavelength that does not overlap significant emission lines of the shield gases or alloy materials used in arc welding. This wavelength seems to be suitable in this respect for all welding experiments done to date. Figure 4 shows GMA welding of heavy-section aluminum alloy plate. In this process the shield gas is pure argon and aluminum wire is used as a consumable electrode in place of the tungsten electrode used in the GTA process. The welding current was $\sim$400 A and the weld bead was $\sim$29 mm wide in a flat-bottom V groove. The typical video picture without laser or shuttering is shown in Figure 4a. Some evidence of the wire electrode can be seen below the rim of the gas cup, but there is little detail regarding the groove. The bright elliptical area represents the arc light surrounded by a rather large depression in the welding pool. A great deal more detail is to be seen with the laser illumination (Figure 4b), although the pictures can be somewhat confusing without the dynamics of the actual video.

The laser pulses are generated at 30 pulses per second with a 10 ns pulse width. The residual welding arc light seen in the video is negligible, even though the welding arc is very much brighter than that of the low-power GTA welding done earlier. The profile and position of the welding groove are important parameters to be measured and are revealed by the thin shadows cast across the groove by the laser light.

The work at INEL is now with GMA and GTA welding of stainless and carbon steels. The current sensor hardware is developed well enough to produce the required imagery on pool and groove geometry for a variety of welding conditions. The emphasis now is on development of techniques for real-time computer processing of the video to extract the required data.

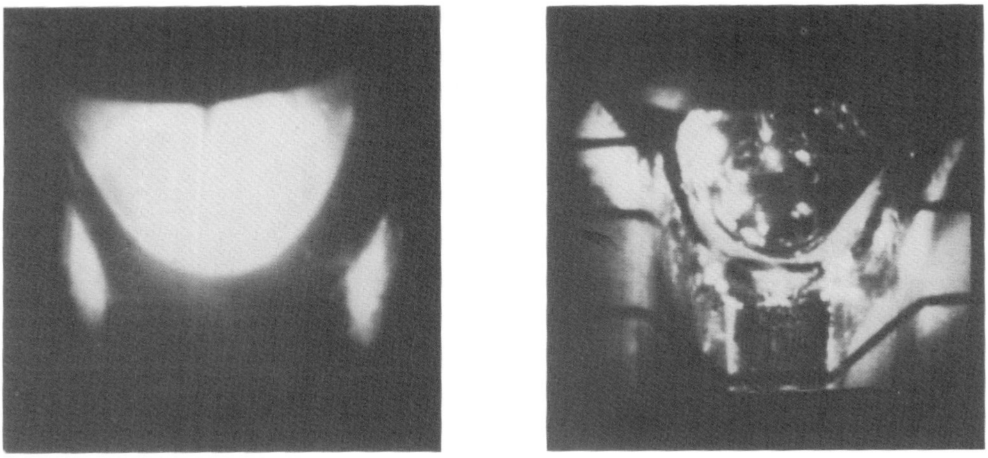

a) Without laser illumination

b) Using laser illumination and shadows to reveal groove profile.

Fig. 4. GMA Welding of Aluminum.

ULTRASONIC SENSING OF WELD POOL PENETRATION

In contrast to sensing the surface of the weld pool using electro-optical methods, ultrasonic sensing uses sound waves to "look at" the interior of materials that are opaque to light. A transducer operating in the pulse-echo mode converts electrical signals into high-frequency sound waves that travel to the area of the weld pool. When the sound wave hits the metal/molten metal interface, a portion of the energy in the wave is reflected. Some of the reflected sound energy may find its way back to the transducer, which converts the reflected sound into an electrical signal that can be observed on an oscilloscope or digitized by an analog-to-digital converter for computer analysis. This ultrasonic echo provides information about the location of the interface.

Feasibility work for detecting the molten pool started in 1977.[4 5] The proof-of-principle for a topside inspection system was achieved in the concurrent inspection system in 1984. The concurrent welding research involves positioning a transducer on the topside of a weld sample, ultrasonically sensing defects 330 mm behind the welding torch in the solidified weld metal, and analyzing the digitized data using pattern-recognition techniques.[6]

The current real-time monitoring system for detecting the molten weld pool uses a 5 MHz contact transducer mounted on various lucite wedges to generate refracted ultrasonic beams in carbon steel. A fixture design allows the transducer, mounted on the lucite wedge, to move along the plate parallel to the weld preparation in alignment with the welding torch (Figure 5). Carbon steel 25.4 mm thick is used in two weld geometries - a single bevel V-groove having a 30° included angle, a 4.76 mm root opening, and a 6.35 mm backup bar; and a V-groove having a 60° included angle, a 4.76 mm root opening, and a 6.35 mm backup bar.

Two methods are utilized to collect ultrasonic data from the molten weld pool. The first method is a video system which includes a camera, monitor, and a video recorder to record the real-time ultrasonic signal displayed on the oscilloscope screen. The video system provides a convenient method to review a welding run quickly. The second data

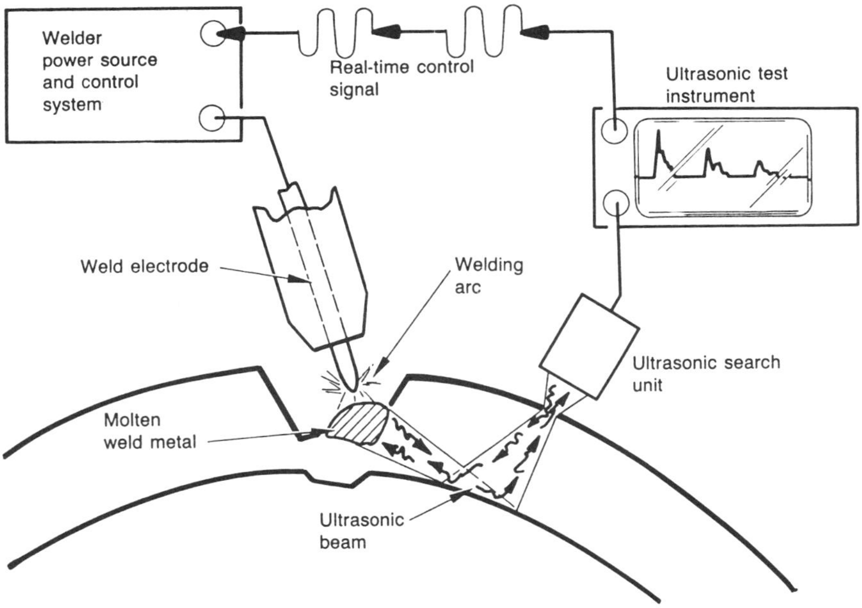

Fig. 5. Ultrasound weld monitoring schematic.

acquisition method uses a DEC LSI 11-23 computer to digitize and store the amplified ultrasonic signal. The data acquisition software digitally records the same A scan (the voltage output of the transducer as a function of ultrasonic transit time) displayed on the oscilloscope. Position data are acquired from an encoder mounted on the side-beam welder and the software allows these position data to be entered in the file header of each A scan in the file.

Using the transducer fixture and the encoder mounted on the side beam welder, alignment is established between the torch and transducer (Figure 6). Data have been acquired using various lucite wedges which provide different angles of refraction and modes of sound propagation in the weld sample to sense the molten weld pool. A 45° refracted longitudinal sound wave allows sensing of the molten weld pool interior since longitudinal sound waves propagate in the liquid metal. This sensing technique can detect good weld, porosity in the pool, and lack of side wall penetration on the side of the weld nearest the transducer. A 60° refracted shear wave on a 30° bevel plate allows sensing of the molten-solid interface (shear waves are not supported in liquid) and amount of side wall penetration. The 60° refracted angle is not optimum for sound reflection at the molten/solid interface. The real value of sensing with a 60° refracted shear wave is in determining if there is full side wall penetration. There is a specular reflection pathway when the pool does not penetrated the side wall which results in a large echo returning to the transducer. Using a 45° refracted shear wave allows for enhanced detection of the molten/solid interface because the reflection coefficient is greater than for the 60° refracted shear wave. The reflected signal is of sufficient amplitude and resolution to provide information about the region around the molten/solid interface.

An evaluation process is currently underway using the 45° refracted shear wave data. Reflectors seen before, during, and after welding are evaluated using machined samples to simulate these three unique points of

Fig. 6. Weld fixture used for transducer and torch alignment.

the welding process. Initial correlations between the machined samples and the acquired data indicate that the geometries of the forming molten/solid interfaces can be detected and grouped into several basic geometry types. This information provides the potential of monitoring the interface and providing control input which can assure an optimum interface geometry throughout a welding pass. Ray-tracing field codes are being developed which will allow for the validation of data acquired during the welding pass. These field codes consider the effects that thermal boundaries, plate geometry, and transducer sound fields have on received ultrasonic echoes during a weld pass. But, for the ultrasonic sensing to be implemented in an industrial application, research efforts need to be directed to establishing a non-contact method of sound inducement and detection. One method that is currently being investigated is laser-induced sound in the weld pool coupled with a contact transducer for sound detection.

CONCLUSION

Process sensing and control of automated welding is possible for the selected parameters discussed above. The success of this integrated control scheme can result in improved weld quality, increased productivity, and energy cost savings in appropriate welding applications.

ACKNOWLEDGMENTS

This work is supported by the U. S. Department of Energy, Office of Energy Research, Office of Basic Engineering Sciences under DOE Contract No. DE-AC07-76ID01570. Appreciation is expressed to U. S. Wallace, C. L. Shull, and C. Stander for technical assistance.

REFERENCES

1. H. B. Smartt, C. J. Einerson, and A. D. Watkins, "Modelling and Control of Gas Metal Arc Welding," in preparation, to be submitted to Welding Journal.

2. W. E. Lukens and R. A. Morris, "Infrared Temperature Sensing of Cooling Rates for Arc Welding Control," Welding Journal, 61, 1, January 1982, pp. 27-33.

3. H. B. Smartt and J. F. Key, "An Investigation of Factors Controlling GTA Weld Bead Geometry," Proceedings, ASM Conference on Trends in Welding Research in the United States, November 1981, New Orleans, LA.

4. L. A. Lott, J. A. Johnson, and H. B. Smartt, "Real-Time Ultrasonic Sensing of Arc Welding Processes," Proceedings, 1983 Symposium on Nondestructive Evaluation Applications and Materials Processing, p. 13-22, American Society for Metals, Metals Park, OH.

5. L. A. Lott, "Ultrasonic Detection of Molten/Solid Interfaces in Weld Pools," Materials Evaluation, 42:337-341 (1984).

6. J. A. Johnson and N. M. Carlson, "Weld Energy Reduction by Using Concurrent Nondestructive Evaluation," to be published, NDT International, June 1986.

ULTRASONIC CHARACTERIZATION OF SPOT WELDS

F. Nadeau, C. Néron and J.F. Bussière

Industrial Materials Research Institute
National Research Council Canada
75 De Mortagne, Boucherville (Quebec) Canada J4B 6Y4

INTRODUCTION

Resistance welding is one of the most widely used joining processes in industry. Basically, it uses mechanical pressure and high intensity current to hold and weld sheet metal assemblies. The bond is assured by the presence of a "weld nugget", a volume common to both sheets which melts and resolidifies during the operation. The most common application is automotive engineering where many of the sheet steel components are spot welded[1]. It is also used in other fields such as the aerospace industry and with other materials such as aluminum or magnesium alloys, copper, nickel, etc.[2]. As any other industrial process, a whole area of technologies has evolved along with resistance welding to meet requirements of quality assurance via either process control or product testing. The major breakthroughs are on the side of process control where much of the research efforts have been concentrated. Parameters such as current and time can now be dynamically controlled through adaptive feedback mechanisms using sensors that measure electrical resistance,[3,4] displacement,[5] infrared radiation,[6] ultrasonic transmission[7] and even acoustic emission.[8] As for testing, most of it is still done destructively. A common method is the peel test where a sample weld is simply pulled apart: if the weld is good, one of the sheets will tear around the nugget whose width can then be measured. Of course, process performance and weld quality can only be statistically estimated.

By contrast, the application of non destructive testing technologies to resistance welding has progressed slowly when compared to the advances in process control. One reason is probably because resistance welding is a cheap process and NDT is often an expensive proposition (this is not true for certain applications such as in the aerospace industry). Another reason is that assessing the quality of any type of bond including spot welds, is often a major technical challenge. For instance, the number one NDT technique, X-rays, is not reliable[10] at showing lack of fusion, one of the commonest flaws in spot welds. Other techniques have been investigated such as eddy currents,[9] electrical resistivity measurements[10] which is showing some promise of success, and also ultrasonic inspection. It has been shown[11,12] that the nugget can be detected and even sized by measuring the attenuation of bulk waves transmitted through the weld.

Lamb waves[13] have also been successfully used to size the total area of the weld. The technique does not however differentiate the nugget from the surrounding weakly bonded "stick zone" and therefore can be misleading in assessing weld strength. Images of spot welds produced with a scanning laser acoustic microscope (SLAM)[14] have also been shown to reveal the presence of a nugget in mild steel.

In the present study, high resolution ultrasonic C-scan imaging was investigated as a technique to characterize spot welds in jet engine components, an application where advanced NDT procedures are economically feasible. We find that the method can be both efficient and reliable but is material dependent because it relies on microstructural contrast between parent and nugget material.

SAMPLE

The samples studied were made from two different alloys: Inconel (AMS 5540), composed of 75% nickel, 15% chromium and 7% iron, and stainless steel (AMS 5504). The Inconel has a single phase (austenite) solid solution microstructure with a carbide dispersion. The material is annealed and cold worked, causing twinning in the austenitic grains. Each Inconel sample consisted of a pair of identical 2 x 7 cm strips of cold rolled sheets. The two strips were joined by five or six spot welds performed with the same welding parameters. Two different sheet thicknesses were used, $t = 2.1$ mm and $t = 1$ mm. The welding parameters were changed from one sample to another to vary the penetration. The resulting penetrations (defined as the ratio between the thickness of the nugget and the combined thickness of the two plates) were measured from macrographs of the end spot weld on each sample. They are: 0% and 50% for the thicker strips ($t = 2.1$ mm) with a spot diameter of about 10 mm, and 0%, 8%, 27%, 59%, 65% for the others ($t = 1$ mm) with a spot diameter of 6 mm.

AMS 5504 stainless steel is a hardenable martensitic stainless steel. In the sheet form (cold worked and annealed) the microstructure consists of a number of chromium carbides distributed in a ferrite matrix. The amount of martensite depends on the heat treatment, cooling rates, etc. The stainless steel samples consisted of seam welded strips of unequal thicknesses (1,2 mm and 0,9 mm). The seam width was around 6 mm. Again, various degrees of penetration were obtained: 0%, 37%, 52%, 61% and 76%.

ULTRASONICS

Ultrasonic C scans were performed on all samples with a high precision computerized immersion system. Images were produced from either the interface echo (reflexion from the bottom of the top plate) or the bottom echo (reflexion from the bottom of the bottom plate). Two immersion type focussed ultrasonic transducers were used: a 6.35 mm diameter 25 MHz and a 10 MHz (19 mm dia.) transducer, each having a focal length of 25 mm. Adequate focussing was obtained by adjusting the transducer to sample distance to maximize the amplitude of the selected echo (interface or bottom) which could then be gated out for imaging. Micrographs of selected samples were also produced using both a conventional optical microscope and an acoustic microscope. The acoustic microscope is similar to that described in ref. 15 and was operated at 450 MHz.

RESULTS

Figure 1 shows a C-scan performed with the 25 MHz transducer and using the interface echo. The samples in this case are the 2 mm thick Inconel ones. We see that the interface echo disappears over the entire surface of the weld, producing the dark areas on the image. In the 50% penetration weld, the outline of molten metal ejected from the nugget is clearly visible. This last characteristic however is the only indication of any difference in penetration between the two welds. Poorly bonded regions such as over the entire area of the 0% penetration sample or near the edge of the 50% sample do not cause any more of an interface echo that the heavily bonded nugget area. Therefore, the interface echo can not be used to directly assess the thickness of the nugget.

Although lack of fusion is the main source of defects in spot welds, other types of defects such as solidification cracking, inclusions, voids, etc. may occur. Some of these are ultrasonic scatterers that will be visible on an interface echo C-scan. Such is the case for the 50% penetration weld of Figure 1 where the small white spot inside the dark area corresponds to such a defect. This sample was subsequently cut through the defect which turned out to be a large void as shown in the macrography of Figure 2.

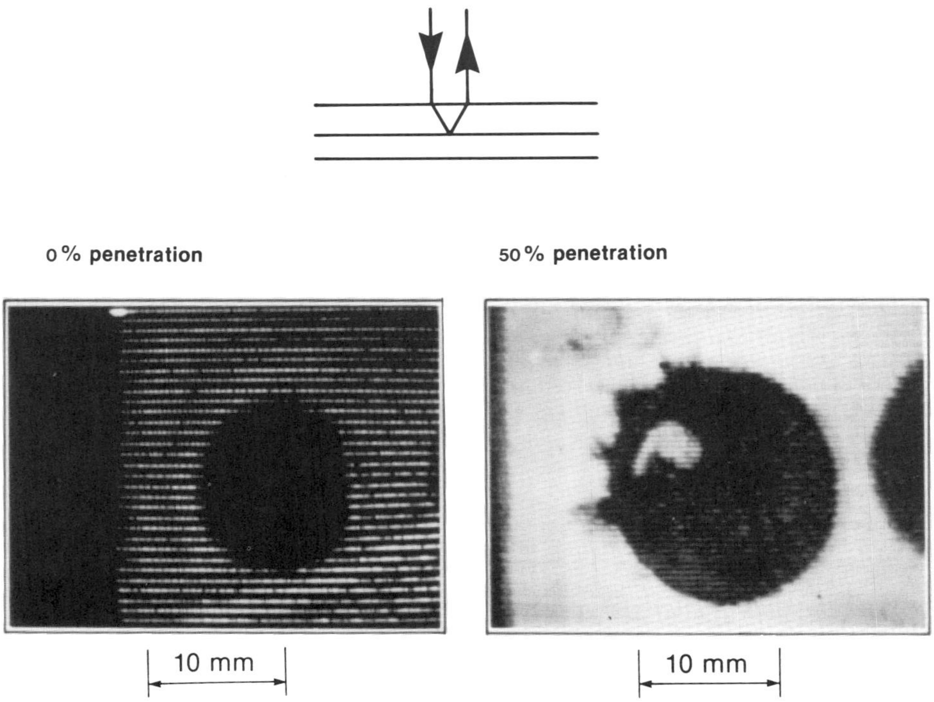

Fig. 1. C-scan images of spot welds in 2.1 mm thick Inconel plates using the interface echo. Note the absence of echo over the entire area except for a small defect in the 50% penetration weld.

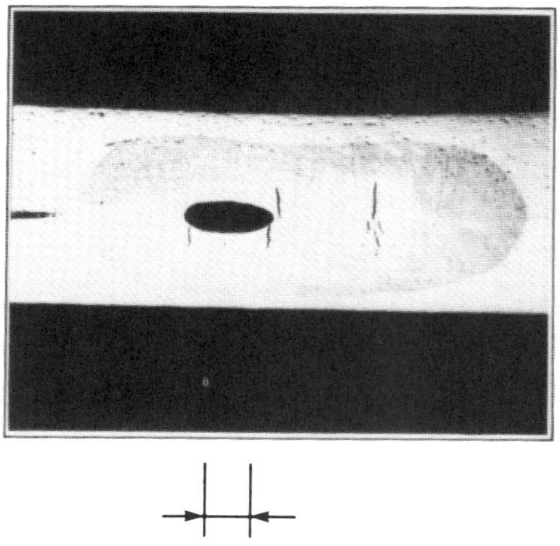

Fig. 2. Macrograph showing a void in the highly penetrated spot weld (50% penetration) of Figure 1.

Figure 3 illustrates the same two welds scanned with the same 25 MHz transducer but imaged with the bottom echo. Here the difference between the two samples is striking as the more penetrated weld produces a dark contrast inside the weld contour. We know that the amplitude of the bottom echo does not depend solely on the reflective properties of the interface but on the total through transmission coefficient of the sample. We also know from Figure 1 that, essentially, there is no interface reflexion within the weld perimeter. This constrast can therefore be attributed only to either attenuation of ultrasound as it propagates through the bulk of the weld or variations in the transmissive and reflective properties of, respectively, the top and bottom surfaces of the sample. This last hypothesis is based on the fact that the hotter welding parameters used to achieve the high penetration of the second weld induce greater deformation of these surfaces. However, after polishing one or both of these surfaces, virtually the same images were obtained. Thus, the contrast is caused by attenuation in the weld itself.

One would expect the area outside the perimeter of the weld to be dark since the interface between the two plates is highly reflective (almost 100%) and thus the transmission coefficient of the sample is close to zero. This is not true in this case because, the two plates being of equal thickness, the second interface echo falls in the same time gate as the bottom echo. Interestingly, near the edge of the weld, those two echoes interfere, producing the dark contour that outlines the weld.

An additional set of Inconel samples was studied. These were made from thinner 1 mm plates welded to various well controlled levels of penetration. Figure 4 shows the bottom echo C-scans obtained for some of these samples, along with the amplitude v/s position profile of a single line scan through the center of the welds. A very distinctive and progressive rise of the attenuation with increasing penetration is observed. The very same measurements were performed on the stainless steel samples. The long wavy shape typical of seam welds was clearly

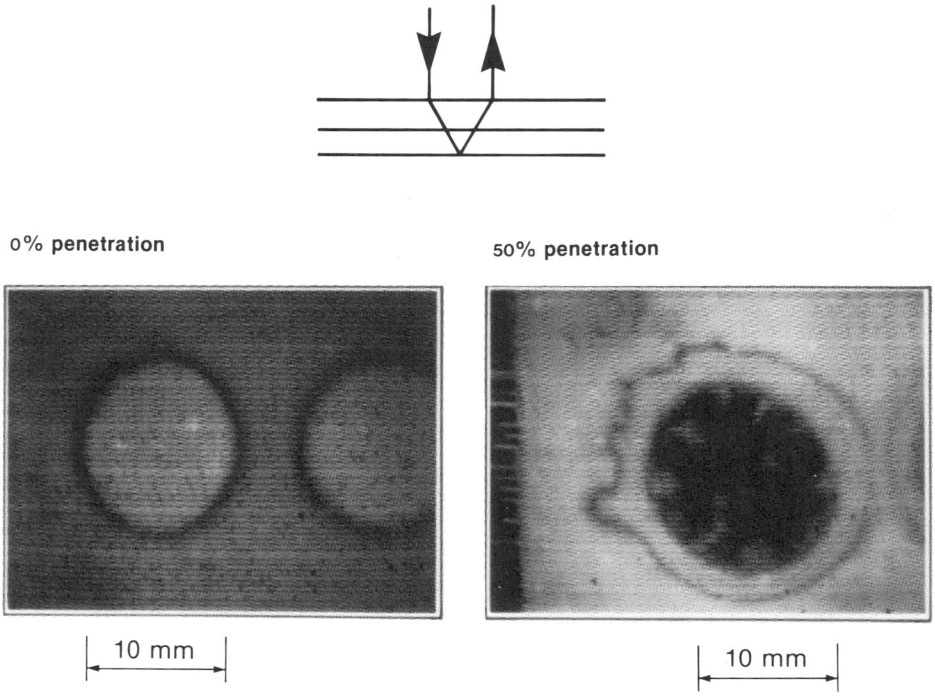

Fig. 3. C-scan images of spot welds in 2.1 mm thick Inconel plates using the bottom echo. Note the larger attenuation (darker area) at the center of the thicker nugget (50% penetration).

outlined in the images. However, no significant or consistent differences in attenuation could be found between the various levels of penetration. The average attenuation measured in this material was also quite low (less than 0.3 db/mm).

DISCUSSION

The microstructures observed in a typical spot weld can be numerous (parent metal, heat affected zone, nugget edge and center) and each attenuates ultrasound differently but the most significant difference is expected to be between the nugget and the unmolten material. Assuming two uniform attenuation coefficients, one in the nugget and one outside, and using the data from the 0% penetration sample to estimate the latter, one can plot the attenuation in the nugget as a function of twice the nugget thickness and obtain, for the Inconel samples, the curve illustrated in Figure 5. The straight line represents the theoretical values for a coefficient of 2.1 db/mm. Agreement between this simplified model and the actual data points is good.

Assuming that most of the attenuation is due to scattering at the grain boundaries and that the wavelength of the 10 MH_Z ultrasonic pulse is much greater than the average grain size, we know[16] that the attenuation coefficient, α, is proportional to the grain size, D, to the third power:

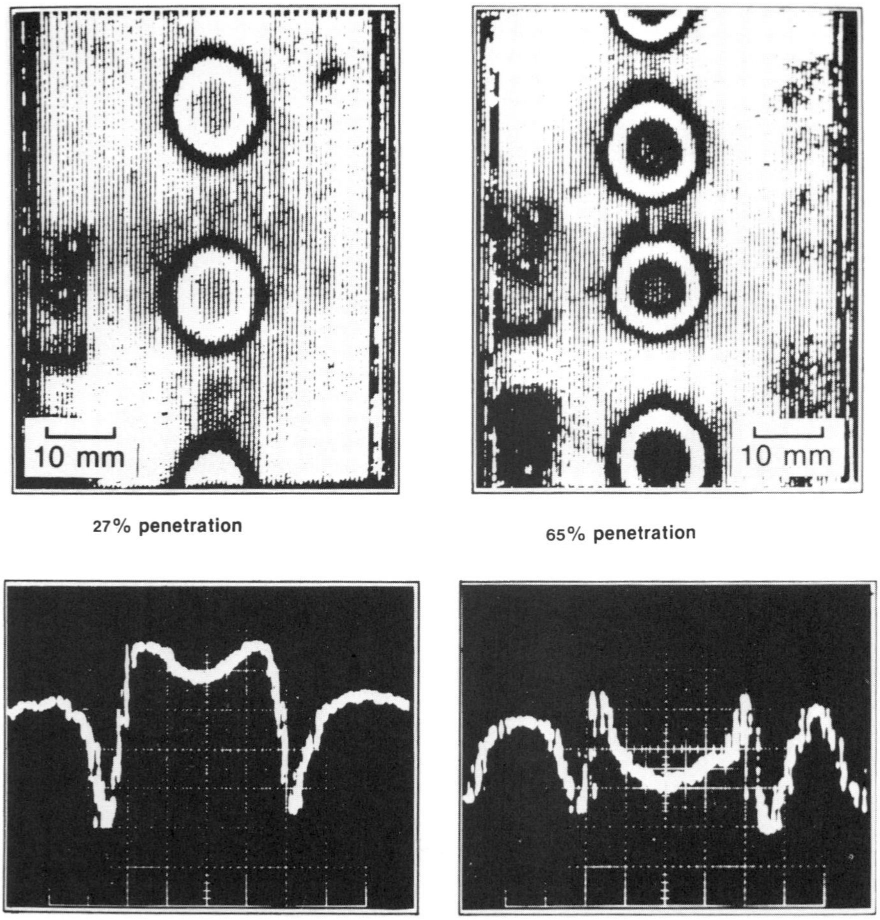

Fig. 4. C-scan images (bottom echo) of spot welds of various penetrations in 1 mm thick Inconel plates (top figures). Note the increased attenuation in the center of the nugget for 65% penetration compared to 27%. The bottom figures are line scans of rectified amplitude across the center of each weld.

$$\alpha = kD^3$$

$$\frac{d\alpha}{\alpha} = \frac{3dD}{D} \qquad (1)$$

According to (1), the relative change in attenuation is equal to three times the relative change in grain size. Since the attenuation in the nugget is about fifteen times that in the parent metal, a grain size five times larger is to be expected. This corresponds to what was observed as illustrated in Figure 6. The area shown in that figure is at the edge of the nugget in the 4 mm thick Inconel sample penetrated at 50%. It was imaged both optically and acoustically. These images reveal an enormous difference in microstructure between the parent metal, where the grain size averages about 75 μm and the typical cast structure

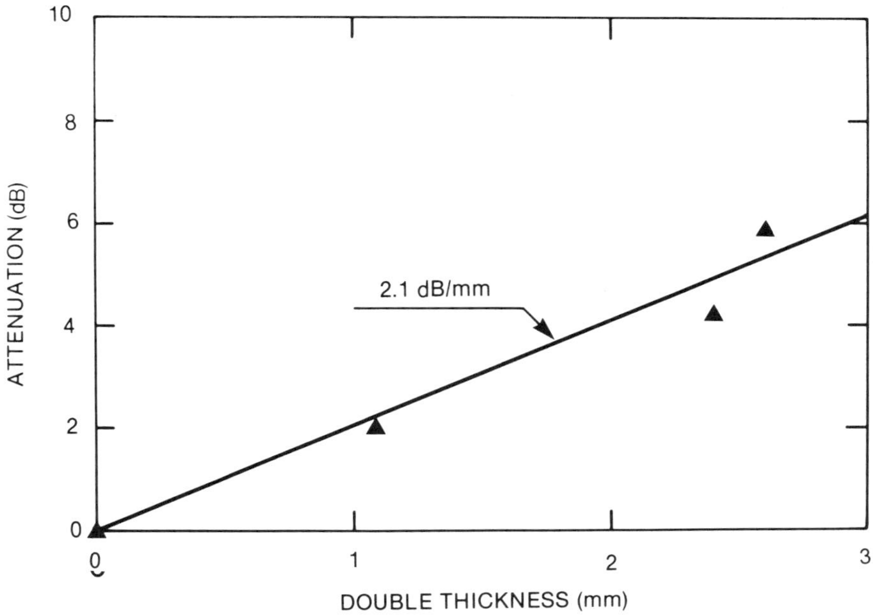

Fig. 5. Attenuation in the nugget (Inconel) as a function of nugget thickness

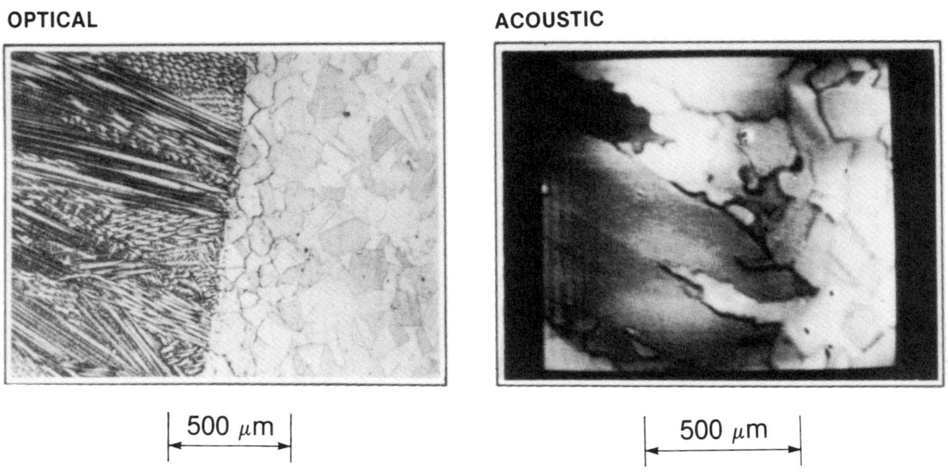

Fig. 6. Optical and acoustical micrographs of a spot weld nugget edge (Inconel). The acoustic micrograph was obtained with a frequency of 450 MHz. The parent metal in each case is on the right hand side.

of the nugget, where columnar austenitic grains as long as 500 μm can be observed. Note that the grain boundaries in the nugget appear much more clearly in the acoustic micrograph (which was performed on an un-etched sample) than in the optical one where they are burried in the fine substructure of the columnar dendrites.

By contrast, the microstructure of the stainless steel samples consisted of almost pure martensite in the nugget and a mixture of martensite and ferrite in the parent metal. It has been shown[16] conclusively that "the martensitic transformation lowers the scattering contribution to the attenuation in polycrystalline metals". It is therefore not surprising that the technique could not detect even the presence of a nugget in these samples.

CONCLUSION

It has been shown that ultrasonic C-scan imaging of the amplitude of the echo transmitted through a spot weld can be used to detect and size the nugget of the weld. The technique however only works on alloys yielding nugget microstructures that attenuate ultrasound much more than the parent metal. Image contrast from the nugget arises because of increased attenuation associated with larger grain size in the nugget as compared to the parent metal. A good quantitative relationship was found to exist between nugget penetration and attenuation. However, because this quantitative relationship is dependent on the evolution of microstructure in the nugget during solidification, such a correlation will be material dependent.

The technique was shown to work well for Inconel, an austenitic superalloy, but no contrast associated with the nugget could be observed for a martensitic stainless steel. The acoustic microscope was a valuable tool in this study, producing micrographs where the true ultrasonic scattering boundaries are revealed.

ACKNOWLEDGEMENT

The authors would like to acknowledge the precious contribution of Mr. Reno Burlone of Pratt & Whitney who supplied the samples and information for this work.

REFERENCES

1- Dickinson, D.W., "Welding in the Automative Industry - State of the Art". American Iron and Steel Institute, Washington, Research Report #SG-81-5 (August 1981).

2- "Resistance and Solid-State Welding and Other Joining Processes", A.W.S. Welding Handbook, 7th Edition, Vol. 3, pp. 14-15, W.H. Kearns Editor.

3- Dickinson, D.W., Franklin, J.E. and Stanya, A., "Characterization of Spot Welding Behavior of Dynamic Electrical Parameter Monitoring", Welding Journal Research Supplement (June 1980), pp. 170-176.

4- Schumacher, B.W., Cooper, J.G. and Dilay, W., "Resistance Spot Welding Control That Automatically Selects the Welding Schedule for Different Types of Steel", S.A.E. Technical Paper #850407, International Congress & Exposition, Detroit (March 1985).

5- Roden, W.A., "Evaluation of Resistance Welding In-Process Monitors", Welding Journal Research Supplement, (November 1968), pp. 515-S - 521-S.

6- Brown, B. and Bawgs, E., "The Measurement and Monitoring of Resistance Spot Welds Using Infrared Thermography", SPIE Vol. 581, Thermosense VIII (1986), pp. 57-69.
7- Burbank, G.E. & Taylor, W.D., "Ultrasonic In-Process Inspection of Resistance Spot Welds", Welding Journal Research Suppl. (May 1965), p. 193s.
8- Havens, J.R., "Controlling Spot Welding Quality and Expulsion", S.M.E. Paper #AD76-279 (1976).
9- Fastritsky, V.S., Fishkin, P.S. and Rybalkin, E.P., "Method and Apparatus for Non-Destructive Quality Testing of Spot Welds", U.S. Patent #4 287 474 (Sept. 1981).
10- Cohen, R.L. and West, K.W., "Spot Weld Strength Determined from Simple Electrical Measurements", Welding Journal Dec. 1984), pp. 17-23.
11- Papadakis, E.P., "Ultrasonic Velocity and Attenuation: Measurement Methods with Scientific and Industrial Applications", Physical Acoustics, Vol. 12, Chapter 5, pp. 343-348, W.P. Mason & R.N. Thurston Editors, Academic Press (1976).
12- Dubetz, M. and Bilge, J.U., "Ultrasonic Method of Inspecting Spot Welds", U.S. Patent #4 265 119 (May 1981).
13- Fendel, F., Peretz, M. and Rokhlin, S.I., "Ultrasonic Lamb Wave Method for the Sizing of Spot Welds", Ultrasonics (March 1984), pp. 78-84.
14- Adams, T., "Nondestructive Evaluation of Resistance Spot Welding Variables Using Ultrasound", Welding Journal (June 1985), pp. 27-28.
15- Quate, L., Atalar, A., and Wickramasinghe, H.K., Proc. IEEE (1979), Vol. 67, pp. 1092-1104.
16- Papadakis,, E.P., "Scattering in Polycrystalline Media", Methods of Experimental Physics: Ultrasonics, Vol. 19, P.D. Edmons Editor, Academic Press (1981), pp. 237-278.

AN NDE SYSTEM FOR THE DETECTION OF EARLY DAMAGE IN HIGH-TEMPERATURE ROTORS

T. Goto, Y. Kadoya, T. Takigawa[+] and K. Kawamoto[+]

Takasago Technical Institute
Mitsubishi Heavy Industries, Ltd., Takasago, Japan
[+]Turbine Engineering Department
Mitsubishi Heavy Industries, Ltd., Tokyo, Japan

INTRODUCTION

The assessment and extension of the life of fossil power plants is now a matter of prime importance for the electricity industry.[1] Consequently, in recent years, advances in life prediction methods, especially for high-temperature components, have become a frequent, and closely followed, topic at materials-related conferences. One of the conclusions obtained from research work to date on material degradation and damage in long-service components is that the development of non-destructive examination techniques to detect early damage, that is, damage before crack initiation, is urgently needed. Crack initiation life, as compared with crack propagation life, has the advantage of making it possible to formulate a life extension program with more options.

The high-temperature rotor is one of the key components in a high-pressure turbine, and in service is subjected to high centrifugal and thermal transient stresses. For this reason, in-service inspection is carried out using non-destructive tests. These tests can detect both cracks and flaws. However, they provide no information on rotor life till cracks initiate.

We, therefore, developed an inspection system, the MACH NDE SYSTEM, standing for Material Characteristics Non-Destructive Evaluation system, for the high-temperature rotor. The critical locations of the rotor are the inner bore region, where creep-related damage is accumulated, and the periphry grooves, such as blade grooves, where the fatigue damage is accumulated. Our aim was to detect such damage by observing material properties change before crack initiation in order to predict the crack initiation life. This paper presents the results of the studies conducted for the development of the system and introduces the system itself.

MATERIAL PROPERTIES CHANGES IN ROTOR MATERIAL DURING CREEP

Changes in the material properties were investigated in the laboratory by applying non-destructive tests to Cr-Mo-V rotor steel specimens subjected to long-term heating and creep in order to find clues to promising examination methods for damage detection and an analytical method for non-destructive life prediction.

Long-term heating

The non-destructive tests used were X-ray diffraction broadening analysis, eddy current testing (ECT), electro resistance measurement, ultrasonic velocity measurement, ultrasonic attenuation measurement and hardness test.

As examples, the results of X-ray, ECT and hardness tests are shown in Fig.1, with the readings plotted against the time-temperature parameter (Larson-Miller parameter) of

$$P = T(20 + \log t) \qquad (1)$$

where T is the heating temperature (K) and t is the heating period (hrs).

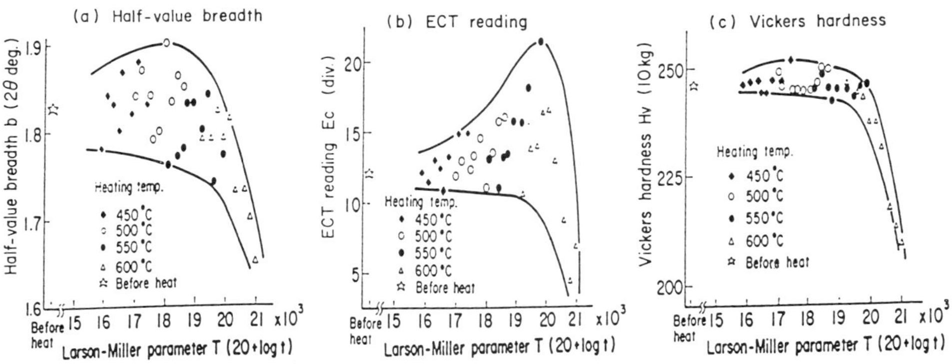

Fig.1 Plotting of half-value breadth (a), ECT reading (b) and Vickers hardness (c) against Larson-Miller parameter for long-term heating

It is clear from Fig.1 (b) that the material property studied using ECT has a mode of change during long-term heating at constant temperature which depends on the heating temperature. At 550°C, after a short period in which it decreases, the ECT reading increases with heating time. On the other hand, the reading at 600°C begins to increase till a peak value is reached, and then decreases with heating time. It seems difficult, therefore, to express the changes in ECT reading simply using a time-temperature parameter.

If it is possible to express them using a time-temperature parameter, one may expect the material property test to yield information on service temperature or service hours, though one or other would have to be known before the test is carried out. Comparison of the test results suggests that the hardness test is the most useful, since a simple curve approximation of material property change against the time-temperature parameter can be most easily obtained using the hardness test.

It is clear that the approximation curve of the hardness change during long-term heating must depend on the temper history adopted in material production, because Eq. (1) does not include the effect of the history. The changes may also depend on the chemical composition of the Cr-Mo-V rotor. In order to reduce the influence of these factors, the hardness divided by the value before heating was plotted to give the hardness change expression. The result is shown in Fig.2, where the data from two other Cr-Mo-V charges found in the literatures[2,3] are included. This figure indicates that a simple curve approximation of hardness change against the time-temperature parameter may be possible for the high-temperature rotor, if the hardness-initial value ratio is used. The authors call this curve the "softening curve".

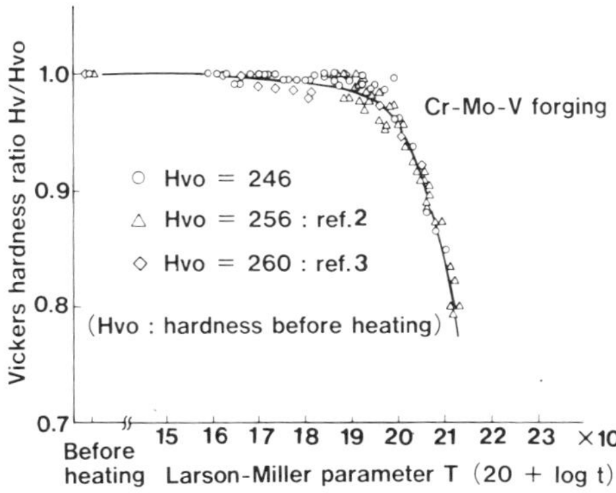

Fig.2 Plotting of hardness ratio to the value before heating against Larson-Miller parameter for long-term heating of three charges of Cr-Mo-V steel

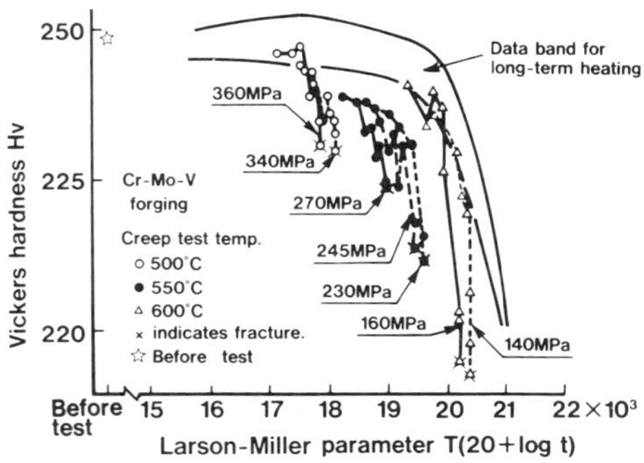

Fig.3 Plotting of hardness against Larson-Miller parameter for creep in Cr-Mo-V steel

Softening due to creep

A relatively smooth curve for change during creep was obtained with the hardness test. Fig.3 shows the plotting of hardness of creep specimens against Eq. (1), with T being creep test temperature and t loading hours. The data obtained by long-term heating are also shown for comparison. From Fig.3, the following may be pointed out.

(i) The change in hardness indicates the dependence on stress at a given temperature. The lower the stress, the more the curve for hardness change during creep tends to move to the right side of the graph and the closer to the softening curve obtained from long-term heating tests.
(ii) The hardness value at rupture also shows the dependence on stress at a given temperature. Therefore, it is impossible to define the rupture or creep damage by the hardness test alone.

VanLeeuwen[4] successfully used Manson's minimum-commitment method to analyze the effect of stress, temperature and time on change in tensile strength during long-term heating and creep for aluminium. One of the present authors,[5] therefore, used this method in this study.

Let the phenomenon depending on temperature T and time t be λ and λ be expressible by the function of G as

$$\lambda = f(G) \qquad (2).$$

Typical examples of G are the Dorn parameter and the Larson Miller parameter expressed by Eq. (1). From these parameters, it is evident that it is possible to express G by the addition of a temperature term H(T) and a time term F(t) as

$$G = H(T) + F(t) \qquad (3).$$

That is, the phenomenon occurring during long-term heating is a function of Eq. (3). Where stress is added, the phenomenon shows some shift ΔG in G, but let ΔG be a function of stress only.

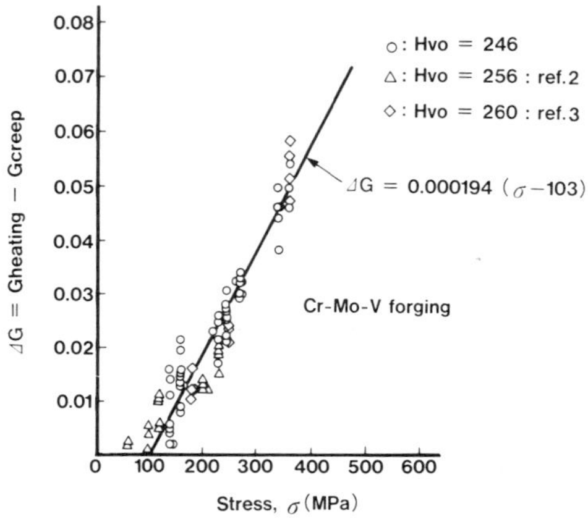

Fig.4 Plotting of ΔG against creep stress for Cr-Mo-V steel

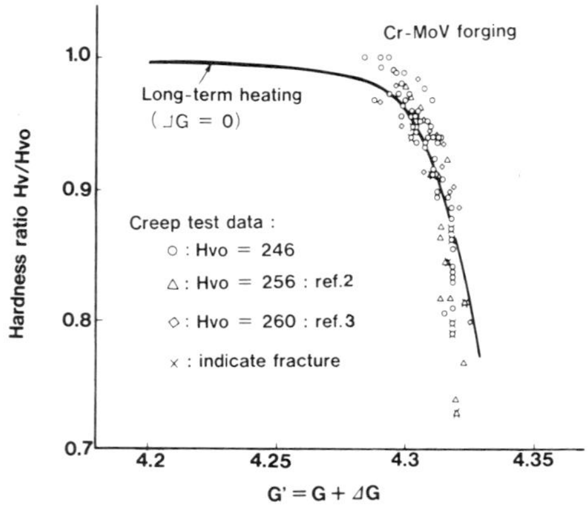

Fig.5 Plotting of hardness ratio against parameter G' for creep in Cr-Mo-V steel

The Larson-Miller parameter was used to define the form of G, and the hardness change in the form of hardness ratio Hv/Hvo was defined as λ. For long-term heating data, the softening curve was used instead of actual experimental data. ΔG for each creep specimen was read from the Hv/Hvo-G diagram and plotted against the creep stress in Fig.4, where the data from two other Cr-Mo-V charges found in the literatures[2,3] are included.

Fig.4 shows a good correlation existing between ΔG and σ (MPa), and the relation can be expressed by

$$\Delta G = 0.000194\ (\sigma - 103) \quad (4).$$

The hardness of a creep specimen can, therefore, be expressed using the parameter of G',

$$G' = \log T + \log\ (20 + \log t) + \Delta G \quad (5),$$

and can be correlated with the long-term heating data. Fig.5 shows the plotting of creep data against G', and good correlation between creep data and softening curve is obtained.

Estimation of residual creep life using hardness test

From the experimental study described so far, a method of estimating the residual creep life using the hardness test was proposed.[8] The method uses the distribution of temperature and/or stress in a machine part and the creep rupture diagram.

Fig.6 shows the method used for the case in which, though the service temperature and time are known, the initial hardness and critical stress are not known. It should be noted that this method is not effective for a machine part subjected to a stress below a threshold value. For a Cr-Mo-V steel, the value is about 100 MPa and may be higher than the level of creep stress after relaxation on the bore surface of rotor, which is the location at which we want to know when the cracks initiate.

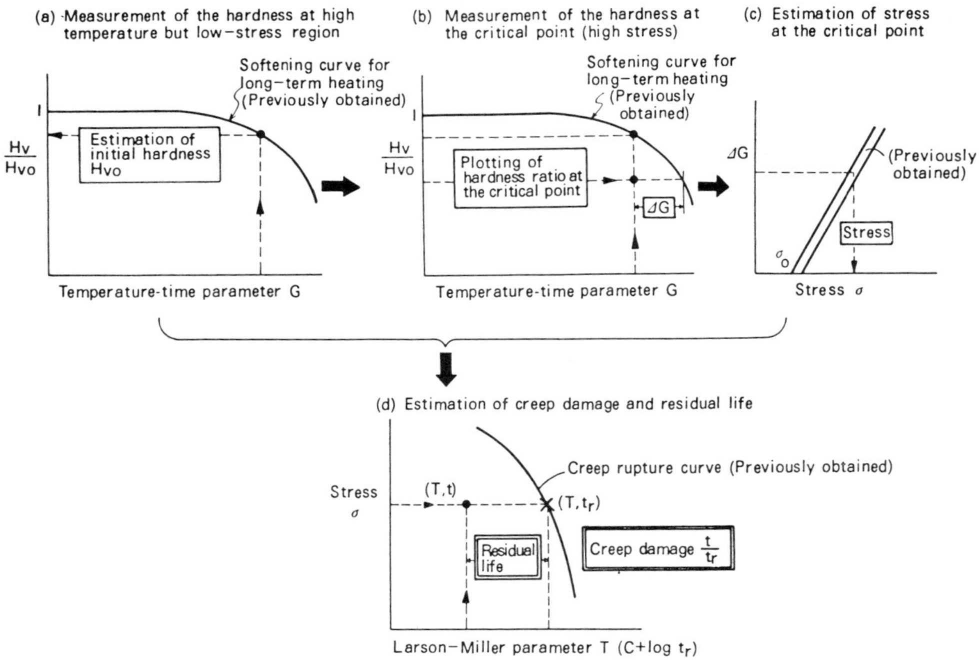

Fig.6 Illustration of a method to detect creep damage and estimate residual life for cases where service temperature and time are known

It is, however, important to notice here that the point at which softening is observed on the rotor bore surface is most probably the site of concentration of severest damage. It is also pointed out that the stress during relaxation period and the transient thermal stress may increase such a possibility, though their effect on the softening behavior of bore surface is not yet clear.

Cavitation due to creep

The physical damage which we can most easily understand is the cavitation phenomenon occurring during creep. ECT, ERT and UT methods were tried to detect the creep voids existing in creep specimens. However, no definite conclusion as to whether these methods can detect this phenomenon or not has yet been reached.

Therefore, replication observation was adopted in this study. It seems that the evaluation of cavitation for residual life prediction may be possible using three methods. They are (i) the calculation of creep strain accumulation[6] using a kachanov-type damage model, (ii) the classification of the cavitation situation[7] and (iii) the empirical correlation between cavitation and creep parameters.[8]

For method (ii), many examples of actual rotor failure would be needed. Therefore, we adopt methods (i) and (iii). Sinya et al.,[9] carried out density measurement of creep specimens, and found the following relation for Cr-Mo-V steel.

$$-\Delta D/D = 2.5\times10^3\, \varepsilon t \sigma^3 \exp\,(-234000/RT) \qquad (6)$$

where $\Delta D/D$ is density change, ε creep strain (%), t exposure time (hr), σ stress (kg/mm^2), R gas constant and T temperature (K). They also give a creep life prediction map, shown in Fig.7, derived from Eq. (6).

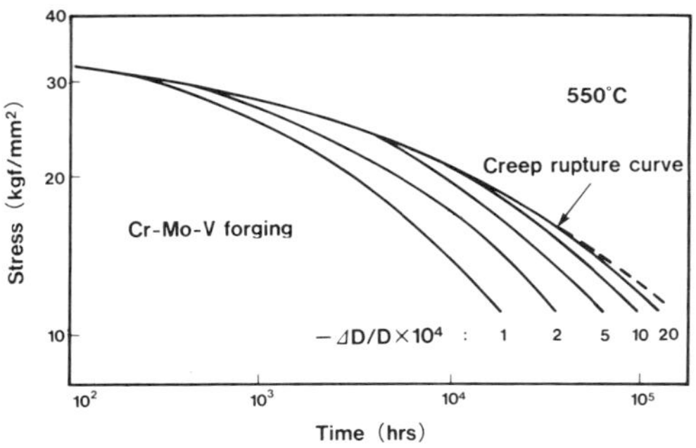

Fig.7 Diagram of iso-change in density for creep in Cr-Mo-V steel[9]

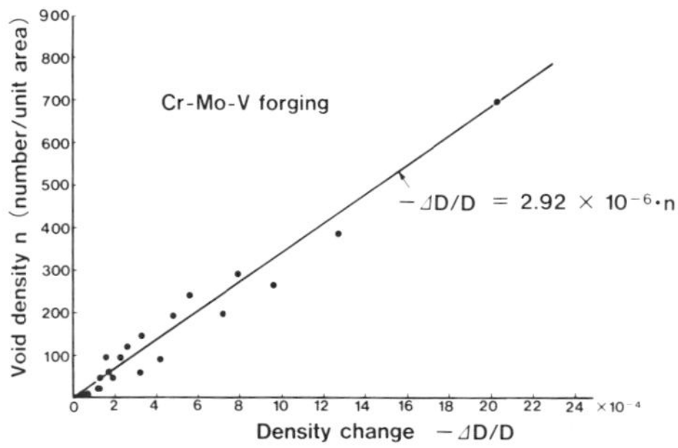

Fig.8 Plotting of void density against density change for creep specimen

Although these researchers used density measurement, they also suggested that there will be a correlation between the degree of creep voids and density change and that the replication technique is promising.

Therefore, various parameters, such as averages of void size, void number in unit area, and so forth, obtained by replication observation were investigated in terms of correlation to change in density. It was found, as shown in Fig.8, that the best parameter is the void number in unit area.

Fig.9 shows an example of void observed by the replication method and by the direct observation of the same void.

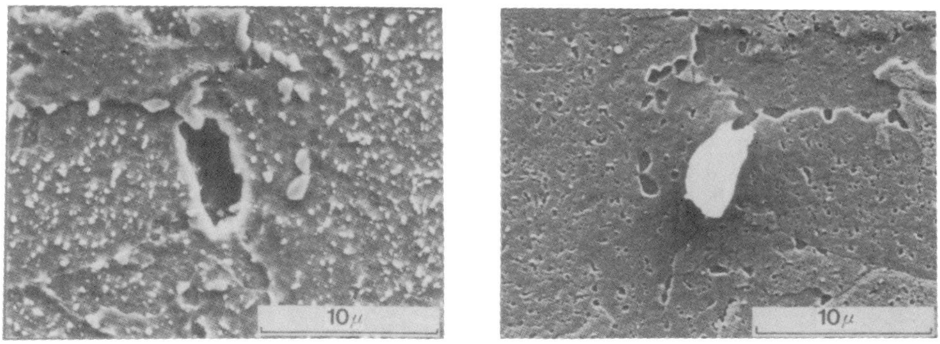

Direct method **Replica method**

Fig.9 Creep cavity observed by replica method and direct method (Ruptured specimen; 2,742 hours, 600°C, 196 MPa)

The calculated rotor creep stress shows a peak slightly below the bore surface. Therefore, it is to be expected that observation of bore surface may yield data for phenomena occurring somewhat later than the occurrence of critical damage at the location of highest stress. However, what is said in our section on estimation of residual creep life by hardness test may hold true here, too. We should also point out that the number of fossil power plants which are cyclically operated is increasing. The role of bore surface inspection would, therefore, clearly appear to be increasingly important.

THE MACH NDE SYSTEM FOR ROTOR BORE

The typical size of rotor bore is 150mm in diameter and 10m in length. The apparatus developed is able to carry out the complete series of inspection processes, grinding and polishing, hardness test, etching and replication at any position inside the bore. The apparatus comprises the four working units, which carry out the various processes, the control box and the data acquisition box. Fig.10 shows the whole of the apparatus.

Fig.10 MACH NDE SYSTEM for rotor bore

Figs.11 and 12 show, respectively, the hardness test unit and the replication unit working in a rotor bore. During the inspection, the distribution of hardness in the longitudinal direction is first obtained and compared with the distribution of calculated temperature, stress and creep damage. From this comparison, the locations for replication are decided and the replication process is then carried out.

Fig.11 Hardness test unit

Fig.12 Replication unit

PERIPHERY GROOVE FATIGUE DAMAGE

The high-temperature rotor has grooves cut circumferentially on its periphery, where fatigue cracks due to transient thermal stress and the grooves' stress concentration have sometimes been observed.

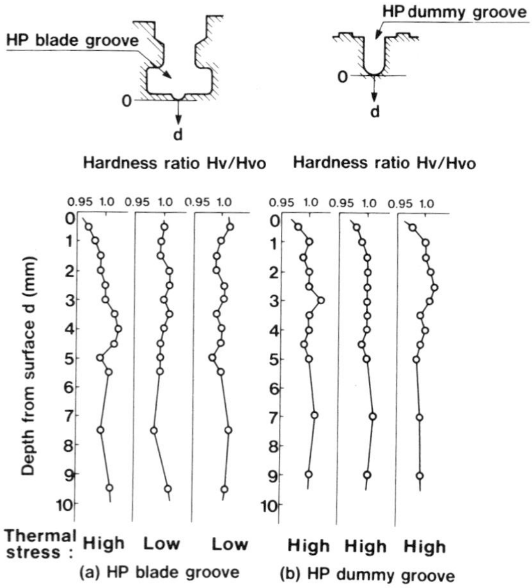

Fig.13 Hardness distributions in periphery grooves

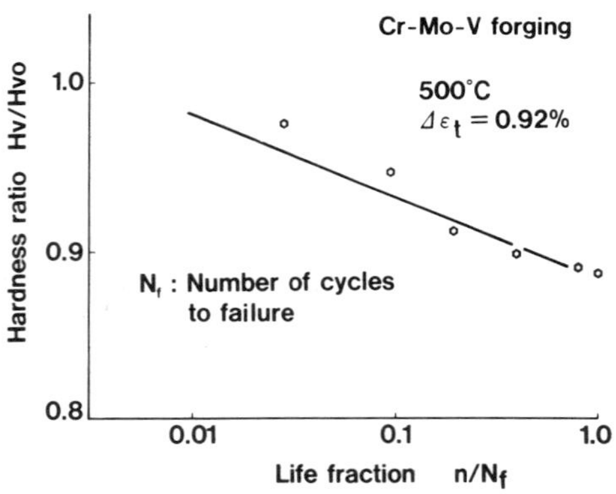

Fig.14 Changes in hardness during high-temperature, low-cycle fatigue

Fig.13 shows the Vickers hardness distribution at the dummy groove roots and the blade groove roots from surface to inside observed in a discarded long-term-use HP·IP rotor. The groove roots, where higher thermal stresses were estimatd to exist in service, are seen to have been softened, although the decreases in hardness is restricted in the surface region to no more than 2 mm in depth. Fatigue tests conducted using specimens sampled from the same rotor revealed that crack initiation life at the surface of high-thermal-stress grooves is short, but that recovery of fatigue life is possible if skin peeling of the softened surface region is carried out.

Fig.14 shows the change in hardness during high-temperature, low-cycle fatigue, and indicates that the repetition of stress (strain) may be the cause of the softening. The changes in hardness may be represented by an equation such as:

$$H_V/H_{VO} = A + B \log \frac{n}{N_f} \qquad (7)$$

where H_{VO} is Vickers hardness before fatigue test, n is the number of cycles, N_f is the number of cycles to failure, and A and B are constants.

Fig.15 shows the relationship existing between the hardness change at failure and the plastic strain range, regardless of test temperature. From these test results, it was found that the constants A and B may be expressed as a function of the plastic strain range.

The breadth of X-ray diffraction can also give the material property changes during high-temperature, low-cycle fatigue. Fig.16 shows the changes in breadth obtained from the same specimens for which the hardnesses are shown in Fig.14.

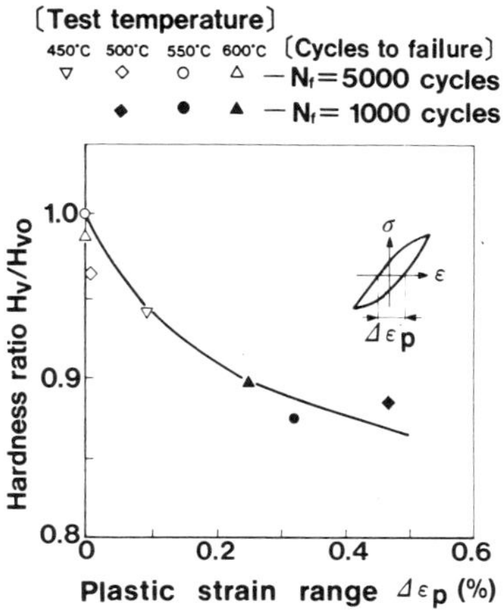

Fig.15 Softening at failure against plastic strain range

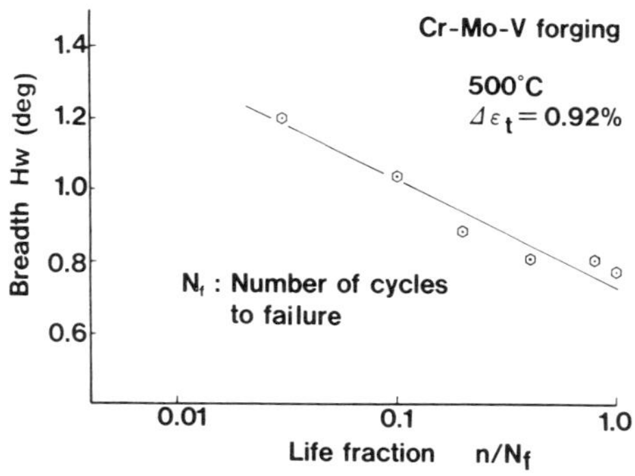

Fig.16 Change in breadth of X-ray diffraction during a high-temperature, low-cycle fatigue

For a turbine, there are many types of start up, cold-start, warm-start and hot start. During continuous operation, load changes often occur. Therefore, the change in thermal stress, that is, peak-to-peak stress during one cycle of operation, is not uniform. Therefore, further investigation will be needed before we can have an entirely adequate method of residual life prediction using observation of the changes in hardness and/or X-ray diffraction breadth. However, we believe that the depth at which decrease in hardness and/or X-ray breadth is found should indicate the depth at which the fatigue-damaged layer is. Therefore, if softening is found at the bottom of the periphery groove, it is recommended that skin peeling be carried out till hardness and/or X-ray breadth is confirmed to be recovered in order to prevent crack initiation.

THE MACH NDE SYSTEM FOR PERIPHERY GROOVES

The typical size of dummy grooves is 3mm in radius and 15mm in depth. An apparatus enabling us to obtain the distribution of hardness and/or X-ray breadth at surface and various depths at the bottom of such grooves, in in-service rotors was developed. For detection, the X-ray diffraction method was adopted because the accuracy of the hardness test at such groove roots is inadequate.

Fig.17 MACH NDE SYSTEM for periphery grooves

The apparatus, which is shown in Fig.17, comprises the unit, which can carry out surface grinding, electrical polishing and inspection of the rotor, and the control and data acquisition box. Fig.18 shows the X-ray inspection unit working on the rotor. Rotor surface preparation before X-ray measurement is carried out by a grinder and an electric polisher to make a small and shallow chord area along the groove shown in Fig.19.

Fig.18 X-ray inspection unit on rotor

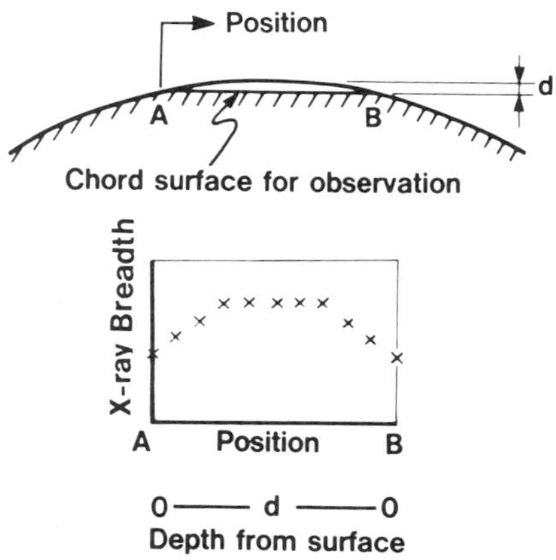

Fig.19 Surface for X-ray observation and inspection method

Automated repetition of step-wise movement and X-ray exosure along the chord yields the distribution of the diffraction breadth along the chord. Since the two ends of the chord correspond to the surface and the center of the chord corresponds to the maximum depth from the groove root, the distibution represents the distribution of the breadth in the direction of depth from surface, as illustrated by the lower figure of Fig.19. If flat distribution of breadth at around the center of chord is not obtained, then the depth d of chord surface should be increased.

In the apparatus, a PSPC (Position Sensitive Proportional Counter) is used as the X-ray detector to give increased accuracy and to reduce the inspection time. With the PSPC, it is possible to obtain the entire diffracted X-ray distribution without the need for the detector to pan.

CONCLUSIONS

A non-destructive evaluation system to detect the creep and fatigue damage in high-temperature rotors was developed. This system gives the greatest sensitivity and accuracy at present obtainable. It will contribute greatly to assessment and extension of the life of high-temperature rotors, by giving clear information only obtainable by direct observation.

It will, however, not be possible to reach final conclusions regarding the effectiveness of the system until enough field performance data have become available.

ACKNOWLEDGEMENT

The authors wish to express their gratitude to the staff of the Kansai Electric Company for the cooperation and fruitful discussions which assisted them so considerably in their research. The data of density changes and the test specimens in Fig.8 were kindly provided by National Research Institute for Metals, Japan.

REFERENCES

1. Conference on Life Extension and Assessment of Fossil Plants, EPRI, Washington, D.C. (1986).

2. Yamada, M., Watanabe, O., Komatsu, S. and Nakamura, S., The Change of Mechanical Properties and Microstructure of Cr-Mo-V Steel During High Temperature Damage, 123rd Committee on Heat-Resisting Metals and Alloys, Japan 22, 1 (1981).

3. Kirihara, S., Shiga, M., Sukekawa, M., Yoshioka, T. and Asano, C., Fundamental Study on Non-Destructive Detection of Creep Damage for Low Alloy Steel, Journal of the Society of Materials Science, Japan 33:1097 (1984).

4. VanLeeuwen, H.P., and Schra, L., Parametric Analysis of the Effects of Prolonged Thermal Exposure on Material Strength, International Conference on Creep and Fatigue in Elevated Temperature Applications, C137/73 (1973).

5. Goto, T., Study on Residual Creep Life Estimation Using Non-Destructive Material Properties Tests, Proc. 2nd International Conference on Creep and Fracture of Engineering Materials and Structures, Pineridge Press, 1135 (1984).

6. Cane, B.J., Aplin, P.F. and Brear, J.M., A Mechanistic Approach to Remanent Creep Life Assessment of Low Alloy Ferritic Components Based on Hardness Measurements, Journal of Pressure Vessel Technology 107:295

7. Neubauer, B. and Wedel, U., Restlife Estimation of Creeping Components by Means of Replicas, Proc. of ASME Int. Conf. on Advances in Life Prediction Method, 307 (1983).

8. Woodford, D.A., A Parametric Approach to Creep Damage, Metal Science 3:50 (1969).

9. Sinya, N., Kyono, J. and Yokoi, S., An Assessment of Creep Damage by Density Change Measurement for Cr-Mo-V Steel, Tetsu-to-Hagane 70:573 (1984).

NONDESTRUCTIVE DEGRADATION EVALUATION OF FERRITIC STEELS FOR HIGH TEMPERATURE APPLICATIONS

W. Lempp[+], N. Kasik[+], and U. Feller[*]

[+]BBC Brown Boveri and Cie, Baden, Switzerland
[*]Federal Office of Metrology, Wabern-Bern, Switzerland

ABSTRACT

The assessment of deterioration of ferritic steels as employed for power plant components with respect to creep, softening and embrittlement is achieved mainly by observation of changes in microstructure. The experience gathered in laboratory investigations is outlined. The obtained knowledge can be applied best in field work if metallography by replication techniques is performed. As a valuable supplement hardness measurements can and should be untertaken. Results from measurements of changes of other physical properties are as yet unable to adequately differentiate changes. Preliminary tests in a static magnetic field indicate some potential for a method based on measurements of reversible permeability.

INTRODUCTION

Assuming that during an overhaul of a power plant examinations of components concerning deformations, presence of cracks, wear and erosion have shown satisfactory results, the problem of material degradation still needs to be resolved. Any quantitative nondestructive determination of degradation of materials of components exposed to elevated temperatures would be an important contribution towards the assessment of safety and estimation of life extension.

Three aspects of degradation are of major concern

- progress of creep towards stress rupture
- softening below specified strength values
- embrittlement below acceptable limits.

Several investigations have been performed and results published on the behaviour of CrMo-steels as used for pipes and tubes. The judgement of the stage of creep reached is mainly based on the observation of the coagulation of the carbide precipitates and the presence of creep cavities which already appear during the secondary stage of creep (Fig.1) and may be used as a guideline for the evaluation, embrittlement being of minor relevance in this type of steel. This type of analysis leaves the impression of assessment being an easy task. The following discussion shall show that quite a number of parameters are to be considered, the situation being more complex with MoV-, CrMoV- and 12 Cr-steels, and support by experience from laboratory work is essential. The discussion is limited to material properties and methods of

NDE which have been investigated in the laboratory and for which the results obtained can be used at present as a basis for field work or show some potential for future application.

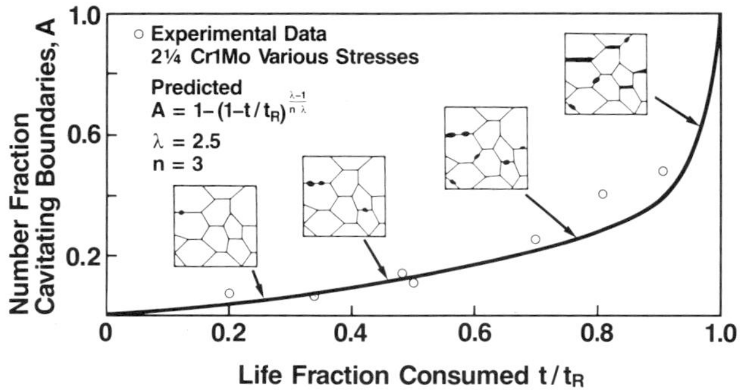

Fig.1: Growth of creep cavities in CrMo-steel (after /1/)

BASIC INFORMATION FROM LABORATORY TESTS

Materials

Investigations were carried out on

- CrMo-steel as applied for piping operated at elevated temperatures
- CrMoV-steel forgings and castings as used for turbine rotors, blades and casings
- MoV-steel which was used in the past for turbine rotors
- 12 Cr-steel (Class II) employed for turbine rotors and blades.

Limits of chemical composition and characteristic tensile values are stated in Table 1. Work was performed on specimens from stress rupture tests (some of them with running times of more than 100'000 hrs) and turbine components retired after more than 100'000 hrs of operation.

Table 1: Composition and tensile properties of investigated materials

Steel	Composition %							Tensile properties		
	C	Si	Mn	Cr	Mo	V	W	R_m MPa	$R_{p0,2}$ MPa	A_5 %
1Cr $\frac{1}{2}$ Mo	0,10 0,18	0,10 0,35	0,4 0,7	0,8 1,2	0,45 0,65			430 550	295	24
2$\frac{1}{4}$ Cr 1Mo	0,15	0,15 0,50	0,4 0,7	2,0 2,5	0,9 1,1			440 590	265	20
MoV	0,20	0,5	0,6	0,3 0,5	0,70	0,30		700	600 700	15
CrMoV	0,17 0,25	0,6	0,3 0,5	1,2 1,5	0,7 1,2	0,25 0,35		690 830	590	16
12 Cr	0,18 0,24	0,5	0,3 0,8	11,0 12,5	0,8 1,2	0,25 0,35	0,6	850 1050	700	13

Initial state of microstructure

During examination one problem encountered is that the initial microstructure even for the same steel quality may well be different due to variation in chemical composition, degree of forging and heat treatment. In addition the microstructure in large forgings shows a considerable change from the outer sections towards the centre as a result of varying cooling rates when quenching. Accordingly, changes in properties of the microstructures in the course of time will be different and any conclusions from examinations performed on its surface only will need appropriate backing up from experiments relevant for the microstructure in the interior.

Carbide precipitation and coagulation

The steels under investigation show characteristic patterns of carbide precipitation and coagulation when exposed to temperatures in the range of 500°C for long times.

Fig.2 shows the microstructure found in a CrMo-steel from a live steam pipe retired after 160'000 hrs of operation. Carbide precipitation (Mo_2C) in the ferrite grains, coagulation of Fe_3C in the pearlite and coagulation of $M_{23}C_6$-carbides along grain boundaries associated with precipitation free zones are observed. In MoV-steels typical features are the appearance of hairlike carbides of type VC and equiaxial carbides of type Mo_2C in the ferrite as well as comparatively coarser carbides of type $M_{23}C_6$ on the former austenite grain boundaries (Fig.3). Similar processes occur in the CrMoV-steel, whereby larger amounts of intergranular Cr-rich carbides

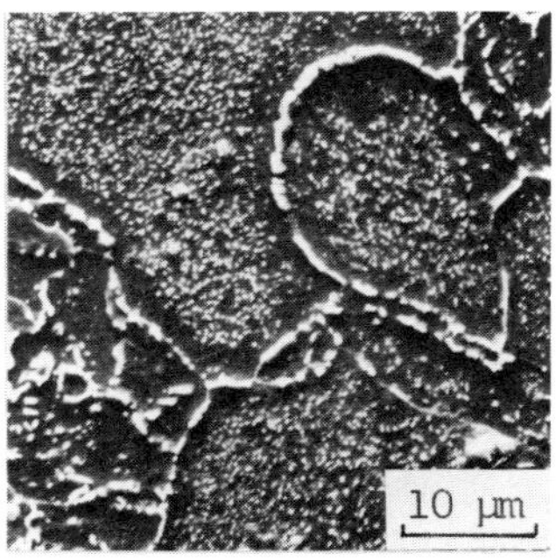

Fig.2: REM-micrograph of CrMo-live steam pipe after 160'000 hrs

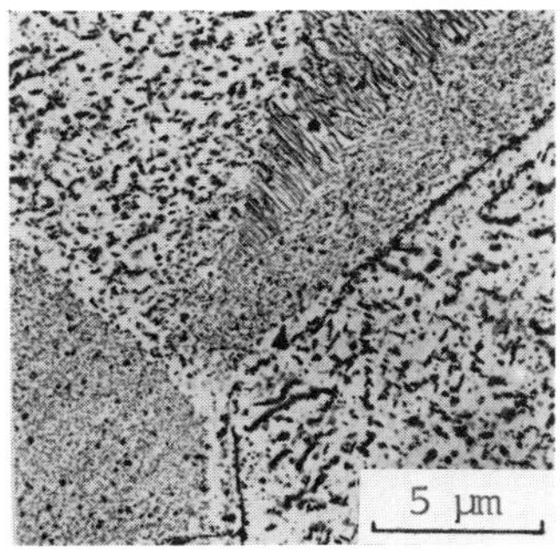

Fig.3: TEM-micrograph of MoV-steel after 130'000 hrs

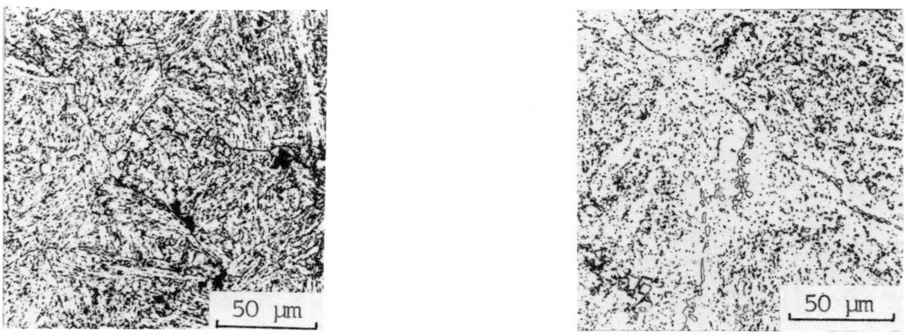

a) 550°C, 230 MPa, 20'000 hrs   b) 650°C, 50 MPa, 25'000 hrs

Fig.4: Carbide morphology in 12%Cr-steel

$M_{23}C_6$ can be observed than in the MoV-steel. The initial microstructure of the 12 Cr-steel mainly consists of tempered martensite. Some δ-ferrite may be observed. Carbide development has been studied extensively / 2,3 / and one can expect that the majority of carbides after long time exposure at elevated temperatures will be of the type $M_{23}C_6$. The coarsening of the carbides on the austenite grain boundaries (Fig.4) can be related to the Larson-Miller-parameter (LMP) - which can represent the combined influence of temperature and time - as shown in Fig.5. Such a correlation theoretically would allow determination of residual life of a component loaded by mainly static primary stresses. However, the influence of initial microstructure on both carbide morphology and the stress rupture curve gives rise to considerable scatter.

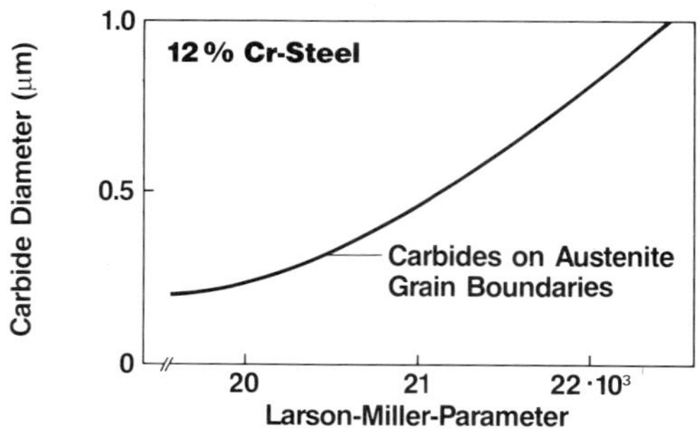

Fig.5: Coarsening of carbides vs. LMP=T(20+logt)

Creep Cavities

Creep cavities in heat resistant steels predominantly occur in the form of circular voids after some time of exposure to stress at elevated temperatures.

In the CrMoV-steels cavities will be found located at bainite grain boundaries, presumably the prior austenite grain boundaries, whereas in the CrMo-steels cavities are observed in both ferrite/ferrite and ferrite/bainite boundaries. Progress of creep can be monitored either on the basis of density of cavities or by the proportion of cavitated (and possibly cracked) grain boundaries.

However, a threshold value in terms of accumulated creep strain exists, before cavities become visible. Apart from the resolution of the method of detection it depends on the type of steel. In CrMo-steels cavities are observed during the second stage of creep, in CrMoV-steels only at the end of the second and beginning of the third stage and in the 12Cr-steels virtually no cavities can be found even in the vicinity of the fracture surface of broken creep specimens.

Even in one class of steels there is no simple relation between rupture life and the extent of cavitation / 4 /, because creep damage is complicated by many factors, e.g. amount of trace elements, initial microstructure and test conditions. Fig.6 shows as an example the observed cavity density during creep tests at 550°C on three melts of CrMoV-steel.

Mechanical Properties

The microstructural changes described above are associated with changes in mechanical properties, some of them considered to be essential for the assessment of a component.

Long term operation of quenched and tempered steel forgings and castings at elevated temperatures is accompanied by a decrease in proof strength, ultimate tensile strength and hardness. Fig. 7a shows the results of hardness measurements vs. LMP of specimens aged at elevated temperatures for long times. A moderate drop in hardness can be observed. When measured on creep specimens the decrease in hardness is more pronounced, as shown in Fig.7b, by the additional influence of the applied stress.

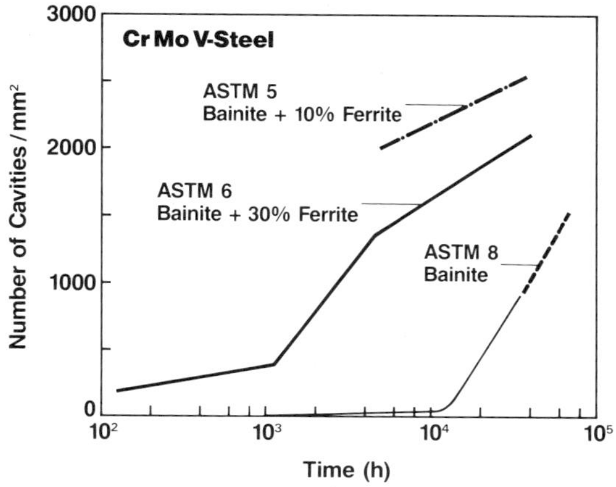

Fig.6: Influence of microstructure on cavity density in 3 melts of CrMoV-steel

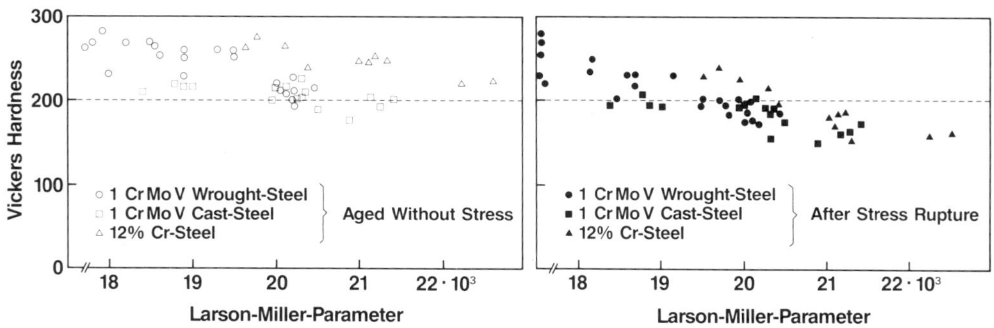

Fig.7: Hardness vs. LMP. a) Aged specimens b) After stress rupture

The correlation of the variation in mechanical properties with observed microstructural changes is still not developed to a stage where reliable quantitative statements can be made, however, some of our own empirical observations may be stated:

- Softening is related to the coagulation of the carbides of the bainite or pearlite and in a more advanced state, to cavity formation. A hardness of 200 HV may be considered as a rough indication of a fairly late stage of secondary creep in MoV and CrMoV-steels, whereas this limit is approximately 220 HV for 12 Cr-steel.

- MoV-steel and CrMoV-forgings and castings embrittle substantially, CrMo-steels to a lesser extent. This is caused by formation of carbide coagulation on the grain boundaries, possibly enhanced by phosphorus segregation, and precipitation of carbides in the ferrite matrix.

- Low cycle fatigue behaviour, when examined on material from retired components, seems only moderately changed when compared with the initial condition.

- The threshold value for crack propagation under cyclic stress appears hardly to be changed. Crack propagation rates were observed to be slightly higher in MoV-steel.

Physical Properties

Microstructural changes in metals result in changes of physical properties. Attempts to utilize measurements of these for life time considerations date back quite some time. Many of these approaches require the measurement of a physical property with a high resolution (like e.g. density, electrical conductivity, velocity of sound) and thus are not easily applicable to field work. At present methods are considered most promising which make use of changes of the magnetic properties of ferritic steels.
In general two methods can be distinguished, based on application of

- alternating magnetic fields
- static magnetic fields.

The former methods are widely used in quality control. The applied magnetizing forces normally are small and changes in magnetic permeability are of predominant influence. Due to the skin effect associated with high test frequencies they are restricted to surface effects. Lowering of the test frequency can improve the situation as has been shown recently / 5 /.

Applying a static magnetic field results in a more complete penetration of the material and thus can reflect its condition more comprehensively. In order to explore the potential of such a method a number of tests on new and crept specimens was performed, measuring the virgin magnetization curve and

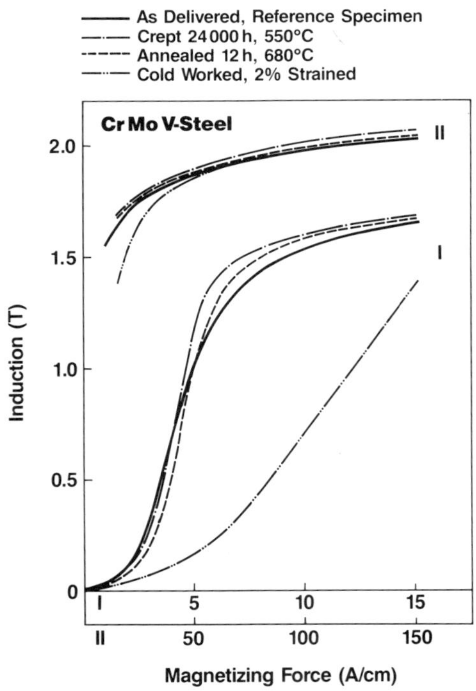

Fig.8: Dc-commutation curves of CrMoV-steels after different treatments

the hysteresis loops by means of a NPL permeameter according to IEC 404-2. Remanent induction and coercive force displayed comparatively small differences for the various material conditions. Further it is known from published measurements of coercive force / 6,7 / that the differences with increasing life time are fading out which is a drawback in life time considerations.

On the basis of these observations the dc-commutation curves in a range of magnetizing force between zero and 200 A/cm were recorded after carefully demagnetizing the specimens under computer control. The specimens were taken from a CrMoV-forging and used in the conditions stated in Fig.8. One specimen was aged at a temperature and for a time which results in the same LMP as for the crept specimen. One specimen was cold worked to a similar amount of strain as the crept specimen. The resulting magnetizing curves are plotted in Fig.8. In the range of intermediate inductions the aged and crept specimen show a definite change in slope, the cold worked specimen falling way off towards lower values as could be expected.

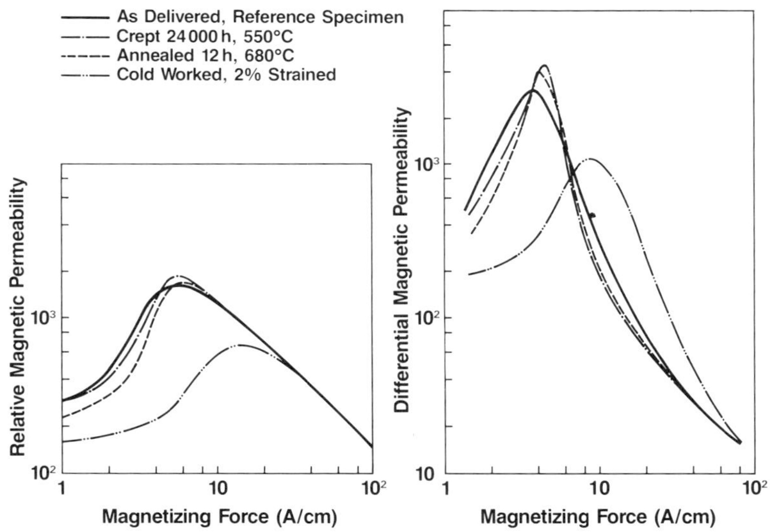

Fig.9: Evaluation of magnetizing curves on the basis of (a) relative, (b) differential magnetic permeability

A quantitative evaluation can be performed on the basis of the relative magnetic permeability $\mu_r = (1/\mu_0)$ B/H and the differential magnetic permeability $\mu_d = (1/\mu_0)$ dB/dH. Figs. 9a and b display the results. The differential permeability gives the most significant differences between the various conditions of the material and still results in distinct values near the transition into the saturation. This range of magnetizing force is much less sensitive to disturbing effects (like magnetic after effects, remanent fields and stress induced fields) and a method operating in this range could offer definite advantages.

Since there are indications that the reversible permeability shows a similar characteristic to the differential permeability / 8 /, a test method could be established on the principle depicted in Fig.10, whereby the signal of the superimposed oscillating current would respond to changes of the reversible permeability.

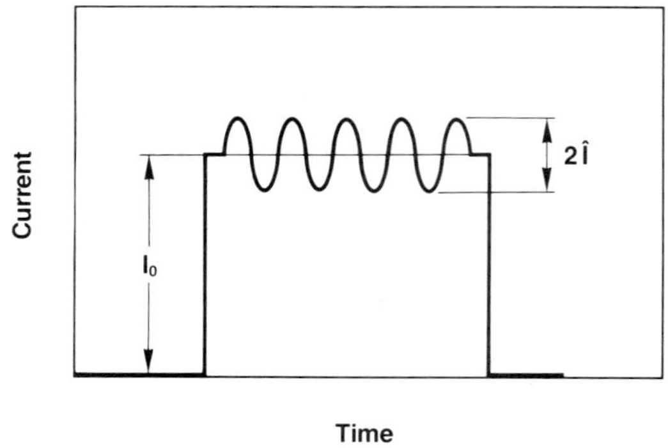

Fig.10: Principle of potential test method

EXPERIENCE FROM FIELD WORK

From a practical point of view the non-destructive assessment of material degradation of a component is till now being approached only indirectly by the observation of microstructure and measurement of hardness. Both methods are restricted in application, since only the accessible surface can be examined. Other methods have as yet not found widespread acceptance.

Surface Metallography

The microstructure of a selected place on the surface can be revealed by the usual methods, i.e. grinding, polishing and etching. Inspection can be done on site where, however, the observation of the important details is limited due to problems of access, time restrictions, lack of sufficient magnification of portable microscopes and difficulties in taking pictures for documentation. Therefore replication techniques have been developed to a high standard and are an estabished method nowadays.

The surface microstructure is replicated by means of a triafol film softened with acetone. After drying the film can be removed and conveyed to the laboratory where it can be observed in the optical microscope and in the scanning electron microscope. When inspecting, the fact that a negative image of the surface is obtained has to be kept in mind. Interpretation requires experience which best can be gained by test series on creep specimens where direct inspection and replica inspection can be compared. Two problems are to be kept in mind. First, the initial state of microstructure usually is not known and has to be deducted. This may be difficult and in some cases almost impossible without other sources of information. Second, the observer's understandable intention of detecting creep cavities must not lead to misinterpretations in cases where there are faulty replicas or casting and welding porosity. In both cases metallurgical experience and the previously described support from laboratory experiments are an absolute prerequisite if reliable conclusions are to be drawn.

Hardness Measurements

Hardness measurements can and should supplement the findings from surface metallography, yielding valuable information as stated above.

Portable diamond pyramid hardness testers are available, but handling is intricate and access limited.

Very convenient and reliable - if properly employed - is a dynamic device which uses a small hard metal ball bouncing against the surface and comparing its velocity when hitting and returning / 9 /. The probe is of the size of a fountain pen, access is not normally a problem and handling is easy. Restrictions arise from the stiffness of the object to be measured. With wall thicknesses below about 15 mm no reliable values can be obtained.

CONCLUSIONS

At present the assessment of degradation of ferritic steels exposed to stresses at elevated temperatures is mainly based on detailed observations of changes in microstructure concerning carbide coagulation, carbide precipitation and creep cavities. However, these phenomena have been shown to depend on a number of parameters like steel type and variation of composition in different melts, initial microstructure, operation time and temperature. Nevertheless some experience has been gathered in interpreting microstructural features with respect to mechanical properties and state of creep after long time of operation.

It is this knowledge which still supports surface metallography as the main non-destructive method of examination. If supplemented by hardness measurements an at least qualitative assessment of the state of material can be made by experienced metallurgists. It is to be made clear at this point that these methods can be applied to selected surface points of a component only and the results serve as a kind of calibration for findings of other disciplines like stress analysis, design and analysis of operational history. The cooperation of experts from all fields concerned is a must for a reliable life time study.

The detailed picture of the state of a material gained by metallography is difficult to obtain by other NDE-methods reflecting the change of selected physical properties. Methods based on changes in magnetic properties are attractive because of their presumably simple application. By consideration of the reversible permeability in the range of intermediate inductions an as yet not explored potential for non-destructive testing can be opened up.

REFERENCES

/1/ Gooch, D.J., Townsend, R.D.: CEGB Remanent Life Assessment Procedures. EPRI, Conf. Life Extension and Assessment of Fossil Plants, 1986

/2/ Hede, A., Aronsson, B.: Microstructure and Creep Properties of some 12%-Chromium Martensitic Steels, J. Iron and Steel Inst. 207 (1969) 1241-1251

/3/ Irvine, K.J., Crowe, D.J., Pickering, F.B.: The Physical Metallurgy of 12%-Chromium Steels. J. Iron and Steel Inst. 195 (1960) 386-404

/4/ Tipler, H.R., Hopkins, B.E.: The Creep Cavitation of Commercial and High-Purity Cr-Mo-V-Steels. Metal Science (1976) 47-56

/5/ Clark, W.G., Metala, M.J.: Eddy Current Determination of Material Properties. EPRI, Conf. Life Extension and Assessment of Fossil Plants, 1986

/6/ Relander, K., Geiger, T.: Gefügeänderungen in den warmfesten Stählen 13CrMo 44 and 10CrMo 9 10 bei langzeitigen Glühungen zwischen 550 und 780°C. Arch. Eisenhüttenwesen 37 (1966) 897-906

/7/ Hartnagel, W. et al.: Determination of Residual Life of Turbine Components. COST 501, Progress Report 1983

/8/ Gonda, P. et al.: Computerized Evaluation of Magnetic Properties. J. Magn. Magn. Mater. 41 (1984) 241-243

/9/ Equotip by PROCEQ, Zurich, Switzerland

NONDESTRUCTIVE EVALUATION OF CREEP DAMAGE IN SERVICE EXPOSED 14 MoV 6 3 STEEL

H. Willems*, W. Bendick**, and H. Weber**

* Fraunhofer-Institut für zerstörungsfreie Prüfverfahren
D-6600 Saarbrücken 11, FRG

** Mannesmann Forschungsinstitut GmbH
D-4100 Duisburg, FRG

ABSTRACT

Ultrasonic techniques as well as electrical resistivity measurements were applied to assess the degree of creep damage in 14 MoV 6 3 steel (0.5 Cr, 0.5 Mo, 0.25 V). Samples were cut from a tube bend which had failed after 130,000 h/535°C service exposure providing us with a complete variety of damage states. Using metallographic methods the microstructural damage was classified in terms of micropore concentration and microcrack formation, respectively. The actual porosity of the material was determined quantitatively by density measurements.

The obtained results show that both ultrasonic velocity (longitudinal waves, shear waves, and surface waves) and electrical resistivity can be used for damage characterization whereas ultrasonic attenuation measurements did not prove to be appropriate in this case. The experimental results are in good agreement with theoretical predictions. Especially ultrasonic velocity measurements seem to have a high potential for practical applications.

INTRODUCTION

Components in power stations are subjected to service conditions under which creep processes take place limiting the component's lifetime. The probability of failure generally increases with operating time. However, the lifetime itself can only be predicted with great uncertainty /1,2/. Thus, the assessment of service-exposed components has become necessary to improve the reliability and safety of power plants. An important part of this procedure is the detection of creep damage. Besides differences in material behavior the typical damage and fracture behavior of different components like e.g. straight pipes, weldments, fittings, and tube bends has to be taken into consideration /3/. Common methods of non-destructive testing are sufficient in cases in which damage starts locally as a crack on the outer surface of the component. This is true for creep damage in weldments and fittings. For tube bends, however, creep damage can develop in form of micropores and microcracks covering large areas of the bend and eventually leading to a catastrophic fracture of the whole component. To prevent this type of failure it is necessary to safely detect micropores of 1 to 3 μm in diameter. In nowadays practice this is done by metallographic investigations using the surface replication technique /4/. However, the main disadvantage of this technique is that it gives only information about the damage state at the surface and not its propagation into the wall, and secondly it can reasonably only be applied to limited areas of the component. Therefore, the development of a non-destructive test method to identify creep damage in form of micropores is of great practical importance.

MATERIAL DAMAGE BY CREEP

The strain-time-behavior of a metallic materials at high temperatures generally is represented by a creep curve with primary, secondary, and tertiary part (Fig. 1). The primary and secondary creep range is characterized by dislocation and phase reactions. At the beginning of tertiary creep material damage occurs causing an increase of creep rate. The interaction of creep and material damage finally causes the fracture of the specimen. In the example of Fig. 1 damage appears in form of micropores at the grain boundaries growing with increasing creep rate. For homogeneously stressed creep specimens fracture will occur by plastic instability at a certain loss of internal cross-section as a result of cavitation. In the case of inhomogeneous stressing different stages of creep damage are passed before final failure occurs.

As an example Fig. 2 shows the damage pattern at the surface of a tube bend which has failed by a leakage after an operation of 130,000 h at 535°C. The damage grade has been classified from 0 to 7 with isolated micropores at low damage grades, chaines of micropores and microcracks in the midrange, and larger disintegrations of microstructure for the highest damage grade. Fig. 2 shows a projection of the tube bend into one plane with the outer vertex line denoted as 12.00-position.

In addition to surface replication, several cuts have been made in order to investigate the damage propagation through the tube wall. As a result high degrees of damage have only been found near the tube's surface, whereas low damage grades are spread through the whole wall thickness.

EXPERIMENTAL RESULTS

After thorough metallographic investigation test specimens with different damage grades (up to 6) have been cut out of the damaged tube bend to study the influence of creep damage on physical properties with the aim of developing a non-destructive test method for creep damage.

Fig. 3 shows the dependence of the electrical resistivity as measured by a potential drop method and the density on the microscopically determined damage grade. A decrease in density corresponds to an increase in electrical resistivity of similar magnitude.

For the ultrasonic measurements rectangular bars with different damage grades were used having sizes of $10x10x100mm^3$. The deviations in the specimen thickness were below $\pm$ $3x10^{-3}$mm. The damage grade of each specimen was determined metallographically at several positions. The ultrasonic velocity of the longitudinal wave (v_L) and the shear wave (v_T) were measured at a frequency of 5 MHz by means of the pulse-echo overlap method /5/. Additionally, v_L was measured continuously as a

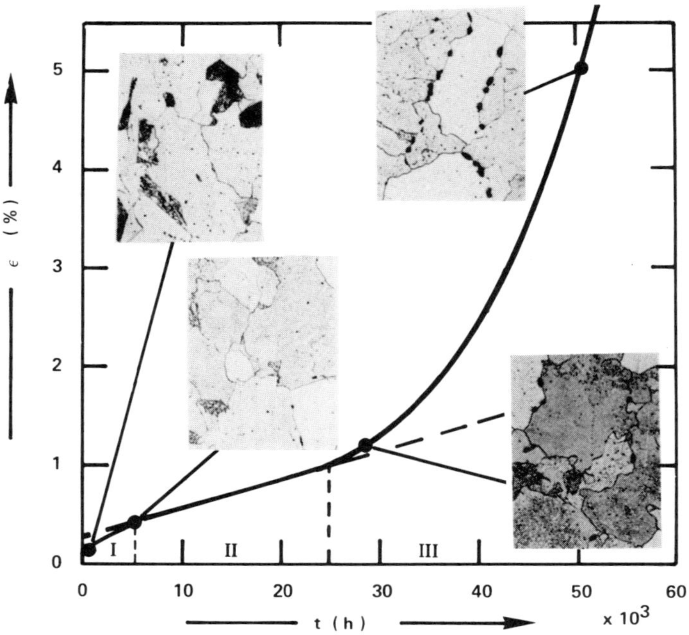

Fig. 1   Development of creep damage in 14 MoV 6 3 (550°C, 98 MPa).

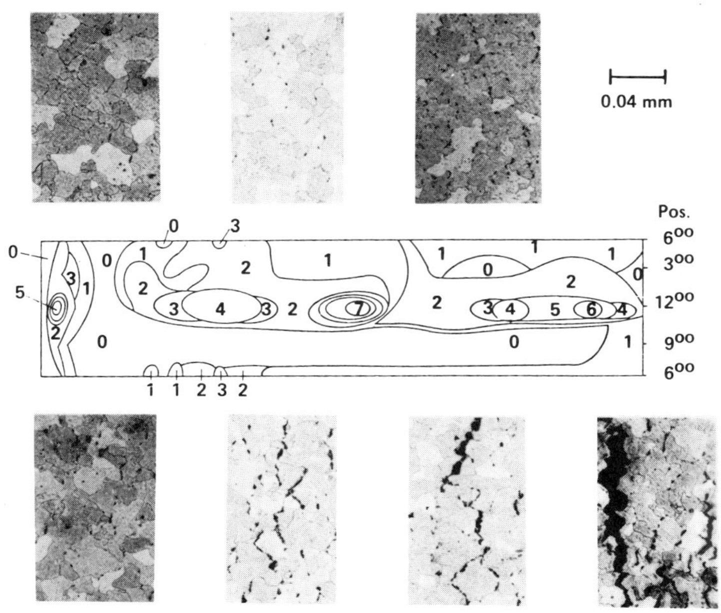

Fig. 2   Damage state at the surface of the tube bend according to the given classification scheme.

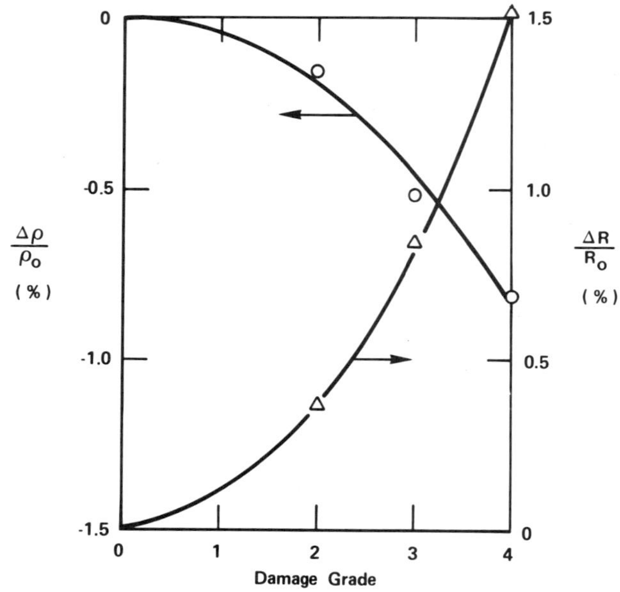

Fig. 3 Change in density ($\Delta\varrho/\varrho_0$) and electrical resistivity ($\Delta R/R_0$) by creep damage (ϱ_0 = 7.843 g/cm^3, R_0 = 0.1905 μΩ m).

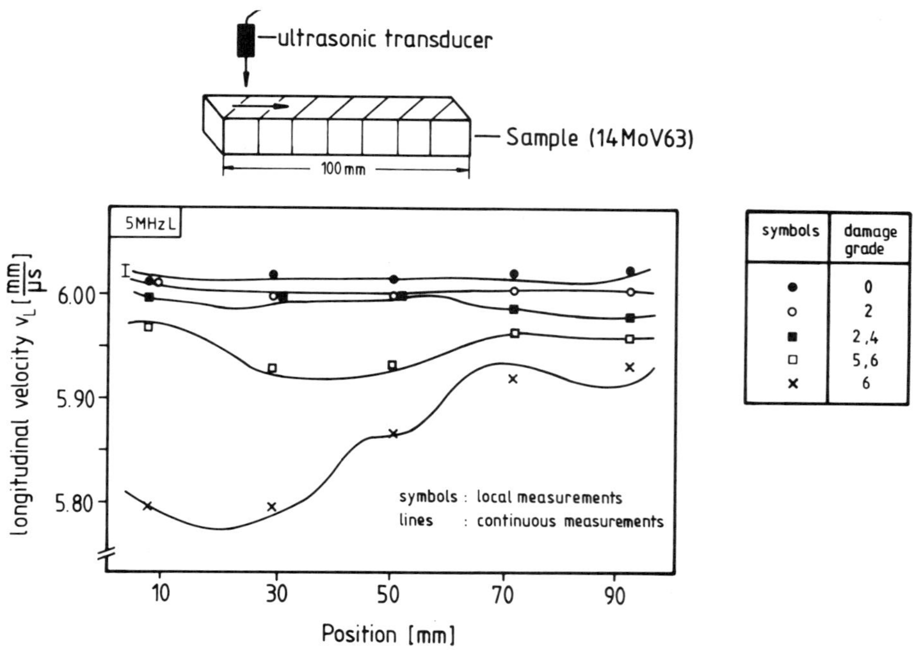

Fig. 4 Longitudinal velocity v_L as a function of position in creep damaged specimens with different damage grades.

function of position (Fig. 4) by recording the time shift /6/ of the first backwall echo with regard to the entrance echo while scanning the probe across the specimen in immersion technique. From Fig. 4 it can be seen that v_L is decreasing with increasing damage grade. Furthermore, one observes a rather inhomogeneous velocity profile at high damage grades indicating an inhomogeneous damage distribution in that case. The velocity v_R of the Rayleigh wave was measured using a pitch-and-catch arrangement with a fixed distance between the ultrasonic transmitter and receiver. Fig. 5 shows the average values of v_L, v_T and v_R versus damage grade normalized to the corresponding velocities measured at the undamaged state. In all cases the reproducibility is better than 10^{-3}. It can be seen that the ultrasonic velocity is decreasing nearly linearly up to damage grade 4. For higher damage grades the velocity drop becomes more pronounced and amounts to 2-3%. Additionally it is found, that the effect is about twice as large for longitudinal waves as for shear waves and surface waves, respectively (see Fig. 5).

Fig. 6 shows the results of ultrasonic attenuation measurements. Here, the amplitude variation of the first backwall echo was recorded as a function of position while the ultrasonic probe was scanned across the specimen in immersion technique. The ultrasonic attenuation coeffients of the different specimens as determined from backwell-echo sequences are also given in Fig. 6. Only in the case of damage grade 6, i.e. mainly microscopic cracks, larger variations of the backwall-echo amplitude were observed. They are associated with scattering losses due to a varying crack density. Here, the largest drop in the ultrasonic amplitude appears at the same position where the largest drop in v_L is observed (see Fig. 4). However, no correlation between the absolut attenuation coefficient α and the damage grade was found (see Fig. 6). For example, one specimen with damage grade 2 exhibited an attenuation which was about two times larger than in the other specimens. Metallographic examinations showed a considerably larger grain size in that specimen thus causing the higher attenuation due to enhanced grain scattering losses. In big components such variations of the grain size are not unusual for the considered steel which has to be taken into account when discussing attenuation measurements.

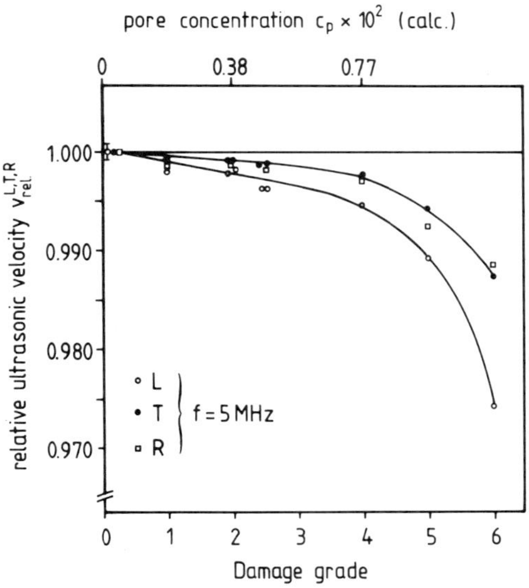

Fig. 5 "Relative ultrasonic velocities v_L, v_T, v_R" vs "damage grade" in creep damaged 14 MoV 6 3.

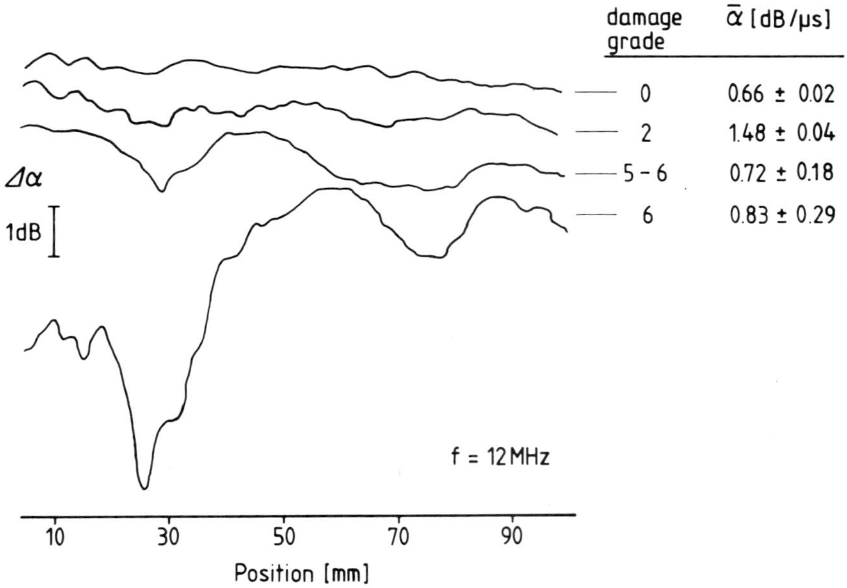

Fig. 6 Variation of ultrasonic attenuation in creep damaged specimens with different damage grades.

DISCUSSION OF RESULTS

Most physical properties are very sensitive to changes on an atomic scale, as e.g. microstructural changes by phase reactions. Therefore, a study of creep damage by measuring physical properties can only be done by comparing damaged and undamaged material of the same microstructure. The service-exposed tube bend is excellently suited for such measurements since the whole bend has attained a uniform microstructure by longtime service-exposure. The different degrees of creep damage result from inhomogeneous stressing which has no influence on the microstructure.

The dependence of density on the damage grade is self-evident. Moreover a quantitative relationship between the density change and the increase of creep rate can be established by model calculations /7/. The density measurement is a simple but very accurate way of determining creep damage in a quantitative manner. However, the method is destructive. Therefore the density measurement can only be regarded as a quantitative scale to compare other test methods.

The experimental results show an almost linear correlation between density and change in electrical resistivity. This dependence can be understood theoretically /8/. Assuming a homogeneous distribution of spherical micropores with a concentration c_p, the resistivity change should amount to:

$$\frac{\Delta R}{R} = (1 - c_p)^{-3/2} - 1 \approx \frac{3}{2} c_p + \ldots \qquad (1)$$

In consideration of the ideal assumptions, the agreement between theory and experimental results is quite good. The experimental values lie slightly above the curve given by equation (1) which might be a result of the very inhomogeneous distribution of micropores. Additionally, it should be mentioned that the resistivity measurements were performed under ideal laboratory conditions using special specimen geometries. Appropriate methods for practical applications have still to be developed.

The discussion of the ultrasonic results is based on the theory of ultrasonic propagation in porous media /9,10,11/. Although the formulation of the theory by Watermann on Truell /9/ is somewhat different from that by Sayers and Smith /11/, the results are the same for the low porosities considered here. Assuming spherical pores in an isotropic matrix with $c_p \leq 0.05$ and $d_p/\lambda \ll 1$ (long wavelength approximation) the following expressions may be derived:

$$v_L = v_L^o \, (1 - k_L \cdot c_p) \tag{2}$$

$$v_T = v_T^o \, (1 - k_T \cdot c_p) \tag{3}$$

Here, d_p is the pore diameter, λ is the ultrasonic wavelength, v_L, v_T and v_L^o, v_T^o are the ultrasonic velocities with and without porosity, and k_L, k_T are constants depending on the material under consideration. Calculated values for k_L and k_T are given in Table 1 which also contains the corresponding experimental values as obtained from a linear regression analysis involving the measured density change of 0.8% for damage grade 4 (see Fig. 3). As can be seen the agreement between calculated and experimental values is very good. The different sensitivity of L-waves and T-waves with regard to porosity is also confirmed by the theory. For damage grades larger than 4 the ultrasonic velocity decreases stronger than linearly (Fig. 5). Because no accurate density measurements for damage grades larger than 4 are available due to the penetration of the measuring liquid into the microscopic cracks, the density change might also be rather non-linear in that range with respect to the damage scale defined here. However, density changes caused by creep damage are not considered to be larger than 2%, i.e. the theory, which assumes spherical pores, seems to underestimate the influence of porosity caused by microscopic cracks.

TABLE 1. Experimental and theoretical results for constants k_L, k_T

	Experiment	Theory /10/	Theory /11/
v_L [mm/µs]	6.013	5.90	6.00
v_T [mm/µs]	3.278	3.23	3.00
k_L	0.66	0.65	0.78
k_T	0.32	/	0.44
k_L/k_T	2.06	/	1.77

Concerning practical applications ultrasonic velocity measurements are considered to be appropriate for surface investigations using surface waves as well as for bulk investigations using L-or T-waves provided that velocity changes measured relative to an undamaged state of the same component can be attributed to porosity. A severe problem in the case of L- or T-waves, however, is the precise determination of the ultrasonic sound-path, e.g. the wall thickness of a tube. In general this will not be possible with sufficient accuracy. However, due to the different sensitivity of L- and T-waves on porosity a sound-path independent expression for the determination of pore concentrations by velocity measurements is readily obtained from equations (2),(3) yielding

$$c_p = \frac{K-1}{k_T \cdot K - k_L} \tag{4}$$

with

$$K = \frac{t_L^{\,o} \cdot t_T}{t_L \cdot t_T^{\,o}} \tag{5}$$

Here, $t_L^{\,o}$, $t_T^{\,o}$, t_L and t_T are the time-of-flights between backwall echoes according to the definitions given above. For each state the time-of-flights of both L- and T-waves have to be measured at the same position in order to eliminate the sound-path.

Concerning ultrasonic attenuation measurements theory /12/ yields for the ultrasonic scattering coefficient α_s^G of longitudinal waves in a polycrystalline material with grain size d_G.

$$\alpha_s^G = S_G \cdot d_G^3 \cdot f^4$$

The scattering coefficient α_s^P for spherical pores in an isotropic matrix is given by /9/

$$\alpha_s^P = c_p \cdot S_p \cdot d_p^3 \cdot f^4 \tag{7}$$

Here, f is the frequency and S_G, S_p are the scattering parameters for grain scattering and pore scattering, respectively. In both cases Rayleigh scattering is assumed ($d_{P,G}/\lambda \ll 1$).

If absorption contributions are neglected the ultrasonic attenuation coefficient in porous, polycrystalline materials might be approximated in the case of small porosities by

$$\alpha = \alpha_s^P + \alpha_s^G = f^4 \left(c_p \cdot S_p \cdot d_p^3 + (1-c_p)\, S_G \cdot d_G^3\right) \tag{8}$$

With $S_G = 0.0035\ (\mathrm{mm}/\mu\mathrm{s})^{-4}$ /12/, $S_p = 0.24\ (\mathrm{mm}/\mu\mathrm{s})^{-4}$ /13/ and reasonable values for c_p, d_p and d_G (see above) it is readily shown that the effect of low concentrations of micropores on the ultrasonic attenuation is far below the attainable accuracy of attenuation measurements. In fact, small variations of grain size will produce larger variations of α than porosity does in the case of creep damage. Therefore, ultrasonic attenuation measurements are not considered to be appropriate for the detection of micropores in creep damaged, polycrystalline steels under practical conditions.

CONCLUSIONS

The obtained results show that both ultrasonic velocity (longitudinal, shear, and surface waves) and electrical resistivity can be used to detect creep damage at the stage of isolated micropores. The experimental results are confirmed by theoretical calculations. In case of electrical resistivity measurements, suitable non-destructive test methods still have to be developed. The next step for the ultrasonic technique is to detect creep damage on real components under service-like conditions. Both activities are planned for the near future.

Our results are considered as a very promising step towards the development of NDT methods for the in-service detection of creep damage on a micropore scale, which is of great importance in the surveillance and maintenance of creep exposed components.

ACKNOWLEDGEMENTS

This work has been financially supported by the "Bundesministerium für Forschung und Technologie" of the FRG within the framework of the European concerted action COST 501 on high temperature materials.

REFERENCES

/1/ Bendick, W. and Weber, H.:
VGB Kraftwerkstechnik 66 (1986) 63-72 and 170-177
/2/ Fabritius, H. and Weber, H.:
Proc. VGB Conf. "Beurteilung von Bauteilen nach Beanspruchung im Kriechbereich", Essen, Feb. 1984, 73-72
/3/ Bruehl, F., Kalwa G., and Weber, H.:
VGB Kraftwerkstechnik 65 (1985) 1059-1068
/4/ Bendick, W., Bruening, B., and Weber, H.:
Sonderbände der Praktischen Metallografie, Nr. 16, 439-450, Dr. Riederer Verlag, Stuttgart 1985
/5/ McSkimin, H.L.:
J. Acoust. Soc. Am. 33 (1961) 12
/6/ Bamberg, J., Schmitt, H., and Arnold, W.:
Rev. Sci. Instrum. 53 (1982) 1613
/7/ Bendick, W., Kalwa, G., and Weber, H.:
Conf. "High Temperature Alloys", Petten, Oct. 15-17, 1985
/8/ Ondracek, G.:
Z. Werkstofftech. 8 (1977) 280-287
/9/ Watermann, P.C. and Truell, R.:
J. Math. Phys. 2 (1961) 512-537
/10/ Hirsekorn, S.:
IzfP-Bericht Nr. 790218-TW, Saarbrücken 1979
/11/ Sayers, C.M. and Smith, R.L.:
Ultrasonics, Sept. 1982, 201
/12/ S. Hirsekorn:
J. Acoust. Soc. Am. 72 (1982) 1021
/13/ S. Hirsekorn:
Unpublished results

A NOVEL METHOD TO MEASURE SENSITIVITY TO AGING IN STEEL SHEETS

Charles Brun, and Pascal Patou

Usinor-Aciers
Research and Development dept.
Montataire, France

INTRODUCTION

Aging of plain carbon steel results from the precipitation of free mobile carbon and nitrogen atoms in the crystal lattice of α-iron.

This precipitation is controlled by the degree of supersaturation of interstitial elements, but also by the diffusion rate of these elements in ferrite and by the nucleus density[1]. The severely strained zones of cold-worked steel present high density of dislocation lines[2]. It is well known that carbon and nitrogen are firmly fixed on dislocation lines and can be even more stable than in the form of carbides and nitrides. Consequently, if the strained metal is not supersaturated with dissolved atoms, the smallest particles redissolve during aging, in order to compensate the migration of dissolved atoms to the dislocation lines, and to maintain the saturation concentration in the matrix[3].

The understanding of the solute carbon and nitrogen precipitation mechanisms depends on two kinds of information :

. initial amounts in interstitial C and N

. kinetics of C and N precipitation.

In the case of aluminum-killed steel, the thermal history of the steel —— from hot-rolling to final annealing after cold-rolling —— induces the precipitation of almost all nitrogen as Al N. The soluble nitrogen content is therefore negligible. In the following, we shall be only concerned by interstitial carbon content of the steel.

INTERNAL FRICTION MEASUREMENT

Whereas the total carbon amount can be easily measured, it is generaly hard to distinguish the carbon in a precipitated form from the one in an interstitial form.

Measuring the damping effect of the unbound carbon atoms has been used for a long time to determine the interstitial carbon amount in iron.

This method - called internal friction measurement - has been improved by many researchers[4, 5, 6]. The Collette torsion pendulum which works at about 1 Hz and 35° C with flat samples, gives very good results for high purity iron[7]. Fig. 1 shows the correlation between internal friction coefficient Q^{-1} and the carbon content, determined by chemical analysis, for solution treated high purity iron, doped with various amounts of carbon.

For industrial steel samples, the distorsion of the Snoek's peak impedes any evaluation of the interstitial carbon content (Fig. 2).

Using a bending pendulum[8] with a natural frequency of about 10 Hz provides better results for industrial steels, but the test needs a highly rarified atmosphere to prevent external friction damping. Indeed, the oscillations amplitude is rather large.

To prevent such inconvenients, we have achieved an internal friction measuring system, which works at high frequency and, therefore, at high temperature.

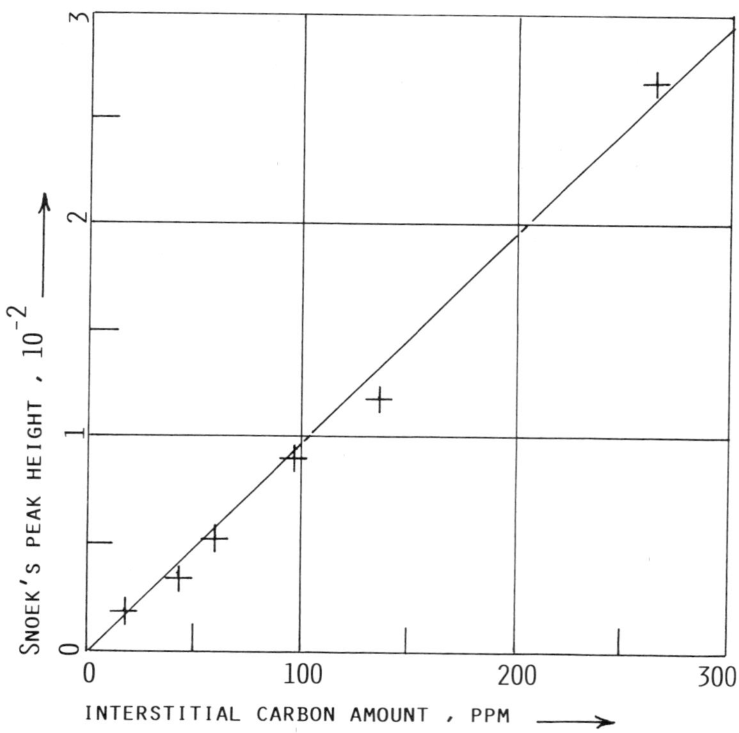

Fig. 1. Correlation between interstitial C amount and Snoek's peak height, as obtained with the torsion pendulum, in high purity iron

DESCRIPTION

In the longitudinal mode, the natural frequency of a 100 mm long steel sample is about 25.5 kHz. In these conditions the Snoek peak occurs at 185° C (365° F). To generate the vibration, a magnetostritive oscillator is used, as shown in Fig. 3. This principle is well known, for it is the main part of the Lankford value "modul-r" system[9]. A 100 x 10 mm^2 sample is introduced in a three coils block. The central coil - with direct current - performs the polarization (north and south poles) of the sample. The drive coil - with alternating current - induces an alternating magnetic field, which makes the sample oscillate, through magnetostriction effect. The pick-up coil gives a potentiel difference which is a function of sample strain.

For internal friction measurement, the coils block is preheated at the chosen temperature. A 1 mm thick sample reaches the right temperature in about 1 minute. Heating system is a convection one, with a preheated gas flux of 20 liters/minute (Fig. 4).

As the strain does not exceed 10^{-6}, the contact of the sample with teflon core or the gas friction do not effect in any way the damping. Fig. 5 shows the influence of the temperature on the resonance frequency ; the rate is about 4.3 Hz per degree C. Therefore, the stabilization of the frequency indicates the stabilization of the temperature.

To measure the damping, the drive coil is cut off and at the same time, the voltage level in the pick-up coil is taken in memory. The count of the free oscillations starts automatically, until the voltage level in

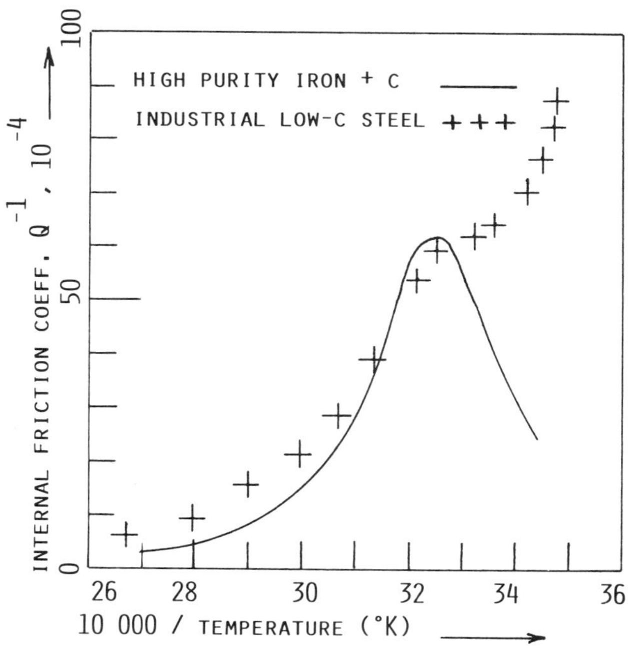

Fig. 2. Snoek's peaks, as obtained with the torsion pendulum

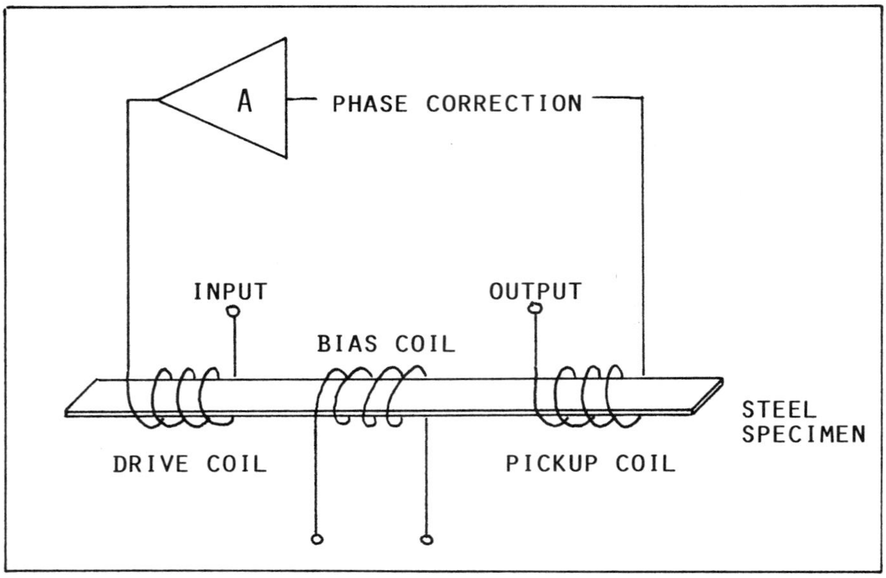

Fig. 3. Schematic view of the magnetostrictive oscillator

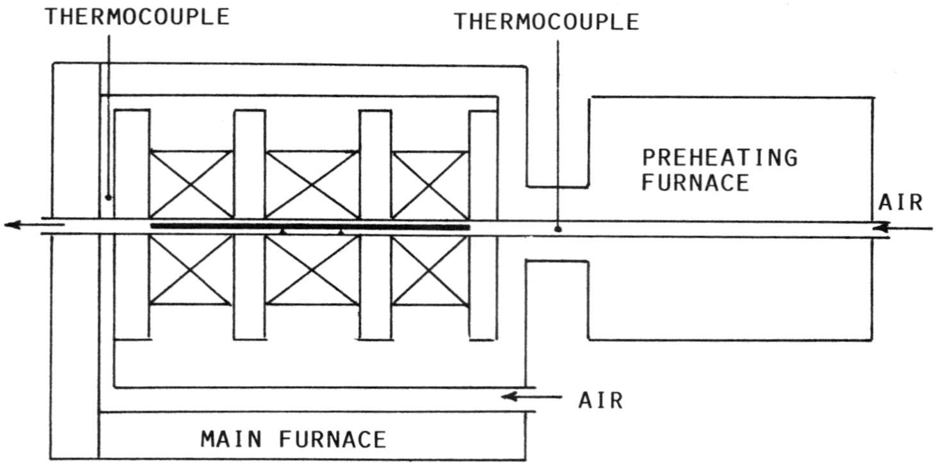

Fig. 4. Schematic view of the cell

the pick-up coil reaches 0.54 time the initial tension, as shown in Fig. 6. The number N of free oscillations is immediately displayed, and the internal friction coefficient Q^{-1} can be calculated with the following relationship :

$$Q^{-1} = \frac{1}{N} \frac{1}{\pi} \operatorname{Ln} \frac{1}{0.54}$$

The N measurement lasts only few milliseconds and it can be easily reproduced every 3 seconds.

VALIDITY OF THE METHOD

Continuous noise

In a Q^{-1} versus temperature diagram, the Snoek peak shows always a bottom noise, wich is rarely negligible (see Fig. 7). That noise proceeds from other causes of damping and it is not due to the Snoek's effect. It is therefore necessary to rectify the maximal Q^{-1} value in order to get the actual Snoek's peak height :

$$\Delta Q^{-1} = Q^{-1}_{max} - Q^{-1}_{noise}$$

Noise measurement can be carried out after a complete precipitation of interstitial carbon, that is to say after an artificial aging treatment, for instance 20 mn at 300° C (572° F)

Carbon precipitation

The measurement of the interstitial carbon amount is achieved at 185° C. It may be expected that the carbon precipitation will occur rapidly at such a high temperature. Fig. 8 shows schematically the Snoek's peak collapse, during measurements at 185° C.

This constitutes an interesting particularity of the present system : when once the temperature is stabilized, it is easy to repeat the damping measurements every three seconds and to follow by this way the collapse of Snoek's peak. This procedure gives therefore valuable information on the carbon precipitation kinetic at 185° C.

Such informations are useful to study aging phenomenons in low carbon steel.

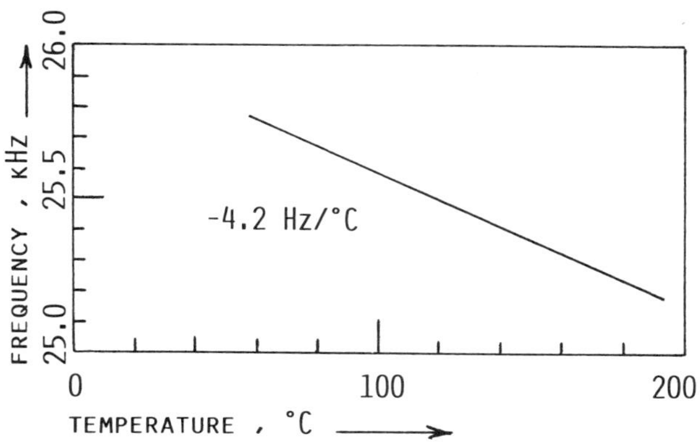

Fig. 5. Influence of temperature on resonance frequency of a 0.1 m long steel specimen

Texture influence

Snoek's peak height can be bound to intertitial carbon amount, by the way of a linear relationship :

$$C_i = k\ \Delta Q^{-1}$$

The factor k depends strongly on the anisotropy of material elastic properties. In the case of a single crystal of pure iron, it is possible to calculate the coefficient k, the value of which is usually between 1.7 and 2.0 as shown in Fig. 9. Furthermore, previous work[11] has shown that the grain size has no influence on the factor k. The curve in Fig. 9 is therefore expected to be available for polycrystalline materials. This has been confirmed by solution treatments at different temperatures and we use now the following relationship :

$$C_i = 1.85\ \Delta Q^{-1}$$

in which C_i stands for interstitial carbon amount (ppm) and ΔQ^{-1} for Snoek's peak height (10^{-4}).

APPLICATIONS

The present system can be used to optimize or control any material or process as regards aging sensitivity. For instance, in the case of cold-rolled and continuously annealed steel sheets, the method is successfully used to optimize the over-aging treatment. Fig. 10 shows the correlation between Snoek's peak height and the traditional aging index. This latter parameter is defined as increase in flow stress at 10 % tensile strain, before and after artificial aging of 4 hours at 100° C (212° F). Fig. 11 et 12 show respectively aging index and Snoek's peak height, as a function of over-aging temperature and time. It appears that the novel method is rather more sensitive.

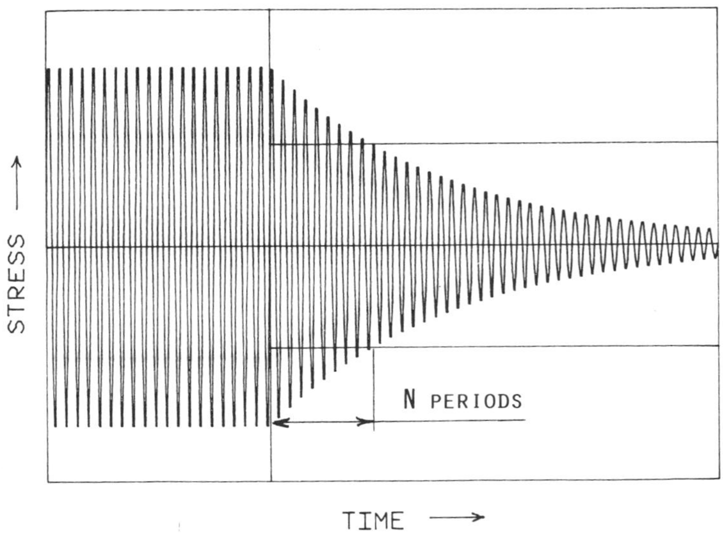

Fig. 6. Evolution of sample stress during measurement

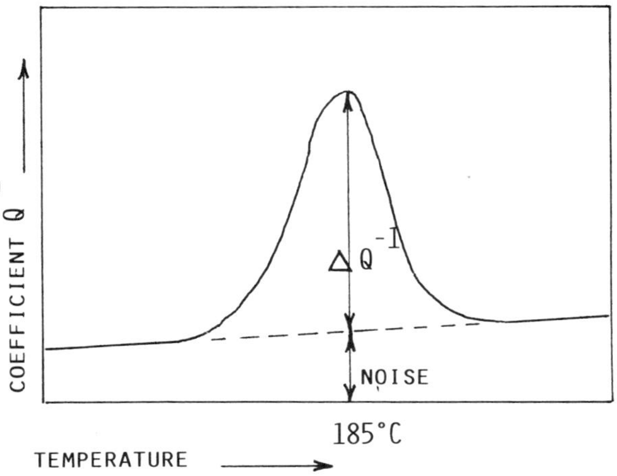

Fig. 7. Schematic Snoek's peak, showing up the noise

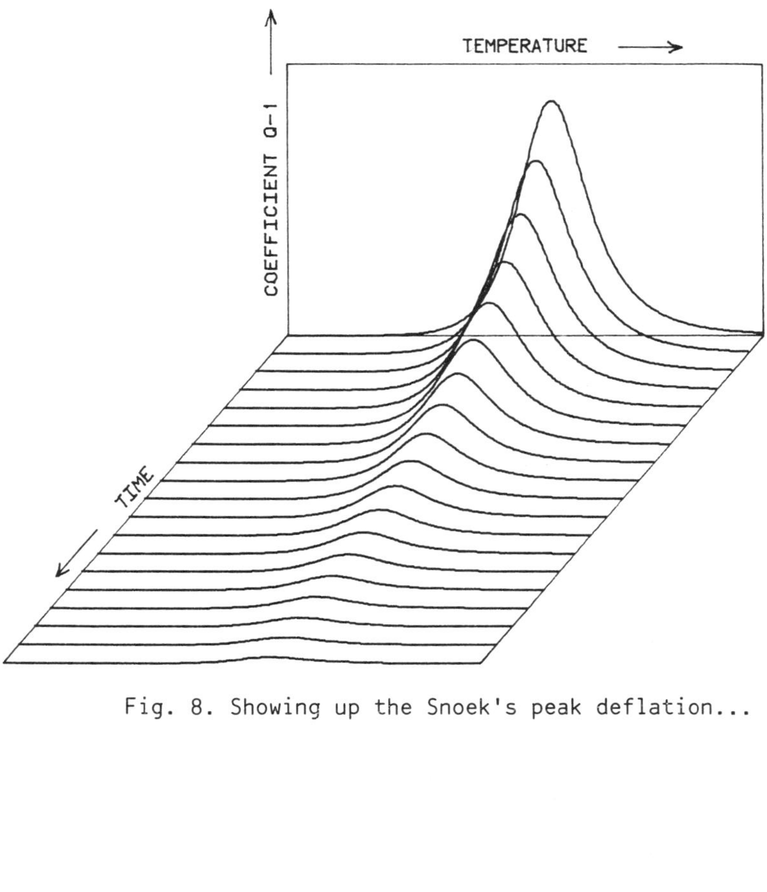

Fig. 8. Showing up the Snoek's peak deflation...

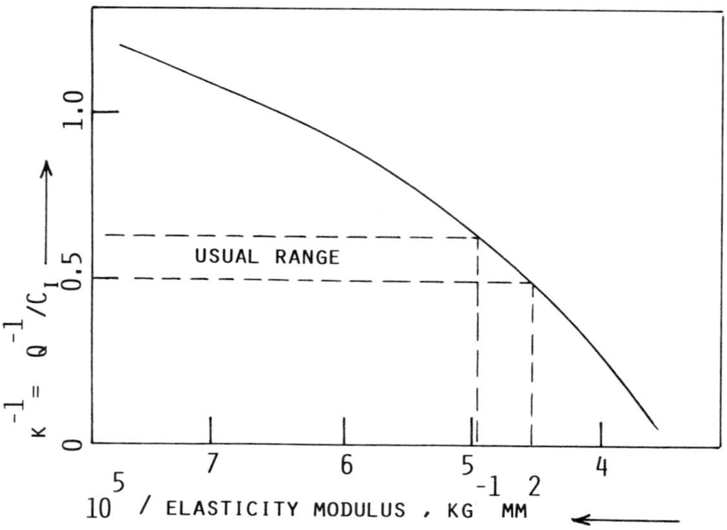

Fig. 9. Theorical evolution of k with Young's modulus, in a monocrystal of pure iron

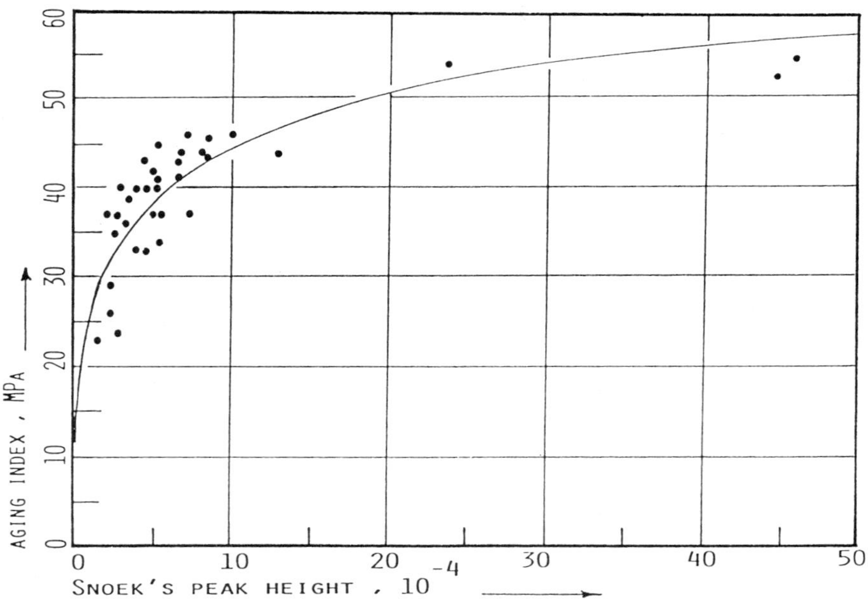

Fig. 10. Correlation between Snoek's peak height and aging index

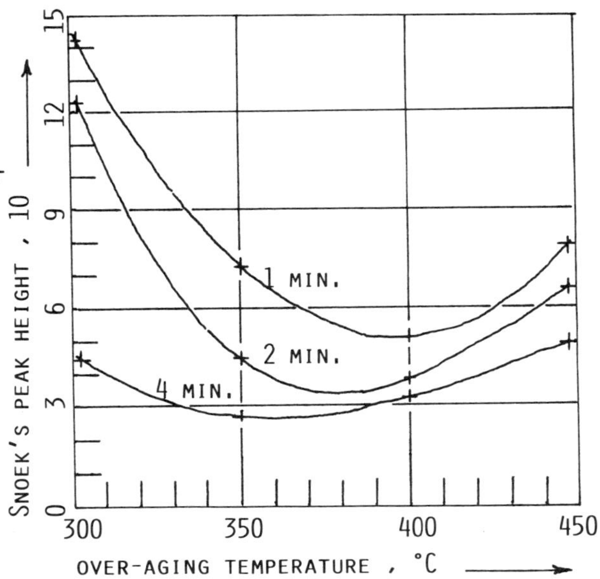

Fig. 11. Influence of over-aging time and temperature on aging-index

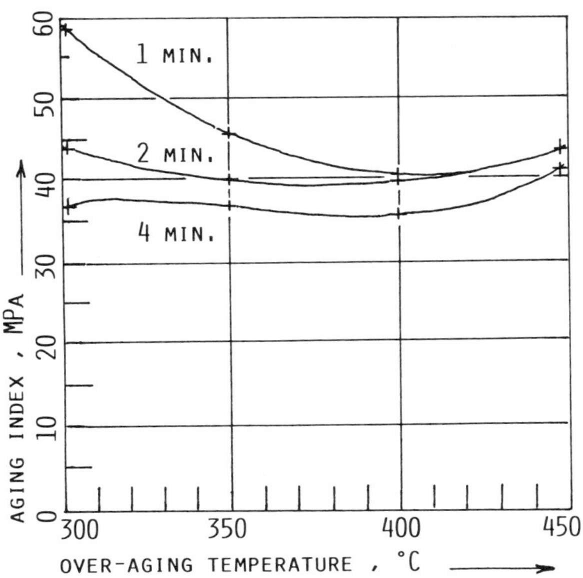

Fig. 12. Influence of over-aging time and temperature on Snoek's peak height

CONCLUSION

The present method allows a rapid and accurate measurement of the damping due to Snoek effect. It is possible to get the interstitial carbon amount, as well as information on the precipitation kinetic at 185° C.

In the case of fundamental research, the method is not so interesting, because of uncertainty on the value of $k = Ci / \Delta Q^{-1}$.

On the other hand, measurements can be easily automated in order to get an actual system for process control.

REFERENCES

1. M. Nacken and W. Heller, Arch. Eisenh., 45 : 54 (1956)
2. W. C. Leslie and A. S. Keh, J.I.S.I., 722 : 728 (1962)
3. W. S. Carswell, Acta Met., 670 : 700 (1961)
4. T. S. Keh, Phys. Rev., 533 (1947)
5. J. L. Snoek, New Developments in Ferromagnetic Materials, in : "Elsevier", Amsterdam (1947)
6. D. Polder, Philips Res. Rep., 5 (1945)
7. G. Collette, C.R. Acad. Sc, 2756 (1958)
8. R. Grossterlinden and U. Lotter, Arch. Eisenh., 317 (1983)
9. G. Blanchard and C. Oliver, Métaux corrosion Industries, 339 (1982)
10. B. Astie, J. Degauque, J. Garigue and J. P. Redoules, INSA TOULOUSE, non published results (1984)
11. J. C. Swartz, Acta Met., 1511 (1969)

INVESTIGATION OF CREEP DAMAGE IN ALLOY 800H

USING ULTRASONIC VELOCITY MEASUREMENTS

H. Willems

Fraunhofer-Institut für zerstörungsfreie Prüfverfahren
Universität, Gebäude 37
D-6600 Saarbrücken 11, FRG

ABSTRACT

Results of ultrasonic velocity measurements on creep tested specimens of austenitic Alloy 800H are presented. It is shown that different influences have to be taken into account in order to correctly describe the observed behavior of the velocity during creep. At high stresses resulting in short lifetimes, the measured velocity changes can be attributed mainly to the plastic deformation of the material whereas at lower stresses the influence of physical discontinuities (micropores, microcracks) becomes dominant. Additionally, a purely thermal ageing effect due to carbide precipitations is noticed. Damage patterns are demonstrated using microradiography as well as metallography. From the results obtained it is suggested that the ultrasonic velocity becomes sensitive to microstructural damage at the transition range from secondary to tertiary creep. Concerning practical applications, this could allow the nondestructive detection of creep damage at a relatively early stage.

INTRODUCTION

The occurrence of creep damage limits the lifetime of components that are exposed to stresses at temperatures higher than approximately half the melting temperature. Such conditions are generally met by a lot of structural components especially in power plants (pipes, turbines etc.). According to conventional safety rules critical parts are usually exchanged long before any failure has to be expected. This procedure is based on statistics drawn from material tests by standardized methods rather than on the actual state of the component concerned. During the last years an increasing need can be stated to develop NDE methods for the detection of early damage stages in order to improve the reliability and safety of components. Basically, techniques are required being sensitive to either small strains or better to small concentrations of micropores and microcracks, respectively. With regard to in-field applications, only replica techniques are used successfully for that purpose up to now /1,2/. These metallographic techniques are by nature restricted to surfaces where appropriate spots have to be selected and to be prepared carefully. In this work the influence of creep damage on the ultrasonic velocity has been investigated on a representative high-temperature alloy for tube components, i.e. Alloy 800H (X10 NiCr AlTi 32 20).

Table 1. Chemical composition (wt %) and solution treatment of investigated Alloy 800H

C	Cr	Ni	Ti	Al	N	Fe
0.070	20.26	31.11	0.31	0.34	0.009	bal.
1130°C/30min ⟶		< 2 min ⟶		~ 830°C ⟶		water

The high-temperature properties of this material are extensively studied by several laboratories within the framework of the European COST 501 action on high-temperature materials.

EXPERIMENTAL

Material

Hot-rolled rod material of X10 NiCrAlTi 32 20 (Alloy 800H) of the same cast was supplied by VEW-Austria to the laboratories concerned. The chemical composition and the solution treatment of the materials are given in Table 1 /3/. The mean grain size was 80 μm with a half-width of the distribution of about ± 50 μm /3/.

Specimens

Creep-tested specimens have been supplied to the author by several laboratories (see Acknowledgements). An overview on the testing conditions is given in Table 2. Except for one, all tests were performed at 800°C. Most tests were run under constant load conditions up to rupture, some tests have been interrupted after the steady state stage. Four specimens were run with a constant strain rate until steady state creep was reached. The cylindrical creep specimens had diameters between 6 mm and 10 mm. In order to enable ultrasonic velocity measurements, plane

Table 2. Creep testing conditions of the investigated specimens

Testing procedure	Temp. (°C)	Stress (MPa)	Strain rate (sec^{-1})	Number of specimens	Time to fracture (h) (order of magnitude)
		initial stress σ_0			
Constant load test	700	57		1	10^4
	800	45		5	10^3
	↓	60		5	10^3
	↓	70-80		6	10^2
	↓	98		2	10^1
	↓	130		2	10^0
	↓	150		1	10^{-1}
	↓	saturation stress			
Constant strain rate test	↓	48	10^{-8}	1	Tests interrupted at beginning of steady state creep /4/
	↓	62	10^{-7}	1	
	↓	77	10^{-6}	1	
	↓	95	10^{-5}	1	

parallel samples had to be machined out of the creep-tested specimens. Therefore, from each specimen several disks were prepared having thicknesses of about 3 mm with an accuracy of $\pm$ 2x10^{-3} mm. From the reduction in diameter the local creep strain of each disk was determined. These samples were used to measure the ultrasonic velocity parallel to the load direction. In some cases samples were also prepared to allow velocity measurements perpendicular to the load direction.

Velocity mesurements

Ultrasonic velocity measurements were performed by means of the pulse-echo overlap method /5/. The absolute measuring error

$$|\Delta v/v| = |\Delta d/d| + |\Delta t/t|$$

was approximately $2 \cdot 10^{-3}$. Here, d is the specimen thickness and t is the time-of-flight between two backwall echoes. The frequencies used were 5 MHz for shear (T) waves and 10 MHz for longitudinal (L) waves. To minimize dispersion effects, narrow band pulses were applied by means of burst excitation. If not mentioned otherwise, the measured velocities were normalized to the velocities obtained at the heads of the specimens, v_H, in order to separate purely thermal effects from load induced effects.

RESULTS AND DISCUSSION

Ageing effects

According to the heat treatment of the material (see Table 1) most of the carbon is in solution for the as-delivered condition. The amount of carbides comes to about 0.2 wt % /6/. Ageing at 800°C leads to the precipitation of carbides at grain boundaries, twin boundaries, dislocation pile-ups etc. As can be seen from Fig. 1a, the carbide content reaches a saturation level already after approximately 15 h ageing. This correlates well with an increase of the ultrasonic velocities both v_L and v_T during the first 16 hours and then remaining constant within the measuring error (Fig. 1b). Concerning the creep tests referred to in what follows the specimens were usually preaged 16h/800°C before testing in order to ensure stable starting conditions.

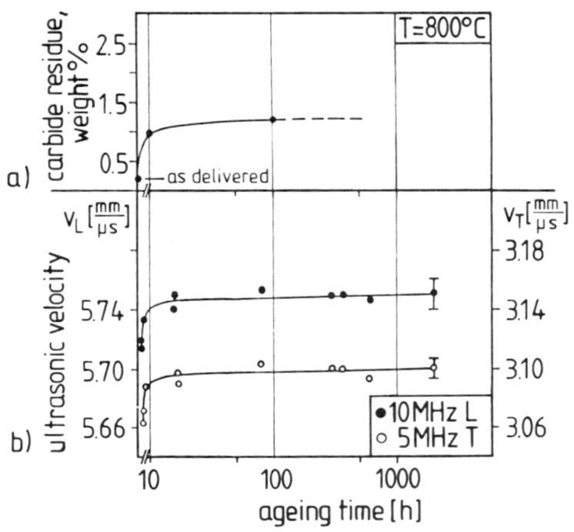

Fig. 1 a) Amount of carbide residue in Alloy 800H as a function of ageing time (after /6/)
b) ultrasonic velocities v_L,v_T as a function of ageing time.

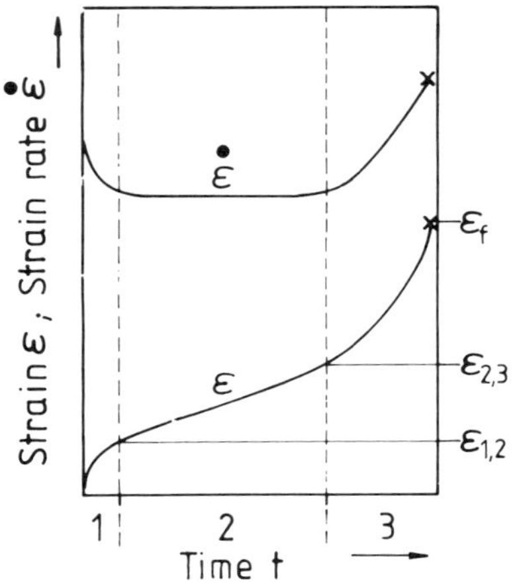

Fig. 2 Creep curve showing strain ε and strain rate $\dot{\varepsilon}$ as a function of time (schematically).

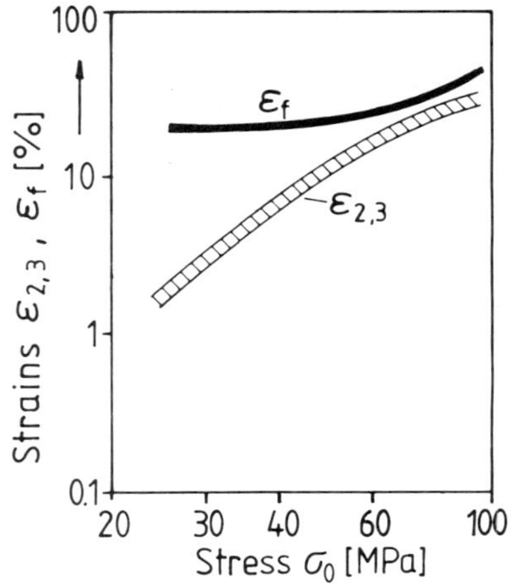

Fig. 3 Strains $\varepsilon_{2,3}$ and ε_f for Alloy 800H (preaged 16h/800°C) as a function of stress σ_0 (results taken from Ref. /3/).

Creep

The time-dependent deformation that occurs in (metallic) materials subjected to a constant load at high temperatures is called creep. A creep curve, i.e. strain ε or strain rate $\dot{\varepsilon}$ as a function of time t, is shown schematically in Fig. 2. One usually observes three different creep stages (Fig. 2): Primary creep is usually characterized by a decreasing strain rate which indicates hardening of the material. During secondary creep (also called steady state creep) the strain rate is rather constant. Finally $\dot{\varepsilon}$ increases (tertiary creep) which ends in fracture. The transitions between the different stages can be characterized by the corresponding strains $\varepsilon_{1,2}$ and $\varepsilon_{2,3}$ (see Fig. 2). The strains $\varepsilon_{2,3}$ and the fracture strain ε_f for the investigated Alloy 800H (preaged state) are shown in Fig. 3 as a function of initial stress σ_0 based on results of Degischer et al. /3/.

Primary creep stage

During primary creep dislocation mechanisms can lead to the formation of subgrain structures consisting of dislocation networks. For stresses above 40 MPa such structures have been found in Alloy 800H by TEM investigations before the secondary creep stage was reached /7/. Ultrasonic measurements on specimens from constant strain rate tests interrupted at the beginning of steady state creep (see Table 2), i.e. end of primary creep when compared to a constant load test, did not show significant changes of ultrasonic velocity as could eventually be expected according to the Granato-Lücke model of dislocation damping which yields /8/

$$\Delta v/v_0 \sim \Lambda \cdot L^2$$

for the low frequency range. Here, v_0 is the velocity without dislocations, Λ is the dislocation density, and L is the dislocation loop length. However, when measured during plastic deformation the effect on ultrasonic velocity due to changes in the dislocation structure is usually well below 10^{-2} (see /9/). Thus, the measuring accuracy was probably not sufficient in our case. In contrary, ultrasonic absorption proved to be sensitive to changes in the dislocation structure /10/.

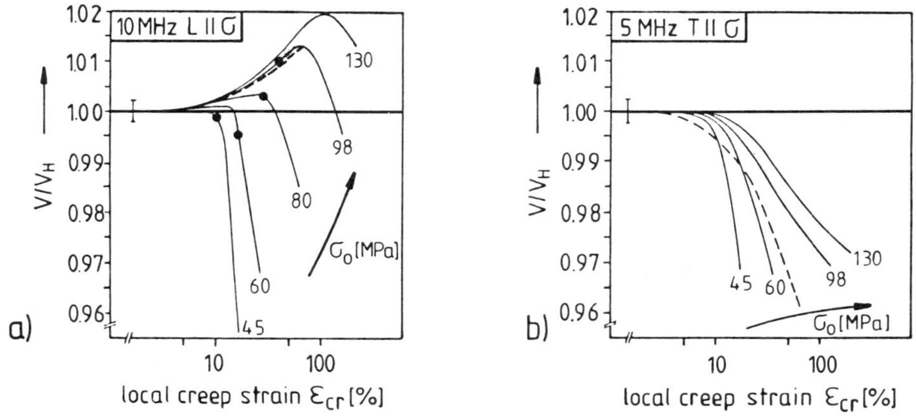

Fig. 4 Normalized ultrasonic velocity in Alloy 800H as a function of creep strain for different initial stresses σ_0 (dashed lines: measurements as a function of plastic deformation at room temperature). Dots corresponds to $\varepsilon_{2,3}(\sigma_0)$ (see Fig. 3).
a) longitudinal waves b) shear waves

Secondary and tertiary creep stage

Secondary creep is generally described in terms of balanced strain hardening and recovery. Towards the end of secondary creep the nucleation of cavities due to vacancy condensation and grain boundary sliding usually takes place. The growth of the vacancies leads to a loss in internal cross section, and an increasing strain rate is observed characterizing the beginning of tertiary creep. Finally, fracture will occur by plastic instability at a certain loss of cross section. For a complete description of the processes involved a lot of microstructural features have to be taken into account which are in many cases not fully understood presently in terms of basic principles (see for example /11/).

Specimens originating from creep tests interrupted at the end of secondary creep as well as ruptured specimens have been investigated according to the procedure described above. The results are summarized in Fig. 4a for L-waves and in Fig. 4b for T-waves, respectively. The normalized velocities are plotted as a function of the local creep strain with initial stress σ_0 as parameter. For σ_0 > 80 MPa v_L measured parallel to the stress direction increases with increasing creep strain, reaches a maximum and then decreases near the fracture area (Fig. 4a). For stresses between 45 MPa and 60 MPa only a decrease of v_L is measured which starts at strains above 10% (Fig. 4a). Concerning T-waves the velocity decreases at strains larger than 10% for all stresses applied (Fig. 4b). Below strains of 10% the data are in the measuring error. Some examples of the variation of both v_L and v_T from the specimen heads across the gauge length are illustrated in Fig. 5. Fig. 6 shows the velocity ratio v_F/v_H (v_F-velocity near the fracture zone) as a function of fracture time t_f. For large times t_f (low stresses) the velocity v_F of L-waves as well as T-waves decreases similarly whereas for short t_f (high stresses) the change of v_L and v_T is opposite (Fig. 6). Referring to the behavior of v_L (Fig. 4,5,6) it is obvious that two different influences are operating. These are considered to be plastic deformation and porosity as will be shown below.

Plastic deformation

The change of ultrasonic velocity as a function of plastic deformation at room temperature (RT) is depicted in Fig. 7. Here, the measuring

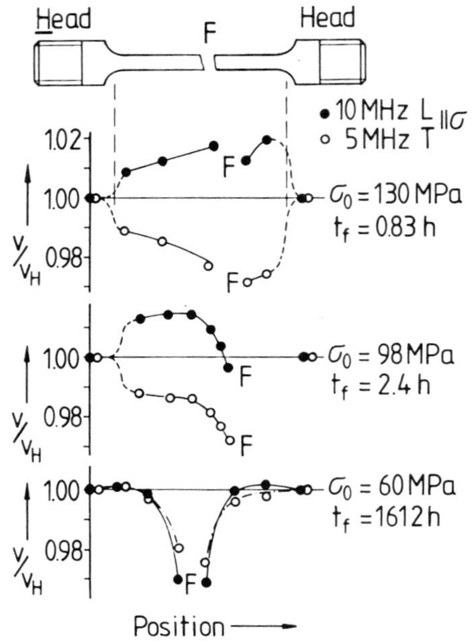

Fig. 5 Variation of ultrasonic velocity across the gauge length of creep damaged specimens (Alloy 800H, 800°C).

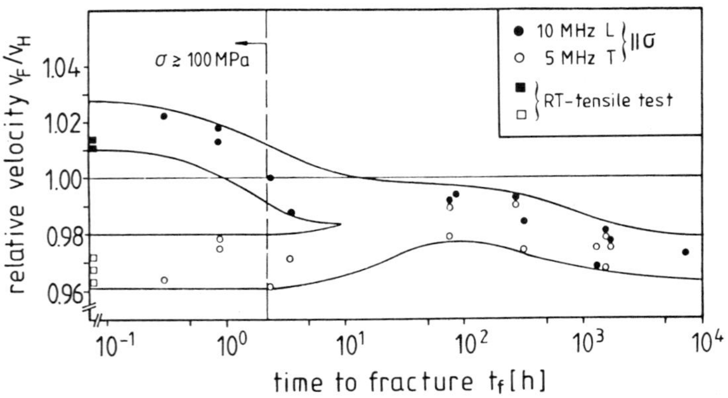

Fig. 6 Relative ultrasonic velocity v_F/v_H near the fracture area of creep damaged specimens as a function of time to fracture.

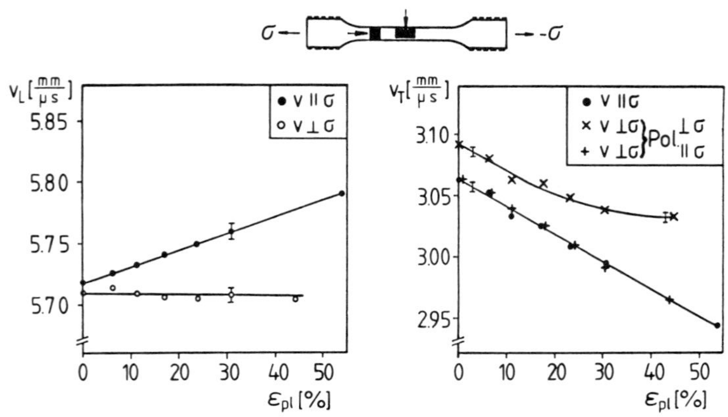

Fig. 7 Ultrasonic velocities v_L, v_T in Alloy 800H as a function of plastic deformation at room temperature.

error $\Delta v/v$ was $\pm 10^{-3}$. It is found that v_L perpendicular to the stress direction is rather independent of the plastic deformation, whereas v_L parallel to the stress direction increases linearly. The transverse wave velocities v_T both parallel and perpendicular to the load direction (the latter one with polarization in the load direction ($v_T^{\parallel}$)) show the same linear decrease up to rupture whereas $v_T^{\perp}$ with both propagation direction and polarization perpendicular to the load levels off at plastic deformations larger than about 30%. At this strain, the onset of contraction was observed. The difference between $v_T^{\parallel}$ and $v_T^{\perp}$ in the undeformed state indicates a texture due to the production process of the material. The uniaxial deformation applied here leads to a pronounced increase of the <111>-texture in the load direction. This is a well-known effect for fcc-materials /12/. Because v_L (v_T) in the <111>-direction is larger (smaller) compared to the isotropic case, the behavior of v_L and v_T parallel to the load direction is readily explained in terms of texture. This should also hold for the propagation perpendicular to the load direction but the situation is more complex in that case.

Comparing now the results measured at room temperature with the results measured on the creep specimens, a rather good agreement can be stated between the plastic strain dependence of v_L at room temperature (dashed line in Fig. 4a) and the creep strain dependence of v_L at stresses above 100 MPa disregarding the near-fracture zone. Concerning shear waves (Fig. 4b) the effect of plastic deformation is larger at room temperature than after the creep tests if the 130 MPa-specimen is assumed to show the pure deformation effect for strains below 100%. Measurements of v_L perpendicular to the stress direction did not show a dependence on the creep strain in agreement with the RT-measurements (Fig. 7). Concerning in service conditions, the applied stresses (and hence strains) are well below the stresses used here. In that case the influence of plastic deformation seems to be negligible.

Porosity

In contrast to the RT-measurements the longitudinal velocity v_L drops if one approaches the fracture zone. Microstructural examinations of these zones exhibit cavities (micropores, microcracks). Some examples obtained by microradiography as well as by metallography are shown in Fig. 8. It is found that the cavity density in a given specimen is the larger the larger the strain. Additionally, for a given strain the cavity density is increasing with increasing time to fracture. It should be mentioned that the microradiographic images show the projection of the whole sample (3 mm thickness) on the film plane. According to Fig. 4a, the change of velocity at stresses below 60 MPa should be mainly correlated with the porosity whereas for higher stresses the plastic deformation must be taken into account. The influence of porosity on the ultrasonic velocity can be calculated by ultrasonic propagation theory assuming spherical pores in an isotropic matrix /9,13,14/. The following expressions can be derived for v_L and v_T as a function of porosity c_p (volume fraction) assuming $c_p < 0.05$ and $d/\lambda \ll 1$:

$$v_L = v_L^{o}\,(1 - k_L \cdot c_p) \quad \text{and} \quad v_T = v_T^{o}\,(1 - k_T \cdot c_p).$$

Here, c_p is the pore concentration, v_L^{o}, v_T^{o} are the velocities without porosity, d is the pore diameter and λ is the ultrasonic wavelength. Using $v_L = 6\,\text{mm}/\mu\text{s}$ and $v_T = 3\,\text{mm}/\mu\text{s}$ one obtains from /14/: $k_L = 0.78$ and $k_T = 0.44$, i.e. the longitudinal wave velocity is nearly twice as sensitive with regard to porosity than the shear wave velocity. These theoretical results are in good agreement with experimental values obtained in service-exposed material as long as micropores are prevailing /15/. In the presence of microcracks the effect was even

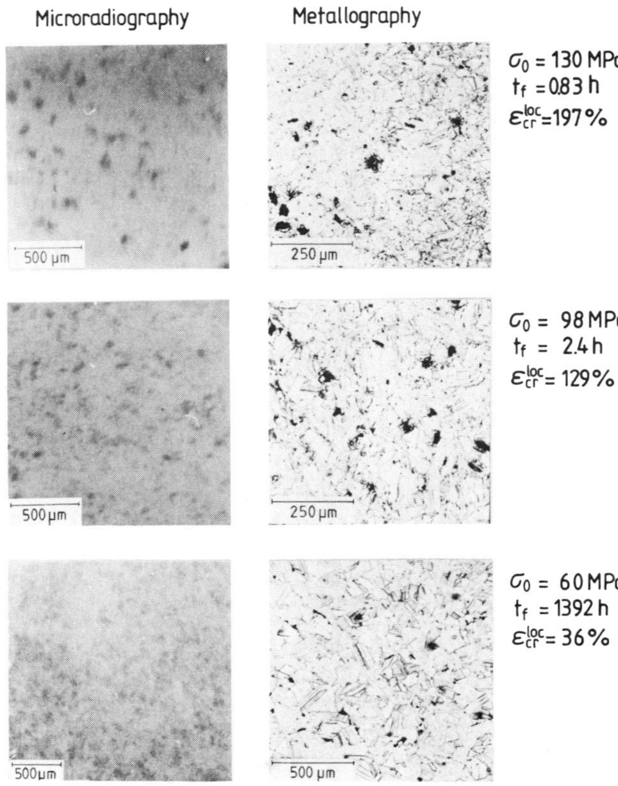

Fig. 8 Examples of creep damage in Alloy 800H.

larger than expected from the values given above. In /15/ the actual porosity has been determined by density measurements. If one relates the theoretical results to the velocity changes as measured on the creep specimens tested at stresses below 60 MPa, one may estimate porosities up to 5% at the fracture zone (Fig. 4). Because in that zone the damage patterns show mainly microcracks (see Fig. 8) this might be a factor 2 too large according to the results mentioned above, i.e. the theory which assumes spherical pores seems to underestimate the influence of porosity produced by microcracks. On the other hand high precision velocity measurements should enable the detection of pore concentrations below 10^{-3}. If one relates the creep strains $\varepsilon_{23}(\sigma)$ at the transition from secondary to tertiary creep (see Fig. 3) to the corresponding velocity measurements, one obtains the values indicated in Fig. 4a by dots. One may recognize that the dropping of v_L starts around these strains because the velocity becomes sensitive to the increasing porosity.

CONCLUSIONS

During high-temperature creep the ultrasonic velocity in Alloy 800H is mainly influenced by microstructural changes, plastic deformation and porosity. Microstructural influences (mainly precipitation processes) can be separated from damage influences by comparing material of the same microstructure, i.e. damaged and undamaged parts of a component submitted to the same temperature. The influence of plastic deformation on the ultrasonic velocity is large in the case of high stresses. Concerning service stresses, the effect is assumed to be negligible, whereas the influence of porosity becomes dominant in that case. Here, ultrasonic velocity measurements seem to be appropriate to detect small concentrations of micropores. Based on the creep properties of the considered material, it is suggested that ultrasonic velocity in Alloy 800H becomes sensitive to porosity at the transition from secondary to tertiary creep assuming reasonable measuring accuracy.

ACKNOWLEDGEMENTS

This work was performed within the European concerted action COST 501 on high temperature materials. The autor wishes to thank Dr. R. Danzer (Montanuniversität Leoben, Austria), Dr. H.P. Degischer (Österreichisches Forschungszentrum Seibersdorf, Austria), Dr. U. Hildebrandt (BBC Mannheim, FRG), and Dr. M. Steen (University of Gent, Belgium) for supplying the creep-tested specimens. Financial support is provided by the "Bundesministerium für Forschung und Technologie" of the FRG.

REFERENCES

1. Neubauer, B., Wedel, U., Restlife estimation of creeping components by means of replicas, ASME International Conference on Advances in Life Prediction Methods, Albany, 1983, pp. 307-313
2. Auerkari, P., Remanent creep life estimation of old power plant steam piping systems, ASME International Conference on Advances in Life Prediction Methods, Albany, 1983, pp. 353-356
3. Degischer, H.P., Aigner, H. Danzer, R., Mitter, W., Evaluation and qualification of uniaxial creep of X10 NiCr AlTi 32 20, paper submitted to: COST 50/COST 501 Conference "High Temperature Alloys for Gas Turbines and other Applications 1986", to be held in Liège, Oct. 6-9, 1986
4. Steen, M., LCF and creep life determination of Alloy 800H and a 9-12% Cr-steel, 1st annual report, COST 501 project B5 (1984)
5. McSkimin, H.L., Pulse superposition method for measuring ultrasonic velocity in solids, J. Acoust. Soc. Am. 33, 12 (1981)
6. Stickler, R., Weidlich, G., Prakt. Metallographie 20, 174 (1983)
7. Guttmann, V., Timm, J., Creep and fracture behavior of high temperature alloys in environments related to energy conversion processes, 2nd progress report, COST 501 project CCR 3 (1985)
8. Granato, A. and Lücke, K., Theory of mechanical damping due to dislocations, J. Appl. Phys. 27, 583 (1956)
9. Truell, R., Elbaum, C., Chick, B.B., Ultrasonic Methods in Solid State Physics. Academic Press, New York and London (1969)
10. Willems H., A new method for the measurement of ultrasonic absorption in polycristalline materials, paper submitted to: Review of Progress in QNDE, University of California - San Diego, Aug. 3-8, (1986)
11. Bressers, J. (Ed.), Creep and Fatigue in High Temperature Alloys. Applied Science Publishers LTD, London (1981)
12. Wassermann, G. and Grewen, J., Texturen metallischer Werkstoffe, Springer Verlag, Berlin/Göttingen/Heidelberg (1962)
13. Hirsekorn, S., Streuung von ebenen Ultraschallwellen an kugelförmigen, isotropen Einschlüssen in einem isotropen Medium unter Berücksichtigung der Mehrfachstreuung, IzfP-Bericht Nr. 790218-TW, Saarbrücken (1979)
14. Sayers, C.M. and Smith, R.L., The propagation of ultrasound in porous media, Ultrasonics, Sept. 1982, 201
15. Willems, H., Bendick, W., and Weber, H., Nondestructive evaluation of creep damage in service exposed 14 MoV 6 3 steel, paper submitted to: Symposium on the Nondestructive Characterization of Materials, Montreal, July 21-23, 1986

DETECTION OF HEAT TREAT DEFECTS AND GRINDING BURNS BY MEASUREMENT OF BARKHAUSEN NOISE

Kirsti M. Tiitto and Richard J. Pro

American Stress Technologies
515 Hollydale Drive
Pittsburgh, PA 15102

INTRODUCTION

Grinding and heat treating processes are integral steps in the industrial production of most finished metallic parts. Among the parts whose properties may be adversely affected by improper grinding and/or heat treatment are camshaft lobes, bearing races, bearings, machine bolts, automotive U-joint trunnions, piston pins, gears, shafts, pipes, and many others. The benefits associated with proper grinding and heat treatment are well-known[1,2]. Included among these benefits are controlled hardness, improved fracture toughness, and reliable dimensional control. Improper grinding may result in numerous surface defects, e.g., retempered areas, rehardened areas, and cracks. Heat treatment errors often result in regions of decreased hardness, i.e., soft spots, as well as decarburizing, denitriding, or overtempering. If left undetected, each of these defects may lead to catastrophic failure of a part operating under load. As a result, it is necessary to detect grinding and heat treat defects as they occur so that steps may be taken to correct the process immediately and prevent the incorporation of faulty parts into their end-products.

The most common grinding burn detection method, nital etching[3], is difficult to use as a production monitoring tool. There are many reasons for this including the amount of time and degree of care necessary to accurately locate grinding burns; the inability to monitor 100% of parts due to the etching of the finished surface, i.e., it is not truly non-destructive; and the difficulties associated with handling and disposal of the required chemicals. As a result of these difficulties, it is usually necessary to conduct periodic audits of a small number of parts with the hope that evidence of grinding burns will be obtained before a large number of burned parts has been produced.

An alternative method for in-process detection of grinding burns and heat treat defects now exists commercially. This method is based upon the detection of Barkhausen noise. Barkhausen noise results from the magnetoelastic interaction which occurs in ferromagnetic materials subjected to an applied magnetic field. This technique is well-suited for 100% inspection of ground or heat treated parts since it is completely non-destructive. A description of the Barkhausen noise measurement principle is presented below. The effects of grinding burns and heat treat defects on Barkhausen noise are outlined and examples of defect detection are provided.

BARKHAUSEN NOISE PRINCIPLE

The magnetic microstructure of ferromagnetic materials is known to consist of an aggregate of regions termed magnetic domains. A magnetic domain is defined as a region in which all magnetic dipole moments are aligned parallel to one another; the magnitude of the vector sum of the dipole moments determines the magnitude of the magnetization of the domain, and the direction of the magnetization is that along which the dipole moments are aligned. Magnetic domains are separated from one another by domain walls. The domains are magnetized to saturation along $\langle 100 \rangle$, which are the directions of easy magnetization in bcc materials. The net magnetization of the bulk material is given by the average of the magnetization of all domains. In the demagnetized state, magnetic domains will be randomly oriented throughout the bulk and the net magnetization of the material will be zero.

If a magnetic domain wall is made to move, the magnetization within the area swept by the wall will change. This changing magnetization will generate an electric pulse. When all pulses generated by wall movements throughout the bulk are counted, a noise-like signal, known as Barkhausen noise, results.

Magnetic domain walls can be made to move by a number of means. If an a.c. magnetic field is applied to the sample, the domain walls will move back-and-forth as the magnetic induction of the material changes along the hysteresis loop. The maximum of the Barkhausen noise signal occurs where the slope of the hysteresis loop is steepest. Therefore, the steeper the slope of the hysteresis loop, the greater is the Barkhausen noise amplitude.

The movement of domain walls and, therefore, the amplitude of the Barkhausen noise signal under an applied a.c. magnetic field, is determined by the state of stress and the microstructure of the material. These effects may be understood as follows.

Effect of Stress

Figure 1 illustrates the effects of variations in the state of stress of a material on the Barkhausen noise amplitude. Figure 1 (a) schematically illustrates a material which contains 4 magnetic domains; the arrows within each domain represent the magnitude and direction of the magnetization for that domain. The net magnetization of the material is zero. The material is demagnetized and is under no applied stress. A typical Barkhausen noise burst for this material is represented next to the sample. In figure 1 (b) a tensile stress has been applied to the sample along the axis shown. Magnetic domains with magnetization aligned with the tensile axis grow at the expense of non-aligned domains. An increase in the Barkhausen noise amplitude, measured along the tensile axis, results due to the change of magnetization in the area swept out by the moving domain walls.

The converse occurs when the material is placed in compression along the measurement direction. The subsequent change in magnetization away from the measurement direction results in a decrease of the Barkhausen noise amplitude with respect to the unstressed state. Application of an external magnetic field in the measurement direction has an effect similar to applied tension with an increase in the Barkhausen noise level with respect to the demagnetized state. By combining an external field with tensile stress, one may generate a large Barkhausen noise signal. Compressive stress acts in an opposite manner than the external field and, therefore, the Barkhausen noise level is decreased under this combination.

On the basis of the above interaction between stress and magnetic field one may determine the stress level of a ferromagnetic sample by applying a

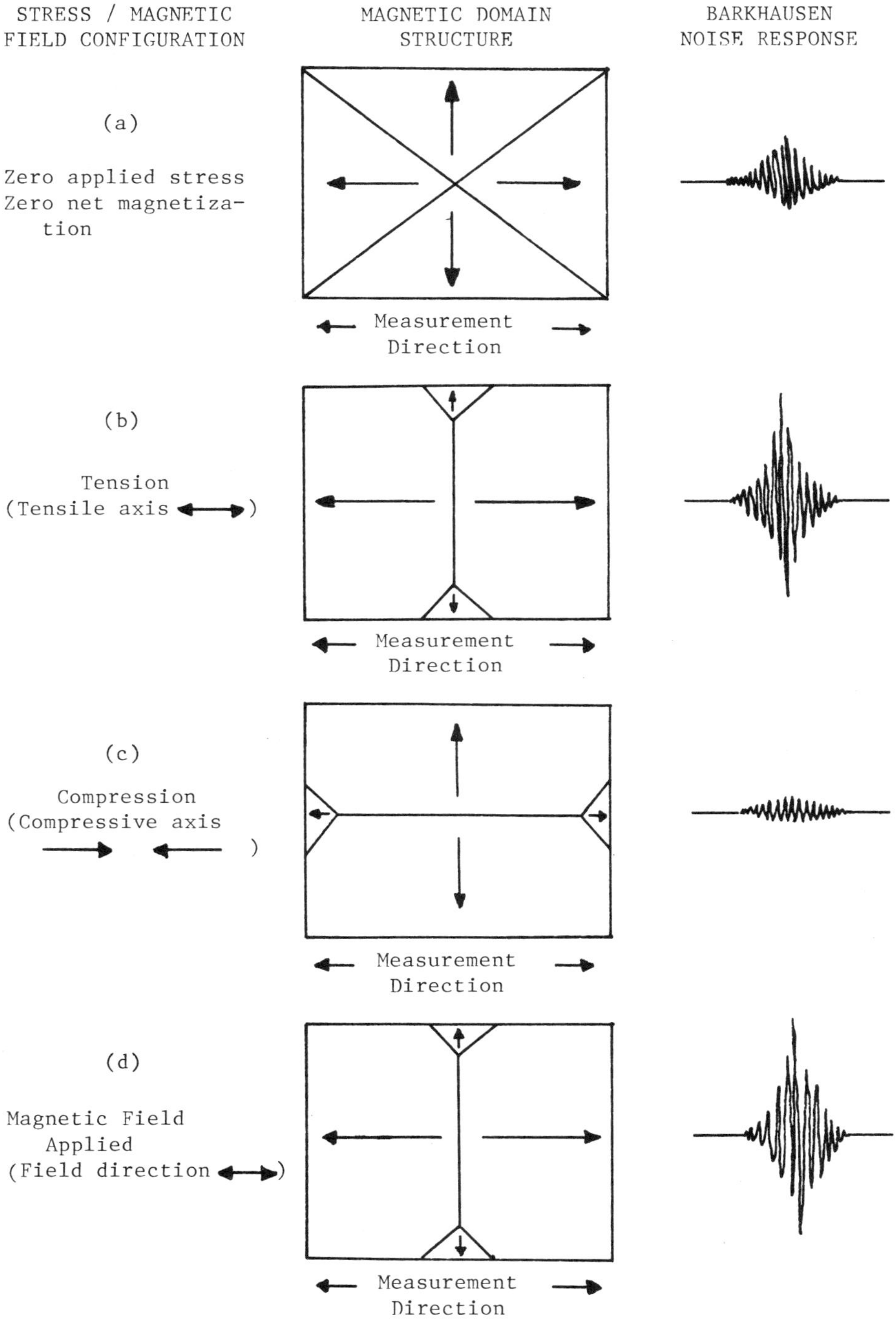

FIGURE 1 Effects of stress and applied magnetic field on Barkhausen noise amplitude: (a) System of 4 magnetic domains configured to give zero net magnetization under zero applied stress. (b) Tensile stress causes domains aligned with the tensile axis to grow at the expense of non-aligned domains. Barkhausen noise amplitude, measured along the tensile axis, increases. (c) In compression, domains aligned with the compressive axis decrease in size. Barkhausen noise amplitude decreases. (d) An applied magnetic field acts like applied tension; Barkhausen noise amplitude increases.

known a.c. magnetic field and measuring the Barkhausen noise amplitude. Absolute stress values may be obtained by using this technique in conjunction with, e.g., a cantilever beam test, to generate calibration curves.

Effect of Microstructure

Hardness changes represent the primary microstructural source of variations in Barkhausen noise amplitude and will, therefore, be used to illustrate this point. High hardness materials (e.g., quenched, quenched and tempered, carburized, or nitrided parts) are associated with slowly-varying hysteresis loops and, therefore, present low levels of Barkhausen noise. Decreases in hardness due to over-tempering, decarburizing, or denitriding result in an increase in steepness of the hysteresis loops with an associated increase in Barkhausen noise. Figure 2 illustrates the continuous change in Barkhausen noise amplitude with respect to hardness variations.

DEFECT DETECTION PRINCIPLES

The influence of the state of stress and microstructure of a material on Barkhausen noise has the following implications:

1. Barkhausen noise can be used to evaluate the state of residual stress when microstructure is controlled.
2. If residual stress is controlled, Barkhausen noise gives valuable information on hardness.
3. Barkhausen noise can be used to detect material defects which include simultaneous changes in stress and microstructure such as heat treat defects and grinding burns.

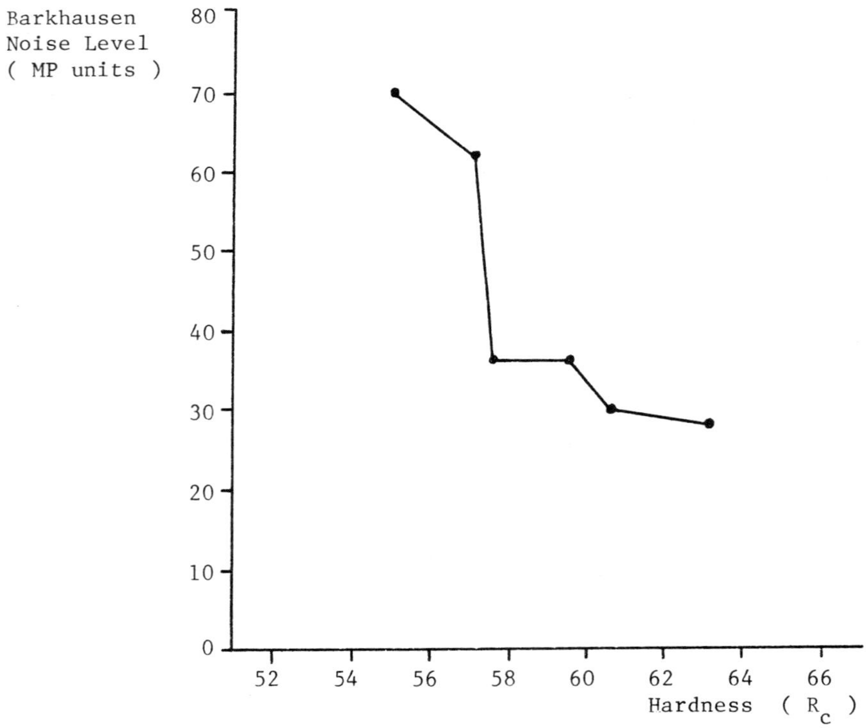

FIGURE 2 Effect of hardness on the Barkhausen noise level from gears. The change in hardness was obtained by tempering case-hardened gears at different temperatures.

Abusive grinding causes microstructural and/or stress changes in hardened surfaces due to the generation of excessive heat. Two distinct classes of grinding burns may be defined. If the excess heat generated by grinding does not raise the surface temperature of a hardened steel above the ferrite/martensite transformation temperature, then the surface is locally over-tempered and a retempering burn is formed; this burn will be referred to as Type I for convenience.

Raising the surface temperature above the transformation temperature results in the formation of brittle, untempered martensite due to quenching by the cooling water. This type of grinding burn, which results in the formation of white martensite, will be referred to as Type II, or a rehardening burn.

Type I grinding burns result in an increase in the Barkhausen noise level due to local decreases in hardness with respect to properly ground surrounding surfaces. Type II burns, with the formation of hard, untempered martensite, cause localized decreases in Barkhausen noise levels (see Fig. 2)

In carburizing, nitriding, and other hardening processes such as induction hardening, some areas may be left unhardened so that soft spots or edges are formed. While the hardened areas have martensitic microstructure, the unhardened spots or edges are more ferritic and have lower hardness. Since the unhardened areas take less volume than the surrounding hardened surface, the residual stress in these two areas will be different and tensile stresses may be generated. Both decreased hardness and decreased compressive stress will increase the Barkhausen noise amplitude in the area of heat treat defects.

EXPERIMENTAL PROCEDURE

A wide variety of parts exhibiting grinding damage or heat treat defects have been tested using commercially available equipment. For the purpose of this paper, data obtained from testing a gear and a camshaft lobe will be presented.

An American Stress Technologies Rollscan 100-2 central unit was used as the Barkhausen noise processing unit (Figure 3). Barkhausen noise sensors suitable for use on gear teeth and camshaft lobes were used in conjunction with the central unit. The sensors apply an a.c. magnetic field to the sample via two magnetizing pole pieces and detect the Barkhausen noise with a third, pick-up pole piece. The measurement direction is defined along the line formed by the pole pieces. The central unit automatically evaluates the Barkhausen noise and displays the measurement result.

When evaluating grinding burns and/or heat treat defects, the difference in Barkhausen noise output between good and defective parts is first assessed and rejection limits established. The Barkhausen noise levels from unknown parts to be inspected are compared to the reference data and any parts which show unacceptable noise levels are rejected. This procedure may take place either manually or automatically. Both static and dynamic evaluations are possible. In dynamic evaluations, speeds up to 5 m/sec (16.4 ft/sec) can be tolerated.

RESULTS AND DISCUSSION

Figure 4 shows data[4] which is typical of grinding burn evaluations on gear teeth, as well as a photo of a typical gear tooth sensor for the Rollscan 100-2. In this application, the sensor was held in contact with the gear manually; fixturing for semi-automated and automated gear tooth

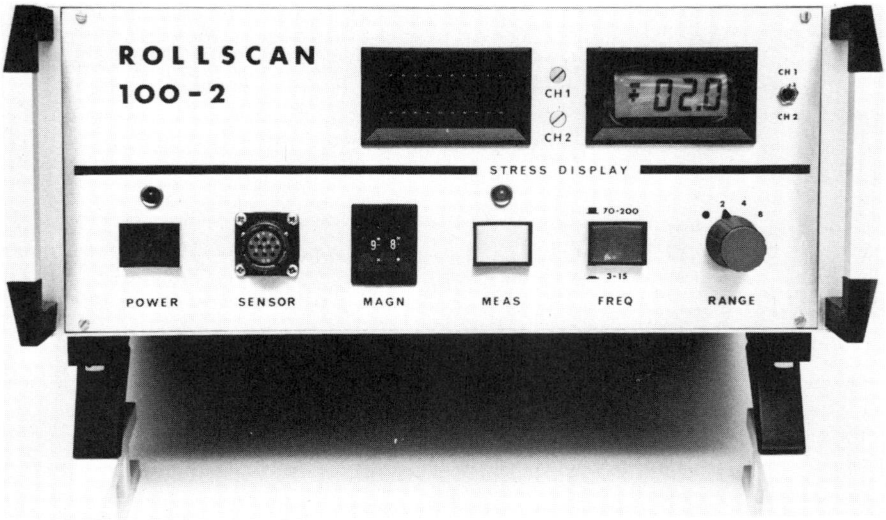

FIGURE 3 American Stress Technologies' Rollscan 100-2
This unit can accept inputs from two sensors simultaneously. The amplitude of the magnetizing current is set by a front-panel dial. The depth of measurement, determined by the measurement frequency band, is selected on the front panel (depth = 0.02 mm or 0.2 mm).

evaluation are also available. Each tooth flank was evaluated separately, and the noise was recorded and plotted against tooth number. Teeth numbered 8 through 16 exhibited grinding burns based on studies done on known good and bad gear samples made with the same material (and, therefore, having the same calibration curve) and using the Barkhausen noise technique. The principle comparison standard for location of grinding burns was the nital etch technique. Verification studies were performed using the nital etch technique; correlation between nital etching and Barkhausen noise detection of grinding burns was excellent. Note that up to four times greater Barkhausen noise levels were obtained from burned teeth than from sound teeth.

Results typical of dynamic evaluation of steel camshaft valve lobes for grinding burns are illustrated in figure 5.[5] Also shown in figure 5 is a typical semi-automated sensor system designed for camshaft evaluation. An operator places the camshaft to be inspected on rubberized end-rollers. The end-rollers are motor driven so as to rotate the camshaft at a pre-set speed. The operator then lowers the sensor assembly onto a particular camshaft valve lobe; stops are provided for rapid location of each lobe. The Rollscan 100-2 unit then evaluates the Barkhausen noise level as the camshaft is rotated through several revolutions. For production purposes an LED display and audible alarm signal the operator when the experimentally determined rejection limit has been exceeded.

Barkhausen noise levels in figure 5 (a) exceed the established rejectio limit over approximately 85% of the lobe surface in question. The size of the contact area between the pick-up pole piece and the sample determines the spatial resolution of the sensor. By minimizing the size of the pick-up pole piece one may sub-divide a part such as a camshaft lobe into many narrow measurement bands, e.g., 1.5 mm (0.06 in.) in width. This principle accounts for the substantial variations observed over the width of the

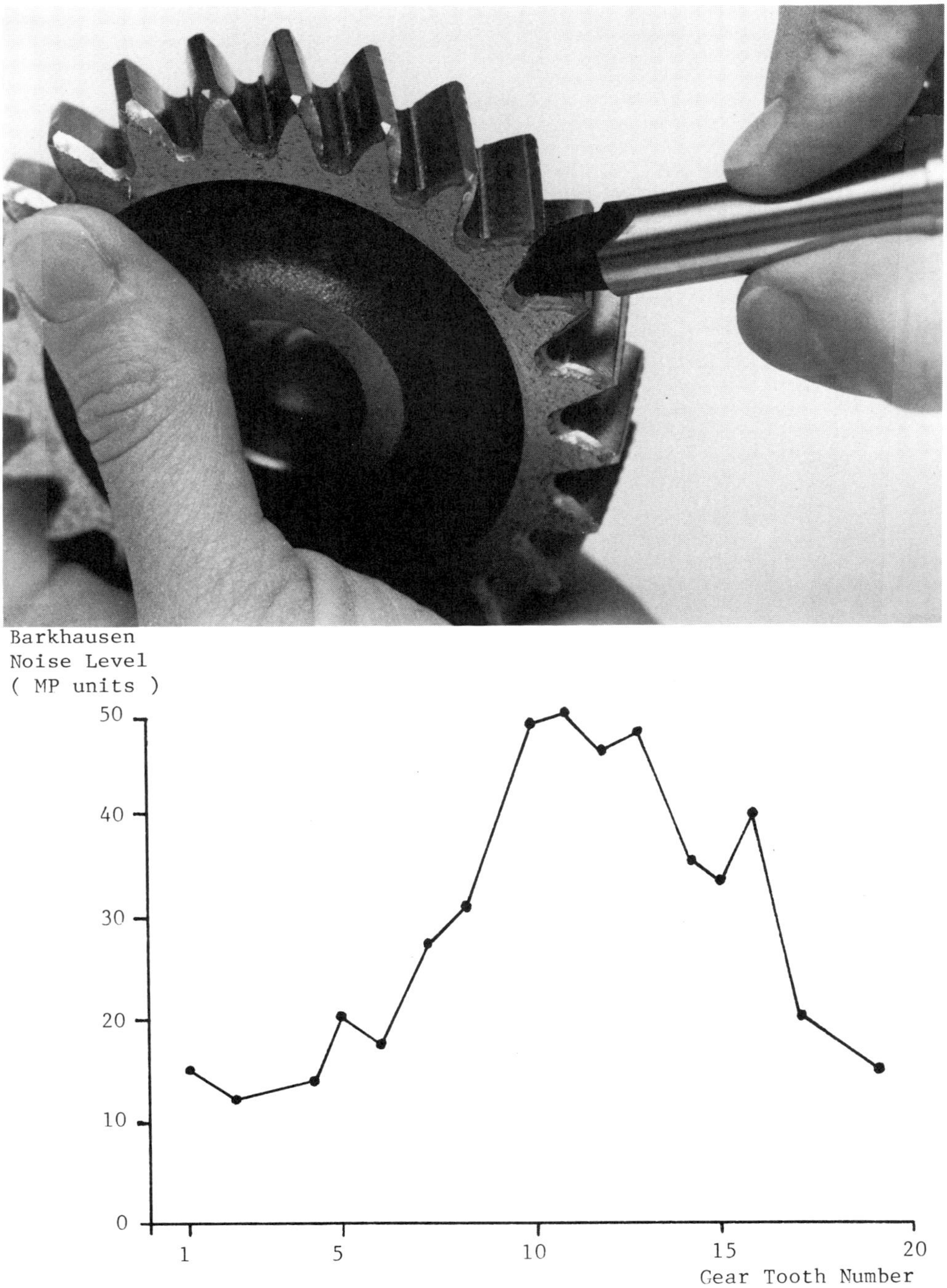

FIGURE 4 Effect of abusive grinding on Barkhausen noise from gear tooth flanks. Gear teeth numbered 8 through 16 exhibit grinding burns as determined by comparison tests with gears which were tested by the Barkhausen noise technique and also nital etched. Up to 4 times greater noise amplitudes are exhibited by burned teeth than by sound teeth. A manually-operated gear tooth sensor is shown in the accompanying photograph.

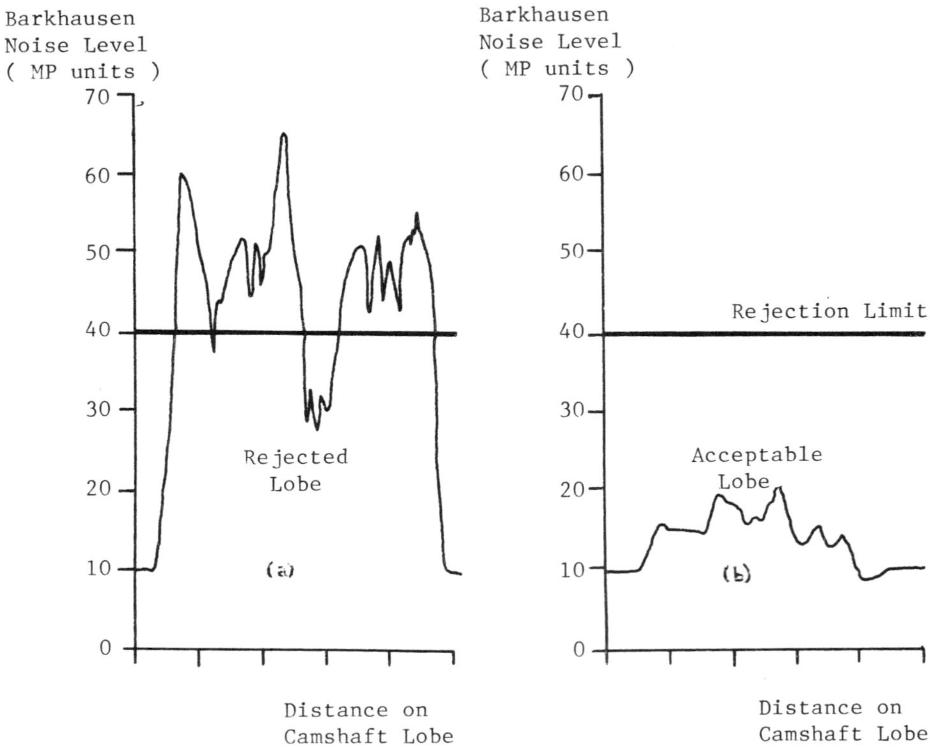

FIGURE 5 Effect of grinding burns on Barkhausen noise from steel camshaft valve lobes. (a) Heavy grinding burn damage results in noise signals above the rejection limit of 40. Note that 85% of the valve lobe surface has been burned by abusive grinding. (b) A sound camshaft lobe exhibits consistently low levels of Barkhausen noise, well below the rejection limit. A semi-automated camshaft inspection system is shown in the accompanying photograph.

of the camshaft lobe evaluated in figure 5 (a).

An acceptable camshaft lobe is shown in figure 5 (b). The Barkhausen noise levels are well below the rejection limit over the entire width of the lobe.

Heat treat defects, which lead to a softening of the microstructure as well as a decrease in compressive stress, cause localized large increases in the Barkhausen noise level with respect to areas which have received proper heat treatment. This effect has been observed in carburized automotive U-joint trunnions, unhardened edges in induction-hardened camshaft valve lobes, and decarburized edges on bearing races, for example.

CONCLUSIONS

Evaluation of a wide variety of ferromagnetic parts for grinding burn and heat treat defects proceeds in a similar manner as the above examples. The Barkhausen noise technique, with its' high level of sensitivity to stress and microstructural changes, has proven to be at least as sensitive to grinding burn and heat treat defects as other, more cumbersome techniques such as nital etching.

It is observed that a change from high compressive stress to zero stress and, further, to tensile stress systematically increases the Barkhausen noise amplitude. In addition, a decrease in hardness of hardened parts is accompanied by an increase in Barkhausen noise amplitude.

Included among the defects which the Barkhausen noise technique is sensitive to are retempering burns, rehardening burns, unhardened soft spots, decarburization, and denitriding.

The non-destructive nature and dynamic sensing capabilities of this technique make it equally suited for automated, 100%-inspection production processes, manual audit inspections, or process control research.

REFERENCES

1. American Society for Metals, "Source Book on Heat Treating, vol. I & II",American Society for Metals, Metals Park (1975).
2. G. Bellows, "Low Stress Grinding for Quality Production", Metcut Research Associates Inc., Cincinnati (1983).
3. United States Air Force, Military Standard 867A Temper Etch Inspection, Department of Defense, Washington, D.C. (1969).
4. K. Tiitto, unpublished.
5. K. Tiitto, unpublished.

INTERNAL STRESS EVALUATION BY ATTENUATION MEASUREMENTS DURING ROOM TEMPERATURE CREEP IN 5N ALUMINIUM

A. Vincent+, S. Djeroud, R. Fougeres

Groupes d'Etudes de Métallurgie Physique et de Physique des Matériaux (UA CNRS 341)

+ and Laboratoire de Traitement du Signal et d'Ultrasons

Bât. 502 - INSA DE LYON 69621 Villeurbanne Cedex France

INTRODUCTION

It is generally well established that the resolved shear stress, due to an applied tensile stress which is large enough in order to induce long-range dislocation movements, can be divided in two components : the first one, $\tau_{eff.}$, originates from short-range stress obstacles which can be overcome with the help of thermal energy and the second one is due to long-range stress obstacles for which thermal activation is not effective ; this latter component leads to the so called "internal stress" (τ_i). In the past, from relaxation /1/ or creep /2/ tests, various methods have been proposed in order to determine this internal stress. More recently, some evaluations of τ_i have been carried out by measuring the dislocation loop radius of curvature during T.E.M. observations /3/.

In this latter case, the question about the relation between the actual internal stress acting during plastic deformation in the bulk of a material and that measured in the T.E.M. specimen arises. In the former case, the methods are based on the fact that an adequate decrement of the applied stress leads to an inverted creep strain : then, τ_i is deduced from the applied stress leading to a zero deformation or relaxation rate. These methods have been criticized /4,5,6/ and corrections have been proposed /7/ in order to take into account the anelasticity phenomenon which occurs when the deformation rate approachs zero.

The aim of this paper is to obtain non destructive experimental data about this anelasticity phenomenon by measuring and analysing the ultrasonic attenuation changes induced by successive decrements of the creep stress. Moreover, this analysis allows us to propose a new method for the evaluation of internal stresses.

EXPERIMENTAL RESULTS

This study was carried out on pure polycrystalline aluminium (99.999%) obtained by an extrusion followed by machining and then recrystallized at 450°C for one hour ; the grain size thus obtained is of order of 1 mm. The diameter of the useful part or a specimen is 9 mm ; the

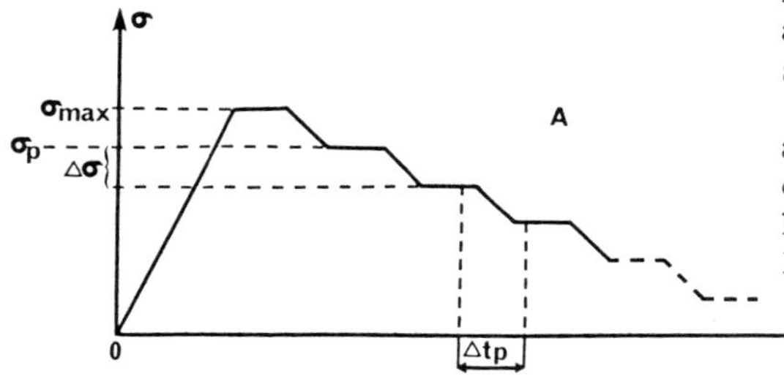

Fig. 1. A : Schematic procedure and definitions used in a step by step unloading.
B : Example of recording attenuation change versus time during the whole experiment. Plastic deformation induced by loading ε_p =0.2%.

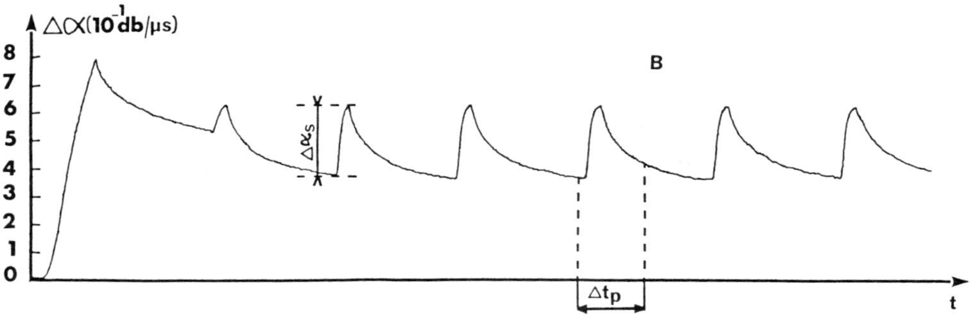

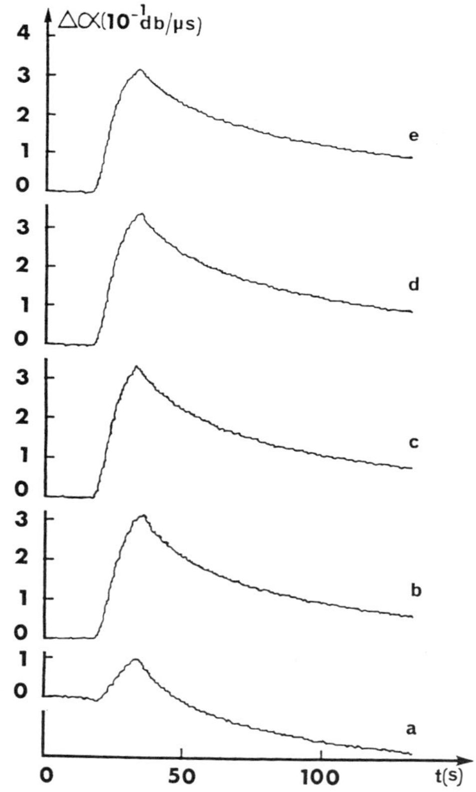

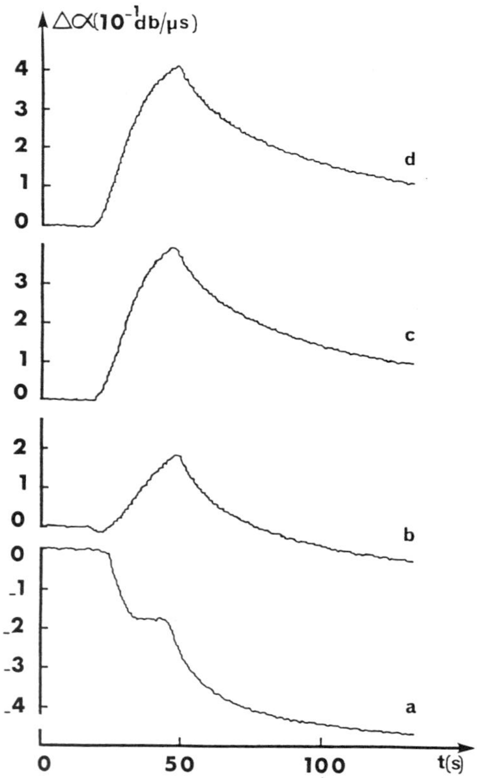

Fig. 2. Recordings $\Delta\alpha$ - t for : ε_p = 0.2%, T = 300 K, ν = 16.9 Mhz, σ_{max} = 6.5 MPa and $\Delta\sigma$= 0.3 MPa. During : a) 1st stress decrement ; b) 2nd ; c) 3rd ; d) 4th ; e) 8th.

Fig. 3. Recordings $\Delta\alpha$- t for : σ_{max} = 21 MPa and $\Delta\sigma$= 0.6 MPa. During : a) 1st stress decrement ; b) 2nd ; c) 4th ; d) 5th.

useful length, 40 mm long, is terminated at both ends by a screwed head sized 12 mm in diameter. The specimen is submitted to a tensile stress through an hydraulic creep machine specially designed for low temperature tests of pure materials. Strain is measured by means of an extensometer (M.T.S. Type 632.26C) leading, via an appropriate commercial amplifier, to a resolution of 10^{-6}. Simultaneously with the creep test, an ultrasonic longitudinal pulsed wave, which frequency ν is about 16.5 Mhz, is sent periodically through the specimen with a propagation direction parallel to the tensile axis ; the emission and the reception of the waves are obtained by means of two quartz transducers glued on the free, flat, parallel sides of each head of the specimen. Changes in the amplitude of the first transmitted echo are followed by a system developed in our laboratory /8/ and stored in a computer memory. Then, the corresponding attenuation variations $\Delta\alpha$ are computed from these data with reference to the attenuation α at the beginning of the recording ; an overall resolution of about 0.001 db/μs is obtained.

RESULTS

Following the application of the initial load, the stress σ is alternatively kept constant and decreased (Fig. 1 A) ; the interval of time (t_p) elapsing between two unloadings as the load decrement itself ($\Delta\sigma$) are fixed for a given overall test. The typical behaviour of ultrasonic attenuation during such a test is shown in Fig. 1 B : a large increase in attenuation is observed during the plastic deformation induced by the initial loading ; then, the attenuation decreases at constant stress and generally it increases during each unloading. The magnitude of this increase arising at each stress reduction will be hereafter labelled $\Delta\alpha_s$. Moreover, more detailed recordings, corresponding each to an interval of time (Δt_p) including an unloading time, are reported in Fig. 2 to 4 :

- the recordings of $\Delta\alpha = f(t)$ are given in Fig. 2 for five stress reductions carried out from a maximum stress σ_{max} = 6.5 MPa which induced initially a plastic deformation ε_p amounting to 0.2%. The most important features of these data are that an attenuation increase is observed from the first load decrement and $\Delta\alpha_s$ goes through a maximum as a function of the number of the load decrement.
- the recordings drawn in Fig. 3 are obtained similarly to the preceding ones but from $\sigma_{max.}$ = 21 MPa and ε_p = 2.4 % : on the contrary to that was observed from ε_p = 0.2%, the attenuation decreases steeply during the first stress reduction ; then, during the subsequent unloadings the classical attenuation increase is observed again and $\Delta\alpha_s$ goes through a maximum.
- in order to show the temperature effect, three attenuation recordings, obtained in same conditions but at T=325K, T=271K and T=219K, are presented for equivalent unloadings in Fig. 4 : it can be seen that the magnitude of the attenuation jump diminishes as the temperature is decreased, to vanish almost completely at 219 K.

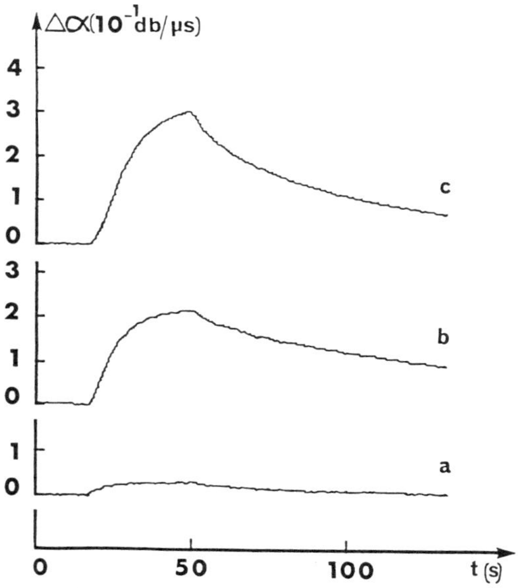

Fig. 4. Effect of temperature on the attenuation increase during the 4th stress decrement ($\varepsilon_p \simeq 1.4$ %) = 0.6 MPa) a) T = 219 K, ν = 16.6 Mhz ; b) T = 271 K, ν = 16.2 Mhz ; c) T = 325 K, ν = 16.1 Mhz.

Moreover, various experiments, not reported in details there, have been carried out in order to study the influence of other experimental conditions upon the attenuation variations : especially, the interval of time (t_p) between two unloadings, the stress decrement ($\Delta\sigma$), and the repetition of the procedure effects have been investigated. If, $\Delta\alpha_s$ depends slightly on these parameters, especially on the magnitude of the stress reduction, the general behaviour of the attenuation evolution remains qualitatively identical to that above described.

Finally, in order to determine the creep mechanical characteristics some experiments have been carried out without any stress reduction subsequent to the initial loading : the creep curves thus obtained are shown in Fig. 5.

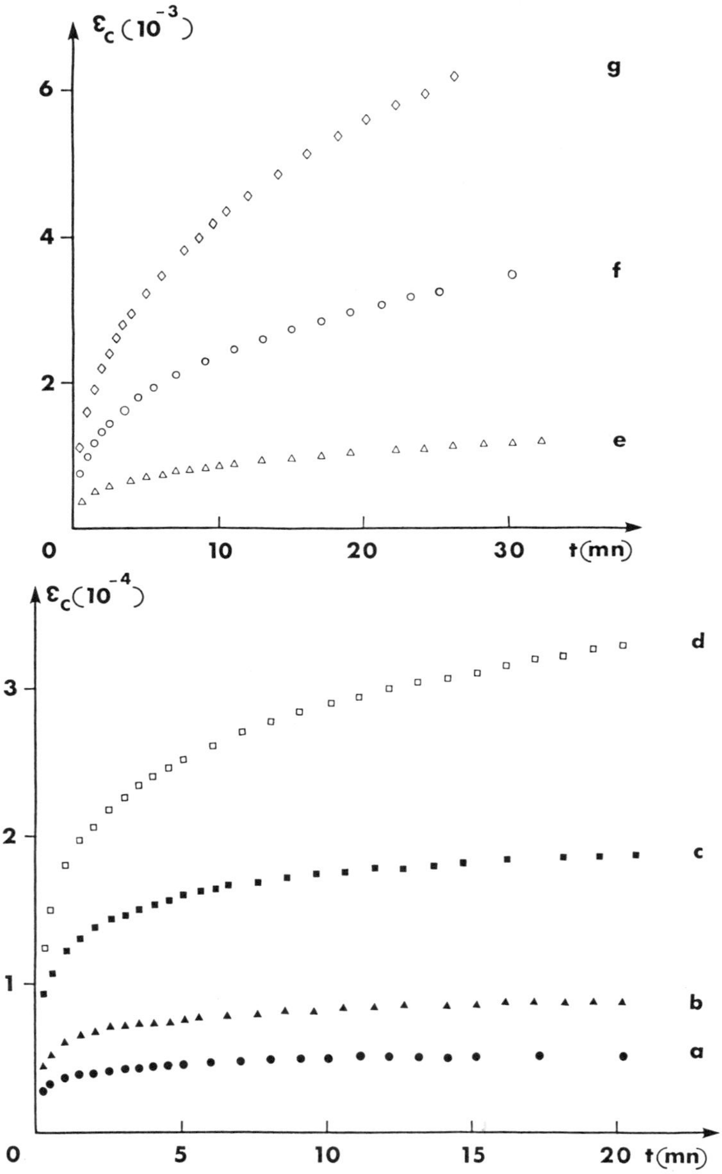

Fig. 5. Creep strain vs time for various initial plastic deformation : a) ε_p = 0.1 % ; b) 0.2 % ; c) 0.4 % ; d) 0.7 % ; e) 1.4 % ; f) 2.4 % : g) 4.8 % (a,b,c pure logarithmic creep ; d,e,f,g primary stage of recovery creep).

DISCUSSION

Outline of the Model

In order to provide a basis for the discussion of the present results, we outline a simple model able to explain the attenuation increase observed during each stress decrement.

In the temperature range investigated in this work, the creep phenomenon is generally ascribed to dislocation movements /9/ ; then, we consider the well known dislocation-string model in order to explain the attenuation evolutions : from the work of GRANATO and LUCKE /10/, in the low megahertz range, the attenuation α can be roughly expressed as : $\alpha \simeq K_{\alpha} \Lambda L^4$ where K_{α} is a factor depending upon temperature and ultrasonic frequency, Λ is the density of mobile dislocations and L is the average free loop length.

Thus, the first attenuation increase which is observed at the beginning of the experiment (Fig. 1 B) is mainly due to the increase in Λ induced by the initial plastic deformation.

In other respects, as recalled in the introductive section, the effective stress τ_{eff} acting on the dislocation in its glide plane can be expressed as $\tau_{eff} = \sigma.f - \tau_i$ (1) where τ_i is the long-range internal stress opposite to the applied tensile stress (σ) and f is the Schmid factor. Moreover, the dislocation movement is not continuous but is of a jerky nature : this is due to the fact that, except at very high strain rates, the dislocation motion is controlled by the thermally assisted jumping over some hard obstacles /11/.

So, during the constant stress time, the dislocations are immobilized before overcoming these obstacles. That enables such dislocation loops to be reached by some point defects which are mobile in aluminium at room temperature /12/ ; this mechanism is schematized in Fig. 6 (A→B). Then, during a stress decrement the line tension tends to move the dislocation loop backwards so that three events may occur for a dislocation loop pinned in the above way by weakly interactive point defects :

i) for a small stress decrement and at low temperature the dislocation loop remains pinned by the immobile point defects.

ii) for a suitable stress decrement and an intermediate temperature, the breakaway of the loop from pinning defects occurs (Fig. 6 B→C).

iii) at higher temperature, the pinning defects are mobile enough to be dragged transversely by the dislocation during its movement induced by the stress decrement.

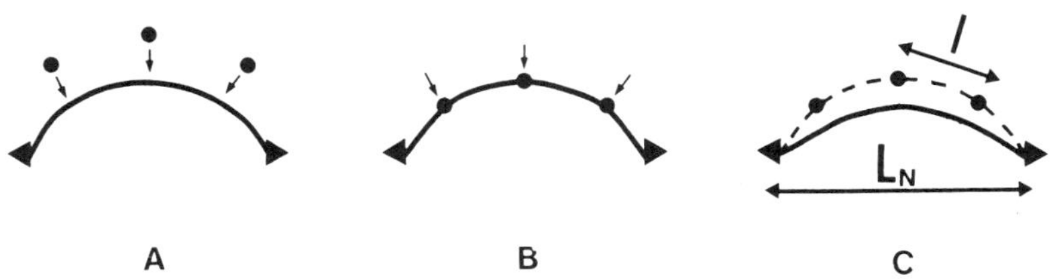

Fig. 6. Dislocation pinning and depinning. A : a free dislocation loop at the end of plastic loading ; B : pinned state at the end of constant stress time ; C : depinning induced by the stress decrement.

In the case of aluminium at room temperature, it has been established in previous studies /13/ that, among these mechanisms, the breakaway is the only one that leads to an increase in ultrasonic attenuation ; then, the attenuation change can be written as : $\Delta\alpha \simeq K_{\alpha} \Lambda_y (L_N^4 - l^4)$ (2) where Λ_y is the dislocation line density concerned with the breakaway (one has to distinguish Λ_y from Λ because of dislocations that are not reached by point defects and because of those concerned with the above situations (i) and (iii)), L_N is the loop length between hard obstacles (nodes or jogs) and l is the free loop length between smooth pinning points (point defects).

Therefore, Fig. 3 (b,c,d) allows us to conclude firmly that dislocation-point defect breakaways occur during the successive unloadings of the specimen.

On the evolution of the attenuation change ($\Delta\alpha_s$).

As soon as a dislocation major loop has been pinned by a first point defect, the relation $L_N^4 \gg l^4$ is satisfied. Then, from our model (relation (2)) it follows that $\Delta\alpha_s$ is dependent mainly on Λ_y. For a given initial plastic deformation (ε_p) this dislocation density concerned with the breakaway depends itself upon :

i) the probability that a motionless dislocation loop will be reached by a mobile point defect (Fig. 6 B) : on one hand this probability should be decreased by an increase in the delay time (t_m) necessary for a point defect to reach the loop by a diffusion process ; on the other hand this probability should be increased by an increase in the waiting time (t_w) of a dislocation line in front of a discrete hard obstacle. Moreover, because of the thermally-assisted character of both processes, t_w and t_m are decreased as the temperature is increased ; in the case of t_m, this effect is clearly shown in Fig. 4 : indeed, from the temperature T = 219 K at which point defects associated with impurities are still immobile in 99.999% pure aluminium /12/ to the temperature T = 325 K, $\Delta\alpha_s$ is multiplied by a factor 3.5 whereas K_α is varied only in the ratio of one to 1.5 /14/. Hence, an increase of Λ_y in the ratio of one to 2.3 is necessary in order to explain this evolution of $\Delta\alpha_s$ from 219 K to 325 K : this effect is mainly due to an increase in the bulk mobility of point defects thus making faster their arrival on dislocations.

In other respects, the waiting time t_w can be expressed as :

$$t_w = (1/\nu_o) \exp (\Delta G_o - \tau_{eff} \cdot b \cdot \Delta A/kT) \qquad (3)$$

where ν_o is the attempt frequency, ΔG_o is the activation energy to overcome the barrier, ΔA is the activation area and k is the Boltzmann constant.

Moreover, it must be noticed, that some influence upon ΔG_o is expected from the point defects that have reached the motionless dislocation. This complex question has not been solved exactly up to now. Nevertheless, additional pinning points will undoubtly lead to an overall stronger anchoring of the loop, so that one can expect longer waiting times, t_w, for smaller migration times.

ii) the probability that the stress decrement induces the breakaway of these dislocations : in the temperature range investigated in this study (i.e. around room temperature) a competition arises between the breakaway and dragging mechanisms /13/ ; thus, the breakaway probability appears to be a complicated function of T and magnitude of stress variation /14/.

To sum up, the density of dislocations concerned with the depinning process (Λ_y), as the associated attenuation increase ($\Delta\alpha_s$), are complicated functions of temperature (T), effective stress (τ_{eff}), magnitude of stress decrement ($\Delta\sigma$) interval of time t_p and initial plastic deformation (ε_p).

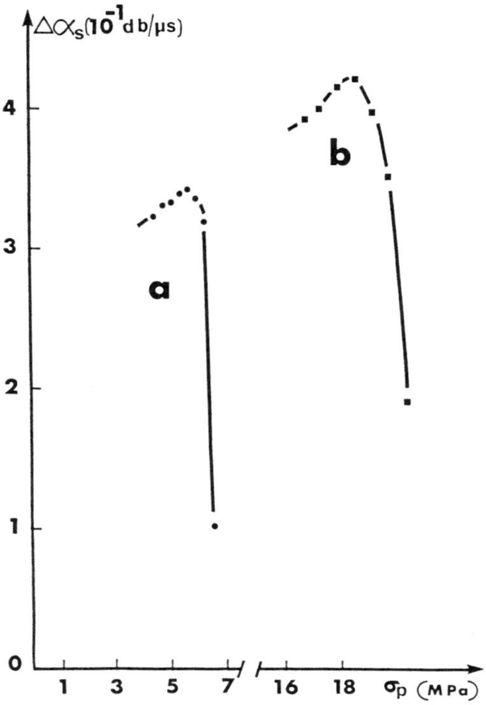

Fig. 7. Magnitude of attenuation increase vs stress at T = 300 K ; a) ε_p = 0.2 % B) ε_p = 2.4 %

In spite of this complexity, as a first approach, the evolution of $\Delta\alpha_s$ reported in Fig. 7 as a function of the stress σ_p, from which a stress reduction is applied (see Fig. 1), can be easily understood on the following basis : each test was carried out at constant temperature with identical successive stress decrements and intervals of time t_p between these ; moreover, in the temperature range which has been investigated, the dislocation microstructure induced by the initial plastic deformation is not expected to be profoundly affected during the test : indeed, the creep process is noticeable only during the short interval of time t_p elapsing between the end of plastic loading and the first unloading step. This latter assumption allows us to conclude that the evolution of $\Delta\alpha_s$ is mainly governed by the variation of the effective stress which remains the only one parameter varying strongly during such a test. Now, it must be considered that t_m and t_w are distributed each around an average value $\overline{t_m}$ and $\overline{t_w}$: although at present, we are not able to precise the laws governing these distributions, it can be expected that the more highly the relation $\overline{t_m} > \overline{t_w}$ will be satisfied, the lower the density Λ_y will be ; on the contrary, if $\overline{t_w}$ becomes larger than $\overline{t_m}$ most of the dislocation loops will be pinned by at least one point defect during their motionless waiting time on hard obstacles, so that Λ_y will tend towards the total density of mobile dislocations. From these arguments, it can be deduced, at least approximatively, that Λ_y will be maximum as $\overline{t_w}$ will be also maximum. Finally, let us consider relation (3) showing that t_a will take its maximum value as τ_{eff} = 0.

Thus from all these arguments it can be concluded that the maximum of $\Delta\alpha_s$ plotted as a function of σ_p (Fig. 7) arises from this process : that is to say the stress σ_p, from which the unloading jump leads to the maximum value of $\Delta\alpha_s$, is roughly equal to the average internal stress $\overline{\sigma_i}$ = $\overline{\tau_i/f}$ acting backwards on the dislocation movement. This conclusion is in good accordance with previous experiments carried out by a classical dip test method on the same aluminium at room temperature /15/ : it was established the occurence of a backward strain as the creep stress is reduced below a critical value (i.e. approximatively the internal stress). Besides the distribution functions of f, ΔG_o and ΔA, a further more accurate analysis of the internal stress should include the influence of the stress decrement characteristics (magnitude and rate) in order to take into account the fact that at lower unloading rates the dragging mechanism should compete more and more efficiently with the breakaway process : in the present experimental conditions, our previous work on breakaway and dragging mechanisms /13/ allows us to estimate that about three quarters of all the pinned loops are concerned with the breakaway mechanism, and the remaining part with the dragging mechanism.

Internal stresses

The values of $\overline{\sigma_i}$ obtained through the method above described are $\overline{\sigma_i}$ = 4.4 MPa ; 5.6 MPa ; 7.3 MPa and 18.6 MPa respectively for ε_p = 0.1 % ; 0.2 % ; 0.4 % and 2.4 % of plastic deformation induced by the initial loading. The direct comparison of these values with those obtained through other methods is difficult because the experimental conditions are noticeably different (torsional creep and dip test method in reference /15/, single crystals and stress relaxation in reference /16/. Nevertheless, let us consider the ratio $\overline{\sigma_i}/\sigma_{max}$ (where σ_{max} is the applied stress that induces the initial plastic deformation (ε_p) of the specimen) : then we obtain ratios around 0.85, sligthly lower but in accordance with those deduced from references /15/ an /16/ respectively about 0.95 and 0.9.

Anomalous attenuation change

An anomalous attenuation decrease induced by the unloading has been reported in Fig. 3 : with our experimental conditions this phenomenon arises only during the first stress decrement and if the amount of initial plastic deformation exceeds about 0.4 %. Moreover the magnitude of this phenomenon is increased as a function of plastic deformation. These latter features lead us to analyse the creep curves given in Fig. 5 : this was done through the differential method proposed by CRUSSARD /17/ assuming a creep law $\varepsilon_{creep} = \beta t^m$.

For $\varepsilon_p \lesssim 0.4$ % the value m = o is obtained, thus indicating that the creep is purely logarithmic.

Above $\varepsilon_p \simeq 0.4$ %, m is observed to increase graduately up to the value $m \simeq 0.33$ which is reached for $\varepsilon_p \simeq 2$ %. This correlation between the appearance of an anomalous attenuation decrease and a transition from logarithmic to power law creep suggests that this anomalous behaviour could be associated to the occurence of new mechanisms for dislocation motions ; then two kinds of mechanisms could explain this attenuation decrease :

(i) during the first creep time preceding the first stress decrement, the mobility of dislocations becomes so high that $\overline{t_w} \ll \overline{t_m}$, then Λ_y should nearly equal zero ; consequently during the first stress decrement $\overline{t_w}$ could increase significantly, thus allowing some fixed dislocations to be pinned by point defects during the unloading itself.

(ii) if it is assumed, as often proposed in the review by OIKAWA and LANGDON /18/, that the primary creep stage is associated to a competition between creation and annihilation of dislocations, the stress reduction could stop rapidly the creation process whereas annihilations could go on : in this case the attenuation decrease would be associated to a dislocation density lowering.

A better identification of these mechanisms which contribute to the observed anomalous attenuation decrease requires further experimental data providing some additional characteristics of this phenomenon : this study is in the course.

REFERENCES

/1/ A.A. SOLOMON, New techniques and apparatus for examining the elevated temperature deformation of Metals, Rev. Sci. Instrum. 40, 1025, (1969).

/2/ C.N. AHLQUIST and W.D. NIX, A technique for measuring mean internal stress during high temperature creep, Scripta Met., 3, 679 (1969).

/3/ J. LEPINOUX and L.P. KUBIN, In situ T.E.M. observations of the cyclic dislocation behaviour in persistent slip bands of copper single crystals, Phil. Mag. A, 51, 5, 675 (1985).

/4/ G.J. LYOYD and R.J. Mc ELROY, On the anelastic contribution to creep, Acta Met., 22, 339 (1974).

/5/ P. PAHUTOVA, M. CADECK and P. RYS, Anelasticity and measured internal stress in high temperature creep, Scripta Met., 11, 1061, (1977).

/6/ F. DOBES, The influence of anelasticity on the measurement of internal stress in creep, Scripta Met., 15, 215 (1980).

/7/ J. CHICOIS, A. HAMEL, R. FOUGERES et J. PEREZ, Estimation expérimentale de l'influence de l'anélasticité sur la mesure des contraintes internes en fluage à basse temperature, Scripta. Met., 15, 599 (1981).

/8/ A. VINCENT, J.L. BOUVIER VOLAILLE et P. FLEISCHMANN, Utilisation d'un microordinateur pour l'étude des variations rapides de l'atténuation et de la vitesse des onde ultrasonores. J. Phys. E, 15, 765 (1982).

/9/ O.D. SHERBY, J.L. LYTTON and J.E. DORN, Activation energies for creep of high-purity aluminium, Acta Met., 5, 219 (1957).

/10/ A. GRANATO and K. LUCKE, Theory of mechanical damping due to dislocations, J. Appl. Phys. 27, 6, 583 (1956).

/11/ G.B. GIBBS, The thermodynamics of thermally-activated dislocation glide, Phys. Stat. Sol., 10, 507 (1965).

/12/ A. VINCENT, S.M. SEYED REIHANI and J. PEREZ, The nature of dislocation-pinning defects in cold-worked aluminium studied by ultrasonic waves. Phys. Stat. Sol. (a), 39, 651 (1977).

/13/ A. VINCENT et J. PEREZ, Etude de l'interaction dislocation-défauts ponctuels par méthode ultrasonore sous contrainte quasi-statique, Phil. Mag. A, 40, 3, 377 (1979).

/14/ G. GREMAUD, Complex interaction mechanisms between dislocations and point defects involving simultaneously depinning-repinning and dragging processes, J. de Physique C9, 12, 44, 607 (1983).

/15/ J. CHICOIS, A. HAMEL, R. FOUGERES, C. ESNOUF, G. FANTOZZI and J. PEREZ, Internal Stress measurement in aluminium by DIP TEST method and correlation with the Bordoni relaxation evolution, J. de Physique, C5, 10, 42, 169 (1981).

/16/ A.A. ALY, T.G. ABDEL-MALIK, A.M. ABDEEN and H.M. ELLABANY, Stress relaxation in single crystals of copper and aluminium, Mat. Sci. and Engng 62, 181 (1984).

/17/ C. CRUSSARD, Etude rhéologique du fluage et de l'hystérésis mécanique des matériaux, Métaux, Corrosion, Industrie, 299 (1963-1964).

/18/ H. OIKAWA and T; LANGDON, The creep characteristics of pure metals and metallic solid solution alloys, in : "Creep behaviour of crystalline solids", B. WILSHIRE and R.W. EVANS ed., Pineridge Press, Swansea (1985).

ND DETERMINATION OF GRINDING DEFECTS

USING THE 3MA PROTOTYPE DEVICE

Reinhard Conrad, Rudi Jonck, Werner Theiner, and
Bernd Reimringer
Robert Bosch GmbH, Stuttgart, FRG
IzfP Fraunhofer Institut, Saarbrücken, FRG

INTRODUCTION

Full automatic inspection in production line is already a dream for many years. Especially for production engineers, who are engaged with components characterized by great accuracy of manufacture and by high stress in operation.
On Fig. 1 and Fig. 2 you can see an example:

Fig. 1: Distributor injection pump

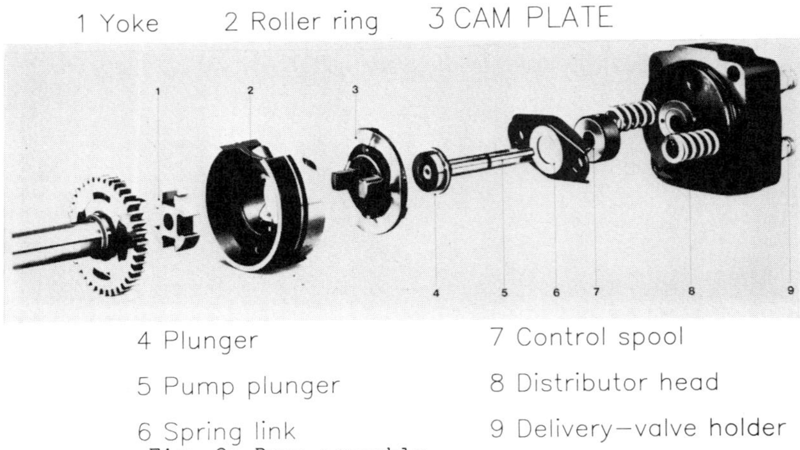

Fig. 2: Pump assembly

In Fig. 1 is shown a distributor injection pump, which is used in passengers cars, trucks and other engines. In Fig. 2 is shown the corresponding pump assembly. Our testing results are related to part number 3, the cam-plate. The lobes on the surface of the cam-plate convert the input-shaft rotation into pump-plunger reciprocation.
You must know that all these parts are ground and that grinding is one of the most relevant machining processes for surface finishing. If one of these machining parameters is miscarried, the microstructure and the residual stress states are changed in near surface zones. By the new nondestructive (nd) testing method, described and used in this paper, it is possible to evaluate microstructure gradients in the range of 3 µm up to 200 µm with a spatial resolution of 500 µm.

MICROSTRUCTURE IN HEAT-AFFECTED ZONES

In Fig. 3 is shown the metallographical examination of a ground workpiece and the corresponding hardness profile. The grinding conditions were not fulfilled. In near surface zones one can see a new hardened

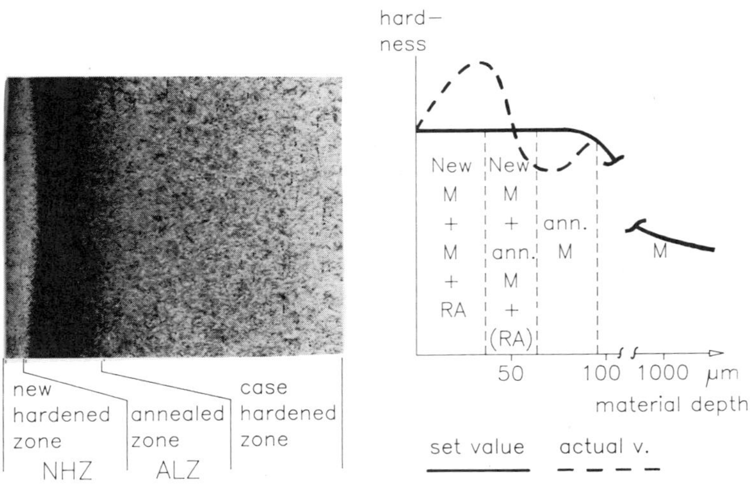

Fig. 3: Metallographic section and hardness profile of a case-hardened and ground 16 MnCr 5 part

zone (NHZ: white region), in deeper sections an annealed zone (ALZ: darker metallographic section), whereas the case hardened microstructure states extend up to more than 1000 µm. The hardness profile is schematically shown on the right side. The broken line means the actual shape caused by grinding defects and the other one the nominal shape. Moreover one can see the different microstructure states: martensite, annealed martensite and retained austenite. Our new ndt-method, the incremental permeability μ_Δ, was applied to evaluate these grinding defects. It was developed in co-operation of the Fraunhofer Institut fuer zerstoerungsfreie Pruefverfahren (IzfP) and the Robert Bosch GmbH.

DESCRIPTION OF THE EQUIPMENT

The incremental permeability can be measured with a conventional eddy current equipment with and without a magnetic field excitation. According to Fig. 4 the workpiece undergoes cyclic magnetization. This magnetization is superimposed an alternating magnetic field in a wide frequency range from 10 kHz up to 35 MHz. By measuring the probe impedance variations of the magnetic permeability of the test object can be recorded and correlated with the impedance during the magnetic excitaton. For more information about theory and so on I refer to the contribution of Dr. Theiner: "Determination of sub-surface microstructure states by micromagnetic ndtechniques" elsewhere in these proceedings.

Our measuring quantities can be deduced from the μ_Δ (H)-curves:

the maximal value	$\mu_{\Delta max}$
the coercive force from the $\mu_{\Delta max}$ position	$H_{c\mu}$

Both quantities are measured over the tangential magnetic field strength H, which can be detected by Hall-probes.

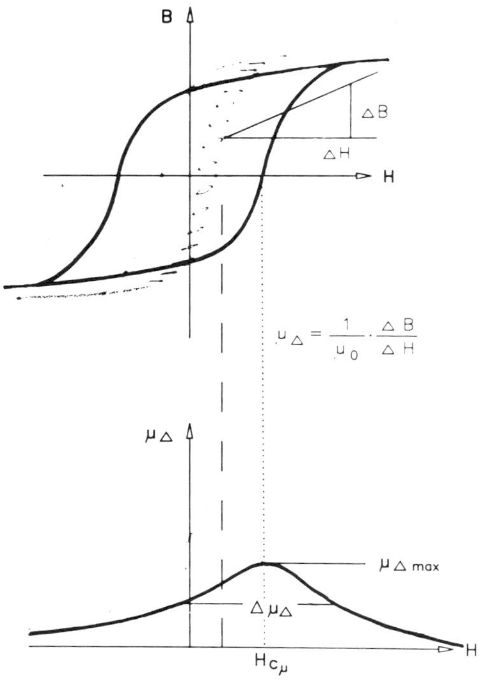

Fig. 4: Incremantal permeability μ_Δ (H)

Fig. 5: Outlay of the prototype

On Fig. 5 one can see the prototype. On the bottom there is the power supply and over that the master unit with the operator control panel. For operation and controlling there are only 6 softkeys. The handling is menue-guided, so no special training of the operator is necessary.

On Fig. 6 is shown the sensor manipulation built in the magnet-yoke.The sensor tests the lobeslide-way of the cam-plate and after rotation of 90 degrees the 4 dog-surfaces too. The number of lobes depends on the engine type. By a switch one can select a 4 or 6 cylinder cam-plate.

Fig. 6: Outlay of the sensor

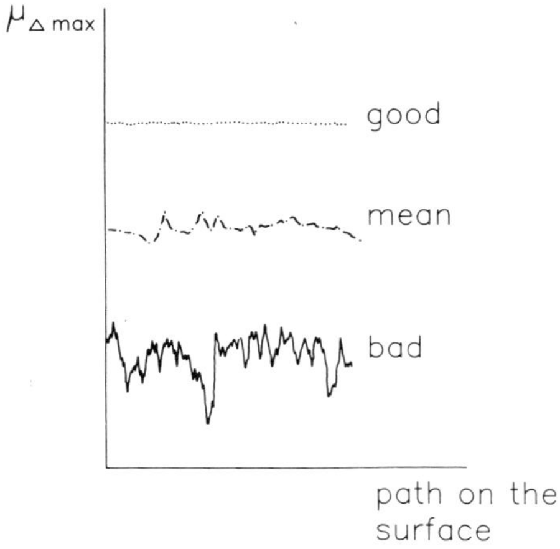

Fig. 7: Examination of grinding states, step 1

MEASURING METHOD

Fig. 7 and 8 are schematic views of our measuring method. In a first step we detect variations of the absolute value of the μ_{Δ}-measuring quantity as a function of the path on the workpiece surface. From the extent of the variations we get an information about the quality of our part. In a second step we evaluate the local depth of the heat affected zone from the $H_{c\mu}$-profile curve. The depth of the local new hardened zone (NHZ) is correlated with the maximum of the $H_{c\mu}$-profile curve. We get the depth of the annealed zone (ALZ) by using the point of intersection of a $H_{c\mu}$ = const line with a profile curve. The most important measuring variable is the frequency f_{Δ} of the superimposed alternating field, which is strictly correlated with the information about the depth of the heat-affected zone.

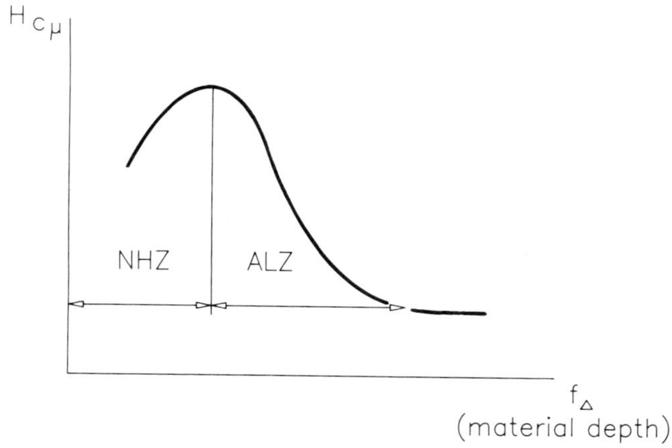

Fig. 8: Examination of grinding states, step 2

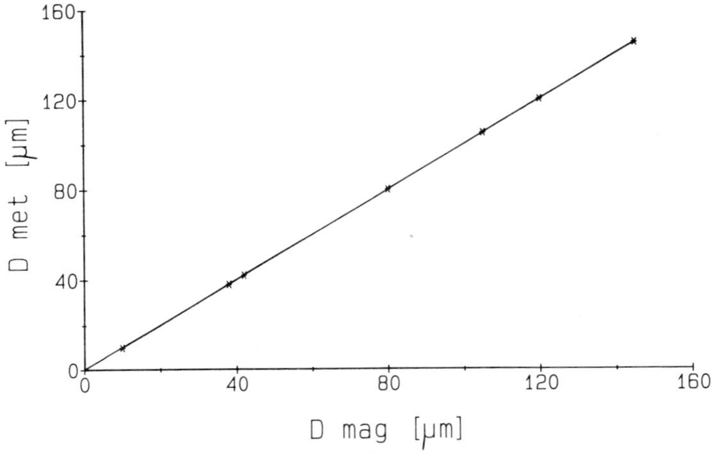

Fig. 9: Grinding defect depth D: metallographic D_{met} versus magnetic D_{mag}

Using this procedure one see in Fig. 9 that the magnetic evaluated grinding defect depth D_{mag} correlates quite resonably with the metallographically determined values D_{met}:

MEASURING RESULTS

The incremental permeability testing results of three different grinding states are shown in Fig. 10. The $H_{c\mu}$-profiles demonstrate that the incremental permeability values of the coercitivity $H_{c\mu}$ are sensitive to microstructure profiles up to 200 µm. "B" denotes a grinding state with a NHZ ~ 60 µm and an ALZ ~ 160 µm, "C" with a NHZ ~ 25 µm and an ALZ ~ 100 µm, and "A" is characterized by soft ALZ - microstructure states up to the surface.

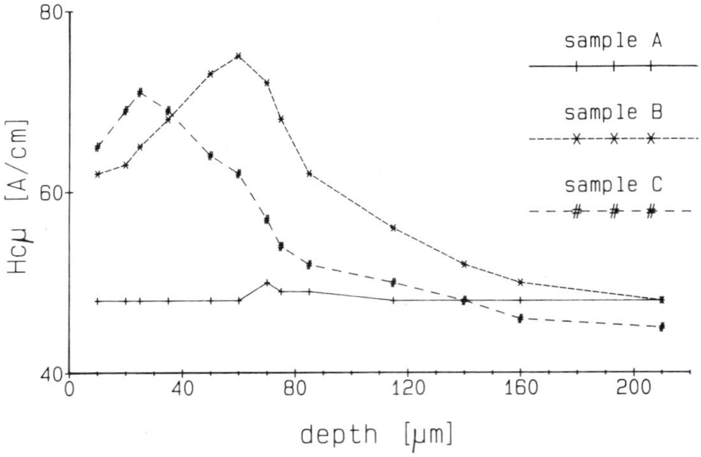

Fig. 10: Magnetically evaluated hardness profiles of 3 different grinding states

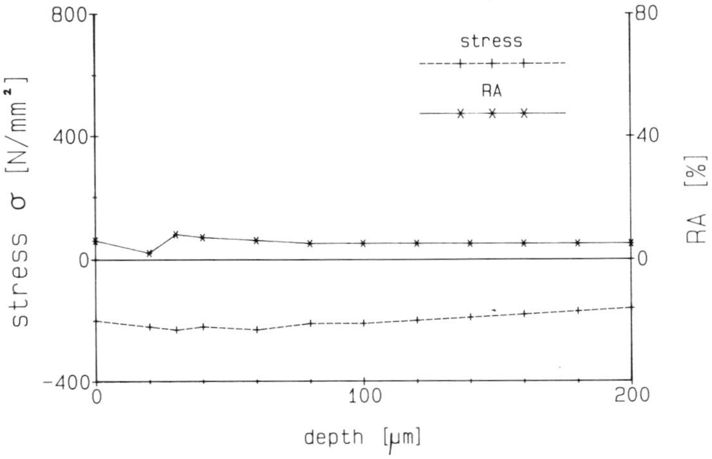

Fig. 11: Retained austenite RA and residual stress before grinding

Besides these local extensions of grinding defects our testing aim is the residual stress and the content of microstructure states. In Fig. 11 and 12 one can see the content of retained austenite and the residual stress distribution before and after grinding.
Before grinding there is a nearly constant part of retained austenite and a mean value of ~ 200 N/mm² stress from a depth of 0 µm up to 200 µm. After grinding one can see great variations. Especially the amount of retained austenite goes up to 50 % in surface near zone. By measuring the mean value of the incremental permeability as a function of the grinding defect depth it seems possible to separate workpieces with a different content of retained austenite as one can see in Fig. 13.

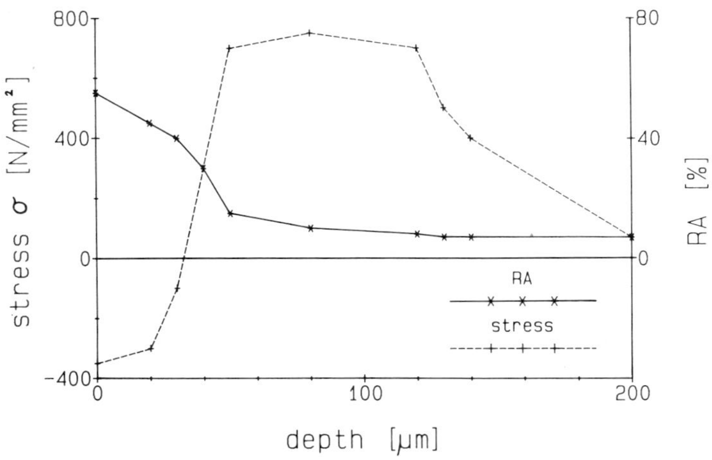

Fig. 12: Retained austenite RA and residual stress after grinding

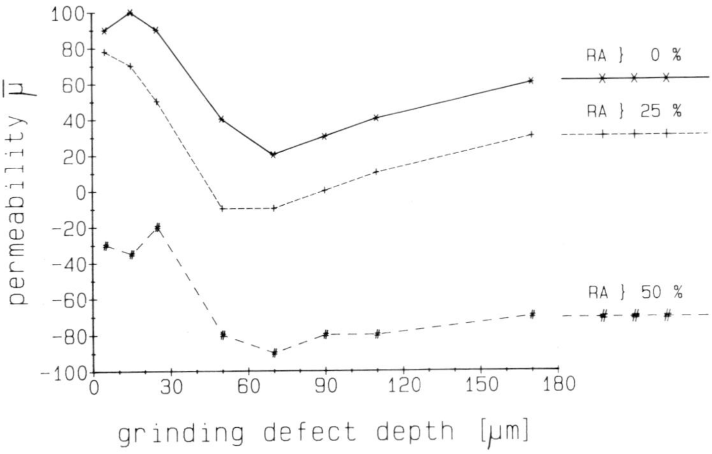

Fig. 13: Permeability, grinding defect depth and retained austenite

But you must know that these results are very preliminary, we have to prove this by statistics.

CONCLUSIONS

It has been demonstrated that the new nd-testing method, the incremental permeability, can be used to detect grinding defects. We are hopeful to have soon a complete equipment in order to solve the problems of nd-testing of microstructure and stress. At last I will give you the meaning of the word 3 MA. It comes from the 3 initial letters M and the initial letter A in the name of our future equipment: Micromagnetic Mulitparameter Microstructure and residual stress Analysis.

REFERENCES

1. R. CONRAD, R. JONCK, W.A. THEINER, Zerstörungsfreie Ermittlung von wärmebeeinflußten Randschichten und deren Dicke, Journal HTM, 4: 213 - 217 (1986).

2. R. CONRAD, R. KERN, W.A. THEINER, Determination of Surface Integrities by Ferromagnetic Quantities, Conference contribution to: "First International conference on surface engineering", Brighton (Great Britain), 25. - 28. June 1985.

3. E. BRINKSMEIER, E. SCHNEIDER, W.A. THEINER, K.H. TÖNSHOFF, Nondestructive Testing For Evaluating Surface Integrity, Annals of the CIRP, 33 (2): 489 - 509 (1984).

4. I. ALTPETER, W.A. THEINER, Determination of Residual Stresses Using Micromagnetic Parameters, in: "New Procedures in NDT", P. Höller, ed., Springer-Verlag, Berlin (1983).

EVALUATION OF CABLE AGING BY WATER EXTRACTION OF SOLUBLE IONS

J.C. Garcia, J.P. Crine, R. Gilbert,
E. Sacher[+] and J.C. Portal[x]

Institut de recherche d'Hydro-Québec
[+]Varennes, Canada
[x]Ecole Polytechnique, Montréal, Canada
Institut National des Sciences Appliquées, Toulouse, France

INTRODUCTION

The major concern for all dielectric users is the aging of electrical insulation. This is especially true for high-voltage cable insulation, where aging results in changes in the material properties. A specific problem encountered with high-voltage cables is a degradation phenomenon known as treeing which is particularly detrimental to cable lifetime in the presence of water and impurities[1]. It is also obvious that the presence of impurities in an insulation will reduce its electric breakdown strength. In a previous study[1], it was shown that cable insulations contain several ppm of various impurities and additives, some of which were ionized (see Table 1). It was also observed that aging generally leads to an increase in the impurity contents[2] caused by contaminant diffusion from the heavily contaminated semiconducting shields. Ions are certainly not welcome in any insulation system and it is of the utmost importance to know and control their contents.

Until recently relatively low ionic contents were difficult to evaluate but with the advent of ion chromatographs, this is now a much more easier task. These instruments were originally designed to measure ions in aqueous solutions but water-extractable ions (i.e. those easily solubilized by water) in solids can be estimated as well. Since our preliminary measurements have shown that ions can indeed be detected in cable insulations, we decided to perform more detailed ionic measurements on the base polymer resins and on the additives used by cable manufacturers. The main objective pursued in this research was to evaluate ion contents in cable insulation and their evolution with time in order to be able in the future to predict cable lifetime under service conditions.

PRINCIPLES OF IONIC CHROMATOGRAPHY

Basically, ionic chromatography (IC) combines the separation ability of ion-exchange resins with the evaluation of the content of these ions from their electrical conductivity (see Fig. 1). The aqueous sample is eluted through a column filled with a resin whose ion-exchange reaction consists in substituting an ion of the resin (stationary phase) with an ion in solution (mobile phase). The various ions remain bonded to the ion-exchange sites for a given time, the so-called retention time, which differs for each ion. This allows an easy identification of the nature of each ion in a chromatogram, such as the one shown in Fig. 1.

TABLE 1

CONTAMINANTS IN XLPE RESINS (ppm)

RESINS	Aℓ+Si	Cℓ	Na	Ca	S
XLPE uncured*	0.2	1.0	—	—	—
XLPE fully cured	0.93	10.0	7	—	260
25% carbon black XLPE uncured	65	55	315	240	6560
" fully cured	69	70	—	—	—

* without peroxide

IONS (ppm) IN CABLES AND RESINS

	CABLES		XLPE RESINS	
	#1	#5	Pure	25% carbon black
Fluoride	0.05	0.1	1.27	1.28
Chloride	1.84	0.62	0.65	0.62
Sodium	0.81	0.48	0.03	0.34
Potassium	0.75	0.45	—	0.15

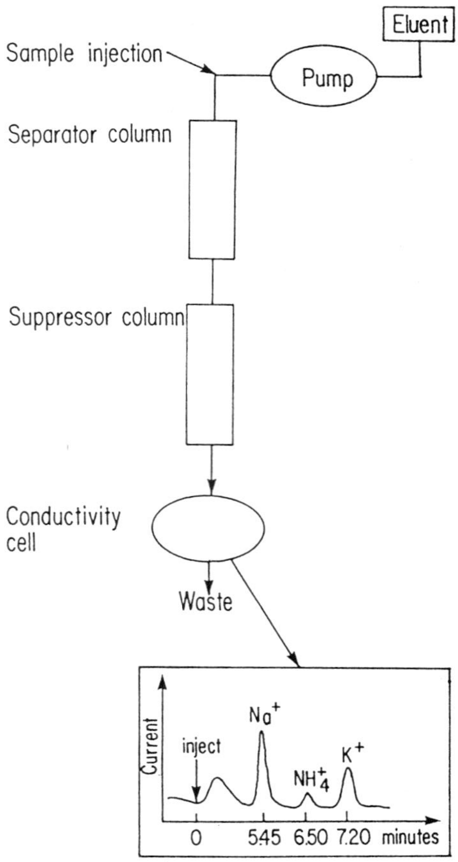

Figure 1 : Schematic diagram of an ion chromatograph and a typical chromatogram obtained with a PE resin.

The ion chromatograph used was a Dionex 16 equipped with standard cationic and anionic columns as well as with Ice columns for the identification of organic ions. Cleaning and column-reactivation procedures suggested by Dionex were strictly followed. The water used for extraction was triply distilled (conductivity $\simeq 10^{-6}$ $(\Omega cm)^{-1}$) and it contained trace ions of the order of 1 ppt (ng/ℓ).

SAMPLES PREPARATION

Commercially available additive-free polyethylene (PE) and crosslinkable PE (XLPE) resins were used. The latter resin contains about 4000 ppm of an antioxidant compound and 2% of a crosslinking agent. In order to evaluate the influence of the antioxidant, another PE sample was prepared with 1500 ppm of a sulfur-containing antioxidant (Irganox 1035). Soluble ions were extracted by soaking the samples (in pellets form) in triple distilled water at ≃ 50°C for ≃ 2 weeks. This relatively mild temperature proved to be high enough to significantly increase the rate of ion extraction without thermally degrading neither the samples nor the containers. The samples and the water were kept in linear PE-bottles, which are known to generate few ions[3]. The contamination of water by the bottles was estimated from blanks bottles filled with water and kept in the same conditions than the other bottles.

Elemental impurity contents in the solid samples were determined by standard neutron activation analysis (NAA) techniques before and after water extraction. In addition, some measurements were performed by plasma-coupled emission spectroscopy (PES) on the water used for extraction; it was used especially for elements easily detected by NAA (such as silicon or iron).

EXPERIMENTAL DIFFICULTIES

Efficiency of water extraction

Although the simple procedure used here (i.e. ion-extraction by soaking samples in water) has been shown to be surprisingly effective, it is clear that only a fraction of the ionic content is detected. The results of Table 2 indicate that after 2 weeks in water the samples have lost a significant amount of impurities; the only exception is Aℓ whose content increases after extraction, which is likely to be due to uncontrolled external contamination. Only a limited fraction of metals and silicon are extracted; identical results were obtained with 3 cables soaked in water at 50°C for 2 weeks (Table 3). Alkalis, being highly soluble are most easily extracted. The results of Tables 1 and 2 suggest that other impurities may be grafted or their movement is restricted by the PE chains. An efficient water extraction of impurities therefore implies long extraction times. Using high pressure could possibly significantly improve ion extraction. High temperature could also improve extraction but could easily cause oxidation and therefore formate and propiorate ions, which is not desirable. Another method to improve the extraction is to use samples with the largest possible A/V ratio (where A is the exposed area and V is the volume). For cable measurements, it is then suggested to tool-cut a ribbon (≃ 100 μm thick) from the isolation. This thin and long sample will be very well suited for water extraction.

Contamination by containers

Since samples are kept for some time in the water contained in bottles, it is obvious that water could be contaminated by these containers. It was shown elsewhere[3] that the bottles generating the lowest ionic contents are polyethylene (PE) and polypropylene (PP) bottles. Specially treated borosilicate glass bottles may be used but Pyrex and Teflon-FEP bottles should be rejected.

TABLE 2
IMPURITIES MEASURED IN VARIOUS SAMPLES (BY NAA) BEFORE AND AFTER SOAKING IN WATER AT 50°C FOR 2 WEEKS AND IONS CONTENT IN WATER (BY PES)

		ADDITIVE-FREE PE	PE + 1500 PPM A.O.	XLPE
AL	Before	3.5 ppm	6.8 ppm	5 ppm
	After	9.1 ppm	7.5 ppm	-
CU	Before	4.8 ppm	1.7 ppm	3 ppm
	After	2.7 ppm	1.1 ppm	-
CA	Before	8.4 ppm	16.4 ppm	6 ppm
	After	5.7 ppm	7.6 ppm	-
NA	Before	6.9 ppm	14.2 ppm	5.5 ppm
	After	4.1 ppm	9.9 ppm	-
	Ions	90 ppb	675 ppb	10 ppb
K	Before	2.9 ppm	7.8 ppm	2.5 ppm
	After	3.4 ppm	7.3 ppm	-
	Ions	90 ppb	545 ppb	10 ppb
CL	Before	9.3 ppm	21 ppm	10 ppm
	After	8.5 ppm	16.5 ppm	-
	Ions	89 ppb	900 ppb	625 ppb
S	Before	<30 ppm	130 $\pm$ 46 ppm	230 $\pm$ 40 ppm
	After	<30 ppm	120 $\pm$ 45 ppm	-
	Ions	150 ppb	250 ppb	500 ppb
ACETATE		?	?	1500 ppb
FORMATE		15 ppb	100 ppb	115 ppb

TABLE 3
COMPARISON OF SOME IMPURITY CONTENTS IN CABLES MEASURED IN SOLID SAMPLES (BY NAA) AND IN THE EXTRACTION WATER (BY PES)

		CABLE 1	CABLE 2	CABLE 3
AL	± Solid	1 ppm	3 ppm	1.5 ppm
	Water	6 ppb	5 ppb	3.0 ppb
SI *	Solid	1320 ppm	-	200 to 300 ppm
	Water	10 ppb	-	40 ppb

±: measured in middle of insulation
* : in insulation shield.

A major problem with water extraction is that impurities extracted from the sample can themselves well be adsorbed by the container's walls (and vice versa). This means that the contents (elemental or ionic) measured can be possibly seriously in error if the containers adsorbs or desorbs ions with a kinetic rate faster than the impurity desorption rate of the sample. In other words, the values reported here are well possibly erroneous and well below the true ionic contents.

It has often been suggested to clean the container's walls by an acid treatment that removes metallic ions[4]. It has been shown that this introduces more ions (especially with HCL cleaning) and it is not a recommended practice for alkali-ion extraction in polymer samples[3].

Preparation of standard solutions

The ion-content estimation requires that a calibration curve be established with accurately known ionic concentrations. With inorganic ions this is straightforward and solutions in the ppm range can be prepared and kept a few days in PE bottles without inducing significant errors (due to adsorption, for example). However, with organics ions, adsorption into the bottles walls is very significant, as recently shown by Gilbert[5]. In fact, solutions in PE or borosilicate glass bottles and containing originally ≈ 150 ppb of formate or propionate ions have lost all organic ions in less than two weeks[5]. When the ion contents are in the ppm range, the change of ion content is much less important albeit not negligible. It is therefore recommended to prepare standard solution (from a master batch of ≈ 1000 ppm) every two or three (at most) days in order to maintain reliable and reproducible results.

Another difficulty with organic ions is their identification since the columns used for that purpose are still in the development stage. A great improvement has been the recent advent of UV detectors coupled to the conductivity cell. In any case, the identification of organic ions still requires a complex analysis.

RESULTS OBTAINED WITH CABLE-GRADE BASE RESINS

The results obtained with additive-free PE, PE+ antioxidant and XLPE resins are shown in Table 2. As already pointed out, the values there are possibly lower than the real ionic contents of the resins due to the slow extraction and to some adsorption into the container. Note that after 2 weeks at ≈ 50°C the water contained in the blanks bottles had the following ions: $C\ell^-$ = 118 ppb, SO_4^{--} = 35 ppb, Na^+ = 60 ppb, K^+ = 50 ppb, formate = 337 ppb, proprionate = 42 ppb. These ion contents were substracted from the values measured for the different samples (i.e. values of Table 2 represent only ions in the samples). It is possible that different bottles could be more or less contaminated and the above values should then be considered as only first approximations.

As already discussed, water easily extracts impurities from PE or XLPE. It may be deduced from Tables 1 to 3 that the ion contents in cable-grade resins or in cables are sufficiently great to have a detrimental influence on the electrical properties of dielectric materials. The sample with 1500 ppm antioxidant contains much more impurities than the two other samples and it is likely to be due to some external contamination during sample preparation. The crosslinking agent does not bring a significantly larger contamination with the exception of 1500 ppb acetate. Since there are no acetate ions found in other samples, they probably come from the crosslinking agent. The larger sulfate ion contents in XLPE arises from the fact that this sample contains more antioxidant, and therefore more sulfur. Interestingly, traces of oxidation by-products (formate and propionate ions) are detected in unaged resins. This suggests that the high temperature required for extrusion causes some oxidation. It is also possible that the antioxidant contains some of these ions as trace contaminants. From a practical point of view, this means that aging (e.g. thermal oxidation) of cables could a priori be evaluated from the time-evolution of their organic and inorganic ion contents. This is in agreement with the results shown in Table 1 where cable #1 was an unaged cable whereas cable #5 was the same cable thermally-aged. It is obvious that aging has led to some variations in ion content and it was suspected at the time that the observed decrease could be due to diffusion of some impurities out of the samples during laboratory aging.

CONCLUSIONS

It has been shown that organic and inorganic ions in PE and XLPE can be partially extracted by water and thus, detected by ionic chromatography. A significative ionic evaluation requires long extraction times at temperatures above 20°C, contaminant-free sample containers, highly purified water and calibration standards prepared every few days.

Only a limited number of metallic (and heavy elements) ions is extracted by water with the simple experimental procedure described here. An improved extraction efficiency would require some modifications, especially in preventing the reabsorption into the container of ions extracted from the samples.

The additives used in cable-grade PE resins contain sizeable amounts of highly soluble impurities. Since aging appears to induce an increase in the concentration of other ions, ionic chromatography could be useful to evaluate cable aging. This will be studied in more details in an EPRI project (RP 7897-2) on cable characterization.

REFERENCES

1. J.P. Crine, and J. Lanteigne, IEEE Trans. Elec. Insul. 19, 220 (1984).
2. S. Sapieha and J.P. Crine, 1980 Proc. of Conf. Elec. Insul. Diel. Phenom. (1980), p. 69.
3. J.P. Crine, to be published in Polym. Eng. Sci. (1986).
4. D.P.M. Laxen and R.H. Harrison, Analyt. Chem. 53, 395 (1981).
5. R. Gilbert, submitted to Can. J. Chem. Eng. (1986).

\PPLICATION OF POLARIZED SHEAR WAVES TO EVALUATE ANISOTROPY)F STEEL SHEETS

J.C. Albert, O. Cassier, B. Chamont, M. Arminjon,
and F. Goncalves

IRSID
Institut de Recherches de la Sidérurgie Française
78105 Saint German en Laye Cedex, France

INTRODUCTION

The use of heavy plates in mechanical construction, and steel sheets in deep drawing, requires homogeneous properties. These properties must be uniform along a same plate or from one sheet to another. In steel plants, the need to control these properties increases with the requirement of product quality. Generally, destructive techniques such as tensile tests, are used for quality control. The use of ultrasonics to test these properties offers great potentiality and the advantages of being nondestructive. In fact, classical destructive testing requires samples to be cut off from steel products. This kind of quality control presents many drawbacks:

- The cost is generally higher than the cost of nondestructive testing due to the need to cut off samples and the time spent for the preparation and the test.

- Destructive control proceeds by sampling. The assessment of quality can't be obtained for the entire product.

A nondestructive measurement based on ultrasonic techniques may provide 100% control of the product, during the process (on-line control), and this leads to a reduced cost in use. This paper presents experiments on different steel products. The use of velocity measurements of ultrasonic waves is shown to be useful in assessing mechanical properties on cold and hot rolled sheets.

PRINCIPLE

Ultrasonic velocities are directly obtained from the stiffness matrix of materials[1]. This relationship has been completed by Hughes and Kelly with the third order elastic constants in 1953. Here, the problem is limited to the evaluation of the homogeneity of characteristics between the rolling and the transverse directions of steel sheets. The steel samples which are studied here present texture symmetries which allow to simplify measurements for the evaluation of materials anisotropy (fig. 1).

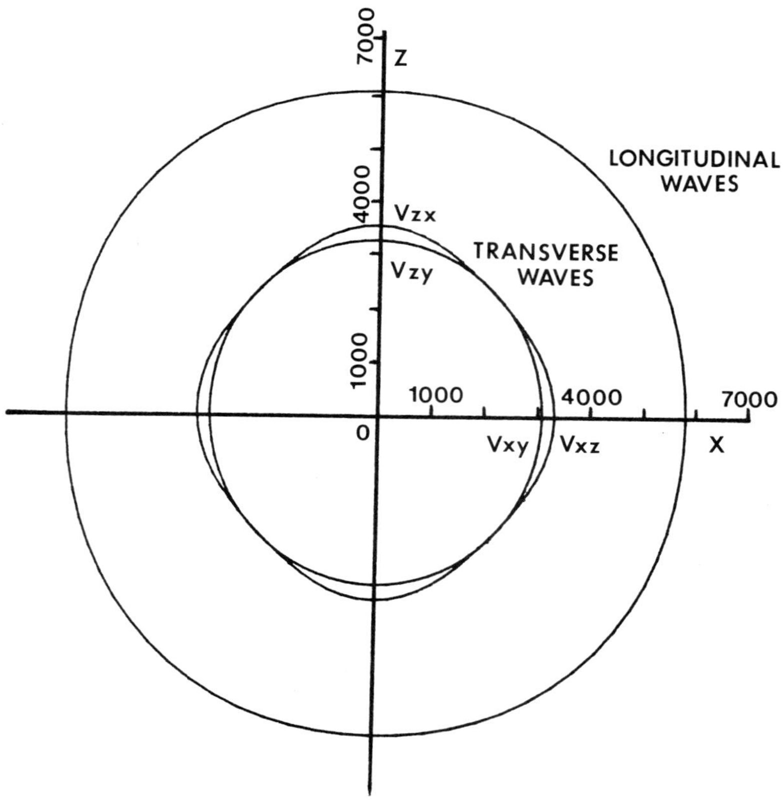

Fig. 1: Phase velocities in xoz plane

We define the z-axis as the direction parallel to one of the thickness, the x-axis as the rolling direction and the y-axis as the transverse direction. We assume that a simple relationship between the shear wave velocities and the elastic constants of the stiffness matrix are sufficient to characterize the state of anisotropy of steel sheets. We neglect scattering, which provides less velocity variations that anisotropy. It is possible to use the acoustic birefringence for the evaluation of materials anisotropy. So, this kind of measurement is often used for the residual stress determination to remove the effect of anisotropy on the measurement[2,3].

The acoustic birefringence may be defined by the following coefficient:

$$Bo = 2\,(Vzx - Vzy) \,/\, (Vzx + Vzy) \qquad (1)$$

Where Vzx is the velocity of the shear wave propagating along the z-axis polarized along the x-axis. The measurement of velocity requires measurements of time of flight and thickness of sample. The coefficient of acoustic birefringence Bo may be expressed versus only the time of flight of two waves. These measurements are made using the principle of pulse echo overlap method. The time of flight is the delay for the maximum of cross-correlation coefficient between two successive echoes.

RESULTS

Some experiments have been made on heavy hot-rolled plate samples with different start-rolling and finish-rolling temperatures (considering the last stage of rolling). Here, the aim is to evaluate the rate of rolling in $\alpha + \gamma$ phase. The rolling of heavy plates is often terminated in the $\alpha + \gamma$ phase to improve the mechanical properties of plates. The deformation of ferrite increases the yield strength and the ultimate strength. The improvement depends on the plastic strain ratio and on the temperature of steel when the strain is applied. For this kind of process, the strain ratio is known (it is obtained by the thickness reduction), but the temperature is not continuously known. In fig. 2, we show the relationship between the acoustic birefringence and a coefficient T, function of the rolling temperatures and the $\gamma \rightarrow \alpha$ transformation temperature Ar3:

$$T = (Ar3 - TER) / (TSR - TER) \quad (2)$$

Where TER is the finish-rolling temperature and TSR is the start-rolling temperature. We obtain a good correlation between T and Bo for slightly different compositions of steel which are sufficient to change the transformation temperature. The start-rolling temperatures may be higher or lower than the different transformation temperatures. These acoustic measurements give us a certain classification of samples, the same as the one resulting from X-ray-diffraction method.

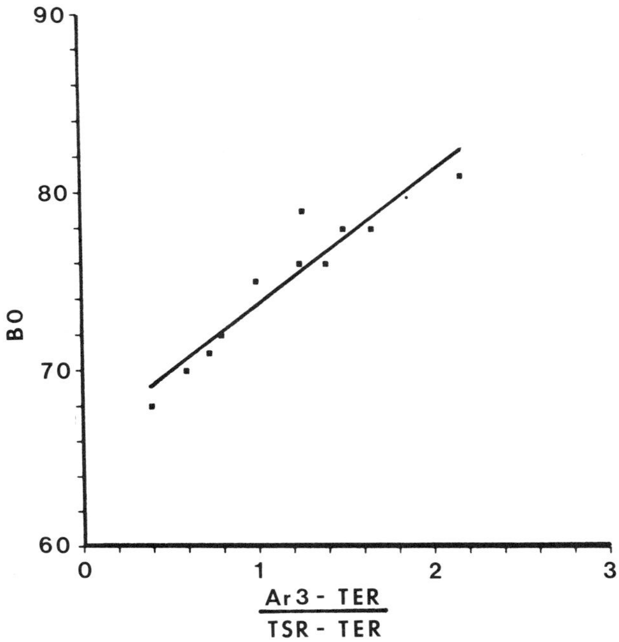

Fig. 2: Acoustic birefringence coefficient as a function of rolling temperature coefficient. Correlation coefficient: 0,94.

The same kind of measurements have been made on other samples cut off from a plate rolled with a temperature gradient on its top and bottom. This plate was rolled in the $\alpha + \gamma$ phase. Convection and radiation phenomena on extremities were sufficient to decrease the temperature by about 50 to 100°C. Consequently, the intercritical strain ratio is increasing on the top and bottom of the plate and the mechanical characteristics vary such as shown in fig. 3. The polycrystal anisotropy is due to the grain elongation induced by the ferrite deformation. This may be detected by the velocity measurements of shear waves (fig. 4). In steel plants, the slabs, which are reheated in furnaces, may present heterogeneous temperature distributions due to the skids in the furnace. The skid marks may induce a larger localized anisotropy on the cold plate. The ultrasonic velocity measurements allow to detect these skid marks as shown in Fig. 4. So, we have found a good correlation between the acoustic birefringence coefficient and the variations of yield strength between the rolling and transverse directions (fig. 5).

The shear wave velocities have also been measured on samples of deep drawing sheets. Usually, the anisotropy coefficients $\bar{r}$ or Δr of this kind of sheets are obtained by the classical destructive tensile tests. These coefficients are calculated with measurements done in three directions with respect to the rolling direction:

$$\bar{r} = (r_0 + 2r_{45} + r_{90})/4 \tag{3}$$

$$\Delta r = (r_0 - 2r_{45} + r_{90})/2 \tag{4}$$

These three coefficients r_0, r_{45}, r_{90} are defined as the ratio of strains:

$$r_\Theta = (\alpha W/W)/(\alpha T/T) \tag{5}$$

Where Θ is the angle between the axis of the tensile sample and the rolling direction, W and T are the width and thickness respectively of tensile sample.

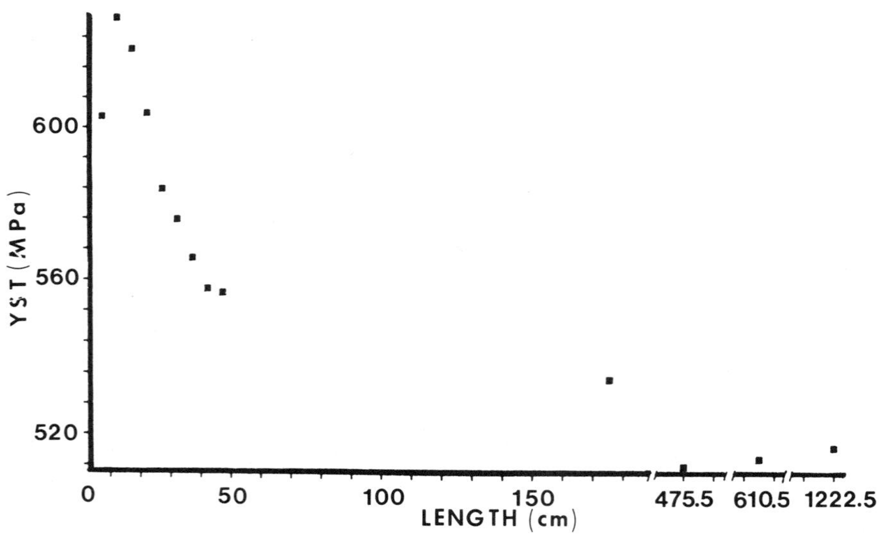

Fig. 3: Yield strength in transverse direction as a function of length.

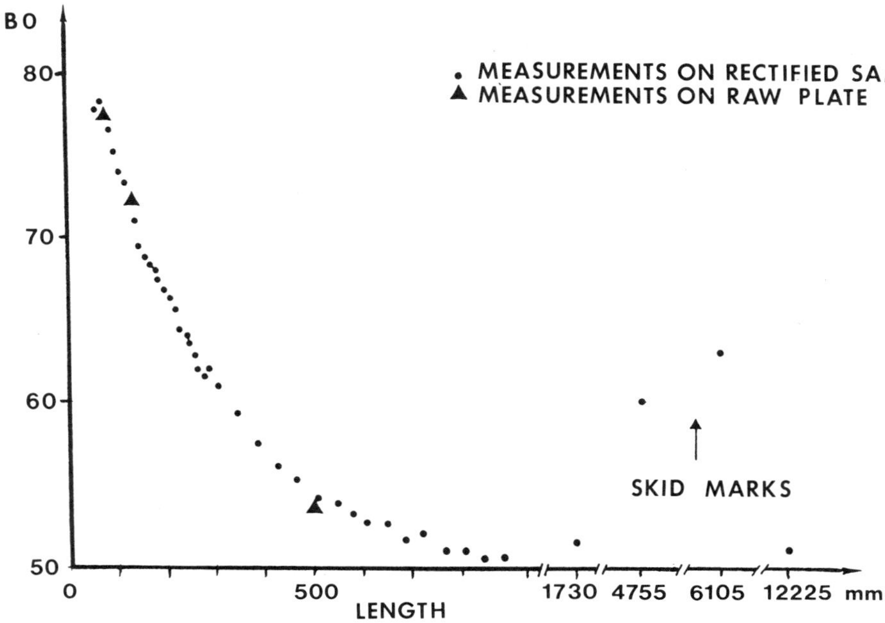

Fig. 4: Acoustic birefringence coefficient as a function of length.

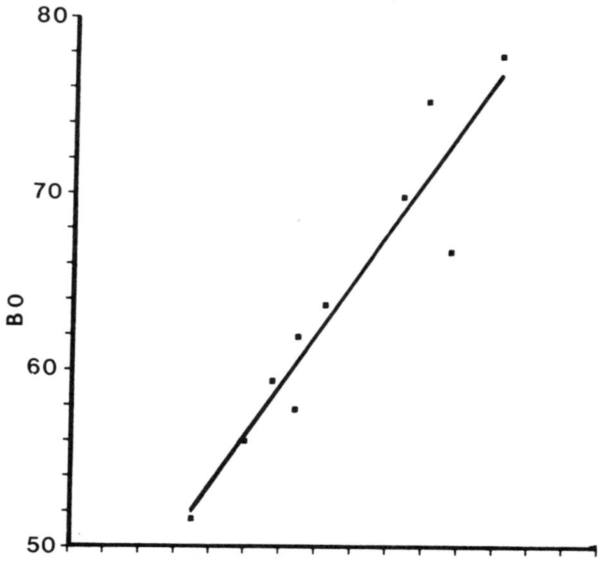

Fig. 5: Acoustic birefringence coefficient as a function of the yield strength coefficient.
Correlation coefficient: 0,94
CYS = (YSR - YST) / (YSR + YST)
YSR : yield strength in rolling direction
YST : yield strength in transverse direction.

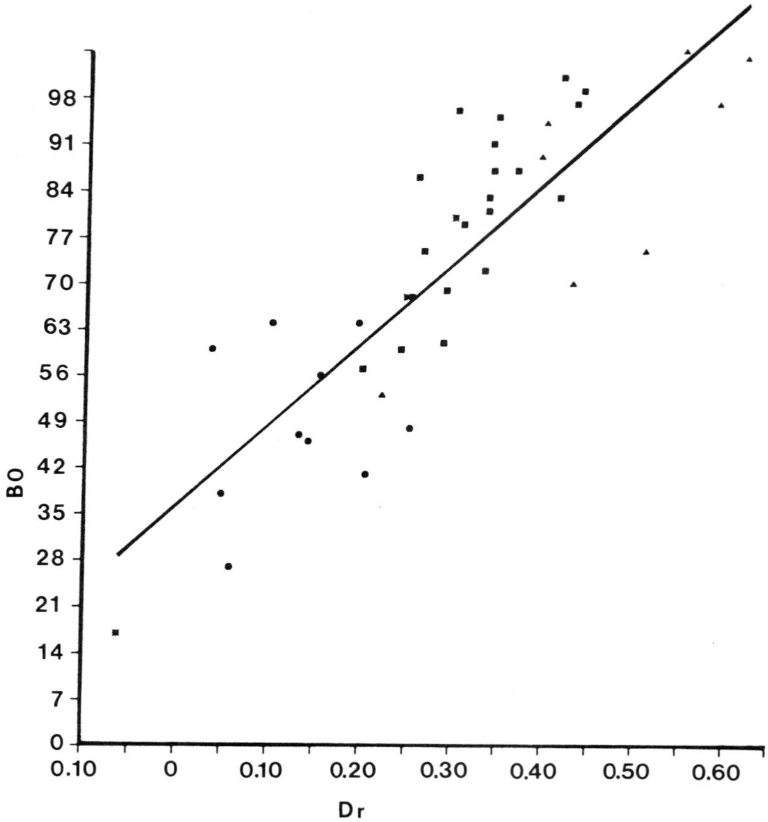

Fig. 6: Acoustic birefringence coefficient function of anisotropy coefficient measured by tensile testing. Correlation coefficient: 0,84.

The very principle of ultrasonic measurement implies the nonexistence of a direct relationship between the classical coefficients $\bar{r}$ and Δr involving the strain ratio at 45°, and the acoustic coefficient Bo. Nevertheless, the acoustic measurements are correlated to the rolling and transverse direction characteristics. Fig. 6 shows the existing relation between the acoustic birefringence coefficient and a coefficient named Dr, expressed by:

$$Dr = (r_0 - r_{90}) / (r_0 + r_{90}) \tag{6}$$

The measurements have been made on 41 samples with different compositions (and two kinds of texture). The residual standard deviation has the same magnitude as the errors on acoustic measurements and tensile measurements.

CONCLUSIONS

The correlations between the ultrasonic measurements and the process parameters or process control parameters are indirect. There is not always a physical and a theoretical relationship between the acoustic measurements which are measurements of the elastic coefficients, and the rolling temperatures or the deep-drawing characteristics. It is obvious that there is a relation between the elastic and plastic characteristics of materials[4,5,6]. It appears that the measurements of the elastic coefficients are not sufficient for a complete characterization of the plastic behaviour of materials[7]. Here, we have shown some possible applications of this kind of measurement. In the aim of industrialization, these measurements may be improved by using the noncontact ultrasonic techniques as EMAT or Laser. The application of this kind of measurement requires a learning or a calibration on a large number of samples for which the state of anisotropy or the characteristics to be evaluated are known. This kind of measurement, which is principally founded upon empirical correlations, may be applied as a relative measurement to different controls:

- Control of the state of anisotropy of the product with respect to a reference state with the intention to:

  - Controlling the process deviations.
  - Optimizing the use of products with regard to their applications.
  - Optimizing crop-head.

- Evaluating the product in permissible deviation range in order to decrease the number of complete characterization destructive tests.

REFERENCES

1. B. Auld, Acoustic Fields and Waves in Solids, Sec 6A, Wiley, New York (1973).
2. A.V. Clark, R.B. Mignogna and R.J. Sanford, Acousto elastic measurement of stress and stress intensity factors around crack tips. Ultrasonics, 21 (1983) 57-64.
3. A. Arora, Effect of texture and stress on acoustic birefringence - Scripta Metallurgica 18 (1984) 763-766.
4. W. Voigt, Lehrbuch der Kristallphysik: Leipzig Teubner 1928.
5. A.Z. Reuss, angew - Math - Mech - 9 (1929) 55.
6. S. Nagashima, Relation between texture and r value in steel sheets. Proceedings Inter. Symp. on textures in Research and Practice (1968) 444-463.
7. P. Parniere, IRSID Internal Report RPC 39 (1978).
8. B.E. Droney, Use of ultrasonic techniques to assess the mechanical properties of steels, Nondestructive Methods for Material Property Determination, ed by C.O. Ruud and R.E. Green, Plenum Press, pp. 237-248, (1984).
9. K. Goebbels, H.J. Salzluerger, Zerstörungsfreies Merssen der plastischen, Anisotropie von Blechen mit Ultraschall, Material prüfung 25 (1983) 279-281.
10. H. Warlimont, G. Hausch, Elastische anisotropie von Metallen und legierungen in Mechanische Anisotropie H.P. Stuwe, Springer Verlag (1974) 35-621.

CORRELATION BETWEEN ELASTIC AND PLASTIC ANISOTROPY IN ROLLED METAL PLATES

J.F. Bussière and C.K. Jen

Industrial Materials Research Institute
National Research Council of Canada
75 de Mortagne Blvd., Boucherville, Québec, J4B 6Y4

I. Makarow, B. Bacroix, Ph. Lequeu and J.J. Jonas

Dept. of Mining and Metallurgical Engineering
McGill University, Montréal, Québec, H3A 2A7

INTRODUCTION

Because of its relative strength and ability to be formed into various shapes, sheet metal is one of the most widely used industrial materials. Forming processes can be classified into several categories including bending, stretching and drawing. To insure high quality in the final product and/or efficiency in processing, material properties should be consistent and optimized according to the different forming processes which involve different strain distributions. These material properties are controlled in the mill by adjustments in chemical composition and thermomechanical processing, resulting in certain desirable textures and microstructures. For deep drawing applications (e.g. Aℓ beverage cans and stainless steel kitchen sinks) texture or preferred crystallographic orientation has a strong influence on the resistance of the material to thinning during a draw and therefore on the maximum attainable depth of draw. It also determines the presence or absence of earing, i.e. nonuniformity in the depth, which can occur, for instance, when drawing a cup (see e.g. fig. 1). Although textures can be characterized by X-ray pole figures, for engineering purposes the most relevant material parameters are obtained from uniaxial tensile measurements. In particular, depth of draw and earing have been shown to be closely related respectively to the average value and the angular variation of the plastic strain ratio, in the plane of the metal sheet. The plastic strain ratio, $R(\alpha)$, pertaining to a direction at an angle α to the rolling direction is defined as the ratio of the true strain along the width of the specimen, $\varepsilon_{yy}(\alpha)$, to that along the thickness $\varepsilon_{zz}(\alpha)$, measured during uniaxial tension in a specimen cut at an angle α to the rolling direction (see fig. 2).

$$R(\alpha) = \varepsilon_{yy}(\alpha) / \varepsilon_{zz}(\alpha) \qquad (1)$$

The plastic anisotropy of a sheet can then be characterized in terms of the average strain ratio $\bar{R}$ and the planar strain ratio ΔR where

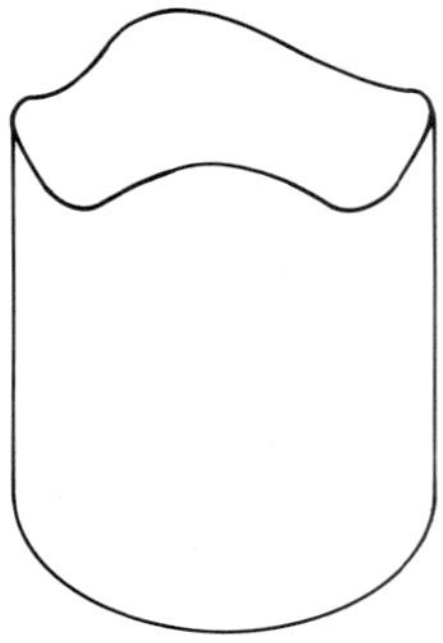

Fig. 1: Drawn cup with ears in the directions of high R value.

$$\bar{R} = [R(0) + 2R(45) + R(90)]/4 \qquad (2)$$

$$\Delta R = [R(0) + R(90) - 2R(45)]/2 \qquad (3)$$

and R(0), R(90) and R(45) are respectively the values of R measured in directions parallel, transverse and at 45° to the rolling direction.

The characterization of plastic anisotropy by uniaxial tensile measurements, although informative, is time consuming, costly and can only be performed on a small fraction of the production. The development of an on-line sensor for measuring the plastic anisotropy of rolled metal sheet would permit the minimization of both material and time losses during subsequent processing and insure, through feedback control, consistency of material properties.

In this paper the possibility of predicting plastic behavior from elastic measurements is explored theoretically for iron (mild steel) and aluminum. Analytic expressions for Young's modulus and bulk acoustic wave velocities are developed as a function of material preferred orientation and measurement direction using the crystallite orientation distribution function (CODF) method[1]. The plastic properties $R(\alpha)$, $\bar{R}$ and ΔR are calculated using the continuum mechanics of textured polycrystals (CMTP) method reported elsewhere[2]. The calculated acoustic parameters and plastic properties are then compared and correlated for several of the preferred orientations commonly observed during the rolling of fcc (e.g. aluminum) and bcc (e.g. iron) metals. Throughout this work, only single ideal orientations are considered with the additional constraint of overall orthorhombic macroscopic symmetry corresponding to that of rolled plates.

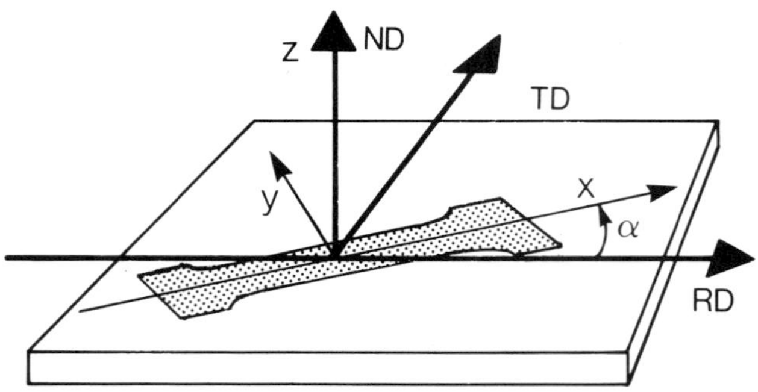

Fig. 2: Tensile specimen cut at an angle α to the rolling direction for the measurement of $R(\alpha)$.

II THEORY

a) Elastic properties

Although only single ideal orientations are considered in this paper, each textured material considered includes a combination of an ideal orientation and its mirror images with respect to the principal deformation axes; these are needed to achieve the overall macroscopic orthorhombic symmetry corresponding to that of rolled plates. For this reason each single textured plate could not, in general, be treated as a single crystal and was analysed within the more general framework of crystallographic orientation distribution functions (CODF). The CODF, f(g), is a statistical distribution which describes the volume fraction of individual crystallites in a polycrystalline sample as a function of their orientations, neglecting their positions, shapes and sizes within the aggregate. In the present case, g is the rotation relating the crystallographic and sample axes. Each crystallite may therefore be identified by an orientation g_i and a volume fraction V_i. A simple definition of the CODF is thus:

$$dV/V = f(g)\, dg \tag{4}$$

where dV/V is the volume fraction of all crystallites having orientations varying from g to g + Δg. This function f(g) may be developed into a sum of independent spherical harmonics $\dot{\dot{T}}^{\mu\nu*}$, each term of the expansion separately fulfilling all the symmetry requirements:

$$f(g) = \sum_{\ell=o}^{\infty} \sum_{\mu=1}^{M(\ell)} \sum_{\nu=1}^{N(\ell)} C_\ell^{\mu\nu} \dot{\dot{T}}_\ell^{\mu\nu*}(g) \tag{5}$$

$M(\ell)$ and $N(\ell)$ are the number of values of μ and ν, determined by the crystal and sample symmetries[3]. In the most general case, f(g) may vary over an infinite number of possible crystal orientations. The determination of the C_ℓ^{mn} coefficients for each generalized spherical harmonic is then very complicated. However, when it is assumed that the polycrystal is composed of crystallites oriented in a limited number of specific ways (including all the possible symmetrical orientations), the coefficients of Eq. 3 may be calculated exactly, as follows:

$$C_\ell^{\mu\nu} = (2\ell + 1) \sum_i V_i \dot{\dot{T}}_\ell^{\mu\nu*}(g_i) / \sum_i V_i \tag{6}$$

Here the V_i are the volume fractions of all the ideal orientations present in the aggregate. The symmetrical generalized spherical harmonic functions $\dot{\dot{T}}_\ell^{\mu\nu*}(g_i)$ can be calculated for all ℓ, μ and ν. Expressions for these functions, for the particular case of cubic crystals and orthorhombic sample symmetries, are listed in reference 4.

Once the CODF is obtained, it is possible to deduce the effective elastic constants of the polycrystalline aggregate using techniques such as those described by Bunge[1], Morris[5] or Sayers[6]. In the present paper, Bunge's method was used because of the intrinsic inclusion of symmetry considerations making it easier to treat single textures of cubic materials with imposed orthorhombic symmetry (i.e. rolled plates).

Different assumptions can be made when averaging elastic constants. Of these, the most common are the Voigt and Reuss approximations, which assume respectively uniform strain and stress in all the crystallites. The Hill approximation is the arithmetic average of the Reuss and Voigt approximations and is often closest to the measured values of polycrystal elastic constants.

Single crystal elastic constants, C_{ijkl}, and compliances, S_{ijkl}, may conveniently be decomposed into isotropic and anisotropic parts and averaged in order to obtain respectively the Voigt and Reuss averages C_{ijkl} and S_{ijkl}.

$$C_{ijkl} = C^0_{ijkl} + C_a(\bar{t}_{ijkl} - t_{ijkl}) \tag{7}$$

$$S_{ijkl} = S^0_{ijkl} + S_a(\bar{t}_{ijkl} - t_{ijkl}) \tag{8}$$

where t_{ijkl} is the anisotropy tensor for cubic crystals (7) and t_{ijkl} is the sum of the CODF coefficients as described below.

$$\bar{t}_{ijkl} = \bar{a}_0^{11}(ijkl) + \bar{a}_4^{11}(ijkl)C_4^{11} + \bar{a}_4^{12}(ijkl)C_4^{12} + \bar{a}_4^{13}(ijkl)C_4^{13} \tag{9}$$

The $\bar{a}_\lambda^{\mu\nu}(ijkl)$, listed in Reference 1, are mathematical constants. The independent coefficients $C_\ell^{\mu\nu}$ introduce the crystal cubic and sample orthorhombic symmetries into the averaging process. Only the zeroth and fourth order $C_\ell^{\mu\nu}$ coefficients are required in Eq. (6) because the compliance and stiffness form fourth rank tensors[8]. In the equations which average elastic properties using the CODF expressions, the value of ℓ is always 4.

For convenience, bulk ultrasonic velocities were calculated using averaged elastic constants (Voigt average) and standard equations relating velocities to the orthorhombic elastic constants (see e.g. ref. 9).

Young's modulus E is simply the inverse of the 1111 component of the oriented and averaged compliance ($\bar{S}'_{ijkl}$) tensor[1,10]. The particular case of rotations in the rolling plane around the normal direction yields

$$\frac{1}{E(\alpha)} = \bar{S}_{1111}\cos^4\alpha + \bar{S}_{2222}\sin^4\alpha + (\bar{S}_{1212} + \frac{1}{2}\bar{S}_{1122})\sin^2 2\alpha \tag{10}$$

where α is the angle between the rolling and measurement directions.

b) Plastic strain ratio

The plastic anisotropy of rolled sheet, as expressed by the R value, has generally been calculated either by crystallographic[11,12] or by continuum[13,14] methods. Here, an alternative model, which combines the best features of both approaches and is known as CMTP(the continuum mechanics of textured polycrystals)[2,15-17] is used to predict the strain ratio R(α) as a function of the angle α of tensile testing in the sheet plane. For this purpose, a Hill type of yield criterion is fitted to the crystallographic yield surface pertaining to a cubic single crystal displaying {111}<110> (fcc) or {110} <111> (bcc) slip. This yield function, which is expressed in the 100 axes of the texture component of interest, is then reoriented as required in the specimen reference frame. By this means, complete analytic expressions of the plastic properties (i.e. stress and strain rate characteristics of rolled sheets) can be readily obtained as a function of the ideal orientation of interest {hkl}<uvw> and of the angle α. When several texture components are present, the overall properties can be obtained by averaging on a volume fraction basis[2,16,17]. This leads, finally, to the R coefficient for the material.

TABLE I - TEXTURES

Texture		Weighting Factor		
ND	RD	ALUMINUM		IRON
		Rolled	Recrystallized	
0 0 1	1 1 0	0	0	1
0 1 3	1 0 0	1	1	0
0 1 3	4 3 -1	1	1	0
1 0 0	0 0 1	1	17	0
1 0 0	0 1 1	1	1	0
1 0 0	0 1 2	1	1	1
1 1 0	0 0 1	1	1	1
1 1 0	-1 1 2	4	1	1
1 1 0	-1 1 4	0	0	1
1 1 0	3 -3 2	1	1	1
-1 1 0	1 1 1	0	0	1
1 1 1	-1 1 0	1	1	1
1 1 1	1 1 -2	0	0	1
1 1 2	-1 1 0	1	1	1
1 1 2	1 1 -1	1	1	0
1 2 3	4 1 -2	1	1	0
1 2 3	6 3 -4	1	1	0
1 3 2	1 1 -2	0	0	1
1 3 5	2 1 -1	4	1	0
1 4 6	2 1 -1	1	1	0
2 1 1	-1 1 1	0	0	1
-3 2 1	1 1 1	0	0	1
3 3 2	-1 1 0	0	0	1
4 1 1	0 1 1	0	0	1
4 1 1	1 4 -8	0	0	1
4 4 11	11 11 -8	4	1	0
5 5 2	1 1 -5	1	1	0
5 5 4	2 2 -5	1	1	0

c) Correlations

In order to simulate the texture variations occurring during the manufacture of aluminum or low carbon steels, a list of the most commonly occurring single textures was established for these materials (see table I). In the case of aluminum, different weighting factors were given to each texture to take into account the difference between recrystallized and cold rolled material[18]. In the case of iron, realistic weighting factors were not available and each allowable bcc texture component was given equal weight. For each texture listed in table I, the angular variation of R, R(α), as well as $\bar{R}$ and ΔR were computed using the CMTP method. These quantities were then expressed as linear expansions of elastic parameters. In order to establish the choice of the relevant elastic parameters, use was made of empirically established correlations published in the literature. Thus, parameters such as Young's modulus[19,20], E, and the through thickness square of the longitudinal velocity[21], V^2_{LN}, were included. In the Roe CODF notation[22], the effects of texture on the elastic properties are completely described by the three fourth order coefficients[6], W_{400}, W_{420} and W_{440} (which are similar to the $C_\ell^{\mu\nu}$ of equation 5). These coefficients, which should relate respectively to the depth of the draw ($\bar{R}$), the tendency to form two ears, and the tendency to form four ears[23], can be expressed in terms of bulk ultrasonic velocities using the theory developed by Sayers[6]. The velocity V_{LN} depends only on W_{400}. The following two parameters depend respectively on W_{420} and W_{440}:

$$dV^2_{SN} = (V^2_{S1N} - V^2_{S2N})/(V^2_{LN} + V^2_{S1N} + V^2_{S2N}) \quad (11)$$

$$dV_{SH}(\alpha) = [V_{SH}(\alpha) - V_{SH}(0)]/V_{SH}(0) \quad (12)$$

where V_{S1N} and V_{S2N} are the velocities of shear waves propagating in the normal direction and polarized in the rolling and transverse directions, respectively.

In addition to the parameters defined by equations (11) and (12), bulk longitudinal, vertically polarized and horizontally polarized shear wave velocities, $V_L(\alpha)$, $V_{SV}(\alpha)$ and $V_{SH}(\alpha)$, propagating in the plane of the plate at an angle α to the rolling direction were also included. Although only bulk waves are considered here, it is also possible to relate texture to the propagation velocities of Lamb[24] and Rayleigh waves[25].

For industrial purposes, the average value of $R(\alpha)$, $\bar{R}$, and the planar variation of R, ΔR, are only defined in terms of the angles 0, 45 and 90° (equations (2) and (3)). For the present calculations, the elastic parameters of interest were defined in an analogous manner; e.g. only the ultrasonic velocities for propagation at 0, 45 and 90 degrees to the rolling direction were taken into consideration. These are $\bar{E}$, $\bar{V}_L$, $\bar{V}_{SH}$, V_{SV}, ΔE, ΔV_L, and ΔV_{SH}. Because ΔV_{SV} tends to have values very close to zero, the following parameter was defined instead.

$$\delta V_{SV} = V_{SV}(90°) - V_{SV}(0°) \quad (13)$$

The above parameters were used to establish correlations with $\bar{R}$ and ΔR. To reproduce the variation of R with angle, $R(\alpha)$, the following equation was used.

$$R(\alpha) = a\, V_L(\alpha) + b\, V_{SH}(\alpha) + c\, V_{SV}(\alpha) + d\, E(\alpha) + e\, V_{LN}^2 + f\, dV_{SN}^2 + g\, dV_{SH}(\alpha) + h \quad (14)$$

where $R(\alpha)$ is obtained from the CMTP method and the letters a to h represent adjustable constants

For correlations relating to $\bar{R}$ and ΔR, which are of more practical interest, the following regression equations were used, respectively:

$$\bar{R} = a'\,\bar{V}_L + b'\bar{V}_{SH} + c'\bar{\bar{V}}_{SV} + d'\bar{E} + e'V^2_{LN} + f' \quad (15)$$

and

$$\Delta R = a''\Delta V_L + b''\Delta V_{SH} + c''\delta V_{SV} + d''\Delta E + e''dV^2_{SN} + f''V^2_{LN} + g'' \quad (16)$$

where again $\bar{R}$ and ΔR are calculated by means of the CMTP method and a' to f' and a" to d" are constants.

$\bar{R}$ and ΔR were also obtained by initially determining the best fit to the entire angular dependence of R, $R(\alpha)_{FIT}$ and subsequently calculating $\bar{R}$ and ΔR, using equations (2) and (3).

It should be noted that in some cases some of the constants in equation (15) can be intentionally set to zero. For instance, although Young's modulus can be measured acoustically (see section 3.3), it is not possible to do it on-line; thus d, d' or d" may be set to zero. With few exceptions, all the regressions are based on values of $\bar{R}$, ΔR and the elastic properties at only three angles, 0°, 45° and 90°.

In this work only the aluminum (fcc) and iron (bcc) polycrystalline plates are selected for the demonstration, although the programs and algorithms developed can be used for other metals with cubic crystal symmetry and orthotropic plate symmetry, such as rolled copper and brass.

RESULTS & DISCUSSION

To predict the detailed variation of R with angle, $R(\alpha)$, equation (14) was used for seven equally spaced angles from 0 to 90°, for each texture listed in table I. Best fits were obtained when either a or c was set equal to zero. Using c = 0, the following multiple correlation coefficients (R^2) and standard deviations, σ, were obtained for the aluminum rolling and recrystallization textures, respectively: $R^2_{rolling} = 0.85$, $R^2_{recryst} = 0.89$, $\sigma_{rolling} = 0.24$, $\sigma_{recryst} = 0.21$. Typical results are shown in figure 3, where R (α), computed using the CMTP method, is compared to the elastic regression approximation (eq. (14)) for two different rolling textures composed of single ideal orientations in $A\ell$. In general, the agreement is good. Similar results were obtained for the recrystallized textures and for the iron.

The correlation results obtained for $\bar{R}$ with the parameters of equation (15) are listed in table II, where R^2 is the multilinear correlation coefficient, and σ the standard deviation. The use of Young's modulus alone is seen to give values of R^2 between 0.71 and 0.86 whereas

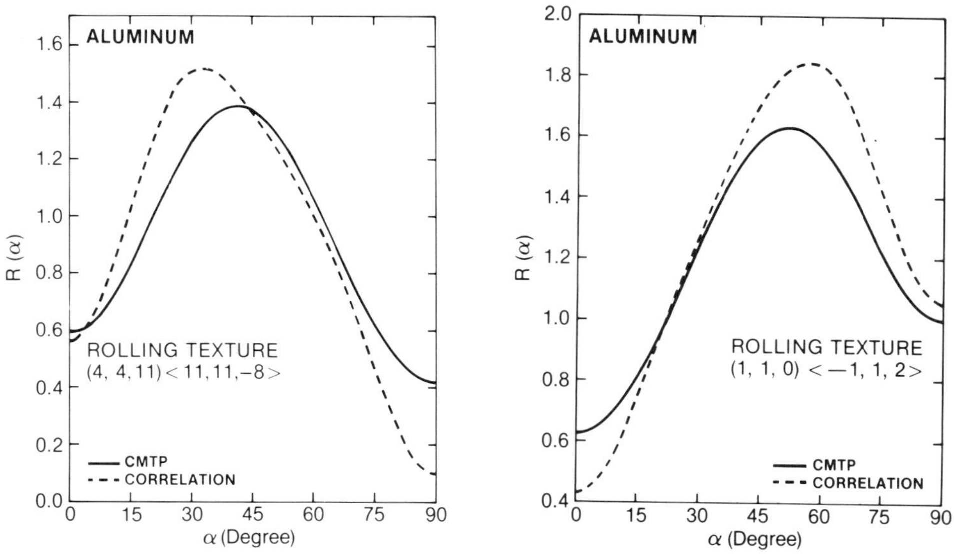

Fig. 3: Comparison of $R(\alpha)$ calculated from the plastic theory (CMTP) and deduced from the elastic properties (eq. (14)) for rolled aluminum with rolling textures a) (4,4,11) $\langle 11, 11, -8 \rangle$ and b) (1, 1, 0) $\langle -1, 1, 2 \rangle$.

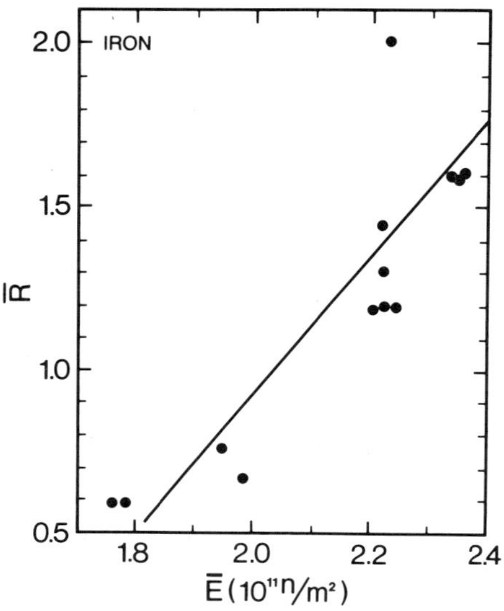

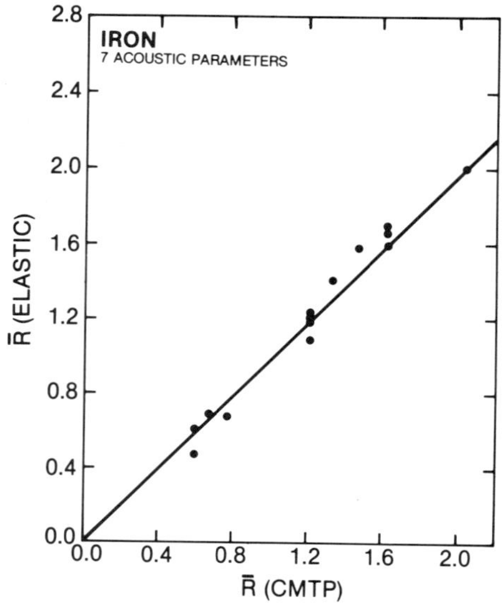

Fig. 4: Comparison between calculated values of $\bar{R}$ and the average Young's modulus $\bar{E}$ for pure iron.

Fig. 5: Correlation between $\bar{R}$ calculated using the CMTP method and predicted using seven elastic parameters (eq. (14)).

the use of several parameters can significantly improve accuracy. For instance, combining V^2_{LN} and V_{SH} for the case of iron yields R^2 = 0.95 and σ = 0.10. In general, the best results are obtained using R_{FIT}, which is deduced by calculating R using the best fit to $R(\alpha)$ using equation (14) at angles of 0, 45 and 90°. For illustration purposes, figures 4 and 5 show the correlations obtained respectively with $\bar{E}$ only and with $\bar{R}_{FIT}$ for the case of iron. The use of $\bar{E}$ only, which is the basis of a commercial instrument[26], is seen to be considerably improved when all the elastic parameters are included.

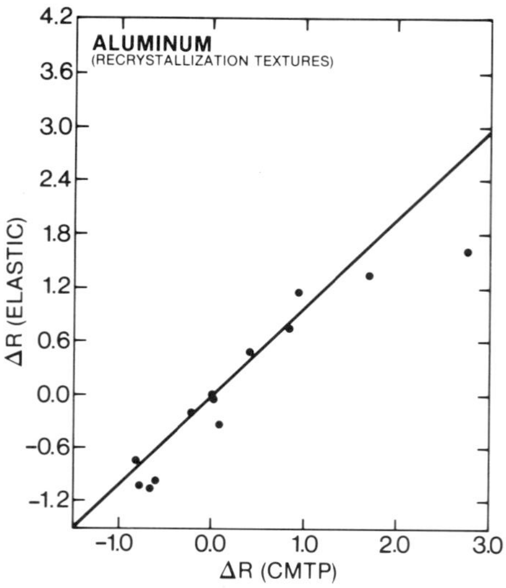

Fig. 6: Correlation between ΔR calculated from the CMTP method and using the best elastic fit to $R(\alpha)$ (eq. (14)).

TABLE II - CORRELATION RESULTS FOR $\bar{R}$

PARAMETERS INCLUDED						RESULTS					
E	V^2_{LN}	V_{SH}	V_{SV}	V_L	R_{FIT}	ALUMINUM Rolling		ALUMINUM Recryst		IRON	
						R^2	σ	R^2	σ	R^2	σ
X						0.71	0.19	0.86	0.15	0.76	0.20
	X					0.71	0.19	0.86	0.15	0.77	0.19
		X				0.69	0.19	0.85	0.16	0.67	0.23
			X			0.72	0.18	0.86	0.15	0.80	0.18
				X		0.71	0.19	0.86	0.15	0.77	0.20
			X	X		0.72	0.18	0.86	0.15	0.91	0.13
	X		X			0.72	0.18	0.86	0.15	0.95	0.10
		X	X			0.79	0.16	0.90	0.13	0.84	0.17
	X	X				0.74	0.18	0.88	0.14	0.80	0.19
					X	0.90	0.11	0.95	0.09	0.98	0.06

TABLE III - CORRELATION RESULTS FOR ΔR

ΔE	ΔV_{SH}	δV_{SV}	ΔV_L	V^2_{LN}	dV^2_{SN}	ΔR_{FIT}	ALUMINUM Rolling		ALUMINUM Recryst		IRON	
							R^2	σ	R^2	σ	R^2	σ
X							0.82	0.39	0.87	0.31	0.85	0.35
	X						0.80	0.41	0.74	0.44	0.75	0.45
		X					0.34	0.74	0.18	0.77	0.14	0.84
			X				0.80	0.40	0.74	0.44	0.77	0.44
	X	X					0.84	0.37	0.86	0.32	0.79	0.42
X		X			X		0.97	0.17	0.98	0.13	0.89	0.32
						X	0.81	0.39	0.92	0.24	0.90	0.28

The correlations obtained for ΔR using equation (16) are listed in table III, where ΔR_{FIT} was obtained using the best fit to $R(\alpha)$ from equation (14) using the angles 0, 45, and 90°. In this case, it is seen that improvements over the use of ΔE can be obtained with multiple parameters, although the improvement is more modest than for the case of $\bar{R}$. Figure 6 illustrates results using ΔR_{FIT} for recrystallization textures in pure aluminum.

CONCLUSION

Using a simplified approach to calculate the plastic strain ratio, $R(\alpha)$ and the CODF formalism to calculate the elastic properties, the possibility of predicting $R(\alpha)$ values from Young's modulus and ultrasonic velocities was explored theoretically for rolled plates of pure aluminum and iron. The results confirm empirically established correlations between Young's modulus and plastic strain ratio observed in mild steel which have resulted in the production of a commercial instrument. These correlations were also found to exist in aluminum. In addition, it was shown that considerable improvements in accuracy can be obtained when more than one parameter (i.e. several ultrasonic velocities) are included simultaneously. In the present work, only single ideal textures and bulk wave velocities were considered. Further work will include more complex textures and take into account velocities characteristic of thin plates.

REFERENCES

1. H. J. Bunge, "Texture analysis in material science, mathematical methods", Butterworths, London 1982.
2. Ph. Lequeu, F. Montheillet and J.J. Jonas, "A simplified method for the prediction of plastic anisotropy in rolled sheet", Proc. AIME Symposium on Textures in Nonferrous Metals, Detroit, Sept. 1984, pp. 189-212.
3. H. J. Bunge, "Texture analysis in material science, mathematical methods", Butterworths, London, p. 52, fig. 44, 1982.
4. H. J. Bunge, "Texture analysis in material science, mathematical methods", Butterworths, London, p. 389, 1982.
5 P.R. Morris, "Averaging fourth-rank tensors with weight functions", J. Appl. Phys., Vol. 40, No. 2, Feb. 1969, pp. 447-448.
6. C.M. Sayers, "Ultrasonic velocities in anisotropic polycrystalline aggregates", J. Phys. D. Appl. Phys., Vol. 15, pp. 2157-2167, 1982.
7. H. J. Bunge, "Texture analysis in material science, mathematical methods", Butterworths, London, p. 323, 1982.
8. H.J. Bunge, "The effective elastic constants of textured polycrystals in second order approximation", Kristall and Technik, Vol. 9, pp. 413-424, 1974.
9. B.A. Auld "Acoustic fields and waves in solids", John Wiley & Sons, N.Y., Vols. I and II, 1973.
10. J. F. Nye, Physical Properties of Crystals, Clarendon Press, Oxford, 1957.
11. J.F.W. Bishop and R. Hill, "A thoery of the plastic distorsion of a polycrystalline aggregate under combined stress", Phil Mag., 42, 414 1951.
12. J.F.W. Bishop and R. Hill, "A theoretical deviation of the plastic properties of a polycrystalline face centered metal", Phil. Mag., 42, 1298, 1951
13. R. Hill, "A theory of the yielding and plastic flow of anisotropic materials", Proc. Roy. Soc. London, A193, 281, 1948
14. R. Hill, "Theoretical plasticity of textured aggregates", Math. Proc. Camb., Phil. Soc., 85, 179, 1979

15. F. Montheillet, P. Gilormini and J.J. Jonas, "Relation between axial stress and texture development during torsion testing: A simplified theory", Acta Metall., 33, 705, 1985
16. Ph. Lequeu, F. Montheillet and J.J. Jonas; "Computer modelling of the plastic properties of textured polycrystals", Ottawa, Canada, May 1986, in press.
17. Ph. Lequeu, P. Gilormini, F. Montheillet, B. Bacroix and J.J. Jonas, "Yield surfaces for textured polycrystals - I. Crystallo graphic Approach", Acta Metall. 1987, 35, pp. 439-451.
18. M. Bull, Alcan Int. Ltd. private communication
19. C.A. Stickels and P.R. Mould, "The use of Young's modulus for predicting the plastic-strain ratio of low carbon steel sheets", Metallurgical Trans., Vol. 1, pp. 1030-1312, 1970.
20. P.R. Mould and T.E. Johnson, Jr., "Rapid assessment of drawability of cold-rolled low carbon steel sheets", Sheet Metal Industries, pp. 328-348, June 1973.
21. H. Kitagawa, "Analysis of elastic anisotropy in cold rolled steel sheet", Proc. of 6th Int'l Conf. on Textures in Metals, Tokyo, pp. 1166-1178, 1981.
22. R.J. Roe, "Description of crystallite orientation in polycrystalline materials. III General solution to pole figure inversion", J. Appl. Phys. 36, 2024, 1965.
23. G.J. Davies, D.J. Goodwill and J.S. Kallend, "Elastic and plastic anisotropy in sheets of cubic metals", Metallurgical Trans., Vol. 3, pp. 1627-1631, 1972.
24. S.S. Lee, J.F. Smith and R.B. Thompson, "Inference of crystallite orientation distribution function from the velocities of ultrasonic plate modes", these proceedings.
25. P.P. Delsanto, R.B. Mignona and A.V. Clark, "Ultrasonic texture analysis for polycrystalline aggregates of cubic materials displaying orthorhombic symmetry", these proceedings.
26. B. Engart, "A drawability tester for steel users and producers", Metal Progress, p. 47, June 1985.

ULTRASONIC TEXTURE ANALYSIS FOR POLYCRYSTALLINE AGGREGATES

OF CUBIC MATERIALS DISPLAYING ORTHOTROPIC SYMMETRY*

P. P. Delsanto, R. B. Mignogna
Naval Research Laboratory, Washington, DC 20375-5000
A. V. Clark, Jr.
Fracture and Deformation Division
National Bureau of Standards, Boulder, CO 80303

INTRODUCTION

The study of the applications of the acoustoelastic effect, i.e. the stress-dependence of the propagation velocity of ultrasonic waves in deformed elastic media, has undergone considerable progress in recent years.[1] Techniques for the determination of applied and residual stresses have been proposed both for bulk[2-5] and for surface[6-9] ultrasonic waves. However, for the practical application of these techniques to fabricated materials, the difficulty of separating the often competing effects of stress and texture remains a vexing problem.

To attack this problem it is necessary first to have a complete understanding of texture effects in the absence of stresses. Therefore in this paper we investigate the propagation of Rayleigh waves (RW) on the surface of unstressed materials, characterized as having an orthotropic distribution of cubic crystallites. Examples are rolled plates of alloys of aluminum, copper and iron. The texture, or preferential alignment of the crystallographic axes, can be conveniently described in terms of the orientation distribution function,[10,11] which, together with the elastic constants of the crystallites, defines the material anisotropy.

Assuming that the material is not strongly anisotropic, we can study the propagation of RW's with the perturbation formalism, which we have recently proposed for deformed orthotropic materials.[12] The formalism is briefly reviewed in Sec. I, where the formula for the change in RW phase velocity due to a slight anisotropy is reported. This general formula is then applied, in Sec. II, to the special case of an orthotropic distribution of cubic crystallites. The corresponding change in RW velocity is given in function of the orientation distribution function (W_{lmn} coefficients) and of the elastic constants of the crystallites.

In Sec. III we proceed to the determination of the texture (W_{lmn} coefficients). Two cases are discussed: in the first case, reliable absolute

* Contribution of the Naval Research Laboratory and of the National Bureau of Standards; not subject to copyright.

measurements of the propagation velocity are assumed to be possible; in the second, only relative measurement (i.e. the difference between measurements at different angles) are considered. In Sec. IV, the correlation coefficients between the W_{lmn} coefficients and the RW velocities are plotted, together with other quantities of interest, and an error analysis is performed.

I. THE PERTURBATION FORMALISM

We give here a brief review of the perturbation formalism for the propagation of Rayleigh waves on the surface of an unstressed orthotropic material plate. Complete details and the discussion of stress effects can be found in ref. 12.

Let us assume that the material occupies the half space $x_3 \geqslant 0$, with vacuum in $x_3 < 0$. Since RW's penetrate into the material only up to a depth of approximately one wavelength λ_0, the finite depth d of the material plate does not appreciably affect our results, as long as $d >> \lambda_0$, as is usually the case. We choose the x_2-axis in the direction of propagation of the wave and the x_1-axis in the transverse direction. We also assume that two material symmetry axes, which we call $\tilde{x}_1$ and $\tilde{x}_2$, lie on the plate surface with an angle ϕ between x_1 and $\tilde{x}_1$.

In the $\tilde{x}_k$-system (with $\tilde{x}_3 \equiv x_3$) the only non-zero second order elastic constants are, for orthotropic materials, $\overline{C}_{kl}$ $(k,l = 1,2,3)$ and $\overline{C}_{kk}$ $(k = 4,5,6)$, where Voigt's abbreviated notation is used. Since the material is supposed to be only slightly anisotropic, we can assume a fictitious material as its isotropic approximation or "unperturbed" case. The Lamé constants λ and μ of this fictitious isotropic material are chosen in such a way as to minimize the "perturbative corrections":

$$\tilde{C}'_{km} = \overline{C}_{km} - C_{km} \tag{1}$$

where

$$C_{km} = \begin{pmatrix} \lambda + 2\mu & \lambda & \lambda & & & \\ \lambda & \lambda + 2\mu & \lambda & & & \\ \lambda & \lambda & \lambda + 2\mu & & & \\ & & & \mu & 0 & 0 \\ & & & 0 & \mu & 0 \\ & & & 0 & 0 & \mu \end{pmatrix} \tag{2}$$

We can then write the equation of motion for the infinitesimal displacements $\overline{w}_k$ caused by the passing RW

$$\overline{S}_{klmn}\, \overline{w}_{m,ln} = \frac{1}{\mu}\, \rho\, \ddot{\overline{w}}_k \qquad (k,l,m,n = 1,2,3) \tag{3}$$

where

$$\overline{S}_{klmn} = S_{klmn} + S'_{klmn} \tag{4}$$

$$S_{klmn} = \frac{\lambda}{\mu}\, \delta_{kl}\, \delta_{mn} + \delta_{km}\, \delta_{ln} + \delta_{kn}\, \delta_{lm} \tag{5}$$

$$S'_{klmn} = \frac{1}{\mu}\, C'_{klmn}\, . \tag{6}$$

Here S_{klmn} is the effective stiffness tensor for the fictitious isotropic material, divided by μ to make it nondimensional. S'_{klmn} are the corresponding perturbative corrections. C'_{klmn} are given by the perturbative corrections (in Voigt's notation) $\tilde{C}'_{pq}$ after a rotation from the $\tilde{x}$-system to the x-system. ρ is the material density and the notation $A_{i,k} \equiv \partial A_i / \partial x_k$ is used.

The equation of motion for the unperturbed case

$$S_{klmn}\ w_{m,ln} = \frac{1}{\mu}\ \rho\ \ddot{w}_k \qquad (k,l,m,n = 1,2,3) \tag{7}$$

can be easily solved by assuming a plane wave solution

$$w_j = W_j \exp\left[ik\ (n_l x_l - \mathrm{v}t)\right] \qquad (j = 1,2,3) \tag{8}$$

where the wave number k is assumed to be real, while the direction cosines n_l may be complex. Due to our assumptions about the geometry of the system $n_1 = 0$, $n_2 = 1$. n_3 and the phase velocity v are the unknowns of the problem. They can be both determined by substituting Eq. (8) into Eq. (7) and solving the corresponding system together with the system of boundary conditions

$$T_{3k} \equiv \mu\ S_{k3mn}\ w_{m,n} = 0 \qquad (k = 1,2,3) \tag{9}$$

which require that the surface $x_3 = 0$ be traction-free.

As a result we obtain the Rayleigh wave equation

$$[(\mathrm{v}/\mathrm{v}_T)^2 - 2]^2 = 4\ [1 - (\mathrm{v}/\mathrm{v}_L)^2]^{1/2}\ [1 - (\mathrm{v}/\mathrm{v}_T)^2]^{1/2} \tag{10}$$

where $\mathrm{v}_T = \sqrt{\mu/\rho}$ and $\mathrm{v}_L = \sqrt{(\lambda + 2\mu)/\rho}$ are the transverse (shear) and longitudinal wave velocity, respectively.

One of the solutions of Eq. (10) is the RW phase velocity v, which is approximately given by[14]

$$\mathrm{v} = \frac{0.87 + 1.12\ \nu}{1 + \nu}\ \mathrm{v}_T \tag{11}$$

where ν is the Poisson's ratio

$$\nu = \lambda / [2(\lambda + \mu)]\ . \tag{12}$$

For physical values of ν $(0 < \nu < 0.5)$ one finds $\mathrm{v} < \mathrm{v}_T < \mathrm{v}_L$. One also finds that n_3 is pure imaginary, i.e. RW's decay exponentially with x_3 (depth direction).

In our treatment we assume that the "perturbative corrections" $\tilde{C}'_{km}$ are small, compared to C_{km}, and therefore can be treated at the lowest order of perturbation. Consequently the corrections v', w'_j, $W'_j \ldots$ to v, w_j, $W_j \ldots$ are also small. The treatment follows the same procedure as for the "unperturbed" case, keeping at any step only first or second order terms, as needed. The result is a linearized formula in which the change in RW phase velocity v' due to the anisotropy is given by

$$\begin{aligned}\frac{\mathrm{v}'}{\mathrm{v}} = \frac{1}{2\mu}\ \{ & A_{2222}\ [\ \sin^4\phi\ \tilde{C}'_{11} + \cos^4\phi\ \tilde{C}'_{22} + 2\sin^2\phi\ \cos^2\phi\ (\tilde{C}'_{12} + 2\tilde{C}'_{66})] \\ & + A_{2233}\ (\sin^2\phi\ \tilde{C}'_{13} + \cos^2\phi\ \tilde{C}'_{23}) \\ & + A_{23}\ (\sin^2\phi\ \tilde{C}'_{55} + \cos^2\phi\ \tilde{C}'_{44}) + A_{3333}\ C'_{33} \}\end{aligned} \tag{13}$$

where

$$A_{2222} = \alpha\, [(up + u - 2p)\, \beta - u(1-p)/2]/(p-u)$$
$$A_{2233} = \alpha[2(2-u)\beta - u] \tag{14}$$
$$A_{23} = \alpha(2-u)\, [4\beta - u/(2-2u)]$$
$$A_{3333} = A_{2222}\, (1 - u/p)$$

and

$$\alpha = \frac{2(u-1)\,(u-p)}{u(2u^2 - 3up + u + 2p - 2)} \tag{15}$$
$$\beta = p/(2-2p) + 1/u \tag{16}$$
$$p = \lambda/\mu + 2 \tag{17}$$
$$u = \rho v^2/\mu\,. \tag{18}$$

We note that u is a function only of p and consequently that the coefficients A_{2222}, A_{2233}, A_{23} and A_{3333} depend only on λ/μ or, equivalently, on the Poisson's ratio ν.

II. CHANGE IN RW PHASE VELOCITY DUE TO AN ORTHOTROPIC DISTRIBUTION OF CUBIC CRYSTALLITES

We assume now that the material contains an orthotropic distribution of cubic crystallites. In the reference system of cubic crystals there are only three independent second order elastic constants: $C^o_{11} = C^o_{22} = C^o_{33}$, C^o_{12} $C^o_{13} = C^o_{23}$ and $C^o_{44} = C^o_{55} = C^o_{66}$. We define, as a measure of the anisotropy:

$$c = C^o_{11} - C^o_{12} - 2C^o_{44}\,. \tag{19}$$

For an isotropic material $c = 0$. We note that other authors give a different definition of the anisotropy, e.g.[15]

$$c' = 2\, C^o_{44}/(C^o_{11} - C^o_{12})\,. \tag{20}$$

In this case $c' = 1$ indicates isotropy. We then assume

$$\lambda = C^o_{12} + c/5$$
$$\mu = C^o_{44} + c/5\,. \tag{21}$$

We need now to derive the resulting elastic constants for the orthotropic distribution. From Eqs. (1), (19), (21) and Eq. (4) of ref. 10, it follows

$$\tilde{C}'_{kk} \equiv \bar{C}_{kk} - \lambda - 2\mu = c(0.4 - 2 <r_{kk}>) \qquad (k = 1,2,3)$$
$$\tilde{C}'_{km} = \bar{C}_{km} - \lambda = c\,(<r_{km}> - 0.2) \qquad (k \neq m = 1,2,3) \tag{22}$$
$$\tilde{C}_{kk} \equiv \bar{C}_{kk} - \mu = c\,(<r_{kk}> - 0.2) \qquad (k = 4,5,6)$$

where

$$r_{kk} = \sum_{i<j} a^2_{ki}\, a^2_{kj} \qquad (k = 1,2,3)$$
$$r_{km} = \sum_{i} a^2_{ki}\, a^2_{mi} \qquad (k \neq m = 1,2,3) \tag{23}$$
$$r_{44} = r_{23},\ r_{55} = r_{13},\ r_{66} = r_{12}$$

and a_{ij} is the rotation matrix between the reference frames of the crystallite and of the orthotropic symmetry axes ($\tilde{x}_k$-system).

The rotation between the two reference frames can also be defined by three Euler angles θ, ψ and Φ and the orientation distribution of the crystallites by a normalized probability function $w(\cos\theta, \psi, \Phi)$. Following ref. 10, we proceed with the expansion:

$$w(\cos\theta, \psi, \Phi) = \sum_{l=0}^{\infty} \sum_{m=-l}^{l} \sum_{n=-l}^{l} W_{lmn}\, Z_{lmn}(\cos\theta) \exp(-im\psi) \exp(-in\Phi) \qquad (24)$$

where Z_{lmn} are generalized Legendre functions and are defined in ref. 16.

Because of symmetry arguments, the number of nonzero independent coefficients W_{lmn} is reduced, for the case under consideration, to only three: W_{400}, W_{420} and W_{440}. The $<r_{km}>$ can be explicitly written in terms of the W_{lmn}.

$$\begin{aligned}
<r_{11}> &= 0.2 - a\,(3\,W_{400} - 2\sqrt{10}\,W_{420} + \sqrt{70}\,W_{440}) \\
<r_{22}> &= 0.2 - a\,(3\,W_{400} + 2\sqrt{10}\,W_{420} + \sqrt{70}\,W_{440}) \\
<r_{33}> &= 0.2 - 8a\,W_{400} \\
<r_{44}> &= <r_{23}> = 0.2 - 4a\,(2\,W_{400} + \sqrt{10}\,W_{420}) \\
<r_{55}> &= <r_{13}> = 0.2 - 4a\,(2\,W_{400} - \sqrt{10}\,W_{420}) \\
<r_{66}> &= <r_{12}> = 0.2 + 2a\,(W_{400} - \sqrt{70}\,W_{440})\,.
\end{aligned} \qquad (25)$$

From Eqs. (13), (22) and (25) it finally follows

$$\frac{v'}{v} = \frac{c}{\mu}\,(\alpha_0\,W_{400} + \alpha_1\,W_{420}\cos 2\phi + \alpha_2\,W_{440}\cos 4\phi) \qquad (26)$$

where

$$\begin{aligned}
\alpha_0 &= a\,(3A_{2222} + 8A_{3333} - 4A_{2233} - 4A_{23}) \\
\alpha_1 &= 2a\sqrt{10}\,(A_{2222} - A_{2233} - A_{23}) \\
\alpha_2 &= a\sqrt{70}\,A_{2222}\,. \\
a &= 2\sqrt{2}\pi^2/35
\end{aligned} \qquad (27)$$

III. DETERMINATION OF THE TEXTURE COEFFICIENTS

We now apply Eq. (26) to the inverse problem of determining the texture coefficients W_{400}, W_{420} and W_{440} by measuring the RW propagation velocity $\bar{v}$ at several angles ϕ. The experimental apparatus and technique for the measurement of $\bar{v}$ are described in an accompanying paper.[13] We only note here that absolute measurements of $\bar{v}$ (ϕ) may be more difficult (or less reliable) than relative measurements of $\bar{v}$ (ϕ) minus $\bar{v}$ at some fixed angle ϕ_0. We, therefore, consider separately the two cases of absolute and relative measurements of the RW velocity.

Absolute measurements. We assume in this subsection that absolute measurements of $\bar{v}$ (ϕ) are possible and sufficiently reliable. Subtracting from $\bar{v}$ the isotropic velocity v, given by Eq. (10), we obtain v' (ϕ). From Eq. (26) it follows

$$v'(0^o) = \frac{cv}{\mu} (\alpha_0 W_{400} + \alpha_1 W_{420} + \alpha_2 W_{440})$$
$$v'(45^o) = \frac{cv}{\mu} (\alpha_0 W_{400} - \alpha_2 W_{440}) \tag{28}$$
$$v'(90^o) = \frac{cv}{\mu} (\alpha_o W_{400} - \alpha_1 W_{420} + \alpha_2 W_{440})$$

The system of Eqs. (28) can be easily solved

$$W_{400} = a_o g_o$$
$$W_{420} = a_1 g_1 \tag{29}$$
$$W_{440} = a_2 g_2$$

where

$$g_o = v' (0^o) + 2v' (45^o) + v'(90^o)$$
$$g_1 = v' (0^o) - v'(90^o) \tag{30}$$
$$g_2 = v' (0^o) - 2v' (45^o) + v' (90^o)$$

and

$$a_k = \frac{\mu(1 + \delta_{k1})}{4cv\alpha_k} . \tag{31}$$

Relative measurements. If only differences of RW velocities at different angles can be measured reliably, we choose as fixed angle $\phi_o = 0^o$ and define

$$h(\phi) \equiv \bar{v} (0^o) - \bar{v} (\phi) = v' (0^o) - v' (\phi) . \tag{32}$$

We note that, in this case, it is not necessary to subtract the isotropic velocity v. From Eq. (26)

$$h(\phi) = \frac{2cv}{\mu} (\alpha_1 W_{420} \sin^2 \phi + \alpha_2 W_{440} \sin^2 2\phi) \tag{33}$$

from which it immediately follows

$$W_{420} = a_1 h(90^o)$$
$$W_{440} = a_2 [2h(45^o) - h(90^o)] . \tag{34}$$

We note that, with only relative measurements of the RW velocities, the determination of W_{400} is not possible.

Error analysis. Eqs. (29) and (34) make the error analysis very easy. In fact, e.g.

$$\Delta W_{400}/ W_{400} = \sqrt{(\Delta a_o/a_o)^2 + (\Delta g_o/g_o)^2} \tag{35}$$

with similar expressions for W_{420} and W_{440} .

The errors Δg_k due to experimental inaccuracies in the measurement of the RW velocities are discussed elsewhere.[13] We simply note here that they can be greatly reduced if, instead of only three measurements at 0°, 45° and 90°, more measurements at other angles and a least square fit are performed. The errors Δa_k are due to inaccuracy in the determination of the elastic constants and will be discussed in the next section.

IV. NUMERICAL RESULTS

As we have seen in Sec. III, the texture coefficients W_{400}, W_{420} and W_{440} can be easily correlated to the RW phase velocities at 0°, 45° and 90°. The calculation of the correlation coefficients a_k (Eq. 31) is, however, rather cumbersome, since it requires a numerical solution of Eq. (10) and the evaluation of the A_{klmn} coefficients (Eqs. 14-18) and of the α_k coefficients (Eq. 27). Therefore we plot here the a_k and α_k coefficients for an extended range of values of the elastic constants, which includes several examples of cubic crystals: Al, Fe, Cu and NaCl.

In Table 1 we report, from ref. 15, the values of the elastic constants of these crystals, in units of 10^{10} Pascal. Since all the quantities of interest can be expressed as functions of only the two ratios C^o_{11}/C^o_{44} and C^o_{12}/C^o_{44}, we also report these ratios and the corresponding anisotropy c, as defined in Eq. (19), but divided by μ to make it nondimensional.

Table 1. Elastic constants and anisotropy

	C^o_{11} [a]	C^o_{12} [a]	C^o_{44} [a]	C^o_{11}/C^o_{44}	C^o_{12}/C^o_{44}	c/μ [b]
Al	10.82	6.13	2.85	3.8	2.15	−0.38
Cu	17.02	12.3	7.51	2.23	1.64	−1.89
Fe	23.7	14.1	11.6	2.04	1.22	−1.53
NaCl	0.49	0.124	0.126	3.89	0.98	0.766

[a] Units: 10^{10} Pascal

[b] The anisotropy c is defined in Eq. (19)

Figures (1) and (2) show the plots of c/μ vs. the two ratios C^o_{11}/C^o_{44} and C^o_{12}/C^o_{44}, respectively. We can see that the dependence of c/μ over these ratios is very strong. Figures (3) and (4) show the plots of α_o vs. C^o_{11}/C^o_{44} and C^o_{12}/C^o_{44}, respectively. Plots of α_1 and α_2 are similar. The dependence is very weak. The coefficients a_k depend (see Eq. 31) on c/μ, v and α_k. Since $v = \sqrt{\mu u/\rho}$, there is an explicit dependence of a_k on the density ρ and, through μ on C^o_{44}. Therefore we cannot plot, in a single representation, the coefficients a_k for different crystals. In Figs. 5 and 6 we show the plots of a_o vs. C^o_{11}/C^o_{44} and C^o_{12}/C^o_{44} for values of ρ and C^o_{44} corresponding to *Al*. Plots of a_1 and a_2 are similar. Plots for other crystals (Fe, Cu and NaCl) look very similar, but the points corresponding to the physical values of the elastic constants are further away from the asymptota then in *Al*. All these plots show a very strong dependence on C^o_{11}/C^o_{44} or C^o_{12}/C^o_{44}, due to the presence of c in the denominator of Eq. (31), with asymptota corresponding to zeroes of c and therefore to isotropy of the material. Around these points our technique for the determination of the texture coefficients is no longer applicable because of the large errors associated with any uncertainty of the elastic constants. On the other side, the coefficients W_{lmn} become, in this case, meaningless. In fact, if the crystallites are isotropic ($c = 0$), any orientation distribution is acoustically equivalent.

We can also use Figs. (1)-(6) for evaluating the term $\Delta a_k/a_k$ in the error analysis of Sec. III. We can see that, as observed above, any small uncertainty in the elastic constants affects seriously the reliability of our technique in the proximity of the asymptota. Therefore a better precision in the determination of the elastic constants is required, for Al than for Fe, Cu or NaCl. Since v and α_k are

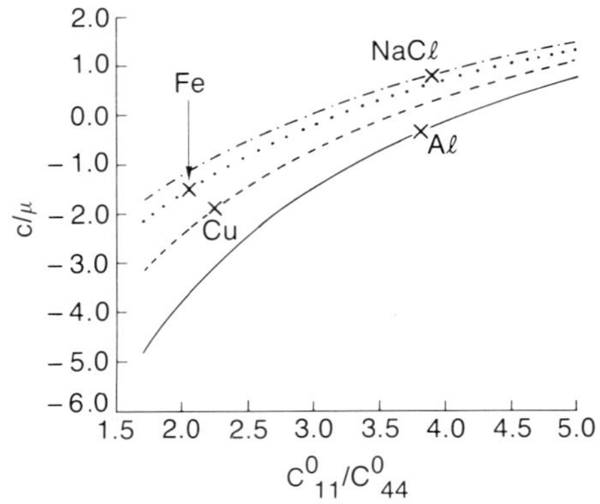

Fig. 1 — Anisotropy c/μ vs. C_{11}^{o}/C_{44}^{o} for several values of C_{12}^{o}/C_{44}^{o}: 2.15 (solid line), 1.64 (dashed line), 1.22 (dotted line) and 0.98 (chain-dashed line).

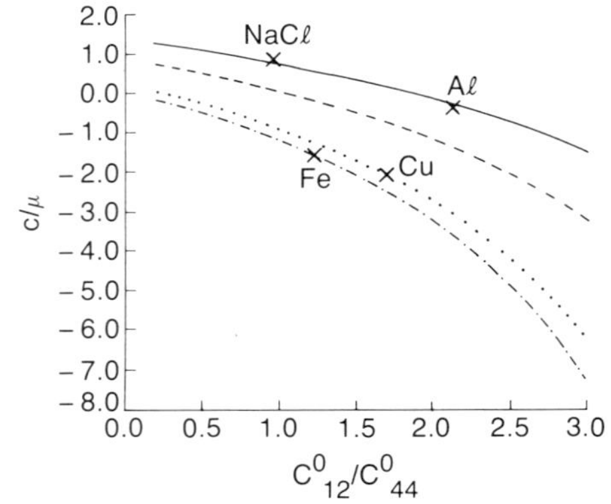

Fig. 2 — Anisotropy c/μ vs. C_{12}^{o}/C_{44}^{o} for several values of C_{11}^{o}/C_{44}^{o}: 3.85 (solid line), 3.05 (dashed line), 2.23 (dotted line) and 2.04 (chain-dashed line).

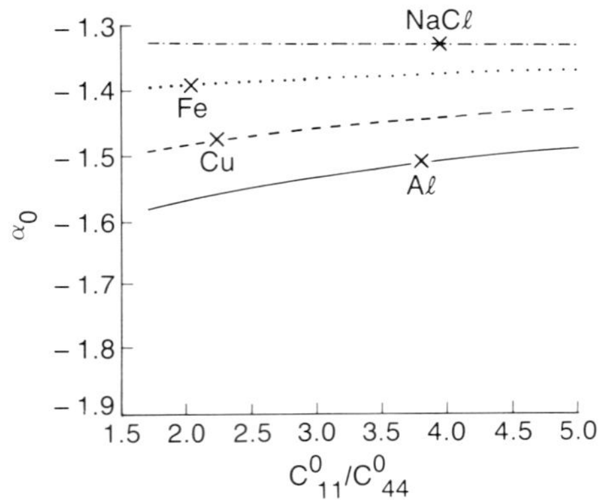

Fig. 3 — Coefficient α_o vs. C_{11}^{o}/C_{44}^{o} for several values of C_{12}^{o}/C_{44}^{o}: 2.15 (solid line), 1.64 (dashed line), 1.22 (dotted line) and 0.98 (chain-dashed line).

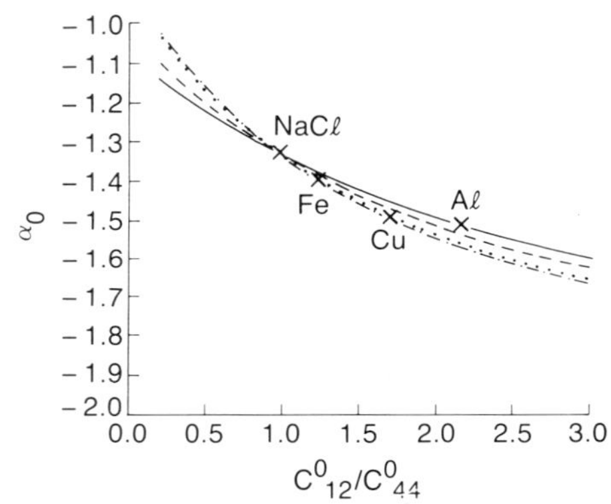

Fig. 4 — Coefficient α_o vs. C_{12}^{o}/C_{44}^{o} for several values of C_{11}^{o}/C_{44}^{o}: 3.85 (solid line), 3.05 (dashed line), 2.23 (dotted line) and 2.04 (chain-dashed line).

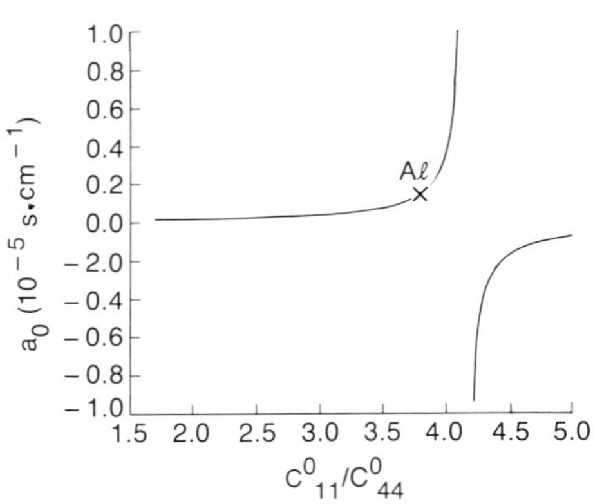

Fig. 5 — Coefficient a_o vs C_{11}^{o}/C_{44}^{o} for values of C_{12}^{o} and C_{44}^{o} corresponding to *Al*.

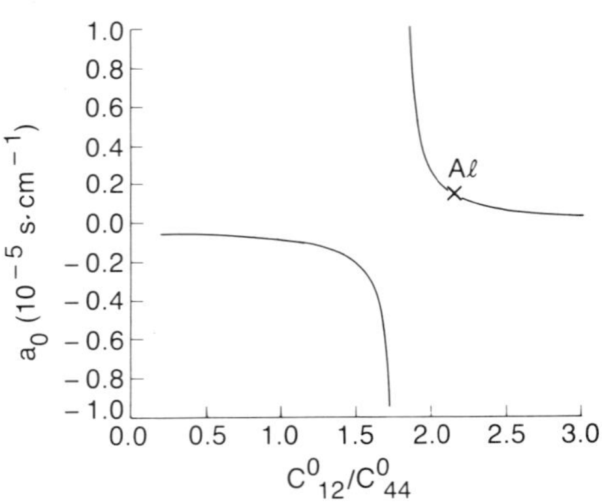

Fig. 6 — Coefficient a_o vs. C_{12}^{o}/C_{44}^{o} for values of C_{11}^{o} and C_{44}^{o} corresponding to *Al*.

relatively insensitive to small variations in the elastic constants, any corresponding error $\Delta v/v$ or $\Delta\alpha_k/\alpha_k$ can be neglected, compared with $\Delta(c/\mu)/(c/\mu)$. Therefore

$$\frac{\Delta a_k}{a_k} \simeq \frac{\Delta(c/\mu)}{c/\mu} \approx \frac{\Delta c}{c} \qquad (36)$$

i.e. the error in the determination of the W_{lmn} coefficients due to uncertainties in the elastic constants of the material is approximately equal to the uncertainty in the anisotropy and can be immediately evaluated from Figs. (1) and (2).

CONCLUSION

We have presented a perturbation formalism for the calculation of the change in Rayleigh wave velocity due to an orthotropic distribution of cubic crystallites in an unstressed homogeneous plate. A technique, based on this formalism, is then proposed for the determination of the texture coefficients, by means of several measurements of RW velocity at different propagation angles. Simple formulas are derived, which correlate the W_{400}, W_{420} and W_{440} coefficients, which characterize the material texture, with the measured RW velocities. The correlation coefficients are also plotted and an error analysis is performed to evaluate the error in the texture coefficients, due to uncertainties in the elastic constants of the cubic crystallites. Our results are used in the calculation of the texture coefficients for a rolled 2025-T351 *Al* plate.[13]

References

1. D. S. Hughes and J. L. Kelly, Phys. Rev. 92, 1145 (1953)
2. R. B. King and C. M. Fortunko, J. Appl. Phys. 54, 3027 (1983)
3. R. B. King and C. M. Fortunko, NBSIR 84-3002, National Bureau of Standards, Boulder, Colorado (1984)
4. A. V. Clark and R. B. Mignogna, Ultrasonics 22, 205 (1984)
5. A. V. Clark and R. B. Mignogna, in Rev. Progress in Quantitative NDE, Vol. 4B, D. O. Thompson and D. E. Chimenti, Eds. (Plenum Press, New York, 1985), p. 1095
6. M. Hirao, H. Fukuoka and K. Hori, J. Appl. Mech. 48, 119 (1981)
7. P. P. Delsanto and A. V. Clark, in Rev. Progress in Quantitative NDE, Vol. 5B, D. O. Thompson and D. E. Chimenti, Eds. (Plenum Press, New York, 1986), p. 1407
8. G. T. Mase and G. C. Johnson, "Surface Waves in Anisotropic Materials," to appear in Rev. Progress in Quantitative NDE, Vol. 5.
9. A. Zeiger and K. Jassby, J. of Nondestr. Ev. 3, 115 (1982)
10. C. M. Sayers, J. Phys. D 15, 2157 (1982)
11. D. R. Allen, R. Langman and C. M. Sayers, Ultrasonics 23, 215 (1985)
12. P. P. Delsanto and A.V. Clark, "Rayleigh Wave Propagation in Deformed Orthotropic Materials," submitted to J. Acoust. Soc. Am.
13. R. B. Mignogna, P. P. Delsanto, B. B. Rath, C. L. Vold and A. V. Clark, "Ultrasonic Measurements on Textured Materials," to be presented at the 2nd Int. Symposium on the Nondestructive Characterization of Materials, Montreal, July 21-3, 1986
14. I. A. Viktorov, Rayleigh and Lamb Waves (Plenum Press, N.Y., 1967)
15. W. P. Mason, Physical Acoustics and the Properties of Solids (D. Van Nostrand Co., N.Y., 1958)
16. R. J. Roe, J. Appl. Phys. 37, 2069 (1966)

ULTRASONIC MEASUREMENTS ON TEXTURED MATERIALS

R. B. Mignogna, P. P. Delsanto, A. V. Clark Jr.,*
B. B. Rath, and C. L. Vold
Naval Research Laboratory
Washington, D.C. 20375-5000

INTRODUCTION

Over the past 30 years the study of acoustoelasticity has gone through various stages of development. Hughes and Kelly reported their initial work on acoustoelasticity in 1953.[1] Theoretically they considered materials to be homogeneous and isotropic but found limited experimental agreement with theory due to initial anisotropy. Early on, it was recognized that most materials display an initial anisotropy in regards to acoustic and acoustoelastic measurements. This was attributed to preferential grain orientation (texture). Numerous works considering ultrasonic texture measurement prior to 1973 are referenced by Green.[2] Not until twenty years after Hughes and Kelly, in 1973, did Iwashimizu and Kubomura present a theory to account for both initial anisotropy and stress. The theory they proposed assumes that a homogeneous orthotropic symmetry exists in rolled plates. Their theory was found to agree rather well with experimental data and spurred renewed interest in the potential of acoustoelasticity. A number of refinements have been made using slight orthotropic symmetry as a basis. However, all of these theories consider only the macroscopic symmetry displayed by the material.[3,4]

More recent work by Allen and Sayers[5–7] begins to view the effects of texture on acoustics and acoustoelasticity in a more fundamental way and theoretically takes into account that metals are composed of grains or crystallites. Each of these grains has its own symmetry axes. The alignment of these axes may range from a random to a highly ordered distribution. Ordered alignment of the grains usually results in the material displaying some type of macroscopic symmetry such as the orthotropic symmetry of rolled plate mentioned earlier and has been described in terms of crystallite orientation distribution functions (CODF). This type of description was initially developed for the analysis of x-ray diffraction data and is now being used in conjunction with acoustoelasticity.[5–8] Using CODF's for acoustoelastic applications, both the symmetry of the crystallites and the macroscopic texture induced symmetry are considered.

A major assumption in most acoustoelastic developments is that the texture is considered homogeneous through the thickness of the sample. A homogeneous texture through the thickness is, in general, not found in most fabricated

* Current address: NBS, Boulder, Co. 80303

materials. It is the purpose of this paper to describe ultrasonic measurements of texture in conjunction with theoretical results reported earlier in this conference for the case of Rayleigh waves incorporating crystallite orientation distribution functions.[8] We also present acoustic birefringence measurements to support the surface wave results showing texture variations through the plate thickness. Discussion is also presented in regards to comparing results of various measurement techniques, i.e. surface waves to bulk or plate waves in the presence of texture and stress inhomogeneity.

EXPERIMENTAL CONSIDERATIONS AND RESULTS

The two most commonly used materials for acoustoelastic studies appear to be rolled plates of aluminum alloys and iron alloys. Most of these products have two important features in common. The basic crystalline structure of these materials is some form of cubic symmetry and the rolled products display orthotropic symmetry. However, they also have one unfortunate feature in common, their sensitivity to the rolling processes such as amount of reduction and thermal history during and after processing. Because of this, various types and degrees of texture can be found for the same alloy depending upon the mechanical and thermal processing. We have shown previously that commercially obtained plates of the same alloy and temper designation but various thicknesses can have both a different texture as shown by x-ray pole figures and a different initial acoustic birefringence.[9] In that report we also showed a difference in the x-ray pole figures taken at the surface of the plate and those taken from the center-plane of the same plate.

The material chosen for this example was a 19.0 mm rolled plate of 2024-T351 aluminum. The 2024 alloy is an aluminum-copper-magnesium alloy, the T351 temper designation indicates solution heat treatment, natural aging and stress relief by stretching (0.5 to 3% permanent set). Materials not stress relieved may have substantial residual stresses. An example of this is the case of 7075 aluminum, with a T6 temper designation (solution heat treated and artificially aged) residual stress may range to 26000 PSI. However, with a T651 temper designation (solution heat treated, artificially aged and stress relieved by stretching) residual stress ranges only to 2000 PSI.[10] These stresses (and also texture) are inhomogeneous through the thickness and usually symmetric about the center plane depending upon the stress relief process. In many cases it is assumed little or no residual stress exists in the materials used for study. However, with the possibility of such large residual stress present in commercially available materials, it is evident that a great deal of care is necessary to match materials and assumptions.

The microstructure of the 2024-T351 plate shows a grain size variation through the thickness that is symmetric about the center-plane. Taking the fabrication processes into consideration we believe it is quite reasonable to assume that most properties of the plate are symmetric about the center plane of the plate. Figure 1 shows photomicrographs of the plate edge, parallel and perpendicular to the rolling direction. Only photomicrographs of the surface to a depth of 1.6 mm and +0.8 mm about the center plane are presented. However a nonuniform variation exists from the top surface to the center of the plate. It should also be noted that the grains are extremely elongated along the rolling direction: up to 5 to 10 times their width (perpendicular to the rolling direction) as can be seen in the photomicrographs.

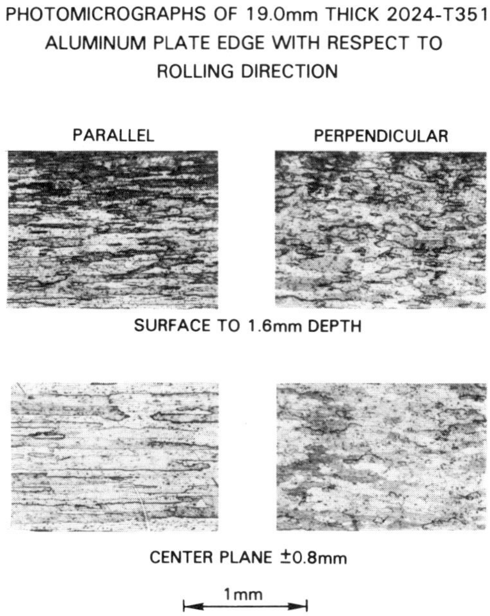

Fig. 1 — Photomicrographs of plate edge, parallel and perpendicular to rolling direction.

Figure 2 shows [111] x-ray pole figures, labeled A through D, corresponding to various depths in the plate. Specimens were prepared by removing material to expose these depths, where *A* is the top surface, *B* is 0.4 mm below the surface, *C* is 3.3 mm below the surface and *D* is 9.5 mm below the surface (center plane of plate). The peak diffracted x-ray intensity for *A* through *D* was 1.49, 1.60, 3.04, 3.61 respectively. The pole figures are presented for qualitative comparison to the ultrasonic measurements.

A specimen was machined from the 19.0 mm plate for surface wave measurements. The thickness of the original 19.0 mm plate was reduced by machining from one side only followed by polishing of the machined surface to 600 grit. The specimen thus had one original surface and the center plane of the plate exposed as the second surface. The specimen dimensions were 200 by 200 by 10 mm.

A device was constructed to measure the time of flight of Rayleigh waves (RW) as a function of angle on the surfaces of the specimen. The device consisted of a commercial wedge RW transducer as the transmitter and two conical PZT elements as receivers. The frequency used for this work was 4.5 MHz. The wedge of the transmitter was modified to reduce the area of contact. The contact area was reduced from 25 by 19 mm to 5 by 19 mm in order to minimize couplant variation and enhance the devices capability of having all three elements (transmitter and two receivers) contact the plate in a repeatable fashion. The conical receivers were in dry contact with the plate surface. All three transducers were mounted in a semi-rigid fixture. The two conical elements were cut from a 6.35 mm thick poled PZT disk. The elements were initially machined as cylinders 6.35 mm in diameter and then a cone having a height of 3.2 mm was machined on one end of each elements, the total height remaining 6.35 mm.

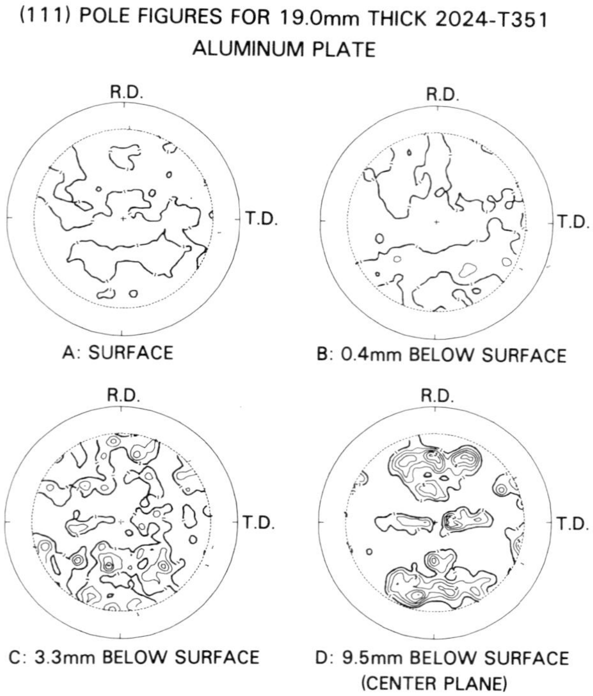

Fig. 2 — [111] x-ray pole figures corresponding to various depths in the plate.

The conical receivers were then rigidly mounted on a 6.35 by 9.5 by 31.3 mm fused quartz bar, with a spacing of 27.5 mm. The receiver elements were mounted on fused quartz for both mechanical and thermal stability. Fused quartz has a very small coefficient of thermal expansion, 0.5×10^{-6} mm/mm/C°, thus enabling any temperature change of the transducer device, i.e., any thermal expansion of the fused quartz bar, due to handling during the measurements to be neglected. The quartz bar was designed to rotate about the center of its length when mounted in the semi-rigid fixture. The ability to rotate insures that the two receivers and the transmitter will contact the plate whether the plate is flat or has some curvature. However, the fixture maintains all three transducers rigidly in line on the plate surface. A schematic diagram of the device is shown in Fig. 3.

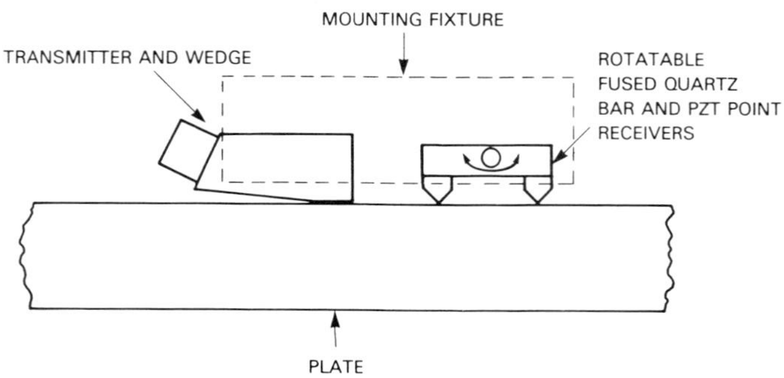

Fig. 3 — Schematic diagram of surface wave device consisting of a transmitter and two PZT point receivers.

The outputs of the receiver transducers were connected in parallel to an amplifier. A pulsed oscillator was used to drive the transmitter and give a quasi-cw pulse. The RW time of flight was measured from the difference in arrival times of the wave at the two receiver transducers. This was done by considering the two signals as "echoes" and using a modified pulse-echo-overlap[11] techniques (P.E.O.).

Two sources of experimental error were considered in the measurements. The first is due to the P.E.O. technique itself and the second is due to the finite size of the two receiver transducers in contact with the surface of the plate. The second source of error is seen when attempting to make repeated measurements, i.e., lifting the entire device from the plate surface and setting it down at a slightly different position. The finite size of the receiver transducers manifests itself as an "uncertainty" in the otherwise fixed spacing between the two caused by slightly different areas of the receiver transducers coming into contact with the plate surface. In order to evaluate the errors, repeatability measurements were made on a front surface mirror, considering the aluminized glass to be an ideal material and surface (homogeneous, isotropic and having minimal surface roughness). A set of forty measurements were made on the front surface mirror by lifting the entire device from the surface and then placing it back on the surface at a slightly different position and remeasuring the RW travel time. The average travel time was approximately 3 μsec with a standard deviation of $\pm 0.9 \times 10^{-4}$ or ± 0.3 nsec.

RW time of flight measurements were made on both surfaces of the 200 by 200 by 10 mm plate at 15° intervals with respect to the rolling direction. At each angle 13 or more time of flight measurements were made. The standard deviation of these measurements was substantially higher than those on the front surface mirror to $\pm 7 \times 10^{-4}$ or ± 2.4 nsec. Increasing the number of measurements did not significantly reduce the standard deviation of the measurements. We believe the increase in the standard deviation is due to surface roughness, however further investigation to increase the repeatability of the measurements is in progress. The maximum differences in the time of flight occurred at 0° and 90° for both sides of the plate. However, the normalized difference in time of flight measurements on the exposed center plane $|[t(0)-t(90)]/\langle t \rangle|$, where t is the time of flight, were much larger than those for the surface, 8.5×10^{-3} versus 1.6×10^{-3}.

Analysis of the RW wave data was accomplished using the method proposed by Delsanto *et al.*[8]. Although the time of flight data was averaged over a number of measurements for each angle, further averaging was accomplished by using a least squares routine for the data set as a function of angle. The least squares results were then used for calculation of the texture coefficients, W_{400}, W_{420} and W_{440}.

Figure 4 shows the experimental data and the least squares fit for the absolute measurement method described by Delsanto *et al.*[8] in terms of $V'(\phi) = V(\phi) - V$ as a function of the angle ϕ. $V(\phi)$ is the measured velocity and V is the calculated isotropic velocity. The error bar shown is the standard deviation from the least squares fit. Both the data from the top surface and the center plane of the plate are shown on the same graph. The texture coefficients for the surface plane are: $W_{400} = 0.0087$, $W_{420} = 0.0006$ and $W_{440} = -0.0017$. The coefficients for the center plane are: $W_{400} = 0.0087$, $W_{420} = -0.0037$ and $W_{440} = -0.0025$.

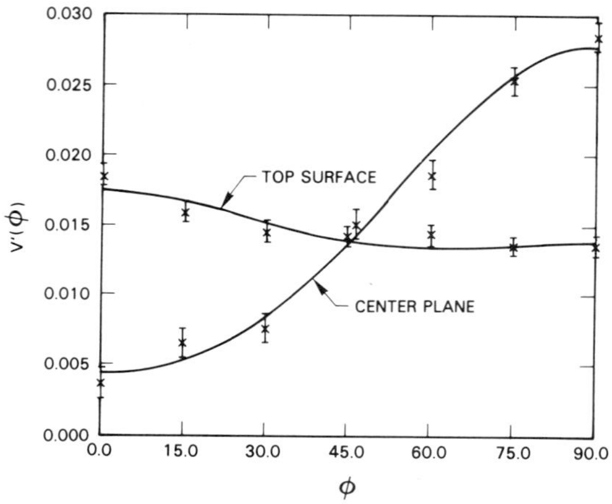

Fig. 4 — RW velocity data and least squares fit as a function of angle for both the top surface and center plane of the plate using the absolute measurement method. The texture coefficient for the surface plane are: $W_{400} = 0.0087$, $W_{420} = 0.0006$ and $W_{440} = -0.0017$. The coefficients for the center plane are: $W_{400} = 0.0087$, $W_{420} = -0.0037$ and $W_{440} = -0.0025$.

The relative measurement method presented by Delsanto[8] defines the function $h(\phi)$ as the difference of the RW wave velocity at various angles from a fixed angle ϕ. Figure 5 shows the experimental data and the least squares fit in terms of $h(\phi)$ as a function of the angle ϕ. The error bar shown is the standard deviation from the least squares. Both the data from the top surface and the center plane of the plate are shown on the same graph. The texture coefficients for the surface plane are: $W_{420} = 0.0007$ and $W_{440} = -0.0025$. The coefficients for the center plane are: $W_{420} = -0.0038$ and $W_{440} = -0.0025$. W_{400} is not determined using the relative measurement method. An advantage of the relative method is, however, the fact that only two unknowns, W_{420} and W_{400} are used in the least squares determination. Also the isotropic velocity need not be subtracted from the measured velocity as is necessary in the absolute measurement method.

Four samples, 25 by 25 mm square by approximately 5 mm thick, were cut from the plate. The samples were cut at various depths from the surface so that x-ray pole figures could be made at those depths. Through thickness birefringence measurements were also made on these same specimens. The birefringences being defined as $B = (V_{S\parallel} - V_{S\perp})/\langle V \rangle$ where $V_{S\parallel}$ and $V_{S\perp}$ are the horizontally polarized shear (SH) wave velocities propagating normal to the plate and polarized parallel and perpendicular to the rolling direction, respectively. $\langle V \rangle$ is the average SH wave velocity. The positions from which the four specimens (labeled A,B,C,D) were cut are shown pictorially along with the corresponding birefringence measurements in Fig. 6. Also indicated in Fig. 6 is the depth where the x-ray pole figures were made. Although some of the specimens overlap depth wise and are not discrete layer by layer, the birefringence was found to vary by more than a factor of two, from 4.37×10^{-3} to 9.20×10^{-3}. It should be noted that the acoustic axes did not rotate from the rolling direction. The birefringence measurements were made as described in a previous report.[12] The initial acoustic birefringence through the entire thickness of plate was found to be 6.80×10^{-3}.

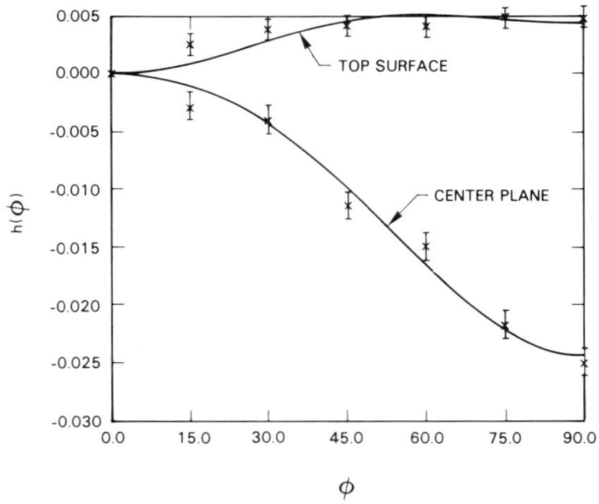

Fig. 5 — RW velocity data and least squares fit as a function of angle for both the top surface and center plane of the plate using the relative measurement method. The texture coefficient for the surface plane are: $W_{420} = 0.0007$ and $W_{440} = -0.0025$. The coefficients for the center plane are: $W_{420} = -0.0038$ and $W_{440} = -0.0025$.

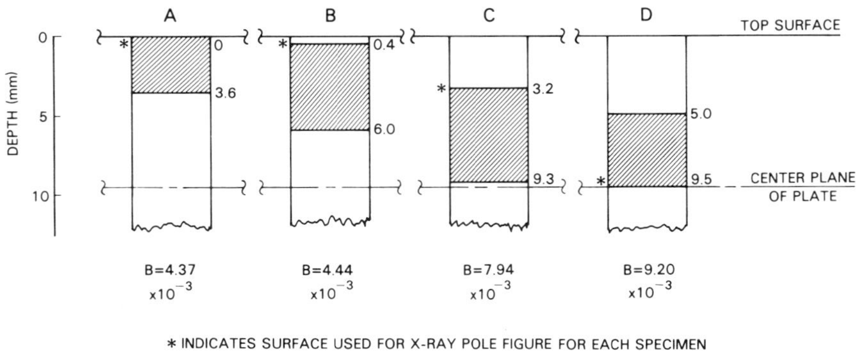

Fig. 6 — Positions relative to surface and center plane of plate of birefringence specimens.

Allen *et al.* recently reported results for an ultrasonic velocity combination technique using bulk waves [6]. Using Allen's results relating W_{420} to the difference of the velocity squared of plane polarized shear waves propagating normal to the plate and polarized parallel and perpendicular to the rolling directions $V_{S\parallel}^2 - V_{S\perp}^2$, we can write

$$W_{420} = 7\sqrt{5}\,\rho\,(V_{S\parallel}^2 - V_{S\perp}^2)\,/\,32\,\pi^2\,c \tag{8}$$

where c is defined in terms of the single crystal elastic constants as

$$c = c_{11} - c_{12} - 2c_{44}$$

and ρ is the density. By defining the $V_{S\parallel}$ and $V_{S\perp}$ as

$$V_{S\parallel} = \langle V \rangle + \Delta V/2$$

and

$$V_{S\perp} = <V> - \Delta V/2$$

the birefringence can be written as

$$B = \Delta V / <V>$$

and Eq. as

$$W_{420} = \frac{7\sqrt{5}\, B\rho <V>^2}{16\, \pi^2 c} \tag{9}$$

Using Eq. 0 and the birefringence data for the four samples A, B, C, and D, the values of W_{420} for each of the samples are −0.0011, −0.0011, −0.0020 and −0.0023 respectively.

CONCLUSIONS AND DISCUSSION

The RW data have been analyzed assuming no stress is present in the plate. The data from the top surface of the specimen show a negative trend as a function of angle in respect to the center plane data. A very slight amount of warping occurred during machining of the specimen indicating that some residual stress was contained in the plate and is now redistributed due to reducing the thickness of the plate. Although the RW wave data display this negative trend, birefringence measurements did not display a 90° rotation of the acoustic axis as would be expected if the RW results were due to a homogeneous stress or different texture. Initial RW wave measurements on the surface of a full thickness specimen indicate the opposite trend but same magnitude as compared to the 10 mm thick specimen (top surface). Although some residual stress appears to be present we expect them to be small but not negligible. Because of these tentative results and the qualitative agreement of the x-ray pole figures we must also conclude that the surface texture is substantially weaker than that of the center plane with W_{420} of the order of −0.0001 for the surface. Further measurements and analysis are in progress to verify these conclusions.

As can be seen, the W coefficients calculated from the birefringence data agree with those calculated from the surface wave measurements that a gradient exists in the plate. However, the birefringence measurement is an average through the thickness whereas the RW wave measurement is more descrete or layer-like, averaging only to a depth of approximately one wave length (0.7 mm). If the birefringence samples were of the order of the RW wave length, the expected agreement of W_{420} would be better for each "layer."

The analysis of the RW wave data indicate that W_{400} and W_{440} are relatively insensitive to the texture variation in this particular plate. For this particular case, by examination of Sayer's results[5] and Eq. 30 of Delsanto[8], some physical significance of W_{400}, W_{420} and W_{440} can be seen. W_{400} is a measure of the anisotropy along the plate normal with respect to complete isotropy and a dc shift of the RW velocity data at all angles will occur for a different W_{400}. W_{400} is also more sensitive to any deviation of the elastic constants due to alloying elements, assuming single crystal values for the pure materials are used in the calculation of the isotropic velocity. W_{440} describes the asymmetry of the RW velocity data in the plane of the plate with respect to 45°. W_{420} is a measure of the magnitude and sign (with respect to the rolling direction) of the anisotropy in the plane of the plate, i.e. proportional to the birefringence and difference between the RW velocity at 0° and 90°. (A more detailed discussion will be presented at a later date.) Further more, the values obtained for these coefficients are subject to averaging by the wave mode used to measure them and the type of texture and stress variations that may be found in the plate. This is evident for the results of W_{420} calculated from the birefringence measurements and the surface wave measurements discussed previously. However, measurement of the W coefficients will also be quite sensitive to any stress present in the plate, especially W_{420}. Birefringence measurements of W_{420} will be sensitive only to stress that does not average to zero but the surface wave measurements will be sensitive to any surface stress. It may also be of interest to consider the possible effects of substantial through thickness texture and stress gradients on the propagation of plate wave modes, i.e. what type of averaging occurs. The distribution of texture and stress through the plate thickness may be such that the effects are additive. For this case, the wave cannot be considered to pass through one "layer" at a time as with bulk or surface waves. Even surface waves will require similar consideration depending upon the frequency of the waves, i.e., the depth of penetration. Some of the stress and texture variations may become negligible for very thin or very thick materials, but at the same time the gradients of these variations may be much greater for think plates and smaller in thick plates.

ACKNOWLEDGMENTS

The research reported in this paper was conducted at the Naval Research Laboratory under sponsorship of the Office of Naval Research.

REFERENCES

1. D.S. Hughes and J.L. Kelly, Phys. Rev. 92, 1145 (1953)
2. R.E. Green, Jr., "Treatise on Material Science and Technology," Vol. 3: Ultrasonic Investigation of Mechanical Properties. Academic Press, New York, 1973.
3. K. Okada, J. Acoust. Soc. Jpn. (E) *1*, No. 3, 193-200 (1980).
4. Clark, A.V. and Mignogna, R.B., Ultrasonics, *21*, No 5 217-225 (1983).
5. C.M. Sayers, J. Phys. D *15*, 2157 (1982).
6. D.R. Allen and C.M. Sayers, Ultrasonics, *22*, 179 (1984).
7. D.R. Allen, R. Langman and C.M. Sayers, Ultrasonics *23*, 215 (1985)
8. P.P. Delsanto, R.B. Mignogna and A.V. Clark, Jr., "Ultrasonic Texture Analysis for Polycrystalline Aggregates of Cubic Materials Displaying Orthotropic Symmetry," to be presented at the 2nd Int. Symposium on the Nondestructive Characterization of Materials, Montreal, July 21-23, 1986.
9. R.B. Mignogna, A.V. Clark, B.B. Rath and C.L. Vold, in Nondestructive Methods for Material Property Determination, C. Olaf and R.E. Green, Jr., Eds. (Plenum Press, New York, 1984), p. 339.
10. "Aluminum," Vol. III: Fabrication and Finishing. American Society for Metals, Metals Park, Ohio, 1967.
11. D.H. Chung, D.J. Silversmith, and B.B. Chick, Rev. Sci. Instrum. *40,* 718-720 (1969).
12. A.V. Clark, and R.B. Mignogna, Ultrasonics *21*, 57-64 No. 2, (1983).

INFERENCE OF CRYSTALLITE ORIENTATION DISTRIBUTION FUNCTION FROM THE VELOCITY OF ULTRASONIC PLATE MODES

S. S. Lee, J. F. Smith, and R. B. Thompson

Ames Laboratory
Iowa State University
Ames, Iowa 50011

INTRODUCTION

Techniques have recently been reported for the ultrasonic measurement of stress.[1-3] In these, the effects of texture are suppressed by comparing the velocity of two shear waves whose directions of propagation and polarization have been interchanged. Here, an alternate goal is described. It is desired to enhance, rather than suppress, the textural effects so that the preferred orientation of crystallites can be determined.

The degree or type of texture can be quantitatively described by a crystallite orientation distribution function (CODF) which gives the probability that a given crystallite in the sample has a specified orientation with respect to the sample axes. From the work of Bunge[4] and Roe[5], the crystallite orientation distribution function can be expanded as a series of generalized Legendre functions with the expansion coefficients $W_{\ell mn}$. These coefficients can be directly determined from the analysis of X-ray or neutron pole figures, or indirectly inferred from physical property measurements. For example, when symmetries are taken into account, the Voigt-average elastic constants of a rolled plate of cubic crystallites are functions of the lowest coefficients W_{400}, W_{420}, and W_{440}.[4-6] This paper presents techniques to infer these coefficients from the velocities of ultrasonic plate modes, such as can be measured by EMATs placed on the plate surfaces.

THEORY

The theory for the propagation of ultrasonic plane waves in a biaxially stressed, orthorhombic continuum has been presented by Thompson *et al.*[7] In the theory, the x_1, x_2, x_3 axes may be thought of as coinciding with the rolling, transverse, and thickness directions of a rolled plate, respectively. Wave propagation has been considered in the 1-2 plane, corresponding to the plane of the plate and the effects of a biaxial stress in that plane on the wave propagation have been taken into account. The final expressions for propagation of the plane longitudinal and transverse waves, to first order in elastic anisotropy, were found to be

$$\rho V_L^2 = C_L + \frac{\alpha}{2} C_L \cos 2\theta - \frac{\beta}{2} C_T(1-\cos 4\theta)$$
$$+ (\overline{C}_{16} + \overline{C}_{26}) \sin 2\theta + \frac{1}{2} (\overline{C}_{16} - \overline{C}_{26}) \sin 4\theta$$
$$+ \frac{1}{2} (\sigma_a + \sigma_b) + \frac{1}{2} (\sigma_a - \sigma_b) \cos 2 (\Omega-\theta) \quad (1)$$

$$\rho V_T^2 = C_T + \frac{\beta}{2} C_T (1-\cos 4\theta) - \frac{1}{2} (\overline{C}_{16} - \overline{C}_{26}) \sin 4\theta$$
$$+ \frac{1}{2} (\sigma_a + \sigma_b) + \frac{1}{2} (\sigma_a - \sigma_b) \cos 2(\Omega-\theta) \quad . \quad (2)$$

Here, ρ is density, $V_{L,T}$ is the wave velocity of longitudinal and transverse plane waves, θ is the angle of propagation with respect to rolling direction, σ_a and σ_b are principal stress values with orientation angle Ω and $(\Omega+90°)$ with respect to rolling direction. $C_T = \overline{C}_{66}$, $C_L = (\overline{C}_{11} + \overline{C}_{22})/2$, $\alpha = (\overline{C}_{11} - \overline{C}_{22})/C_L$, $\beta = [(C_L - C_{12})/2 - C_T]/C_T$. The latter two parameters are measures of longitudinal and shear wave velocity anisotropies, respectively. The $\overline{C}_{ij}$ are elastic constants which are modified by stress.

In many applications, determination of texture would be done in the unstressed state and the stress-related terms in the above equations would vanish in this case. This simplifies the expressions to the following form:

$$\rho V_L^2 = C_L + \frac{\alpha}{2} C_L \cos 2\theta - \frac{\beta}{2} C_T (1-\cos 4\theta) \quad (3)$$

$$\rho V_T^2 = C_T + \frac{\beta}{2} C_T (1 - \cos 4\theta). \quad (4).$$

These plane wave solutions for an unbound medium must be modified if they are to be applied to wave propagation in a plate geometry. The previously proposed isotropic correction[3] relating the plate wave velocities to plane wave velocities may not be correct for strongly textured materials. Accordingly, the derivation of a more general correction is presented below.

When the wave length is large with respect to the plate thickness, the dynamic stress component, σ_{3i}, can be assumed to be approximately zero throughout the plate thickness. Then, the equations of motion for the lowest order symmetric Lamb (S_o) and horizontally polarized shear (SH_o) modes of the plate[8] can be put in the same form as the equation of motion for plane waves propagating and polarized in the 1-2 plane of an unbounded medium by introducing the effective elastic constants,

$$\hat{C}_{11} = C_{11} - C_{13}^2/C_{33} \quad (5a)$$

$$\hat{C}_{12} = C_{12} - C_{13}C_{23}/C_{33} \quad (5b)$$

$$\hat{C}_{22} = C_{22} - C_{23}^2/C_{33}. \quad (5c)$$

These are each reduced because the unconstrained Poisson's effect in the thin plate reduces the stress required to produce a given strain. The elastic constant C_{66} is not changed since shearing stresses in the 1-2 plane are uninfluenced by the plate faces.

The long wavelength limit of the velocities of the S_o and SH_o plate modes may then be determined as solutions of the reduced Christoffel equations[9]:

$$\begin{vmatrix} \Gamma_{11} - \rho V^2 & \Gamma_{12} \\ \Gamma_{12} & \Gamma_{22} - \rho V^2 \end{vmatrix} = 0 \qquad (6)$$

where

$$\Gamma_{11} = \hat{C}_{11} P_1^2 + C_{66} P_2^2,$$

$$\Gamma_{12} = (\hat{C}_{12} + C_{66}) P_1 P_2, \text{ and}$$

$$\Gamma_{22} = C_{66} P_1^2 + \hat{C}_{22} P_2^2 \quad .$$

In these expressions, P_1 and P_2 are direction cosines of the wave normal with respect to rolling direction. The solutions of the above equation have the same form as the plane wave solutions in Eq. (3) and Eq. (4) if one corrects the elastic constant parameters by adding "$\wedge$".

Hence, a new set of parameters is obtained:

$$\hat{C}_L = (\hat{C}_{11} + \hat{C}_{22})/2 \qquad (7a)$$

$$\hat{\alpha} = (\hat{C}_{11} - \hat{C}_{22})/\hat{C}_L \qquad (7b)$$

$$\hat{\beta} = [(\hat{C}_L - \hat{C}_{12})/2 - \hat{C}_T]/\hat{C}_T \qquad (7c)$$

$$\hat{C}_T = C_{66}, \text{ same as } C_T \text{ in plane wave} \quad . \qquad (7d)$$

Now, the angular dependence of the solutions for S_o SH_o mode propagation in the plane of the plate, to lowest order anisotropy, is obtained:

$$\rho V_{S_o}{}^2 = \hat{C}_L + \frac{\hat{\alpha}}{2} \hat{C}_L \cos 2\theta - \frac{\hat{\beta}}{2} \hat{C}_T (1-\cos 4\theta) \qquad (8)$$

$$\rho V_{SH_o}{}^2 = \hat{C}_T + \frac{\hat{\beta}}{2} \hat{C}_T (1-\cos 4\theta) \quad . \qquad (9)$$

Here, V_{SH_o} is the velocity of the SH_o mode[8] of the plate and V_{S_o} is the velocity of the fundamental symmetric Lamb mode[8].

If one restricts attention to the velocities measured at 0°, 45°. and 90°, one finds

$$\rho V_{SH_o}{}^2(0°) = \hat{C}_T \qquad (10a)$$

$$\rho V_{SH_o}{}^2(45°) = \hat{C}_T + \hat{\beta}\hat{C}_T \qquad (10b)$$

$$\rho V_{S_o}{}^2(0°) = \hat{C}_L + \frac{\hat{\alpha}}{2} \hat{C}_L \qquad (10c)$$

$$\rho V_{S_o}{}^2(45°) = \hat{C}_L - \hat{\beta} \hat{C}_T \qquad (10d)$$

$$\rho V_{S_o}{}^2(90°) = \hat{C}_L - \frac{\hat{\alpha}}{2} \hat{C}_L \qquad (10e)$$

In order to quantify the relationship between the velocities, and hence, the elastic constant parameters $\hat{\alpha}$, $\hat{\beta}$, $\hat{C}_L$, and $\hat{C}_T$, to the texture, it is useful to introduce the coefficients $W_{\ell mn}$ as defined by the CODF expansion[5]:

$$w(\xi, \psi, \Phi) = \sum_{\ell=0}^{\infty} \sum_{m=-\ell}^{\ell} \sum_{n=-\ell}^{\ell} W_{\ell mn} Z_{\ell mn}(\xi) e^{-im\psi} e^{-in\Phi} \tag{11}$$

where w is the CODF expressed in terms of the Euler angles θ, ψ, ϕ, $\xi = \cos\theta$, and $Z_{\ell mn}$ are generalized Legendre functions. The details of the definition of the angles are presented by Sayers in Ref. 10. Therein, Sayers[10] has derived the relationship between the elastic constants of an orthorhombic aggregate of cubic crystallite and the $W_{\ell mn}$ using the method of Voigt to compute the polycrystalline average. Following Bunge[4], Sayers concludes that only three independent coefficients, W_{400}, W_{420}, and W_{440}, need to be considered.

Inserting Eq. (5) into Eq. (7), using the relationship between elastic constants and $W_{\ell mn}$, and considering only terms which vary to first order in the elastic anisotropy, $C° = C_{11}{}^0 - C_{12}{}^0 - 2C_{44}{}^0$, one concludes:

$$C_T = C_{44}{}^0 + C^0\{\frac{1}{5} + \frac{4\sqrt{2}}{35}\pi^2 (W_{400} - \sqrt{70}\, W_{440})\} \tag{12a}$$

$$\hat{\hat{\beta C}}_T = \beta C_T = \frac{16\sqrt{35}}{35}\pi^2 C^0 W_{440} \tag{12b}$$

$$\hat{C}_L = C_{11}{}^0 - \frac{C_{12}{}^{0^2}}{C_{11}{}^0} - \frac{2}{5} C^0(1 + \frac{C_{12}{}^0}{C_{11}{}^0} + \frac{C_{12}{}^{0^2}}{C_{11}{}^{0^2}})$$
$$+ \frac{4\sqrt{2}}{35}\pi^2 C^0(3 + 8\frac{C_{12}{}^0}{C_{11}{}^0} + 8\frac{C_{12}{}^{0^2}}{C_{11}{}^{0^2}}) W_{400}$$
$$+ \frac{8\sqrt{35}}{35}\pi^2 C^0 W_{440} \tag{12c}$$

$$\hat{\alpha}\hat{C}_L = -\frac{32\sqrt{5}}{35}\pi^2 C^0(1 + 2\frac{C_{12}{}^0}{C_{11}{}^0}) W_{420} \tag{12d}$$

where $C_{11}{}^0$, $C_{12}{}^0$, and $C_{44}{}^0$ are single crystal elastic constants.

Inserting Eqs. (12a-d) into Eqs. (10a-e) and seeking a stable data reduction scheme to deduce the coefficients of the CODF from velocity data, one concludes:

for W_{400}

$$\frac{1}{2}\{\rho V_{SH_0}{}^2(45°) + \rho V_{SH_0}{}^2(0°)\} = C_{44}{}^0 + C^0(\frac{1}{5} + \frac{4\sqrt{2}}{35}\pi^2 W_{400}) \tag{13a}$$

$$\frac{1}{4}\{\rho V_{S_0}{}^2(0°) + \rho V_{S_0}{}^2(90°) + 2\rho V_{S_0}{}^2(45°)\} =$$
$$C_{11}{}^0 - \frac{C_{12}{}^{0^2}}{C_{11}{}^0} - \frac{2}{5} C^0(1 + \frac{C_{12}{}^0}{C_{11}{}^0} + \frac{C_{12}{}^{0^2}}{C_{11}{}^{0^2}})$$
$$+ \frac{4\sqrt{2}}{35}\pi^2 C^0(3 + 8\frac{C_{12}{}^0}{C_{11}{}^0} + 8\frac{C_{12}{}^{0^2}}{C_{11}{}^{0^2}}) W_{400} \tag{13b}$$

for W_{420}

$$\rho V_{S_0}{}^2(0°) - \rho V_{S_0}{}^2(90°) = \frac{-32\sqrt{5}}{35}\pi^2 C^0(1 + 2\frac{C_{12}{}^0}{C_{11}{}^0}) W_{420} \quad (13c)$$

for W_{440}

$$\rho V_{SH_0}{}^2(45°) - \rho V_{SH_0}{}^2(0°) = \frac{16\sqrt{35}}{35}\pi^2 C° W_{440} \quad (13d)$$

$$\frac{1}{2}\{\rho V_{S_0}{}^2(0°) + \rho V_{S_0}{}^2(90°) - 2\rho V_{S_0}{}^2(45°)\} = \frac{16\sqrt{35}}{35}\pi^2 C^0 W_{440} \quad (13e)$$

There are two possible experimental routes to determine W_{400} and W_{440}. Their relative merits will be discussed in the section on experimental results.

If stress exists, then Eqs. (13a-e) can be generalized by including the stress related terms which are expressed in Eqs. (1) and (2). One concludes:

for W_{400}

$$\frac{1}{2}\{\rho V_{SH_0}{}^2(45°) + \rho V_{SH_0}{}^2(0°) - (\sigma_a - \sigma_b)(\cos 2\Omega + \sin 2\Omega)\}$$

$$= C_{44}{}^0 + C^0(\frac{1}{5} + \frac{4\sqrt{2}}{35}\pi^2 W_{400}) + \frac{(\sigma_a + \sigma_b)}{2} \quad (14a)$$

$$\frac{1}{4}\{\rho V_{S_0}{}^2(0°) + \rho V_{S_0}{}^2(90°) + 2\rho V_{S_0}{}^2(45°) - (\sigma_a - \sigma_b)\sin 2\Omega\}$$

$$= C_{11}{}^0 - \frac{C_{12}{}^{0\,2}}{C_{11}{}^0} - \frac{2}{5}C^0(1 + \frac{C_{12}{}^0}{C_{11}{}^0} + \frac{C_{12}{}^{0\,2}}{C_{11}{}^{0\,2}})$$

$$+ \frac{4\sqrt{2}}{35}\pi^2 C^0(3 + 8\frac{C_{12}{}^0}{C_{11}{}^0} + 8\frac{C_{12}{}^{0\,2}}{C_{11}{}^{0\,2}}) W_{400}$$

$$+ \frac{(\sigma_a + \sigma_b)}{2} + \frac{(\overline{C}_{16} + \overline{C}_{26})}{2} \quad (14b)$$

for W_{420}

$$\rho V_{S_0}{}^2(0°) - \rho V_{S_0}{}^2(90°) - (\sigma_a - \sigma_b)\cos 2\Omega$$

$$= -\frac{32\sqrt{5}}{35}\pi^2 C^0(1 + 2\frac{C_{12}{}^0}{C_{11}{}^0}) W_{420} \quad (14c)$$

for W_{440}

$$\rho V_{SH_0}{}^2(45°) - \rho V_{SH_0}{}^2(0°) + \frac{(\sigma_a - \sigma_b)}{2}(\cos 2\Omega - \sin 2\Omega)$$

$$= \frac{16\sqrt{35}}{35}\pi^2 C^0 W_{440} \quad (14d)$$

$$\frac{1}{2}\{\rho V_{S_0}{}^2(0°) + \rho V_{S_0}{}^2(90°) - 2\rho V_{S_0}{}^2(45°) + (\sigma_a - \sigma_b)\sin 2\Omega\}$$

$$= \frac{16\sqrt{35}}{35}\pi^2 C^0 W_{440} - (\overline{C}_{16} + \overline{C}_{26}) \quad . \qquad (14e)$$

In each of these equations, the experimentally observable velocities and the principal stress orientations and differences, which can be inferred from two shear wave stress measurement techniques[1-3,7], are placed on the left-hand side, and can be considered as known quantities. The unknown $W_{\ell mn}$, as well as single crystal elastic constants, appear on the right hand side. In some of the equations, the unknowns $(\sigma_a + \sigma_b)$ and $(\overline{C}_{16} + \overline{C}_{26})$ also appear on the right hand side. There is no basis for eliminating these. However, it is possible to unambiguously determine the coefficients, W_{420} and W_{440}, from Eqs. (14c) and (14d) since there are no such terms. Determination of W_{400} cannot be made independent of these stress related quantities and also requires an absolute velocity measurement. This will be discussed further below.

It should be noted that the coefficients of the CODF are themselves weakly modified by stress. In general, these are expected to be small effects and hence, they are neglected here.

EXPERIMENTAL PROCEDURE

Initial experiments have been performed on 1/16 inch thick, 1100 H-14 aluminum and 1/24 inch thick commercially pure copper samples obtained from commercial vendors. The samples were placed horizontally on the top of a lab bench and the angular dependence of the velocities was measured for the "as received" condition. The details of the experimental methods and the sample configuration have been described previously.[1]

EXPERIMENTAL RESULTS

Experimental determination of the $W_{\ell mn}$ based on the theory requires measurement of the angular dependence of SH_0 and S_0 mode velocities.

W_{420} was determined from measurements of the angular dependence of the S_0 mode velocity, which were made with meander coil EMATs[11] at a frequency of approximately 500 KHz. Velocity measurements were made for values of $\theta = 0°$ and $\theta = 90°$ with respect to the rolling direction on both samples. W_{420} was then determined from the difference of the above two velocities, as can be seen from Eq. (13c).

W_{440} was determined by SH_0 mode velocity measurement. The waves were excited with periodic permanent magnet EMATs[12] at the same frequency. The values of W_{440} were determined from the difference in velocities at $\theta = 0°$ and $\theta = 45°$, as can be seen from Eq. (13d). The W_{440} values were checked for internal consistency with independent predictions based on the S_0 mode velocity measurements and Eq. (13e). The results are tabulated, along with the values for W_{420} in Table 1. The good agreement between the values for W_{440} inferred independently from the S_0 and SH_0 measurements provides a measure of the self-consistency of this approach.

Equations (13a) and (13b) also predict that determination of W_{400} could be made from the absolute measurement of the velocity of the SH_0 and S_0 mode. Since the texture induced changes in the absolute velocity are relatively small, and since the Voigt averaging technique can introduce absolute errors, a greater error is inherent in the determination of W_{400} than in the other two coefficients. The values obtained for the samples described above do not appear physically meaningful.

Table 1. Coefficients $W_{\ell mn}$, defined in the text, as determined from ultrasonic measurements

Material	W_{420}	W_{440}
Aluminum	-0.000129	-0.00594[a] -0.00583[b]
Copper	+0.000846	-0.00302[a] -0.00288[b]

[a]The values from angular dependence SH_0 mode velocity.
[b]The values from angular dependence S_0 mode velocity.

A complete set of the $W_{\ell mn}$ values are presently being independently determined by X-ray diffraction.[13] Preliminary results indicate that good agreement is obtained for the W_{440} values. However, considerable scatter is found in the x-ray values for W_{420} (which the ultrasound indicated to be relatively small) depending on the details of the data processing. It is premature to make a comparison to the ultrasonic results. It is interesting to note that similar good agreement for W_{440} and scatter in W_{420} was reported by Allen et al.[14] in a comparison of several x-ray and ultrasonic values.

CONCLUSIONS

Expressions for the angular dependence of the velocities of the SH_0 and S_0 plate modes, propagating in the plane of a rolled plate at long wavelength, have been derived for textured cubic polycrystals. From the relationship between elastic constants and the coefficients of a CODF expansion, the values of W_{420} and W_{440} can readily be determined from measurement of differences in velocities without requirement for absolute velocity data. In principle, the absolute values of the velocities should determine W_{400}, but attempts at experimental implementation have not yet been successful. Even should it prove impossible to obtain a satisfactory measurement of the W_{400} coefficient with the present technique, knowledge of the W_{420} and W_{440} values may be sufficient for establishing process control or material acceptance procedures for rolled metal plates.

ACKNOWLEDGEMENT

This work was supported by USDOE, Office of Basic Energy Sciences, Division of Materials Sciences under contract No. W-7405-Eng-82.

REFERENCES

1. R. B. Thompson, S. S. Lee and J. F. Smith, in: "Review of Progress in Quantitative Nondestructive Evaluation 3", D. O. Thompson and D. E. Chimenti, Eds. Plenum Press, New York (1984), 1311-1319.
2. S. S. Lee, J. F. Smith and R. B. Thompson, in: "Review of Progress in Quantitative Nondestructive Evaluation 4," D. O. Thompson and D. E. Chimenti, Eds., Plenum Press, New York (1985), 1061-1069.
3. S. S. Lee, J. F. Smith and R. B. Thompson, in: "Review of Progress in Quantitative Nondestructive Evaluation 5," D. O. Thompson and D. E. Chimenti, Eds., Plenum Press, New York (in press).
4. H. J. Bunge, Krist. Tech. 3, 431 (1968).

5. R. J. Roe, J. Appl. Phys. 36, 2024 (1965).
6. H. Pursey, H. L. Cox, Philos. Mag. 45, 295 (1954).
7. R. B. Thompson, S. S. Lee and J. F. Smith, J. Acoust. Soc. Amer. (in press).
8. B. A. Auld, Volume II, Chapter 10 in: "Acoustic Waves and Fields in Solids," Wiley-Interscience, New York (1973).
9. M. J. Musgrave, Chapter 7 in: "Crystal Acoustics," Holden-Day, San Francisco (1970).
10. C. M. Sayers, J. Phys. D 15, 2157 (1982).
11. R. B. Thompson, IEEE Trans. on Sonics and Ultrasonics SU-20, 340 (1973).
12. C. F. Vasile and R. B. Thompson, J. Appl. Phys. 50, 2583 (1979).
13. G. C. Johnson, private communication.
14. D. R. Allen, R. Langman, and C. M. Sayers, Ultrasonics 23, 215-222 (1985).

APPLICATION OF NEUTRON TOMOGRAPHY TO NON-DESTRUCTIVE TEXTURE GRADIENT ANALYSIS

Tong-Tsung Wang and Brent L. Adams

Mechanical Engineering Department
Brigham Young University
Provo, UT 84602

Steven T. Lawson, Fred K. Ross and Aaron D. Krawitz

University of Missouri
Columbia, MO 65211

INTRODUCTION

Most engineering materials are polycrystalline and very often the single crystals that constitute the polycrystalline solid are not randomly oriented. This preferred orientation, or texture, is described quantitatively by the Crystallite Orientation Distribution Function (CODF), which gives the volume fraction of crystals oriented in a specific orientation with respect to the macroscopic coordinate system. Typically, the CODF is expressed as an infinite series of spherical harmonics(1). It has been found that the magnetic(2), thermal(3), and mechanical(4) properties of a polycrystalline solid are all related to its texture. Thus, texture can be an important factor in the performance of material in various applications.

Conventionally, texture is determined from crystallographic pole-figures measured by X-radiation. For metal alloys, however, the penetration depth of X-ray is only of the order of microns. Thus, the nondestructive measurement of the bulk texture by the X-ray method is not feasible. In the case of neutron diffraction, the penetration power of thermal neutrons is typically two or more orders of magnitude higher than for X-rays; consequently, a true bulk texture can be measured conveniently. The measurement of texture gradients with a collimated neutron beam is a recent topic of interest. For example, Choi, Prask and Trevino(5) successfully separated pole-figures from a two-layer copper sample. Recently, Adams(6) proposed a method for characterizing the three-dimensional texture gradient by the deconvolution of a set of pole-figures determined from overlapping probe regions. The concept was adapted from X-ray (medical) tomographic analysis. Although the same information can be obtained by sampling through the specimen with a highly collimated, narrow beam, the overlapping probe technique has an advantage in neutron economy.

In this paper, we describe experiments applying the overlapping probe technique to a sample with a known texture gradient. Details of the numerical method are provided, and consideration of the resolution of the analysis is discussed in terms of experimental and numerical uncertainties.

MATHEMATICAL METHOD FOR TEXTURE DECONVOLUTION

A general mathematical method for the deconvolution of texture using an overlapping probe technique has been published by Adams[6]; in that work Bunge's formalism for the CODF[7] was used. Roe's formalism[1] is widely used in the literature, and will be followed in the present derivation. (A comparison between the methods of Bunge and Roe is given by Esling et.al.[8]). In Roe's formalism, the CODF can be expressed as an infinite series of the form:

$$f\,(\psi,\theta,\phi) = \sum_{l=0}^{\infty} \sum_{m=-l}^{l} \sum_{n=-l}^{l} W_{lmn}\, Z_{lmn}\,(\cos\theta)\, e^{-im\psi}\, e^{-in\phi} \qquad \text{EQ.(1)}$$

where ψ,θ and ϕ are the Euler angles of the crystal coordinate system defined with respect to the sample coordinate system and $Z_{lmn}(\cos\theta)$ are the augmented Jacobi polynominals. Similarly, the jth normalized pole-figure can be expressed as :

$$q^{j}(\chi,\eta) = \sum_{l=0}^{\infty} \sum_{m=-l}^{l} Q^{j}_{lm}\, P_l^{m}(\cos\chi)\, e^{-im\eta} \qquad \text{EQ.(2)}$$

in which χ and η are the polar and azimuthal angles of the pole of the diffracting plane defined in the sample coordinate system, and $P_l^{m}(\cos\chi)$ are the associated Legendre polynominals. As has been shown by Roe[1], Q^{j}_{lm} is related to W_{lmn} by:

$$Q^{j}_{lm} = 2\pi \sqrt{\frac{2}{2*l+1}} \sum_{n=-l}^{l} W_{lmn} P_l^{n}(\cos \Xi_j)\, e^{in\Phi_j} \qquad \text{EQ.(3)}$$

where Ξ_j and Φ_j are the polar and azimuthal angles of the pole of the jth diffracting plane defined in the crystal coordinate system. The Euler angles (ψ,θ,ϕ) and the two sets of polar angles (χ,η) and (Ξ_j, Φ_j) are defined in accordance with Roe's[1] convention. In practice, we can only measure intensities from a finite set of diffracting planes. Thus, Ξ and Φ are discrete variables. However, mathematically we may treat them as continuous variables. By combining EQS.(1) to (3), we obtain:

$$q(\chi,\eta;\Xi,\Phi) = 2\pi \sum_{l=0}^{\infty} \sqrt{\frac{2}{2*l+1}} \sum_{m=-l}^{l}\sum_{n=-l}^{l} W_{lmn} P_l^m(\cos\chi)\, e^{-im\eta} P_l^n(\cos\Xi)\, e^{in\Phi}$$

EQ.(4)

This function is equivalent to the Axis Distribution Function (ADF) defined by Bunge[7]. If we fix Ξ and Φ and vary χ and η over $0\le\chi\le\pi/2$ and $0\le\eta\le 2\pi$, the ADF represents a pole-figure. Conversely, fixing χ and η and varying Ξ and Φ, the ADF represents the inverse pole-figure. If the probe region covers a volume with uniform texture, the measured intensity is related to the ADF by:

$$I(\chi,\eta;\Xi,\Phi;\mathbf{x}) = R(\Xi,\Phi)\; q(\chi,\eta;\Xi,\Phi) \iiint_{V(\mathbf{x})} \exp[-\Sigma L(\chi,\eta;\Xi,\Phi;\mathbf{x};\mathbf{y})]\; dV$$

EQ.(5)

where **x** is the position vector of the centroid of the probe region and $R(\Xi,\Phi)$ is the reflectivity of the diffracting plane. $L(\chi,\eta;\Xi,\Phi;\mathbf{x},\mathbf{y})$ is the linear path length traversed by the neutron in irradiating the differential volume dV located at **y.** Σ is the macroscopic thermal neutron removal cross-section. The integration is performed over the whole probe region $V(\mathbf{x})$. Here we have made the assumption that the reflectivity depends only on the type of the crystallographic plane diffracting the neutrons. Because texture gradients may be significant in the sample, the probe region might cover a volume with a non-uniform texture. In this case, the measured intensity is related to the ADF by:

$$I(\chi,\eta;\Xi,\Phi;\mathbf{x}) = R(\Xi,\Phi) \iiint_{V(\mathbf{x})} q(\chi,\eta;\Xi,\Phi;\mathbf{y})\; \exp[-\Sigma L(\chi,\eta;\Xi,\Phi;\mathbf{x};\mathbf{y})\; dV$$

EQ.(6)

The essential assumption used in EQ.(6) is that the distance over which significant changes in the texture occur is very much larger than the average grain size. The concept of reconstruction space[9] is now introduced. Reconstruction space is defined in the sample coordinate system and is considered to consist of a compact array of K volume elements which completely encompass the volume occupied by the sample. The elemental volume of the kth element of the reconstruction space, V_k, has a constant value for all elements. The texture inside each volume element is considered to be uniform . Thus, the dimensions of V_k fix the inherent resolution in the problem. Replacing the integration in the previous equation by a summation over the K elements, we obtain

$$I(\chi,\eta;\Xi,\Phi;\mathbf{x}) = R(\Xi,\Phi) \sum_{\kappa=1}^{K} V_k(\chi,\eta;\Xi,\Phi;\mathbf{x})\; \exp[-\Sigma L_k(\chi,\eta;\Xi,\Phi;\mathbf{x})]\; q_k(\chi,\eta;\Xi,\Phi)$$

EQ.(7)

where $V_k(\chi,\eta;\Xi,\Phi;\mathbf{x})$ is the volume of the kth element of the reconstruction space intercepted by the probe specified by χ, η, Ξ, Φ and $\mathbf{x}$. L_k is the average linear path traversed by the neutron in irradiating the kth element. At this point it is more convenient to use indices to represent the discrete geometric parameters of the analysis. Let a be associated with χ, β with η, γ with $\mathbf{x}$ and j with Ξ and Φ. Then EQ.(7) may be written as:

$$I^{\alpha\beta\gamma j} = R^j \sum_{k=1}^{K} V_k^{\alpha\beta\gamma j} q_k^{\alpha\beta j}$$

EQ.(8)

In EQ.(8) the attenuation correction has been combined in the $V_k^{\alpha\beta\gamma j}$ terms. Generally we may truncate the infinite series in the ADF at l=16 without introducing serious truncation error[(6)], hence the number of data points may always be much larger than the number of CODF coefficients for all the volume elements in the reconstruction space to be deconvoluted. For such an over-determined system, we can always find the optimum solution that minimizes the object function Θ, which is defined as :

$$\Theta = \sum_{\alpha=1}^{N_\alpha} \sum_{\beta=1}^{N_\beta} \sum_{\gamma=1}^{N_\gamma} \sum_{j=1}^{N_j} \left\{ \frac{I^{\alpha\beta\gamma j}}{R^j} - \sum_{k=1}^{K} V_k^{\alpha\beta\gamma j} q_k^{\alpha\beta j} \right\}^2$$

EQ.(9)

Denote the CODF coefficients for the kth element as $W_{lmn,k}$. Θ can be minimized with respect to the unknowns R^j and $W_{lmn,k}$ by setting the derivatives equal to zero:

$$\frac{\partial \Theta}{\partial (R^j)^{-1}} = 0$$

$$\frac{\partial \Theta}{\partial W_{lmn,k}} = 0$$

EQ.(10)

EQ. (10) forms a square system of linear equations of order K times the total number of W_{lmn} coefficients, which is readily solved by the standard pivoting strategies.

EXPERIMENTAL METHOD

The experiment was carried out at the University of Missouri Research Reactor (MURR). Two cylindrical samples were constructed for this study by stacking 1/2 inch diameter circular plates punched from a brass sheet. The 0.016" thick 70/30 brass sheet was supplied by The Olin Brass Company. In the first sample the rolling directions were aligned in each layer for the purpose of determining the bulk texture. The second sample was fabricated to possess a known texture gradient through its thickness. This was accomplished by rotating each successive layer three degrees from the previous one. Thus, in 31 layers a 90° rotation through the thickness was achieved.

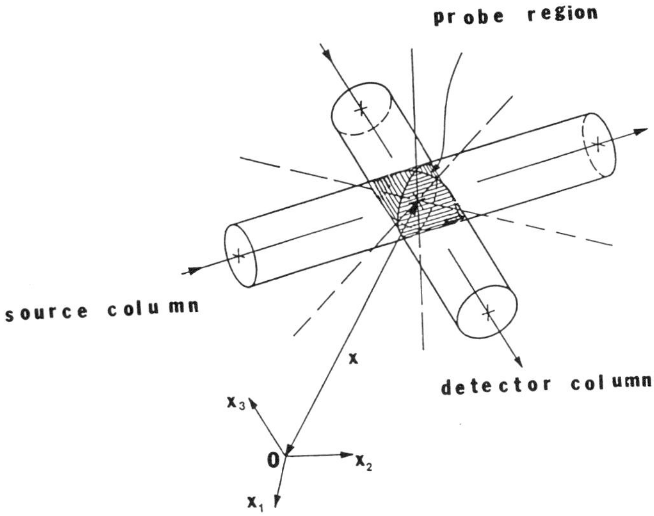

Figure 1: Probe region as defined by the intersection of the source column and the detector column.

The aligned sample was analyzed using bulk texture techniques with a 1cm square incident neutron beam and a 15' Soller collimation at the detector. Three pole figures, (111) (220) and (200), were measured; they constitute the three strongest reflections for brass.

For the second sample, probe regions were created by masking the incident and counted beam with neutron-absorbing plates with 6mm holes drilled in their centers. The intersections of the two created pencil-beams defines a discrete probe region as the intersection of two cylinders (Fig. 1). The probe center can be positioned accurately within the sample using a telescope centered upon the probe region. Data was collected for a series of 15 overlapping probe regions. The first probe was centered upon the top reference layer (with a rotation of 0°) and each consecutive probe was centered two layers further into the sample along the cylindrical axis. Scattered intensities were collected over a hemisphere grid, using the constant area method to choose values of χ and η. Measurement of 1001 points over the hemisphere and a monitor count of 50,000 gave adequate resolution. The computer-controlled diffractometer methodically varied the position of the sample and recorded the scattered intensity for each point on the hemisphere grid along with the corresponding χ and η values for the point. This procedure was peformed for two sets of pole-figures, (111) and (220), for each of the fifteen runs giving 30 data sets in all.

TEXTURE GRADIENT ANALYSIS

The procedures involved in the texture gradient analysis can be summarized in three steps:

1) Evaluation of the volume (corrected by the attenuation effect) of each layer intercepted by the probe region:

The probe region is specified by the measuring conditions χ,Ξ,Φ and $\mathbf{x}$. (Note: Because the sample is cylindrical in shape and η is the rotation angle around the cylindrical axis, the volume of interception is independent of η for the present experiment.) A computer program was developed to evaluate the volume of interception for each layer corresponding to different measuring conditions. In the program the attenuation factors for each layer at a given measuring condition was calculated based on the path traversed by the neutrons in irradiating the center of each layer.

2) Prediction of the texture gradient based on the CODF of the aligned sample:

The traditional series method (taking into account the cubic-orthotropic symmetry the sample possessed) was used to get the CODF coefficients of the unrotated brass sheet up to 20th order based on the three measured pole-figures from the aligned sample. The effect of the symmetry considerations is to reduce the number of independent CODF coefficients. For samples with cubic-orthotropic symmetry Roe[10] has shown that m is restricted to be even, n is restricted to be multiple of 4 and W_{lm0} and W_{lm4} are the only independent CODF coefficients up to l=22 for a given l and m, i.e. :

$$W_{lmn} = G_{ln} W_{lm0} + H_{ln} W_{lm4}$$

EQ.(11)

where G_{ln} and H_{ln} are constants first given by Roe[10]. Combining EQ.(3) with EQ.(11) it can easily be seen that two pole-figures are sufficient to solve the CODF coefficients up to l=22 for samples with cubic-orthotropic symmetry. Explicitly, the set of equations for solving the independent CODF coefficients are given by:

$$2\pi \sqrt{\frac{2}{2*l+1}} \left\{ W_{lm0} \left[\sum_{n=0(4)}^{l} G_{ln} P_l^n(\cos \Xi^j) \cos n\Phi^j \right] + W_{lm4} \left[\sum_{n=0(4)}^{l} H_{ln} P_l^n(\cos \Xi^j) \cos n\Phi^j \right] \right\} = Q^j_{lm}$$

EQ.(12)

Here again because of the symmetry consideration W_{lmn} and Q^j_{lm} are restricted to be real. Applying the orthogonal properties of the spherical harmonics and replacing the integration over the semi-sphere covered by the pole-figure by summations, from EQ.(2) it is easy to obtain:

$$Q^j_{lm} = \sum_{\alpha=1}^{N_\alpha} \sum_{\beta=1}^{N_\beta} \frac{I^j(\chi_\alpha, \eta_\beta)}{4\pi I^j_{av}} P_l^m(\cos \chi_\alpha) \cos m\eta_\beta$$

EQ.(13)

where $I(\chi_\alpha,\eta_\beta)$ is the measured intensity at $(\chi_\alpha, \eta_\beta)$ and $I_{av.}$ is the average intensity of the pole-figure. After solving W_{lm0} and W_{lm4}, all the other CODF coefficients can readily be determined from EQ.(11) and the CODF can be calculated from EQ.(1). Contours plots of the aligned CODF at different θ sections are given in Fig.(2).

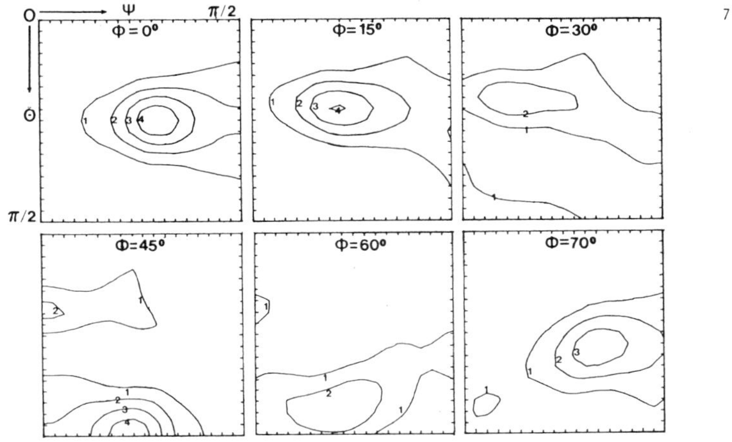

Figure 2: Contour plots of the CODF for the aligned brass sample. (Units of times random)

The CODF coefficients of a rotated sheet-specimen are related to that of the unrotated one by[1]:

$$W'_{lmn} = W_{lmn}\, e^{-im\alpha}$$

EQ.(14)

where α is the angle of rotation around the normal direction. From EQ. (5) , based on the interception volume obtained from step (1) and the CODF coefficients of the aligned sample, a computer program was developed to construct the theoretical pole-figures to simulate the 15 runs. As a comparison between the ideal pole-figures and the measured pole-figures, contour plots of the composite CODF corresponding to the two sets of pole-figures are given in Fig.3 and Fig.4 , in both cases the relative rotation for each successive run is evident. (NOTE: Since the probe region samples different volume in the specimen for a single pole-figure, the composite CODF does not have the same physical interpretation associated with the usual CODF.) For the rotated sample the orthogonal sample symmetry no longer exists thus both Q_{lm} and W_{lmn} are imaginary. Both EQS.(12) and (13) must be modified to include the imaginary parts:

$$2\pi\sqrt{\frac{2}{2*l+1}}\ \Big\{A_{lm0}\Big[\sum_{n=0(4)}^{l} G_{ln}P_l^n(\cos\Xi^j)\cos n\Phi^j\Big] + A_{lm4}\Big[\sum_{n=0(4)}^{l} H_{ln}P_l^n(\cos\Xi^j)\cos n\Phi^j\Big] -$$

$$B_{lm0}\Big[\sum_{n=0(4)}^{l} G_{ln}P_l^n(\cos\Xi^j)\sin n\Phi^j\Big] - B_{lm4}\Big[\sum_{n=0(4)}^{l} H_{ln}P_l^n(\cos\Xi^j)\sin n\Phi^j\Big]\Big\} = \alpha_{lm}^j$$

$$2\pi\sqrt{\frac{2}{2*l+1}}\ \Big\{A_{lm0}\Big[\sum_{n=0(4)}^{l} G_{ln}P_l^n(\cos\Xi^j)\sin n\Phi^j\Big] + A_{lm4}\Big[\sum_{n=0(4)}^{l} H_{ln}P_l^n(\cos\Xi^j)\sin n\Phi^j\Big] -$$

$$B_{lm0}\Big[\sum_{n=0(4)}^{l} G_{ln}P_l^n(\cos\Xi^j)\cos n\Phi^j\Big] - B_{lm4}\Big[\sum_{n=0(4)}^{l} H_{ln}P_l^n(\cos\Xi^j)\cos n\Phi^j\Big]\Big\} = \beta_{lm}^j$$

where EQ.(15)

$W_{lm0} = A_{lm0} + i\,B_{lm0}$ and $W_{lm4} = A_{lm4} + i\,B_{lm4}$

$Q_{lm} = \alpha_{lm} + i\,\beta_{lm}$

The series expansion was truncated at $l=8^{th}$ order.

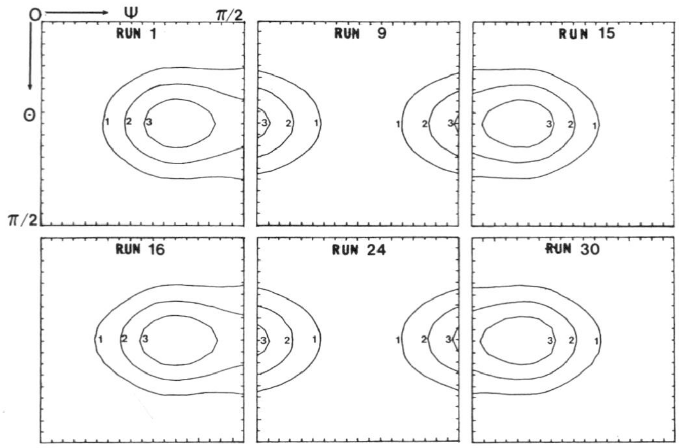

Figure 3: Contour plots of the composite CODFs ($\phi = 0^{o}$ section) corresponding to the ideal pole-figures constructed based on the ideal texture gradient(l truncated at 8th order in the series expansion Eq.(7), units of times random).

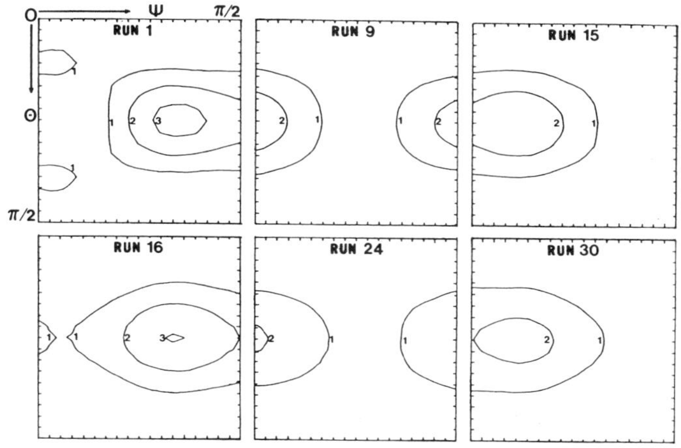

Figure 4: Contour plots of the composite CODFs ($\phi = 0^o$ section) corresponding to the measured pole-figures. (Units of times random)

3) Texture deconvolution of the ideal and experimental pole-figures:

By taking into account the cubic crystal symmetry and the reflection plane parallel to the rolling plane of the rotated brass sample, EQ.(4), the ADF can be rewritten in another form:

$$q_k^{\alpha\beta j} = \frac{1}{4\pi} + 8\pi \sum_{l=4(2)}^{L\leq 22} \sqrt{\frac{2}{2*l+1}} \sum_{m=0(2)}^{l} \left(A_{lm0,k} C_{lm}^{\alpha\beta j} + A_{lm4,k} D_{lm}^{\alpha\beta j} + B_{lm0,k} E_{lm}^{\alpha\beta j} + B_{lm4,k} F_{lm}^{\alpha\beta j} \right)$$

EQ.(16)

where

$$C_{lm}^{\alpha\beta j} = (1+\delta_m^0)^{-1} P_l^m(\cos\chi_\alpha) \cos m\eta_\beta \sum_{n=0(4)}^{l} (1+\delta_n^0)^{-1} G_{ln} P_l^n(\cos\Xi_j) \cos n\Phi_j$$

$$D_{lm}^{\alpha\beta j} = (1+\delta_m^0)^{-1} P_l^m(\cos\chi_\alpha) \cos m\eta_\beta \sum_{n=0(4)}^{l} (1+\delta_n^0) H_{ln} P_l^n(\cos\Xi_j) \cos n\Phi_j$$

$$E_{lm}^{\alpha\beta j} = (1+\delta_m^0)^{-1} P_l^m(\cos\chi_\alpha) \sin m\eta_\beta \sum_{n=0(4)}^{l} (1+\delta_n^0)^{-1} G_{ln} P_l^n(\cos\Xi_j) \cos n\Phi_j$$

$$F_{lm}^{\alpha\beta j} = (1+\delta_m^0)^{-1} P_l^m(\cos\chi_\alpha) \sin m\eta_\beta \sum_{n=0(4)}^{l} (1+\delta_n^0)^{-1} H_{ln} P_l^n(\cos\Xi_j) \cos n\Phi_j$$

In EQ.(16) $A_{lm0,k}$, $A_{lm4,k}$, $B_{lm0,k}$ and $B_{lm4,k}$ are the real and imaginary parts of the independent CODF coefficients of the kth volume element in the reconstruction space. By combining EQ.(16) with EQS. (8) to (10), the CODF coefficients together with the reflectivity of the diffracting planes are readily solvable. For the present study, we truncate the series expansion at $l=8^{th}$ order, and the total number of volume elements is set to be equal to 15 (i.e. two layers per volume element). The total number of unknowns is equal to 317 which includes 180 A_{lm0} , 135 B_{lm0} and two reflectivities R^j. A computer program was established based on the methodology described above.

RESULTS AND DISCUSSION

1) The deconvolution of the ideal pole-figures

To test the computer progam for the texture deconvolution , the set of 30 ideal pole-figures generated with the series expansion truncated at $l=8^{th}$ order was used as input. The deconvoluted textures are in good agreement with the rotated sheet texture used to generate the ideal pole-figures (Fig.5). To further test the sensitivity of the program to small perturbations of the input data, ideal pole-figures generated by truncating the series expansion at $l=12^{th}$ order were used as input. Although only minor differences existed between the two sets of pole-figures, the deconvoluted texture for the latter was found to be very different from that which was expected (Fig. 6.a). From the deconvoluted texture of the volume elements 30 pole-figures were reconstructed (Fig. 6.b). The reconstructed pole-figures agreed well with the original input pole-figures. This result indicates that in order to adequately deconvolute the texture either : (1) some other constraints together with EQ.(10) have to be satisfied,or (2) a higher order of deconvoluted CODF coefficients is necessary. The CODF contour plots of the deconvoluted volume elements (Fig. 6.a) show the existence of regions with large negative values which should never occur in real materials. A possible solution of this problem would be to impose a constraint forcing all values of the CODF to be positive. Practically, this constraint is very difficult to enforce numerically considering that the CODF is a continuous function and the constrained optimization is a very tedious computation process. As for going to higher order, l, of the CODF coefficients, the size of the system of equations to be solved will increase rapidly. For samples with cubic crystal symmetry $l=12$ is an adequate choice. When l is less than 12, as has been shown by Roe[10], W_{lm0} is the only independent CODF coefficient for a given l and m. Thus for l less than 12 the least square fitting is relatively unconstrained.

2) The deconvolution of measured pole-figures

The results of texture deconvolution with the 30 experimental pole-figures as input are presented in Fig. 7.a. Considering : 1). the sensitivity of the software to the quality of the input data as mentioned before, 2). the uncertainties in the intercepted volumes of the volume elements by the probe region because of the very simplified model used in this study, and 3). the noise level of the measured data, the discrepancy between the present result and the predicted result is not unexpected. To reduce the noise level and enforce the sample and crystal symmetry on the measured data, experimental pole-figures were reconstructed from the composite CODF

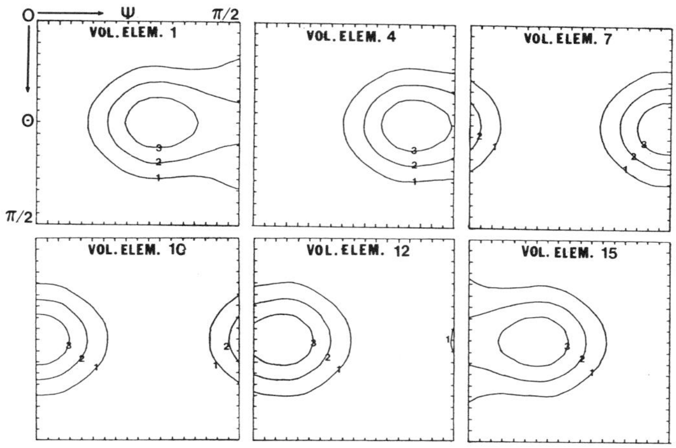

Figure 5: Contour plots ($\phi = 0^o$ section) of the CODFs for several volume elements deconvoluted from the ideal pole-figures (l truncated at 8th order in the series expansion Eq.(7)) . (Units of times random)

coefficients corresponding to each run. No significant improvement over the previous result was achieved for this smoothed set of pole-figures. Pole-figures were then reconstructed based on the deconvoluted texture of each volume element and, like previous case, they agreed well with the measured pole-figures(Fig. 7.b).

CONCLUSIONS

The results of present study indicate:

1) The unconstrained least square fitting method of texture deconvolution can accurately deconvolute idealized texture gradients .

2) The method is very sensitive to the noise in the input data; consequently additional constraints are necessary to adequately deconvolute typical measured texture gradients. The non-negativity of CODF is an obvious constraint that should be imposed.

3) The order, l, of the CODF coefficients strongly affects the resolution of deconvolution in cubic materials; orders higher than l = 8 are desirable.

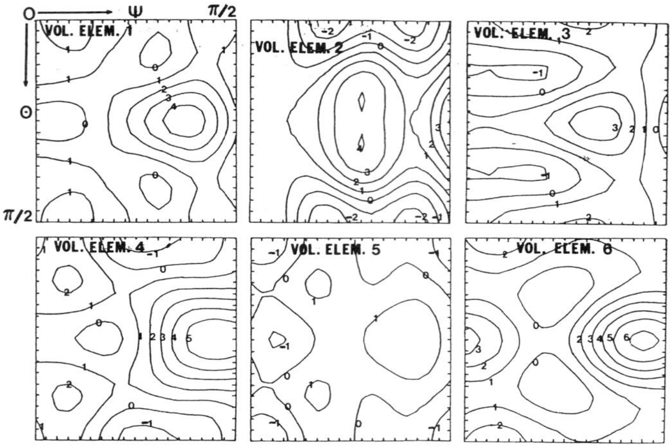

Figure 6(a): Contour plots ($\phi = 0^{o}$ section) of the CODFs for several volume elements deconvoluted from the ideal pole-figures (1 truncated at 12th order in the series expansion in Eq.(7)).(Units of times random)

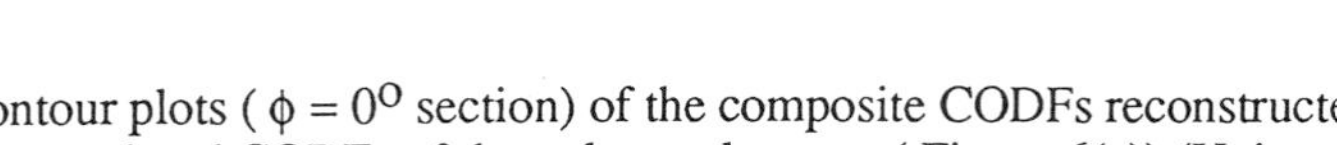

Figure 6(b): Contour plots ($\phi = 0^{o}$ section) of the composite CODFs reconstructed from the deconvoluted CODFs of the volume elements (Figure 6(a)).(Units of times random)

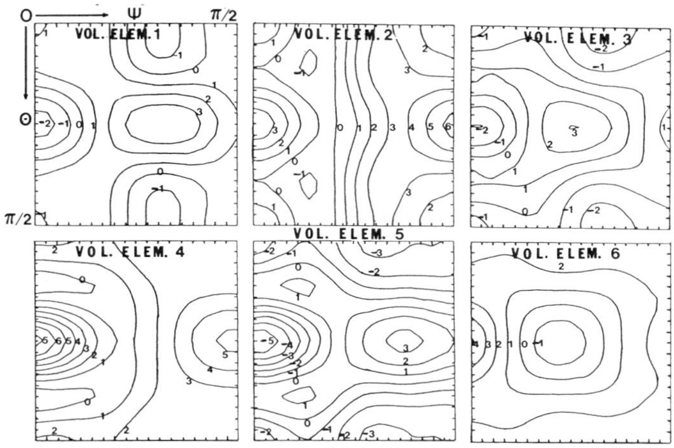

Figure 7(a): Contour plots ($\phi = 0^{o}$ section) of the CODFs for several volume elements deconvoluted from the measured pole-figures. (Units of times random)

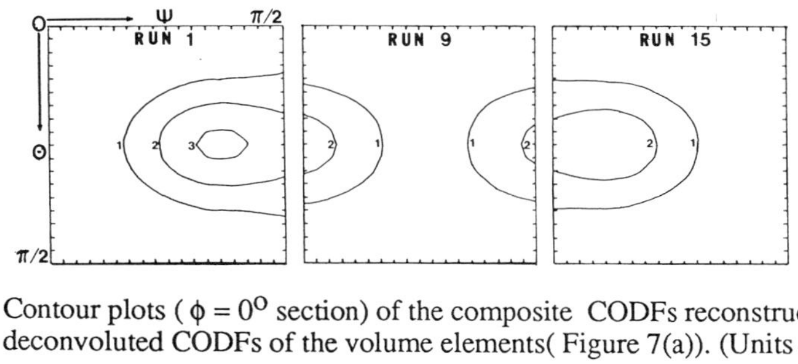

Figure 7(b): Contour plots ($\phi = 0^{o}$ section) of the composite CODFs reconstructed from the deconvoluted CODFs of the volume elements(Figure 7(a)). (Units of times random)

ACKNOWLEDGEMENT

T. T. Wang and B. L. Adams wish to acknowledge the support of the National Science Fundation under grant # DMR-8451907. The authors gratefully acknowledge the provision of beam-time by the University of Missouri Research Reactor (MURR) for the experiments reported in this work.

REFERENCES

1. R. J. Roe, J. Appl. Phys., 36:2024 (1965).
2. P. R. Morris, and J. W. Flowers, Texture of Crystalline Solids, 4:129 (1981).
3. E. F. Sturcken and J. W. Croach, Trans. Met. Soc. A.I.M.E.,227:934 (1963).
4. H. J. Bunge and W. T. Roberts, J. Appl. Cryst., 2:116 (1969).
5. C. S. Choi, H. J. Prask, and J. Trevino, J. Appl. Cryst., 12:327 (1979).
6. B. L. Adams, Scripta Metall., 18:999 (1984)
7. H. J. Bunge, "Texture Analysis in Materials Science", Butterworths, London (1982).
8. C. Esling, E. Bechler-Ferry and H. J. Bunge, in: "Quantitative Texture Analysis" H. J. Bunge and E. Esling, ed., Deutsche Gesellschaft für Metallkunde, Oberursel (1982).
9. M. B. Katz, in: "Lecture Notes in Biomathematics, No.26" S. Levin, ed., Springer Verlag, New York (1978).
10. R. J. Roe, J. Appl. Phys., 37:2069 (1966).

MAGNETOSTRICTION AND THE EFFECT OF STRESS AND TEXTURE*

J.A. Szpunar and D.L. Atherton

Department of Physics, Queen's University
Kingston, Ontario K7L 3N6, Canada

INTRODUCTION

Changes of domain structure during the magnetization of a steel specimen result in magnetostriction. Magnetostrictive strain can be measured and it is well known that there are various structural parameters which are responsible for magnetostrictive changes. In this work we have studied the effect of stress on magnetostriction in constructional steel and the effect of texture on the anisotropy of magnetostriction at saturation in strongly textured Fe-Si.

The problem of changes in magnetoelastic properties at low field strength has been discussed by various authors[1,2]. The problem is extremely complex especially in such materials as steel where inclusions, pearlite and cementite structure and texture contribute to a complexity of the magnetoelastic interactions.

Another parameter which may strongly affect the anisotropy is texture. Texture influences the magnetic properties of soft and hard magnetic materials and these problems have been discussed in previous papers[3,4]. Depending on the statistical orientation distribution of grains in a polycrystalline specimen the magnetization curve and the hysteresis curves of specimens having different texture are different. A well known example of a material where texture has been used to improve the magnetic properties is Fe-Si steel.

In a demagnetized state the domain magnetization directions are parallel to the direction of magnetization and at low field domain wall movement is responsible for an increase in magnetization. Movement of 180° domain walls does not contribute to magnetostriction changes and therefore, at low field the magnetostriction signal does not change much. Total magnetostriction changes which occur between the demagnetized state of the specimen and saturation are called the saturation magnetostriction. Rotational processes are involved in magnetization of the specimen to saturation and therefore the magnetostriction at saturation can be represented as a function of texture.

---

*Research supported by National Research Council (Industrial Materials Institute) and Department of Energy, Mines and Resources (CANMET).

MAGNETOSTRICTION AND STRAIN

In steel the magnetocrystalline anisotropy dominates all other energy terms and magnetization within domains is aligned along the easy axis determined by magnetocrystalline energy. Tension and compression can influence only the distribution of domain magnetizations between various easy directions. A uniform tension of amplitude σ applied to a crystal can only modify the effect of the crystal anisotropy energy. In order to describe these effects we use here the known relation for the energy term depending on the direction of magnetization.

Suppose that a uniform tension is applied to a crystal such that γ_1, γ_2, γ_3 are directional cosines of σ with respect to the crystal axis. The magnetization is directed in the direction $M(\alpha_1, \alpha_2, \alpha_3)$ and the energy is:

$$E = K_1(\alpha_1^2\alpha_2^2 + \alpha_2^2\alpha_3^2 + \alpha_1^2\alpha_3^2) - \frac{3}{2}\lambda_{100}\sigma(\alpha_1^2\gamma_1^1 + \alpha_2^2\gamma_2^2 + \alpha_3^2\gamma_3^2)$$

$$- 3\lambda_{100}\sigma(\alpha_1\alpha_2\gamma_1\gamma_2 + \alpha_2\alpha_3\gamma_2\gamma_3 + \alpha_3\alpha_1\gamma_3\gamma_1)$$

Thus a term known as the magnetoelastic energy is added to the crystal energy. Since this energy is a product of $\lambda\sigma$ we may expect that a crystal which has positive magnetostriction when deformed by tension will behave like a crystal having a negative magnetostriction constant but deformed in compression. In construction steel the magnetocrystalline energy is the dominant part of the total energy. K_1 is high and the easy magnetic direction is <100>. For steel $\lambda\sigma$ is positive for tension and the corresponding energy term will have a minimum for the domain magnetization orientations which are parallel to the direction of stress. If the $\lambda\sigma$ product is negative as for compressive stress the energy minimum will correspond to an orientation of domain magnetization which is normal to the direction of the applied compressive strain. Irrespective of whether tension or compression is applied magnetization at low field takes place through domain wall movements. The effect of tension is to favour domains with magnetization oriented along or opposite to the direction of tensile strain. Thus, magnetization can increase faster for steel under tension than for steel which is under compression. Therefore a significant increase in magnetization is observed at low field strength. Such an increase is often caused by the movement of 180° domain walls which does not produce a magnetostrictive signal.

Compressive strain has the opposite effect on magnetization and magnetostriction. The results of measurements of magnetostriction for various tensile and compressive stress are presented in Figs. 1, 2 3 and 4. Magnetostriction was measured along both the initial and anhysteretic magnetization curves.

Comparing these experimental results we draw the following conclusions. The results obtained along the anhysteretic curve show a stronger variation of magnetostriction than the results obtained along the initial magnetization curve. Tensile stress shifts magnetostriction towards the negative direction and a stress of 400 MPa is enough to suppress the initial positive value of magnetostriction measured. Higher stress is necessary to suppress the initial value of magnetostriction measured along the anhysteretic curve. We also observe that the magnetostrictive maximum is shifted towards lower field by

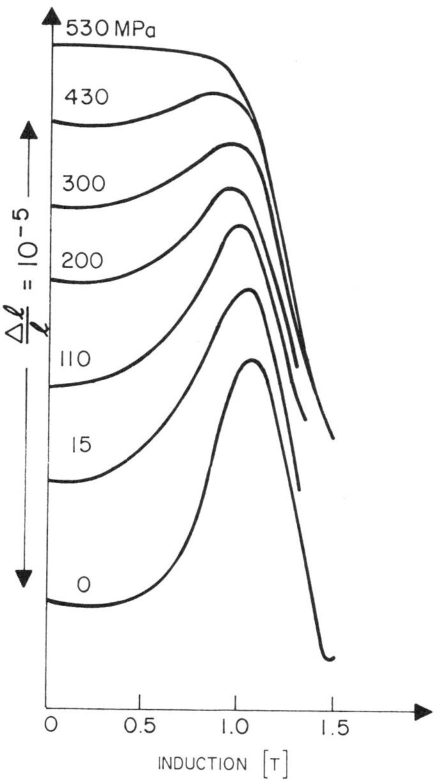

Fig. 1: Magnetostriction measured along the anhysteretic magnetization curve for various tensile isostresses.

tension and towards higher field by compression. All the significant changes in magnetostrictive behaviour are observed at inductions of less than 1 T and can be explained by changes in the existing domain structure. At magnetizations above 1T, where rotational process become significant, the magnetostrictive behaviour is less effected by stress as avident in Figs. 1, 2, 3 and 4.

Without having full information about the structure and texture of the material we can only claim that the observed behaviour agrees with the present understanding of the effect of strain on domain structure. As we have already stated tensile and compressive stresses change the statistical distribution of domain orientations.

Under tensile stress the domain mangetizations are oriented parallel to the strain direction, magnetization increases by 180° domain wall movements and this process does not produce magnetostrictive strain. Rotation of domain vectors from the 100 direction towards the axis of tension is repsonsible for the contraction in the direction of the applied strain. This effect is observed at fields above 1 T.

A compressive strain increases the proportion of the domains oriented perpendicular to the strain direction. If a field is then applied it will remove those domains which cause a large positive magnetostriction. Such an effect is observed for specimens which have not been strained by compression. One possible explanation of this fact is that a residual compressive strain already exists in these specimens.

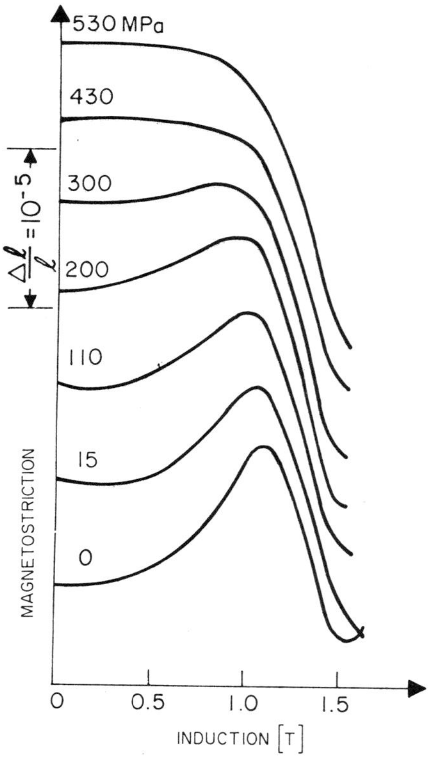

Fig. 2: Magnetostriction measured along the initial magnetization curve for various tensile isostresses.

MAGNETOSTRICTION AND TEXTURE

The saturation magnetostriction of a polycrystalline specimen is characterised by a single constant. It's value depends on the individual magnetostrictive properties of the crystals and on the orientation of these crystals.

It is not entirely clear how the averaging should be done. At saturation each crystal tries to strain magnetostrictively in the direction of the field. Because the crystals are oriented differently the strain in the specified direction is different for each crystal. Usually one of two assumptions can be made namely that the stress is uniform throughout but strain varies or that strain is uniform and stress varies.

Magnetostriction in polycrystalline textured materials is described by a fourth rank tensor defined in the specimen reference frame.

$$\frac{d\ell}{\ell} = \sum b_i b_j \lambda_{ijk\ell}(g) a_k a_\ell$$

where a (a_1, a_2, a_3) is a unit vector in the direction of magnetization and $b(b_1, b_2, b_3)$ is the unit vector in the direction in which $d\ell$ is measured. $\lambda_{ijk\ell}(g)$ is a function of Euler Angles $g \equiv (\Phi_1, \Phi_2, \Phi_3)$ describing the orientatio

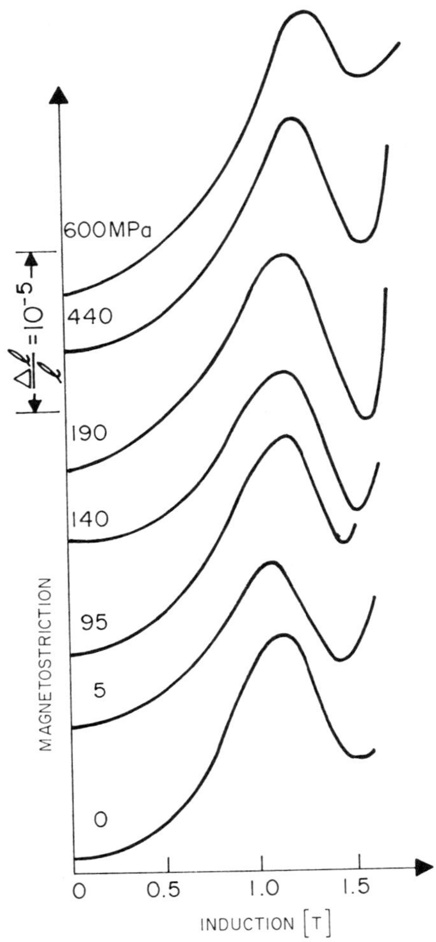

Fig. 3: Magnetostriction measured along the anhysteretic magnetization curve for various compressive isostresses.

of the crystal reference frame in the specimen reference frame. Since the ODF is known, the mean value of λ can be defined as

$$\lambda_{ijk\ell} = \int \lambda_{ijk\ell}(g) f(g) dg$$

For cubic symmetry the following formula has been obtained [5]:

$$\bar{\lambda}_{ijk\ell} = \lambda_{ijk\ell} + \lambda_a \left[a_o(ijkl)(g) - \bar{a}_4^{11}(ijk\ell) C_4^{11} \right]$$

$$+ \bar{a}_4^{12}(ijk\ell)\, C_4^{12} + \bar{a}_4^{13}(ijk\ell)\, C_4^{13}$$

where

$$\lambda_a = \lambda_{1111} - \lambda_{1122} - 2\lambda_{1212}$$

The single crystal tensor for cubic symmetry is taken assuming

$$\lambda_a \times 10^6 = 23.7 \qquad \text{and } \lambda_{111} \times 10^6 = 4.1$$

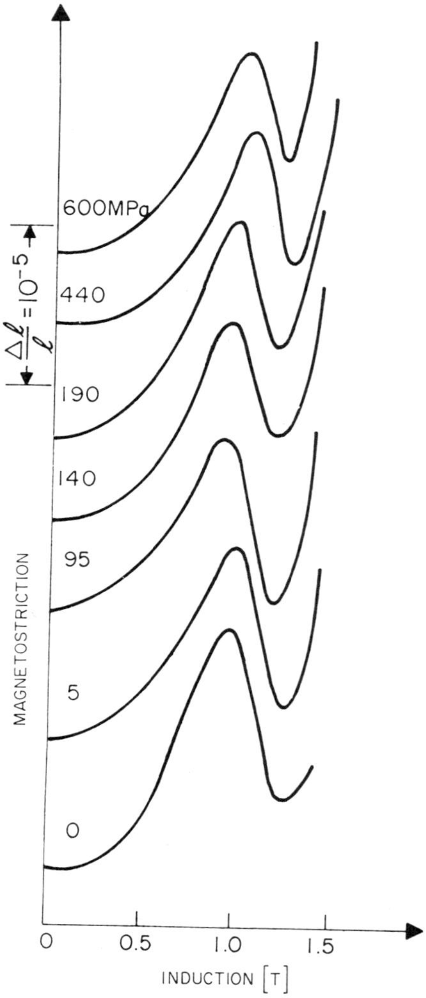

Fig. 4: Magnetostriction measured along the initial magnetization curve for various compressive isostresses.

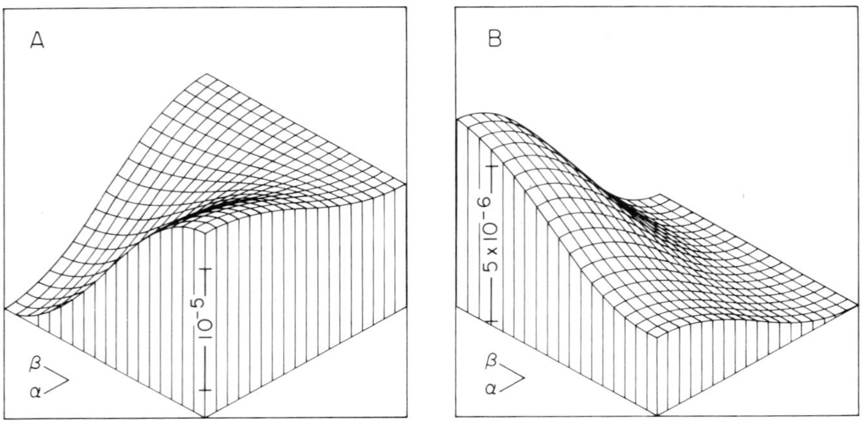

Fig. 5: Graphical representation of the anisotropy of magnetostriction $d\ell/\ell$ $(\alpha\beta)$ for a textured Fe-Si specimen. The direction of the magnetic field is aligned in the following directions: A(90,90) and B(90,45).

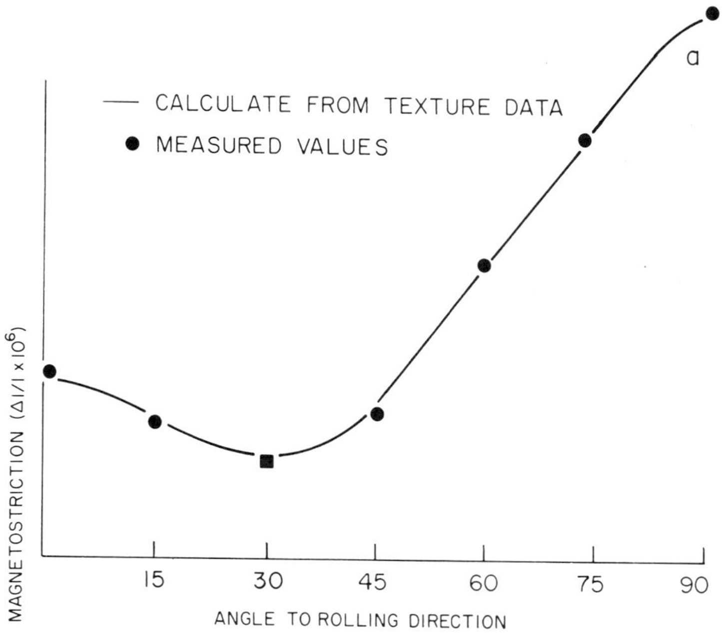

Fig. 6: The changes of magnetostrictive strain calculated in the plane of the Fe-Si sheet. Experimental points are shown for comparison.

The series expansion coefficients of the texture function for the specimen investigated are:

$C_4^{11} = .008, \quad C_4^{12} = 6.1, \quad C_4^{13} = 3.61$

The coefficients a(i,j,k,ℓ) can be found in a paper by Bunge [5]. Texture in the Fe-Si specimen was measured experimentally using the neutron diffraction technique [4] and analysed by means of ODF. The anisotropy of the magnetostrictive strain has been calculated at a field which saturates the specimen. Two-dimensional computer drawings are generated and the difference between the magnetostrictive strain in the direction defined by angular co-ordinates α,β and the direction in which the strain is minimum are displayed. Such graphical representation of the anisotropy of magnetostriction for a specimen which has strong Goss type texture is displayed in Fig. 5 and compared with experiments in Fig. 6.

CONCLUSIONS

We have demonstrated that magnetostriction in constructional steel is strongly affected by external strain. Tensile strain changes the magnetostriction versus field relationship significantly while compressive stress has much less impact on magnetostriction. Magnetostriction versus stress curves show distinctive maxima which disappear only when the specimen is subjected to high tensile stress (higher than 500 MPa). There are only small differences between magnetostriction measured along the initial and the anhysteretic curves. Our experiments agree with our understanding of the influence of stress on magnetic properties and we are working at present on a theoretical model.

Texture affects the saturation magnetostriction in Fe-Si steel and the results of calculations agree reasonably well with experiments. In Fe-Si oriented steel the texture is very strong and dominates other possible reasons for anisotropic properties. The method applied is based on a description of the crystal orientation distribution.

REFERENCES

1. W.F. Brown, Phys. Rev. 75: 147 (1949).
2. L. Brugel et Rimet, J. de Physique. 27: 589 (1966).
3. J.A. Szpunar and M. Ojanen, Met. Trans. 64: 561 (1975).
4. J.A. Szpunar, Atomic Energy Rev. 141: 199 (1976)
5. H.J. Bunge, J. Magn. Magn. Mat. 4: 305 (1977).

MAGNETIZATION AND STRESS EFFECTS IN STEEL*

D.L. Atherton, J.A. Szpunar and B. Szpunar

Department of Physics
Queen's University
Kingston, Ontario K7L 3N6

INTRODUCTION

Stress has long been known to be one of the major factors affecting the magnetic behaviour of ferromagnetic materials. It therefore influences magnetic inspection techniques for steel. It has also been proposed that magnetic measurements can be used to monitor stress[1]. We are particularly concerned with magnetic inspection techniques for pipelines [2]. We investigate stress effects by two complementary approaches, namely by experimental studies of magnetic flux leakage signals using laboratory test rigs to reproduce the behaviour of magnetic inspection tools used for pipelines and by studies of the magnetic behaviour of representative steel specimens under tensile and compressive strains.

When the magnetization of a ferromagnet is changed by alterations in the applied field there are accompanying magnetostrictive dimensional changes. Conversely changes in stress induced strains produce mangetomechanical magnetization changes. It has been presumed that these effects might be correlated, using Le Chatelier's principle, but this is not justifiable since magnetization effects are hysteretic therefore equilibrium thermodynamic relationships cannot be applied. The application of Le Chatelier's principle would suggest that a ferromagnet with positive magnetostriction (i.e. which expands on magnetization) should have its magnetization increased by positive strain (tension) and decreased by compression. Fig. 1 shows an example of the application of both tension and compression cycles applied under the same conditions. Both cause increased magnetization in this case. The changes are hysteretic and part of this magnetization change remains when the stress is relaxed.

We have shown that the first magnetomechanical effect is that the initial application of stress to a magnetized ferromagnet causes a shift in magnetization towards the anhysteretic or equilibrium magnetization [3]. Points on the anhysteretic magnetization curve are obtained, rather similarly to demagnetization, by applying AC of steadily diminishing amplitude but superimposed on a DC bias field. This gives the anhysteretic magnetization for the applied bias field. When considering the magnetomechanical

*Research supported by National Research Council (IMIR) and Department of Energy Mines and Resources (CANMET) and by Natural Sciences and Engineering Research Council.

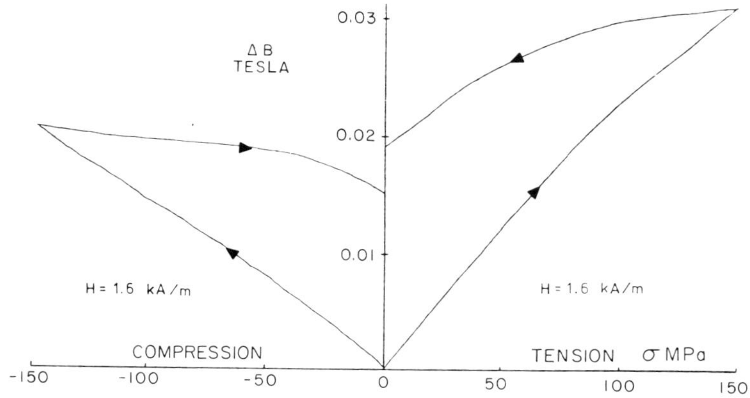

Fig. 1: The changes in magnetization, ΔB, occuring during a tensile and a compressive cycle from the same initial conditions.

effect, stress and magnetization histories are important. Fig. 2 shows the results of applying similar tensile stress cycles at different points on an initial magnetization curve and around a major hysteresis loop.

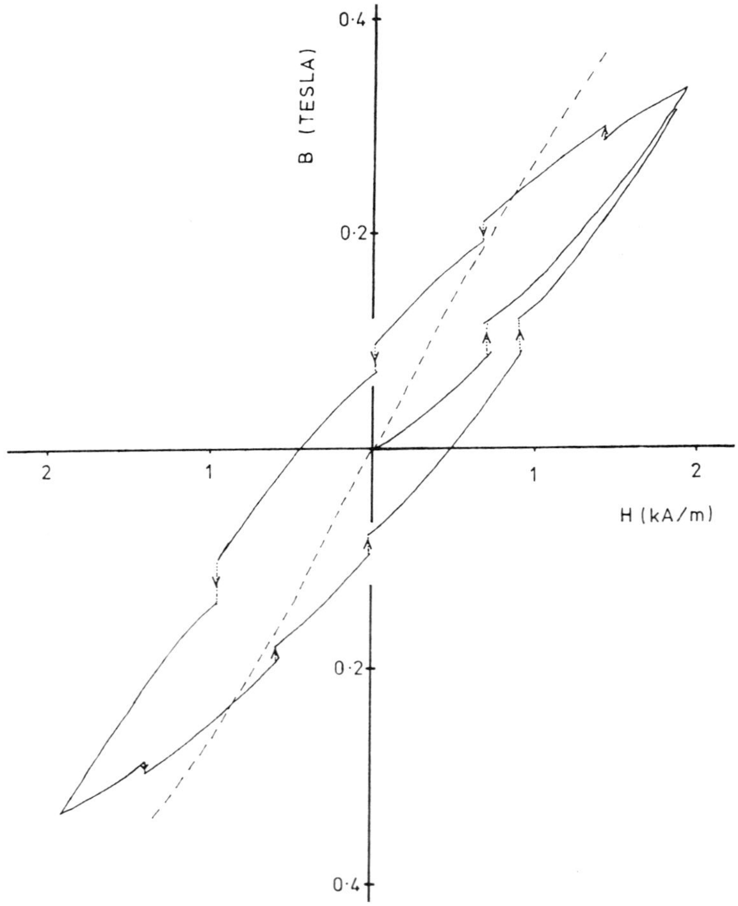

Fig. 2: The effects of identical tensile stress cycles applied at different points on the magnetization curve. The dashed curve is the anhysteretic magnetization curve.

The senses of the irreversible stress induced changes in magnetization depend on the starting positions but are always such as to shift the magnetization toward the anhysteretic (shown as a dotted curve). In addition there are reversible changes which become apparent during subsequent stress cycling. Fig. 3 shows the changes in remanent magnetization over four cycles of tension or compression for a sample of pipeline steel. Fig. 4 shows the effects of repeated stress cycles of initially

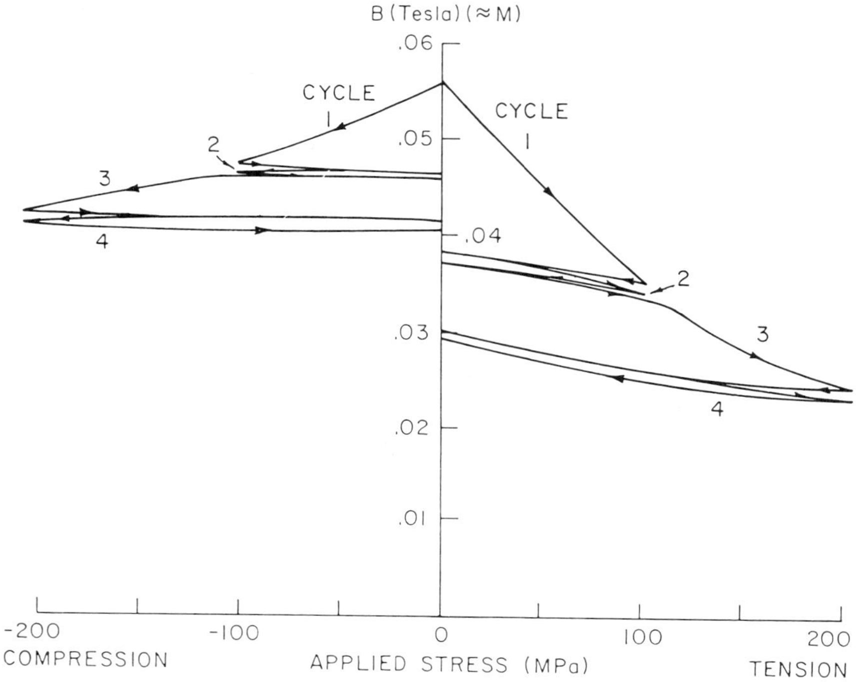

Fig. 3: Changes of remanent magnetization produced over four tension or compression cycles with the stress aligned with the magnetization. The pipeline steel sample was magnetized in a field of 5kA/m before both the series of tensile stress cycles and compressive cycles.

modest amplitude and then greater amplitudes at various fields. In general at the end of the first stress cycle there remains an irreversible shift in magnetization. Subsequent stress cycles cause essentially reversible changes until the previous maximum stress is exceeded when there are again irreversible changes occuring during the first cycle. Fig. 5 gives examples of the reversible, irreversible and total magnetization changes occuring under tensile and compressive stresses at various fields. It is possible that the reversible effects may be partly explicable in terms of the stress dependence of the anhysteretic which we have also measured on small samples of pipeline steels [4].

It may be possible to correlate the reversible magnetomechanical effects described above with magnetostriction, but magnetostriction measurements are customarily made along the initial magnetization curve which is, of course, hysteretic and therefore precludes the application of Le Chatelier's principle. We have therefore made magnetostriction measurements under anhysteretic conditions [5]. Examples of tensile isostress measurements are given in references 5 and 6. They are very similar in form to our measurements made along the initial magnetization curve and are complex. The double inflexion occuring at low stress is more distinct under all compressive stresses and presumably indicates changes in the dominant mechanisms.

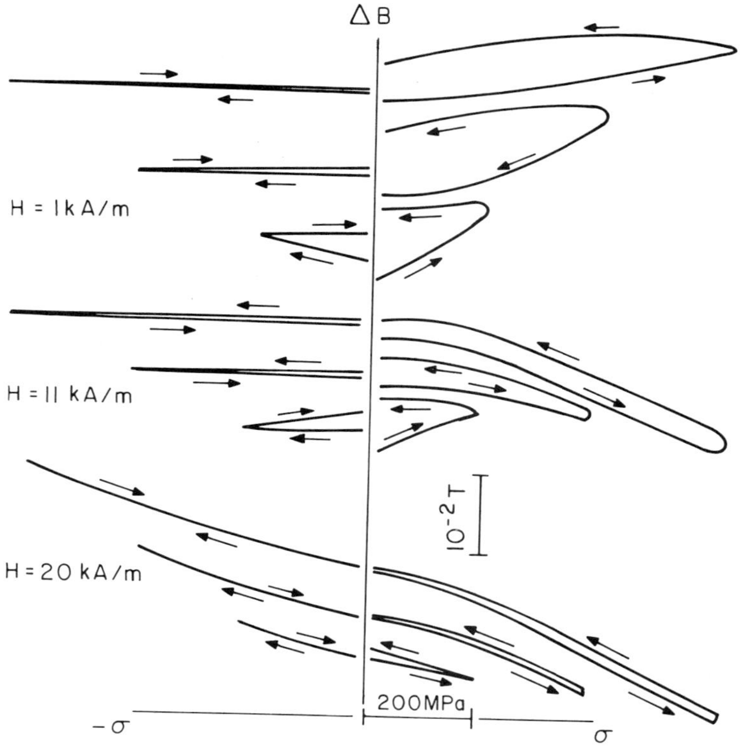

Fig. 4: Changes of magnetization caused by repeated cycles of tension or compression at various fields. The sample has been demagnetized between each stress reversal or change in field.

Recently [5] we have also shown that the magnetization process itself which is known to result from both reversible and irreversible processes, can be split into separately measureable components by applying incremental field reversals during field changes. Fig. 6 gives an example of small field reversals applied along the initial magnetization curve and the subsequent reversible component of magnetization extracted from them. It is possible that magnetostriction can also be separated into reversible and irreversible components and that attempts at thermodynamic correlations should use the reversible components of the magnetomechanical and magnetostrictive effects. It is not obvious if there should be any correlations between irreversible effects, although they may all be attributeable to the same cause, namely the interaction of magnetic domains with pinning sites.

A correlation for the reversible components may be derived from the Gibbs free energy relationship:

$$dG = -SdT + \sigma d\varepsilon + \mu_o MdH$$

With constant elastic modulus and isothermal conditions this becomes:

$$\varepsilon d\sigma + \mu_o MdH = 0$$

therefore $\left(\frac{\partial M}{\partial \sigma}\right)_H = \frac{1}{\mu_o}\left(\frac{\partial \varepsilon}{\partial H}\right)_\sigma = \frac{1}{\mu_o}\left(\frac{\partial \varepsilon}{\partial M}\right)_\sigma\left(\frac{\partial M}{\partial H}\right)_\sigma = \frac{\chi_{\sigma,H}}{\mu_o}\left(\frac{\partial \varepsilon}{\partial M}\right)_\sigma$ where $\left(\frac{\partial M}{\partial \sigma}\right)_H$ and $\left(\frac{\partial \varepsilon}{\partial M}\right)_\sigma$ are the (anhysteretic) magnetomechanical [4] and magneto strictive [6] coefficients and $\chi_{\sigma,H}$ the incremental susceptibility. The

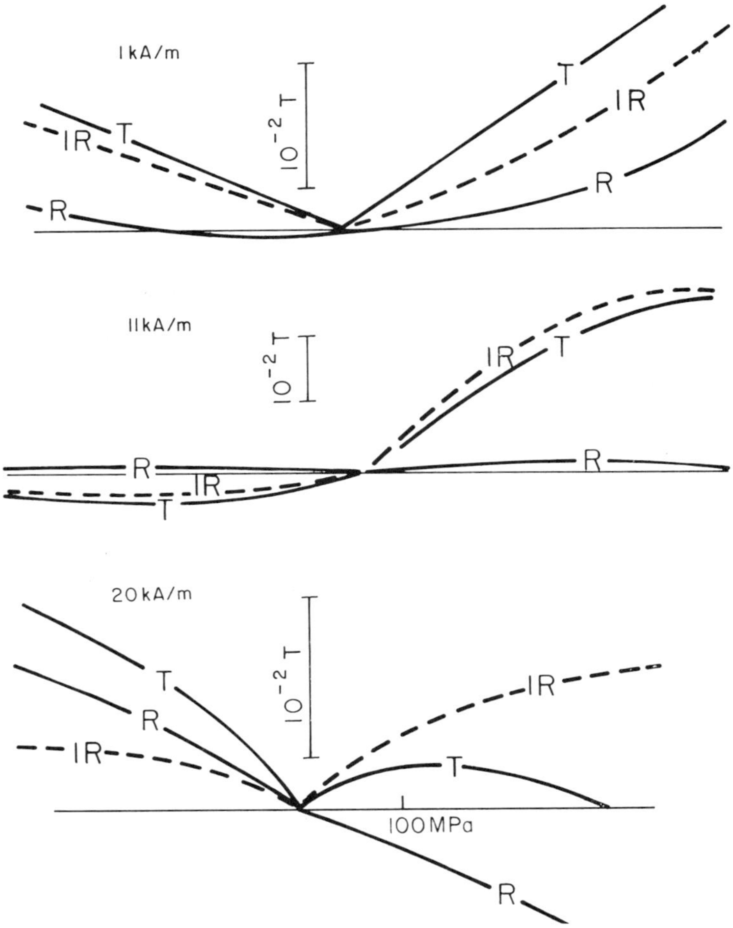

Fig. 5: The total (T), reversible (R) and irreversible (IR) stress induced changes in magnetization at various fields.

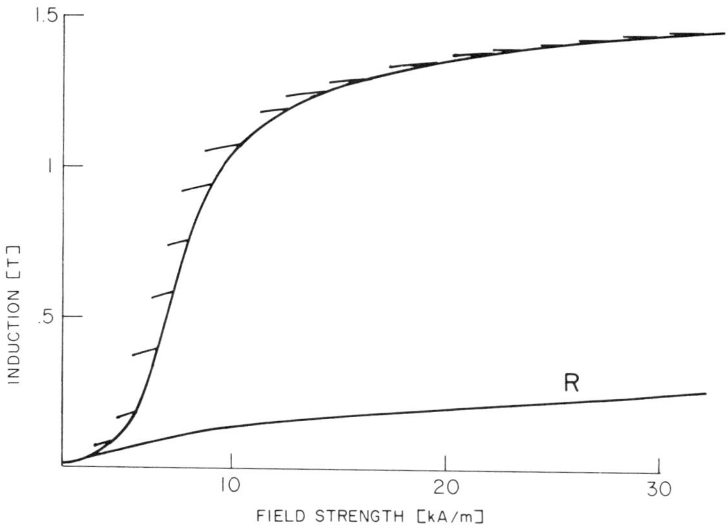

Fig. 6: Initial magnetization curve showing the small field reversals used to separate the reversal component (R).

technique and precision with which these quantities can be measured is not yet adequate to test this relationship conclusively over a representative range of parameters but initial spot checks suggest a quantiative agreement within a factor of two. Further work is needed though.

APPLICATIONS TO PIPELINE MONITORING

We have already shown that above ground magnetometer survey of stress induced magnetization changes over buried pipelines could be used to monitor bending stress anomalies [7]. Our laboratory scale tests [8,9] of the magnetization changes induced in steel pipes by bending stress or internal pressurization have shown that the irreversible magnetization changes might be used to indicate the maximum stress level since the pipeline was last effectively reinitialized by magnetic pigging. The greatest challenge is however to measure absolute stress levels magnetically.

We show here that the magnetic flux leakage signals obtained using pipeline inspection tools vary with wall stress caused by gas line pressure. For our laboratory tests we have used a test rig with a stationary

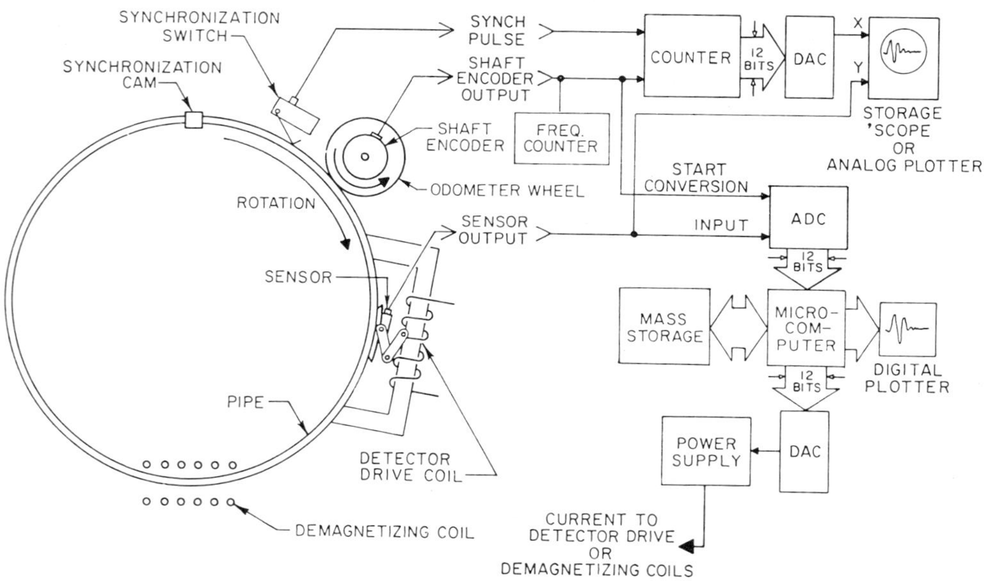

Fig. 7: Schematic of the data acquisition system used with the rotating drum test rig to evaluate the response of magnetic leakage flux detectors.

magnetic flux leakage detector pressed against the outside of a rotating drum [10]. The drum consists of a short section of line pipe sealed with end caps and pressurized by internal hydraulic pressure. It rests on a set of truck wheels which are used to drive it. The microcomputer based data acquisition system used is shown schematically in Fig. 7. Figs. 8 and 9 give plots of the leakage fields recorded over a complete track around the circumference of the drum at 0 and 6.9 MPa gauge pressure corresponding to about 85% of yield strength. Leakage flux anomalies due to the seam weld and simulated near and far side corrosion pits are apparent superimposed on background noise. The background noise shown in these particular records is repeatable and is due entirely to magnetic irregularities in the pipe and is therefore detailed signal rather than "noise". In practice, we use signal processing to enhance the signal to noise ratio and to identify near and far side defects in order to allow their severity to be gauged. It is evident that there are significant changes in both the defect signals and the noise on these traces. In fact the amplitude of leakage flux signals decrease 40% with increasing line pressure then increase until, at high pressure, there has been a 40% overall increase. The background

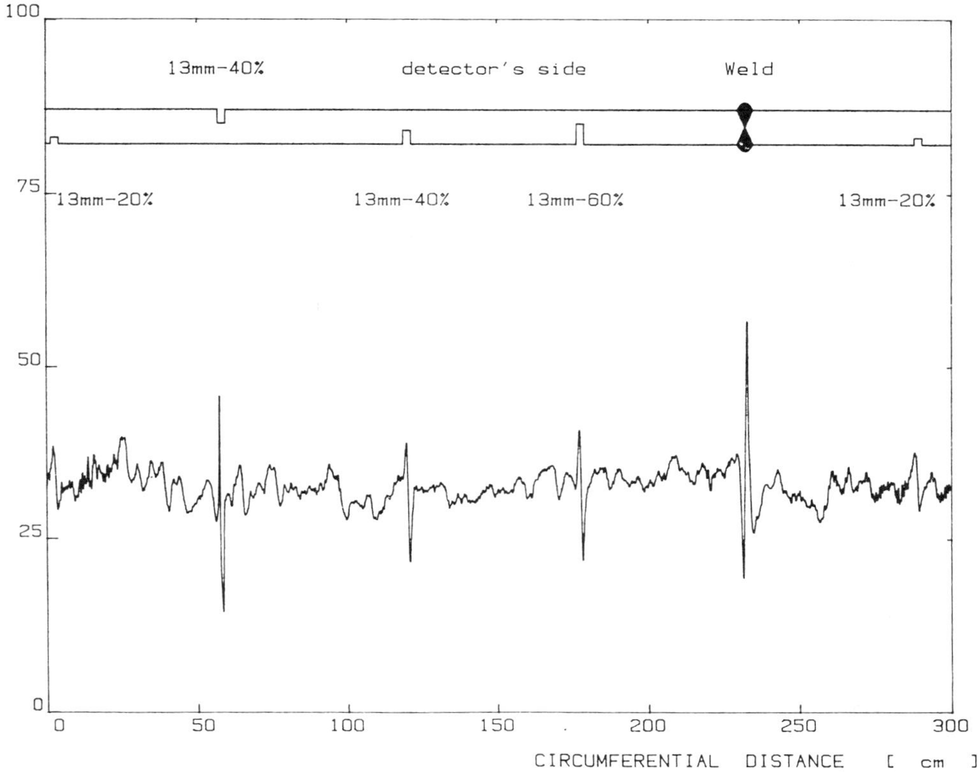

Fig. 8: Plot of the magnetic leakage fields measured around the circumference of an unpressurized section of line pipe. A schematic cross section of the pipewall is shown.

magnetic noise decreases monotonically so that, at high pressure, the signal to noise ratio for the raw data is improved by 80% as shown in Fig. 10. These effects are independent of defect penetration and reverse with decreasing pressure so that they do not appear to be due to local yielding but rather to more general overall stress induced changes in pipe wall incremental permeability. The obvious applications are that, where possible, inspection under maximum line pressure will produce improved signal to noise ratios and, more significantly, where detector response is calibrated, in order to allow corrosion penetration to be assessed, by running tools in test pipes with known defects, then these test pipes must be at the same pressure as the line to be inspected or corrections made.

CONCLUSIONS

Bulk magnetic effects are very complex, particularly in line pipe steels, but they are large enough to give very valuable techniques for nondestructive inspection of steels. Separating reversible and irreversible magnetic effects is important not only because thermodynamic

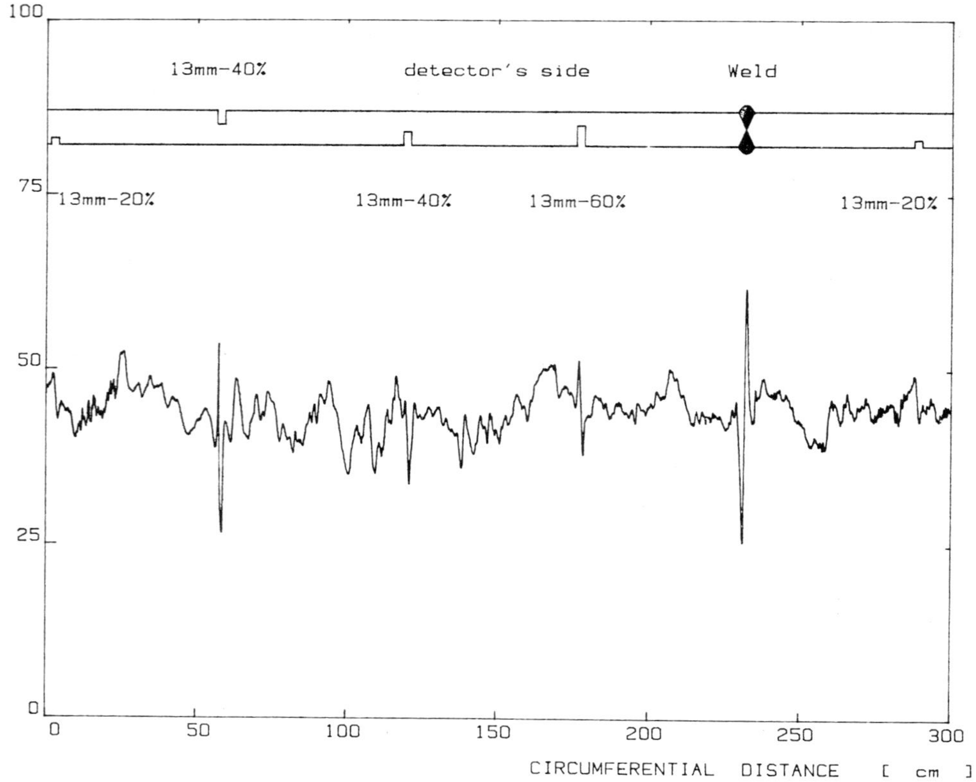

Fig. 9: Plot of the magnetic leakage fields measured around the circumference of a section of line pipe pressurized internally to 6.9 MPa.

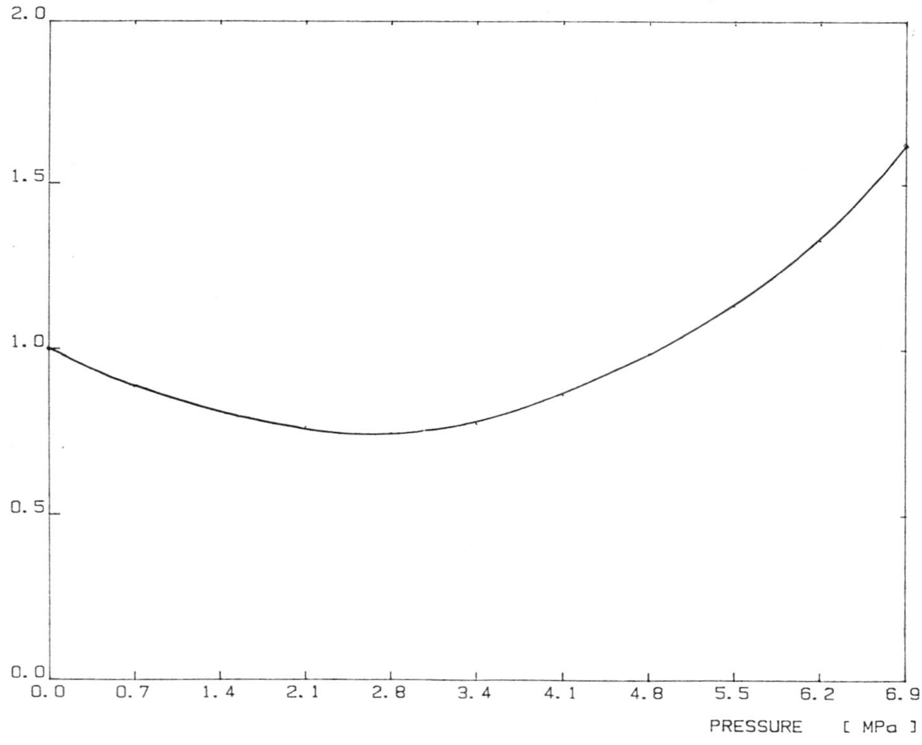

Fig. 10: Normalized signal to noise ratio as a function of internal pressur

correlations may be possible between purely reversible components but also because a clear distinction between them is an essential step in our understanding and hence in the further development of magnetic inspection techniques. There are also many significant details in the operation and interpretation of the results of existing magnetic inspection tools which demand improved understanding. Stress is a very important parameter determining ferromagnetic behaviour. In many cases, such as gas pipelines, the problems associated with defects are not porosity but stress concentration leading to fracture. The long term goal is therefore to inspect for stress anomalies rather than to detect geometric irregularities such as corrosion pits.

REFERENCES

[1] D.L. Atherton and A. Teitsma, "Detection of anomalous stresses in gas pipelines by magnetometer survey", J. Appl. Phys., Vol. 53, No. 11, pp. 8130-8135, 1982.

[2] A. Teitsma, P. Porter, H.A. French and D.L. Atherton, 1983, "Recent developments in corrosion detection", Proc. Int. Conf. on Pipeline Inspection, Edmonton, June 13-16, 1983, pp. 455-470.

[3] D.C. Jiles and D.L. Atherton, "Theory of the magnetisation process in ferromagnets and its application to the magnetomechanical effect", J. Phys. D: Appl. Phys., Vol. 17, pp. 1265-1281,1984.

[4] L.G. Dobranski, D.C. Jiles and D.L. Atherton, "Dependence of the anhysteretic magnetization on uniaxial stress in steel", J. Appl. Phys., Vol. 57, No. 1 pp. 4229-4231, 1985.

[5] D.L. Atherton and J.A. Szpunar, "Effect of stress on magnetization and magnetostriction in pipeline steel", (presented at 1986 Intermag Conf.), to be published in IEEE Trans. on Magnetics, 1986.

[6] J.A. Szpunar and D.L. Atherton, "Magnetostriction and effect of stress and texture" presented at this conference.

[7] D.L. Atherton and D.C. Jiles, "Effects of stress on magnetization", N.D.T. International, Vol. 19, No. 1, pp. 15-20, 1986.

[8] D.L. Atherton, L.W. Coathup, D.C. Jiles, L. Longo, C. Welbourn and A. Teitsma, "Stress Induced Magnetisation Changes of Steel Pipes - Laboratory Tests", IEEE Trans. on Magnetics, Vol MAG-19, No. 4, pp. 1564-1568, 1983.

[9] D.L. Atherton, C. Welbourn, D.C. Jiles, L. Reynolds and J. Scott Thomas, "Stress-Induced Magnetization Changes in Steel Pipes - Laboratory Tests, Part II" IEEE Trans. on Magnetics, MAG-20, Vol. 6, pp. 2129-2136, 1984.

[10] D.L. Atherton and C. Welbourn, "A Rotating Drum Test Rig for the Development of Pipeline Monitoring Tools", C.S.N.D.T. Journal, Vol. 6 No. 8, pp. 51-56, 1985.

EVALUATION OF SURFACE LAYER RESIDUAL STRESSES AND LATTICE DISTORTIONS IN ION IMPLANTED MATERIALS BY X-RAY DIFFRACTION

E. D. Roll, R. N. Pangborn and M. F. Amateau

Department of Engineering Science and Mechanics
The Pennsylvania State University
University Park, PA 16802

ABSTRACT

Shallow incidence X-ray diffraction techniques were employed to investigate the effects of ion implantation in metallic and ceramic materials. In the initial study, X-ray rocking curve profiles were used to evaluate the lattice distortion introduced by implantation of lithium in aluminum single crystals. By preparing asymmetrically cut crystals such that the diffraction condition ranged from glancing incidence to symmetric reflection, a depth profile of the implantation damage could be constructed. In this way, the damage associated with multiple implantations at two different fluences could be compared. In a subsequent investigation, low order diffraction peaks obtained using Cr Kα radiation have also been monitored for the purpose of estimating the residual stress and lattice disorder imposed by implantation of argon in alumina and silicon carbide. The maximum and average compressive stresses in the surface layer of the specimens, implanted at various energies and fluence, were evaluated and correlated with theoretical depth profiles of the implantation damage.

INTRODUCTION

X-ray diffraction techniques have been used extensively to study the lattice distortion and residual stresses introduced in crystalline materials by various types of mechanical deformation [1,2]. They are particularly useful for investigating the effects on the surface of preparations or treatments, such as grinding, rolling, shot peening, and welding, since the surface layer affected by these processes is of the same order in thickness as the depth probed by the X-rays. Implanted ions, however, are typically distributed over depths of 0.1 to 1 μm, a layer depth significantly smaller than the X-ray penetration depth. In this investigation, shallow angle incidence, combined with the use of a relatively long X-ray wavelength, was employed to maximize the contribution of the implanted layer to the total diffracted intensity. In this way, it was demonstrated that X-ray diffraction can be a useful and convenient tool for nondestructive evaluation of ion implantation induced damage and residual stresses.

EXPERIMENTAL PROCEDURE

Specimen Preparation

Aluminum single crystal wafers were wire spark cut from a rectangular-section bar with a [111] axis. Six specimens were prepared with asymmetric orientations, selected to give incidence angles of from 2.3 to 29.3 degrees for the (111) reflection using Cr and Cu Kα radiation. The specimens were subsequently lapped, annealed for one hour at 500°C, and electropolished to ensure that the surfaces were stress and damage free. Half of the surface area of each specimen was implanted with lithium ions at five energy levels (190, 120, 72, 45, and 25 KeV) to produce a surface layer with maximum possible depth and a nearly flat concentration profile. The fluences at each energy (9.2, 7.2, 5.5, 3.8, 3.1 x 10^{16} ions/cm^2) gave an implanted lithium concentration of about 7 atomic percent as measured using ion beam analysis. After evaluation by X-ray diffraction, the implantation was repeated on the same specimens to give a concentration of about 14 atomic percent.

Sintered alumina (Al_2O_3) and hot pressed silicon carbide (SiC) were obtained from vendors as 6.35 x 3.18 x 50.8 mm bars with chamfered edges. All specimens were polished with consecutively finer diamond paste to a 1 µm finish. The specimens were annealed at 1200°C for 24 hours to eliminate any surface stresses and damage introduced during polishing. A final polish with 0.25 µm diamond paste was applied to remove any reaction layer formed during annealing. One specimen of each type was implanted with argon ions under the following conditions: 1.50 MeV, 1.0 x 10^{16} ions/cm^2; 0.75 MeV, 1.0 x 10^{16} ions/cm^2; 0.15 MeV, 3.0 x 10^{17} ions/cm^2. The high energy implantations were conducted to maximize the depth of the surface layer subjected to lattice damage. The high dose introduced at the low (0.15 MeV) energy was intended to concentrate the implantation effects within a shallow surface layer.

X-Ray Diffraction Methods

X-ray double crystal diffractometry was applied to evaluate the lattice distortion associated with implantation of the aluminum single crystals. The shape of rocking curves, obtained by recording diffracted intensity as the crystal is slowly rotated through the Bragg angle for a selected hkl reflection, is highly sensitive to the degree of lattice imperfection [3,4]. The halfwidth (full width at half maximum) or integral breadth (integrated intensity divided by the maximum intensity) of the rocking curve provides a relative measure of the lattice distortion imposed by a deformative process, when compared to the rocking curve width for the virgin sample. In this investigation, rocking curves were recorded for the asymmetrically cut crystals using Cr and Cu Kα radiation. The penetration distances of the X-rays into the specimens were calculated using the standard formulation for photoelectric absorption and a reflected-to-incident intensity ratio of 1/e. Table 1 summarizes the measurements made and approximate depth probed in each case. It should be noted that, in relatively perfect single crystals, the photoelectric absorption can be accompanied by attenuation due to primary and secondary extinction which further reduce the effective penetration depth. Five rocking curves were recorded on strip charts for the implanted and unimplanted halves of each specimen and later transferred to computer files using a digitizing board.

A GE XRD-5 diffractometer fitted with a linear position sensitive proportional counter was used for investigation of implantation effects in the polycrystalline ceramic materials. Although X-ray residual stress measurements are conventionally conducted using high order reflections, the penetration depths of the X-rays are also high for high 2Θ angles. Thus, while the accuracy with which the change in lattice spacing (Δd)

TABLE 1. Diffraction Conditions for Al Single Crystals

Sample #	Angle of Cutψ(deg)	X-Ray Target	Incidence Angle	Penetration(μm)
1	27	Cr	2.3	0.95
2	26	Cr	3.3	1.34
3	25	Cr	4.3	1.71
4	24	Cr	5.3	2.14
5	17	Cu	2.3	2.87
5	17	Cr	12.3	4.10
6	0	Cr	29.3	6.08

can be determined from the angular shift in the diffraction peak position ($\Delta 2\Theta$) is maximized for back reflections, the capability to disclose effects confined to a shallow, near-surface layer is compromised. For this investigation, symmetric reflections from low order crystallographic planes were chosen to maximize the contribution of the ion implanted layer to the total diffracted intensity. Strong reflections at 2Θ angles of 38.42 degrees (012) and 51.78 degrees (101) were used in the analyses of the alumina and silicon carbide, respectively. Since the surface normal strains measured for the implanted specimens were very large, the use of shallow diffraction angles did not appreciably hinder the calculation of the implantation-induced surface layer stress. Furthermore, as a result of the enhanced contribution by the surface layer to the diffracted intensity, a distinct auxiliary peak could be resolved, angularly separated from the diffraction peak associated with the bulk material. Two profiles were collected for each of the implanted and unimplanted specimens. Background subtraction, curve fitting, and stripping of the $K\alpha_1/K\alpha_2$ doublet were performed for each profile. In order to isolate the auxiliary peak corresponding to the implanted surface layer, the profiles for the unimplanted specimens were normalized with respect to the main peak of the profile for the implanted specimen and then subtracted from this latter profile. The angular shift of the auxiliary peak and its angular extent could be used to evaluate the internal stress level and distribution. The changes in the background level and the integral breadth of the residual peak from the bulk material were also used in evaluating the damage due to implantation.

RESULTS AND DISCUSSION

Lithium-Implanted Aluminum Single Crystals

Depth profiles of the ion concentration for lithium-implanted aluminum single crystals are shown in Figure 1. A relatively flat profile extending over a 1 μm depth was obtained, giving maximum concentrations of about 7 and 14 atomic percent lithium. The similar data from ion beam analyses for crystals cut at 27 and 0 degrees to the (111) planes indicate little tendency for ion channelling during implantation.

Figure 2 shows the results of rocking curve analyses for the aluminum single crystals implanted with lithium ions. The rocking curve breadths, normalized with respect to the unimplanted half of each specimen, are plotted as a function of the calculated penetration depth of the X-rays. Data are given for rocking curves acquired using glancing incidence as illustrated by the insert in the figure. It can also be seen that the normalized integral breadths are consistently higher than the normalized halfwidths. The largest breadth increases after the first implantation were obtained for the specimen giving an X-ray penetration depth of 1.71 μm. The second implantation caused a further increase in the broadening of about 20%, and the specimen giving the maximum broadening had an asso-

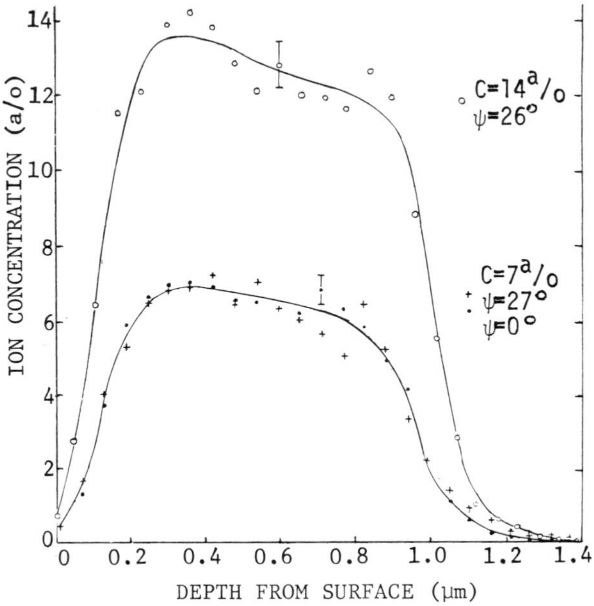

Fig. 1. Concentration profile measured by ion beam analysis for Li^+-implanted aluminum single crystals.

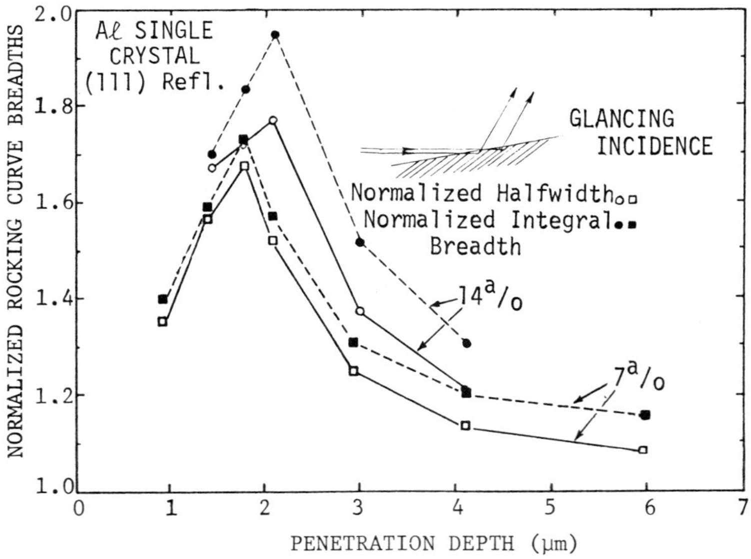

Fig. 2. Depth profiles of normalized rocking curve breadths for 7% Li^+ and 14% Li^+ concentration in Al single crystal.

ciated X-ray penetration depth of just over 2 μm. The larger broadening disclosed by the integral breadths indicates that crystalline defects which contribute to the "tails" of the rocking curves (i.e., point defects such as vacancies and interstitials) constitute a significant proportion of the damage.

In order to construct a profile of the "damage" versus depth from the experimentally measured profile of broadening versus X-ray penetration depth, an algorithm for extracting the local damage from the averaged values was developed. The measured broadening $\beta(d)$, represents an average broadening over the particular penetration depth, d. It may be expressed in terms to the broadening at any depth, x, in the form:

$$\beta(d) = \int_0^\infty \beta(x) e^{-x/d} dx \; / \; \int_0^\infty e^{-x/d} dx \qquad (1)$$

Integrating the denominator results in simply the penetration depth, d, and the resultant expression is similar in type to that of a Laplace transform:

$$f(s) = \{F(t)\} = \int_0^\infty F(t) e^{-st} dt \text{ ; where } t \doteq x \text{ and } s \doteq 1/d. \qquad (2)$$

After substituting an appropriate polynomial expression for the measured depth profile, $\beta(d)$, and using the experimental data to solve for the required coefficients, the inverse Laplace transform of Equation (1) gives the depth profile of interest, $\beta(x)$. The resultant "corrected" profile is shown in conjunction with the experimental profile in Figure 3 for implantation to 7 atomic percent.

Argon-Implanted Alumina and Silicon Carbide

Figures 4(a) and 4(b) give the diffraction profiles for alumina and silicon carbide, respectively, for both the unimplanted specimens and for specimens implanted with argon at 1.50 MeV to a dose of 1.0 x 10^{16} ions/cm^2. Ion implantation is seen to cause four primary modifications of the diffraction profiles: (1) a reduction in the maximum intensity of the main diffraction peak, (2) the formation of a small auxiliary peak to the low angle side of the main peak, (3) an increase in the breadth of the main peak, and (4) an increase in the background intensity about the main peak. These effects

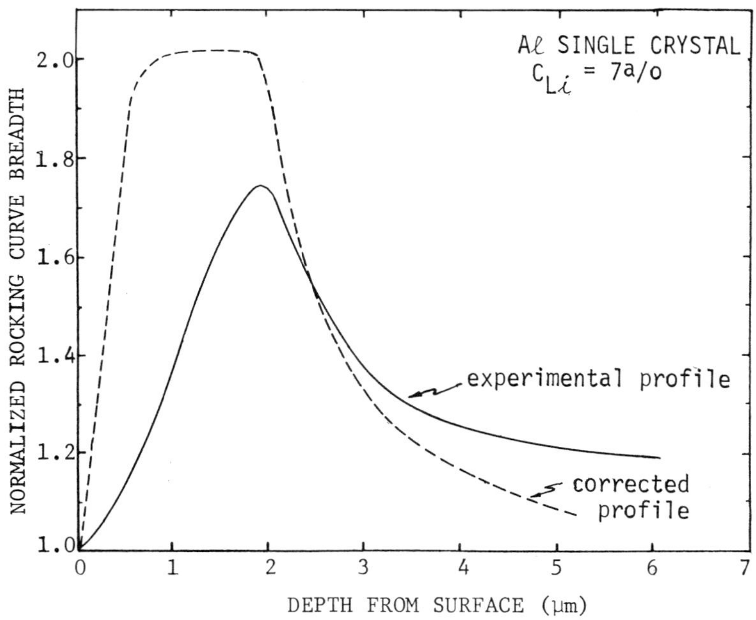

Fig. 3. Experimental and corrected depth profiles for Al single crystals implanted with Li^+ to 7 atomic percent.

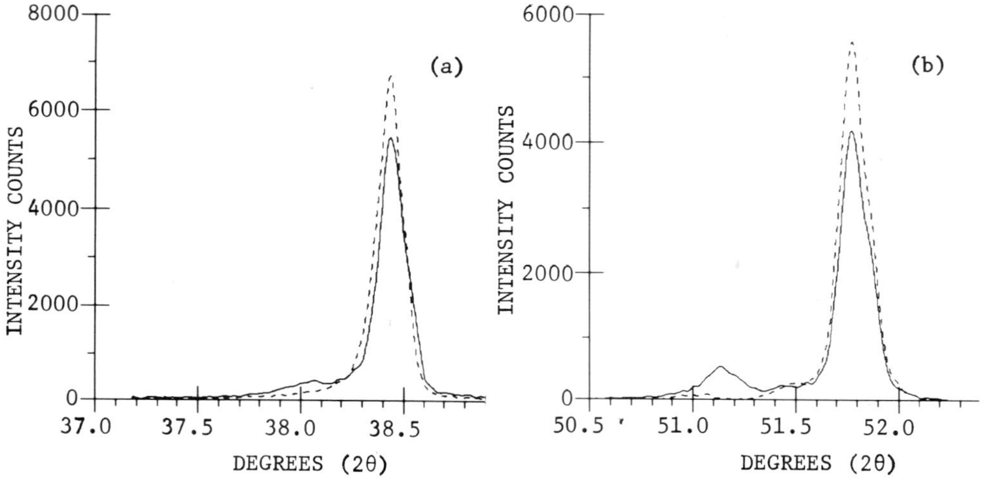

Fig. 4. Diffraction peak profiles for unimplanted (broken curve) and 1×10^{16} Ar^+/cm^2 (1.50MeV) implanted (solid curve) (a) alumina and (b) silicon carbide.

may be explained as follows. The reduction of the maximum intensity of the main peak is associated with the loss in diffracted intensity from the near-surface layer which, on account of the surface layer strains, is manifested by the new, auxiliary peak displaced angularly from the main peak. The broadening of the main peak and increase in the background level are caused by the increased lattice distortion and disorder due to implantation. The broadening may be influenced, as well, by the tensile stress gradient that must exist in the underlying material to balance, or equilibrate, the in-plane compressive stresses introduced at the surface.

For correlation with the X-ray data, depth profiles of the atomic displacements due to implantation were generated using the E-DEP-1 computer code [5] with displacement energy threshold corrections formulated by Robinson and Oen [6]. The depth profiles in Figure 5 show that the maximum damage levels for the same dose at two different energies are, theoretically, nearly equivalent, but that the average damage and depths of the layers differ markedly. The experimentally measured increases in background level for alumina specimens implanted under these conditions (1.50 and 0.75 MeV, 1×10^{16} ions/cm^2) were 29 and 21%, respectively, corroborating the prediction. Implantation at a still lower energy (0.15 MeV), but to a much higher dose (3.0×10^{17} ions/cm^2) gave an equally substantial, 23% increase in the background level, and the results for silicon carbide exhibited a similar trend.

It is to be noted that damage levels of between 100 and 270 displacements per atom (dpa) have been observed for alumina with retention of crystallinity [8, 9]. For covalently bonded solids, the resubstitution of displaced lattice ions occurs less easily than in ionically bonded ceramics, with the consequence that silicon carbide, for instance, is rendered amorphous at a low damage threshold of only 0.2 dpa [9].

The elastic strains introduced by ion implantation, and the material response to them, i.e., the residual stresses associated with the constraint of the lattice expansion, were evaluated by focusing attention on the auxiliary peaks. These satellite peaks represent the diffracted intensity from the near surface layer. In order that they could be studied independently from the remainder of the diffraction profile, they were isolated by subtraction of the peak contributed by underlying material according to the procedure described briefly in the experimental section. Figures

6(a) and 6(b) show example peaks thus obtained for alumina and silicon carbide. The stronger peak obtained for implantation of alumina at 1.50 MeV as compared to 0.75 MeV is the result of the deeper penetration of implanted ions at the higher energy. The surface layer, therefore, makes a greater contribution to the total diffracted intensity (which derives from the full depth probed by the incident X-rays of about 5 μm). The position of the maximum intensity of the peak relative to that for the unimplanted specimen gives a measure of the average strain for crystallographic planes parallel to the surface. The slightly greater displacement of the intensity maximum for the peak from the specimen implanted at 0.75 MeV, as compared to 1.50 MeV, reflects the somewhat greater strain and damage within the shallow surface layer (see also Figure 3).

Elasticity theory was used to derive the expression relating the

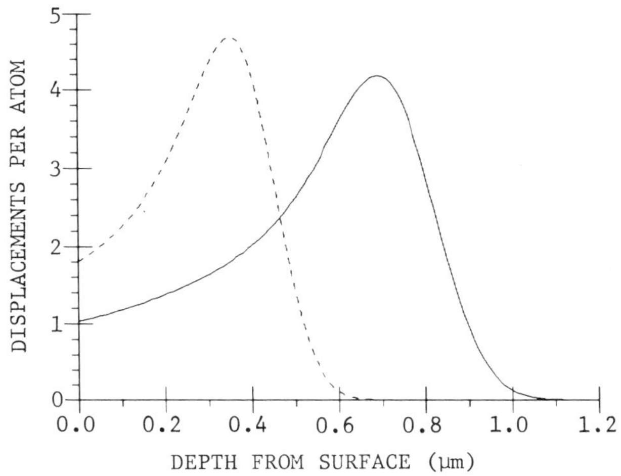

Fig. 5. Theoretical implantation damage depth distributions for 1 x 10^{16} Ar^+/cm^2 (1.50 MeV) implanted (solid curve) and 0.75MeV implanted (broken curve) alumina.

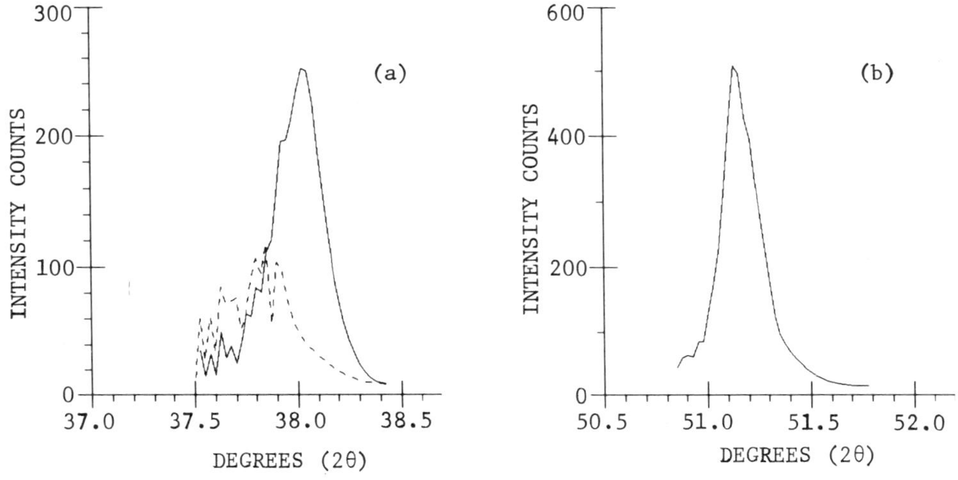

Fig. 6. Auxiliary diffraction peaks from (a) 1 x 10^{16} Ar^+/cm^2 (1.50 MeV (solid curve) and 0.75 MeV (broken curve)) implanted alumina, and (b) 1.0 x 10^{16} Ar^+/cm^2 (1.5 MeV) implanted silicon carbide.

measured strains normal to the surface, ε, to the in-plane stress, σ_R:

$$\sigma_R = -E\ \varepsilon/(1+\nu)\ , \qquad (3)$$

where E is the elastic modulus and ν is Poisson's ratio. For alumina and silicon carbide, values for the average modulus for all directions in the diffraction plane were computed from single crystal stiffness coefficients [10, 11] on account of the anisotropy inherent to the hexagonal structure out of the basal plane. The maximum compressive stresses for alumina specimens implanted at 1.50 and 0.75 MeV were calculated to be 7.93 GPa (1150 Ksi), as determined from the maximum extent of the auxiliary peaks. The average stresses were 3.63 GPa (526 Ksi) and 4.62 GPa (669 Ksi), respectively. Maximum and average stresses for the silicon carbide implanted at 1.50 MeV were 6.59 GPa (955 Ksi) and 4.16 GPa (618 Ksi). The stresses reported by other investigators are consistent with these high values. Krefft and Eernisse [12] employed a deflecting cantilever technique to estimate the average compressive stress in alumina implanted with argon at 0.5 MeV to a dose of 3.0×10^{15} ions/cm^2. Their value of 3.47 GPa (500 Ksi) is consistent with the values reported herein. Page, [13] using the same method, estimated that the average residual stresses were in excess of 7 GPa (1010 Ksi) for Ti^+-implanted alumina.

The observed broadening of the main diffraction peaks could be ascribed to the tensile stress gradient in the material underlying the compressively stressed surface layer. Increases in the integral breadth ranged from 7 to 18% for the various specimens and implantation conditions investigated.

Since no auxiliary peaks were exhibited by specimens of alumina or silicon carbide implanted at low energy (0.15MeV) to a high fluence of 3.0×10^{17} ions/cm^2, either the layer depth was insufficient to generate a detectable auxiliary peak, or sufficient damage was introduced to prevent the stress from being supported by the layer. Under these conditions, the damage level in alumina was over 200 dpa, exceeding the critical damage level of 100 dpa at which amorphization has been reported. For the lower doses at 1.50 and 0.75 MeV, damage levels were considerably less (about 4 dpa), with the consequence that the still-crystalline structure could support a significant internal stress. Similarly, the damage level in the silicon carbide specimens implanted at 0.15 MeV was over 100 dpa for the entire surface layer, again significantly higher than the threshold reported for amorphization of 0.2 dpa. Subsequent indentation hardness measurements made using Knoop and Vickers indenters disclosed a 43% decrease in hardness for these specimens. Unexplained is the result that the damage levels from 1 to 4 dpa for silicon carbide implanted at 1.50 MeV also exceeded the threshold for this material, yet well-resolved auxiliary peaks were nevertheless disclosed prominently by the diffraction profiles from these specimens. Hardness tests revealed a 32% increase in hardness for these implanted specimens, suggesting that crystallinity was, in fact, retained for the most part.

CONCLUSIONS

(1) Rocking curves obtained by double crystal diffractometry could be used to evaluate the depth profile of the lattice damage for ion-implanted aluminum single crystals. The observed broadening of the rocking curves is attributed to the distortions introduced by implantation. No evidence that significant elastic stresses were constrained within the surface layer was found, and the double crystal diffractometer does not, in any event, allow the evaluation of absolute peak positions. Given the

capability for stress transfer exhibited by such ductile materials, the resultant strain/stress gradients would simply contribute to the broadening of the rocking curve profiles and would extend, as found experimentally, over depths larger than those in which implanted ions actually reside. Extension of the technique to the evaluation of implantation damage in common metals and alloys could be accomplished by using a recent modification of the instrumentation for application to polycrystalline materials [14]. This instrument would allow the angle of incidence of the X-rays to be conveniently manipulated so as to minimize the depth probed in conducting the measurements.

(2) Shallow angle diffraction experiments, conducted using a conventional diffractometer, were used to evaluate implantation damage in selected ceramics. The compressive residual stress in the surface layer, degree of induced lattice disorder, and tendency for amorphization could be estimated from the diffraction profiles. In particular, a distinct auxiliary peak was obtained, reflecting the substantial elastic strains supported by the surface layer in specimens which had not been too severely damaged.

REFERENCES

1. X-Ray Studies on Mechanical Behavior of Materials, S. Taira, ed., The Society of Materials Science, Japan, 309 p. (1974).
2. S. Weissmann, "The Application of X-ray Topography to Materials Science," in Nondestructive Evaluation of Materials, J. J. Burke and V. Weiss, eds., Plenum Publ. Corp., 69 (1979).
3. R. W. James, The Optical Principles of the Diffraction of X-Rays, G. Bell and Sons, Ltd., London (1950).
4. A. D. Kurtz, S. A. Kulin and B. L. Averbach, "Effect of Dislocations on the Minority Carrier Lifetime in Semiconductors," Phys. Rev. 101, 4, 1285 (1956).
5. I. Manning and G. P. Mueller, "Depth Distribution of Energy Deposition by Ion Bombardment," Comp. Phys. Comm. 7, 85 (1974).
6. M. T. Robinson and O. S. Oen, "On the Use of Thresholds in Damage Energy Calculations," J. of Nuc. Mater. 102/104, 1315 (1981).
7. T. F. Page and P. J. Burnett, "Criteria for Mechanical Property Modification of Ceramic Surfaces by Ion Implantation," in Proc. Int. Conf. Radiation Effects in Insulators, Guilford, UK, July (1985).
8. C. J. McHargue, Oak Ridge National Laboratory, Personal Communication referenced in M.S. Thesis by S. R. Zimmerman, The Pennsylvania State University, August (1986).
9. J. M. Williams, C. J. McHargue and B. R. Appleton, "Structural Alterations in SiC as a Result of C^+ and N^+ Implantation," Nuc. Inst. and Meth. 209/210, 317 (1983).
10. G. Arlt and G. R. Schodder, "Antiresonance of Conducting Piezoelectric Resonators," Acoust. Soc. of Amer. 37, 151 (1965).
11. R. F. S. Hearmon, "The Elastic Constants of Anisotropic Materials II," Adv. Phys. 5, 323 (1956).
12. G. B. Krefft and E. P. Eernisse, "Volume Expansion and Annealing Compaction of Ion Bombarded Single-Crystal and Polycrystalline α-Al_2O_3," J. Appl. Phys. 49, 2725 (1978).
13. T. F. Page and P. J. Burnett, "Modifying the Tribological Properties of Ceramics by Ion Implanting," Proc. Brittish Ceramic Soc. 34, 65 (1984).
14. R. Yazici, W. Mayo, T. Takemoto and S. Weissmann, "Defect Structure Analysis of Polycrystalline Materials by Computer-Controlled Double-Crystal Diffractometer with Position-Sensitive Detector," J. Appl. Cryst. 16, 89 (1983).

EFFECTS OF GRAIN SIZE AND COOLING RATE ON MAGNETOACOUSTIC STRESS MEASUREMENT IN AISI 4140 STEEL

D. Utrata and M. Namkung*

Association of American Railroads

3140 S. Federal Street, Chicago, IL 60616

ABSTRACT

Effects of microstructures on the uniaxial stress dependence of low-field magnetoacoustic response were studied. Test samples were cut from a single length of AISI 4140 steel bar and heat treated to obtain both coarse- and fine-grained specimens for study. Within these two classifications, the specimen cooling rates included both air- and furnace-cooling to contrast coarse and fine pearlitic development. The specific test procedure involved measurement of the fractional frequency shift of phase-locked ultrasonic waves propagated in a sample simultaneously being magnetized. Curves of frequency shift vs. magnetic induction were then obtained for each specimen at various applied test loads.

The results indicate that both the permeability of the samples and the quantitative behavior of the fractional frequency shift curves are dependent on heat treatment and microstructure. The most pronounced difference in behavior was found between the as-recieved sample and all of the heat treated samples.

Qualitatively, the test results from this study agree with previous investigations of various low- and medium- carbon steels. It is concluded that the stress dependence of acoustic fractional frequency shift exhibited in the magnetoacoustic test method is more dependent on carbon content than on microstructure.

INTRODUCTION

Recent research in the development of the low-field magnetoacoustic test method as a tool for determining residual stress in ferromagnetic materials involved studying the effects of carbon content, heat treatment, and microstructure on test response.

---

* Department of Physics, College of William and Mary, Williamsburg, VA 23185

The ultimate goal of this phase of the work is to determine the range of uncertainty in stress measurement introduced when a variation in micro-structral features is encountered in unknown samples. To that end, different samples obtained from a sigle bar stock were heat treated to introduce a variation in both grain size and interlamellar pearlitic spacing. Further, previous work was performed only on carbon steels, and suggested the existence of a unique test response characteristic only to medium carbon steels. This study introduces medium-carbon alloy steels into the test matrix of samples currently being studied.

EXPERIMENT

Sample Preparation

The material used in this study was 35mm diameter, hot-rolled AISI 4140 steel rod. The chemical composition of the steel is shown in Table 1. Five sample blanks were cut from the same length of rod to minimize

Table 1. Chemical Composition of Steel Samples

Element	C	Mn	P	S	Si	Ni	Cr	Mo	Cu
Weight %	0.42	0.96	.010	.023	0.27	0.17	0.98	0.16	0.12

differences in initial stress states and microstructures. Four of these sample blanks were then heat treated to provide the test matrix shown in Figure 1.

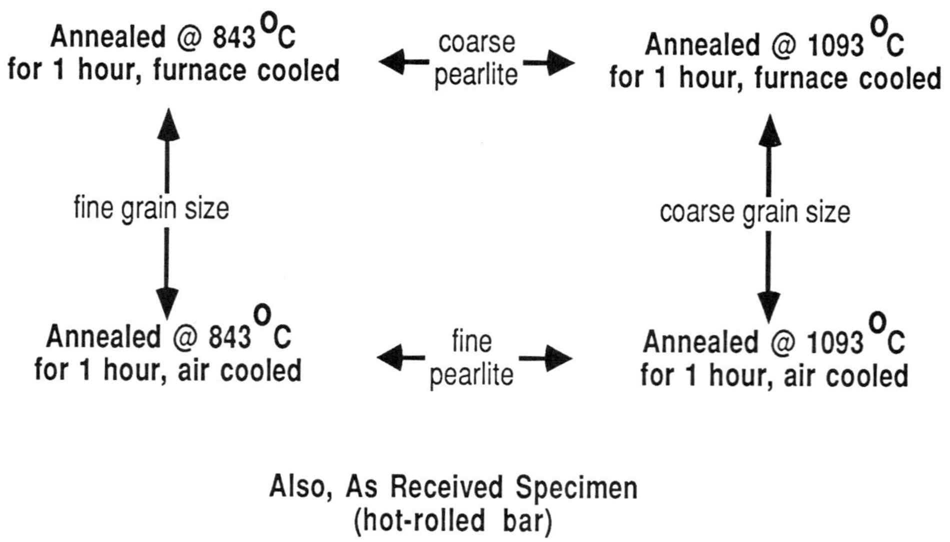

Fig. 1. Test matrix of samples used in this study.

The microstructures of these five samples were then obtained by performing standard metallographic examinations on samples cut from the ends of the sample blanks. The microstructures of the samples are shown in Figures 2-6. As may be seen from the photomicrographs, the desired manipulation of microstructural features was achieved.

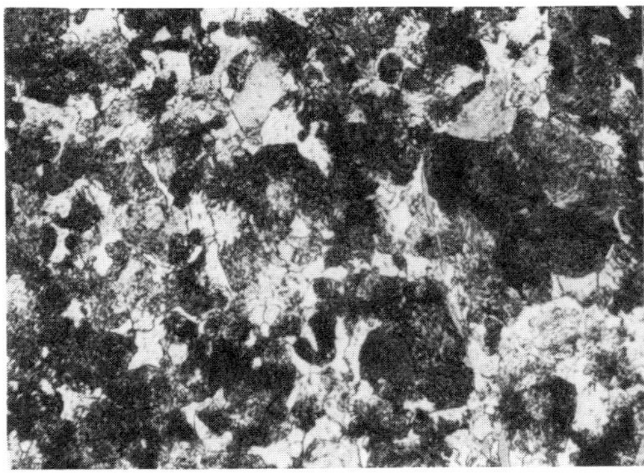

(2)

(3)

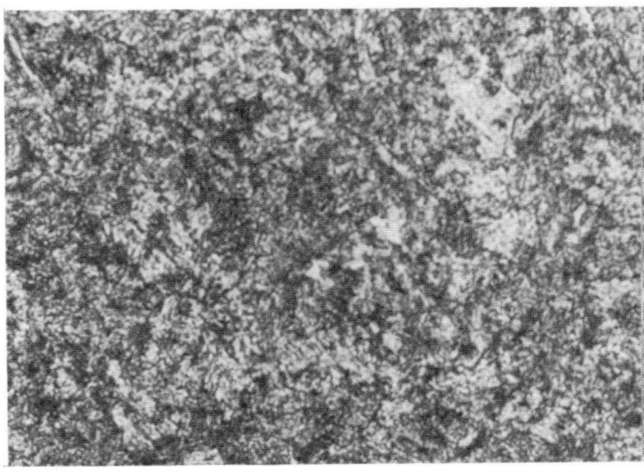

(4)

(5)

(6)

Figs.2-6. Microstructures of steel samples: (2) annealed at 843°C and furnace cooled; (3) annealed at 1093°C and furnace cooled; (4) annealed at 843°C and air cooled; (5) annealed at 1093°C and air cooled; (6) hot-rolled, as-received.

Subsequent to heat treatment, the blanks were machined into final specimen form, shown schematically in Figure 7.

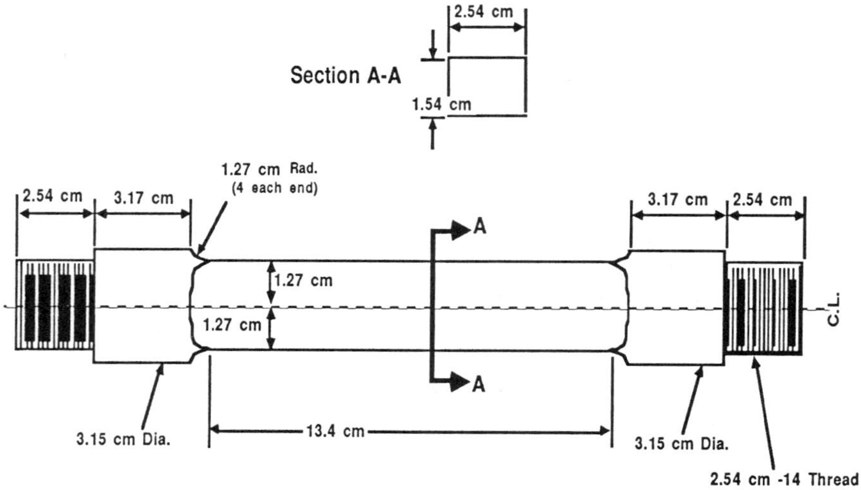

Fig.7. Schematic drawing of test sample.

Test Procedure

The development of the low-field magnetoacoustic test procedure has been previously documented[1,2,3]. In essence, pulsed-phase-locked-loop circuitry is used to accurately monitor the change in ultrasonic wave frequency within a test sample, while a magnetic field is being applied. This fractional frequency change ($\Delta F/F$) is essentially equivalent to the fractional velocity change ($\Delta V/V$) of the ultrasonic wave, which is dependent on the magnetic domain structure in the sample. The initial domain structure has been shown to be related to the state of stress in a ferromagnetic material[4,5]. Therefore, when an applied magnetic field induces a change in the domain structure, the change in phase-locked ultrasonic wave frequency exhibited during the test procedure is related to the state of stress in the test sample.

All of the test samples in this study were evaluated using the same procedure employed in our previously reported research. Namely, a pair of electromagnets was oriented to induce a magnetic field in the sample parallel to the axis of applied stress. The direction of compressional ultrasonic wave propagation was perpendicular to both stress and field. A calibrated pick-up coil was used to measure magnetic induction in the samples. A total of five test runs were performed on each sample at a given stress level; three tests where the applied magnetic field value was steadily increased to a maximum and then decreased back to zero were followed by two tests where the applied magnetic field value was alternately increased incrementally and then returned to zero, up to a maximum level. After each individual test run, the applied magnetic field was cycled about zero by alternately changing polarity and applying decreasing amplitudes of applied field.

In the following report, all of the data presented were obtained from the last test run (the second discontinuous run). Experience in the laboratory has shown this data to be quite reproducible. The data obtained with the continuous-field tests may yield some information regarding the time-response of the various samples, but to date has not been systematically evaluated.

RESULTS

The fractional frequency shift (ΔF/F, in parts per million) as a function of magnetic induction (B, in kilo Gauss), were obtained for each sample at applied loads ranging from 250 MPa in tension to 250 MPa in compression. For clarity, only the response curves for the two extreme loads and the unloaded condition are shown in the data plots. Further, individual data sets are not presented; the emphasis in this investigation was on studying the effect of a change in one microstructural variable. Therefore, the plots stress comparisons between data sets.

It was noted, however, that all samples exhibited behavior similar to that first associated with medium carbon steels[2]. The characteristic trait of these steels was that under compressive loading the initial slope of the ΔF/F vs. B curve was negative, and the zero-stress curve fell between the tensile and compressive curves. In low carbon steels, the unstressed curve was always above both the tensile and compressive loading curves.

Figures 8 to 11 show the results of comparing different cooling rates for fine-grained material, different cooling rates for coarse-grained material, different grain sizes for furnace-cooled material, and different grain sizes for air-cooled material, respectively.

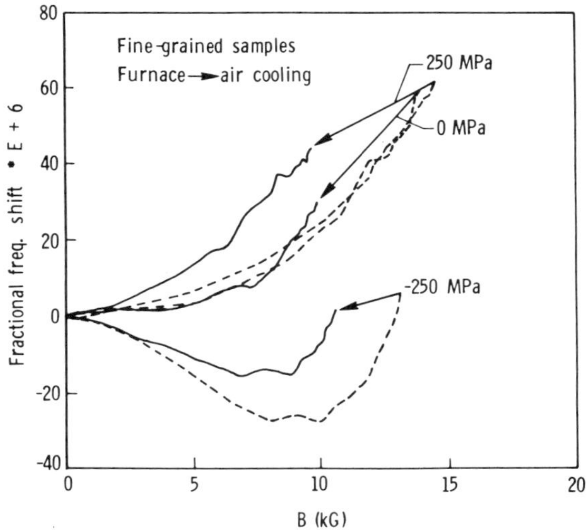

Fig.8. Effect of increased cooling rate for fine-grained samples on magnetoacoustic test response.

From Figure 8, it may be observed that a more rapid cooling rate in fine-grained samples caused a pronounced decrease in permeability, as evidenced by a shift to the left (lower magnetic induction at the maximum applied field value) in the test curves. Also evident is a lower magnitude of fractional frequency shift in both the positive and negative directions. The physical interpretation of this might be that a refinement in the pearlitic structure, as induced by more rapid cooling, inhibits domain wall motion, or even domain rotation, because of an increase in localized residual stresses. Such inhibited motion would serve to reduce the changes in ultrasonic wave velocity, and therefore reduce changes in the measured frequency shift.

The effect of cooling rate is less pronounced in coarse-grained samples, in that only the permeability of the samples appears to be altered. Figure 9 shows that while the response curves have shifted to lower values of applied magnetic field, the magnitude of the fractional frequency shift exhibits only slight change with increasing cooling rates. This suggests that the coarse-grained samples, having larger ratios of grain volume to grain boundary surface area, are more "tolerant" of localized stresses. Further, this suggests that localized stresses in steels exhibit their influence primarily at the grain boundaries.

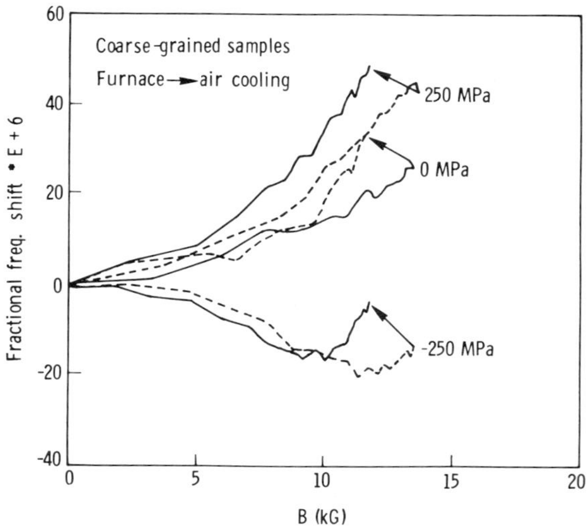

Fig.9. Effect of increased cooling rate for coarse-grained samples on magnetoacoustic test response.

Figure 10 shows the effect of grain refinement for furnace-cooled samples. The permeability of the samples for either grain size appears to be quite similar. The stress effect between samples is quite different; the coarse-grained material exhibits a clear distinction between both applied tension and compression versus the unloaded condition, while the fine-grained material shows a negligible change due to applied tension with respect to the zero-stress case. Because the fine-grained material also exhibits a greater magnitude of frequency shift in both the positive and negative directions than the coarse-grained material, an inhibition of domain wall motion is not surmised to exist; such an effect would diminish the ($\Delta F/F$) value. Rather, the microstructure of the fine-grained, furnace-cooled steel apparently interacts with the domain structure in such a way that the effects

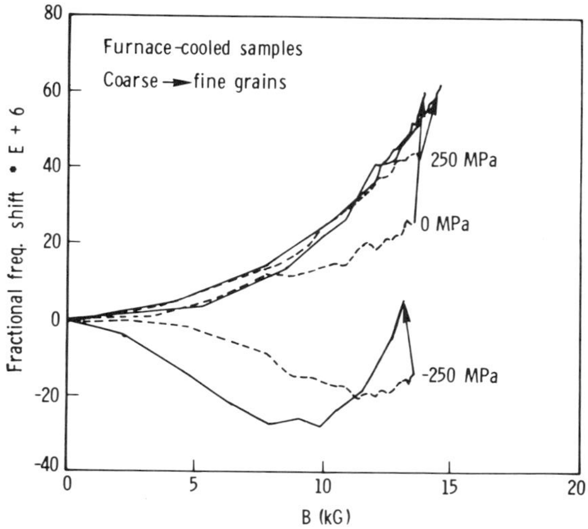

Fig.10. Effect of grain refinenment for furnace cooled samples on magnetoacoustic test response.

of applied tension and compression undergo markedly different behavior. This suggests that there exists a critical cooling rate below which the effects of applied tension will be suppressed due to unique interaction between microstructure and domain structure. A more thorough explanation of this point would require an examination of the actual domain structure and/or development of the behavioral model of acoustic wave/domain structure interaction to discriminate between tension and compression on a microscopic level. Both of these areas were beyond the immediate scope of the project.

The results of air-cooled samples having different grain sizes is shown in Figure 11. Here, grain refinement is seen to lower the permeability of the steel, but produce negligible effect on the fractional frequency shift.

The concept of magnetic domain structure interaction with microstructure is again seen in the response of the as-received, hot-rolled sample, as shown in Figure 12. The heavy lines, representing the as-received sample response at the loading extremes and zero is seen to lie essentially within the spread of all of the other samples (faint lines). The maximum induced magnetic field in this sample is also markedly higher than any of the other samples, and the maximum values of fractional frequency shift are also greater. This suggests that pronounced domain wall motion takes place during the magnetization process, which is logical, considering that an hot-rolled structure is composed of relatively defect-free grains. The fact that the spread between response curves for this sample is the smallest of all the samples tested, however, certainly implies that applied stress will do little to affect domain wall motion in this structure. It is suggested, then, that the microstructural features common to both the hot-rolled

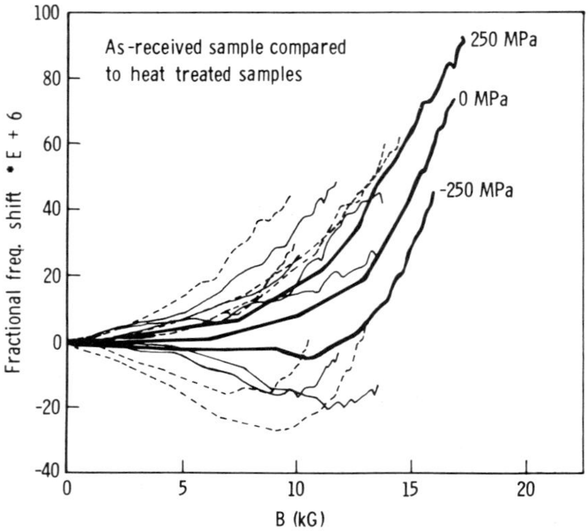

Fig.11. Effect of grain refinement for air cooled samples on magnetoacoustic test response.

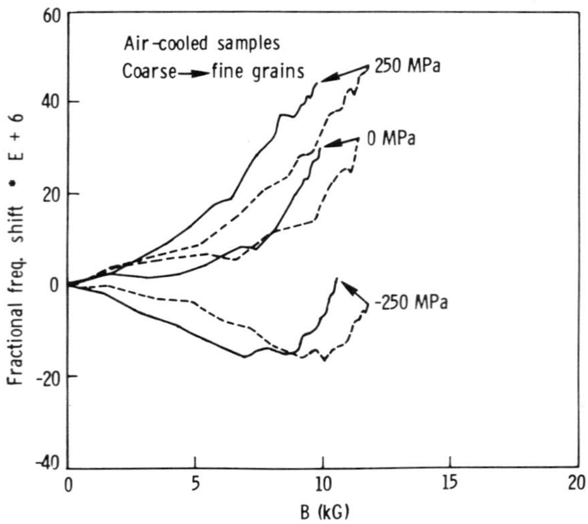

Fig.12. Magnetoacoustic test response of as-received, hot-rolled sample (dark lines) compared to all of the heat treated samples (faint lines).

and the fine-grained, furnace-cooled material enable a distinct difference in interaction between the domain structure and the sample's lattice network to be felt under applied tension and compression. This phenomenon could be related to the existence of an extensive grain boundary network being present in a relatively low-stress matrix.

CONCLUSIONS

Some general observations may be made from these results. A slow cooling rate, such as employed in furnace cooling, apparently emphasizes the distinction between steels having different grain sizes. The slow transformation rate allows an observable effect of an increased grain volume to grain boundary surface area ratio to influence the fractional frequency shift in the magnetoacoustic test.

Further, fine-grained steel samples, with a more extensive network of dislocations and lattice mismatch due to the increased grain boundary surface area present, apparently interacts with the magnetic domain structure to the extent that applied tensile loads will exhibit a ($\Delta F/F$) response essentially similar to the zero-stress test response. The microstructural situation needed for this effect presumably arises when there exists a fine-grained structure with a low-stress matrix. Immediate results would suggest that this situation will arise when the austenitization temperature lies between 843°C and 1093°C, followed by a cooling rate which approaches furnace cooling.

Finally, alloying additions, grain size and cooling rates are quite secondary considerations compared to carbon content for determining the general sample response in the magnetoacoustic test. Within the range of study in this project, the general response characteristics of medium carbon steel, which have been documented in previous work, were found to be present in all microstructures, albeit with occasional slight masking effects.

ACKNOWLEDGEMENTS

This work was performed at NASA Langley Research Center in Hampton, VA, under a project jointly sponsored by the National Areonautics and Space Administration and the Federal Railroad Administration. The authors wish to express their gratitude to these organizations for their support, and also to their own organizations: the Association of American Railroads, Metallurgy Division, and the College of William and Mary, Physics Department.

REFERENCES

1. M. Namkung, D. Utrata, S. G. Allison and J. S. Heyman, Magnetoacoustic Stress Measurement in Railroad Rail Steel, Review of Prog. QNDE, 2;1481 (1986).
2. M. Namkung, D. Utrata, S. G. Allison and J. S. Heyman, Magnetoacoustic Stress Measurements in Steel, Proceedings of IEEE Ultrasonics Symposium, 2;1022 (1986).
3. M. Namkung, D. Utrata, J. S. Heyman and S. G. Allison, Low-Field Magnetoacoustic Residual Stress Measurement in Steel, submitted the Solid Mechanics Reserach for QNDE Symposium, Northwestern University, Evanston, IL (1985).
4. D. M. Bozorth, "Ferromagnetism," Van Nostrand, New York (1951).
5. B. D. Cullity, "Introduction to Magnetic Materials," Addison Wesley, Menlo Park (1972).

IN-SITU STRESS MEASUREMENT WITH THE CANMET PORTABLE X-RAY STRESS DIFFRACTOMETER

R.A. Holt, CANMET, Energy Mines and Resources, Ottawa, Canada

M. Brauss, PROTO Mfg. Ltd., Windsor, Canada

J. Boag, Ontario Hydro Research Division, Toronto, Canada

ABSTRACT

A new instrument has been developed to measure in-situ residual or applied stresses by X-ray diffraction. The instrument has the capability to compensate for texture gradients or coarse grain size, to operate in both single- and multiple exposure modes, to determine surface stress tensors without repositioning and to examine diffraction lines wider than the detector aperture. Laboratory tests on ferritic steels show the instrument to be capable of a standard deviation of 14.6 MPa in the single exposure mode and 5.8 MPa in the multiple exposure mode. The diffractometer was used to measure stress near welds in the Deaerator Storage Tank in Ontario Hydro's Lakeview Thermal Generating Station (T.G.S.) - Unit #7 at Mississauga, Ontario. Large compressive residual stresses were found parallel and perpendicular to the welds by single- and multiple exposure techniques.

INTRODUCTION

The X-ray method of stress measurement applies Bragg's law to measure elastic distention of the crystal lattice at the surface of polycrystalline materials (1).

Recent developments in position sensitive detectors, solid state X-ray power supplies and computer technologies make it possible to design fully automated instruments for in-situ measurement of stress by X-ray diffraction [2-6]. In this paper we describe some results obtained in the laboratory and in the field with a new instrument designed for this purpose.

THE INSTRUMENT

The principles of X-ray stress measurement are reviewed in Appendix A, and the coordinate system used in X-ray stress measurement is shown in Figure 1. System x,y,z is fixed to the specimen, system u, v, w to the instrument. The direction of stress measurement is u, the axis of the instrument v and the specimen normal z. The head of the instrument is shown in Figure 2. It has two position sensitive proportional counters necessary for the single exposure technique (SET).

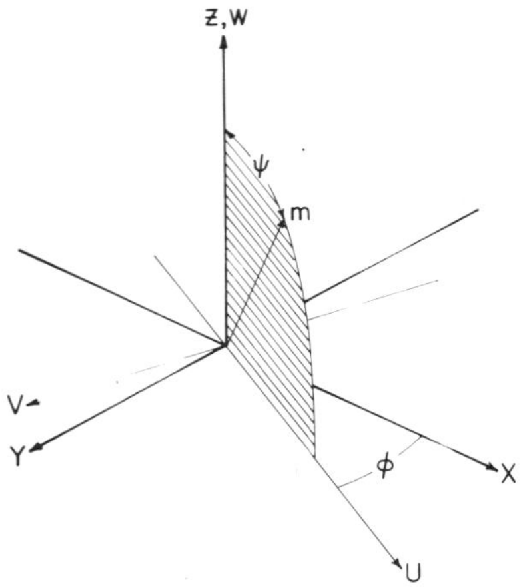

Fig. 1. Coordinate system used in X-ray stress measurement.

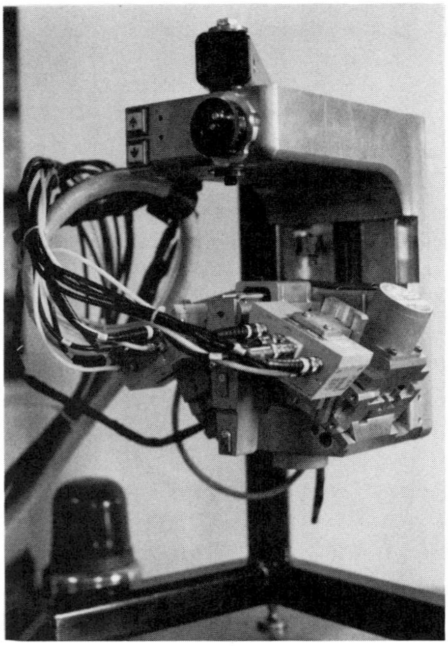

Fig. 2. Head of CANMET Portable Stress Diffractometer. Mostly magnesium alloy, it weights ~15kg.

These simultaneously measure the lattice spacing in two directions m, on each side of the incident beam and in this mode, accurate placement of the instrument relative to the specimen is not necessary (8,9). The orientation of the X-ray source about the axis of the instrument (alpha rotation) is controlled precisely by a stepping motor to compensate for errors from coarse grain structure or texture gradients in the specimen (5,6). These two features are described in detail in a paper which gives the results obtained with an earlier version of the instrument (6). Because the alpha rotation of the source is not coupled to the detectors, the Bragg profiles move in the detectors with the orientation of the incident beam. In this way, complete profiles which are broader than the angular aperture of the detector can be detected.

The X-ray source and detectors can be rotated about the axis of the instrument (omega-rotation). This allows automated measurement by the multiple exposure technique (MET or $\sin^2\psi$ method). The MET is sometimes more accurate than the SET (see Appendix A)

The height of the instrument above the surface to be examined is controlled by buttons at the front of the head. The distance can be adjusted to within +/- 0.05mm by mechanical contact with the specimen of a sensor based on a linear potentiometer. The head can be rotated manually about the z-axis (phi-rotation) to determine stress in several directions and thus extract the surface stress tensor.

The X-ray source is a miniature Kevex tube rated at 30 kV and 6.7 mA. M.Braun detectors and position analysing electonics are used. The instrument is controlled by an IBM PC compatible computer. The control and data reduction software consists of modules to:

- collect and examine profiles for preselected alpha and omega positions to set up the detectors and assess unknown specimens.
- calibrate the angular sensitivity of the detectors.
- record variations in response along the length of both detectors when uniformly illuminated by scatter from a glass slide or radioactive source for subsequent correction of the measured profiles.
- assess the usefulness of a given stress free specimen to a particular stressed specimen and set up conditions for absolute calibration of the detectors (necessary only with the SET).
- perform absolute calibration of the detectors.
- measure stress by either SET or MET (the Bragg profiles can be wider than the detector aperture).
- review raw data, and carry out data reduction with alternative data reduction options; data reduction includes background subtraction, geometric corrections (e.g. polarization, absorption), elimination of profile asymmetry and parabolic fitting to a selected part of the profile.

The electronics are packaged in shock and moisture proof cases, Figure 3, and the cables and head are packed in similar cases for transportation.

LABORATORY MEASUREMENTS

Testing was carried out in the laboratory to determine the "X-ray elastic constants", K_1 of Eq. 5, Appendix A, for three ferritic steels, to assess the reproducibility of the measurements and to assess the error associated with measuring a stress of zero (Figure 2). The {211} Bragg reflection was examined with Cr k-alpha radiation.

Some problems were experienced with drift in the position sensitive proportional counters and their associated position analysing electronics, which could lead to significant errors. Care is required in setting up the detectors, and for SET measurements, frequent calibration is required. For the results reported below, the instrument was calibrated for each SET measurement with a coupon of 1020 steel, stress relieved for 1 h at 950K.

X-ray elastic contants for {211} were determined by measuring "stress" in tapered, constant stress cantilever beams mechanically loaded under the diffractometer. A typical value of K_1 of 178 GPa was used for the preliminary calculation of stress based on the literature survey by Cullity (1). The applied outer fibre strains in the beams were measured with resistance strain gauges and converted to stress with an assumed bulk Young's modulus of 200 GPa. The value of K_1 was then corrected to obtain agreement between measured and applied stresses. Results for Stelco X-65 measured by both the SET and MET are shown in Figure 4. The X-ray elastic constants for the three steels are shown in Table 1. They agree reasonably well with the range reported by Cullity (typically 170-190 GPa with extremes of 162 and 227 GPa).

In 122 SET measurements on the unloaded Stelco X-65 beam over an 18 day period, with an average counting time of 75 s for each measurement, a standard deviation of 14.6 MPa was achieved.

The stress relieved 1020 steel was abandoned as a calibration standard when it was found that some variability occurred in the angles of the Bragg reflections depending on the position of the X-ray spot on

Fig. 3. Electronics for CANMET Portable Stress Diffractometer packaged in shock and moisture resistant cases.

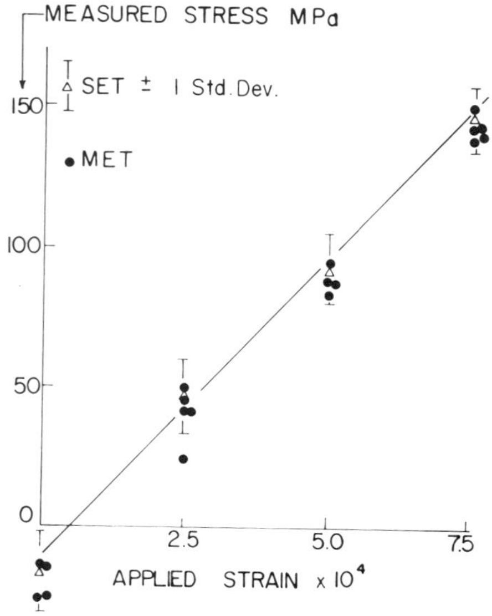

Fig. 4. Measured stress vs applied strain for Stelco X-65 tapered cantilever beam.

the coupon. A finer grained specimen of a microalloyed steel (0.05% C, 0.03% Nb) was prepared by stress relieving at 980 K for 1 h. A series of 39 MET measurements made with a total counting time of 225 s over a 26 day period gave a mean of -18.9 MPa and a standard deviation of 5.8 MPa.

The variation of lattice spacing with $\sin^2(\Psi)$ was approximately linear. SET measurements on the same specimen gave a mean of -6.3 MPa with a standard deviation of 9.6 MPa. The difference in the means is probably due to a systematic "zero" error in the multiple exposure measurements, whereas the stress measured by the SET may have been left by metallographic preparation of the specimen surface.

Table 1. X-ray elastic constants of three ferritic steels

Method Material	X-ray Elastic Constant GPa* SET	MET
SA-106-Gr.B	186(238)	186(238)
A-516-Gr70	171(219)	175(224)
Stelco X-65	167(214)	165(211)

* numbers in brackets are the equivalent value of Young's modulus assuming Poisson's ratio = 0.28

FIELD MEASUREMENTS

The field measurements were made at Ontario Hydro's coal-fired Lakeview T G S - Unit #7 at Mississauga, Ontario. The structure examined was the Deaerator Storage Tank, a low-pressure vessel in which water is collected after removal of gasses between the condenser and boiler. The vessel is horizontal, 3.3 m in diameter and 11.6 m in length, and is fabricated from 16 mm, A212-64 Gr. B. Cracks had been found adjacent to welds inside several similar vessels at Lakeview and during an outage the inside of the Deaerator Storage Tank of Unit 7 had been inspected for similar cracks by a magnetic particle technique. Prior to inspection the areas adjacent to the welds had been cleaned by grit blasting. The object of the X-ray stress measurements was to verify that compressive stresses had been left by the grit blasting.

Areas immediately adjacent to one horizontal and one circumferential weld were selected for stress measurement. The head of the stress diffractometer was mounted on a stand designed for large diameter pipe and cylindrical vessels, Figure 5. The stand was attached to the inside wall of the vessel (approximately 30 degees from horizontal) by permanent magnets at each end of horizontal "feet". Once attached to the vessel, the stand allows four degrees of freedom in manually positioning the head with its vertical axis passing through the spot to be measured. The head is then focused and rotated by the built-in adjustments previously described.

The two spots measured were cleaned by electropolishing "dime"-sized areas for 300 s with methanol-10%perchloric acid at 20V and about 0.5 A. This procedure, which removes about 0.05 mm, was expected to remove any remaining oxide or surface contamination without disturbing stresses left by the grit (approximately 0.25 mm).

Stress was measured by both the SET and MET in directions parallel and perpendicular to each weld from an X-ray spot size at the specimen of about 10mm normal x 2mm parallel to the direction of measurement. The detectors were calibrated in-situ on the stress-relieved micro-alloyed steel specimen previously described. Because a measured value of K_1 was not available, a value of 178 GPa was used. To verify the accuracy of the instrument in each orientation, a MET measurement was made on the standard. The results are summarized in Table 2. The measurements on the standard agreed with those made in the laboratory.

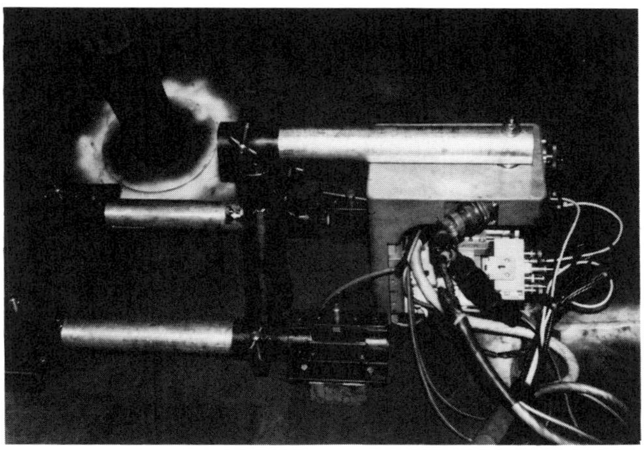

Fig. 5. CANMET Portable Stress Diffractometer, mounted on field stand inside Deaerator Storage Tank at Lakeview T.G.S. - Unit #7.

Table 2. In-situ stresses measured on the inside surface of a Separator Storage Tank at Lakeview G S - Unit #7.

Position	Method	Direction of Measurement	
		Parallel to Vessel MPa	Perpendicular to Vessel MPa
Near Circumferential Weld	SET	-305, -295	-359, -358, -367
	MET	-322, -340	-
Near Horizontal Weld	SET	-328	-415
	MET	-332	-349
On Standard Specimen	MET	-20	-32, -22

High compressive stresses were found in both orientations near both welds. Considering that the X-ray profiles were very broad, reproducibility was good when measurements were repeated. Plots of lattice spacing vs. $\sin^2(\psi)$ were relatively straight, Figure 6, validating the SET, and agreement between single and multiple exposure measurements was also good.

CONCLUSIONS

A new instrument has been developed for in-situ measurement of stress by X-ray diffraction. It has a unique combination of features, including the capabilities of both single and multiple exposure measurement, of compensating for affects of grain size and texture gradients, of rotation about the specimen normal without repositioning and of accessing Bragg profiles broader than the angular aperture of the detectors used.

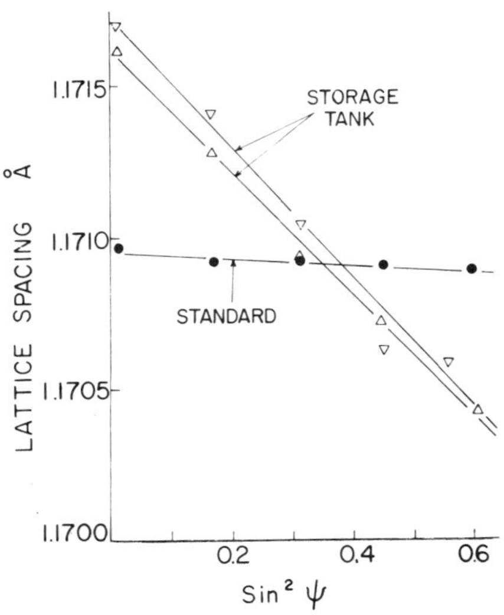

Fig. 6. Plot of lattice spacing vs. $\sin^2$ for some of the MET measurements of Table 2.

During laboratory testing, the instrument made accurate, reproducible measurements by both SET and MET, although detector drift has caused some problems.

The instrument was used to measure stresses near welds on the inside of a Deaerator Storage Tank at Lakeview TGS. High compressive stresses were found to have been introduced by grit blasting to clean the area for magnetic particle inspection. These should help protect the areas from stress corrosion cracking or corrosion fatigue.

ACKNOWLEDGEMENTS

The instrument described is a "commercial prototype" built by PROTO Manufacturing Ltd. under a contribution arrangement with the Program for Industry-Laboratory Projects (PILP). We wish to acknowledge the support and encouragement of D.W.G. White, E. Brauss, A. Hunter and R. Philar, and the technical assistance of H. Wong and N. Shah.

REFERENCES

1. B.D. Cullity, "The Elements of X-ray Diffraction" 2nd Edition, Addison-Wesley, Reading Mass.,1978.
2. M.James and J.B. Cohen, "PARS - A Portable X-ray Analyser for Residual Stress", Journal of Testing and Evaluation, Vol. 6(1978),pp 91-97(See also U.S. patent #4095103).
3. C.O. Ruud, "X-ray Analysis and Advances in Field Instrumentation", Journal of Metals, Vol.13, No.6(1979),pp 10-15.
4. L. Castex, J.M. Sprauel and M. Barral, "A New, In-Situ, Automatic Strain Measuring X-ray Diffraction Apparatus with PSD", Advances in X-ray Analysis, Vol.27(1984), pp 267-272.
5. C.M. Mitchell, "The CANMET Portable Stress Diffractometer", Proc. International Conference on Pipeline Steels, Edmonton, Alberta (1983) Gov't of Canada Publishing Centre, Ottawa(See also U.S. Patent # 4561062)
6. R.A. Holt "A Comprehensive Approach to In-Situ Stress Measurement", Advances in X-ray Analysis, Vol.29(1986), pp 29-35.
7. G. Maeder, J.L. Lebrun and J.M. Sprauel, "Present Possibilities for the X-ray Diffraction Method of Stress measurement", N.D.T. International, October 1981, pp 235-247.
8. J.T. Norton "X-ray Stress Measurement by the Single Exposure Technique", Advances in X-ray Analysis, Vol.11(1968), pp 401-410.
9. C.M. Mitchell, "A Dual Detector Diffractometer for Measurement of Residual Stress", Advances in X-ray Analysis, Vol. 20(1977), pp 379-391.

Appendix A. Principles of X-ray Stress Measurement

Measurements deleted of stress by X-rays is based on Bragg's law which describes the diffraction of monochromatic X-rays by the planes of a crystal, i.e.:

$$\lambda = 2d \sin \Theta \qquad \text{Eq. 1}$$

where:

λ is the wavelength
d is the interplanar spacing of the diffracting planes
Θ is the diffraction angle

We define a set of axes x, y and z in which z is normal to the specimen and x and y are aligned with some axes of symmetry in the specimen (Fig. 1). In practical terms, we can measure the lattice spacing in any direction, m, within a cone 50-60° from the specimen normal. We define the axes u, v and w where w is the projection of m into the specimen surface. The angles ϕ and ψ define the orientation of m, u, v relative to x, y and z.

The elastic strain along m is:

$$\varepsilon_m = \ln \frac{d_m}{d_o} \simeq \frac{d_m - d_o}{d_o} \simeq \frac{d_m - d_o}{d} \qquad \text{Eq. 2}$$

where:

d_m is the interplanar spacing among m,
d_o is the interplanar spacing with no stress, and the change from the exact value d_o to the approximate value d causes only a small error in ε_m.

Then:

$$\varepsilon_m = \varepsilon_w(\cos^2\Psi) + \varepsilon_u(\sin^2\Psi) + \varepsilon_{uw} \cos\Psi \sin\Psi \qquad \text{Eq. 3}$$

and, for material exhibiting isotropic elasticity:

$$\varepsilon_w = \frac{\sigma_w}{E} - \frac{\nu\sigma_u}{E} - \frac{\nu\sigma_v}{E} \qquad \text{Eq. 3a}$$

$$\varepsilon_u = \frac{\sigma_u}{E} - \frac{\nu\sigma_w}{E} - \frac{\nu\sigma_v}{E} \qquad \text{Eq. 3b}$$

$$\varepsilon_{uw} = \frac{T_{uw}}{G} \qquad \text{Eq. 3c}$$

where:

ε_w, $_u$, $_{uw}$ are normal and shear strains in the u, v, w coordinate system.
σ_u, σ_v, σ_w and T_{uw} are normal and shear stresses in the u, v, w coordinate system.
E, ν, G are Young's modulus, Poisson's ratio and the shear modulus.

The X-rays normally sample only a thin surface layer in which stress and structure gradients are small. With no surface tractions

$$\sigma_w = T_{uw} = 0$$

and

$$\varepsilon_m = -\frac{\nu}{E}(\sigma_u + \sigma_v) + \frac{1 + \nu}{E} . \sigma_u . \sin_2\Psi \qquad \text{Eq. 4}$$

Referring to equation 2, σ_u may be found from the slope of a plot of the measured lattice spacing d_m versus $\sin^2\psi$ by applying a factor of:

$$K_1/d = E/(1+\nu)\ d \qquad \text{Eq. 5}$$

If the assumptions of isotropic elasticity and absence of large stress gradients in the measured volume are valid, such plots are linear and only two measurements are necessary to obtain σ_u. If two detectors are used the two measurements can be made simultaneously, intercepting diffracted beams on each side of the incident beam within the u-w plane. If only one detector is used, the X-ray source and detector must be rotated by several tens of degrees relative to the specimen surface to obtain lattice parameter measurements at two values of The two methods are called single exposure and double exposure techniques (SET and DET) respectively. With either technique, measurements in three planes normal to the specimen surface define fully the surface stress tensor.

For the SET the relative positions of the Bragg reflections in both detectors must be calibraed so that a stress free specimen gives a stress of zero. A specimen with Bragg reflections close to those of the unknown, but known to be stress-free, can be used for this purpose with any orientation of the incident X-ray beam. If the specimen contains a small but unknown stress, the detectors can be calibrated accurately with the incident X-ray beam normal to the specimen because Ψ_m and hence d_m will be the same for both detectors.

If the assumption of anisotropic elasticity is invalid, or if stress gradients are sufficiently steep that the stress varies significantly within the layer penetrated by the X-ray beam, then plots of d vs. $\sin^2\Psi$ may not be linear. In this event, both the SET and DET may give incorrect results, and measurements must be made for a number of values of Ψ at each value of ϕ (multiple exposure technique). From such measurements, a triaxial stress tensor can be extracted, or corrections made for effects of anisotropic elasticity (7).

AXIAL STRAINS AT A GIRTH WELD IN A 914 mm LINEPIPE

T.M. Holden, B.M. Powell and S.R. MacEwen

Atomic Energy of Canada Limited, Chalk River Nuclear Laboratories, Chalk River, Ontario K0J 1J0 Canada

R.B. Lazor

The Welding Institute of Canada, 391 Burnhamthorpe Road East, Oakville, Ontario L6J 6C9 Canada

ABSTRACT

High resolution neutron diffraction measurements have been made of axial strains in and adjacent to a multipass girth weld in a complete section of 914 mm (36") linepipe. The experiments were carried out at the NRU reactor, Chalk River, with 1 mm wide apertures before and after the sample to define the location of the strain measurement. The diffraction angle was measured to a precision of ±0.003° leading to an accuracy of 1 in 10^4 in interplanar spacing. The strains were measured in the subset of grains whose [110] direction of the body centred cubic structure was aligned along the axis of the pipe. The strains were measured at ten positions through the 16 mm thickness of the pipe on the weld centre and offset by 4, 8, 20 and 50 mm from the weld centre. In addition, the spatial variation of axial strain as a function of axial distance from the weld centre was mapped out along the mid-thickness, near the root of the girth weld, and at the three-quarter-thickness position. The mid-thickness is in a state of tensile strain with respect to the outside and inside surfaces of the pipe. The effect is strongest at the weld centre and is less, though still present, well away from the weld. The strain peak at mid-thickness extends 5 mm from the weld centre, followed by a minimum at about 9 mm from the weld centre quite close to the edge of the weld. A further slight but extended maximum in strain is centred at 20 mm from the weld centre, well outside the weld. At a position 3 mm from the inside surface of the pipe, a broad strain minimum on the weld centre and a broad maximum centred at 20 mm from the weld centre are observed. The results are compared with strain gauge measurements on the weld surface and through-wall residual stress distributions measured using a layering technique.

INTRODUCTION

Neutron diffraction provides a method for mapping out the variation of the strain within and adjacent to weldments. The ability of thermal neutrons to penetrate up to 30 mm in iron, makes them extremely useful for obtaining diffraction information about volumes inside a sample. The spectrometer resolution is readily adapted to measure interplanar spacings with a precision of $\pm 1 \times 10^{-4}$ Å which is sufficiently good to determine the strains near a weld.

Experimental studies of weldments by neutron diffraction have been reported for welded plates[1] and a tube welded to a plate,[1] but the present measurements are the first ones to be reported on axial strains on a linepipe girth weld. Surface stresses[2] near welds in similar pipes have been deduced from the centre-hole rosette gauge method and through-wall distributions of axial residual stresses[2] have been obtained by the "block removal and layering method". The results of the three methods are compared qualitatively to describe the strains in and adjacent to the girth weld.

EXPERIMENTS

Linepipe

Two short lengths of 914 mm diameter, 16 mm wall thickness, Grade 448 N/mm^2 linepipe were prepared for welding with a V-preparation using a 75° included angle and a 1.6 mm land and a 1.6 mm root face. An internal clamp was used to remove misalignment and to maintain root spacing during deposition of the root pass. Welding was performed vertically down, with the pipe axis in the horizontal position by two welders using conventional stovepipe welding techniques. Six passes were required to complete the girth weld, a typical cross-section of which is shown in Fig.1.

Neutron Diffraction

The experiments were carried out with the N5 spectrometer at the NRU reactor, Chalk River, employing the (331) planes of a squeezed silicon crystal as monochromator to provide a neutron beam of wavelength 2.0411 Å. Soller slit collimators were used to achieve horizontal angular collimation before and after the linepipe of 0.4° and 0.3° respectively. Slits in absorbing cadmium sheet each 1 mm wide and 25 mm high defined the volume in the sample where the diffraction occurred. The diffracting volume is 25 mm high (hoop direction of the pipe) and had a diamond shaped cross-section in the axial-radial plane with the long diagonal (2.0 mm) in the radial direction and the short diagonal (1.2 mm) in the axial direction. Each diffraction measurement gives the average lattice parameter and hence strain at the centroid of the diffracting volume.

The linepipe, weighing 100 kg, was suspended above the spectrometer so that one wall of the pipe on its equatorial diameter was on the centre of the sample table. The linepipe was oriented so that its axis bisected the incident and scattered neutron beams and hence the strains measured were axial. Positioning in the horizontal plane was achieved with a computer controlled XY translator to ± 0.05 mm.

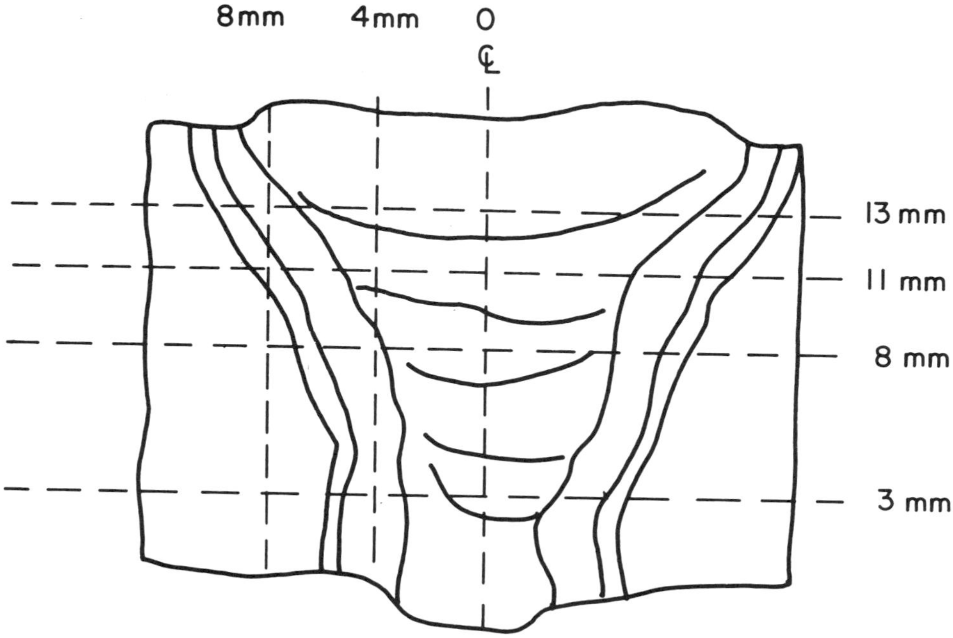

Fig.1. Cross-section of a girth weld in 16 mm line pipe. The inside surface is defined as the zero of position through the wall. The centreline of the weld, designated ℄, is defined as the zero in the axial direction. The dashed lines indicate the loci of neutron scans near the weld.

A typical diffraction peak is shown in Fig.2. For neutron diffraction, the instrumental line-shape is Gaussian so the peak position was obtained by fitting a Gaussian profile on a sloping background to the data. The parameters of importance are the angular position, $2\theta_{hkl}$, of the peak, its full-width at half-maximum and the integrated intensity.

From the angular position, the interplanar spacing d_{hkl} for the planes with Miller indices (h,k,l) may be deduced from Bragg's Law,

$$\lambda = 2d_{hkl}\sin\theta_{hkl}$$

The lattice spacing is for that subset of grains which is aligned along the axis of the pipe. The measurement gives no information about grains oriented, for example, along the radial direction of the pipe.

RESULTS AND DISCUSSION

The interplanar spacing for the (110) planes as a function of distance from the weld-centre at the mid-thickness position and close to the quarter and three-quarter thickness positions (actually 3/16 and 13/16) is shown in Fig.3. The through-wall variation of interplanar spacing at several locations in and adjacent to the weld is shown in Fig.4. The reference lattice spacing was obtained in a subsequent experiment on a cylinder of material 10 mm in diameter cut from the edge of the pipe and annealed for 8 hours at 675°C. The strains calculated with this reference are indicated on the right-hand scales of Figs.3 and 4.

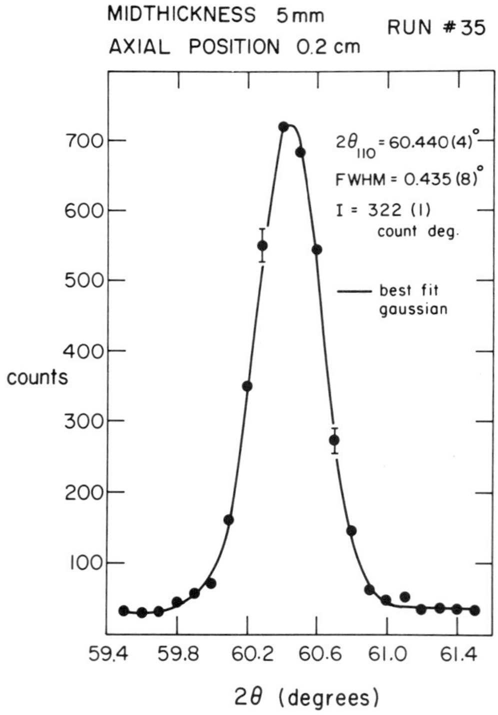

Fig.2. Typical neutron diffraction scan with intensity plotted as a function of scattering angle. The curve through the points represents the best fit to a Gaussian upon a sloping background.

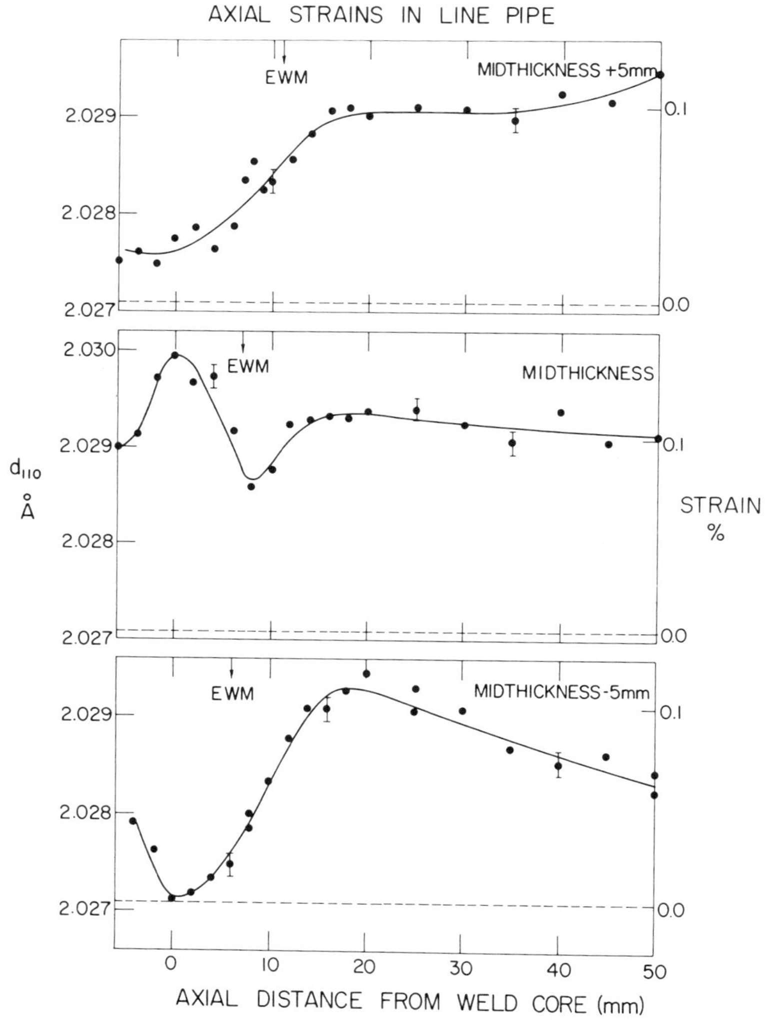

Fig.3. Axial variation of the (110) interplanar spacing of grains oriented with [110] axes along the axis of the linepipe in and adjacent to the girth weld. Measurements were made on the wall mid-thickness and at 3/16 and 13/16 of the wall thickness. The edge of the weld in each scan is denoted by EWM. The right hand scale refers to the axial residual strains deduced with respect to the reference lattice spacing discussed in the text.

The strains deduced from the measurements are expected to be a superposition of the residual strains generated by forming and seam welding the sections of pipe prior to girth-welding as well as the residual strains generated by girth-welding. The results show that the axial lattice strain on the weld centreline is lowest on the inside and outside diameter positions and that there is a high tensile strain in the mid-thickness. This pattern also appears on the through-wall scan off-set from the weld centre by 4 mm but still mostly within the weld. An asymmetric peak is observed at an offset of 8 mm where the loci of scans pass through the parent pipe as well as the weld. On the wall mid-thickness, the high tensile strain evident on the weld centreline in Fig.3 decreases as the edge of the weld is approached and there is a local minimum just outside the weld (9 mm) followed by a broad maximum centred around 16-20 mm from the weld center-line. Girth welding typically induces strains as far as 300 mm to either side of the weld metal. Consequently, the axial strains at 50 mm still contain a component due to the welding operation.

The through-wall variation of axial residual stress deduced from the centre-hole rosette gauge (open circles) and the layering method (full line) on the weld centreline in a similar linepipe is shown in Fig.5. At both surfaces, the stresses are shown to have a range from -100 to +200 N/mm , the result of using different circumferential positions for the two methods. The average surface stresses (rosette gauge) are relatively more tensile at the OD compared to the ID, which is generally true for welds in materials of similar thickness. The layering results show a peak tensile stress at 6.5 mm from the ID, and a similar tensile peak is observed for the diffraction experiments at 9 mm from the ID. Assuming that the value of Young's modulus for iron in the [110] direction is 220 GPa, the stress deduced from the neutron experiments would be 330 MPa, which is 40% larger than the maximum stress obtained by layering. Although there is a difference in the measured stresses near the wall centerline, both methods show a peak tensile axial component at the centre of the weld bead.

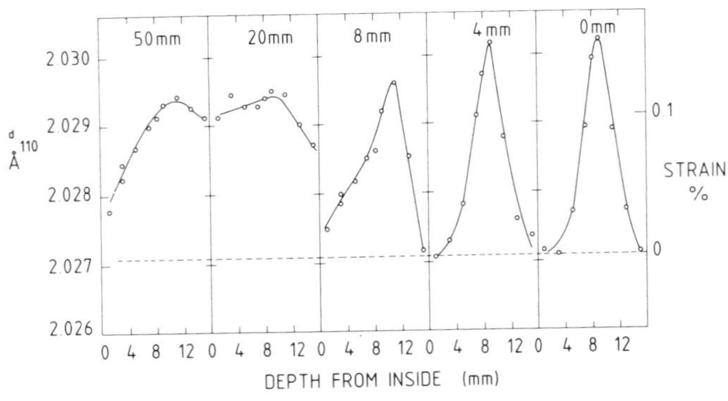

Fig.4. Through-wall variation of the (110) interplanar spacing of grains oriented with [110] axes along the axis of the pipe at several axial locations. The right-hand scale refers to axial residual strains deduced with respect to the reference lattice spacing discussed in the text.

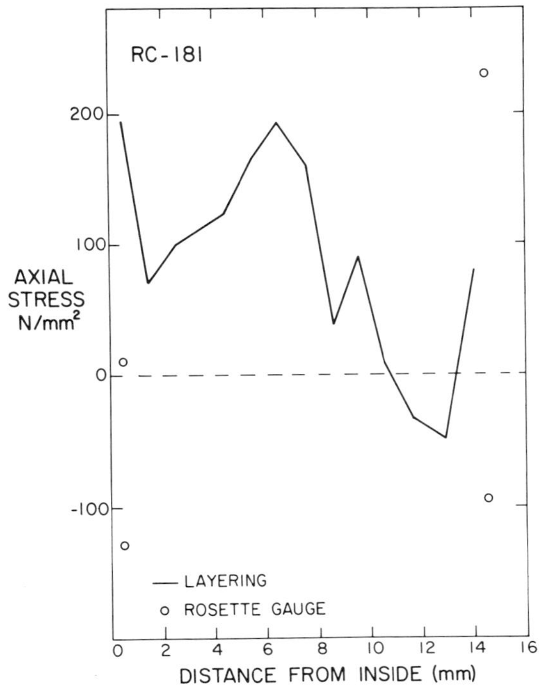

Fig. 5. Through-wall variation of the axial residual stress on the weld centreline of a similar welded linepipe determined by rosette gauge (open circles) and by layering methods (solid lines), from (2).

Finally, the layering results exhibit abrupt changes in stress measurement through the thickness whereas neutron diffraction indicates a smooth distribution of strains, and thus stress, through the grith weld.

As an initial trial to use neutron diffraction techniques to measure the axial residual strains in a linepipe grith weld, these experiments have been extremely encouraging. The results of these trials have indicated that neutron diffraction is a viable nondestructive approach to measuring through-wall residual strains in welded components.

ACKNOWLEDGEMENTS

We acknowledge the expert technical assistance of H.F. Nieman, A.H. Hewitt, M.F. Potter and D. Tennant.

REFERENCES

1. A. J. Allen, M.T. Hutchings, C.G. Windsor and C. Andreani, Advances in Physics 34:445 (1985).
2. R.B. Lazor, R.H. Leggatt and A.G. Glover, AGA PR-140-169-/PR-164-170 (1985).

MEASUREMENT OF STRESS IN STEELS USING MAGNETICALLY INDUCED VELOCITY CHANGES FOR ULTRASONIC WAVES

H. Kwun

Southwest Research Institute
6220 Culebra Road
San Antonio, Texas 78284

ABSTRACT

Because of the magnetoelastic interaction in a ferromagnetic material, the velocity of ultrasonic waves in the material changes upon application of an external magnetic field to the material. Mechanical stresses influence the magnitude of this velocity change as well as the way the velocity change varies as a function of the applied magnetic field. This stress dependence of the magnetically induced velocity change can be used to nondestructively determine applied and residual stresses in ferromagnetic structural steels. In this method, shear or longitudinal waves are used to determine internal stresses while surface waves are used to determine surface stresses. This paper briefly reviews the underlying physics involved in the magnetically induced velocity change phenomenon, and describes an instrumentation system used to measure such velocity changes along with examples of its stress dependence. Measurement results of residual welding stresses and surface stresses determined by using the above described method are presented as well as the capabilities and limitations of the method.

INTRODUCTION

It is well known that the magnetic properties of a ferromagnetic material change upon application of a mechanical stress to the material and, conversely, the elastic properties of the material change upon application of a magnetic field. This bilateral effect is caused by the magnetoelastic interaction between the domain magnetization and the mechanical strain in the lattice.[1,2]

The effect of stress on the magnetic properties has long been used in various electromagnetic methods for nondestructive measurement of stress in ferromagnetic structural steels.[3,4] In these methods, the stress dependence of one or more magnetic parameters such as permeability,[5] coercive force,[6] remanence,[7] and hysteresis loops[8] have been used as well as physical parameters influenced by these magnetic parameters such as eddy current,[9] nonlinear generation of harmonics,[10,11] and Barkhausen noise.[12,13]

In addition, the effect of stress on the changes in the elastic properties of a ferromagnetic material caused by an externally applied magnetic field has been used for stress measurement. Methods based on the magnetostriction[14,15] and the ΔE effect,[16,17] which refers to the elastic moduli change with magnetization, are examples of such uses.

Besides changes in the physical dimension and elastic moduli, attenuation and velocity of ultrasonic waves propagated in a ferromagnetic material change upon application of a magnetic field to the material. This magnetically induced change in the attenuation or the velocity of ultrasonic waves is also dependent on the state of stress and, therefore, has been used for nondestructive measurement of stress.[18-23] The methods based on the magnetically induced attenuation or velocity change for ultrasonic waves are, unlike the other methods, applicable to measurement of bulk stress in the material. Moreover, unlike the ultrasonic methods based on the stress-induced velocity change caused by the nonlinear elasticity of the material (known as the acoustoelastic effect[3,4,24]), the above ultrasonic methods are not sensitive to the variation in nominally the same material. Because of these advantages, the magnetically induced velocity (and attenuation) changes for ultrasonic waves have been actively investigated recently.

This paper briefly describes the physical background on the magnetically induced velocity change (MIVC) phenomenon and the instrumentation system for measuring such velocity changes. Examples of the stress dependence of the MIVC for ultrasonic waves are also given. An application of the MIVC to measurements of residual welding stresses is then described along with the capabilities and the limitations of the method.

PHYSICAL BACKGROUND

When an ultrasonic wave is propagated through a ferromagnetic material, two things occur: one is the ordinary elastic vibrations of atoms caused by the stress of the ultrasonic wave, and the other is the rotational vibrations of domain magnetizations caused by the magnetoelastic interaction between the magnetic domains and the ultrasonic wave.[25,26] Because of the magnetostrictive strain caused by the rotation of domain magnetization,[1,2] the overall strain produced by the stress of the ultrasonic wave is the sum of this magnetostrictive strain ε_m and the ordinary elastic strain ε_e. The general expression for the velocity of sound in the material, $V = (C/\rho)^{\frac{1}{2}}$ where C and ρ are the elastic constant and the density of the material, respectively, can be rewritten as the following:

$$V = (\sigma_u/\rho\,\varepsilon)^{\frac{1}{2}} = [\sigma_u/\rho(\varepsilon_m + \varepsilon_e)]^{\frac{1}{2}} = (\sigma_u/\rho\varepsilon_e)^{\frac{1}{2}}\,(1+\varepsilon_m/\varepsilon_e)^{-\frac{1}{2}} \simeq V_e\,(1-\varepsilon_m/2\varepsilon_e)$$

where σ_u is the stress of the ultrasonic wave, ε is the overall strain due to σ_u, and $V_e = (\sigma_u/\rho\varepsilon_e)^{\frac{1}{2}}$ is the purely elastic value which would be obtained in the absence of the magnetostrictive strain. Therefore, the velocity of sound in a ferromagnetic material at zero magnetic field and zero stress is always smaller than the purely elastic value by approximately $V_e\varepsilon_m/2\varepsilon_e$. In structural steels, the reduction in the velocity due to this magnetostrictive strain term is on the order of 0.01% to 0.1%.

Upon application of a magnetic field to a material that is in a stress free state, the magnetic domains in the material align toward the direction of the magnetic field to minimize the Zeeman magnetic energy. Because of the restraint on the orientation of the domain magnetizations imposed by the applied magnetic field, the extent of the rotational vibration of the domain magnetizations and, therefore, the magnetostrictive strain ε_m, caused by the ultrasonic wave are reduced. Since the restraint on the domain magnetization rotation increases with increasing magnetic field H, ε_m decreases progressively with increasing H and eventually becomes zero as H is further increased (Fig. 1). Accordingly, the velocity of sound increases rapidly initially with increasing H and then slowly approaches the purely elastic value in the high H region, as depicted in Fig. 1 with a solid line.

When a static mechanical stress is present in the material, the magnetoelastic interaction between the static stress and the magnetic domains restricts the rotational vibrational motion of the domain magnetizations caused by the ultrasonic wave. Therefore, the velocity at H = 0 becomes larger than the velocity at zero stress as shown in Fig. 1 (neglecting the acoustoelastic effect[24]). When an external magnetic field is applied to the material, the combined effects of the static stress and the applied magnetic field control the extent of the rotational vibration of the domain magnetizations caused by the ultrasonic wave. In ferromagnetic materials having a positive magnetostriction such as most structural steels, the two effects are additive for $\sigma \| H$ and subtractive for $\sigma \perp H$ when the stress is tensile; when the stress is compressive, the two are additive for $\sigma \perp H$ and subtractive for $\sigma \| H$.[19-22] Consequently, depending on whether the combined effects are additive or subtractive, the velocity can either increase or decrease with increasing H initially (broken lines in Fig. 1). When the effects are additive, the velocity increases with H and gradually approaches the purely elastic value similarly to the stress-free case. When the effects are subtractive, the velocity reaches a minimum where the combined effects are minimum, increases with increasing H as the effect of H becomes dominant over the effect of the static stress, and gradually approaches the purely elastic value as H is further increased. Since the velocity at H = 0 is larger under the stress than at zero stress, the magnitude of the velocity change caused by the applied magnetic field decreases with increasing stress.

As illustrated in the above, the magnitude and shape of the magnetically induced velocity change (MIVC) curve as a function of H are dependent on the magnitude and sign (tensile and compressive) of the stress and the relative orientation of the stress and the magnetic field. These characteristic stress dependences of the MIVC are utilized for determining the magnitude, sign, and direction of an unknown stress.

INSTRUMENTATION

Since the magnitude of the MIVC for ultrasonic waves is only on the order of 0.01% to 0.1% in structural steels, an instrumentation system capable of measuring a velocity change with an accuracy of 1 part per million was required for the investigation. To meet the requirement, a system based on the phase-comparison technique was used.[21]

A block diagram of the system used is illustrated in Fig. 2 (the operating principle of the system can be found in reference 21). The system was applicable to measurement of MIVC for various ultrasonic wave modes such as longitudinal, shear, and surface waves. The system could be computer controlled for automatic data acquisition and analysis.

STRESS DEPENDENCE OF MIVC

Examples of the uniaxial stress dependence of MIVC are shown in Figs. 3 through 6, which were obtained from 0.5-inch thick specimens of ASTM A-514 steel. Fig. 3 was for longitudinal waves, Figs. 4 and 5 for shear waves, and Fig. 6 for surface waves. In these figures V_o is the velocity at H = 0

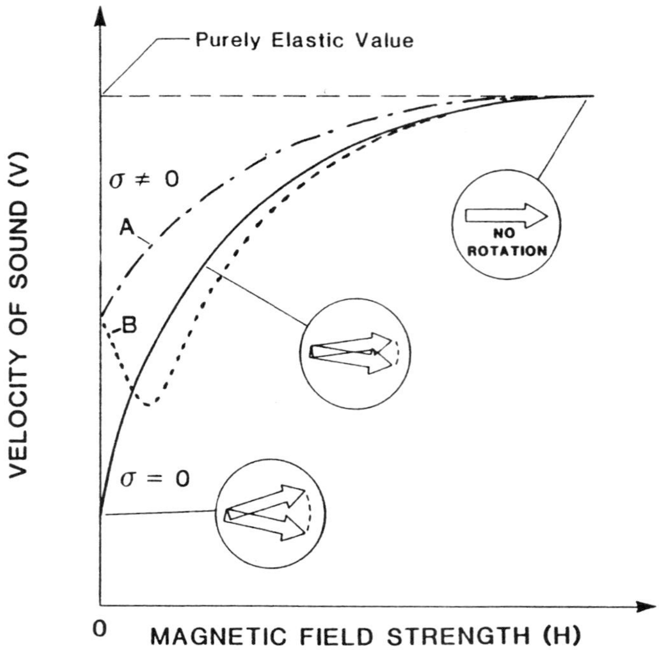

Fig. 1. Effects of a magnetic field on rotational vibration of a domain magnetization and on velocity of an ultrasonic wave. (A is when the effects of stress and magnetic field are additive, and B is when the two effects are subtractive.)

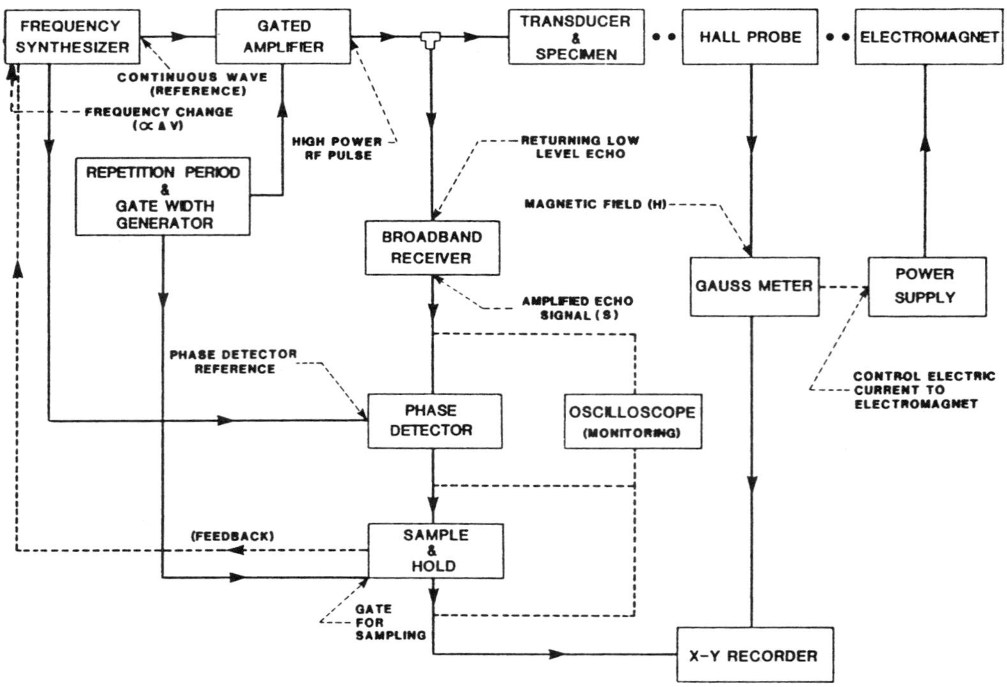

Fig. 2. Block diagram of an instrumentation system for measuring magnetically induced velocity change for ultrasonic waves

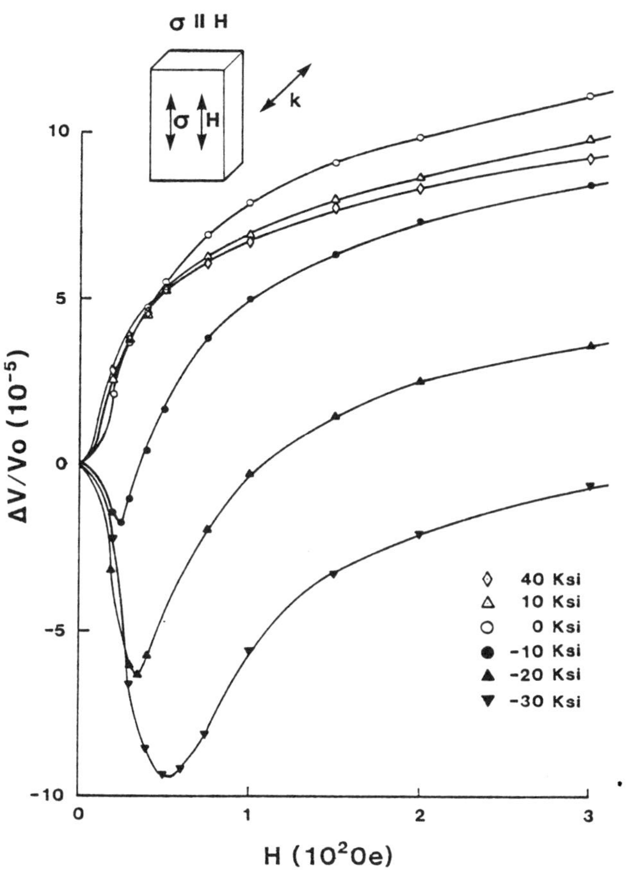

Fig. 3. Stress dependence of MIVC for 10-MHz longitudinal waves in ASTM A-514 steel

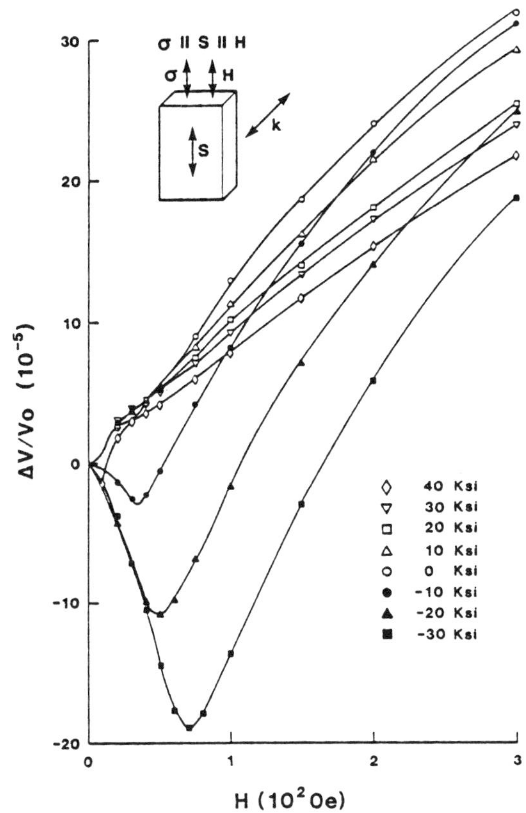

Fig. 4. Stress dependence of MIVC for 5-MHz shear waves in ASTM A-514 steel (S is the polarization direction of shear waves)

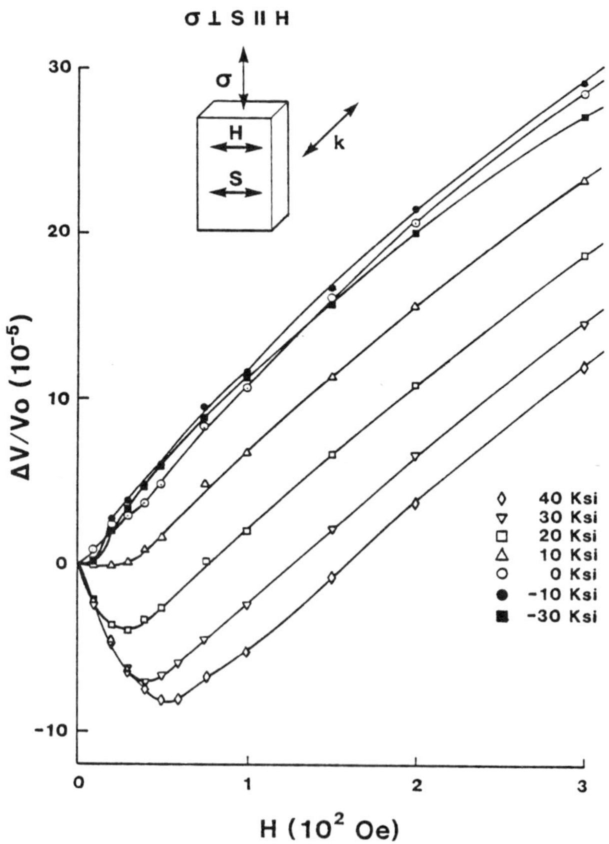

Fig. 5. Stress dependence of MIVC for 5-MHz shear waves in ASTM A-514 steel (S is the polarization direction of shear waves)

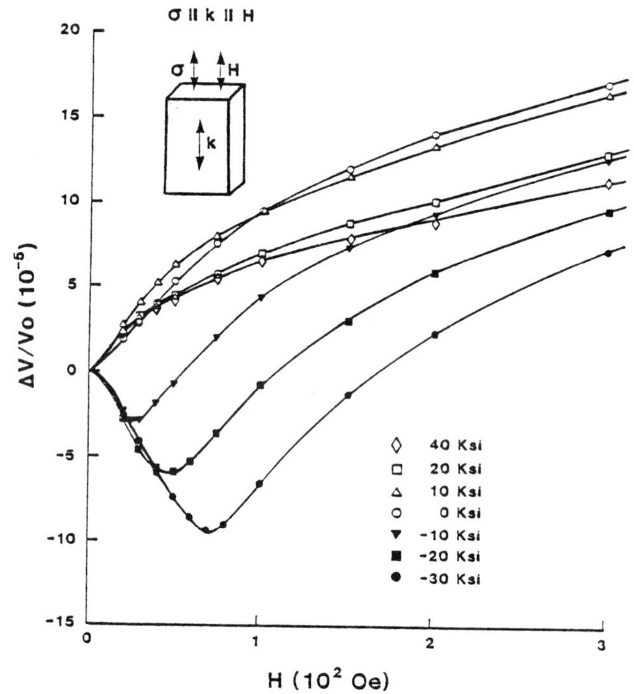

Fig. 6. Stress dependence of MIVC for 5-MHz surface waves in ASTM A-514 steel

and ΔV is the magnetically induced velocity change defined as $\Delta V = V - V_o$ where V is the velocity at H. Additional examples of the stress dependence of MIVC can be found in refs. 19 through 23.

As described in the previous section, MIVC at zero stress generally increased rapidly with H initially and then gradually leveled off toward a saturation value. Upon application of stress to the specimen, both the magnitude and the shape of the MIVC curve as a function of H changed. With increasing stress, the magnitude of the MIVC became smaller. When the relative orientation between the stress and the magnetic field was set so that their effects on the rotational motion of the domain magnetization were subtractive to each other ($\sigma \| H$ for compression and $\sigma \perp H$ for tension), the MIVC curve exhibited the characteristic minimum in the low magnetic field region.

Because both tension and compression decrease the magnitude of MIVC, a state of bending stress through a thickness of a part also affects the MIVC. An example of the effect of bending stress on MIVC for shear waves obtained from the ASTM A-514 steel specimen is shown in Fig. 7. The MIVC obtained under a bending stress seemed to agree with the integrated value over the thickness using the uniaxial stress dependence of the MIVC. Note that the magnitude and the shape of MIVC under bending stresses were quite different from those of MIVC under uniaxial stresses, suggesting the possibility of determining the stress distribution through the thickness. The ultrasonic methods based on the acoustoelastic effect, on the other hand, cannot tell the difference between a zero stress and a bending stress.

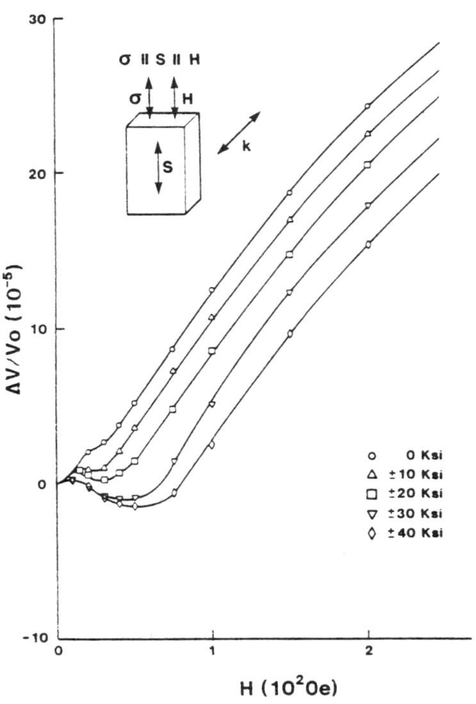

Fig. 7. Effect of bending stress on the MIVC for 5-MHz shear waves in ASTM A-514 steel (S is the polarization of shear waves)

APPLICATIONS

The stress dependence of MIVC was applied to determine the distribution of residual stress around a weld line. Results of a bulk stress measurement through a 0.5-inch thick plate are shown in Figs. 8 and 9. Fig. 8 was obtained from a rectangular butt-welded ASTM A-588 steel specimen and Fig. 9, from the flange of a T-shaped, full penetration-welded ASTM A-514 steel specimen. For comparison, results of destructive stress measurements conducted by using the layer removal method[27] after the MIVC measurement was made are also shown (dotted line in the figures). As shown, the two results agreed fairly well.

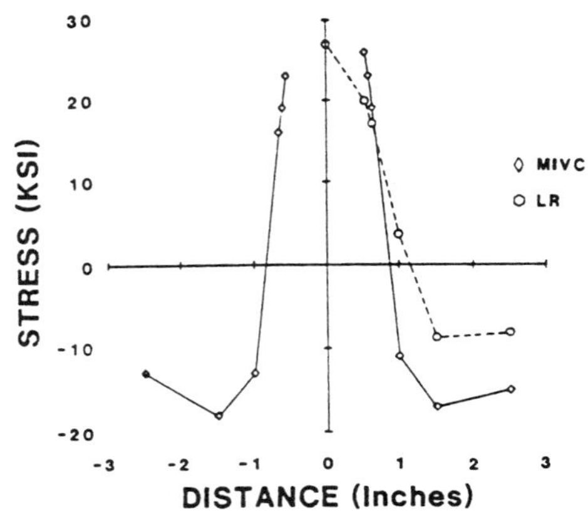

Fig. 8. Distribution of residual bulk welding stresses in a butt-weld specimen of ASTM A-588 steel, determined by the MIVC method and the Layer Removal (LR) method. Stresses are those in the direction parallel to the weld line.

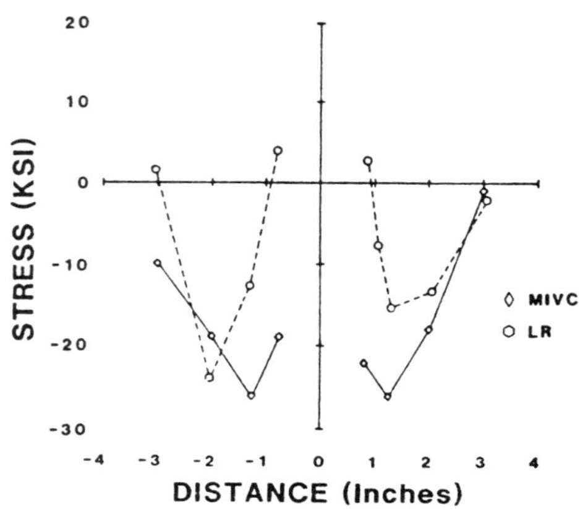

Fig. 9. Distribution of residual bulk welding stresses in the flange of a T-shaped, full-penetration-welded specimen of ASTM A-514 steel, determined by the MIVC method and the Layer Removal (LR) method. Stresses are those in the direction parallel to the weld line.

The MIVC for surface waves was also applied to determine surface stresses in a specimen subjected to a bending load. Results of a surface stress measurement in 0.25-inch thick ASTM A-36 and SAE 4340 steel specimens are tabulated in Table 1. The average stress experienced by the surface waves calculated under the assumption that the penetration depth of surface waves is approximately one wavelength of the wave is also given in the table.

As shown in the table, the MIVC stress indications agreed well with the calculated values. Notice that at 40-KSI bending stress, the MIVC for 2.2-MHz and 5-MHz surface waves indicated different stress values from each other, indicating that a stress gradient could be determined if it were large enough.

Table 1. Results of a Surface Stress Measurement in 0.25-Inch Thick ASTM A-36 and SAE 4340 Steel Specimens

Material	Applied Bending Stress at Specimen Surface (KSI)	MIVC Stress Indication (KSI)		Calculated Average Stress (KSI)	
		2.2 MHz	5 MHz	2.2 MHz	5 MHz
ASTM A-36 Steel	-10	-8	-8	-8	-9
	-20	-16	-16	-16	-18
AISI/SAE 4340 Steel	-20	-16	-16	-16	-18
	-40	-28	-32	-31	-36

CAPABILITIES AND LIMITATIONS

The method based on the stress dependence of the magnetically induced velocity change (MIVC) for ultrasonic waves is capable of measuring both bulk and surface stresses. It can determine the magnitude, sign (tensile or compressive), and direction of an unknown stress. The accuracy of the method is approximately ±5 KSI presently. The method can be applied to measurement of residual welding stresses, bending stresses, and stress gradients at the surface. It also has potential to determine the stress distribution through the thickness of a material.

The applicability of the method is limited only to ferromagnetic materials. It also requires an electromagnet to magnetize a part. Since the required electromagnet is generally heavy and bulky, a mechanical device may be needed to facilitate the positioning of the magnet. Application of the method may also be limited to simple geometry parts because of difficulty in magnetizing complex geometry parts.

ACKNOWLEDGMENT

A part of the results presented in this paper was obtained from the work sponsored by the Federal Highway Administration, Department of Transportation, under Contract No. DTFH 61-83-C-00016.

REFERENCES

1. C. Kittel, Physical Theory of Ferromagnetic Domains, Rev. Mod. Phys., 21:541 (1949).
2. R. M. Bozorth, "Ferromagnetism," Van Nostrand, New York (1951).
3. M. R James and O. Buck, Quantitative Nondestructive Measurements of Residual Stresses, CRC Critical Review in Solid State and Material Sciences, 9:61 (1980).
4. C. O. Ruud, A Review of Selected Nondestructive Methods for Residual Stress Measurement, NDT Int. 15:15 (1982).
5. S. Abuku and B. D. Cullity, A Magnetic Method for the Determination of Residual Stress, Exp. Mech., 11:217 (1971).
6. T. Jakel, Quality Control of Ferromagnetic Components by Coercive Field Strength Measurement, Brit. J. NDT, 26:287 (1984).

7. M. Suzuki, I. Komura, and H. Takahashi, Nondestructive Estimation of Residual Stress in Welded Pressure Vessel Steel by Means of Remanent Magnetization Measurement, Int. J. Pres. Ves. & Piping, 6:87 (1978).
8. R. Hochschild, Electromagnetic Methods of Testing Metals, "Progress in Nondestuctive Testing," Vol. 1, E. G. Stanford and J. H. Fearon, eds., MacMillan, New York, p. 57 (1959).
9. W. G. Clark, Jr., and B. J. Taszarek, Stress Mapping with Eddy Currents, Mat. Eval., 42:1272 (1984).
10. M. M. Shel', Measurement of Stresses in HardMagnetic Steels by Means of the Magnetoelastic Method, Ind. Lab., 33:361 (1967).
11. H. Kwun and G. L. Burkhardt, Effects of Stress on the Harmonic Content of Magnetic Induction in Ferromagnetic Material, in "Proc. 2nd Nat. Seminar Non-Destructive Eval. Ferromagnetic Mat., Dresser Atlas, Houston (1986).
12. R. L. Pasley, Barkhausen Effect - An Indication of Stress, Mat. Eval., 28:157 (1970).
13. G. A. Matzkanin, R. E. Beissner, and C. M. Teller, "The Barkhausen Effect and Its Applications to Nondestructive Evaluation," NTIAC-79-2, Nondestructive Testing Information Analysis Center, Southwest Research Institute, San Antonio (1979).
14. R. B. Thompson, Strain Dependence of Electromagnetic Generation of Ultrasonic Surface Waves in Ferrous Metals, App. Phys. Lett., 28:483 (1976).
15. V. N. Makarov and T. Kh. Biktashev, Effect of Plane Stressed State on Magnitude of Magnetostriction, Sov. J. NDT, 19:487 (1983).
16. M. Kersten, The Temperature Coefficient of the Modulus of Elasticity of Ferromagnetic Substances, Z. Physik, 85:708 (1933).
17. B. S. Berry and W. C. Pritchet, Magnetoelasticity and Internal Friction of an Amorphous Ferromagnetic Alloy, J. App. Phys., 47:3295 (1976).
18. W. J. Bratina and D. Mills, Investigation of Residual Stress in Ferromagnetics Using Ultrasonics, Nondestr. Test., 18:110 (1960).
19. H. Kwun and C. M. Teller, Tensile Stress Dependence of Magnetically Induced Ultrasonic Shear Wave Velocity Change in Polycrystalline A-36 Steel, Appl. Phys. Lett., 41:144 (1982).
20. H. Kwun and C. M. Teller, Stress Dependence of Magnetically Induced Ultrasonic Shear Wave Velocity Change in Polycrystalline A-36 Steel, J. Appl. Phys., 54:4856 (1983).
21. H. Kwun, Effects of Stress on Magnetically Induced Velocity Changes for Ultrasonic Longitudinal Waves in Steels, J. Appl. Phys., 57:1555 (1985).
22. H. Kwun, Effects of Stress on Magnetically Induced Velocity Changes for Surface Waves in Steels, J. Appl. Phys., 58:3921 (1985).
23. M. Namkung and J. S. Heyman, Residual Stress Characterization with an Ultrasonic/Magnetic Technique, Nondestr. Test. Comm., 1:175 (1984).
24. D. S. Hughes and J. L. Kelly, Second-Order Elastic Deformation of Solids, Phys. Rev., 92:1145 (1953).
25. W. P. Mason, Rotational Relaxation in Nickel at High Frequencies, Rev. Mod. Phys., 25:136 (1953).
26. S. B. Grigorev and L. K. Kudryashova, Magnetoelastic Attenuation due to Rotation of Magnetization, Sov. Phys. Solid State, 17:99 (1975).
27. G. Sachs and G. Espey, Measurement of Residual Stress in Metal, Iron Age, 148:63 (Sept. 18, 1941) and 36 (Sept. 26, 1941).

SOME OPTICAL METHODS OF CHARACTERIZING SOLID AND PARTICULATE MATERIALS DURING MANUFACTURING PROCESSES

F Alan Wedgwood

National NDT Centre
AERE Harwell, Didcot
Oxon England

ABSTRACT

Non destructive Testing is usually seen as an end-of-process quality control step. It is becoming apparent however that inspection should be carried out as early as possible in the process. In this paper two methods of non-invasive measurement using lasers will be described. Since they are non-invasive these measurements can be made on hot or fast moving materials and hence at an early stage in some manufacturing processes.

The first method is generation and detection of ultrasound in solids. This can be used for measuring grain scattering and attenuation of ultrasound using lasers of moderate power. Applications include measurement of grain size in metals during rolling operations and monitoring during welding.

The second method is optical sizing of particles or droplets in a gas stream. Instrumentation has been refined so that a wide range of particle sizes and types can be measured. Applications to powder metallurgy, plasma spraying, spray drying and paint spraying will be described. In the last example the philosophy of this sort of measurement is discussed. In application to paint spray quality control it was found that the economical solution in practice is to develop an accurate model using the optical measurements for validation and use the model, with cheaper on-line measurements, for process control.

INTRODUCTION

Optical inspection is by far the most important method in quality control. This is because it can largely be carried out by direct visual methods, and is non-invasive. The fact that it is non-invasive means that it is an attractive option for on line quality control. But for material property inspection its use is far less obvious since light is strongly absorbed by most materials of interest. Even here however due to the advantages for inspecting very hot or delicate materials efforts are being made to develop it. In this conference there are many papers describing methods of material inspection which are to some extent optical.

Roughly speaking optical methods in quality control can be classified as in table I.

TABLE I

METROLOGY	Visual assessment Dimension Edge location Surface contour
SURFACE	Roughness Coatings Oxide layers etc.
BULK PROPERTY	Transparent materials Opaque materials

Most materials of interest are opaque and this review deals with them. In order to interrogate the interior of an opaque material it is necessary to convert the light energy to another form at the surface. For a totally remote method it is also necessary to convert back to light when the interrogating energy is reflected or transmitted by the material. In practice there are only a few methods of doing this. Such methods are summarised in table II.

TABLE II

EXCITATION	INTERROGATION	DETECTION
Light-Heat	Thermal Diffusivity	Heat-light (emission)
Light-U/S	Ultrasonic Wave Scattering	U/S-Light (modulation)
Magnetic energy	Permeability	Magnetic-light (modulation)
Light-Electric energy	Conductivity	Electric-Light (emission)
X-Ray	Density	X-Ray-Light (emission)

The first two lines in the table refer to methods of bulk inspection which work in principal for all materials. In both cases light is absorbed at the surface to give heat. This heat can be used directly to interrogate the material, in which case thermal conductivity and thermal capacity can be measured, generally in terms of their ratio, the thermal diffusivity. The heat signals are generally detected by infra-red emission from the surface which gives a remote inspection method. Other heat sensing methods such as gas expansion are used as in photo-acoustic spectroscopy but these do not give strictly remote inspection methods.

In the second method an intense pulsed light source is used to generate elastic waves, either by sudden expansion of the material surface or at higher energies by sudden surface evaporation of the material. The surface then becomes a source of ultrasonic wave pulses which can be used to interrogate the material in various ways. For detection of ultrasonic waves one has to measure the surface displacements which are very small. This can be done remotely by reflecting light from the surface and using interferometric methods. For some time this possibility of remote ultrasonic inspection has excited research workers but there is a serious problem due to the low sensitivity of the detection method. For example typical contacting sensors such as P.Z.T. probes have a displacement sensitivity of a picometer whereas typical interferometers have a broad-band sensitivity of 100 pm. At Harwell we therefore believe that laser ultrasonic methods are more likely to be applied in material characterization than for defect detection since lower sensitivity is required.

The next two lines in the table refer to magneto-optic and electro-optic conversion. Such methods only apply to special materials such as semi-conductors. Alternatively magneto- or electro-optic coatings can be applied to other materials for NDT purposes, however such methods cannot be considered remote.

The last entry in the table is out of place but is included as a reminder that X-Rays can be used for remote materials characterization. For instance using modern fine-focus sources and projection magnification it is possible to see microstructural details to about ten microns. Such methods are used routinely to check some turbo-engine blades.

A further way of inspecting opaque materials with light is to introduce the light to the material with implanted optical fibres. Obviously this method would be expensive for most materials and one would have to study carefully the effect of the implanted fibres on the material strength. However where glass or carbon fibres are already present as in reinforced plastics the introduction of a small number of optical fibres as an interrogation medium can be envisaged.

The last optical method of quality control for a bulk material considered here is to inspect the material in a finely divided form before it is fabricated. Such a strategy is possible for the following types of materials.

* Powder formed ceramics

* Powder formed metals

* Fibre reinforced materials

* Spray coatings

The sort of useful measurements which can be made optically on such materials, at various stages before fabrication are:

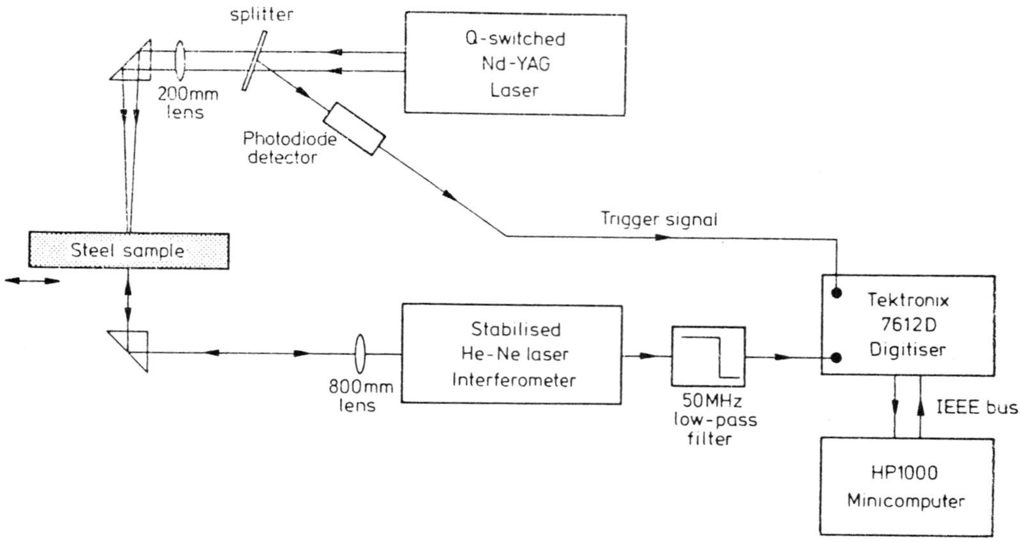

Fig. 1 Apparatus to measure forward ultrasonic scatter from metallic specimen, which can be scanned ultrasonically.

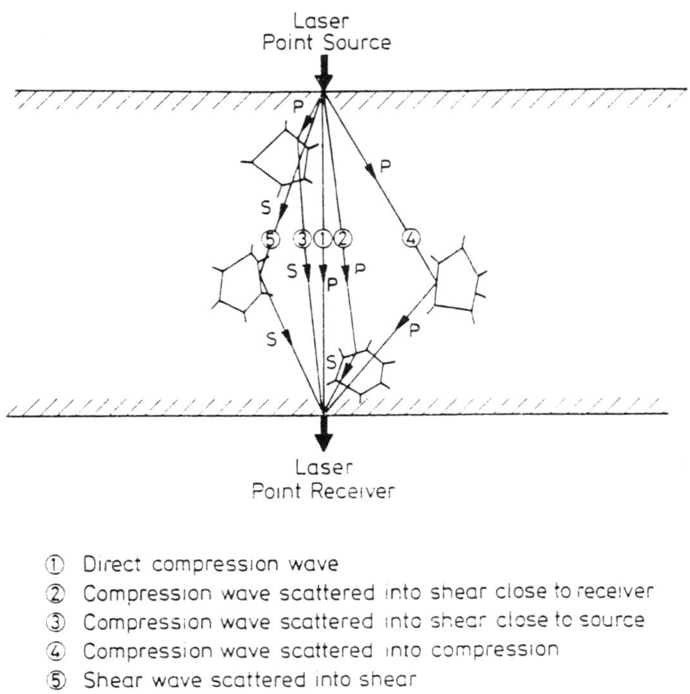

Fig. 2 Schematic diagram to show how P and S waves may be scattered by microstructure to generate noise between main P peak and S edge seen in figures 3 and 4.

* Particle size
* Particle velocity
* Surface state
* Fibre pre-preg quality.

Although some of these properties can be determined by other methods from the bulk powders or fibres before fabrication it can be more convenient to measure them during formation; this is because the measurement can then be used to control or optimise the process. Also in the case of particle size it is notoriously difficult to estimate particle size from a bulk powder where agglomeration can occur. In such cases it is better to measure the true particle size while the particle is being formed. In the case of spray coatings the quality of the final product can depend on drop size and velocity immediately before impact and such measurement can be made optically.

In this review the optical methods introduced above will be illustrated by three examples of work which has been carried out at Harwell:

* Microstructural Monitoring by laser Ultrasonic methods
* Laser Monitoring of size and velocity of Molten Droplets
* Modelling a Paint Spray Process

The third example is not a materials characterization problem since the original objective of the project was to measure the surface quality in a paint spray process. However it illustrates a very important point which arises in actual on-line quality control in manufacturing. That is, for most present industrial processes, on-line sensors have to be reasonably cheap and rugged. In this project, where paint finish could be measured by an optical device with adequate accuracy, the cost and complexity of the device ruled it out for a production line. Instead the device was used in a development laboratory to develop a predictive model of the spray process. This model could then be used in production using readily measured variables such as temperature, spray-gun pressure etc. In the same way the equipment used in the first examples for microstructure and particle monitoring will probably be used for many years for process modelling and process development before they are seen on line in industry.

MICROSTRUCTURAL MONITORING BY LASER ULTRASONIC METHODS

A number of workers have shown that the measurement of ultrasonic attenuation in metals at a range of frequencies can give information about grain size [1,2]. These generally use contacting probes or near contacting probes such as EMATs. If such a measurement is required on line, with a hot material as in a rolling mill, a remote method would be required using lasers both for generation and detection of ultrasound.

Laser generation of ultrasound was reviewed by Scruby[3] et al. in 1982. At low power densities the ultrasound is generated by sudden lateral expansion of the heated surface. In this case much of the forward ultrasonic energy is in shear waves. At higher power densities the surface is evaporated suddenly and the reaction force on the surface gives compression waves in the forward direction. This energy range is to be preferred since the ultrasound is more intense but it does damage

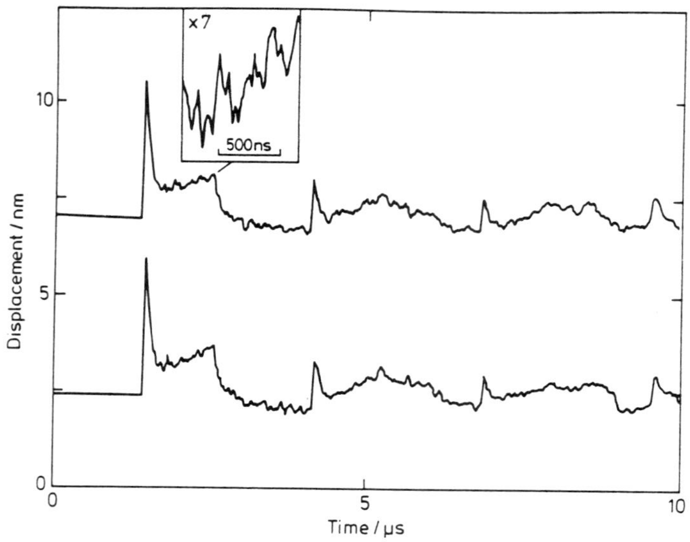

Fig. 3 Displacement waveforms measured at two points about 4 mm apart on the 0.17% carbon (coarse grained) steel sample.

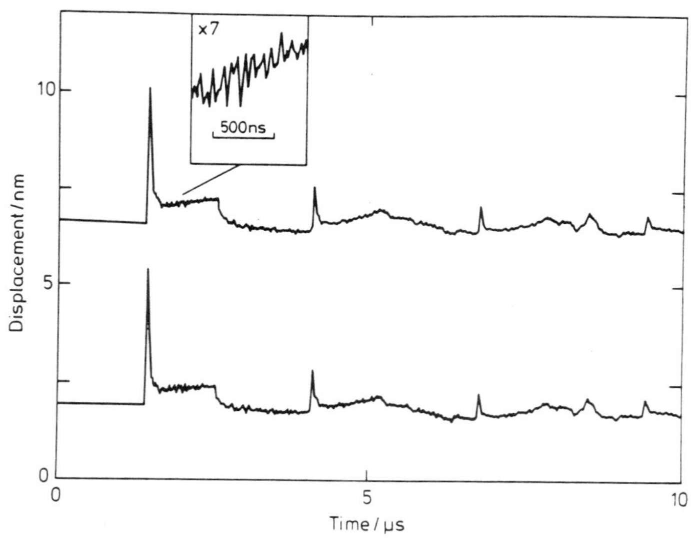

Fig. 4 Same as figure 3 but for the 0.51% (fine grained) steel sample.

the surface. However for hot process inspection this damage is not important. Using a modest pulsed laser in this 'ablation' regime it is possible to produce ultrasonic pulses as large as those produced by contacting transducers. Detection of ultrasound using laser interferometers is much less sensitive than with a contacting transducer particularly if the surface has a low reflectivity. The sensitivity can be increased by putting up the laser power but such lasers are expensive; using a low power HeNe laser the practical limit of sensitivity at broad band is about 100 p.m.

A system using large lasers both for generation and detection of ultrasound was described by Kaule[4]. This was capable of finding laminar defects in red hot steel billets but the overall cost was very high. We believe that a system designed for measuring ultrasonic attenuation needs much less sensitivity than one designed for detecting discrete defects and so is likely to be adopted earlier in rolling mills, particularly in those using controlled rolling to produce carefully graded microstructure.

The present study (Scruby et al.[5]) is a laboratory feasibility study for a laser ultrasound microstructure monitor. It uses a 100 mJ Neodymium - YAG laser to generate the ultrasound and a 2 mW HeNe laser in the detection interferometer in the arrangement shown in figure (1). The specimens were 8 mm thick discs of steel containing 0.17% and 0.51% carbon. Metallographic examination showed that the low carbon content disc had considerably coarser structure than the other.

The received waveforms from the two specimens are shown in figures (3) and (4). The succession of sharp peaks at 2.5 microsecond intervals are multiple thickness reflections of the initial compression wave pulse. In principal since these peaks are taken with a large bandwidth (50 MHz) it is possible to extract frequency dependent attenuation and obtain grain size information as described by Reynolds and Smith[1]. However in this case, due to the small size of the source and detector on the specimen one observes the grain more directly. The mechanism for this is indicated in figure (2) and is thought to be forward scatter and mode conversion from individual grains. This gives the fine structure seen between the compression wave peak and shear wave edge shown in the insets to figures (3) and (4). The fine structure of the low carbon content sample can be seen to be much coarser than for the large carbon content sample. When the waveform is averaged while the sample is moved this fine structure is removed due to spatial averaging whilst the sharp peaks remain. Similar grain noise has been used by Goebbels[2] in back-scatter geometry. We believe that there is an advantage in using the two opposite sides of a plate as it avoids the large air blast signal which accompanies the ultrasound generation process. The method has also been used successfully for non-ferrous metals and alloys with thickness 1 - 40 mm. (unpublished work).

LASER MONITORING OF SIZE AND VELOCITY OF MOLTEN DROPLETS

The method of Laser Doppler Anemometry for measuring the velocity of small particles in a fluid is well known[6] and there are several manufacturers of the necessary equipment. Two narrow beams of light from the same laser are crossed as shown in figure 5. The small ellipsoidal cross-over volume is arranged to be at the point where velocity is to be measured. In this cross-over region there are real interference fringes with a spacing (S) which depends on wavelength and crossing angle. A particle traversing the region scatters a burst of

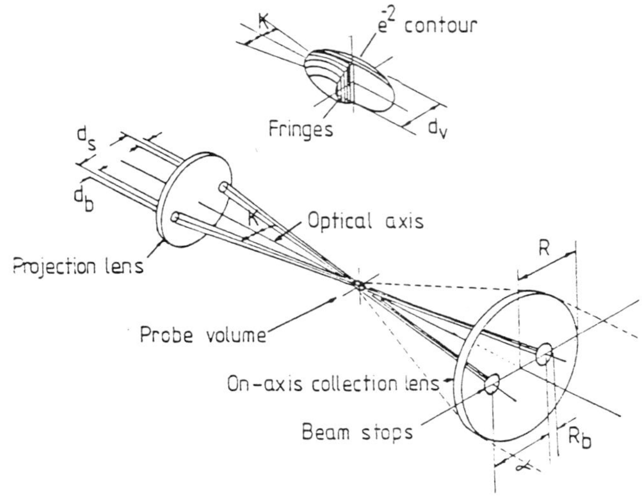

Fig. 5 Probe volume formation at the intersection of two focused laser beams. The inset shows the fringes in the probe volume. The collection aperture is centred on the optical axis.

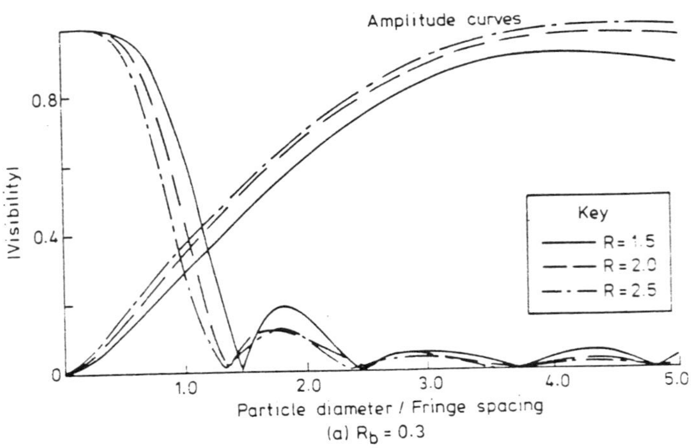

Fig. 6 Theoretical visibility and amplitude curves for diffracting spheres. The parameters R and R_b are given in unit of the beams angular separation.

light which is modulated with a frequency which is proportional to its velocity ($f = V/S$). By measuring these bursts of light and using a counter it is possible to measure a flow pattern non-invasively. Usually the scattered light is collected in the forward direction where the intensity is largest but other geometries can be used.

Measuring drop size at the same time is harder and instrumentation for this is still developing. There are several methods of doing it either using the total intensity of the scattered light or by looking in detail at the phase or amplitude of the Doppler burst. For much of the work at Harwell two methods are used in conjunction in order to increase the dynamic range of the instrument. The first is called the visibility method; particles of different size give different depths of modulation to the Doppler burst. In particular very small particles give full modulation and particles with diameter S give very small modulation. The visibility is simply a measure of this modulation. Such a technique has a useful range of 0.1 S - 0.9 S where S is typically 50 micron. For a wider range the absolute intensity of the burst is measured. For this method to be accurate it is necessary to check whether the particle has passed through the centre of the cross-over region. In order to do this two cross over regions are used one from the blue line of an Argon Ion laser and one from the green line. The blue volume is much smaller than the green and at the centre of it, thus the green signals can be validated by checking whether a blue signal occurs simultaneously. Results of theoretical calculating for fringe visibility and signal amplitude are shown in figure 6. These calculations are carried out using diffraction theory which is a good approximation for the case of opaque spheres such as metal or ceramic droplets. The instrument has been validated using a mixture of bronze spheres with a bimodal distribution of sizes with peaks at 25 and 70 micron diameter. Such bimodal distributions are a severe test of any size distribution measuring device (figure 7).

Using a fast analogue processor it is now possible to measure the visibility, frequency and absolute intensity of Doppler bursts at about 1000 per second. This means that this equipment can now be used in a number of manufacturing process applications where the particle density previously was too high for such measurement. At Harwell studies have been made in spray dryers, liquid metal sprays and flame sprays mainly with a view to understanding the physical processes involved and checking theoretical models.

Flame and Plasma spraying are used for coating materials with either metal or ceramic materials and in some cases plasma spraying is used for direct fabrication of complex shapes. The quality of the material deposited depends on the size, impact velocity and temperature of impacting droplets although no model has yet been published for the process. We have used the Laser Doppler method to measure droplet size and velocity for flame sprayed aluminium oxide[7]. In this work a METCO oxy-acetylene flame gun was employed with METCO 101 powder fed into the oxygen stream. A Harwell two-colour Laser Doppler system was used to monitor drops at a rate of about 200 per second. Measured velocities ranged from 70 m/s near the gun to 20 m/s at 250 mm from the gun, the drop off being due to the slowing down of the hot gas stream from the gun (Figure 8). The visibility signals showed an increase with distance from the gun (Figure 9). In the case of a flame gun the mean drop volume is expected to be nearly constant with distance but the powder is in the form of needle like particles which give a low and variable visibility, when they melt the effective diameter decreases and the visibility increases. Thus the present data is interpreted as particle melting to

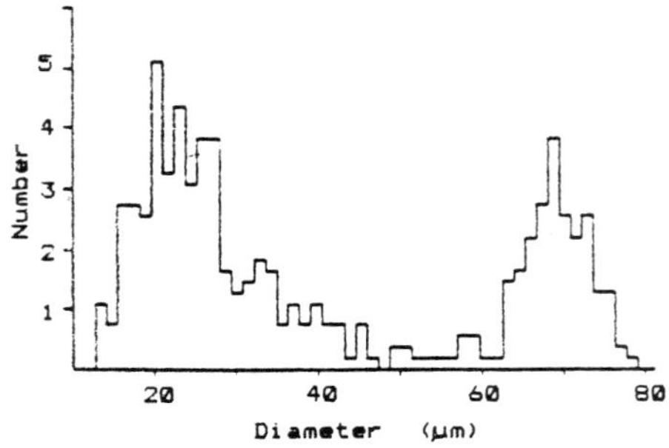

Fig. 7 Validation check of particle sizer using bimodal distribution of bronze spheres.

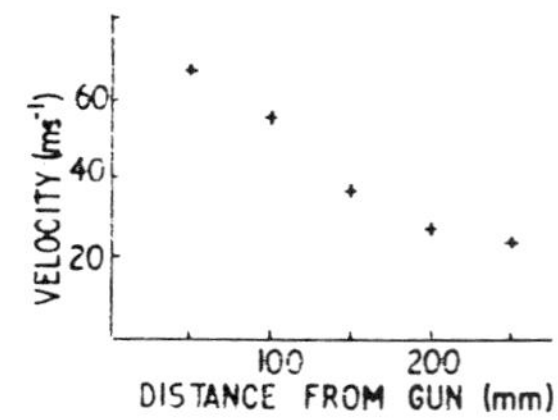

Fig. 8 Axial profile of drop velocity.

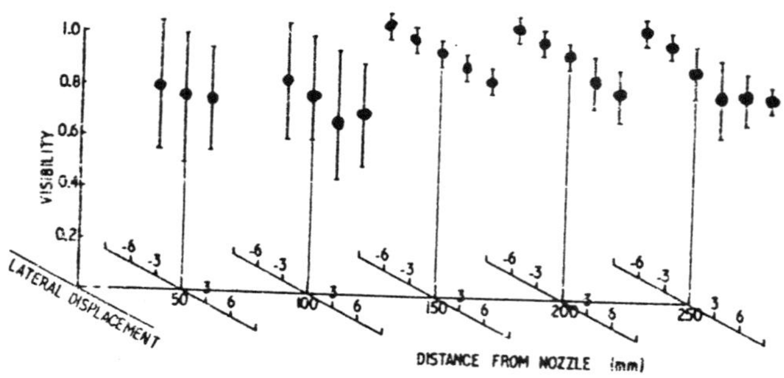

Fig. 9 Axial and radial profiles of the drop sizing visibility parameter.

give spherical drops; this appears to happen between 100 and 150 mm from the gun. Obviously it would be better to get a direct measurement of the drop temperature but light emission methods are difficult due to the high luminosity of the gas stream.

It is planned to use this type of data to correlate the velocity and degree of melting of impacting drops with the porosity and strength of alumina coatings. Work is also starting on an air pressure driven metal sprayer for the manufacture of metal powders as feedstock for powder metal fabrication. At present the velocity of the droplets near the gun have been successfully measured but the spray density is too large for the existing particle sizer.

MODELLING A PAINT SPRAY PROCESS

In automobile manufacture the final coat of paint is now usually applied by automatic spray gun operation, either with air pressure sprays or with electrostatic rotating bell sprays. Manual application of paint is used less and less due both to labour costs and the potential health hazards of working in spray booths. Whilst automatic spray painting is faster and more consistent it is less adaptable and a number of paint finish defects can appear due to changes in the spray booth environment or paint quality. Many of these changes would be detectable at source by a skilled paint sprayer but with automatic spraying the resultant defects only become apparent when the car body emerges from the hardening oven 30-40 minutes after spraying. If defects are detected at that stage on a production line and it is due to mis-setting of the spray guns a large number of cars will need manual repair. It is thus an advantage to inspect the paint quality in the wet state before stoving. Such inspection can be done using a light scattering method however in this project it was considered that the necessary apparatus would be too delicate to operate on the spray line. Instead a mathematical model was developed to predict the final quality.

Harwell was asked to collaborate in a project to control the final coat spraying using both air pressure and electrostatic bell machines. We undertook two tasks:

1) To develop a quantitative method of assessing the finish on painted test panels.

2) To develop a mathematical model which relates the various process parameters to the paint finish.

While the first objective involves optical and image processing methods which are well known the second objective is more complex and involves a wide range of skills including theoretical physics, high speed photography, laser scattering and liquid spray technology.

The most serious defects occurring in spray painting affect the appearance of the finish coat and so an optical method of assessment is appropriate giving images of defect surfaces (Figure 10). These images are then digitised and analysed using an image processor to give parameters which refer to the mean amplitude and wavelength of roughness in the paint finish. The parameters correlate well with expert assessment of the finish as carried out by the paint shop inspectors.

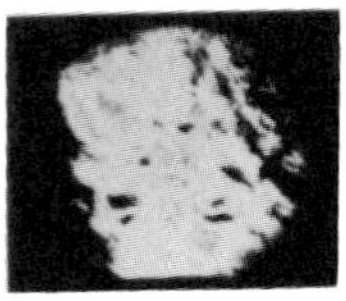

T15 Sags & solvent boll

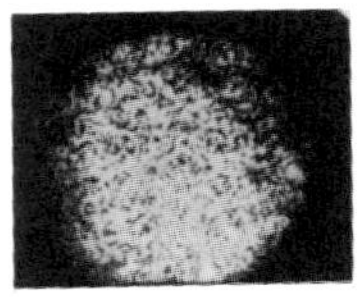

T18 Solvent boll

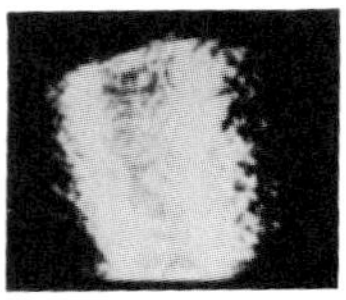

T14 Acceptable

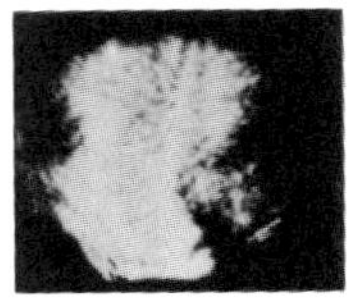

T17 Runs

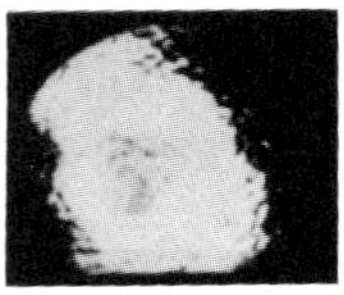

T13 Acceptable peel

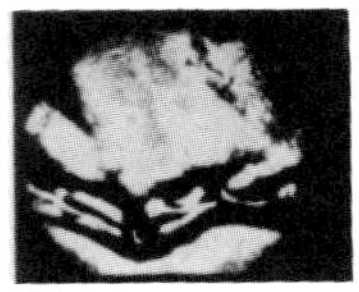

T16 Sags & run

Fig. 10 Typical faults in paint as revealed by optical topography. The surface gradients in the figure are of order 0.1^{o}.

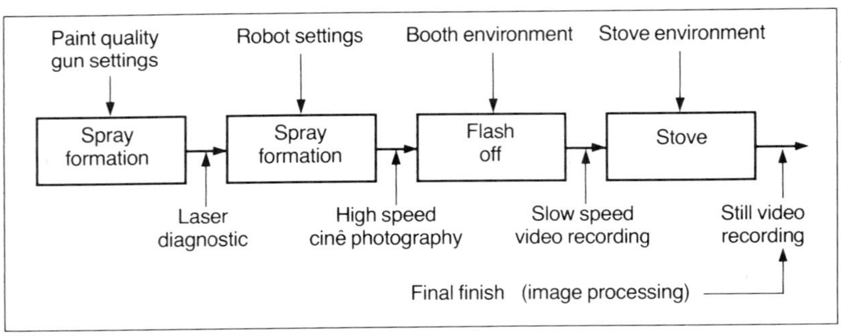

Fig. 11 Flow chart of a paint spray process showing the model stages and experimental validation stages.

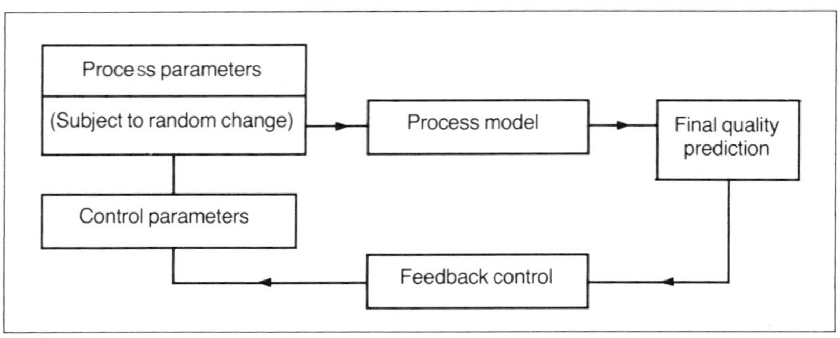

Fig. 12 Control loop for a paint spray process.

In the mathematical model the objective is to predict the finish parameters from given input data on paint quality, spray settings and booth environment. Obviously with a total process as complicated as this the model must be broken down into stages which can be individually tested experimentally. A simple breakdown is given in Figure 11.

Many models of spray formation are available from the literature[8]. Choosing the right one for the job was done by using laser scattering experiments to determine the drop size and velocity distributions and find the best fitting mathematical models. Equations of motion for the drops were then solved numerically. These had to take account of air resistance, evaporation and drop collisions. The mathematical models were checked by laser probing at a variety of positions in the spray 'fan'.

Drop size was measured using a Malvern laser scattering drop sizer and velocities using laser Doppler anemometry.

The deposition of drops on a surface is governed by their incident energy, paint viscosity and surface tension and the nature of the surface, i.e. whether dry or with a pre-existing liquid film. Expressions were derived for the flattening of drops on impact in each case and these were checked by high speed cine photography. Then a statistical model was developed to estimate the initial roughness after deposition. After deposition the paint film relaxes for several minutes during the flash-off period. During this time it is governed by surface tension which causes levelling, gravity which initiates sagging and evaporation which increases viscosity and eventually stops both the previous processes[9]. These processes were modelled using simple approximations to the hydrodynamic[10] and diffusion equations and checked by video photography, image processing and periodic weighting. It was found that the image processing needed in each frame of a video film is a slow and expensive process so special analogue circuitry was designed for on-line readout of surface roughness and mean wavelength.

All the stages of the model calculation were then combined using a programme on an IBM P.C. Validation of the model was carried out by comparing predicted and measured final finish parameters for a wide range of input parameters. Several hundred validation runs were carried out for each spray gun and correlations were between 85% and 95%.

The model is used in a control loop as in figure 12. For example, if a process parameter (such as temperature) changes, the measured change is fed into the process model; this then predicts the change of final quality. The change in parameters, such as gun position, needed to rectify the quality change, is then calculated and fed back to the automatic paint sprayer. Such a system is common in chemical engineering but much less so in other manufacturing processes. It has the advantage that the sensors needed on line are well tried traditional sensors such as thermometers, pressure gauges and position sensors. Sophisticated optical detectors and image processors are not required on-line.

ACKNOWLEDGEMENTS

I am indebted to Chris Scruby and Mike Yeoman for allowing me to use their unpublished work in this paper.

I would also like to thank Gaydon Technology (previously B.L. Technology) who commissioned the work on paint modelling, for permission to describe this work.

REFERENCES

1. W.N. Reynolds and R.L. Smith, Brit. J. NDT Sept, 291 (1985).
2. K. Goebbels, Res. Tech. NDT Vol. IV. ed. R.S. Sharpe, (Academic Press, London) (1982).
3. C.B. Scruby, R.J. Dewhurst, D.A. Hutchins, S.B. Palmer, Res. Tech. in NDT, Vol. V ed. R.S. Sharpe (Academic Press, London) (1982).
4. W. Kaule, 8th World Conf. NDT Cannes, 1-7 (1978).
5. C.B. Scruby, R.L. Smith, B.C. Moss, Int. J. NDT (to be published).
6. L.E. Drain, "The Laser Doppler Technique" (Wiley) (1981).
7. N.P. Smith, J.O.W. Norris, D.J. Hemsley, M.L. Yeoman, Third International Conference on Liquid Atomization and Spray Systems, London (1985).
8. I. Filkova, P. Cedik, Proc. 3rd Int. Drying Symp., U. B'ham (1982).
9. T.C. Patton, "Paint Flow and Pigment Dispersion", Wiley N.Y. 2nd Ed. (1978).
10. H.E. Huppert, Nature 300 p427 (1982).

NOVEL NON-DESTRUCTIVE X-RAY TECHNIQUE FOR NEAR-REAL TIME DEFECT MAPPING

T.S. Anathanarayanan, R.G. Rosemeier, W.E. Mayo* and S. Sacks**

Brimrose Corp. 7720 Belair Rd. Baltimore MD 21236
*Rutgers University Piscataway NJ 08854
**SPAWAR Washington DC 20363

ABSTRACT

A novel x-ray technique is presented which offers the capability for near-real time defect mapping in a variety of electronic materials. The method relies on a recently developed computer controlled x-ray rocking curve analyzer which utilizes localized x-ray line broadening as a means of quantifying the local dislocation density. By the use of electronic x-ray detectors and image processing techniques, a high quality image of the defect distribution can be produced in a fraction of the time required for conventional x-ray topographs. Moreover, the system requires minimum operator intervention due to its microprocessor control, thus making it ideal as a powerful non-destructive tool for both QC and research applications. The operating principle of the system will be presented along with application to the study of defect distributions in a variety of substrate materials and epitaxial thin films. The primary focus of the discussion will be on III-V and II-VI epitaxial films grown by LPE technique. Among the materials to be reported are HgCdTe on CdTe, AlGaAs on GaAs, GaAs and PbSe on BaF_2.

INTRODUCTION

Several x-ray diffraction techniques have been developed over the past four decades for examining surface and sub-surface crystal microstructure [1]. These techniques have been principally limited by the detector technology utilized in their implementation. Photographic film and the scintillation/proportional point counters have been the most popular x-ray detectors used thus far. These detectors require phenomenally long periods of time for data acquisition and have hence been extremely tedious to use.

X-ray detection has been revolutionized with the advent of the unique new high resolution 1-D (linear) and 2-D (spatial) x-ray array detectors in combination with the powerful new micro-computer technology. The 1-D x-ray detector is based on the linear position sensitive detector technology. Recent research has extended the x-ray detectors to the 2-D arrays which are based on the CCD & x-ray image intensifier technology [2]. These 2-D detectors have similar spatial resolution as the 1-D detectors (100-150 microns).

The power and analytical ability of several of these x-ray diffraction techniques have been amply demonstrated by numerous investigators. Some of these techniques include: x-ray diffractometry [1], x-ray topography [3], x-ray rocking curve analysis [4] etc. These techniques can be effectively used to quantitatively measure the micro-lattice strain state in crystalline materials. The focus of the present study is x-ray rocking curve analysis. This technique has been successfully utilized to map micro-plastic and micro-elastic strain in various substrates and epitaxial films namely; GaAs, AlGaAs, HgCdTe, CdTe, PbSe.

Some of the pioneering work in quantitave rocking curve analysis was done by S. Weissmann et. al. [4] at Rutgers University. This group initially utilized photographic film and photodensitometers to implement rocking curve analysis. The film has recently been replaced by a linear position sensitive detector [5]. The state-of-the-art Brimrose 2-D x-ray digital detector has also been used by the authors in this study[6].

THEORY

Rocking curve analysis involves the measurement of the FWHM (full width at half maxima) of individual Bragg diffraction spots. By oscillating the analyzed crystal about the diffractometer axis through its diffracting domain the entire reciprocal volume can be recorded. The FWHM measured thus is a function of the local dislocation density. Hersch et. al. [7] showed this relation to be $\rho = \beta^2/2b$ where, β - FWHM , ρ - dislocation density, b - Bergers vector. Dinan et. al. [8,9] have clearly shown the utility of this technique for the characterization of various substrates and epitaxial films used in micro-electronics applications. They used a point counter in their experiments.

In the present study both 1-D and 2-D x-ray sources were used. Topographic images were obtained by scanning the specimen surface with the 1-D x-ray beam while no such translation was required in case of the 2-D x-ray beam and 2-D detector combination.

EXPERIMENTAL METHODOLOGY

The schematics of the double crystal diffractometer (DCD) [10] used for the study are shown in Figure 1 & 2. This DCD is similar to a conventional diffractometer. The principal distinction between the DCD and the conventional diffractometer is that, the detector in the former is stationary and the specimen is rocked (oscillated), while in the latter both the specimen and the detector rotate (θ - 2θ combination). In the DCD the LPSD is placed on the Ewald sphere represented by the diffractometer circle. The specimen being analyzed is mounted on an automated multi-axis goniometer with numerous degrees of freedom. This goniometer performs both crystal rocking and specimen surface scanning for implimenting the rocking curve analysis in the topographic mode. Several prior investigations have clearly established the efficiency of the rocking curve analysis for quantitative evaluation of micro-lattice strains in crystalline materials [11,12].

There are several factors that effect the resolution of this technique. They include:

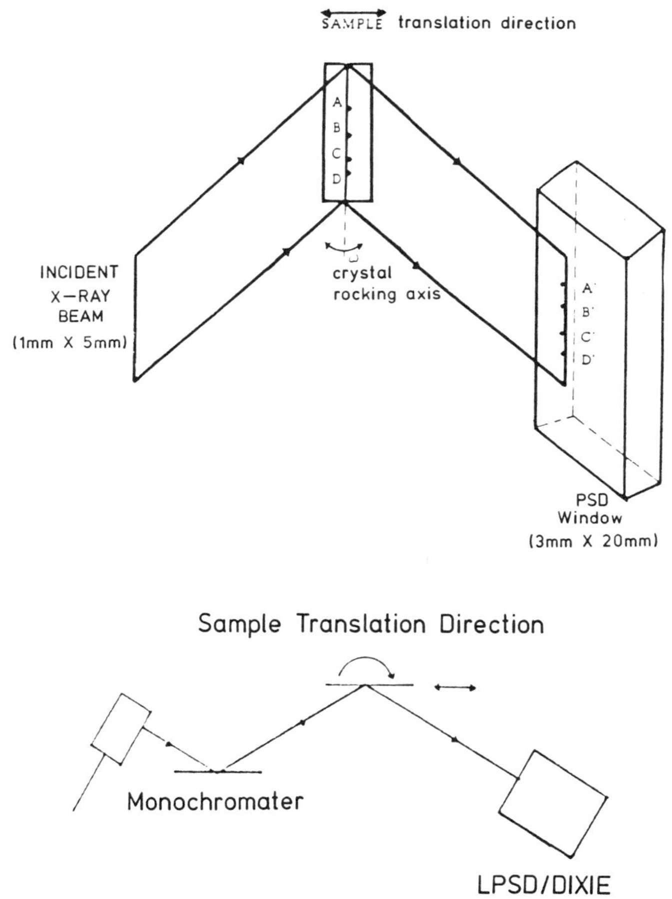

Figure 1: Schematic of Double Crystal Diffractometer

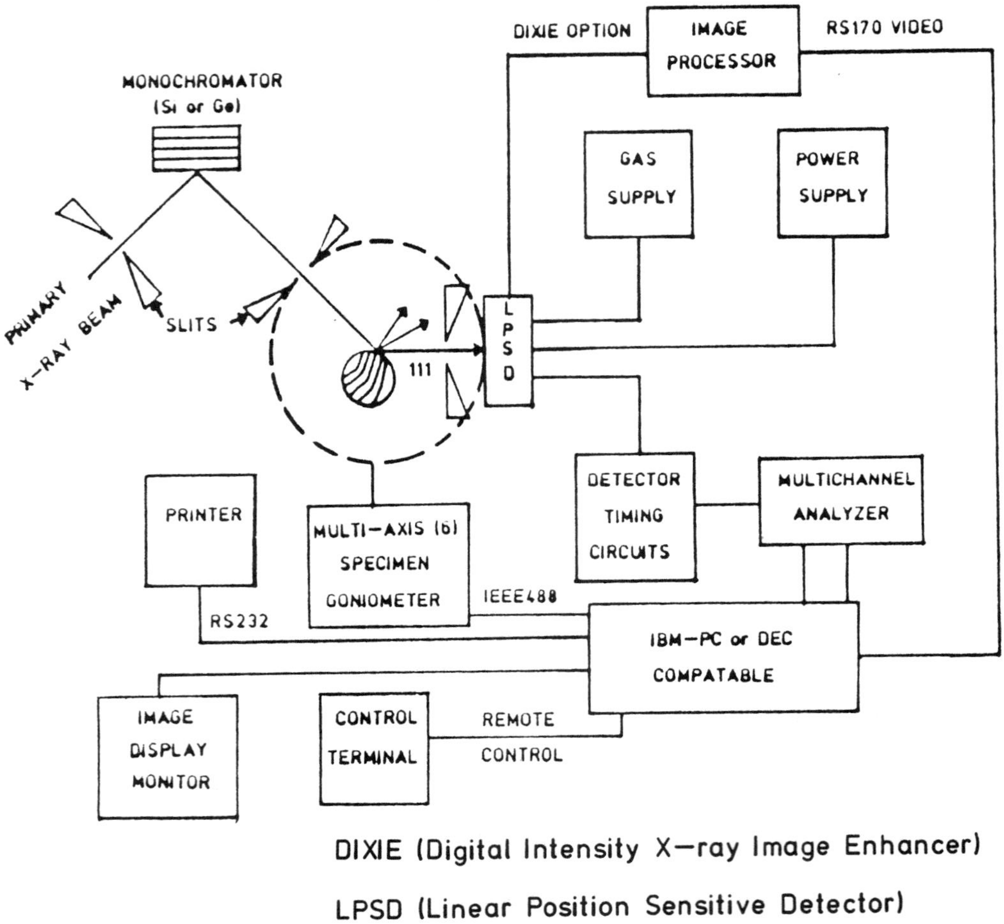

Figure 2: Schematic of Rocking Curve Analyzer

1. Wavelength dispersion of x-ray source i.e., degree of monochromation.

2. Vertical and horizontal divergence i.e., degree of collimation.

3. Spatial resolution of the detectors.

4. Resolution of the translating mechanisms.

The current study used a Si(111) monochromated x-ray beam (Cu K_α radiation). No attempt was made to separate the $K_{\alpha 1}$ & $K_{\alpha 2}$ wavelengths in the x-ray beam. Collimating slits minimized vertical and horizontal divergence in the beam. The ultimate beam size was 0.1mm X 10mm. The spatial resolution of the LPSD used was about 60-100 microns. The specimen goniometer had an angular resolution of 0.1 arc minute and linear resolution of 0.01mm. These resolution limits have since been improved by at least two orders of magnitude. As a matter of fact some recent experiments conducted at Brimrose have already utilized a 2-D detector system replacing the LPSD (1-D). The introduction of the 2-D detector has phenomenally enhanced the rocking curve analysis. The data acquisition time has been reduced by 5 orders of magnitude (810,000 sec to 2 sec) for $1cm^2$ surface. The combination of 2-D detectors, high speed computers and image processors has made it possible to impliment the rocking curve analysis in near real time. Preliminary data obtained with the 2-D detector system shows great promise for on-line process quality control in a variety of applications. The great degree of automation has improved precision and reproductibility of results for experimental analyses.

RESULTS AND DISCUSSIONS

A color rocking curve topograph of a section of a wafer of AlGaAs epitaxial film on a GaAs substrate is shown in Figure 3. The colors represent varying values of rocking curve halfwidths (β). The data clearly indicates a dislocation density gradient towards the edge of the substrate wafer. Epitaxial film quality is closely related to the quality of the substrate it is grown on. The present data tends to support this hypothesis. However, much more experimentation is required to make an unequivocal conclusion.

Figure 4. shows a GaAs substrate rocking curve topograph. One can clearly see pockets of high plastic strain regions due to dislocations.

A rocking curve topographs of HgCdTe (epi-film) grown on CdTe is depicted in Figure 5. These topographs show subgrain structure typical of such multiconstituent materials. By varying the energy of x-rays it is feasible to study the substrate below the epitaxial film as well.

In Figure 6. a near real time rocking curve topograph obtained from a PbSe epitaxial film grown on BaF_2 substrate is illustrated. The x-ray beam was not monochromated in this case. Figure 6a shows the progressive change in the digital real time Berg-Barrett topograph as the crystal is being rocked. Figure 6b is an elastic strain map obtained by tracking Bragg peak for each pixel and then plotting it back as a deviation from the average Bragg angle. The elastic strain

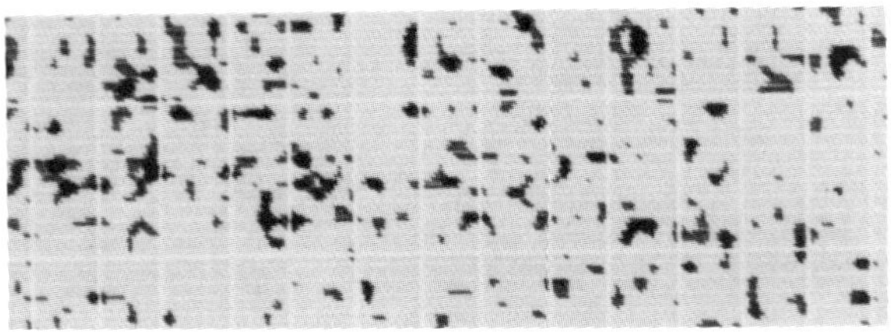

Figure 3: AlGaAs Epi on GaAs (4 Layer Laser).
100 Sample Orientation; 422 Reflection;
Cu K alphal Reflection; Scan Area 1.5x5mm

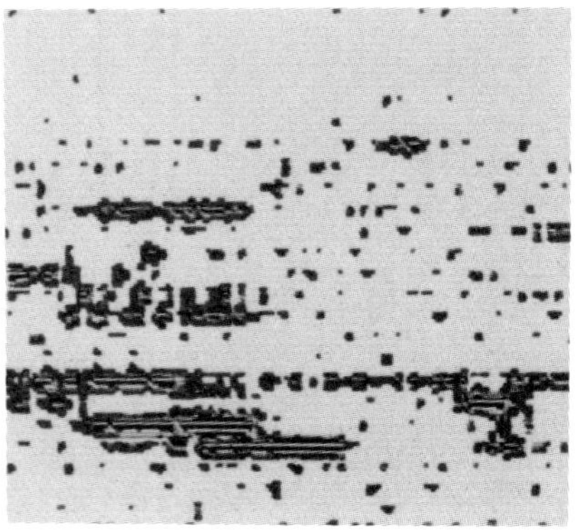

Figure 4: GaAs Wafer. 100 Wafer
422 Reflection;
Cu K alphal radiation;
Scan Area 3x5mm

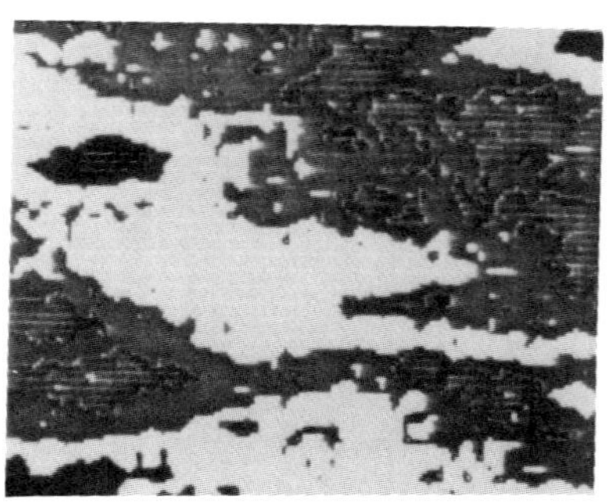

Figure 5: HgCdTe Epi on CdTe
111 Sample Orientation;
422 Reflection;
Cu K alphal Radiation;
Scan Area 4x5mm

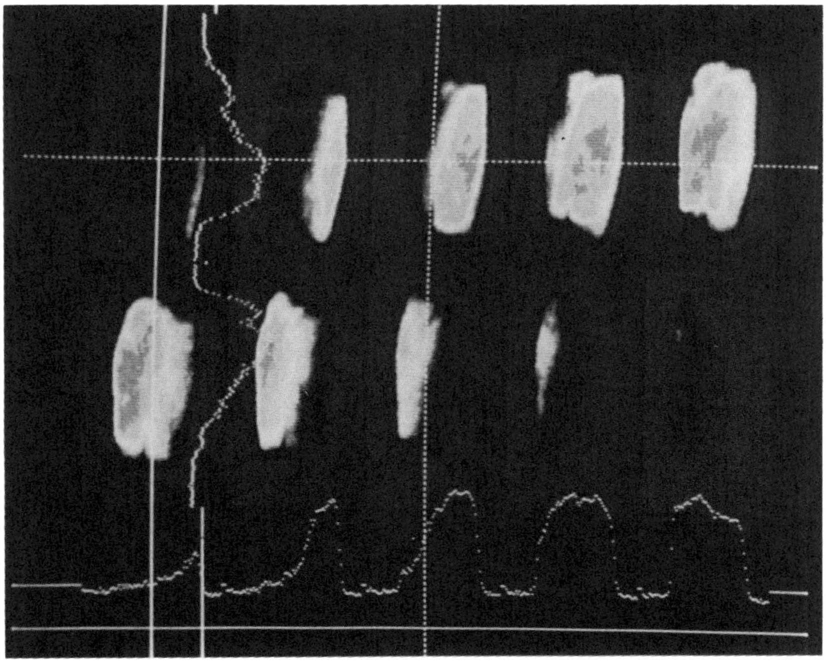

Figure 6a: Successive Digital Real Time Berg-Barrett Topographs of PbSe Epi Film on BaF During Rocking Curve Experiment

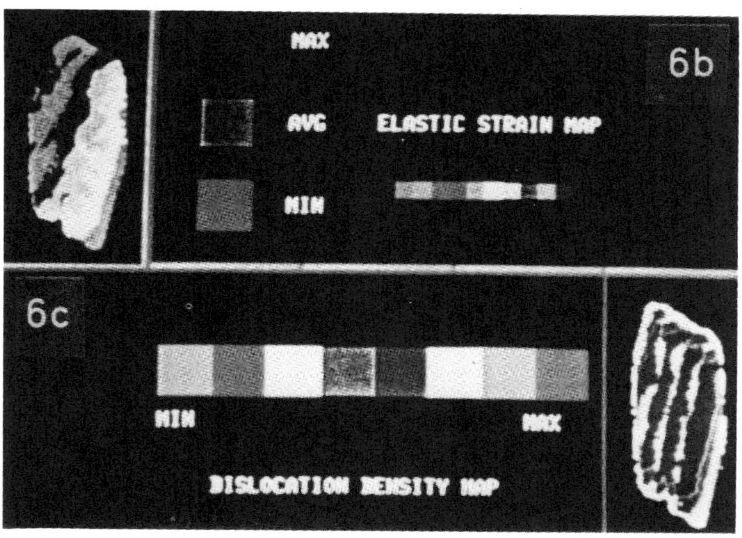

Figure 6b:

Elastic Strain Map

Figure 6c:

Plastic Strain Map

map clearly indicates elastic bending contours across one diagonal of the specimen. The plastic strain map shown in Figure 6c was obtained by tracking the rocking curve halfwidth value of each pixel. This topograph shows bands of high dislocation regions. The origin of such an effect is still under investigation and will be published at a later date.

CONCLUSION

Rocking curve analysis is a powerful technique for characterizing surface and sub-surface micro-lattice strain states. Many different materials have been studied utilizing this technique. The topographic results are extremely informative and relatively simple to interpret. This technique has now been implemented in near real-time and thus offers great potential in the industry oriented production quality control environment or as a research oriented high resolution microstructural characterization tool.

ACKNOWLEDGMENT

The authors wish to acknowledge DARPA, ONR (Chemical Propellants Branch) & SPAWAR of the U.S Department of Defense for supporting this research effort. The authors would like to thank Dr. P. T. Coyne for developing the real time software for rocking curve analysis and A. Wiltrout, K. Aversa and D. Roche for preparation of manuscript.

REFERENCES

1.) Guinier, A., "X-Ray Diffraction," W. H. Freeman & Co., 1963 ed., San Francisco.

2.) Reifsnider, K., and R. E. Green, Jr., "An Image Intensifier System for Dynamic X-ray Diffraction Studies," Rev. Sci. Instr. 39, 1651, (1968).

3.) Weissmann, S., "The Application of X-ray Topography to Materials Science," Non-destructive Evaluation of Materials, Ed. J. J. Burke and V. Weiss, Plenum Publishing Corp., 1979.

4.) Reis,A., J. J. Slade and S. Weissmann, J. Appl. Phys. Vol. 22, p. 655 (1951) J. J. Slade, Jr., and S. Weissmann, J. Appl. Phys. Vol. 23, p. 323 (1952).

5.) Pangborn, R., Ph. D. Thesis, Rutgers University, 1982.

6.) Rosemeier, R. G., T. S. Ananthanarayanan and W. Mayo, "Feasibility Study on Real Time X-ray Topography - Phase I Final Report," DARPA, September 1984 - February 1985.

7.) Hirsch, P. B., P. G. Partridge, and R. L. Segall, Phil. Mag. Vol. 4, p. 721 (1959).

8.) Qadri, S. B., M. Fatemi, and J. H. Dinan, J. Appl. Phys. "Double-crystal X-ray Topographic Studies of Bulk and epitaxially Grown $Zn_xCd_{1-x}Te$," p. 239 (1986).

9.) Qadri, S. B., and J. H. Dinan, J. Appl. Phys. "X-ray Determination of Dislocation Density in Epitaxial ZnCdTe," p. 1066 (1985).

10.) Mayo, W., Ph. D. Thesis, Rutgers University (1982).

11.) Mayo, W. E., J. Chaudhuri and S. Weissmann, "Residual Strain Measurements in Micro-electronic Materials," Am. Soc. for Metals (ASM), 1983 Symposium on Non-destructive Evaluation, Application and Materials Processing, Philadelphia, PA, October 1983, Publ. #8311-005.

12.) Weissman, S., and W. E. Mayo, "Determination of Strain Distributions and Failure Prediction by Novel X-Ray Methods," Am. Soc. for Metals (ASM), 1983 Symposium on Non-destructive Evaluation, Application and Material Processing, Philadelphia, PA, October 1983, Publ. #8311-006.

THE TRACE ELEMENT ANALYSIS FACILITY AT THE UNIVERSITY OF MONTREAL

P.F. Hinrichsen*, A. Houdayer**, and A. Belhadfa

Laboratoire de Physique Nucléaire
Université de Montréal, C.P. 6128, Succursale "A"
Montréal, Québec H3C 3J7, Canada

RESUME

Le montage expérimental ainsi que les performances obtenues à l'aide d'une installation PIXE (Proton Induced X-ray Emission) pour l'analyse quantitative des éléments-traces sont décrits. L'accent est mis sur les charactéristiques qui augmentent le taux de production ainsi que la sensibilité tout en minimisant le coût d'analyse par échantillon.

ABSTRACT

The design and performance of a PIXE (Proton Induced X-ray Emission) facility for trace element analysis is presented, with emphasis on those features which enhance the throughput and sensitivity, and minimize the cost per sample.

INTRODUCTION

The detection of the characteristic X-rays emitted when a target is bombarded with electrons (SEM), X-rays (XRF), or with charged particles (PIXE)[1,2] are well established techniques for elemental analysis. The combination of excitation of X-rays by protons, or other charged particles, and their detection with a Si(Li) detector constitutes a rapid method of multi-element analysis with sensitivity down to the ppm level [1]. The major advantages of PIXE are that 1) only milligram quantities of sample are required, 2) measurements take only minutes, 3) it is inexpensive, and 4) that absolute concentrations at the ppm level can be measured. The system parameters, such as particle type and energy, and selective X-ray absorbers can be varied so as to optimize the sensitivity for the elements of particular interest in a given sample. PIXE is ideally suited to rapid survey or monitoring of many samples. To exploit these advantages, and keep the cost per sample to a minimum, the apparatus was designed for maximum throughput, and for on-line data analysis. A clean laboratory facility is available for target preparation, for evaporating conducting layers of a controlled thickness, or for adding known quantities of a reference element to a sample for internal calibration.

---

* On leave from John Abbott College, Ste Anne de Bellevue, Québec H9X 3L9, Canada
**On leave from CEGEP André-Laurendeau, Montréal, Québec H8N 2J4, Canada

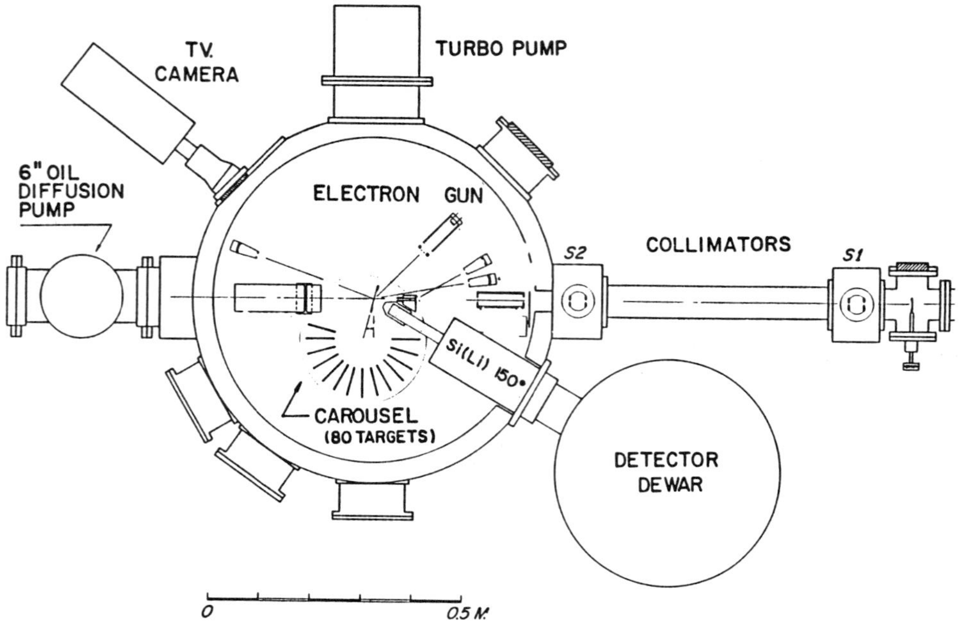

Fig. 1 Diagram of the PIXE scattering chamber

Quantitative results from thin samples, i.e. less than 1 mgm/sq cm, can be rapidly extracted from the X-ray spectra. However, for targets in which the incident particles loose a significant amount of energy, corrections for the variation of cross section with energy as the particle penetrates the target, for absorption of the emitted X-rays as they leave the target, and for fluorescence of the target material by the primary X-rays, are required. A program for the analysis of PIXE data [3] which takes these correction into account, is available in our laboratory.

PIXE should be seen within the context of a broad selection of techniques available at our Laboratory. These include, high energy implantation of a wide variety of ions, RBS (Rutherford Back Scattering), ERD (Elastic Recoil Detection) with time of flight for mass discrimination, and channeling combined with ERD, RBS or PIXE, for the detailed investigation of dopant location and the quality of the crystal structure of substrates. A PIXE-microbeam is also being installed.

EXPERIMENTAL TECHNIQUE

The tandem accelerator produces a proton beam of up to 100 nA in a 4 mm diameter spot on the target in the PIXE chamber, which is shown in fig. 1. We have also obtained beam currents of up to 5 nA through a final aperture of 200 micron diameter, during studies of the spatial variation of sulphur content of polyethylene insulators. For routine measurements the targets are mounted on 5cm x 5cm slide mounts and up to 80 targets can be loaded into the carousel. For measurements on fragile targets, or when the target position or angle are to be varied, the targets are mounted on a 10 target ladder, which can be introduced into the vacuum chamber via an air lock, and can then be manually manipulated. For insulating targets we use an electron flood gun which consists of a heated tungsten filament, at -400 V relative to ground, to reduce the continuous bremsstrahlung background due to electrons which are accelerated into the beam spot.

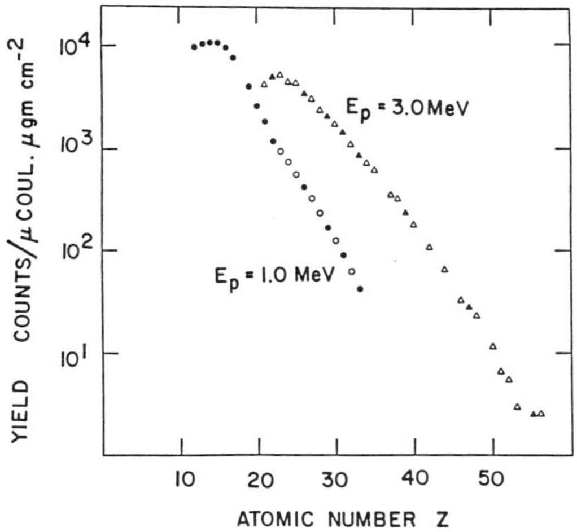

Fig. 2 The absolute calibration of the system at proton energies of 1.0 and 3.0 MeV. The 3.0 MeV data includes the X-ray attenuation of a 37.7 mg/sq cm kapton absorber. The solid points are from standard targets, while relative measurements are represented by open points.

The Ortec Si(Li) X-ray detector is 4 mm in diameter and nominally 4 mm thick, with an energy resolution of FWHM = 153 eV at 5.9 keV. The detector has a 12.7 micron thick beryllium window, is 44 mm from the target, and at an angle of 135° to the beam. A surface barrier detector, mounted at 170° to the beam, is used for simultaneous PIXE-RBS measurements[2].

In many experiments, especially with thick targets, pileup rather than available beam current, or irradiation damage to the target, limits the data acquisition rate, and constitutes a severe constraint on the throughput of samples. We have installed a beam pulsed pileup rejection system similar to that first suggested by Jaklevic[4,5]. The proton beam is pulsed off for the duration of each X-ray pulse. This system significantly improves pileup rejection and is non-paralysable, so high counting rates can be achieved while target heating and radiation damage is reduced to a minimum. Varying the pulse length allows the effective current to be rapidly adjusted without affecting the other beam parameters.

CALIBRATION

The ultimate sensitivity for the detection of an impurity, within a given time, depends on the signal to noise ratio and the counting rate. The background in PIXE is primarily due to electron and particle bremsstrahlung and is typically two orders of magnitude less than for electron excitation in SEM. For targets containing materials of higher Z than the impurities of interest, another source of background is the tail of the detector response[3]. Optimal data can be obtained for a wide range of Z by using two bombarding energies[1], namely 1.0 and 3.0 MeV. 1.0 MeV is ideal for the region Z = 11 to 26, and for $Z > 20$ we chose $E_p = 3.0$ MeV, at which energy the L X-ray production cross sections are high. For the latter a 37.7 mg/sq cm thick kapton absorber is placed in front of the detector to stop backscattered particles, and reduce the bremsstrahlung background. Analysis of data taken at these two energies allows impurities

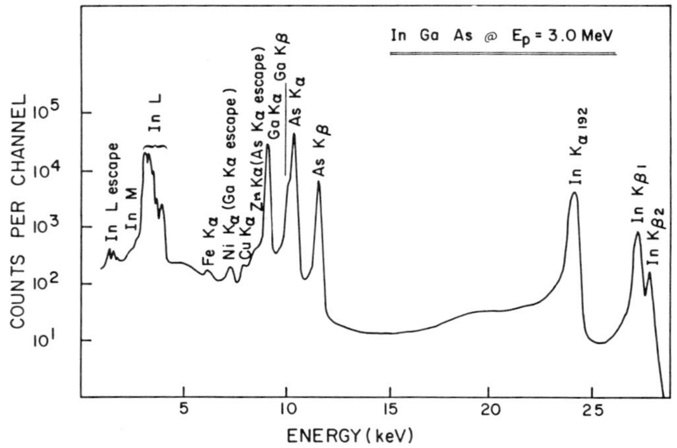

Fig. 3 X-ray spectrum from a thick sample of InGaAs bombarded at E_p = 3.0 MeV. Fe, Ni, Cu, and Zn impurities are clearly evident.

in organic targets to be detected at the few parts per million level, over a wide mass region.

The PIXE system was calibrated at 1.0 and 3.0 MeV, in terms of the total yield in the K_α plus K_β lines, defined as the counts per microcoulomb of protons per microgram/sq cm of each specific target element, using a set of standard thin targets[6]. The absolute calibration curves obtained are shown in fig. 2, and were extended by making relative measurements, using targets prepared from solutions which contained two or more elements in precisely known concentrations. The estimated uncertainty in these calibrations is 10 percent.

TARGET PREPARATION

For thick self supporting targets the risk of contamination is small provided care is taken to keep the surface clean. The latter can be a problem when a microtome is used to slice samples, as the blade can be a source of contamination, or transport impurities from one part of the sample to another. For liquids the backing foil on which the sample is freeze dried can be the source of background peaks. The backing foil should be a strong organic material, and as thin as possible. It should withstand extended irradiation by the beam without damage or mechanical distortion. We have found that polyimide, film available from Du Pont, contain less impurities than the other commonly used target backings. This material, which is used as an insulator in the manufacture of integrated circuits, is available as a liquid, and can therefore also be used as a binder in the preparation of thin targets of powdered materials.

RESULTS

PIXE can be applied to a wide variety of investigations, and the following is a sample of some of our recent measurements. Figure 3 is an X-ray spectrum from an InGaAs wafer bombarded with 3.0 MeV protons. The spectrum reveals the presence of 13 ppm of iron, 36 ppm of copper, as well as nickel and zinc. The background in the 10 to 25 keV region of the

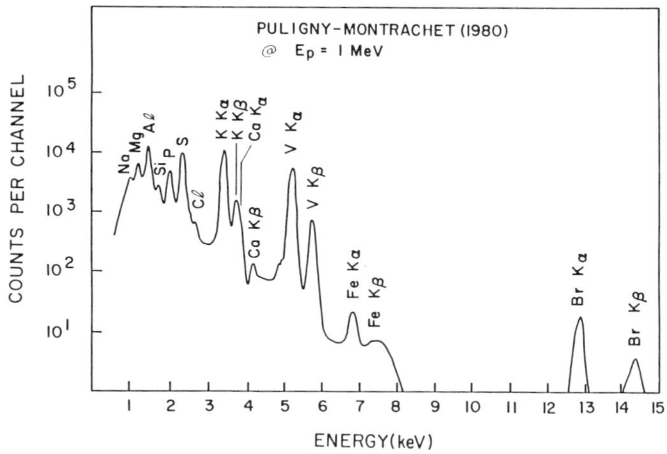

Fig. 4 A typical X-ray spectrum from a sample of polyethylene high voltage cable insulator bombarded at E_p = 1.0 MeV.

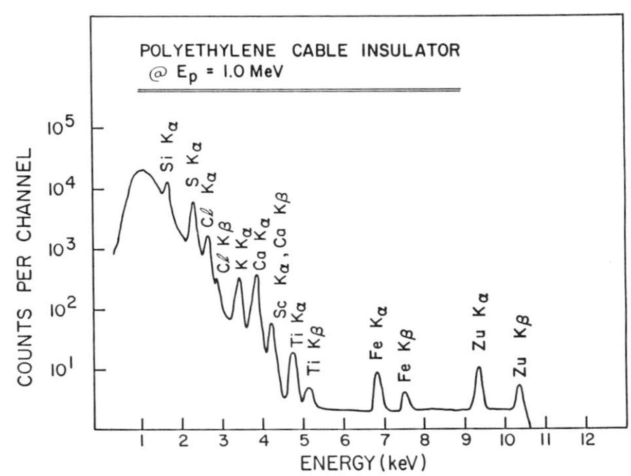

Fig.5 PIXE spectrum from Puligny Montrachet wine spiked with vanadium.

spectrum is due to In K X-ray events for which there was incomplete charge collection in the Si(Li) detector. A further complication is the overlap between the silicon escape peaks of the As and Ga K X-rays and the Ni and Zn peaks. This limits the sensitivity for these elements in this matrix to 50 ppm. These problems can be minimized by the use of selective X-ray filters tailored to the specific measurement[2].

Figure 4 shows the spectrum from a sample of polyethylene high voltage cable insulator, taken at 1.0 MeV proton energy. A wide variety of impurities can be easily identified. Studies of the spacial distribution of these impurities can give valuable information on the effect of additives, such as anti-oxidants, on the properties of the material, and its ageing. Small inclusions during the manufacturing process can become points of initial breakdown, i.e., "Trees". PIXE has the great advantage that the proton beam can be focussed onto these "Trees", which are typically less than 200 microns in size, and reveal their elemental composition, which is a valuable clue to their origin.

A third example of the versatility of PIXE analysis is shown in figure 5. A few microlitres of "Puligny Montrachet" wine, which had been spiked with a standard solution of 1000 ppm of vanadium, were deposited on a nucleopore filter and freeze dried. The spectrum, which was obtained at 1.0 MeV bombarding energy, allows the concentrations of many trace elements to be determined, and it is hoped to be able to correlate this "fingerprint" with that from a sample of the soil from the vinyard of origin. Such studies would allow counterfeit European wines to be rapidly identified on a routine basis.

REFERENCES

1. S.A.E. Johansson, and T.B. Johansson,NIM 137 (1976) 473.
2. I.V. Michell, and J.F. Ziegler, Ch.5, "Ion Beam Handbook for Materials Analysis", Ed. J.W. Mayer, and E. Rimini (Academic Press Inc. N. Y. 1977)
3. J.L. Campbell, W. Maenhaut, E. Bombelka, E. Clayton, K. Malmqvist, J.A. Maxwell, J. Pallon, and J. Vandenhaute, NIM B14 (1986) 204.
4. J.M. Jaklevic, F.S. Goulding, and D.A. Landis, IEEE Trans. Nucl. Sci. NS-19 (1972) 392.
5. K.G. Malmqvist, E. Karlsson, and K.R. Akselsson, NIM 192 (1982) 523.
6. L.M. Heagney, and J.S. Heagney, NIM 167 (1979) 137.

EFFECT OF CAVITIES ON ULTRASONIC ATTENUATION AND VELOCITY

Anmol S. Birring and J.J. Hanley

Nondestructive Evaluation Science and Technology Division
Southwest Research Institute
San Antonio, Texas 78284

ABSTRACT

Cavitation, voids, or porosity in materials can initiate several types of failure mechanisms. One type is creep, which begins with cavitation and eventually leads to microcracking, cracking, crack coalescing, and ultimate failure. Failure from creep can be avoided if cavitation is detected early.

A theoretical investigation of nondestructive ultrasonic methods for detection of creep cavitation in steels was performed. Ultrasonic scattering theories available in the literature were converted into computer software and used to predict the effect of cavities on ultrasonic attenuation and velocity. From the results, we theoretically concluded that measurements of (1) absolute velocity and (2) velocity and attenuation variation with frequency have a strong potential to detect cavitation.

The study also looked at the effects of scattering from grain boundaries and from graphite and carbide inclusions in steels. Grain scattering affects attenuation and velocity, and methods were formulated to compensate for such effects. The scattering from inclusions is expected to be less than that from cavities, as cavities produce a higher impedance mismatch than inclusions. The study concludes that ultrasonic methods will be able to detect creep cavitation in carbon steels used for high-temperature applications.

INTRODUCTION

Nondestructive detection of creep cavitation is extremely useful to assess the integrity of steels in components and structures. Knowing the amount of cavitation, engineers can predict their remaining life.

Work done by ERA Technology[1] showed that for a 1/2Cr 1/2Mo 1/4V steel casing operating at a stress of 84.9 MPa at 550°C, 8 percent of the grain boundaries were cavitated at 30-percent life and 13 percent at 50-percent life. The direct relation of cavitation with remanent life shows a need for a nondestructive evaluation (NDE) method to detect cavitation at grain boundaries.

It is known that cavitation in material affects ultrasonic attenuation and velocity. Thus, ultrasonics was evaluated for detection of cavitation. By measuring the changes in ultrasonic attenuation and velocity, cavitation in the material could be predicted.

During the investigation, ultrasonic scattering theories of Truell et al.[2] were transferred into a software package. The software package proved to be an excellent tool to predict how cavities affect attenuation and velocity of waves. The software also predicted how attenuation and velocity changes with frequency for a fixed cavity size. Additional software was written to predict the effect of grain boundaries and inclusions. Using the software, an NDE method for detection and measurement of creep cavitation in materials can be developed. This paper presents the work done to achieve the theoretical results on the effect of cavities on ultrasonic wave propagation.

ULTRASONIC SCATTERING

Ultrasonic scattering is produced from discontinuities such as cavities, inclusions, or anisotropy in materials, which produce an impedance mismatch. The scattering, in turn, produces changes in attenuation and velocity of ultrasonic waves. A discussion is presented of the theories used to compute attenuation and velocity changes because of scattering from creep cavitation, grain boundaries, and inclusions.

Two-Phase Scattering

The two-phase scattering theory developed by Truell et al.[2] was used to calculate attenuation due to scattering in a material. This theory assumes that individual scatterers are independent of each other so that scattering from a scatterer is not dependent on the scattering from another scatterer.

The two-phase scattering theory relates the size and number of cavities to the attenuation coefficient "α." The attenuation coefficient, an experimentally measurable quantity, can be determined from Eq. (1).

$$A = Ao \exp(-\alpha x) \tag{1}$$

where Ao is the ultrasonic amplitude at x = 0, A is the ultrasonic amplitude at x, and x is the path length.

The two-phase scattering theory determines the attenuation coefficient from the number of scatterers and scattering cross section from Eq. (2) as follows:

$$\alpha = 1/2\ n_o \nu \tag{2}$$

where n_o is the number of scatterers per unit volume and ν is the scattering cross section. The scattering cross section is defined as the ratio of the intensity scattered into unit space angle per unit time to the recurring intensity per unit time and unit area. The scattering intensity is calculated from Eq. (3) as follows:

$$\nu_n = 4 \sum_{0}^{\alpha} (2m + 1)\ [\,|A_m|^2 + m\ (m + 1)\ (k/k_t)\ |\ B_m|^2] \tag{3}$$

where $\nu_n = \nu/\pi a^2$ (a = radius of scatter) is the normalized scattering coefficient. The coefficients in Eq. (3) are evaluated from the boundary conditions of the scatter to matrix. The values of the coefficients are given in Section 32 of Ref. 2. Using these equations, one can calculate attenuation due to cavities and inclusions in an elastic medium. This theory also calculates changes in attenuation with ultrasonic frequency.

Multiple Scattering

The multiple scattering theory is different from the two-phase theory, as it does not assume that the scattering from individual scatterers is independent. The multiple scattering theory computes the complex solution vector ψ of the wave equation.

$$\psi = \exp[i(\beta x + \omega t)] \tag{4}$$

where $\beta = k + i\alpha$, t = time, and ω = frequency. (5)

In Eq. (5), $k = 2\pi/\lambda$, the wave number for longitudinal waves.

Sayers[3-5] has solved Eq. (4) for the case of cavities where the ultrasonic wave propagation is in the Rayleigh Region ($ka < 1$). His solution gives

$$\left(\frac{\beta}{k}^{2}\right) = 1 - \frac{4}{3} \pi a^{3} n_{o} \left(A + Bk^{2}a^{2} - iCk^{3}a^{3}\right) \tag{6}$$

The values of A, B, and C have been solved by Sayers to compute the ultrasonic velocity for low cavity concentrations.

$$v'/v = 1 + (1/2)\,(\Delta V/V)\,(A + B\,k^{2}\,a^{2}) \tag{7}$$

where A and B are solved in Ref. 4, v' = velocity of longitudinal waves when the material has cavities, v = velocity of longitudinal waves when the material has no cavities, and $\Delta V/V = (4/3)\,\pi a^{3} n_{o}$, the fractional volume of cavities.

Besides cavities, grain boundaries also scatter ultrasonic waves when the grains of the materials are anisotropic and randomly placed. In such a case, the impedance mismatch at the grain boundary generates the scattering. Hirsekorn[6] has presented a theory that allows one to calculate the scattering coefficients and velocity of ultrasonic waves in polycrystals of cubic symmetry. The multiple scattering theory of Hirsekorn includes the effect of mode conversion.

Hirsekorn[6] has calculated the value of the complex propagation factor "β" for the wave equation Eq. (4) as follows:

$$\beta = k\left[1 - \frac{1}{(4\pi)^{3}}\left(\frac{A}{\rho_{o} v^{2}}\right)\left(-B_{1} + \frac{k_{t}}{k} B_{2}\right)\right] \tag{8}$$

where $A = C_{11} - C_{12} - 2C_{44}$, the anisotropy factor; k_t = wave number for shear waves; and B_1 and B_2 are solved in Ref. 6. The real and the imaginary parts of Eq. (8) are then used to compute the ultrasonic velocity and attenuation, respectively, using Eq. (5).

ATTENUATION AND VELOCITY VARIATIONS BECAUSE OF SCATTERING

The attenuation and velocity are also affected by the ultrasonic waves scattered by cavities, inclusions, and grains in a medium. From this, a relationship of attenuation to velocity can be obtained. The relationship can then be used to formulate an ultrasonic approach for detection and measurement of cavities. This section presents the affect of cavities, inclusions, and grains on ultrasonic attenuation and velocity.

Attenuation

Changes in attenuation due to ultrasonic scattering from cavities, inclusions, and grain boundaries can be calculated using the two-phase scattering theory. Scattering from each of these three types of scatterers (cavities, inclusions and grains) is discussed as follows.

Cavities Attenuation by cavities was computed using the two-phase theory. Figure 1 is a plot of increased attenuation by the increase in cavity size at 15, 25, and 50 MHz. This computation assumed that the number of cavities was fixed at 500,000/cc and only their size increased. The figure also shows that attenuation increases with frequency. In particular it was found that $d\alpha/df$ depends on the cavity. Thus, a measurement of $d\alpha/df$ will infer cavity size.

Inclusions Inclusions are present in materials for material strengtheners or in the form of impurities. In steel, inclusions can be graphites or carbides (Fe_3C). Scattering is produced by inclusions because of the difference in impedance between the inclusion (C or Fe_3C) and the matrix material (Fe). Ultrasonic scattering was calculated from graphite inclusions by assuming $E = 796$ Kg/mm^2, $\rho = 2.16$ gm/cc, and $\tau = 0.3$ (Poisson's ratio).

Using these values, the attenuation of ultrasonic waves was calculated. Figure 2 shows the increase in attenuation of ultrasonic waves with an increase in the size of graphite inclusions when the ultrasonic frequency is 20 MHz.

Grain Scattering from grains is produced when the crystal structure in the grain is anisotropic. The degree of anisotropy of a crystal A is denoted by $A = C_{11} - C_{12} + 2C_{44}$ where C is the elastic constant.

Using the multiple scattering software,[6] the effect of attenuation on grain size was calculated. Figure 3 is a plot of attenuation versus grain size. This figure shows that the attenuation gradually increases grain size in the Rayleigh region. The non-Rayleigh region has no direct correlation of attenuation.

Velocity

The multiple scattering theory is used to calculate changes on the velocity of ultrasonic waves due to scattering from cavities and grains.

Cavities The effect of cavities on ultrasonic longitudinal wave velocity was calculated. The results determined that a 1-percent volume concentration of distributed cavitation in iron reduced the velocity of 20-MHz ultrasonic waves from 5920 to 5620 m/sec (Figure 4). The velocity also decreased with an increase in ultrasonic frequency (see Figure 5). Cavities in a material can be measured if ultrasonic velocity measurements are taken at several frequencies. Then the slope of velocity versus frequency dv/df decreases with an increase in cavity size (see Figure 5).

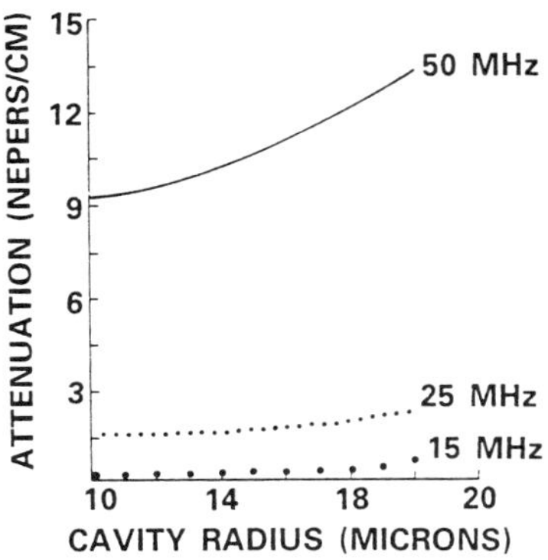

Fig. 1. Attenuation change because of an increase in cavity size. Attenuation changes are calculated for three frequencies 15, 25, and 50 MHz (n_o = 500,000 cavities/cc).

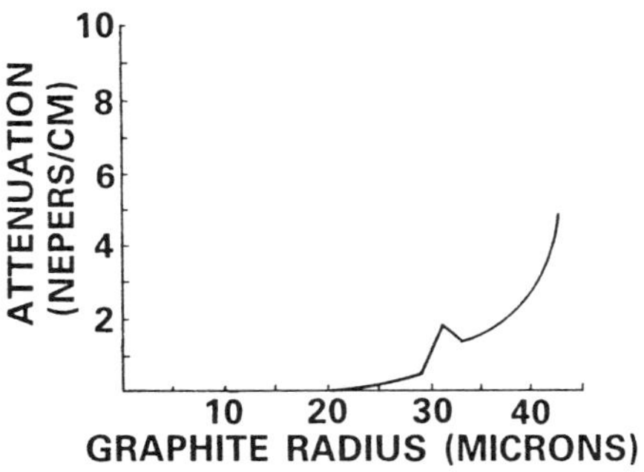

Fig. 2. Attenuation variation with an increase in the size of graphite inclusions at 20 MHz (n_o = 100,000/cc)

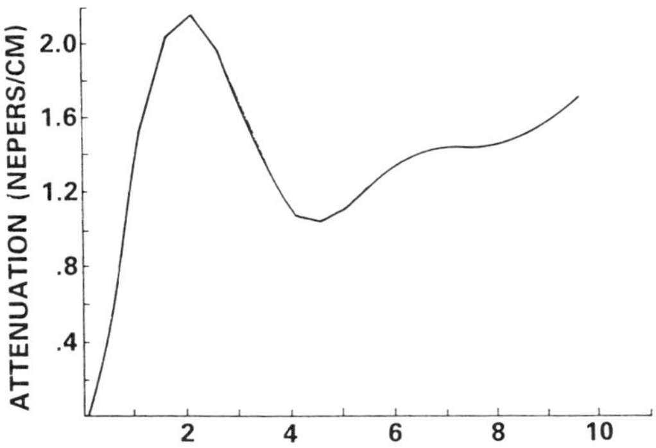

Fig. 3. Attenuation variation because of grain size. The attenuation increases gradually in the Rayleigh region.

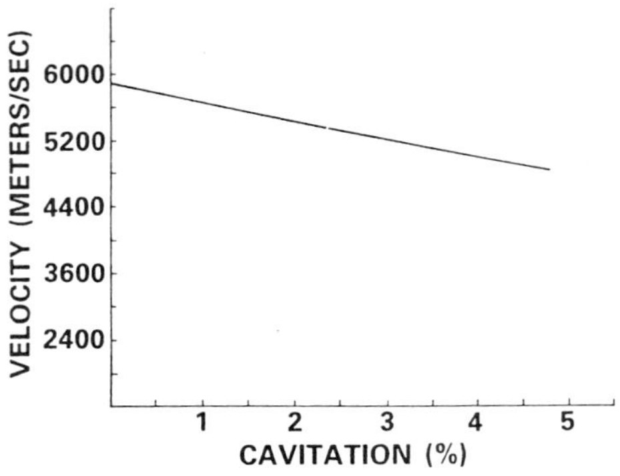

Vol (%)

Fig. 4. Decrease of longitudinal-wave velocity in iron with an increase in cavitation at 20 MHz. A 1-percent cavitation reduces the velocity from 5920 m/sec to 5650 m/sec (4.6-percent reduction).

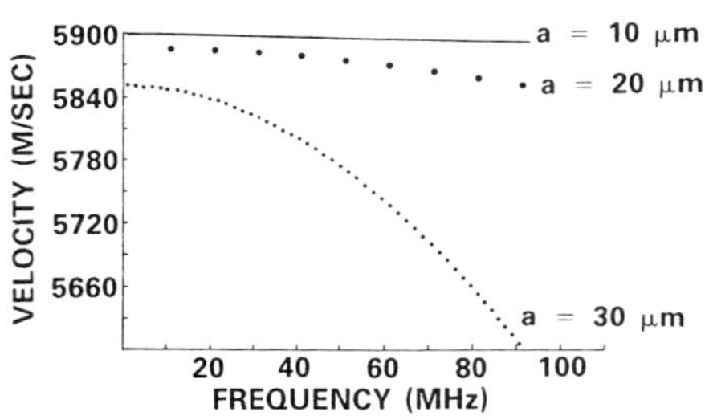

Fig. 5. Variation of ultrasonic velocity with frequency for different cavity sizes. The slopes of these curves dv/df decrease with an increase in cavity size (n_o = 500,000 cavities/cc).

Grains. The multiple scattering software was used to calculate the velocity variation with grain size. The results showed that velocity was not a strong function of velocity for $k_{ag} < 1$ (ag = grain radius). Thus, velocity measurements should be taken in the range of $k_{ag} < 1$; e.g., velocity measurements should be taken in the range of 0 to 47 MHz if ag = 20 μm.

DEVELOPMENT OF AN ULTRASONIC APPROACH TO DETECT CAVITATION

Two ultrasonic properties, attenuation and velocity, were used to detect and measure cavitation in materials. In general, attenuation increased and velocity decreased with an increase in cavitation. Using this approach, an operator can detect cavitation in homogeneous and isotropic elastic medium (e.g., gas bubbles in liquids). Detection of cavitation was not obvious when scattering occurred because of grains and inclusions. The following approaches were considered for compensation of grain and inclusion scattering.

Attenuation Measurement

Individual attenuation variations from cavity, inclusion, and grain size were shown in Figures 1, 2, and 3 respectively. At the present time, however, no multiple scattering theory has been developed in the literature that can predict the total, or combined, scattering. Approximations can be made of the total scattering, however, using two-phase systems consisting of several components. To combine the attenuation from the grains and cavities, an equation for total scattering was written as:

$$\alpha_s = x_1{}^2 \alpha_{s1} + x_2{}^2 \alpha_{s2} + x_1 \alpha_{s12} + x_2 \alpha_{s21} \qquad (9)$$

where x_1 = fraction volume of grains, x_2 = fraction volume of cavities ($x_1 + x_2 = 1$), α_{s1} = scattering within grains (from Eq. 8), $\alpha_{s2} = 0$ = scattering within a cavity, α_{s12} = scattering at the grain-cavity phase boundary (Eq. (2), and $\alpha_{s21} = 0$ = scattering at the cavity-grain boundary.

Using Eq. (9), attenuation was plotted against frequency in Figure 6. This figure shows the attenuation curve for various cavity sizes when the grain size is fixed. If these curves can be experimentally measured, the slope $d\alpha/df$ and shape of the curve can be used to predict the cavity size. Figure 6 shows that for a fixed grain size, the value of $d\alpha/df$ increases with an increase in cavity size.

Graphite and carbide inclusions also affected attenuation. The increase in attenuation because of inclusions is less compared to cavities because the matrix-cavity impedance mismatch is higher than the inclusion matrix. In particular, for ka = 1.5, the normalized scattering cross section for a cavity is 2.0 compared to 0.6 from MnS in Fe.[8]

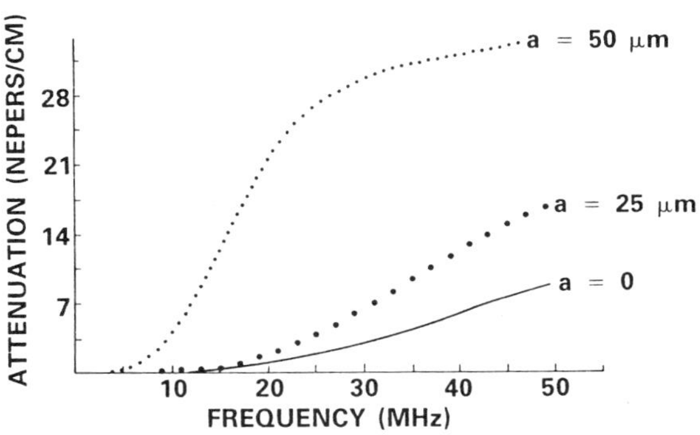

Fig. 6. Attenuation Versus Frequency for a Fixed Grain Size (a_g = 25 μm). Cavity size can be inferred from the value of $d\alpha/df$, the slope of the above curves.

Velocity Measurement

Ultrasonic velocity is affected by cavities, grains, and inclusions. Figures 4 and 5 show a strong dependence of ultrasonic velocity on cavitation. Calculations have also shown that the function of velocity was not strong when $k_{ag} < 1$. Thus, to minimize the effect of grain scattering, velocity measurements to detect cavitation should be performed in a $k_{ag} < 1$ range.

Ultrasonic velocity curves in Figure 5 show the strong dependence of velocity on cavitation and ultrasonic frequency. A reduction in the slope of dv/df will indicate cavitation. This type of measurement can be taken by measuring the velocity for a range of frequencies. The value of dv/df can then be computed and used to detect cavitation.

Inclusions also effect ultrasonic velocity. However, the reduction in velocity because of inclusions is smaller because of the lower inclusion-matrix impedance mismatch compared to the cavity matrix. It was shown that 1-percent MnS only reduced the 20-MHz ultrasonic wave velocity by 1.7 percent[8] compared to a 4.6-percent velocity reduction because of 1-percent cavitation (Figure 5). Since the effect from inclusions is less than from cavities, cavities can be detected in the presence of carbide and graphites in steels.

Thus, attenuation velocity measurements are promising for the detection of cavitation. Both attenuation and velocity measurements should be taken in a range of frequencies. Cavitation increases attenuation and $d\alpha/df$ and decreases velocity and dv/df. To minimize the affect of grain scattering on velocity, the measurements should be taken in the range $k_{ag} < 1$.

CONCLUSIONS

A theoretical investigation was performed to determine the potential of ultrasonic attenuation and velocity measurements to detect cavitation in material. The theoretical study determined the ultrasonic scattering from cavities and its effect on attenuation and velocity. The investigation led to the conclusion that attenuation and velocity measurement and attenuation and velocity as a function of frequency will detect creep cavitation. In particular, it was found that an increase in cavitation increases the attenuation and reduces the velocity. Also, for a fixed cavity size, the ultrasonic velocity decreases with an increase in frequency. The study showed that both attenuation and velocity should be measured for a range of frequencies. Then an increase in $d\alpha/df$ and a decrease in dv/df will indicate cavitation.

The study also included the effects of scattering from grain, graphite, and carbide inclusion in steel and demonstrated that these factors (grains and inclusion) affect ultrasonic scattering. The effect of inclusions on scattering was less than from cavities as cavities produce a high impedance mismatch. The affect of grain scattering on velocity can be minimized if measurements are taken in the range of $k_{ag} < 1$. In this range, the velocity is not strongly dependent on grain scattering. It is concluded, therefore, that creep cavitation is detectable in steels containing carbides and graphite inclusions.

REFERENCES

1. B. J. Cane, P. R. McCarty, and P. F. Aplin, Residual Life Assessment Methods, Progress Report on EPRI RP-2253-4, Palo Alto, CA, June (1985).
2. R. Truell, C. Elbaum, and B. B. Chick, in: "Ultrasonic Methods in Solid State Physics," Academic Press, New York, (1969).
3. C. M. Sayers and R. L. Smith, "The Propagation of Ultrasound in Porous Media, Ultrasonics, 20:201-205 (September 1982).
4. C. M. Sayers, Ultrasonic Velocity Dispension in Porous Materials, Journal of Applied Physics, 14:413-420 (1981).
5. C. M. Sayers, Scattering of Ultrasound by Minority Phases in Polycrystalline Metals, U.K. Atomic Energy Authority Final Report, AERE-R-11162 (January 1984).
6. S. Hirsekorn, The Scattering of Ultrasonic Waves by Polycrystals, Journal of Acoustical Society of America, 72:1021-1031 (September 1982).
7. E. P. Papadakis, The Inverse Problem in Material Characterization Through Ultrasonic Attenuation and Velocity Measurements, in: "Nondestructive Methods for Material Property Determination," C. O. Ruud and R. E. Greed, eds., Plenum Press, New York (1984), pp. 151-160.
8. K. Goebbels, Structure Analysis by Scattered Ultrasound, in: "Research Techniques in Nondestructive Testing," Vol. 4, R. S. Sharpe, ed., Academic Press, London, (1986).

SPATIAL AVERAGING IN POROSITY ASSESSMENT BY ULTRASONIC ATTENUATION SPECTROSCOPY

Peter B. Nagy[a], David V. Rypien and Laszlo Adler

Department of Welding Engineering
The Ohio State University
Columbus, Ohio 43210

The characterization of porosity in cast aluminum via the frequency dependence of the ultrasonic attenuation yields both the average pore radius and the volume fraction[1]. For dilute porosity of less than 6% volume fraction it was found that in case of 50 µm or higher average pore size the experimental measurement of the attenuation could be fit by the theoretical model for single scattering by spherical voids, and the resulting porosity parameters are in good agreement with independent optical and density measurements. Although some deviation between the ultrasonic and optical or density results are accepted as fully justified by the different physical nature of the methods, certain deviations of the ultrasonic results taken under different conditions, e.g. measured on samples of different lengths, show the need for a better understanding of the relation between the measured attenuation and the sought porosity parameters. It is shown that the porosity induced attenuation is smaller than the value predicted by the simplified theoretical model, which can result in consequent discrepancies above the measuring error. Spatial averaging before and after the spectral analysis is found to increase the reliability of the ultrasonic results by effectively rejecting the incoherent components in the attenuated signal.

INTRODUCTION

Porosity in structural materials limits their ultimate strentgh and hence their utility. Recently several studies which discuss the porosity induced ultrasonic attenuation have appeared[1-8]. Porosity assessment by ultrasonic attenuation measurement involves two principal problems: first, how to relate the porosity induced ultrasonic attenuation to porosity parameters such as average pore radius and volume fraction, and second, how to separate the sought porosity induced attenuation from other components contributing to the actually measured total attenuation.

The first question was sufficiently answered by Gubernatis and Domany[4] who have developed formally exact expressions for the plane wave attenuation in porous media. Neglecting correlation between scatterers and presuming relatively weak porosity induced attenuation over a wavelength, their result

---

[a]Permanent Address: Applied Biophysics Laboratory, Technical University Budapest.

simplifies to the following equation which has previously been given by others on physical grounds[2]

$$\alpha(k) = \tfrac{1}{2} n \bar{\gamma}(k), \tag{1}$$

where $\alpha(k)$ is the porosity induced plane wave attenuation coefficient, $\bar{\gamma}(k)$ is the total scattering cross-section of a single scatterer averaged over the whole pore distribution, k is the wave number in the host medium, and n is the number density of the pores. The scattering cross-section can be written as the product of the double geometrical cross-section and the reduced scattering cross-section $\Gamma(ka_p)$

$$\gamma(k) = 2a_p^2 \pi \Gamma(ka_p), \tag{2}$$

where a_p denotes the pore radius. The reduced scattering cross-section $\Gamma(ka_p)$ depends on the host medium only through its Poisson ratio, which makes it possible to determine the pore radius a_p from the shape of the attenuation coefficient versus frequency curve. Subsequently, the pore density can be calculated from the scatteirng induced attenuation by using Eq. 1. Gubernatis and Domany[4] showed that this simple data reduction technique yields accurate porosity parameters for spherical voids of peaked size distribution and less than 5% volume fraction, and a recent comprehensive experimental study[1] confirmed these predictions.

EXPERIMETNAL METHOD

The second problem mentioned in the introduction is how to measure the porosity induced attenuation coefficient $\alpha(k)$. Figure 1 shows the schematic diagram of ultrasonic attenuation measurement by comparison of the front and back echoes from an immersed sample of parallel surfaces. The total attenuation of the sample contains many factors:

$$a = \ln \frac{A_0}{A_1} = a_{imp} + a_{diff} + a_{grain} + \dots + a_p. \tag{3}$$

a_{imp} and a_{diff} are due to the double way through transmission loss at the liquid-solid interface and to the spread of the ultrasonic beam as it propagates forth and back inside the sample, respectively. These two factors can be readily calculated from known parameters or completely eliminated by comparing the backwall echo from the porous sample under study to one from a porosity free, but otherwise similar sample.

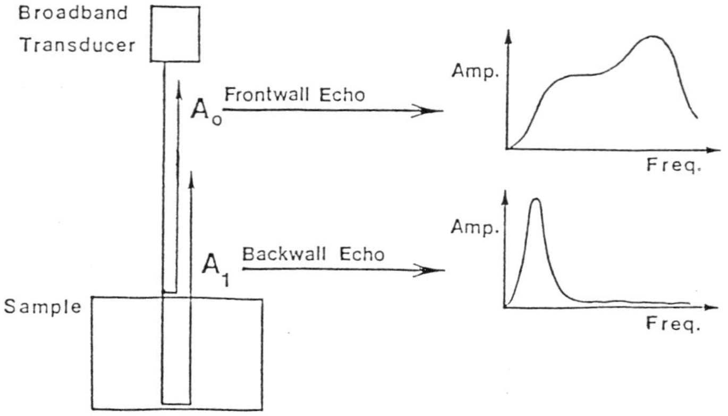

Fig. 1. Schematic diagram of ultrasonic attenuation measurement.

Other contributions in Eq. 3 constitute a more serious problem. The grain scattering induced attenuation a_{grain} depends on the anisotropic elastic properties of the host material and the grain size as well. There seems to be no feasible correction for this term, and the ultrasonic porosity assessment must be limited to applications where the grain scattering induced attenuation is negligible with respect to the porosity induced component a_p. For example, this method is expected to be sensitive to volume fractions on the order of 0.1% in Aluminum when the grains are not bigger than the pores[1].

Other factors not mentioned explicitly in Eq. 3, such as surface roughness induced attenuation can further limit the feasiblitiy of this technique, but for the purposes of this study let us pressume that all of them are either adequately corrected or negligible. The above summarized data reduction is based on the porosity induced attenuation coefficient's being small, but completely disregards the sample thickness on which the total porosity induced attenuation depends. Whenever this attenuation is small enough to let the attenuated coherent signal be measured accurately, the porosity assessment can be made according to the above procedure. As an example, Figure 2 shows the theoretical and experimental frequency-dependent attenuation coefficients in A357 aluminum alloy casting. In certain cases, although the corrected attenuation is entirely due to porosity, it is not necessarily equal to the expected porosity induced plane wave attenuation coefficient times the propagation length. This discrepancy is caused by using a finite ultrasonic beam to interogate the sample instead of an infinite plane wave. The plane wave attenuation is the highest possible scattering induced attenuation, which is based on the total scattering cross-section. As for an infinite plane wave all scattered components regardless of their directions will be completely lost. On the other hand, a finite transducer is somewhat sensitive to the incoherent scattered field as well, i.e. a small part of the scattered energy will appear in the detected signal together with the attenuated coherent signal.

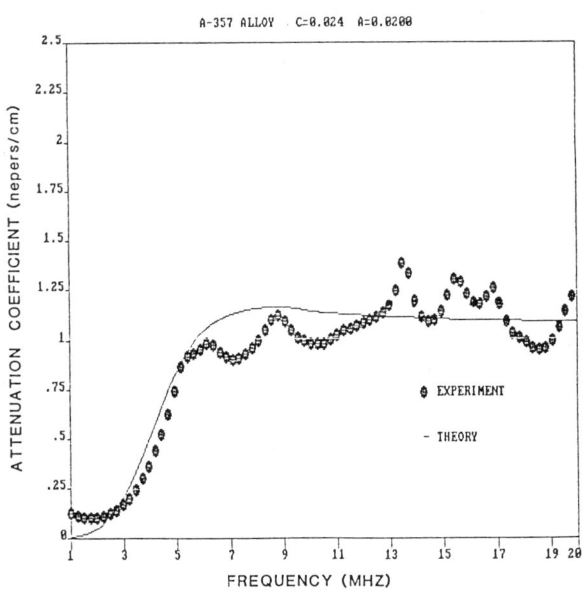

Fig. 2. Comparison of theoretical and experimental attenuation coefficient versus frequency curves for an aluminum sample (20 μm pore radius and 2.4% volume fraction).

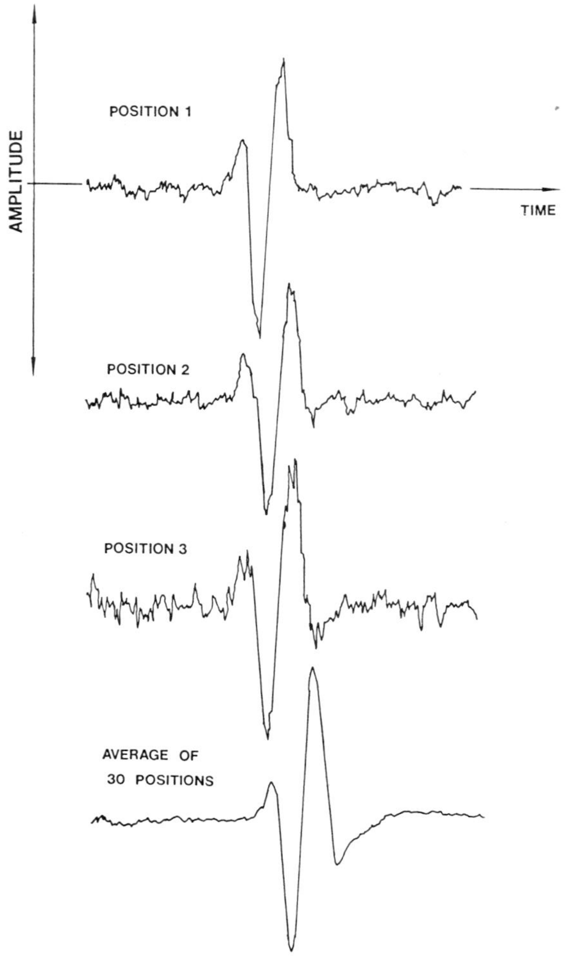

Fig. 3. The effect of spatial averaging on the backwall echo.

SPATIAL AVERAGING

Figure 3 shows the received backwall echoes from a 20 mm thick aluminum sample. The highly attenuated coherent signal is partially submerged in the incoherent backscattered signal. This results in reduced porosity induced attenuation, and consequently in some underestimation of the porosity. Furthermore, the measured porosity induced attenuation will not be linearly proportional to the sample thickness, which will cause disturbing "length dependence" in porosity assessment. In a thicker sample, the saturated high frequency attenuation will be smaller than proportional to the thickness, and the point where the attenuation curve flattens off will shift slightly to lower frequencies. Consequently, the porosity assessment will slightly overestimate the pore radius and considerably underestimate the volume fraction for too thick samples.

A simple solution to this problem is to increase the diameter of the finite beam, i.e. to make the transducer less sensitive to the incoherent signal. In practice this can be easily achieved by laterally moving the transducer and averaging the received time signals. The last picture in

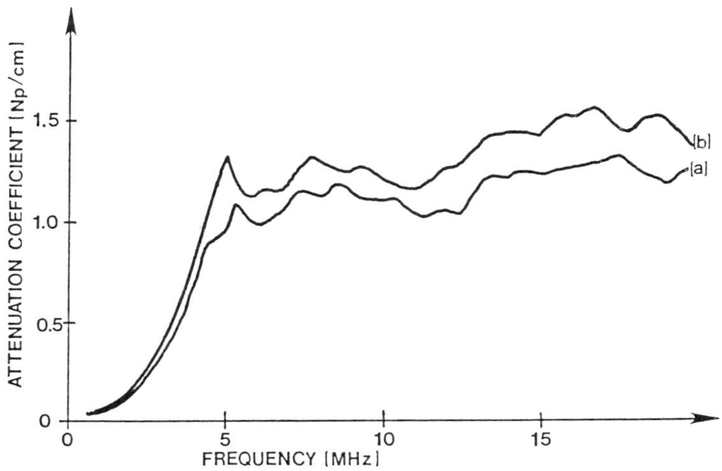

Fig. 4. The effect of spatial averaging in the (a) frequency and (b) time domains on the spectrum of the backwall echo.

Figure 3 shows the enhanced backwall echo after averaging over thirty positions.

Spatial averaging in the time domain will effectively eliminate the incoherent scattering from the received signal thereby resulting in a somewhat higher and more accurate volume fraction. Figure 4 shows the attenuation versus frequency curve for the averaged signal as well as the averaged spectra of thirty different positions. The attenuated coherent echo can be easily recognized in the surrounding backscattered signal in Figure 3 because of its strong low frequency components which are only weakly attenuated by porosity. On the other hand, the substantially attenuated high frequency components are much weaker than the corresponding backscatterd signal, therefore the measured porosity induced attenuation will be lower than the plane wave attenuation presumed in the porosity assessment. Spatial averaging in the time domain will effectively eliminate the backscattered signal while spatial averaging in the frequency domain will similarly reduce the disturbing ripple on the high frequency part of the spectrum, but the averaged spectrum will smoothen to a lower mean value including the averaged backscattered energy and the coherent energy as well.

Both attenuation curves in Figure 4 yield approximately the same 200 μm pore radius, but the calculated volume fractions are 3.3 and 2.8% for spatial averaging in the time and frequency domain, respectively. A subsequent measurement on the two halves of the same sample yielded 3.4 and 3.6% volume fraction indicating only 6% relative "length dependence" in the case of spatial averaging in the time domain versus 22% in the case of spatial averaging in the frequency domain. The 6% relative difference is well within the accuracy of this technique therefore we can conclude that there is no length dependence at all in this case.

CONCLUSION

We showed that a finite diameter transducer will measure a somewhat smaller porosity induced attenuation than the ideal plane wave prediction because its being sensitive to incoherent backscattered terms. The resulting underestimation in porosity volume fraction can be as high as 20-30%. Spatial averaging in the time domain can greatly reduce this effect, but its feasibility is often limited by high gradients in porosity, i.e. it yields unrealistic values whenever the mean porosity parameters change substantially within the region of spatial averaging. This problem can be recognized from the abrupt change of the backwall echo dominated by low frequency components.

REFERENCES

1. L. Adler, J. H. Rose, and C. Mobley, Ultrasonic method to determine gas porosity in aluminum alloy castings: Theory and experiment, J. Appl. Phys. 59:336 (1986).
2. A. G. Evans, B. R. Tittmann, L. Ahlberg, B. T. Khuri-Yakub, and G. S. Kino, Ultrasonic attenuation in ceramics, J. Appl. Phys. 49:2669 (1978).
3. D. O. Thompson, S. J. Wormley, J. H. Rose, and R. B. Thompson, Elastic wave scattering from multiple voids (porosity), in: "Review of Progress in Quantitative Nondestructive Evaluation," Vol. 2A, D. O. Thompson and D. E. Chimenti, eds., Plenum, New York (1983).
4. J. E. Gubernatis and E. Domany, Effects of microstructure on the speed of elastic waves: Formal theory and simple approximations, in: "Review of Progress in Quantitative Nondestructive Evaluation," Vol. 2A, D. O. Thompson and D. E. Chimenti, eds., Plenum, New York (1983).
5. V. K. Varadan, V. V. Varadan, and L. Adler, Nondestructive evaluation of inhomogeneities in welds, in: "Nondestructive Methods in Materials Evaluation," C. O. Ruud and R. E. Green, eds., Plenum, New York (1984).
6. V. K. Varadan, Y. Ma, and V. V. Varadan, Multiple scattering theory for elastic wave propagation in discrete random media, J. Acoust. Soc. Am. 77:375 (1985).
7. J. H. Rose, Ultrasonic characterization of porosity: Theory, in: "Review of Progress in Quantitative Nondestructive Evaluation", Vol. 4B, D. O. Thompson and D. E. Chimenti, eds., Plenum, New York (1985).
8. S. W. Wang, A. Csakany, L. Adler, and C. Mobley, Ultrasonic determination of porosity in cast aluminum, in: "Review of Progress in Quantitative Nondestructive Evalaution," Vol. 4B, D. O. Thompson and D. E. Chimenti, eds., Plenum, New York (1985).

This work was sponsored by the Center for Advanced Nondestructive Evaluation, operated by the Ames Laboratory, USDOE, for the Air Force Wright Aeronautical Laboratories/Materials Laboratory under Contract No. W-7405-ENG-82 with Iowa State University.

INDICATIONS OF MATERIAL CHARACTER FROM THE BEHAVIOR OF DIFFUSE ULTRASONIC FIELDS

R. L. Weaver

Department of Theoretical and Applied Mechanics
University of Illinois at Urbana-Champaign
Urbana, IL 61801

ABSTRACT

While concert and lecture halls have long been studied for their reverberant sound field behaviors, similar studies have begun only recently in solid bodies. Modern instrumentation is allowing the study of incoherent spectral ultrasonic energy densities in solids as they evolve in time and space. In principle this evolution can provide many indications of volume averaged material properties. These include measures of absorption (equivalent to viscoelastic moduli or other sources of internal friction), of mean free path (equivalent to attenuation due to scattering from material inhomogeneities) and of spectral densities of normal modes. It is particularly worth noting that absorption can be measured at low levels and high accuracies far beyond the sensitivities and capabilities of conventional techniques. Furthermore, attenuation as quantified by mean free paths is measurable in a high attenuation domain not otherwise accessible. The concepts, problems and potential of the technique are reviewed and a few laboratory examples presented.

INTRODUCTION

At times long after an ultrasonic source has acted in a finite solid body, long that is compared to an acoustic transit time access the body or between reflections, the ultrasonic field has lost most of its phase coherence. In all but the very simplest of geometries the field after many reflections is essentially incalculable by the theorist. These late time fields will give rise to received signals which appear incoherent, and empty of information. In a recent series of theoretical and experimental papers [1-7] however, it has been shown that the incoherent spectral energy density E(f), which has dimensions of energy per volume per frequency interval, carries information about the source and about the medium. Much of this information is not recoverable from conventional ultrasonic signals.

Ultrasonic diffuse field analysis considers concepts akin to those of statistical mechanics, of architectural acoustics, and of statistical energy analysis (SEA) [8]. The applications and parameter ranges are often very different, though, and ultrasonic diffuse wave analysis is an independent field of study.

SOURCE CHARACTERIZATION

In an early theoretical paper [3] it was suggested that source character (orientation, position, duration) can be recovered by measurement of spectral energy density. This follows from the mechanical impedance $Z(\omega)$ presented to the source being dependent on these parameters. A later study [4] substantiated the suggestion by showing detailed experimental agreement with the predictions of reference [3]. See Figure 1. The study, however, also underlined many of the practical difficulties in measuring diffuse fields. The most significant was due to the inherently stochastic, noisy, nature of the signal: High accuracies in a measured spectral energy density E(f) requires either large bandwidth Δf (and hence poor frequency resolution) or long duration Δt signals (and hence poor time resolution) and the complexities of correcting for ultrasonic absorption and signal decay. Predicted variation in E(f) due to source orientation and position are small. The accuracies necessary to unambiguously resolve these variations were judged impractical outside a laboratory [4].

The strong E(f) variations due to source duration and strength are, however, readily recoverable without prohibitive constraints on bandwidth and signal duration [4]. Work is continuing on the practical application of diffuse field analysis to acoustic emission source characterization.

MATERIAL CHARACTERIZATION BY ULTRASONIC ABSORPTION

Internal Friction (IF) has long been known to characterize materials, especially plastic damage and chemical state. At high frequencies IF is generally measured by means of ultrasonic attenuation. In heterogeneous materials like polycrystals and composites, however, attenuation contains contributions from incoherent scattering from internal surfaces as well as from the IF. The time decay of a diffuse energy field, though, as opposed to the spatial decay of an ultrasonic beam, is not affected by scattering. Thus diffuse field decay rates are true measures of IF. As probes of IF diffuse field decay rates are also, unlike methods of ultrasonic attenuation, insensitive to transducer contact efficiencies, to geometric beam spreading, and to specimen geometry. Inasmuch as energy losses into specimen supports and transducers can be minimized or quantified, non relative absolute measures of absorption can be made. The very long path lengths of typical diffuse field rays (limited essentially by the absorption itself) can allow IF measurements at low absorption levels difficult to attain with conventional ultrasonic attenuation techniques.

The use of diffuse field decay rates as measures of material damage in machined 6061 aluminum is illustrated in reference [5]. Figure 2 shows a typical ultrasonic signal as created and received with broadband piezoelectric transducers. Note the apparent incoherence, note the decay and note the time scale of many milliseconds - long compared to the typical acoustic transit times of order 50 microseconds.

These signals are Fourier analyzed and the evolving power spectral density studied as a function of time. The usual caveats regarding nonstationary processes do apply, but the decay rate is slow, of order inverse milliseconds, compared to the frequency resolutions employed (many kHz) and the process can be treated as piecewise stationary. The power spectral density, at fixed frequency, is generally found to decay exponentially in time like $\exp(-2\sigma(f)t)$ (see Figure 3). The decay rate σ is source independent and taken to indicate material character. Figure 4 suggests that the decay rate is sensitive to material damage.

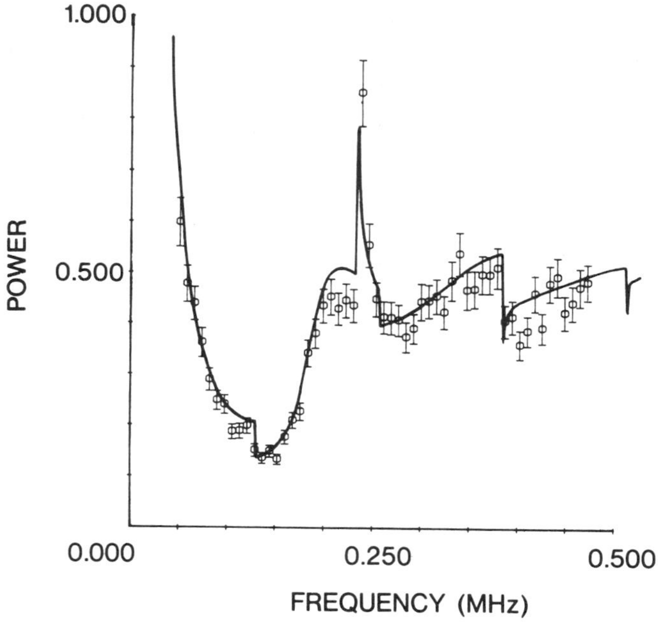

Figure 1 Comparison of the predicted (bold line) and the experimentally determined values for the power spectral density of the surface response of a plate to a step normal point force. The sharp frequency dependencies are characteristic of the position and orientation of the source. [Reference 4]

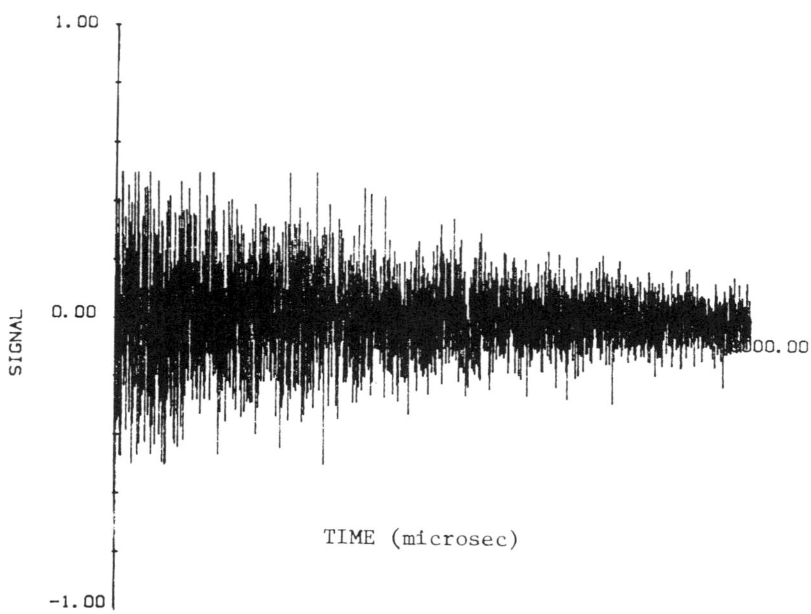

Figure 2 A typical diffuse field signal containing frequencies up to 1 MHz, observed over 8 msec of its evolution. Note the apparently random behavior under the decaying envelope. [Reference 5]

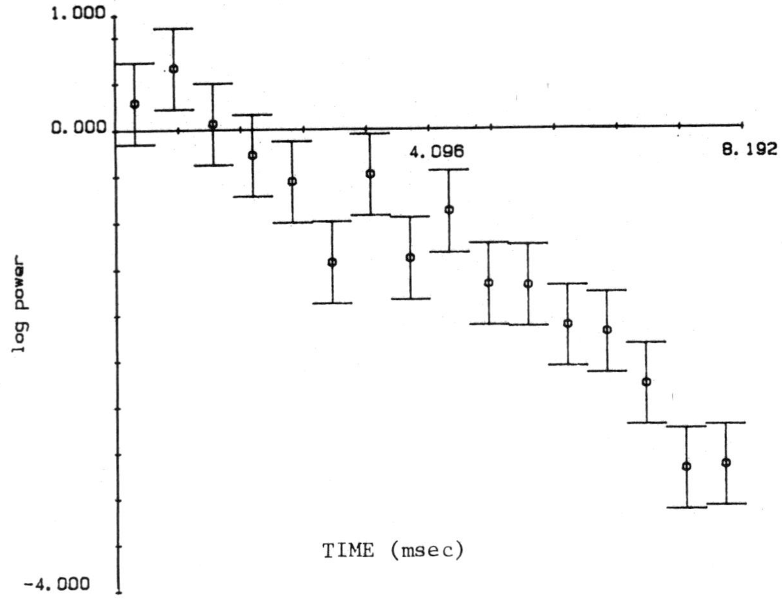

Figure 3 The power spectral density in a narrow frequency band in a signal like that of figure 2 is studied as a function of time. In this example the behavior fits well to an exponential decaying at 0.25 nepers/msec. [Reference 5]

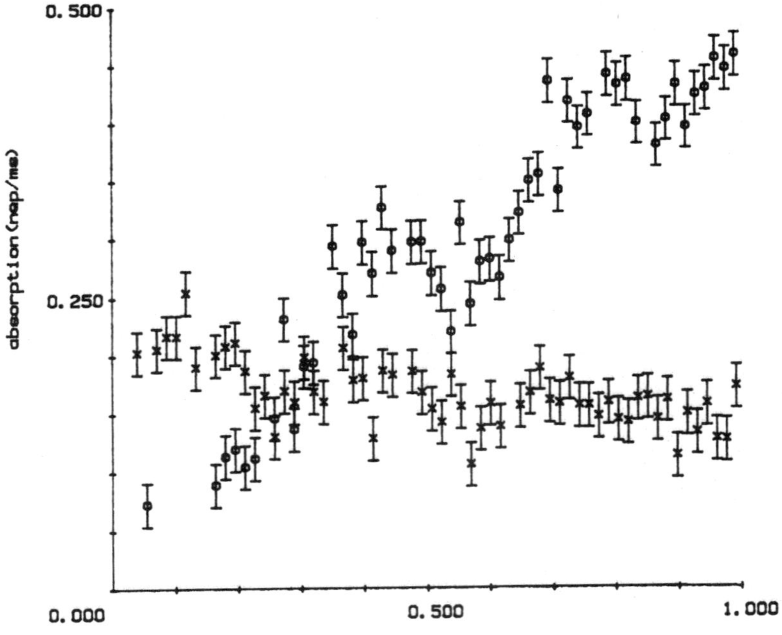

Figure 4 The recovered decay rates in two different samples of 6061-T6 aluminum as received (x's) and after machining (o's). Note that the machining has introduced substantial internal friction. Polishing restores the specimen to a nearly as-received state. Heat treatments had no effect. [Reference 5]

Note that the decay rates are assessed in this example with accuracies of order ± 0.01 nepers/msec, equivalent to accuracies of ± 0.0003 db/cm when translated into conventional attenuation units. This is hundreds of times more accurate than usual quotes for attenuation. The high accuracy is largely due to the long path lengths.

Much work remains to be done along these lines. There is a wealth of applications to be carried out in exploring the relationship between damage and absorption and in exploring how that relationship differs in different materials. Heretofore high frequency IF studies were not possible in heterogeneous materials. There is also work to be done in exploring deviations from simple exponential decay. Such deviations could be caused by at least two phenomena. Nonlinear mechanisms could so act, especially inasmuch as they might transport energy between frequencies. Even in a strictly linear viscoelastic system, however, spectral energy density can decay nonexponentially. If the modes within a narrow frequency band have widely varying individual decay rates, a stochastic superposition of the modes will decay at a rate which itself appears to decrease in time.

MATERIAL CHARACTERIZATION BY INCOHERENT TRANSPORT

In extremely heterogeneous media, in which ultrasound suffers from strong and densely packed scatterings [e.g., from grain boundaries, cracks, voids and inclusions depending on the wavelengths] acoustic energy will not be coherently transported across a body. When attenuation is too high, no direct coherent signal is received. However, one might expect incoherent transport of energy. As a first approximation one would imagine a random walk process with a mean free path of the order of the inverse of the beam attenuation due to scattering. One would then expect the spectral energy density E(f) to be governed by a heat equation as it evolves in time and space. The diffusion constant associated with this heat equation will be a measure of mean free path and thus in a sense, of scattering loss attenuation and scatterer density.

This idea was studied in a small aluminum plate (15x15x0.6 cm) cut with about 100 randomly placed deep scalloped slits of 2 cm lengths. Broadband ultrasound pulses were applied in one corner of the plate and received at two places: adjacent to the source, and in the opposite corner. The resulting time histories of the energy in a band of width 100 kHz centered on 650 kHz are plotted in Figure 5. The evolution fits well to the predictions of the heat equation model, as modified by the internal friction. The mean free path as determined from Figure 5 compares well with the centimeter interslit spacings.

When, however, the same program is carried out in a plate with additional slits and in a band centered on 250 kHz, Figure 6 is obtained. The figures are similar except that in the latter figure the transport process appears to conclude before equipartition is achieved. At late times the region adjacent to the source has 150 times the energy density of the region opposite the source; there appears to be no further transport down this energy density gradient. Behavior of this sort is predicted by the concept of normal mode localization [9]. Whether that phenomenon is in fact responsible for figure 6 is unclear. Work is ongoing.

Transport could also be discussed across a joint between acoustic subsystems, the conductance of the joint characterizing the joint in analogy to SEA's [8] "coupling loss factors".

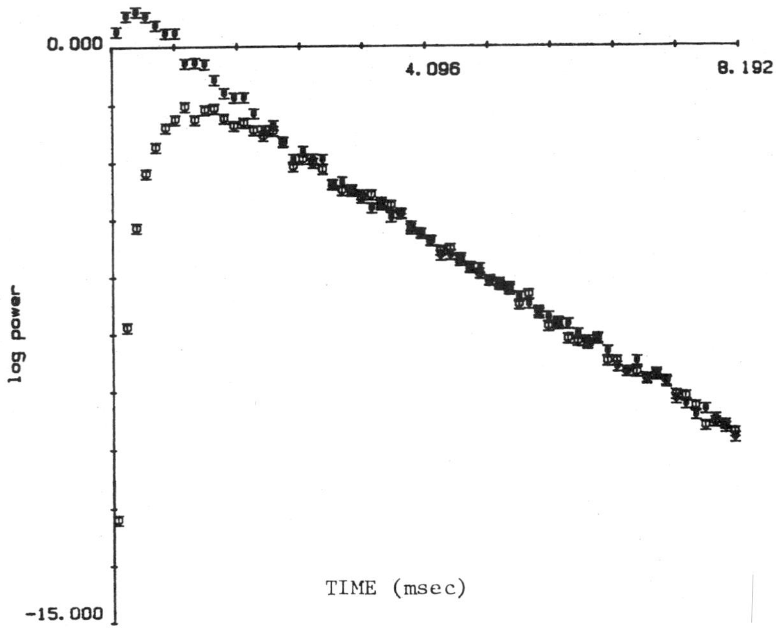

Figure 5 The power spectral densities are plotted as a function of time for the signals received adjacent (upper curve) to the impulsive source and opposite (lower curve) to the source in a highly irregular plate. Note that each curve decays exponentially due to the internal friction. Note also that the region opposite the source sees a diffusive-like arrival of the energy, on a time scale of order milliseconds, long compared to a direct acoustic ray travel time of 75 microseconds.

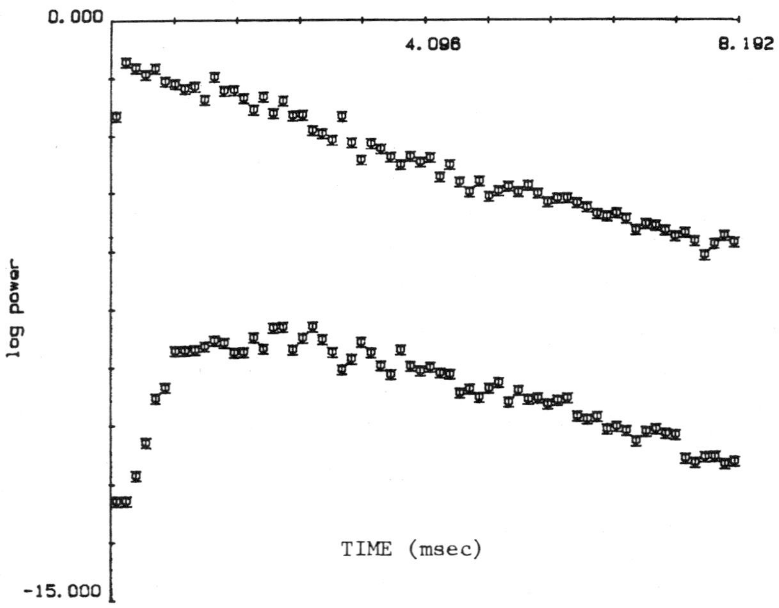

Figure 6 Like Figure 5, but in a different frequency range and in a more highly irregular plate. At late times the adjacent energy density is 150 times stronger than the opposite energy density. This may be due to Anderson localization of the normal modes.

MATERIAL CHARACTERIZATION FROM MODE DENSITY MEASUREMENTS

Diffuse field studies also lend themselves to measures of spectral mode density. This is done by means of the central assumption, like that of SEA, that within a narrow frequency band, and statistically speaking, every mode of vibration shares equally in the available acoustic energy [2,3]. Thus any subsystem of a structure should carry an energy in proportion to its local density of modes. The assumption was crucial to the arguments of reference [3], corroborated in reference [4]. The assumption makes a specific prediction regarding the response of a structure composed of equal volumes of steel and brass: that the brass should hold about 80% of the acoustic energy. This prediction has been substantiated also.

It is suggested here that spectral mode density measurements could be used for materials characterization. In homogeneous media mode densities are approximately proportional to the inverse third power of the shear wave velocity. The concept of mode density, though, has a robust validity even in a heterogeneous medium for which a dynamic modulus is a problematic concept. Such measurements could thus provide a probe of a high frequency effective shear modulus in a regime which is inaccessible to conventional ultrasonics.

Much work, in application, development, and in fundamentals remains to be done. Nevertheless, diffuse field analysis promises to provide nondestructive evaluations not available by means of conventional ultrasonics.

ACKNOWLEDGEMENT

This work was supported by the National Science Foundation Solid Mechanics Program, grant number MSM-8412178.

REFERENCES

1. D. M. Egle, Diffuse Waves in Solid Media, J. Acoust. Soc. Am. 70, 476-480 (1981).
2. R. L. Weaver, On Diffuse Waves in Solid Media, J. Acoust. Soc. Am. 71, 1608-1609 (1982).
3. R. L. Weaver, Diffuse Wave in Finite Plates, J. Sound Vib. 94, 319-335 (1984).
4. R. L. Weaver, Laboratory Studies of Diffuse Waves in Plates, J. Acoust. Soc. Am. 79, 919-923 (1986).
5. R. L. Weaver, Diffuse Field Decay Rates for Material Characterization, in "Solid Mechanics Research for QNDE," from the Proceedings of the ONR Symposium, Evanston, IL, September 1985 (Nijhof, Dordrecht, The Netherlands) (in press).
6. R. L. Weaver, Diffuse Elastic Waves at a Free Surface, J. Acoust. Soc. Am. 78, 131-136 (1985).
7. R. L. Weaver, On the Time and Geometry Independence of Elastodynamic Spectral Energy Density, J. Acoust. Soc. Am. (in press).
8. R. H. Lyon, "Statistical Energy Analysis of Dynamical Systems," MIT Press, Cambridge, MA (1975).
9. E. Abrahamson, P. W. Anderson, D. C. Licciardello, T. D. Ramakrishnan, Scaling Theory of Localization: Absence of Quantum Diffusion in Two Dimensions, Phys. Rev. Lett. 42, 673 (1979).

COUPLING FILM THICKNESS EFFECT ON ABSOLUTE ULTRASONIC VELOCITY MEASUREMENT IN SOLIDS

Alain Vincent

Groupe d'Etude de Métallurgie Physique et de Physique des Matériaux (UA CNRS 341) et Laboratoire de Traitement du Signal et Ultrasons (UA CNRS 1216)
INSA, Bt 502, 69621 Villeurbanne cedex, France

INTRODUCTION

It is now well known that an accurate knowledge of the ultrasonic velocity in solids is very useful. Indeed, in addition to yielding the elastic moduli of solids, it has been observed that this technique could well be used to study the various physical phenomena which profoundly influence the modulus of elasticity.

Thus, during these last thirty years, many methods of measurement of the ultrasonic velocity have been described : most of them have been reviewed by PAPADAKIS [1]. The methods could be broadly classified to fall under two categories : (i) those that enable one to obtain an absolute measure of the ultrasonic velocity and (ii) those developed in order to follow small velocity changes induced by external parameters (e.g. static stress, magnetic field etc.. see for example the review by ALERS [2])

In the case of absolute measurements, the importance of transducer and bond phase shifts (as principal sources of error in velocity measurements) was widely recognized by some pioneers of these techniques like WILLIAMS and LAMB [3] and Mc SKIMIN et al [4]. Indeed, in the case of multiple echo methods, a spurious phase shift occurs between successive echos, when they are reflected for an additional round trip, at the specimen - film - transducer interface. Thus, in the past, the effect of the bond phase shift at reflection has been widely investigated in the case of resonant piezoelectric transducers [1,3 -5]. On the contrary, to our knowledge, nothing has been done in the case of highly damped piezoelectric transducers which are now commercially available. The aim of this work is to contribute to the understanding of spurious bond phenomena occuring during ultrasonic velocity measurements carried out using this kind of transducers.

EXPERIMENTAL INFLUENCE OF THE COUPLING FILM THICKNESS

Principle of the method

In this work, the apparatus "echometre" developed by ODRU and al [6] on the basis of pulse superposition method proposed earlier by Mc SKIMIN and al [4] has been used. A single transducer acts alternatively as emitter and as receiver. The first pulsed wave train emitted leads to successive echos E_1, E_2... due to multiple reflections of the ultrasonic

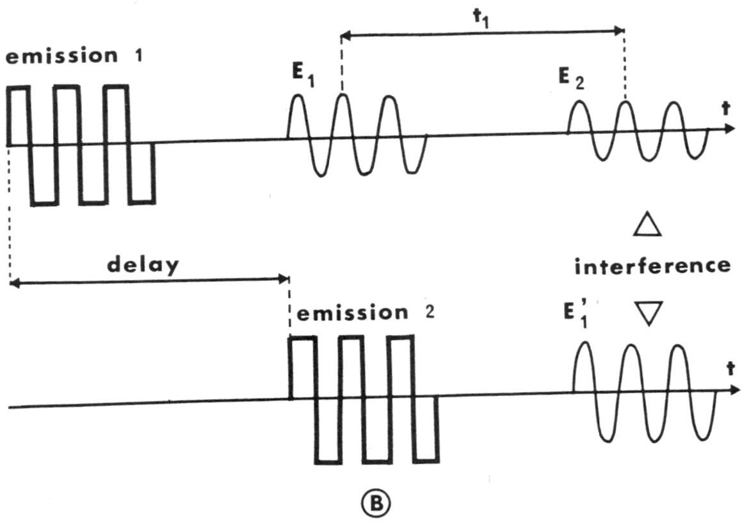

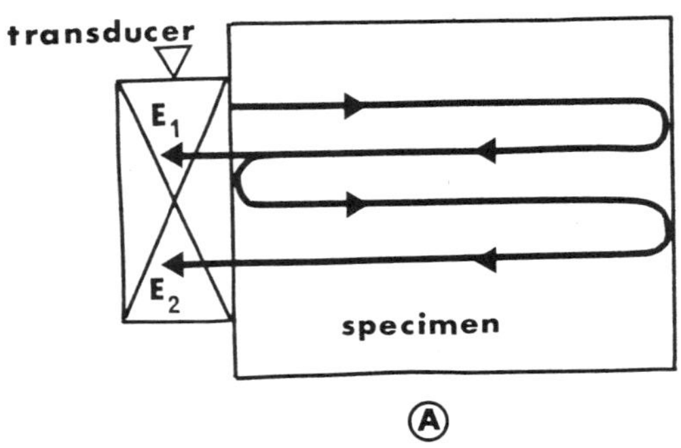

Figure 1 : A) Echo paths in the sample

B) Timing sequence for the double pulse method.

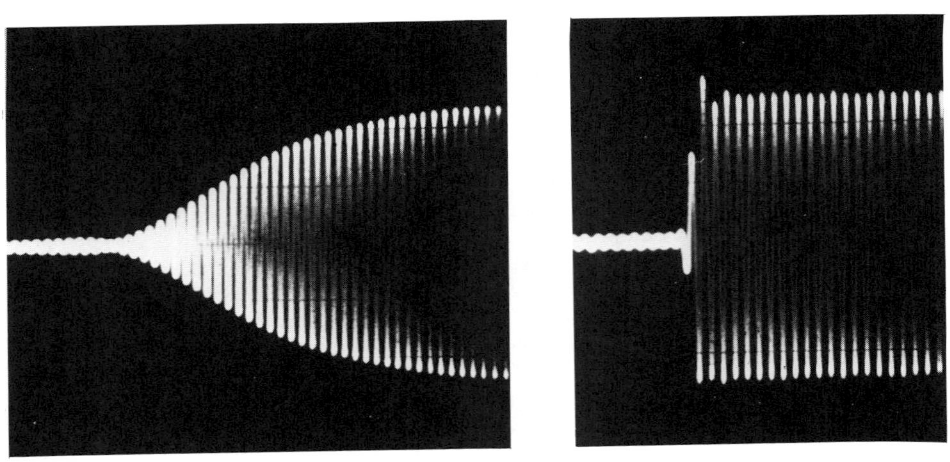

A B

Figure 2 : Oscillograph displays of the rising part of an echo delivered A) by an unbacked quartz transducer (10 Mhz) ; B) by a highly damped transducer (PANAMETRICS - VIDEOSCAN - 10 Mhz).

wave between the parallel sides of the specimen (Fig.1A). A second twin pulsed wave train delayed from the initial one is next applied to the transducer ; the delay can be adjusted so that the first echo of the second train E'_1 coincides and interferes with the second echo E_2 of the first train (Fig.1B) : when the coïncidence is correctly adjusted, the delay time equals the apparent round trip echo travel time t_1 in the specimen. The overall accuracy of the method is mainly determined by : (i) the accuracy of the adjustment which is improved by increasing the ultrasonic frequency, (ii) the possible mismatch error of n cycles which can be made if the coïncidence between E_2 and E'_1 is not chosen rightly, (iii) the phase lag, γ, of echo E_2 which occurs at the reflection on the side of the specimen coupled to the transducer. Thus, from (ii) and (iii), the apparent delay time t_1 is related to the true round trip travel time t_0 by :

$$t_1 = t_0 - 2\pi n/\omega - \gamma/\omega + \pi/\omega \qquad (1)$$

where $\omega = 2\pi f$ (f = frequency of the wave train) and π/ω is the phase inversion occuring at the reflection of E_2 on the rear side of the specimen.

As recalled in introduction , in the past, unbacked quartz transducers were often used : as shown in Fig.2A, their resonant character leads to a very slowly increasing amplitude of the oscillation in the echo. In this case, matching cycle to cycle the echos E'_1 and E_2 is rather uncertain so that frequently n = 1 or 2. Thus specific procedures implying measurements at different frequencies has been proposed by Mc SKIMIN [4] and JACKSON [5]. Now, with highly damped transducers that are commercially available n can be minimized (i.e. n = 0) in a more direct way : indeed, as shown on Fig.2B, they lead to a very well defined rising front of the echos so that matching them cycle to cycle could be reliably done. The matching procedure itself is detailed elsewhere [7]. Moreover, it must be emphasized that in any case the first cycles are only used in order to minimize n as described above : the accurate adjustment of the phase coïncidence or the opposite phase condition is performed on the top, flat, part of the wave train so that the continuous wave analysis may be applied with good approximation in section III.

Experimental procedure and results

The influence of thickness of a coupling layer of Nonaq stopcock grease was studied in the following conditions : the transducer is of the Videoscan Contact series of Panametrics with a nominal frequency of 2.25 Mhz and a diameter of 13 mm. The specimens have smooth plane parallel faces with a thickness of about 2 cm and a diameter much larger than that of the transducer ; one is made of steel (35CD4) and the other of duraluminum (AU4G).

The variation of the coupling film thickness was achieved using the device schematized in Fig. 3 which enables one to maintain a perfect parallelism of the layer.

The changes Δt_1, in the apparent delay time t_1, versus coupling thickness e, obtained using this micrometer, are shown in Fig.4 : the thickness has been varied in the range 0 - 10^{-4} m within which most practical situations would fall. In fact, the residual ruggedness and the unevenness in the flatness of the transducer as well as specimen faces lead to a minimum value of e which is presumably of order of few microns. In the case of the duraluminum specimen (curve a), an increase of 260 ns is observed over the whole range of thickness investigated. A larger increase (370 ns) is obtained in the case of the steel sample ; moreover any accurate measurement becomes impossible between the contact condition and e = 1.5 10^{-5}m since in this domain the amplitude of echo E_2 is strongly reduced in comparison with that of E'_1.

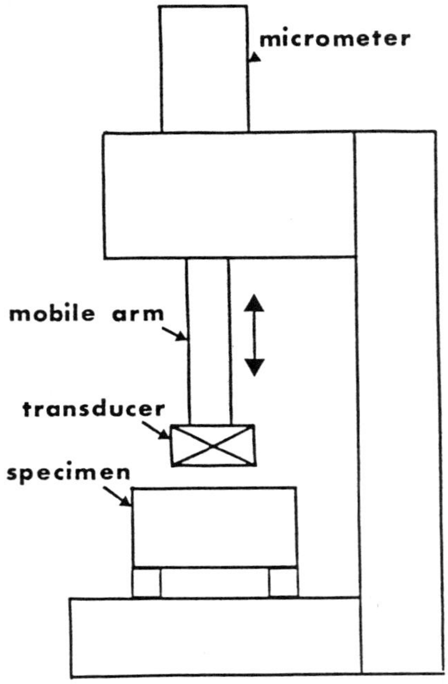

Figure 3 : Experimental device for varying coupling thickness.

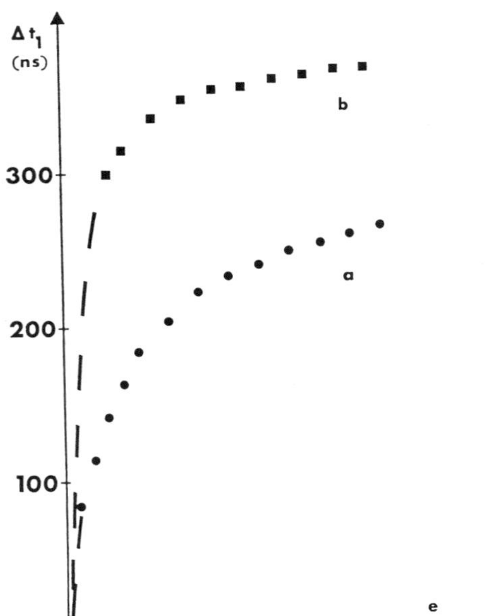

Figure 4 : Changes (Δt_1) in the apparent round trip travel echo time as function of coupling thickness e ; frequency 2.15 Mhz ; plot a : duraluminum specimen - plot b : steel specimen.

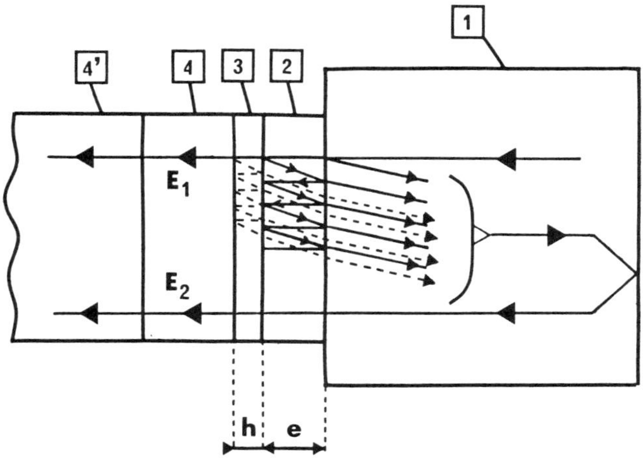

Fig. 5 : Schematic representation of the multilayer model used for calculation of phase shift between echos E_1 and E_2 : 1 specimen ; 2 coupling layer ; 3 transducer wearplate ; 4 piezoelectric plate ; 4' transducer backing.

THEORY AND DISCUSSION

Theory of the reflection on the specimen-transducer interface

In order to determine the phase shift, γ, occuring due to the reflection of the ultrasonic wave at the specimen - transducer interface, the behaviour of the multilayered model shown on Fig.5 has been investigated : layer 2 (thickness e) is the coupling film introduced between the specimen (1) and the transducer ; layer 3 (thickness h) is the wearplate of the transducer ; layer 4 is the piezoelectric plate which is assumed to be perfectly coupled to a backing medium (4') having the same acoustic impedance. In our treatment, the ultrasonic waves have been assumed to be plane and the diffraction phenomenon occuring during multiple reflections inside layers 2 and 3 have been ignored. We have also neglected, the attenuation in all the media.
According to the remark of previous section, the continuous wave analysis can be applied with good approximation. This was done by inserting in a result of SCOTT and GORDON [8], concerning a single layer, the complex impedance of the transducer taking into account its wearplate and deduced from transmission line theory.

Thus, the phase shift γ has been calculated as function of the coupling thickness e, up to $\lambda_2/8$ (λ_2 is the wavelength in the coupling layer) , for various wearplate thicknesses chosen in the range $0 < h < \lambda_3/4$ (λ_3 is the wavelength in the wearplate). The calculation has been restricted to these ranges because most practical configurations will fall within these. The values of the characteristic impedances used in the numerical calculations are : Steel : 45.6, Duraluminum : 17.6, Nonaq : 2.3, Wearplate : 43, Piezoelectric plate : 21 (in 10^6 kg.m^{-2}.s^{-1} ; values for wearplate and piezoelectric medium are those communicated by the manufacturer).

The results are shown in Fig.6 for a duraluminum specimen and in Fig.7 for a steel specimen. The upper part of each figure shows the phase shift γ which is plotted as function of the normalized thickness (e/λ_2), whereas the lower part gives the modulus of the reflection coefficient ($|r|$) also versus (e/λ_2).

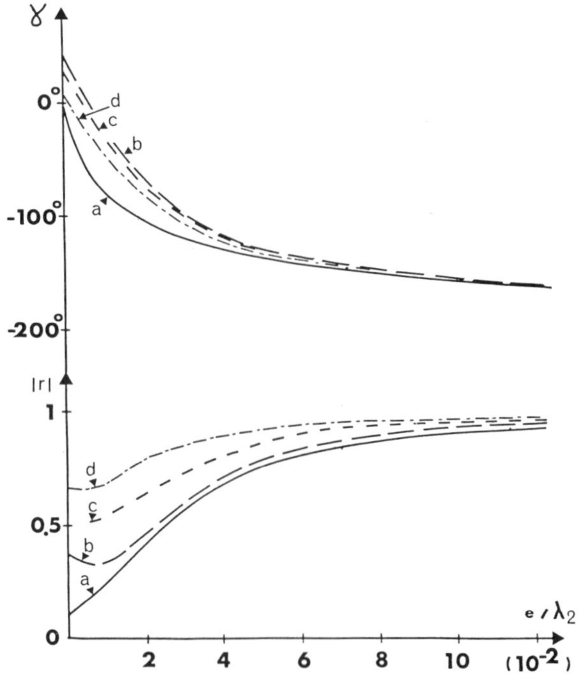

Figure 6 : Phase shift γ and modulus $|r|$ of reflection factor as function of normalized coupling thickness e/λ_2 ; for various wearplate thicknesses h in the case of a duraluminum specimen . Curve a : h = 0 ; b = h $=\lambda_3/13.5$; c : h $=\lambda_3/8$; d : h $=\lambda_3/4.6$.

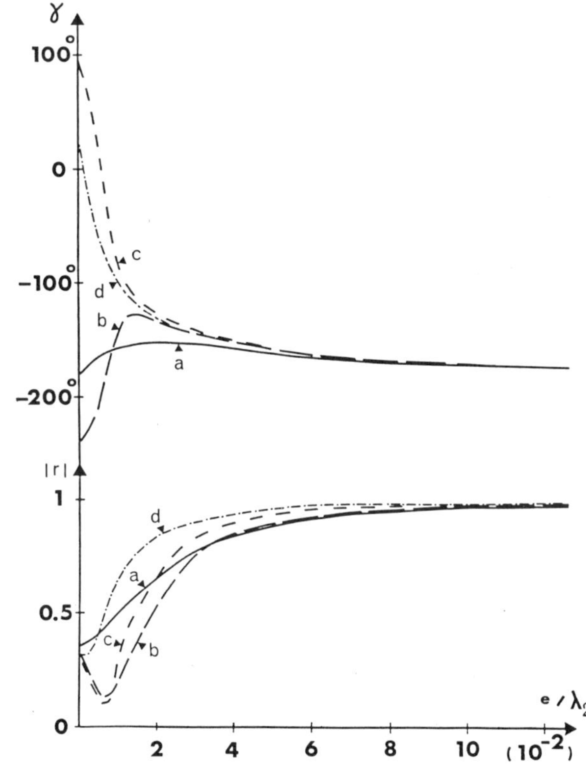

Figure 7 : Same as Fig.6 but for a steel specimen.

The range $0 < e < \lambda_2 /50$ appears to be the most sensitive one to the coupling thickness evolution ; this range is also the one that is mostly influenced by a change in the wearplate thickness.

For a duraluminum specimen, the phase shift γ decreases monotonously when the coupling thickness increases, for all the wearplate thicknesses considered ; the value of γ at $e = 0$ is zero, when no wearplate is introduced in the model (curve a), which is in agreement with what occurs at the reflection of a wave propagating from a medium characterized by a lower acoustic impedance than that of the medium beyond the interface.

For a steel sample (Fig.8), the behaviour of γ is more complex : without wearplate, γ (e=0) takes the value -180°, which is in agreement with a reflection of a wave from a medium characterized by a higher acoustic impedance than that of the medium beyond the interface ; then γ goes through a maximum and finally tends again towards the value -180°. The influence of the wearplate has to be pointed out because at a thickness such that $h > \lambda_3/10$ (curves c and d), the maximum of γ disappears leading to a variation of γ quite similar to that observed with the duraluminum.

A few comments could also be made from the results regarding the modulus of the reflection coefficient : without wearplate, $|r|$ increases monotonously up to its maximum value which should be reached at $e = \lambda_2/4$. The introduction of the wearplate does not change completely this kind of behaviour : nevertheless, in the case of a duraluminum sample, it has to be noticed that the starting value γ (e = 0) is apparently raised due to the wearplate ; in the case of a steel sample, the wearplate leads to a deep minimum in the plot $|r|$ versus e, which is the most marked for $h \simeq \lambda_3/10$.

Application to experimental results

The thickness e of the transducer wearplate used in our experimental work is unknown (not given by the manufacturer). Thus, we searched for the value of h leading to the best fit between the theory described above and the experimental results of Fig.4 : this was achieved by dropping $\gamma(e)$ in relation (1) (with $n = 0$) and using a least square method in order to determine the suitable value of h. This treatment was carried out for the two sets of experimental data of Fig.5 leading in both cases to the common value $h = 0.12\ \lambda_3$. This best accordance between experimental data and theory is shown in Fig.8 and Fig.9, respectively for duraluminum and steel.

This treatment further allows us to determine the true round trip time t_0 which is the value corresponding to $\gamma = 0$ on the theoretical curves of Fig.8 and Fig. 9 : $t_{OD} = 7396$ ns and $t_{OS} = 6663$ ns are thus obtained respectively for duraluminum and steel. It has to be pointed out that the rough measurement of the round trip time in usual direct contact conditions would have led, without any wearplate and coupling corrections, to an error of 58 ns and 137 ns respectively for duraluminum and steel samples. The true value of the ultrasonic velocity in our specimens can be deduced from t_{OD} and t_{OS} after a final correction taking into account the diffraction phase shift occuring with finite beam diameter ; from KHIMUNIN tables [9], the corrections on t_{OD} and t_{OS} are estimated at + 30 ns and + 32 ns respectively ; finally the ultrasonic velocities at 20°C are found to be 6372 ± 6m/s for the duraluminum specimen (AU4G) and 5819 ± 6 m/s for the steel specimen (35CD4).

In addition, the knowledge of h and the theoretical behaviour of $|r|$ allow us to understand the difficulty encountered in measuring t_1 in the case of the steel specimen and with a bonding thickness between $e = 1.5\ 10^{-5}$ m and the contact situation. Indeed, from Fig.7 it can be seen that if a wearplate of about $\lambda_3/10$ is used, there is a deep minimum in the curve $|r|$ versus e, just situated at about 5.10^{-5} m : taking $|r| = 0.1$ at

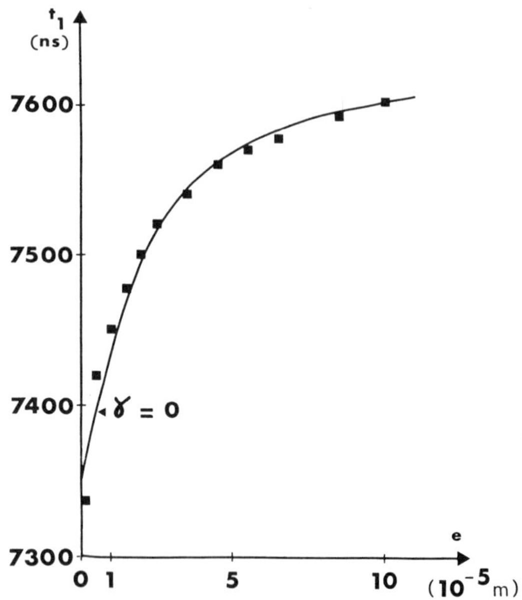

Figure 8 : Best fit between the calculated variation and the experimental data t_1 as function of coupling thickness e for a duraluminum specimen.

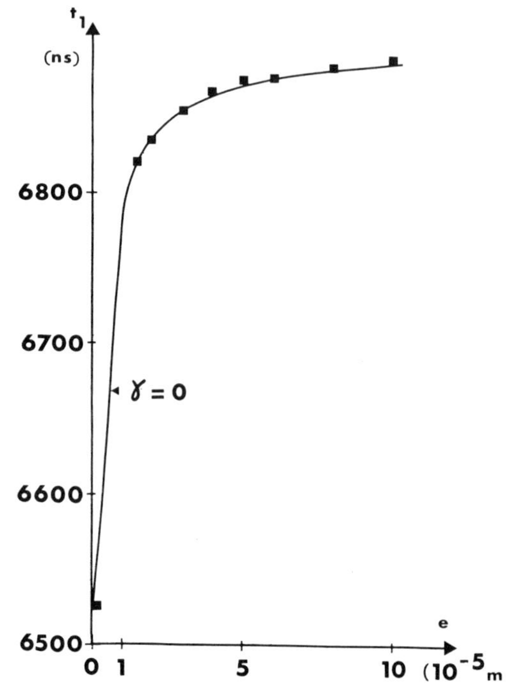

Figure 9 : Same as Fig.8 but for a steel specimen.

the minimum, this explains why the amplitude of echo E_2 is less that 1/10 of E_1, thus leading to a very bad sensitivity for adjusting the coïncidence between E'_1 and E_2.

CONCLUSION

The experimental results of the apparent round trip travel time dependence on the coupling thickness are well explained by the model taking into account the wearplate of the commercially available highly damped transducer used in our experiments. Moreover, from this study, it has to be underlined that using such a transducer : (i) on one hand is much appreciable in order to choose rightly the good coïncidence between successive echos in pulse superposition methods (or pulse overlap methods) ; (ii) on the other hand it is much critical because this apparent round trip time is most sensitive to a coupling thickness uncertainty for the thickness range in which most practical situations would fall.

Finally, it has to be noticed that the critical range of coupling thickness will be shifted towards smaller and smaller values as the frequency of measurements is increased. For example at 20 Mhz this critical range limited by $\lambda_2/50$ will be shortened to $e < 2\ 10^{-6}$ m ; then, with usual surface preparation, e will no longer fall within this critical range. However, the correction for phase shift (γ) will have to be necessarily applied.

ACKNOWLEDGEMENTS

Discussions with Prof. PERDRIX and coworkers about highly damped transducers are gratefully acknowledged.

REFERENCES

1. PAPADAKIS F.P., "The measurement of small changes in ultrasonic velocity and attenuation", Critical Reviews in Solid State Sciences C.R.C., (1973), 373-418.
2. ALERS G.A., "The measurement of very small sound velocity changes and their use in the study of solids", Physical Acoustics, (Ed. W.P. Mason), vol.4, Part A, Academic Press, New York (1966), 277-297.
3. WILLIAMS J. and LAMB J., "On the measurement of ultrasonic velocity in solids", J. Acoust. Soc. Am., 30, 4, (1958), 308-313.
4. Mc SKIMIN H.J. and ANDREATCH P., "Analysis of the pulse superposition method for measuring ultrasonic wave velocities as a function of temperature and pressure", J. Acoust. Soc. Am. 34, (1962), 609.
5. JACKSON I., NIESLER H. and WEIDNER D.J., "Explicit correction of ultrasonically determined elastic wave velocities for transducer-bond phase shifts", J. of Geo. Res. 86, 85 (1981), 3736-3748.
6. ODRU R., RIOU C., VACHER J., DETERRE P., PEGUIN P. and VANONI F., "New instrument for continuous and simultaneous recording of changes in ultrasonic attenuation and velocity", Rev. Sci. Instrum. 49 (2) (1978), 238-241.
7. VINCENT A., "Influence of coupling thickness on ultrasonic velocity measurement", to be published.
8. SCOTT W.R. and GORDON P.F., "Ultrasonic spectrum analysis for non-destructive testing of layered composite media", J. Acoust. Soc. Am., 62, 1 (1977), 108-116.
9. KHIMUNIN A.S., "Numerical calculation of the diffraction corrections for the precise measurement of ultrasound phase velocity", ACUSTICA 32 (1975), 192-200.

POINT-SOURCE/POINT-RECEIVER MATERIALS TESTING

Wolfgang Sachse and Kwang Yul Kim

Department of Theoretical and Applied Mechanics
Cornell University, Ithaca, New York - 14853

INTRODUCTION

Conventional measurements in the ultrasonic testing of materials, when used as the basis of a materials characterization procedure, typically rely on one or two piezoelectric transducers operating as source and receiver, attached to a specimen to launch and detect ultrasonic waves in the object to be characterized. Measurements of signal arrival time (or velocity) and amplitude (or attenuation), possibly as a function of frequency, are then correlated with the composition and the macro- and micro-structure of the material, which may include voids, flaws and inclusions distributed through a region of the material. While relative measurements of the time-of- flight and ultrasonic amplitudes do not present extraordinary measurement challenges, absolute measurements do. It is unfortunate that absolute quantities are often required since they are difficult to obtain reliably with a conventional piezoelectric transducer-based ultrasonic system. For this reason, a considerable effort over the past decade has been undertaken to develop and improve non-contact methods for generating and detecting ultrasonic signals in materials. However, a limiting factor of all the existing non-contact measurement systems is the care required for their use and their reduced sensitivity in comparison to those utilizing piezoelectric transducers.

Over the past several years notable advances have been made in the development of quantitative acoustic emission methods in which the signals emitted by a source in or on a structure are detected by one or more receiving transducers at the surface of the structure. Knowing the transfer characteristics of the sensors and the appropriate dynamic Green's functions of the structure, the temporal and spatial characteristics of the emission source can be recovered from the detected signals by signal processing techniques [1, 2]. An essential assumption in such a measurement procedure is that the material encompassing the source and receiver points is homogeneous, isotropic, elastic and non-attenuative. While the Green's functions for an anisotropic material can, in principle, be computed, none appear to have been published. It is known that the propagation of acoustic emission signals from a source to a receiver point is influenced not only by the geometry of the specimen, but also by the material's macrostructure specified in terms of its size, shape and for composite materials, the ply configuration. In addition, the wave propagation is strongly affected by the specimen material's microstructure, including its anisotropy, heterogeneity, elastic, inelastic and viscous properties, and its wave attenuation characteristics. To determine these variables it is proposed to utilize an ultrasonic testing system analogous to that used in exploration geophysics, consisting of a well-characterized point source and point receiver. This is shown schematically in Figure 1. If both the source and receiver possess known temporal characteristics, then the principal advantage of a point-source, point-receiver system over a conventional ultrasonic system is that quantitative ultrasonic measurements are possible. Although the geometric characteristics of the wave propagation may be more complex than those for the case in which plane waves are used, these effects can be accounted for in the detected signals to recover the material-related wavespeeds and attenuation properties. Because the excitation and detection regions are small, the required specimen surface preparation is minimal and furthermore, specimens of arbitrary geometry can be easily tested, in particular those which are neither planar nor flat.

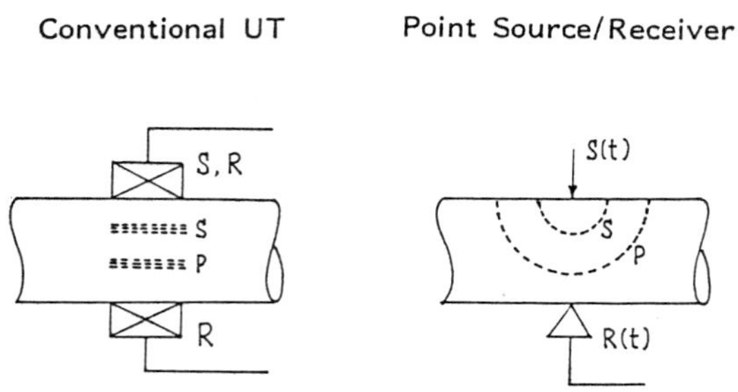

Figure 1 - Conventional ultrasonic and point-source/point-receiver testing configurations.

WAVE PROPAGATION, SYSTEM CHARACTERISTICS AND SIGNAL ANALYSIS

The basis of a point-source/point-receiver testing system is a source and a receiver whose temporal and spatial transfer characteristics are known and a theoretical basis by which the measured signals can be identified, interpreted and processed to recover the characterisitcs of the propagating medium. The term "point" refers ideally to a signal source or detection region whose lateral dimension is much smaller than the effective wavelength of the highest frequency component of interest in the measured signal. This wavelength will also be much shorter than any dominant dimension of the specimen.

Wave Propagation

The analysis of transient elastic waves between a point source and a point receiver in a bounded structure is discussed in several papers [c.f. 3]. The displacement signals u_k detected at a receiver location, $\underset{\sim}{x}$, in a structure from an arbitrary source $f(\underset{\sim}{x}',t)$ located at $\underset{\sim}{x}'$ having source volume V can be written as a sum of contributions due to a monopole, dipole and higher-order terms. The monopole source contribution to the signal can be rewritten compactly as

$$u_k^{(m)}(\underset{\sim}{x},t) = F_j(t) * G_{jk}(\underset{\sim}{x}/\underset{\sim}{x}',t) \tag{1}$$

where $F_j(t) = \int_V f_j(\underset{\sim}{x}',t)\,dV$ is the total force acting throughout the source volume and $G_{jk}(\underset{\sim}{x}/\underset{\sim}{x}',t)$ is the dynamic Green's function of the structure. It is a function of time only for a point source. For all the monopolar sources utilized in this paper, the force is always a force normal to the surface of the specimen, i.e., F_z. The dipole contribution to the signal can be rewritten compactly as

$$u_k^{(d)}(\underset{\sim}{x},t) = M_{ij}(t) * G_{jk,i}(\underset{\sim}{x}/\underset{\sim}{x}',t) \tag{2}$$

where $M_{ij}(t) = \int_V \underset{\sim}{x}'_i f_j(\underset{\sim}{x}',t)\,dV$ represents the moment tensor of the source and $G_{jk,i}(\underset{\sim}{x}/\underset{\sim}{x}',t)$ is the spatial derivative of the Green's function. This representation is used to model many thermoelastic sources [4].

A solution to the forward problem can be readily computed for a specimen of flat plate-like geometry. That is, given the thickness of the specimen, the material's longitudinal to shear wavespeed ratio and the source/receiver separation, the dynamic Green's functions appearing in Eqs. (1) and (2) can be found using one of the available algorithms [3, 5]. If the measurement system includes a point receiving transducer whose output voltage signal is related to the input displacement signal by the sensor's transfer function, R (t), then the results given by Eqs. (1) and (2) must be further convolved with this transfer function to obtain the output voltage signals of the system corresponding to each of the excitations.

An important example is shown in Figure 2. This is the normal displacement signal computed for a vertical force source acting on a plate specimen and detected at the epicenter point on the back surface of the plate directly under the source point. This example was obtained for a 7090 Al/SiC composite 1.884 cm thick whose longitudinal and shear wavespeeds were 0.708 cm/μsec and 0.368 cm/μsec, respectively, with zero wave attenuation. Thus, this result represents the behavior expected for an ideal material. The signals corresponding to other types of source and receiver configurations can be computed similarly. It is seen that the arrivals of both the P- and S-waves can be easily identified. Hence, even though the force was applied normal to the specimen surface, both longitudinal and shear wave modes are excited. The identification of the wave arrivals is possible, provided that the source/receiver separation is not larger than about 10h - 15h, where h is the thickness of the plate. It is also clear from the waveforms shown in the figure that while the wave arrivals are readily identifiable, the signals characteristically possess a "tail", that is, each wave is geometrically dispersed as it propagates through the specimen. This dispersion is unrelated to the properties of the medium and its presence in a signal must be removed if the correct material-related frequency characteristics of a measured signal are to be determined.

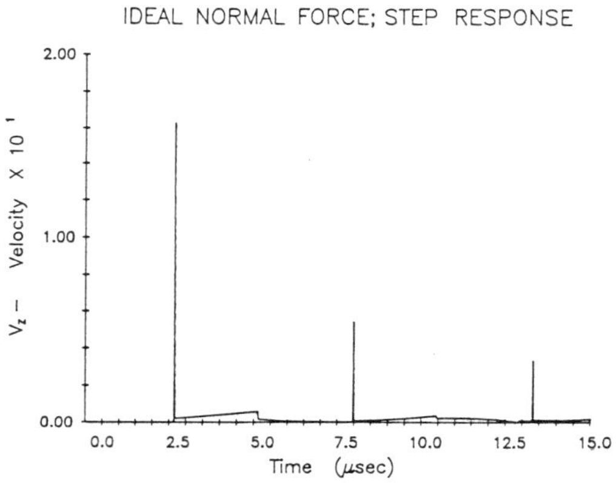

Figure 2 - Ideal epicentral velocity signal; Step normal force excitation; Zero attenuation

Additional insight is gained by considering the evaluation of the displacement signals in the frequency domain. This has recently been carried out for the case of a viscoelastic medium [6]. It is found that the Fourier phase function of the P-wave arrival of the velocity signal is given by

$$\phi(\omega) = kL + A/kL \qquad (3)$$

where k is the wavenumber, L the source/receiver separation and A is a correction factor approximately equal to 2. The first term of the phase function is identical to that derived for plane waves [7]. The amplitude corresponding to the first shear wave arrival can be processed similarly and the phase functions of other source types are expected to exhibit a similar form [8].

Source and Receiver Characteristics

The ideal point source and point receiver with perfect impulse response can only be approximated with real sources and receivers. However, in order to achieve acceptable signal-to-noise ratios, it is possible to use transducers with a finite aperture, provided that the generated and detected acoustic fields in the specimen are uniform and resemble those expected from a point source and point receiver. Equally important is that the transduction characteristics of the source and receiver are known. This includes both the primary and secondary quantities being generated and detected. The temporal transfer characteristics of both source and receiver must be known a priori or determined in a calibration experiment; they must possess an appropriate frequency response relative to the material property being investigated. There are, however, many measurement situations in materials whose viscous dispersion and wave attenuation is sufficiently low so that only a measurement of the arrival of a signal is required and a complete characterization of the sensor is not needed. In these cases, conventional piezoelectric point sensors sensitive to the wave motions normal to the surface of the specimen can be used. With such a sensor the measurement is simplified since no special surface preparation of the specimen or critical transducer alignment with the specimen surface are required. Obviously, in all cases the source and the receiver must possess an adequate signal-to-noise ratio to permit signal identification and subsequent processing operations to be performed reliably. A discussion of various point-sources and receivers and their operating characteristics is given in Ref. 9.

Signal Identification and Waveform Analysis

According to the convolution equations (Eqs. (1) and (2)) for a source of known type and time function and for a specified source-receiver separation, only the time-dependence of the input source function and output displacement signals is required to invert these equations to recover the dynamic Green's function corresponding to the particular testing geometry and specimen material [10]. If the measurement system consists of a source whose excitation is an impulse or a Heaviside step and the receiver is a high-fidelity displacement or velocity sensor then the detected signals will correspond directly to the dynamic Green's function of the specimen, thus requiring no further signal processing. This observation emphasizes the advantage of using a source and a receiver possessing ideal characteristics.

Once the Green's function has been determined, the ultrasonic wavespeeds can be recovered by identifying the arrival times of the P- and S-wave signals. If the instant of excitation is known, only the first arrivals of these signals need to be determined. When this is not known, the arrival of other signals propagating through the specimen are needed. The 3P signal corresponding to the longitudinal wave propagating three times through the specimen is identified in the actual waveform shown in Figure 3. The longitudinal and shear wavespeeds of the material can be recovered from the measured arrival times according to the formulae shown in the figure.

The frequency-dependence of a particular wave arrival is determined by properly windowing that portion of the waveform containing the signal amplitude and transforming this amplitude to obtain the Fourier phase function of the signal. Once this function is found, it is substituted into Eq. (3) which, in turn, is solved numerically to obtain a solution for the dispersion relation of the wave in the material. Once the dispersion relation has been found, the phase and group velocities, c ($=\omega/k$) and v ($=\partial\omega/\partial k$) can be evaluated.

Since the computed, ideal waveform corresponds to the propagation in a non-attenuative material, the attenuation of either the P- or the S-wave amplitude in a real material can be determined by making a comparison of the measured and computed waveform amplitudes of the corresponding wave arrivals. It follows that the frequency dependence of the attenuation can be determined by processing the windowed signal amplitude in the frequency domain to form its magnitude spectrum V (f) and evaluating

$$\alpha(f) = 20 \log_{10} [V(f) / V_{ref}(f)] \qquad (4)$$

where $V_{ref}(f)$ refers to the magnitude spectrum of the wave velocity amplitude of the non-attenuated, ideal signal.

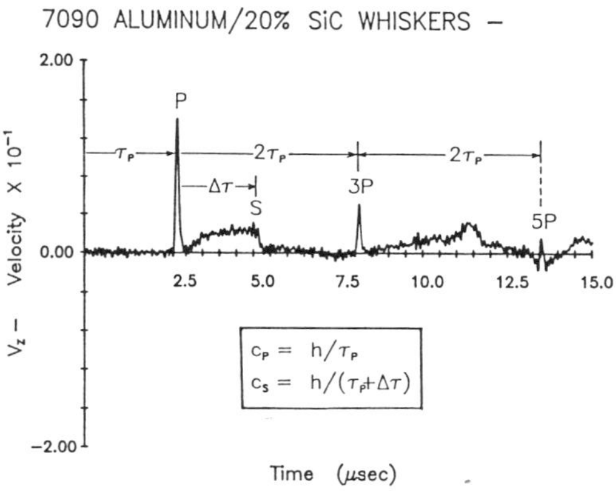

Figure 3 - Measured epicentral velocity signal; Step source. (Wave arrivals are indicated)

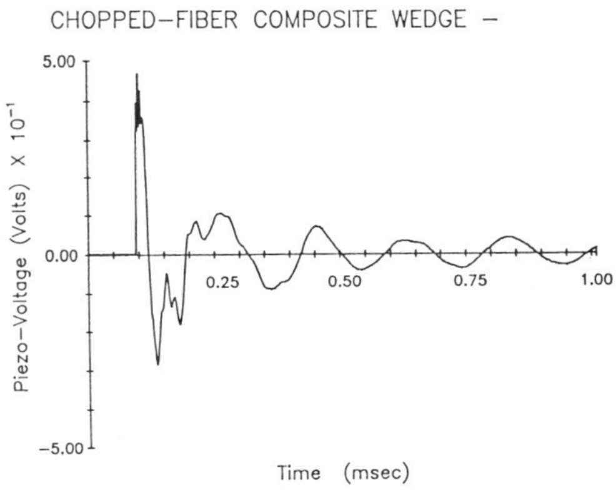

Figure 4 - Ultra-absorptive chopped fiber composite; Step source; Piezoelectric transducer detection.

MEASUREMENTS

The measurement system utilizing a point-source/receiver resembles a system used to make quantitative acoustic emission studies. The exception is the presence of the input source element which may or may not have a sensor attached to it in order to generate a synchronization pulse with the excitation signal. The normal velocity signal at epicenter corresponding to a step excitation on a specimen of a 7090 Al/SiC metal-matrix composite was shown in Figure 3. Measurement of the arrivals of the P- and S-wave amplitudes leads at once to the recovery of the longitudinal and shear wavespeed values, $c_p = 0.700$ cm/μsec; $c_s = 0.380$ cm/μsec. These are within 3% of the values determined in a conventional ultrasonic measurement.

Since in these measurements only the time of arrival of a particular wave is required, an uncalibrated piezoelectric point transducer can often be used. The waveform obtained in a highly attenuative, chopped fiber/epoxy, wedge-shaped specimen having a non-uniform layer of another material on one side is shown in Figure 4. The detection region of the specimen was left unprepared and hence the piezoelectric transducer with the excess couplant exhibited considerable ringing. However, even for this unfavorable testing situation, the first P-wave arrival is easily detected and can be used to determine an effective longitudinal wavespeed value for this material.

To determine the orientation dependence of wavespeeds in a sample, an array of transducers is required. In the simplest configuration, the receiving elements are located equi-distant about the source point. The sensors may be on either side of the sample, but they should be within 10h - 15h of the source so that the first P- and S-wave arrivals can be clearly identified in the detected signals. An example of the results of waveform measurements made in a specimen of graphite-epoxy comprised of 32 plies whose layup was at (+/- 45°) is given in Figure 5. Shown are the wavespeeds of the P-wave in various directions of the material. To obtain this result, the fracture of a capillary was used as a monopolar source with eight point piezoelectric sensors placed at various angles about it. The time of the first arrival was measured in each of the detected waveforms and the wavespeed was computed by dividing the arrival time into the source/receiver separation.

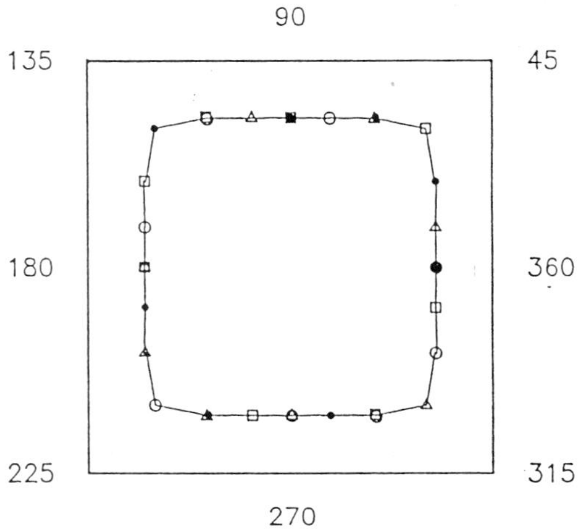

Figure 5 - Wave velocity surface in 32-ply Graphite/Epoxy; Layup: (±45°).

The graphite-epoxy specimen possesses a four-fold symmetry. This can be determined from inspection of the detected signals or from knowledge of the material's fabrication. Recognizing this symmetry, it is possible to generate additional pseudo-points by projecting each of the measured data values in directions oriented at 180, 90, -90 degrees to those measured. As the results of Figure 5 demonstrate, the twenty-four additional points all lie on the same wavespeed surface. This finding verifies the consistency of the measurement results.

Application of the Fourier phase analysis method for determining the dispersion relation and the frequency-dependent phase velocity of the longitudinal wave in a 6061 Al/SiC metal-matrix composite specimen is shown next. The signal resulting from a capillary fracture source detected at epicenter with a piezoelectric point transducer whose response approximated a velocity sensor is shown in Figure 6(a). Also indicated is the windowed, first arrival of the P-wave signal. From the magnitude spectrum it is found that the signal contains little energy above 8 MHz reflecting the frequency response of the transducer and amplifier used to detect the signals. Because the low-frequency correction is only significant at frequencies below 0.5 MHz, it is omitted from the dispersion relation of the derived phase velocity shown in Figure 6(b). It is seen from the latter that the phase velocity between 3 and 10 MHz is nearly constant at 0.690 cm/μsec. At lower frequencies there is a decrease to lower wavespeed values which is due principally to the response of the piezoelectric transducer used to make the measurements and the omission of the low-frequency correction.

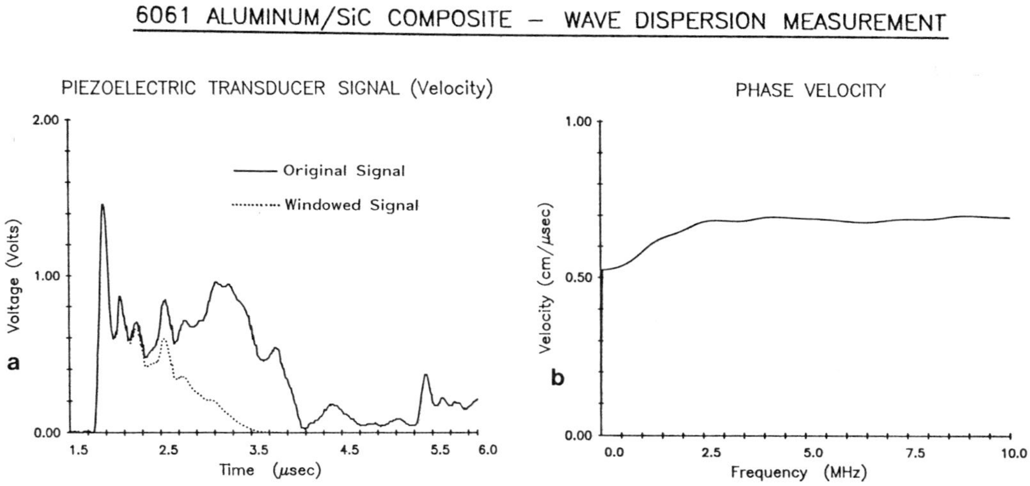

Figure 6 - Wave dispersion measurement in a metal-matrix composite (a) - Original, P-wave windowed signal; (b) - Derived phase velocity

An example of an attenuation measurement is shown in Figures 7(a)-(b). The velocity signal resulting from a step force applied in a 7090 Al/SiC metal-matrix composite specimen was shown in Figure 3. In the procedure, the first P-wave is windowed and compared to the computed response for the ideal case of a non-attenuating material shown in Figure 2. The Fourier magnitude spectra of the measured and ideal P-wave amplitudes are shown in Figure 7(a), while the result obtained from

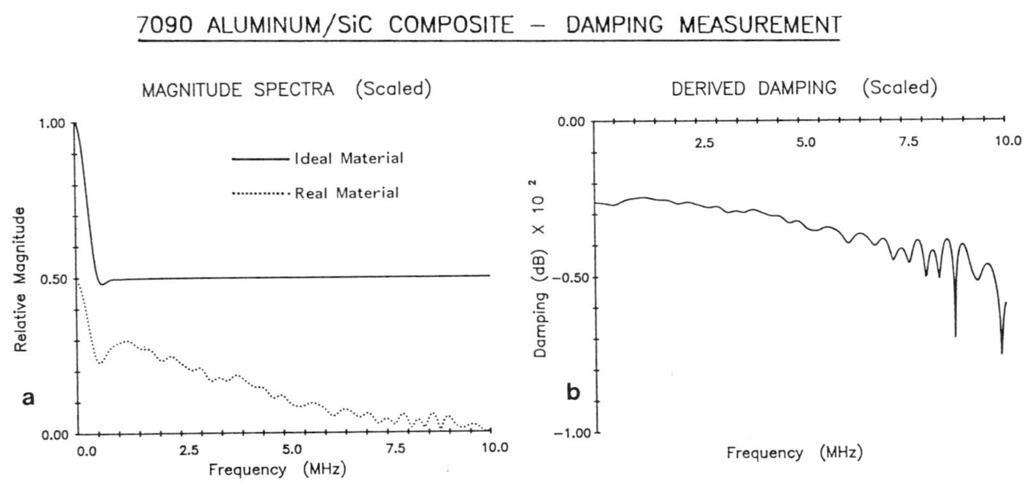

Figure 7 - Wave attenuation measurement in a metal-matrix composite (a) - Magnitude spectra of ideal, real signals; (b) - Derived attenuation

applying Eq. (4) is shown in Figure 7(b). In this example, only a relative measure of the attenuation of the longitudinal wave is determined since the vertical scale in Figure 3 was not calibrated absolutely and the magnitude of the force drop of the source used to generate the signal in this experiment was not measured.

In the waveforms detected in extremely absorptive materials, only the lowest frequencies are able to propagate and, hence, an unambiguous identification of the particular wave arrivals may be difficult. In such cases it may be advisable to choose an epicentral testing configuration with a sufficiently thick specimen so that the separation of the P- and S-wave arrivals is distinct in the detected signals. In cases in which only the lowest frequency components of the signal are propagated, it may also be necessary to consider other sensors to detect the signals.

The few examples shown here were used to illustrate the various signal analysis procedures described in the previous section. Numerous additional examples obtained in a variety of different materials are contained in a full length paper [9].

CONCLUSIONS

The components and characteristics of a point-source/point-receiver material testing system have been described by which the ultrasonic wavespeeds and attenuation can also be determined as a function of frequency in a variety of materials. The method utilizes a source and receiver whose transduction characteristics are known or can be determined in a calibration experiment. The measurements require a minimal amount of surface preparation and they can be made on specimens which are neither planar nor flat. Information regarding the propagation characteristics of both longitudinal and shear wave components is possible from a single waveform. It is also possible to select an excitation source whose time characteristics result in high energies at low frequencies which facilitates measurements in ultra-absorptive materials. It was demonstrated that while the characteristics of the wave propagation are more complex than those for plane waves, the existence of a theory of transient elastic waves permits a proper interpretation of the detected signals, provided that appropriate and calibrated point sources and receivers are utilized to make the measurements and the source/receiver separation is known. Results of several experiments were shown in which a composite material's longitudinal wavespeed can be recovered from the detected waveforms. It was demonstrated that by using an array of sensors, the wave velocity surface of a material can also be determined.

A procedure was also described by which the frequency-dependent wavespeeds and attenuation can be determined from the detected signals. The wavespeeds are recovered from an analysis of the Fourier phase functions of the normal velocity amplitudes corresponding to the arrival of either the longitudinal or shear wave signals. An analysis of the Fourier magnitude spectra of these signal amplitudes are compared to the magnitude spectra of the corresponding wave amplitudes for an ideal, non-attenuating specimen. With the continued development of non-contact point-sources and receivers, this measurement technique shows great promise as a powerful tool for characterizing micro- as well as macro-structural features of a large number of materials under a variety of measurement conditions.

ACKNOWLEDGEMENTS

We acknowledge the valuable discussions we have had with R. L. Weaver and the specimen materials we have received from D. Divecha. This work was supported in part by the Mechanics Division (Dr. Y. Rajapakse) of the Office of Naval Research and by the Solid and Geo-Mechanics Program (Dr. K. Thirumalai) of the National Science Foundation. Use of the facilities of the Materials Science Center at Cornell University which is supported by a grant from the National Science Foundation is also acknowledged.

REFERENCES

1. N. N. Hsu, J. A. Simmons and S. C. Hardy, "An Approach to Acoustic Emission Signal Analysis - Theory and Experiment", Materials Evaluation, Vol. 35, No. 10, 100-106 (1977).

2. Y. H. Pao, "Theory of Acoustic Emission", in Elastic Waves and Non-destructive Testing of Materials, Y. H. Pao, Ed., AMD-Vol. 29, Am. Soc. Mech. Engrs., New York (1978), pp. 107-128.

3. A. N. Ceranoglu and Y. H. Pao, "Propagation of Elastic Pulses and Acoustic Emission in a Plate: Part I. Theory; Part II. Epicentral Response; Part III. General Responses", ASME J. Appl. Mech. 48, 125-147 (1981).

4. C. B. Scruby, "Laser Generation of Ultrasound in Metals", in Research Techniques in Nondestructive Testing, Vol. 5, R. S. Sharpe, Ed., Academic Press (1985), pp. 281-327.

5. N. N. Hsu, "Dynamic Green's Functions of an Infinite Plate - A Computer Program", Report NBSIR 85-3234, National Bureau of Standards, Gaithersburg, MD (November 1985).

6. R. L. Weaver, "Frequency Dependence of Generalized Ray Arrivals at Epicenter in a Viscoelastic Plate". In preparation.

7. Y. H. Pao and W. Sachse, "On the Determination of Phase and Group Velocities of Dispersive Waves in Solids", J. Appl. Phys., 48, 4320-4327 (1978).

8. R. Weaver and W. Sachse, "Viscoelastic Generalized Rays: Theory and Experiment". In Preparation.

9. W. Sachse and K. Y. Kim, "Applications of Ultrasonic Point-Source/Point-Receiver Measurements". Submitted for publication.

10. J. E. Michaels, T. E. Michaels and W. Sachse, "Applications of Deconvolution to Acoustic Emission Signal Analysis", Materials Evaluation, 39, No. 11, 1032-1036 (1981).

LASER-ULTRASONIC DETERMINATION OF ELASTIC CONSTANTS AT AMBIENT AND ELEVATED TEMPERATURES+

Jean-Pierre Monchalin*, René Héon*, Jean F. Bussière** and Bahram Farahbakhsh***

*Physical Metallurgy Research Laboratories, CANMET
Energy, Mines and Resources, Canada
**Industrial Materials Research Institute
National Research Council Canada
75 De Mortagne Blvd, Boucherville, Québec, J4B 6Y4
***Alcan International Ltd
Research and Development Centre
Kingston, Ontario, K7L 4Z4 Canada

INTRODUCTION

It is well known that the elastic constants of an isotropic solid (the bulk and shear moduli or the Young's modulus and the Poisson's ratio) can be determined ultrasonically when both longitudinal and shear wave velocities are measured. At high temperature, traditional ultrasonic techniques are difficult to apply because they require a coupling medium operating in the same temperature range. However, some results have been previously reported using momentary contact[1-2],but the application remains difficult, especially for shear wave coupling, and above 1000°C.Obviously, an ultrasonic technique where ultrasound is generated and detected without contact can avoid such problems.

Contactless generation and detection of ultrasound is possible with the laser-ultrasonic technique. In this technique, ultrasound is generated by the absorption at the surface of the sample of a short and intense laser pulse[3]. The ultrasonic deformation is detected after propagation inside the specimen by a second laser (continuous or long pulse) coupled to an optical interferometer[4]. This technique permits the generation and detection of both longitudinal and shear waves, as required for the determination of the two independent elastic constants encountered in isotropic solids. It also enables the generation and detection over a broad band of ultrasonic frequencies and in particular, at low frequencies, which permit adequate penetration in coarse grained materials[5]. Furthermore, it can be used on specimens which have not been machined to close tolerances and can even be of awkward shapes.

+ Work performed in the laboratories of the Industrial Materials Research Institute

We present below an experimental setup which allows the noncontact measurement of both shear and longitudinal velocities in materials heated up to 1000°C. Examples of results are presented for metal/ceramic composites and ceramics.

EXPERIMENTAL APPARATUS

The experimental setup is sketched in Fig. 1. Generation and detection are performed on opposite sides of the sample, more conveniently shaped as a parallel plate. Although single-side generation and detection would be generally preferred for industrial inspection, operation on different sides is permissible in a laboratory apparatus and avoids the perturbations caused by the heated air or the generated hot plasma existing in single sided operation. The sample is located inside an evacuated quartz tube (pressure ≈ a few 10^{-6} torr) which fits inside a tubular oven. Temperatures as high as 1000°C can be reached, limited in this case only by softening of the quartz tube. Higher temperatures would require a tube made of a different material. The tube is sealed by optical windows pushed against O-rings that are located sufficiently far away from the oven to prevent their overheating. The sample is supported by a holder made of steel located at the center of the oven and which can be slided in and out of the tube. Vacuum below 10^{-5} torr and continuous pumping was sufficient to prevent oxydation of all the samples studied so far.

The laser used for generation is a Q-switched Nd-YAG which provides 8ns pulses with a maximum energy of 0.75J at a repetition rate of 10Hz. The receiving probe is the heterodyne interferometer which has been described earlier[6-7] and which uses in the present setup a 1W single mode argon laser instead of the few mW He-Ne laser used previously. The use of an argon laser gives a better sensitivity (more than an order of magnitude) and permits to probe samples with scattering surfaces. The

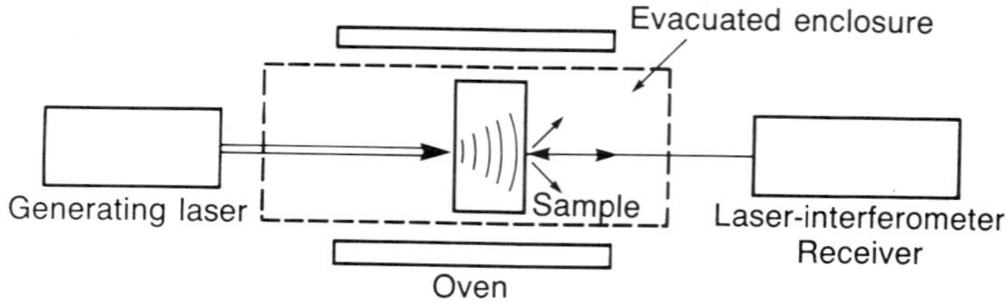

Fig. 1: Schematic of the experimental apparatus.

bandwidth of this detecting system extends from 250kHz to 35MHz, limited by the width of the RF filter used in the demodulation circuit, the Bragg cell frequency and the bandwidth of the gain-controlled amplifier[7]. The demodulation scheme ensures the necessary immunity to ambient vibrations, since a very stable mounting of the sample with respect to the other elements of the interferometer is hardly possible in this case. The signal is recorded with a 5ns resolution transient digitizer and then stored in a computer for further analysis.

EXPERIMENTAL PROCEDURE

As mentioned above, it is necessary to observe the arrivals of both longitudinal and shear wave pulses. This can be done by operating at epicentre (detection and generation at opposite locations) in the thermoelastic regime[3] or at the onset of ablation[5]. In this latter case, a sharp feature corresponding to the longitudinal arrival is observed. A much stronger longitudinal signal is seen when working in the marked ablation regime. However in this case, the arrival of the shear pulse is broad and cannot be accurately timed. A method of obtaining shear wave velocity which has been previously reported[8] but is restricted to specimens shaped as long rods, is based on mode conversion phenomena between longitudinal and shear waves. For samples which are plate-like, another solution, as we found, is to perform detection off-epicentre. This is demonstrated in Fig. 2 which shows the ultrasonic pulses observed on an aluminum alloy plate when the generating and detection spots are moved further and further away from each other. It is seen that a sharp shear feature starts to appear away from epicentre.Even further away, at an angle larger than the limit angle of conversion of shear into longitudinal, one notices the arrival of a head wave[9].

The time intervals measured between the generating laser pulse and the front edge of the longitudinal pulse at epicentre and the front edge of the shear pulse off epicentre were used to determine the longitudinal and shear velocities. Time arrival of the head wave was not used for determination of the velocities, but it was verified that it is in agreement with the values of the velocities deduced from the longitudinal and shear arrivals. The elastic constants (bulk modulus B, shear modulus G, Young's modulus E, Poisson's ratio σ) were determined from the velocities (V_L, V_S) and the density ρ using the standard formulas:

$$B = \rho\ [V_L{}^2 - (4/3)\ V_S{}^2] \tag{1}$$

$$G = \rho\ V_S{}^2 \tag{2}$$

$$E = \rho\ V_S{}^2\ (3\ V_L{}^2 - 4\ V_S{}^2)/(V_L{}^2 - V_S{}^2) \tag{3}$$

$$\sigma = 0.5\ [1 - V_S{}^2/(V_L{}^2 - V_S{}^2)] \tag{4}$$

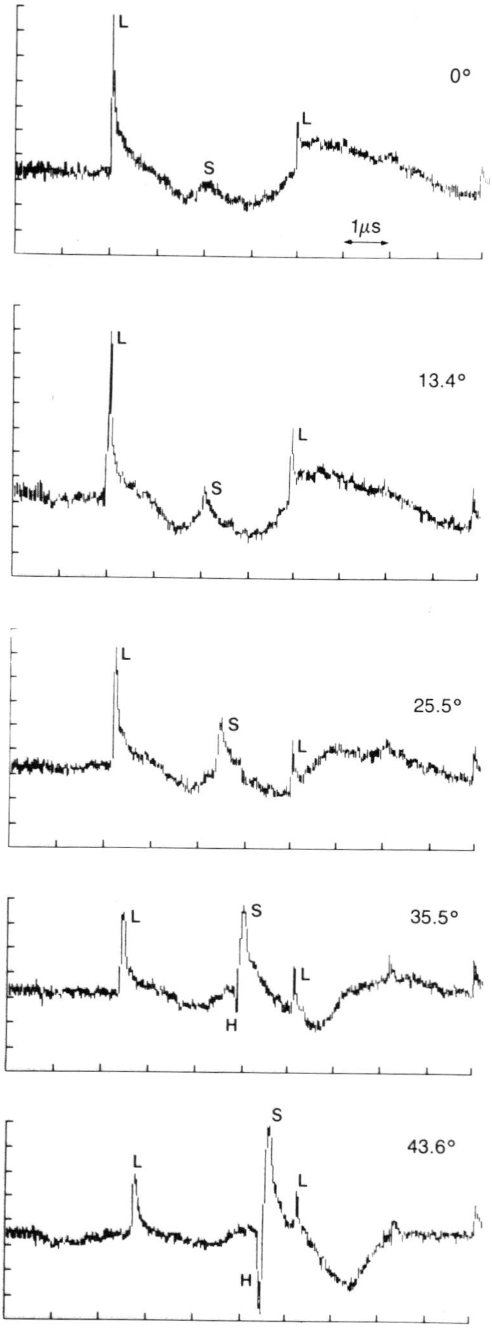

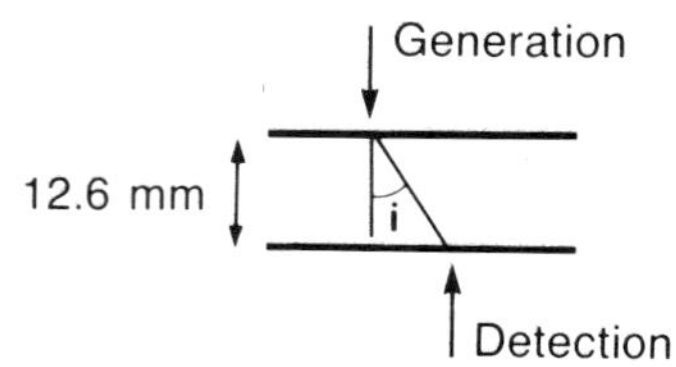

Fig. 2: Signals observed on an aluminum alloy (6061 - T6) plate when the detection spot is at epicentre (i = 0°) and moved away from epicentre. The geometry of generation and detection is indicated above. The values of i are shown on each graph as well as the identification of the longitudinal wave (L) and shear wave (S) pulses. When i is larger than the limit angle (≈ 29.6° in this case), a head wave (H) is produced.

EXAMPLES OF APPLICATIONS

The system was applied to the measurement of the elastic constants of various ceramics and metal/ceramics composites. Fig. 3 shows a result obtained on a hot pressed cermet composed of titanium diborate, alumina and aluminum. This figure shows the large decrease of the shear modulus when the material is heated up to the melting point of aluminum (≈ 660°C). Above this temperature, this modulus vanishes and the material loses cohesion.

Fig. 4 shows the behavior of the longitudinal velocity versus temperature of two PZT-4 specimens. One of them was unpoled and the other one was poled under high electric field for piezoelectric ultrasonic transduction. It is noted, as expected, that poling stiffens the material and increases its velocity. The Curie temperature is revealed by the point where the velocities of the poled and the unpoled materials become identical (≈ 300°C). Above this temperature, one notices a large change of velocity. Later work has shown that this change actually levels off at higher temperature and that it corresponds to a phase transition from a tetragonal to a cubic structure[10].

The last example (Fig. 5) shows the plots of the longitudinal and shear velocities of a material made of a titanium diboride porous skeleton impregnated with aluminum. This material is being developed for new electrolytic cells for producing aluminum and is used at temperatures ranging from 500°C to 950°C. A discontinuity of velocities is noticed at the melting point of aluminum. Below this point the elastic constants can be derived readily by using eq. 1-4. Above the melting point, the material consists of a solid filled with liquid and, according to Biot's theory[11], one should observe the arrival of two longitudinal waves, a slow wave and a fast wave. The slow wave has not been observed so far, which has prevented complete derivation of elastic constants

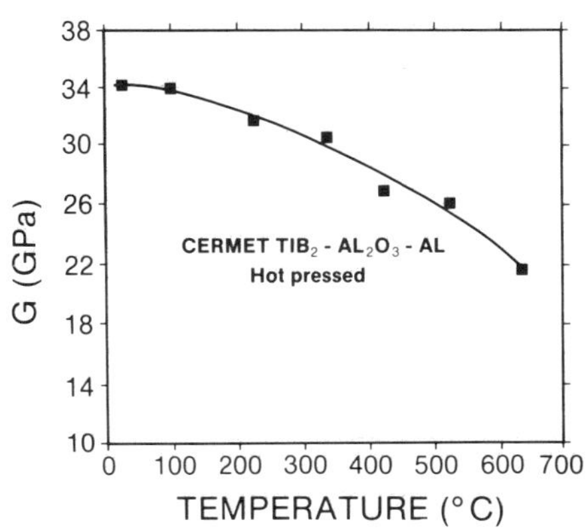

Fig. 3: Shear modulus of a hot pressed cermet made of T_iB_2, Al_2O_3 and Al

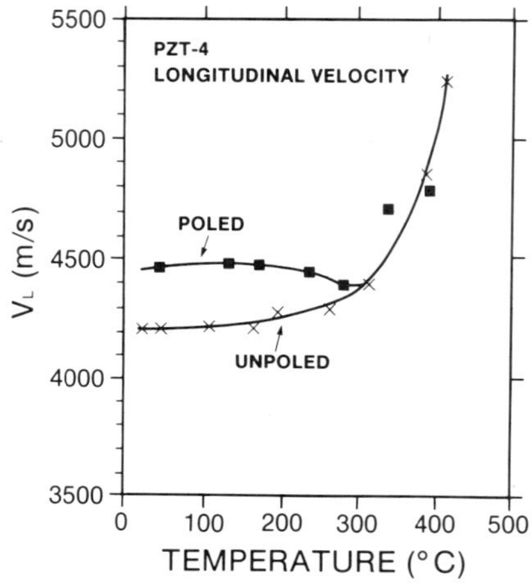

Fig. 4: Velocity of poled and unpoled PZT-4 samples versus temperature

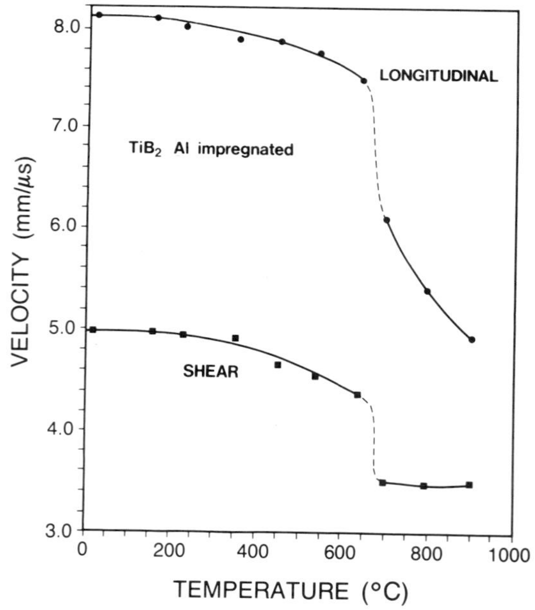

Fig. 5: Longitudinal and shear velocities of a material made of porous TiB_2 impregnated by aluminum.

from ultrasonic data. Work is being pursued to observe the slow wave and to compare the impregnated material with the corresponding empty skeleton.

CONCLUSION

We have developed a technique and laboratory system which allow ultrasonic measurements on materials at elevated temperatures up to 1000°C. The system has been used so far only for measuring ultrasonic velocities and for the determination of elastic constants. Since ultrasonic attenuation can be measured as well at the same time, the system is a powerful characterization means for materials at high temperature, capable of revealing phase and microstructural changes as well as loss of cohesion. Future developments include modelling the evolution of the ultrasonic source which will permit more precise velocity measurements and modification of the system for operation at temperatures higher than 1000°C.

ACKNOWLEDGEMENTS

The help of Mr. D. Pascale for initial mounting of the system is acknowledged, as well as Dr. C.K. Jen for providing the PZT-4 sample and helpful discussions.

REFERENCES

1. E.H. Carnevale, L.C. Lynnworth and G.S. Larson, "Ultrasonic measurement of elastic moduli at elevated temperatures using momentary contact", J. Acoust. Soc. Am., vol. 36, p. 1678-1684, (1964).
2. E.P. Papadakis, L.C. Lynnworth, K.A. Fowler and E.H. Carnevale, "Ultrasonic attenuation and velocity in hot specimens by the momentary contact method with pressure coupling and some results on steel to 1200°C", J. Acoust. Soc. Am., vol 52, p. 850-857, (1972).
3. C.B. Scruby, R.J. Dewhurst, D.A. Hutchins and S.B. Palmer, "Laser generation of ultrasound in metals", in Res. Techniques on Nondestructive Testing (vol. 5), R.S. Sharpe, Ed., Academic Press, New York, p. 281-327, (1982).
4. J.-P. Monchalin, "Optical detection of ultrasound", IEEE Trans. Ultrason., Ferroelectrics, Freq. Contr., vol. UFFC-33, p. 485-499 (1986).
5. L. Piché, B. Champagne and J.-P. Monchalin, "Laser ultrasonics measurement of elastic constants of composites", Materials Evaluation, p. 74-79, vol. 45, no. 1, 1987.
6. J.-P. Monchalin, "Heterodyne interferometric laser probe to measure continuous ultrasonic displacements", Rev. Scien. Instrum., vol. 56, p. 543-546 (1985).
7. J.-P. Monchalin, R. Héon and N. Muzak, "Evaluation of ultrasonic inspection procedures by field mapping with an optical probe", Canadian Metallurgical Quaterly, to be published.
8. C.A. Calder and W.W. Wilcox, "Noncontact material testing using laser energy deposition and interferometry", Materials Evaluation, p. 86-96, January 1980.
9. J.D. Achenbach, Wave propagation in elastic solids, North Holland, New York, 1973, see chap. 7.11.
10. B. Jaffe, W.R. Cook and H. Jaffe, Piezoelectric ceramics, Academic Press, New York, 1971, see chap. 7.
11. M.A. Biot, "Theory of propagation of elastic waves in a fluid-saturated porous solid", J. Acoust. Soc. of Am., vol. 28, p. 168-191 (1956).

LASER/EMAT SYSTEMS FOR THE ULTRASONIC INSPECTION OF ALUMINUM

D.A. Hutchins and F. Hauser

Physics Department, Queen's University
Kingston, Ontario, Canada K7L 3N6

INTRODUCTION

The metals manufacturing industry makes routine use of ultrasonic non-destructive testing techniques for the detection of defects such as cracking. It is becoming increasingly clear, however, that more advanced non-contact techniques are required for the testing of materials which are either in motion or at an elevated temperature. An example is in the continuous casting of aluminum, where differential cooling across the material can lead to stresses which, in turn, can result in porosity or cracks. In this case, an on-line defect detection system is desirable, to detect the onset of manufacturing problems. Hence, the aim is not to detect all defects, but to give an indication of problems in the solidification process.

The research to be described below was initiated to investigate non-contact ultrasonic inspection techniques for possible application to the metals industry. The approach is based on ultrasonic generation by pulsed lasers, and detection by electromagnetic acoustic transducers (EMATs). By means of a brief introduction to these methods of transduction, an outline of their characteristics is now presented.

Ultrasonic generation by pulsed lasers

Several reviews of this subject exist in the literature[1,2], but for the purposes of this paper two principal mechanisms are of interest. The first is thermoelastic expansion, following absorption of optical energy at the metal surface. The resultant acoustic source is dominated by a horizontal force dipole, Fig. 1(a), located just below the metal surface, due to thermal diffusion, and hence also contains a small vertical dipole component. The time dependence of this source is the integral of the laser pulse shape, i.e., close to a step function. Assuming such a source, wave propagation theory may be used to predict the resultant normal displacement waveform at the far side of a metal plate (on epicenter, i.e. on a line passing through the source, perpendicular to the surface of generation). The result is shown in Fig. 1(b). A Poisson's ratio $\nu = 1/3$ has been assumed, which is close to that of aluminum. As can be seen, the waveform contains a large shear (S) signal, a factor of ~ 4 larger than that of the longitudinal mode (L). The corresponding

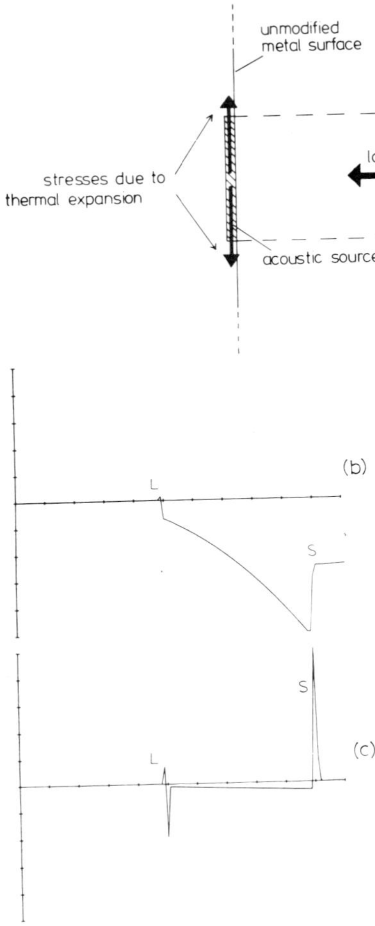

Fig. 1: Thermoelastic generation, showing (a) acoustic source, (b) epicentral displacement waveform, (c) corresponding velocity waveform.

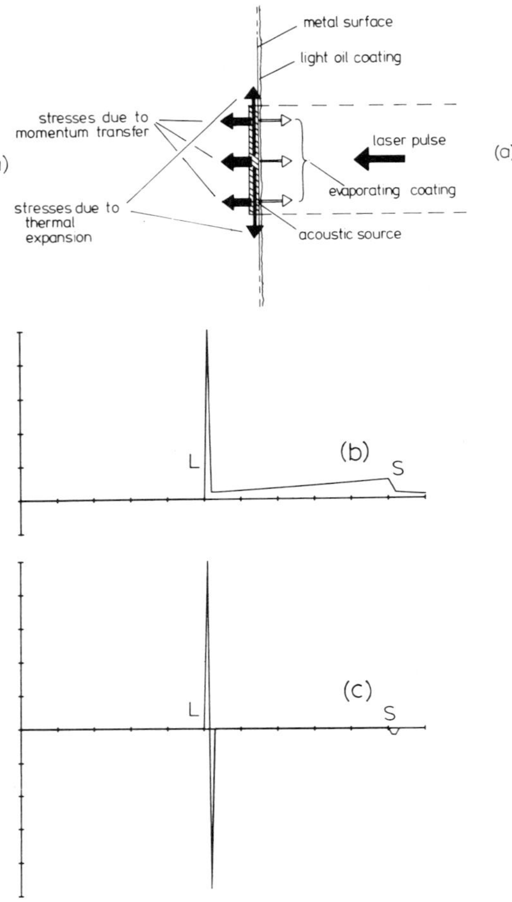

Fig. 2: Generation by liquid evaporation; (a) - (c) as Fig. 1.

normal velocity waveform, obtained by differentiating the displacement waveform of Fig. 1(b), is shown in Fig. 1(c). It contains a prominent monopolar shear pulse and a smaller longitudinal signal which is dipolar in nature.

A second type of acoustic source may be formed by coating the metal surface with a suitable liquid, e.g., oil acetone, water, etc. Under such conditions, the laser energy tends to be absorbed within the liquid, causing its evaporation. The recoil force from this process causes the resultant acoustic source to be dominated by forces normal to the surface, as shown in Fig. 2(a). These forces will have a pulselike time dependence and although thermal forces will still exist, they will tend to be negligible in comparison. The theoretical displacement waveform for this case is shown in Fig. 2(b), again for $\nu = 1/3$. The longitudinal signal is now a prominent positive pulse, with the shear mode now being small in comparison. The corresponding velocity waveform, Fig. 2(c), is dominated by a prominent dipolar longitudinal pulse.

Generation may also be accomplished by ablation of metallic material, at high optical power densities, again leading to a source with normal forces being dominant. It should be noted, however, the surface damage results, and hence thermoelastic and liquid evaporation sources are preferred.

Electromagnetic acoustic transducer (EMAT) detection.

Although EMATs have found limited application to materials testing[3], their low efficiency precludes their use in some situations. For example, the author has found it difficult to design EMATs for use at frequencies in excess of 5 MHz, due to lift-off phenomena. EMATs may be used for both generation and detection, and hence may be used in a conventional pulse-echo or pitch-catch configurations.

In the generation mode, a current pulse is passed through a coil positioned close to the metal surface. Induced eddy currents interact with an applied magnetic field to produce a Lorentz force, and hence ultrasonic generation occurs. In detection, motion of the surface due to ultrasonic pulses caused eddy current generation, which the coil detects. However, the EMAT detection sensitivity depends on the magnetic field direction and the coil geometry. In general, if the magnetic field B is normal to the surface, the EMAT will be sensitive principally to shear waves; conversely, with B parallel to the surface, the longitudinal mode is favoured.

There are a range of EMAT coil designs of use in this investigation, some of which are shown in Fig. 3. Of particular interest is the elongated pancake coil, Fig. 3(c), which can be made sensitive to either shear or longitudinal waves by a single change in the magnetic field direction.

In the results to follow, a range of EMAT designs have been examined for use in conjunction with pulsed laser generation, to investigate the characteristics of such a hybrid system. Both solid state and electromagnets have been utilized, as will now be described. The discussion here will be limited to bulk ultrasonic waves, although it should be noted that in other work, Rayleigh waves have been investigated[4].

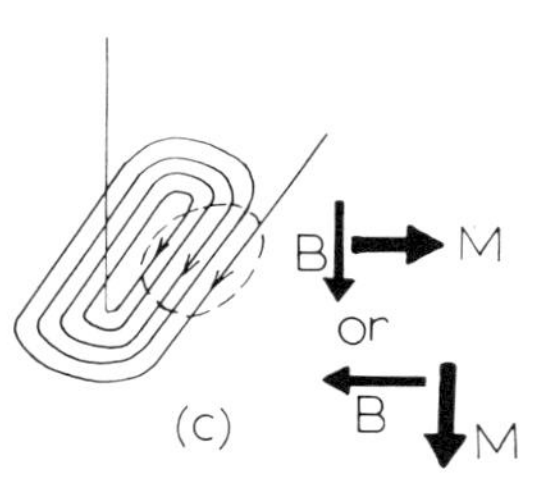

Fig. 3: Schematic diagrams of EMATs used for ultrasonic detection in aluminum.

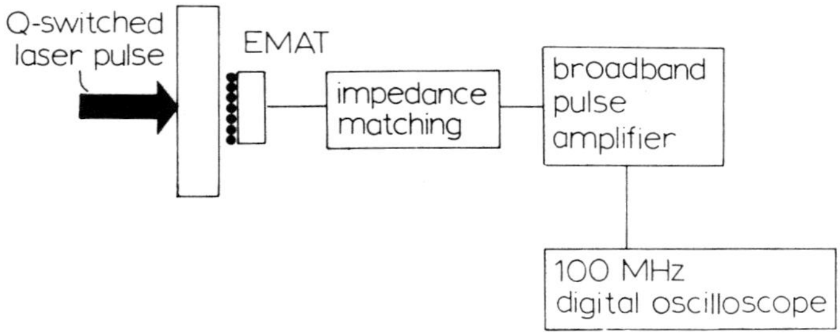

Fig. 4: Schematic diagram of apparatus.

APPARATUS

A schematic diagram of the apparatus used in this investigation is shown in Fig. 4. A frequency-doubled, Q-switched ruby laser provided single pulses of 30 ns half-width duration, at an optical wavelength of 347 nm in the near UV. Using a multimode configuration, this laser supplied pulse energies of ≦ 225 mJ, with the beam being approximately rectangular in cross section and of approximate dimension 6 x 4 mm. The laser beam was incident normally onto one face of parallel-sided metallic plates, of either aluminum or steel, and of 25.4 or 50.8 mm thickness.

The longitudinal and shear waves generated by the laser source propagated through the plate, and were detected on the far side by a variety of EMAT designs such as those of Fig. 3. In this set of experiments, most waveforms were obtained with a small (∿ 0.1 mm) gap between the EMAT coil and the surface, achieved by contacting a brass facing plate to the surface with the coil mounted on its opposite side. This plate contained apertures of various shapes and sizes, to expose the required area of coil to the surface, and was rigidly attached to a brass housing containing the coil and any permanent magnet. The brass housing was sealed to provide minimum noise pickup, and was fitted with a standard BNC connector in each case.

Experiments were conducted with static magnetic fields for the EMATs provided by both permanent magnet and dc electromagnets. In the former

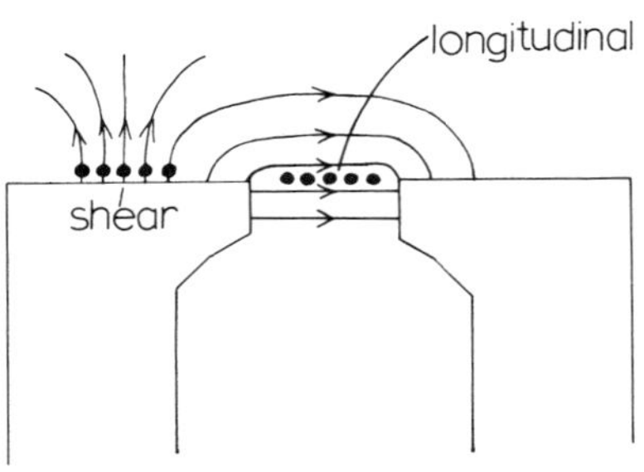

Fig. 5: Electromagnet configuration showing coil positions for maximum sensitivity for a given ultrasonic mode.

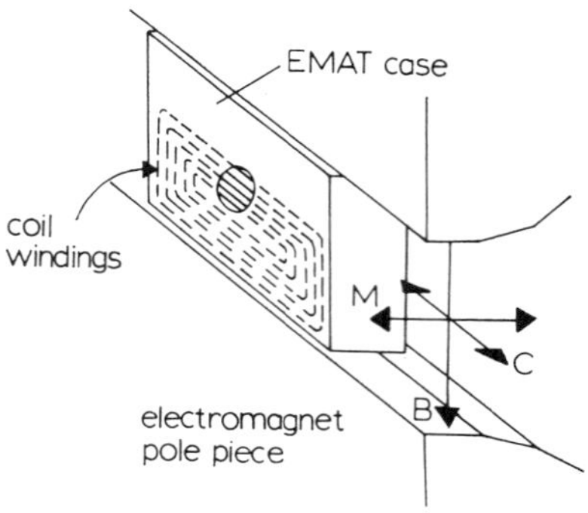

Fig. 6: Electromagnet EMAT sensitive to normal motion.

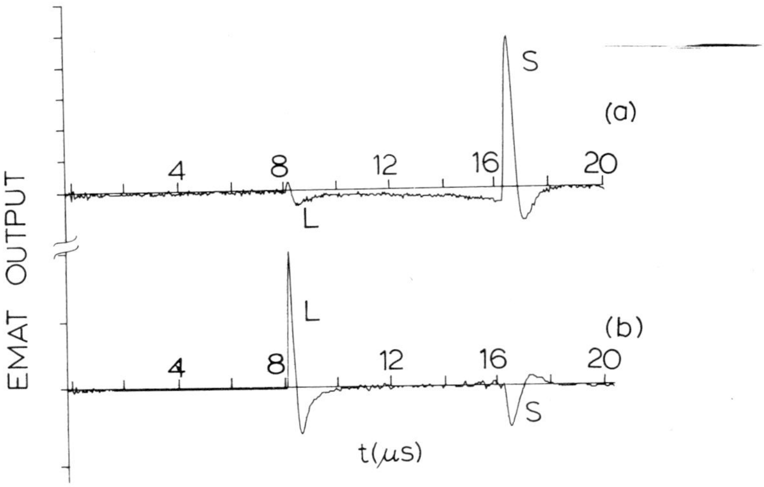

Fig. 7: Experimental waveform in 50.8 mm thick aluminum, following detection by a spiral coil, permanent magnet EMAT (Fig. 3(a)). (a) thermoelastic generation, (b) oil evaporation.

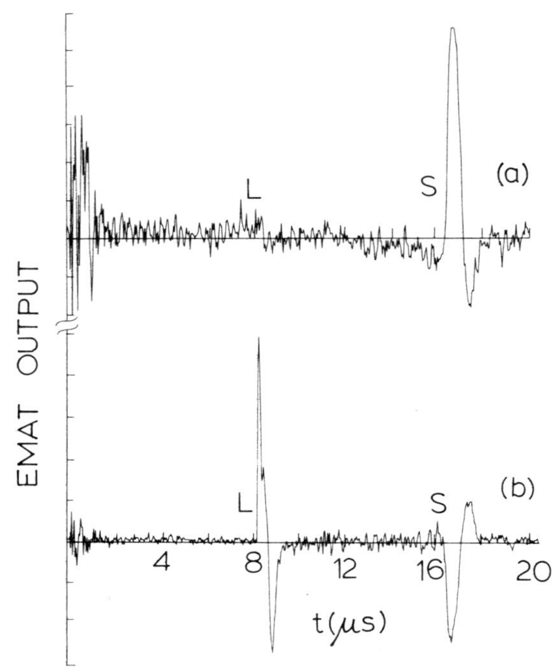

Fig. 8: As Fig. 7, detection by a permanent magnet EMAT of the type shown in Fig. 3(b).

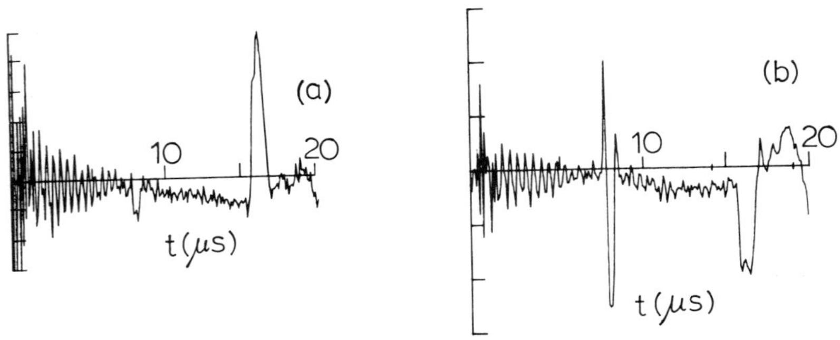

Fig. 9: As Fig. 7, detection by an electromagnet EMAT sensitive to normal motion (Fig. 6).

case, Co-Sm rare earth magnets were used, because of their high fields and light weight. These could easily be incorporated within the EMATs brass housing, being 12.5 mm in thickness. Both rectangular and cylindrical magnets were used, of 25-mm width, with the B field normal to their flat faces. The electromagnet was specifically chosen to allow the same EMAT coil configurations to be inserted within the pole pieces, or at other locations to provide known magnetic fields in specific directions. For this purpose, the Co-Sm magnets were removed from the brass housings. Fig. 5 shows how the shape of the electromagnet pole piece allowed fields, either perpendicular or parallel to the surface of a metallic plate, to be established over the surface area of each EMAT coil. Using a dc power supply, rated at 25 V/250 A, fields of ≦ 0.8T could be produced within the gap between the poles. Fig. 6 shows how a coil such as that shown in Fig. 3(c) could be positioned between the poles of the electromagnet for the detection of normal motion.

RESULTS AND DISCUSSION

(a) Laser-EMAT waveforms in defect-free aluminum

Using the apparatus of Fig. 4, a series of waveforms have been recorded

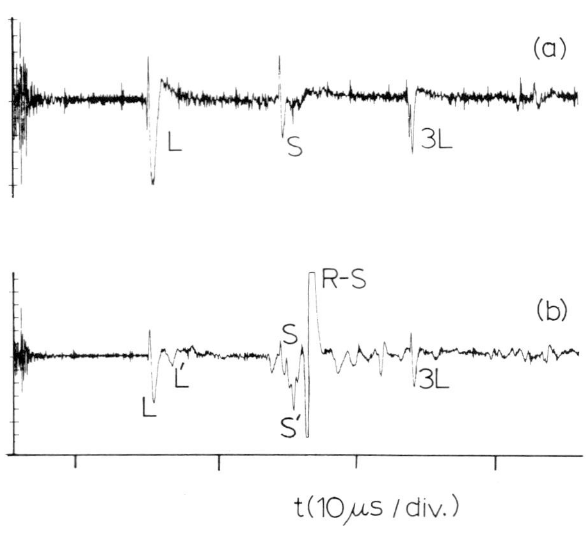

Fig. 10: Experimental waveforms from experiment to simulate detection of surface-breaking crack in a 60 mm aluminum plate. (a) no defect, (b) 7 mm deep slot on generation surface.

for a range of EMAT configurations, to investigate the optimum design for a given application. The first type of EMAT examined was of the type shown in Fig. 3(a), with a spiral coil of 25.4 mm diameter, mounted against a Co-Sm magnet of the same diameter to provide a normal magnetic field B. The results are shown in Fig.7. For thermoelastic generation, Fig. 7(a), a large shear (S) signal resulted with a smaller longitudinal (L) component. Using an oil evaporation source, Fig. 7(b), an increased longitudinal signal resulted, as expected from the discussion above. Use of the EMAT design shown in Fig. 3(b), using a 25.4 mm square Co-Sm magnet, led to the waveforms of Figs. 8(a) and (b) for thermoelastic and oil evaporation sources respectively. Note certain similarities with those of Fig. 7.

It is interesting to note that the waveforms of Figs. 7 and 8 show similarities to those of normal velocity predicted by wave propagation theory (Figs. 1(c) and 2(c)). This occured despite the fact that the nominal design was for sensitivity to horizontal motions. It should be noted, however, that theoretically there is no horizontal motion on-epicentre for the two types of source discussed. Hence the agreement may be interpreted as a degree of sensitivity to normal motion, caused by finite EMAT size and magnetic field curvature from the Co-Sm magnets. Another point of interest, is that the bandwidth of received signals is reasonable, with none of the resonances associated with piezoelectric transducers; hence these detectors would be of use for defect detection at temperatures below $\sim 350^{\circ}C$, the Curie temperatures for the Co-Sm magnets.

Inspection at a higher temperature requires the use of heat shielding or active cooling of solid state magnets, or the use of electromagnets. An electromagnet EMAT of the design shown in Figs. 5 and 6 has thus been used to investigate a possible design for use at elevated temperatures. An additional advantage was that a uniform, horizontal magnetic field could be supplied, leading to sensitivity to normal motion. Resulting waveforms are shown in Fig. 9 for (a) thermoelastic generation and (b) following evaporation of an oil coating. Note, as expected, that the waveforms show reasonable agreement with those of Figs. 1(a) and 2(a).

(b) Results for simulated defects.

The results shown above have been based on the assumption that the EMAT is optimized for the detection of laser-generated ultrasonic waveforms, and this approach is reasonable for the measurement of material thickness, elastic constants and degree of porosity. Consider, however, surface-breaking cracks. Here, it might be expected that a surface-wave EMAT detector would be required. In the present experiments, we have used a line source to generate Rayleigh waves, with a shear wave EMAT on-epicentre with respect to the source, on the far side of an aluminum plate. With no defect present, the waveform of Fig. 10(a) resulted from oil evaporation, with L, S and 3L (multiply-reflected longitudinal) signals being evident. An experiment was then undertaken with a 7 mm deep slot present on the surface of generation to simulate a surface-breaking defect. The resulting waveform contained a large arrival (R-S), which has been interpreted as being due to Rayleigh to shear mode conversion at the base of the slot.

The above preliminary experiment indicates that surface defects may radiate considerable amounts of energy as shear waves, following interaction with laser-generated Rayleigh waves. A shear wave EMAT is a good, non-contact method for the detection of these mode-converted shear waves, and future research will investigate this phenomenon in greater detail.

ACKNOWLEDGEMENTS

This work was funded via a Strategic grant from NSERC, and Contract from the Department of Supply and Services Canada, for which thanks are due to W.R. Sturrock.

REFERENCES

1. D.A. Hutchins, Ultrasonic generation by pulsed laser, to be published in _Physical Acoustics_, Vol. XVIII (Academic Press, W.P. Mason and R.N. Thursten, Eds., 1986).
2. C.B. Scruby, R.J. Dewhurst, D.A. Hutchins and S.B. Palmer, Laser-generated ultrasound in metals, in _Research Techniques in Nondestructive Testing_, Vol. V (Academic Press, R. Sharpe Ed., 1982).
3. H.M. Frost, Electromagnetic-ultrasound transducers: Principles, practice and applications, in _Physical Acoustics_, Vol. XIV (Academic Press, W.P. Mason and R.N. Thurston, Eds., 1979).
4. D.A. Hutchins, T. Goetz and F. Hauser, "Surface wave using laser generation and EMAT detection", to be published in _IEEE Trans. Ultras. Ferr. Freq. Control_, (September 1986).

OPTICS-BASED TECHNIQUES FOR THE CHARACTERIZATION OF COMPOSITES AND CERAMICS

Paolo Cielo, Xavier Maldague, Jean-Claude Krapez
and Richard Lewak*

Industrial Materials Research Institute
National Research Council of Canada
75 de Mortagne Blvd., Boucherville, Québec J4B 6Y4

* Tecrad Inc.
Ancienne-Lorette, Québec

INTRODUCTION

The needs for on-line and unsupervised industrial inspection for quality control and process monitoring are rapidly increasing with the accelerated trend toward factory automation. Optical techniques are particularly well suited for industrial inspection needs because of their noncontact nature and imaging capabilities.

Applications of optical inspection techniques capitalizing on the recent technological advances in the electrooptical technology are certainly going to increase substantially in the years to come. This paper presents a comparative review of a number of optical techniques which have been recently developed at our institute for the characterization of nonmetallic materials. Both thermal and nonthermal methods will be described and their applicability to a number of industrial materials will be discussed.

OPTICAL CHARACTERIZATION OF SINTERED CERAMICS: A COMPARISON OF DIFFERENT TECHNIQUES

For historical reasons, optical techniques for nondestructive material characterization[1] have grown out of similar techniques which were originally directed to nondestructive testing applications. Opto-acoustic and optothermal techniques for the detection of surface cracks or subsurface delaminations have thus been adapted to the measurement of material properties. In some cases, an unbiased new approach may however result in much simpler material characterization methods. One such example is shown here in relation with the characterization of translucent ceramic materials.

The problem which is addressed is the development of a nondestructive technique for the evaluation of the degree of porosity in a ceramic part. The ceramic samples used in this investigation were obtained from Y_2O_3-stabilized ZrO_2 powder fired at temperatures ranging from 1300°C to 1550°C. As a result of the different firing temepratures, the mass

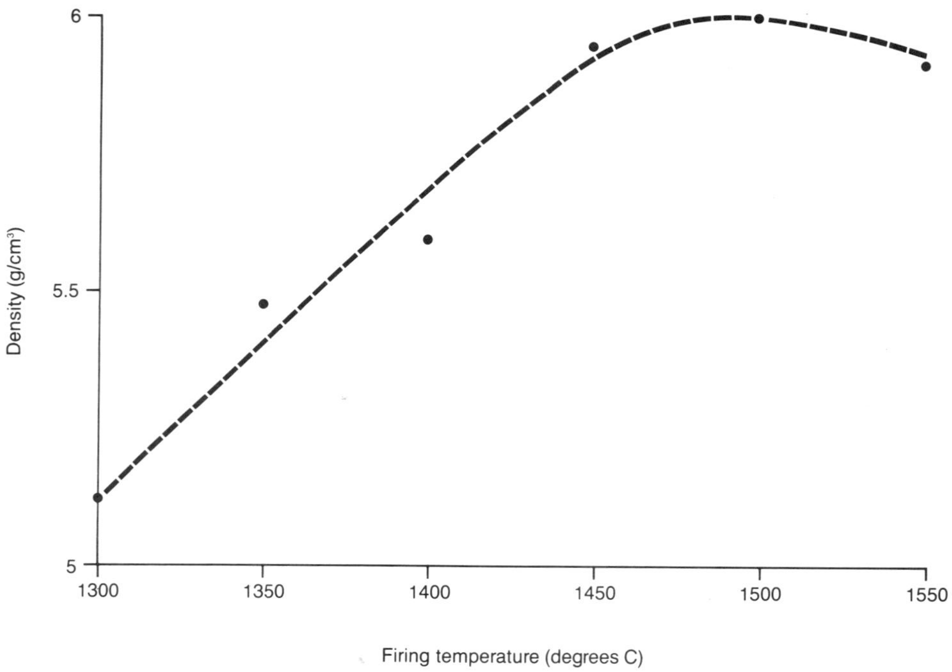

Figure 1: Mass density of the zirconia ceramic samples for different firing temperatures.

density of the samples varied as shown in Figure 1[2]. The increase in density is mainly due to better sintering at higher temperature. Three different laser-based approaches (see figure 2) were analyzed as possible candidates for such a porosity-evaluation problem.

1. Optoacoustic method

Ultrasonic velocity and attenuation measurements provide information on mechanical properties and grain size as well as on the density and distribution of pores in fired ceramics[3]. The need for liquid coupling tends however to restrict the on-line applicability of ultrasonic methods, while liquid-infiltration problems are encountered when analyzing porous materials by such mathods[4]. Dry-coupling laser ultrasound generation was thus considered as a possible approach for non-contact ultrasonic testing of sintered ceramic parts.

The experimental apparatus is shown in figure 2a. A pulsed laser beam is focused to a ring-shaped pattern on the surface of the ceramic pellet to produce a converging surface wave[5] which is detected by a laser interferometer focused in the center of the ring. The time-of-flight measurement yielded surface-acoustic-wave values showing a good correlation with the mass density of the ceramic samples, as figure 2a shows. It must however be mentioned that the laser-irradiated area on the samples had to be coated with a black paint to produce surface absorption of the laser pulse. Moreover, the required laser power is high (typically 0.5J, 15 ns pulses) and the interferometric probe is relatively complex, leading to a rather expensive system.

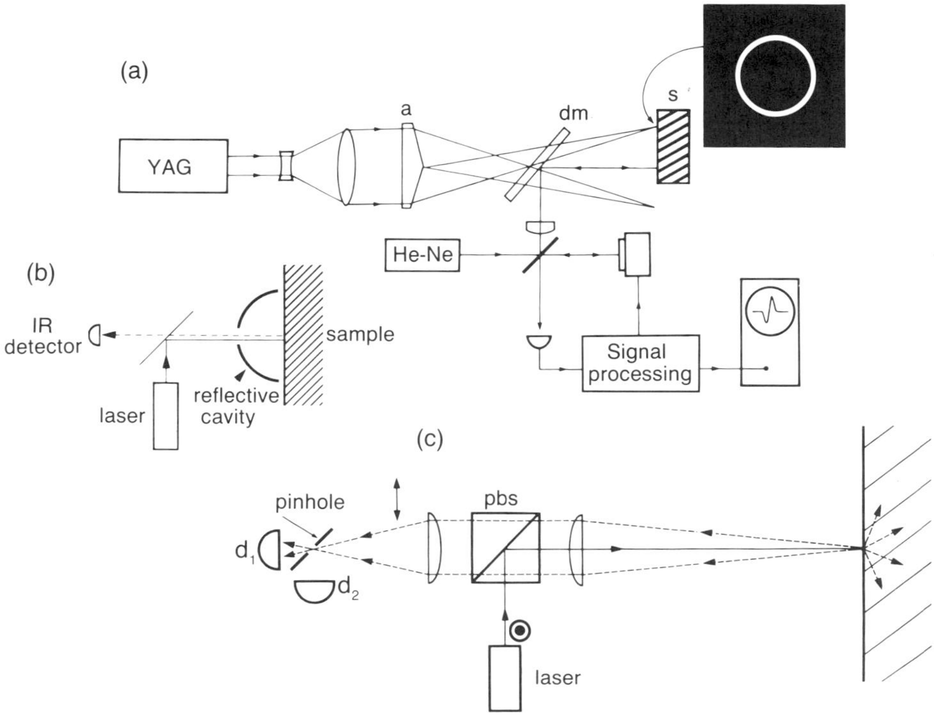

FIGURE 2: Different optical techniques investigated for the characterization of the sintered-ceramic samples: (a) laser-generation of convergent SAW's; (b) monitoring of thermal properties; (c) monitoring of the light-scattering properties. dm: dichroic mirror; s: sample; pbs: polarizing beam-splitter; d_1 and d_2: optical detectors.

2. Optothermal technique

The second method (figure 2b) is based on monitoring the sample thermal properties, which are strongly related to the porosity of ceramic materials[6,7]. A typical approach consists in pulse-heating one side of the sample by a heat source such as a laser beam and measuring the thermal propagation time to the opposite side of the piece to determine the thermal diffusivity. An alternative one-side approach is shown in figure 2b[8]. A laser beam is used to heat the surface to a few degrees C and the surface temperature evolution is monitored by a simple infrared detector. A gold-plated hemispheric cavity is used to obtain a large effective surface absorptivity and emissivity for absolute temperature measurement. It can be shown that such a technique provides an evaluation of the thermal inertia (or equivalently of its square root, the effusivity) and that such a parameter is better related to the material porosity as compared to the thermal diffusivity[8].

The correlation of the measured thermal inertia with the density of the zirconia samples is quite satisfactory, as figure 3b shows. Again,

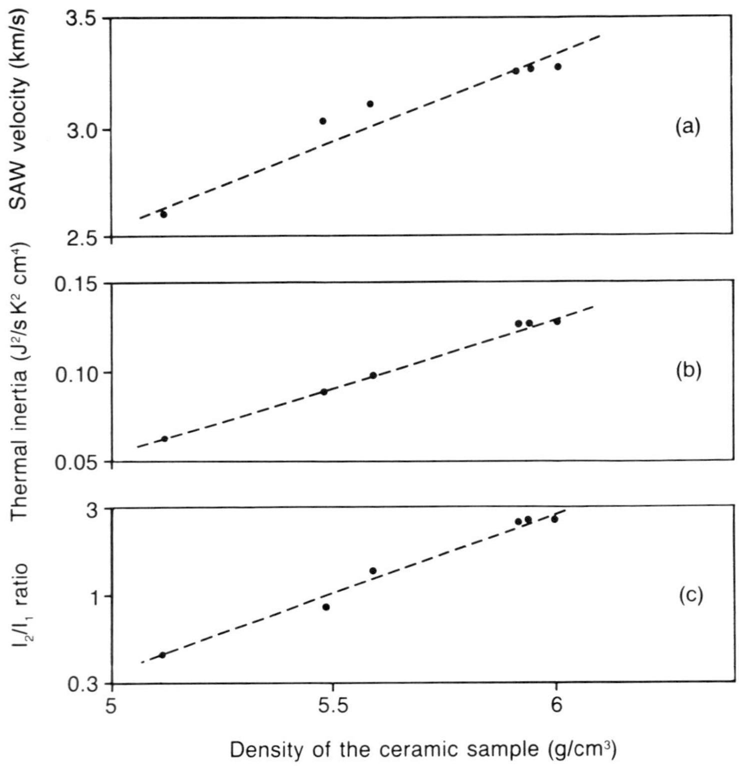

FIGURE 3: Results obtained with the (a) optoacoustic, (b) optothermal and (c) light-scattering methods. In all cases a satisfactory correlation is obtained between the ceramic porosity and the measured parameter.

the surface must be blackened unless the optical penetration depth is smaller than the thermal propagation depth during the observation time. As compared to the optoacoustic approach, the required surface temperature is smaller by two or three orders of magnitude so that thermal stresses are minimized. Moreover, an infrared detector is much simpler and less expensive as compared to an interferometric probe.

3. Light-scattering approach

When the inspected piece is transparent or translucent, such as zirconia and alumina ceramics or most polymers, light-scattering methods may be more conveniently used to characterize inhomogeneous materials. One such approach is shown in figure 2c as applied to the characterization of the porous zirconia pellets. A He-Ne laser beam is projected through a polarizing beam-splitter and focusing optics on the surface of the translucent material to be inspected, where it produces a spot of the order of 0.1 mm in diameter. The laser beam penetrates the porous material and is scattered at each interface, the penetration range being shorter for small-grained or for absorbing materials[9]. Such a scattered beam produces a visible halo whose diameter varies from nearly 0.5 mm to 6 mm for the zirconia samples of lowest and highest density, respectively. A pinhole situated in the image plane of the irradiated surface separates the immediately backscattered radiation, which is focused in the pinhole and detected by detector 1, from the radiation scattered through a larger range, which is mostly collected by detector 2. The use of polarized light eliminates unwanted surface reflections from glazed ceramic samples. As shown in figure 3c, the ratio between the signals I_2 and I_1 obtained from detectors d_2 and d_1 correlates well with the density of the zirconia pellets.

Light-scattering methods require no heating of the piece, so that low-power light sources can be used. A He-Ne laser of a few mW could be used in this case, as compared to a laser of a few watts for the optothermal approach and of several megawatts of peak power for the laser-ultrasound method. Moreover, the visible or near-IR optical components and detectors which can be used for the light-scattering probe are more rugged and less expensive than the medium-IR components required in the optothermal system. In conclusion, the light-scattering approach appears to be the most convenient of the three optical methods considered above for the characterization of translucent zirconia ceramics.

THERMAL CHARACTERIZATION OF COMPOSITES

When the materials to be inspected are opaque, such as graphite-epoxy composites, light-scattering methods cannot in general be used. Some optothermal techniques were thus investigated for the characterization of such materials.

One interesting possibility of thermal methods is the evaluation of phase or molecular orientation in stretched polymer films or in composites by an analysis of the thermal propagation pattern[10]. A typical configuration is shown in figure 4. The inspected part is spot-heated by a narrow laser beam or other point heat source, and the heat-propagation pattern is analyzed by an IR camera. If the material is oriented, such as a unidirectional graphite-epoxy sheet, an elliptical thermal pattern will be observed, with the ratio between the two principal axes (b/a) being related to the square root of the thermal diffusivities in the longitudinal and transverse directions. This may be used to evaluate the orientation of extruded or molded parts, or the relative thermal conductivities of the filler and the matrix in a composite.

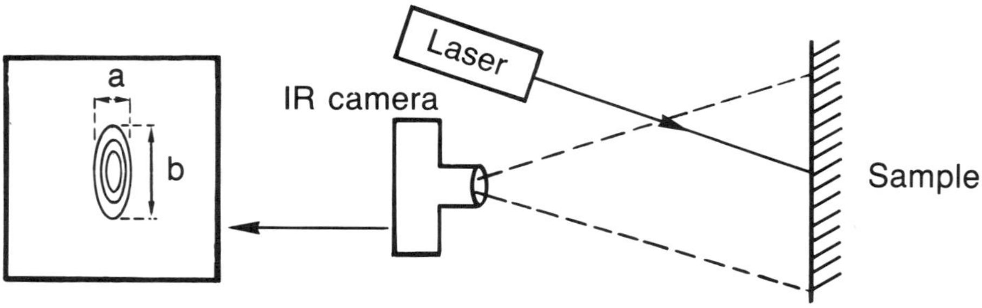

FIGURE 4: Thermographic analysis of phase orientation in composite materials.

Analysis of graphite-epoxy composites

This thermal approach was evaluated for possible application to a typical quality control requirement for graphite-epoxy materials, the evaluation of the filler-matrix proportions in the cured part. Inadequate temperature or part-assembly conditions may produce excessive squeezing of the epoxy material out of the composite, resulting in epoxy-starved regions whose mechanical properties may be affected.

A finite-difference simulation of thermal propagation in the graphite-epoxy material was performed assuming uniform fiber-to-fiber spacing to determine the suitability of thermographic techniques to the detection of epoxy-starved conditions. The eccentricity (b/a) of the simulated thermal pattern was calculated for different graphite-fiber volume concentrations and for different ratios of the fiber thermal conductivity K_f over the matrix conductivity K_m, as shown in figure 5. Isotropic graphite fibers were assumed in the model. As we can see, the eccentricity is a relatively unsensitive parameter for the evaluation of fiber contents when the fiber concentration is in the 30 to 50% range in volume, which corresponds to typical fiber-concentration values of graphite-epoxy materials.

Much more sensitive to fiber concentration is the absolute value of the thermal propagation distance in the transient thermal field, as figure 6 shows. The evaluation of such a parameter may be performed independently of variations in the absorbed heat power by thermal imaging the spot-heated area under transient heating conditions and comparing the recorded thermograph with a reference thermograph obtained on a sample with known fiber concentration.

Such an approach is illustrated in figure 7. The thermographs (a) and (b) show the thermal pattern after a heating period of 20 seconds with a 0.5 W laser of an 8-ply unidirectional NARMCO 5217 sheet cured under normal pressure (690 KPa, pattern a) and under a reduced pressure (69 KPa, pattern b). A sufficient amount of pressure must normally be applied during curing to insure epoxy flow leading to good ply-to-ply bonding. An exceedingly high pressure level generates however excessive overflow of the epoxy leaving an epoxy-starved material of large thermal conductivity. This can be observed in figure 7: the epoxy-poor sheet (a) gives a larger thermal pattern because of the higher conductivity,

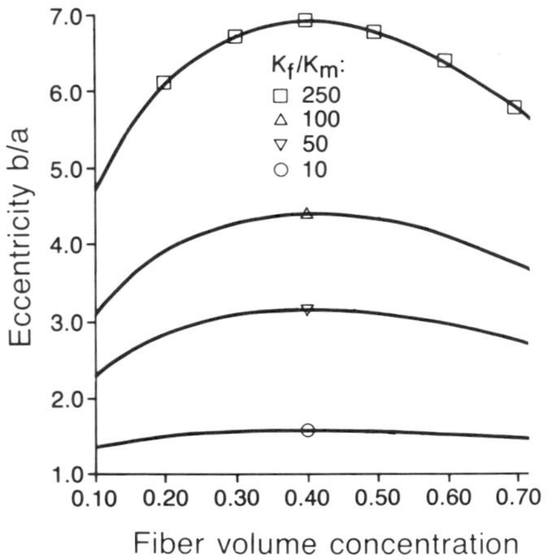

FIGURE 5:
Eccentricity of the elliptical thermal pattern vs. fiber concentration in a unidirectional graphite-epoxy composite as obtained from a numerical model for different ratios of the fiber thermal conductivity K_f and the matrix conductivity K_m.

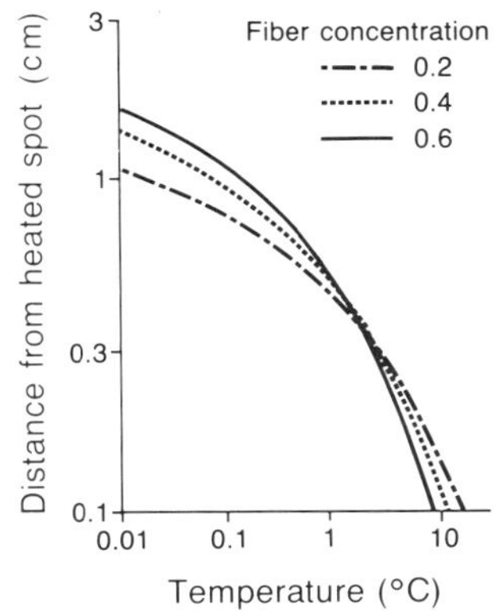

FIGURE 6:
Longitudinal thermal propagation after a 10 ms spot-heating time period obtained for different fiber concentrations with the numerical model assuming K_f = 0.5 and K_m = 0.002 W/cm K.

while the epoxy-rich sheet results in a pattern (b) of smaller extension but with a higher temperature at the center. The difference (c) between the two thermographs is a pattern which is positive along an elongated annular region and negative at the center. Such a temperature inversion of the difference pattern is typical of a thermal conductivity variation, independently of the heat source level and of the surface absorptivity and emissivity. It should also be mentioned that this method is applicable as well to multidirectional sheets, and that the infrared camera could be replaced by a less expensive single-spot infrared detector operating either in the diverging or the converging[11] time-resolved configuration for specific routine tests.

Analysis of steel-polypropylene sheets

Filler-orientation analysis is important for materials containing a dispersed phase such as short fibers in a polymeric matrix. Liquid flow in the extrusion or molding processes may result in local orientation or flocculation of otherwise randomly oriented particles, with substantial effects in the mechanical properties of the product[12,13]. Thermal techniques may be used for the evaluation of filler distribution close to the surface, provided that the thermal conductivity of the filler is substantially different from the thermal conductivity of the matrix.

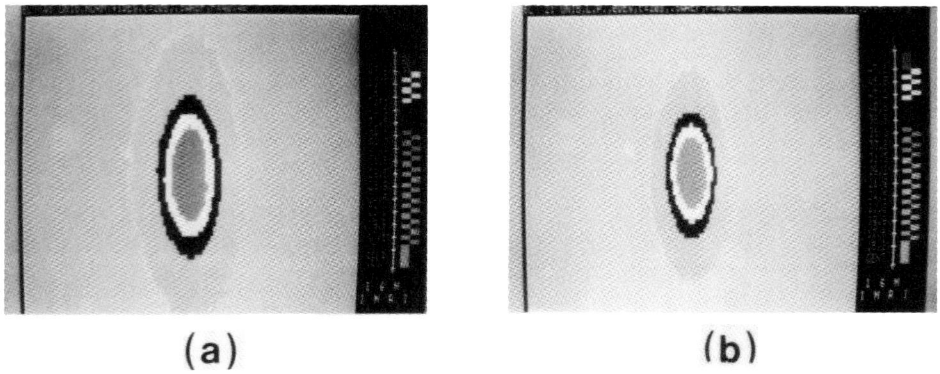

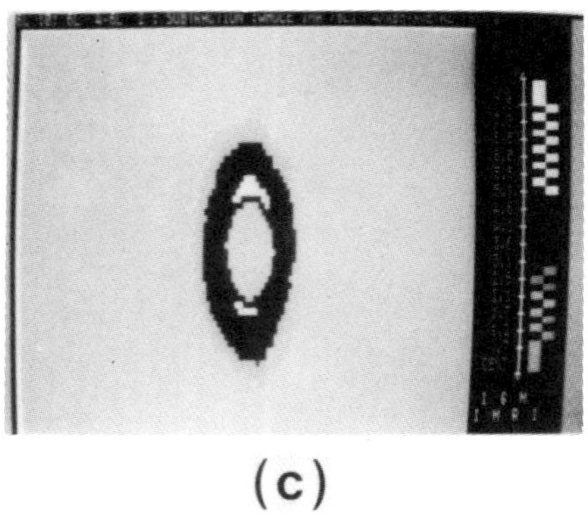

FIGURE 7: Elliptical thermal patterns obtained by laser spot heating a graphite-epoxy sheet cured under (a) normal and (b) low pressure. Thermograph (c) shows the difference of the two patterns. The imaged area is 25 cm^2.

Such an approach was investigated in the case of steel-fiber-filled polypropylene sheets with different steel concentration and orientation. Steel fibers of nearly 1 mm in length are incorporated in the polymer matrix to obtain good thermal and electrical conductivities for specific applications[11]. Strong anisotropy may affect the conductivity properties if the fibers are not randomly oriented. By observing the ellipticity of spot-heated samples by methods of the kind shown in figures 6 and 7, the degree of filler orientation may be evaluated. Such an approach is valid for any composite containing phases with substantially different thermal conductivities.

OPTICAL CHARACTERIZATION OF POLYMER-BASED MATERIALS

As we have mentioned above, nonthermal optical methods have the advantage over optothermal or optoacoustic techniques of not requiring any heating of the sample surface, so that low-power, eye-safe light sources can be used. Nonthermal optical techniques should thus be preferred when possible, i.e. for translucent materials. This section will review some applications of such techniques to polymer-based materials.

Optical spectroscopy analysis of thermosetting polymers

Spectroscopic techniques are widely used for the investigation of many properties of polymeric materials, such as composition, degradation or hydrolysis consequent to moisture infiltration[14,15]. One of the application fields where quality-assurance industrial needs are particularly strong is the nondestructive characterization of epoxy-based products, and particularly of structural composites used in the aerospace industry. Material characterization needs such as the evaluation of the degree of cure before and after autoclave processing or the measurement of absorbed moisture are of primary importance for the integrity assurance of the assembled parts.

Medium-IR spectra provide detailed molecular information on the polymeric resin material. The relatively large molar absorption values in this part of the spectrum require however the use of very thin films, typically 10 μm-thick, for transmission spectroscopy. Spectroscopic inspection of bulk materials in the liquid state, such as molten polymers in an extruder, may be performed by internal-reflection techniques[16]. Another approach, which is however rather qualitative, is diffuse-reflectance spectroscopy, which is applicable to surface characterization of a variety of materials including opaque media such as graphite-epoxy composites[17]. Another approach for polymer analysis is Raman spectroscopy[18]. In this latter case one can work in the visible spectral range where inexpensive optical components and rugged detectors can be used, but the signal level is usually very low.

A relatively little used spectral range for infrared spectroscopy is the near-IR from 0.7 to 2.5 μm[19]. This spectral range is particularly convenient for industrial applications because of the availability of strong wide-bandwidth sources, sensitive detectors and rugged optical components. Moreover, absorption coefficients tend to be much weaker than in the mid-IR, so that thicker material films can be analyzed when the level of impurities is low. A certain amount of effort is being devoted to near-IR spectroscopy at our institute, both for material characterization and for on-line extruded sheet thickness metrology[20].

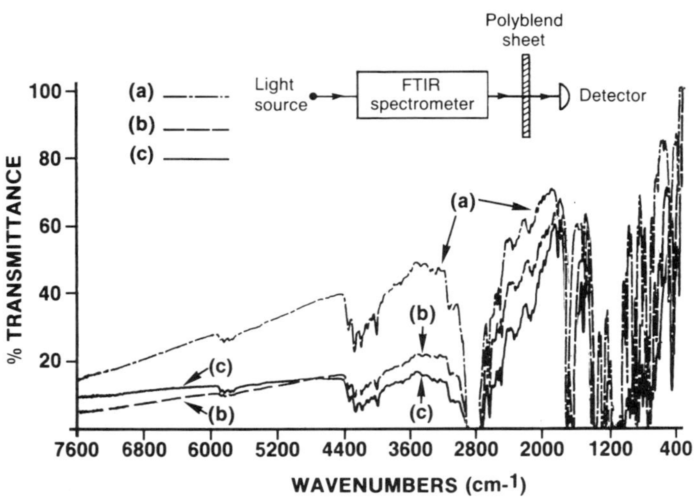

FIGURE 8: Comparative transmission spectra for 90% polypropylene, 10% polycarbonate polymer blends with disperse spherical particles of typical diameters (a) 1-2 μm, (b) 4-10 μm and (c) 10-20 μm.

Tests performed on epoxy sheets showed that polymer curing can be quantitatively followed by monitoring the epoxy-ring peak at 2.2 μm, while water absorption can be evaluated from the absorption band at 1.9 μm.

Spectral-turbidity analysis of polyblends

In addition to molecular absorption analysis, spectral techniques may also be used for size evaluation of material inhomogeneities when such a size is of the order of magnitude of a light wavelength. This approach was recently considered for the evaluation of the average particle dimension in immiscible polymer blends, a parameter which strongly affects the mechanical properties of such materials[20]. 90%-polypropylene, 10%-polycarbonate sheets were prepared under different processing conditions resulting in an average size of the polycarbonate particles of (a) 1.6 μm, (b) 9.8 μm and (c) 18.9 μm, respectively for the three sheets.

Figure 8 shows the transmittance spectra of such films[20]. It can be seen that, apart from the absorption peaks of constant relative depth, the average light-transmission level is lower at shorter wavelengths. This is related to the attenuation produced by light scattering from the dispersed polycarbonate phase, which is strongly related to the ratio between the particle diameter and the wavelength[21]. Consequently, the attenuation at short wavelengths is much weaker for samples containing small dispersed particles (curve a in figure 8) than for samples containing particles larger than the wavelength (curves b and c). The fact that curves b and c cross each other in the short-wavelength spectral range can be explained in terms of the decreasing number of interfaces for an increasing particle size at constant concentration in the geometrical-optics limit[20].

The spectral turbidity approach thus appears to be appropriate for a rapid and noncontact evaluation of the average particle size in polyblends by sampling the transmission spectrum. Such an approach appears to be much less affected by multiple scattering than the small-angle light scattering (SALS) approach where the angular pattern is observed under polarized-light conditions[22,23]. Combined spectral and angular scanning might however prove useful for the evaluation of polyblends with fibrillar morphology.

Far-IR inspection of filler orientation in reinforced polymers

Light-scattering techniques, eventually coupled to appropriate spectral filtering, may also be useful to evaluate the filler dispersion uniformity by line-scanning transmission tests or the filler orientation by angularly scanning the light transmitted through relatively clear polymeric sheets. Backscattering techniques may alternatively be used for near-surface filler evaluation in thick parts.

When the material to be inspected is not sufficiently clear in the visible or mid-IR range, as it is the case for parts containing large concentrations of pigments or of dispersed gaseous particles, radiation of longer wavelength such as microwaves may be used at the expense of a reduced spatial resolution. An intermediate approach is the use of far-IR lasers, in the 50 to 500 μm range, which have recently become commercially available. Far-IR radiation offers a better penetration depth in inhomogeneous media as compared to mid-IR, while focusing capabilities and spatial resolution are better than in the case of microwaves. A demonstration of far-IR possibilities for reinforced-polymer inspection was recently reported[24].

CONCLUSION AND ACKNOWLEDGEMENTS

Optical techniques have found until now relatively few applications in the characterization of industrial materials. When analyzing opaque materials such as metals, optical methods are used mainly for morphological inspection of the part geometry but they are of little use for material characterization except for instrumental applications in combination with other techniques such as ultrasound generation or laser-induced breakdown spectroscopy. When analyzing translucent materials such as polymers or ceramics, optical methods may however more conveniently be directly applied to material characterization by optical-spectroscopy or light-scattering techniques. Significant developments in on-line inspection using such methods are expected in the future, taking advantage of the recent availability of rugged and inexpensive optoelectronic devices as well as of the possibilities offered by optics in terms of remote-inspection capabilities.

The authors wish to acknowledge the contributions of many colleagues, in particular K. Cole of IMRI for the spectral analysis, B. Favis and T. Vu-Khanh of IMRI for the preparation of the polymer-based samples and several discussions, S. Johar and P. Lauzon of the Ontario Research Foundation for the supply of the zirconia ceramic samples, G. Dziub of Canadair for supplying some graphite-epoxy materials, as well as M. Lamontagne of IMRI for his valuable technical assistance.

REFERENCES

1. G. Birnbaum and G.S. White "Laser techniques in NDE" in : "Research Techniques in NDT", R.S. Sharpe ed., 7: 259 (1984).
2. P. Cielo, X. Maldague, S. Johar and B. Lauzon "Some laser-based techniques for the characterization of sintered ceramics", Mater. Eval. 44: 770 (1986).
3. A.G. Evans, B.R. Tittman, L. Ahlberg, B.T. Khuri-Yakub and G.S. Kino "Ultrasonic attenuation in ceramics" J. Appl. Phys. 49: 2669 (1978).
4. D.S. Kupperman and H.B. Karplus "Ultrasonic wave propagation characteristics of green ceramics" Am. Cer. Soc. Bull. 63: 1505 (1984).
5. P. Cielo, F. Nadeau and M. Lamontagne "Laser generation of convergent acoustic waves for materials inspection" Ultrasonics 23: 55 (1985).
6. S.K. Rhee "Porosity-thermal conductivity correlations for ceramic materials" Mater. Sci. Eng. 20: 89 (1985).
7. G. Ziegler and D.P.H. Hasselman "Effect of phase composition and microstructure on the thermal diffusivity of silicon nitride" J. Mater. Sci. 16: 495 (1981).
8. P. Cielo, S. Dallaire, G. Lamonde and S. Johar "Measurement of thermal inertia by the reflective-cavity method". To appear in Can. J. Phys. (Sept. 1986).
9. G. Kortüm "Reflectance Spectroscopy" Springer-Verlag, New York (1969).
10. M.A. Berrie, K.E. Puttick, J.G. Rider, M. Rudman and R.D. Whitehead "Thermal probe analysis of orientation in polymers and composites" Plast. Rubber Proc. Appl. 1: 129 (1981).
11. P. Cielo, L.A. Utracki and M. Lamontagne "Thermal diffusivity measurements by the converging-thermal-wave technique" To appear in Can. J. Phys. (Sept. 1986).
12. R.P. Hegler and G. Mennig "Phase separation effects in processing of glass-bead and glass-fiber-filled by injection molding" Polym. Eng. Sci., 25: 395 (1985).

13. T. Vu-Khanh, B. Sanschagrin and B. Fisa "Fracture of mica-reinforced polypropylene: mica concentration effects" Polym. Compos. 6: 249 (1985).
14. Y.T. Liao and J.L. Koenig "Applications of FTIR spectroscopy to the study of fiber-resin composites" Dev. Reinf. Plast. 4: 31.
15. J.M. Chalmers and M.W. MacKenzie "Some industrial applications of FTIR diffuse reflectance spectroscopy" Appl. Spectr. 39: 634.
16. N.J. Harrick, "Internal Reflection Spectroscopy" Harrick Scientific Corp., 1979.
17. K.C. Cole, C. Lehto and M. Yuhasz "Detection of mold release agent contamination on the surface of epoxy-based composites by diffuse-reflectance FTIR spectroscopy" in: "Optical Techniques for Industrial Inspection", P. Cielo ed., Proc. SPIE vol. 665 (1986).
18. B.J. Bulkin, F. DeBlase and M. Lewin "Raman spectroscopy applied to polymer analysis: low frequency spectra of polyester fibers" in: Optical Techniques for Industrial Inspection", P. Cielo ed., Proc. SPIE vol 665 (1986).
19. D.L. Wetzel "Near-IR reflectance analysis, sleeper among spectroscopic techniques" Anal. Chem. 55: 1165A (1983).
20. P. Cielo, K. Cole and B. Favis "Optical inspection for industrial quality and process control" 6th Int. Conf. on Instrumentation and Automation in the Paper, Rubber, Plastics and Polymerization Industries" Akron, OH, Oct. 27-29, 1986.
21. H.C. Van de Hulst "Light Scattering by Small Particles" John Wiley, London, 1957.
22. R.J. Samuels "SALS from optically anisotropic spheres and disks. Theory and experimental verification" J. Pol. Sci., 9: 2165.
23. R.S. Stein "Optical behavior of polymer blends" in: "Polymer Blends", D.R. Paul and S. Newmann eds., pp. 393-443. Academic Press, New York, 1978.
24. R. Boulay, P. Dubé, P.A. Bélanger, T. Vu-Khanh and P. Cielo, "Analysis of filler concentration and orientation in reinforced polymers by far-IR techniques" in: "Optical Techniques for Industrial Inspection", P. Cielo ed., proc. SPIE vol. 665 (1986).

MECHANICAL CHARACTERIZATION OF PAPER

BY LASER SPECKLE INTERFEROMETRY

B. Castagnede

State University of New York
College of Environmental Science and Forestry
Syracuse, New York 13210

ABSTRACT

A laser speckle interferometry method applied to the determination of strains in paper is described. We are able to characterize the two-dimensional strain field of the viewed surface for different test configurations (tension, compression). The method is shown to provide detailed strain information all along the stress-strain curve until failure.

Paper materials have certain unique characteristics related to their fibrous network composition. Internal light reflections and refractions on numerous surfaces in the translucent structure create a decorrelation phenomenon. This problem is avoided by using a thin coating technique.

The strain field is shown to be inhomogeneous. A technique for strain determinations over a square grid (7 columns, 22 rows) with a 3 mm spacing is described. Our method allows a determination of differences in properties for machine-made papers (which are anisotropic) between the machine direction (MD) and the cross-machine direction (CD). Additionally, a method to compute the in-plane Poisson ratios is proposed.

Details concerning the optical filtering technique used for information processing are given. This semi-automatic system provides a high level of confidence in the measurements and permits the detection of very fine effects.

INTRODUCTION

The measurement of strain in paper has proven to be one of the most intractable problems in the testing of this material. Several techniques that use various mechanical devices [1]- generally extensometers - exist. Some important limitations exist with these methods. First, since they are contact techniques, the measured deformation is inherently perturbed by the mounted device. Secondly, the strain that is measured is an average response over the gage length of the extensometer; point-to-point variations in strain cannot be detected. Mechanical test devices are intensively used in the paper industry for routine measurements. Nevertheless, they are not sufficiently accurate for good understanding of the micromechanics of paper. Acoustical methods, which require a coupling medium, are inadequate for paper materials. For instance, porosity and thickness variations drastically limit the application of most of the acoustical techniques.

The most promising methods for paper strain measurements appear to be optical. Holographic interferometry has been used to determine strain in the thickness direction [2]. Two limitations to such a method are that (a), it has high sensitivity to vibrations, and (b), the analysis of holograms is generally intricate. An attempt to apply laser speckle interferometry to paper was made for a uniaxial tensile test [3]. This work is mainly qualitative; it needs to be extended with a new approach that considers the whole two-dimensional strain field instead of a mere determination of elongations in particular locations. In the present paper we report some new progress along that direction and outline additional work that can be done in the future.

EXPERIMENTAL PROCEDURES

Single-beam laser speckle interferometry is a very simple experimental method for measuring very small in-plane displacements (in the range from a few μm to 0.1 mm) [4,5]. The recording of the displacement field is made on a high resolution photographic plate (i.e. Agfa 10E75 holographic film). Processing of the displacement field information can be done later by spatial filtering (illustrated in Figure 1). Neglecting the contribution of the image of the point source located at the focus (point O in Figure 1), the diffraction pattern consists of a series of straight parallel fringes modulated by a halo function characteristic of the optical set-up used during the recording. Usually this is a Gaussian function with a circular shape. In the particular case where there is also some out of plane movement, this halo function may be modified and exhibit an elliptical shape, or even secondary Newton's rings [6].

The orientation of the fringes θ, is perpendicular to the displacement vector d whose magnitude is related to the interfringe distance S by the simple relationship:

$$d = K/S \quad , \quad \text{where} \quad K = \lambda z/M \quad . \tag{1}$$

In relation (1), M is the magnification factor during the recording, z is the distance between the specklegram and the transform plane, and λ is the wavelength (0.6328 μm).

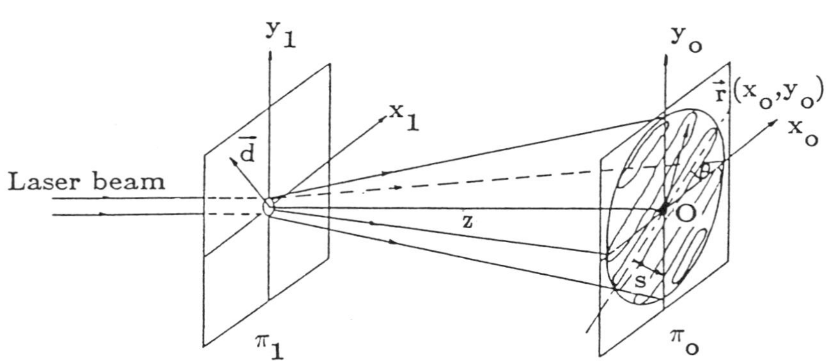

π_1: Specklegram (Photographic plate double exposed)
π_o: Transform plane (fringes pattern)

Fig. 1. Pointwise Filtering Set-Up.

The quality of the diffraction pattern may be adversely affected by several different causes. It is important that the speckle pattern on the sheet remain quite stable during the course of loading. This guarantees that only structural changes will lead to shifts of individual speckle positions. During processing of the specklegrams, the region under illumination should have had as homogeneous a deformation as possible. For paper materials, these two assumptions have not been conclusively verified. As a result, decorrelation occurs, limiting the accuracy of reading displacement information. It is not easy to get a clear explanation of that phenomenon. Recently some theoretical work was done [7,8], describing the origin of speckle decorrelation. Unfortunately these studies, which are limited to the optical set-up, do not include the effect of the material being tested. Uncoated papers exhibit substantial decorrelation, which is related to their translucent character and fibrous structure. The incident light penetrates to some extent before being scattered back, and an average response through the thickness is recorded. Each layer contributes individually, with each being different from one stratum to the next one. The result is a loss of contrast in the fringe pattern obtained after filtering. To avoid or at least limit this problem, here different coatings were used. One was a very thin coating made by a uniform vacuum-metallic deposition of Au/Pd with a thickness of approximately 50 Angstrom. The other one was a high-gloss, fast-drying spray paint. Gold, bronze, and aluminum paints were used for their good reflection properties. The specimens tested were cut from 240 g/m^2 sheets made on a 48-inch, 200 fpm Fourdrinier machine. The pulp used was bleached kraft containing 80 % southern pine and 20 % mixed hardwood.

During processing, the interferogram (or specklegram) is scanned two-dimensionally over a square grid, shown in Figure 2. For each location (i,j), the diffraction pattern is projected on a digitizer tablet. The two items of rough data, the rotation angle θ and the interfringe distance S, are calculated using a least square fitting algorithm for different pairs

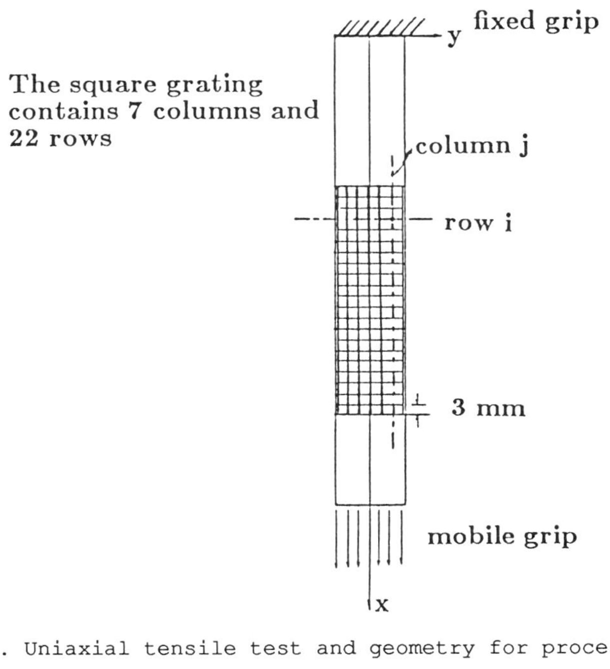

Fig. 2. Uniaxial tensile test and geometry for processing.

a)

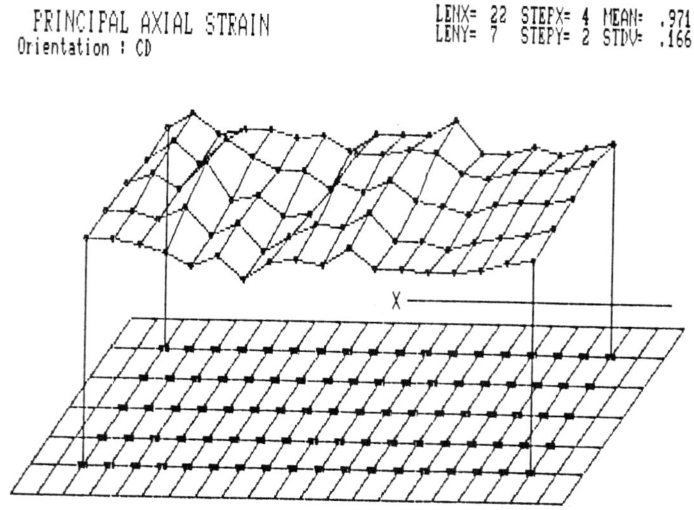

b)

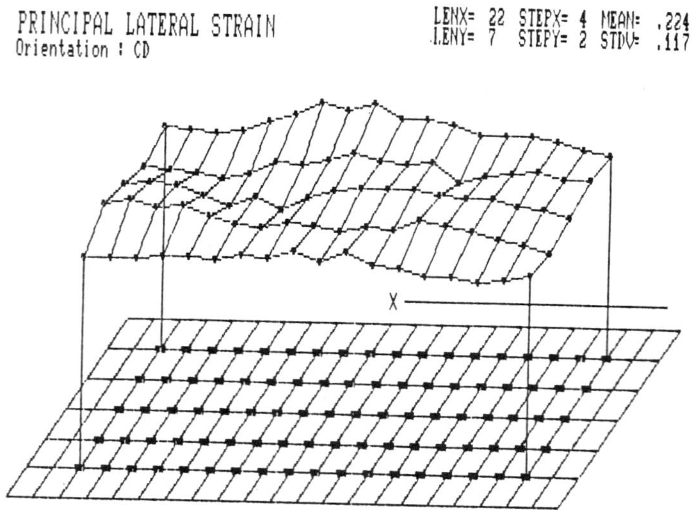

Fig. 3. Distribution of principal strain components.
a) Axial component b) Lateral component
The fixed grip is beyond the right part of the plot.
The scale is given by the coefficient of variation.

a)

POISSON RATIOS

Orientation : CD

LENX= 22 STEPX= 4 MEAN: .214
LENY= 7 STEPY= 2 STDV: .112

.088	.111	.055	-.001	.054
.147	.125	.124	.067	.021
.235	.190	.159	.125	-.020
.262	.187	.232	.230	.045
.318	.160	.276	.327	.033
.299	.073	.271	.330	.136
.277	.227	.301	.234	.142
.363	.188	.221	.178	.215
.369	.232	.269	.164	.172
.507	.274	.217	.206	.262
.589	.389	.238	.279	.198
.543	.238	.102	.297	.318
.420	.259	.187	.209	.153
.320	.109	.099	.338	.224
.223	.047	.125	.296	.188
.223	.129	.260	.339	.139
.181	.112	.243	.362	.137
.242	.162	.257	.324	.129

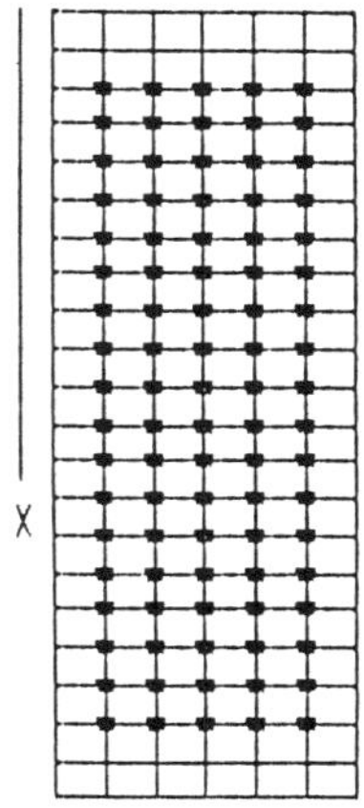

b)

POISSON RATIOS

Orientation : MD

LENX= 22 STEPX= 4 MEAN: .535
LENY= 7 STEPY= 2 STDV: .193

.551	.837	.635	.726	.553
.617	.940	.704	.663	.616
.691	.813	.650	.740	.633
.770	.822	.650	.598	.484
.830	.814	.723	.719	.436
.780	.680	.462	.563	.422
.736	.698	.329	.467	.264
.583	.584	.396	.461	.180
.695	.618	.418	.401	.174
.569	.679	.522	.664	.265
.533	.631	.574	.751	.431
.422	.620	.549	.823	.493
.427	.551	.558	.970	.594
.418	.483	.553	.802	.403
.300	.376	.595	.719	.346
.271	.306	.511	.467	.272
.201	.227	.389	.355	.235
.178	.274	.280	.241	.170

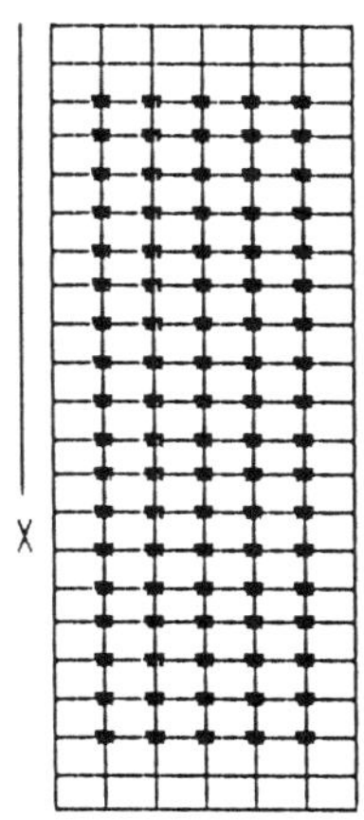

Fig. 4. Distribution of in-plane Poisson ratios.
a) CD orientation b) MD orientation
The fixed grip is beyond the upper part of the plot.

Table. Summary statistics for 2-D strain analysis.
STEPX=4 ; STEPY=2

	Orientation: MD		Orientation: CD	
	Mean	Coef. var.(in %)	Mean	Coef. var.(in %)
Axial	1.000	21	1.000	25
Lateral	0.568	46	0.286	75
Shear	0.089	107	0.202	63
Pr. Axial	0.951	12	0.971	17
Pr. Lateral	0.496	32	0.206	52
Poisson rat.	0.535	36	0.214	52

of fringes. As the constant K in equation (1) is only related to some fixed geometrical parameters, one needs to perform a preliminary calibration. After the calibration, a two-dimensional strain analysis using a finite difference technique can proceed. Principal strain components are calculated using a Mohr's circle procedure.

RESULTS AND DISCUSSION

A limited amount of data is presented here. We believe that it clearly demonstrates the method's versatility as a metrological tool. Tests were performed in tension for two samples (one MD and one CD coupons). The initial strain was around 0.1 % (beginning of the linear range). For both orientations the incremental strain was 0.05 % . During the strain analysis, the average parameters STEPX and STEPY must be fixed. For that purpose, a simulation procedure using perturbation techniques was developed. This calculation permits the definition of a compromise minimizing the influence of the random errors due to the processing of the diffraction patterns. Estimations for the standard deviations are 0.5 % for the interfringe distances, and 15 minutes of arc for the rotation angles.

The statistics summarized in the Table show consistently high values for standard deviations. This indicates that the strain field is inhomogeneous. For axial and lateral strain components, the coefficients of variation are higher for CD than for MD orientation. Even when initial or incremental strain is varied, this trend persists. This feature can be directly related to fiber orientation distribution (during the papermaking process, fibers tend to align with the machine direction). Three-dimensional normalized plots shown in Figure 3 indicate that some spatial pseudo-periodicity in the strain field exists. It is certainly related to variations in paper structural parameters such as basis weight fluctuations, fiber orientation, thickness variations.

In-plane Poisson ratios are calculated as the ratio of the two principal strain components. There is substantial variation in these values represented in Figure 4. The Poisson ratios are distributed in discrete areas of low and high values. The mean value is significant in terms of paper mechanics; each individual value is accurately determined. The following concept must be emphasized. Each 2-D map is like a "static image" corresponding to a given pair of parameters for initial and incremental strain. For a given sample, the change of one or both parameters should entirely modify the stucture of the new map. This behavior is the subject of ongoing studies. Additionally, the overall behavior of paper is presently being studied in terms of variations of the in-plane Poisson ratios along the stress-strain curve. Some theoretical developments [9] predict a decrease in the range of 10 to 20 % as one proceeds from the beginning to the end of the curve.

CONCLUSIONS

Some of the unique possibilities of laser speckle interferometry to characterize the micromechanics of paper materials have been described. Results were presented using a 2-D approach that is appropriate for the description of paper structure (at least as a first approximation). Some limitations remain, particularly in relation to the processing of the specklegrams. At present, this step is tedious and time-consuming. To attempt some systematic studies, a fully automatic video acquisition system, including programmable stages for the 2-D scanning, is needed.

ACKNOWLEDGEMENTS

The author wishes to gratefully acknowledge Dr. R.E. Mark from the College of Environmental Science and Forestry (State University of New York at Syracuse) and Dr. R.W. Perkins of the Department of Mechanical and Aerospace Engineering (Syracuse University) for their encouragment and support. Gratitude is also due to Messrs. A.R.K Eusufzai and C.M. Crosby for their technical assistance.

REFERENCES

[1] V.C. Setterholm and D.E. Gunderson, Observations on Load-Deformation Testing, in : "Handbook of Physical and Mechanical Testing of Paper and Paperboard", vol.1, R.E. Mark, ed., Marcel Dekker, New York, (1983).
[2] D.R. Axelrad and K. Rezai, Deformation Measurement of Thin Foils by Holographic Interferometry. Proc. Conf. Holographic Interferometry and Speckle Metrology, Am. Optic. Soc.,(1980).
[3] M.B. Lyne and H. Bjelkhagen, Pulp and Paper Mag. Canada, 7:29 (1981).
[4] R.P. Khetan and F.P. Chiang, Appl. Optics, 15:2205 (1976).
[5] D.W. Li, J.B. Chen and F.P. Chiang, J. Opt. Soc. Am. A, 2:657(1985).
[6] M. Francon, Information Processing Using Speckle Patterns, in : "Laser Speckle and Related Phenomena", J.C. Dainty, ed., Springer-Verlag, Berlin, (1984).
[7] I. Yamaguchi, Optica Acta, 28:1359 (1981).
[8] I. Yamaguchi, J. Opt. Soc. Am. A, 1:81 (1984).
[9] R.W. Perkins and R.E. Mark, "Micromechanics Constitutive Model for Ribbon-like Fiber Non-wovens.", presented to Penn State University Oct. 7-9 (1985).

OPTICAL JOINT FOURIER TRANSFORM CORRELATION FOR PHASE OBJECT RECOGNITION

Paolo Sirotti

Dipartimento di Elettrotecnica Elettronica Informatica
University of Trieste
Trieste, Italy

INTRODUCTION

Several methods can be used for visualizing phase objects, i.e. transparent planar objects whose index of refraction is a function of two spatial coordinates x, y and refractive properties are negligible[1]. Examples of phase objects are thin biological sections, flames and hot gases fluxes, pellicles, ultrasonic fields ... In this paper, we present a Joint Fourier Transform correlation method[2] to recognize or classify automatically without visualization phase objects. The method is applied to the recognition of different strain states in thin plastic samples (polymethyl methacrylate). We also mention the results obtained on rapidly self modifying liquid pellicles which are other phase objects prepared by smearing microscopy cover glasses with a fast evaporating solvent.

JOINT FOURIER TRANSFORM CORRELATION

Two transparent objects g(x,y), h(x,y) are located at a fore plane of a convergent lens and transilluminated by coherent light (fig. 1) from a common He-Ne laser (λ = 632.8 nm, output power 10 mW). On the back focal plane of the lens (f = focal distance) the Joint Fourier Transform (JFT) of input objects is formed. In order to have exact Fourier Transform relationship between back focal plane and input plane, the input plane should be the front focal plane: if distance = f a quadratic phase error affects the transform, however in JFT correlation this error is irrelevant, provided that it is the same for each single transform interfering on the back focal plane. For the same reason JFT correlation has no vibration isolation problems: the only important parameter is the relative position of the images to be correlated. To simplify, we assume that g(x,y) and h(x,y) are symmetrically located at a distance a from each other along the x-axis.

At the lens back focal plane, the JFT of the two objects can be written as follows:

$$E(u,v) = H(u,v)\, e^{\frac{j\pi a}{\lambda f} u} + G(u,v)\, e^{\frac{-j\pi a}{\lambda f} u} \quad (1)$$

Capital letters indicate Fourier transform, u, v (mm) are the distances measured on the transform plane; they are related to the spatial frequency coordinates p, q (lines/mm) by:

$$p = u/\lambda f \qquad q = v/\lambda f \tag{2}$$

Let us suppose that quadratic recording of JFT is now performed, for example on a photographic film. This recording produces an image which can be expressed as follows:

$$|E(u,v)|^2 = |H|^2 + |G|^2 + G^*H\, e^{\frac{j2\pi a}{\lambda f}u} + GH^*\, e^{\frac{-j2\pi a}{\lambda f}u} \tag{3}$$

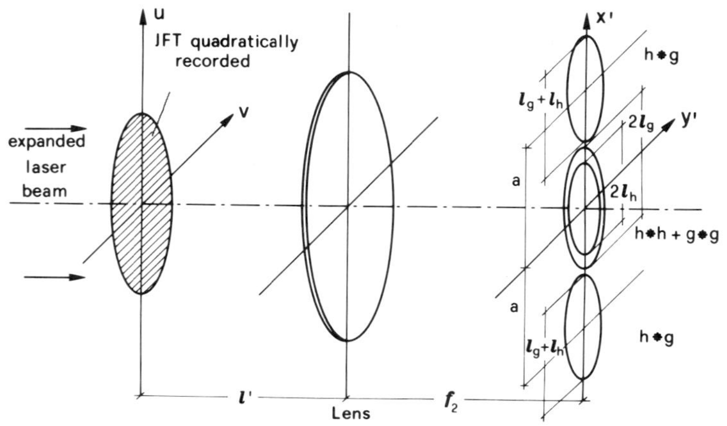

Fig. 1: JFT recording

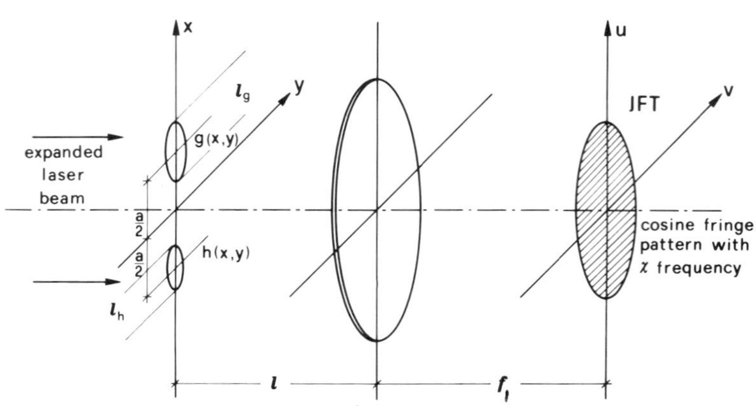

Fig. 2: JFT retransformed to produce the correlation pattern.

A second Fourier transform of this last image, obtained by placing it on the lens input plane again (fig. 2), produces the output pattern:

$$f_o(x',y') = g(x',y')*g(x',y')+h(x',y')*h(x',y')+ \\ g(x',y')*h(x'-a,y')+g(x',y')*h(x'+a,y') \qquad (4)$$

where the symbol * denotes correlation.

The first two terms of equation (4) are the autocorrelation functions of h(x,y) and g(x,y) and are centered on the optical axis: the third and fourth terms are the cross-correlation functions of g(x,y) and h(x,y) and centered at a distance a from the optical axis, on the x'axis.

To avoid overlap of the output terms, the following condition must be satisfied:

$$a - (\iota g + \iota h)/2 \geq \iota g \qquad (5)$$

where ι_g and ι_h are the spatial extensions of g(x,y) and h(x,y), and we have assumed that $\iota g \geq \iota h$.

Representing H(u,v) and G(u,v) by

$$\begin{aligned} H(u,v) &= |H(u,v)| e^{j\Phi(u,v)} \\ G(u,v) &= |G(u,v)| e^{j\Phi(u,v)} \end{aligned} \qquad (6)$$

eq. (3) can be rewritten:

$$|E|^2 = |H|^2 + |G|^2 + |G|\ |H| \cos\left(\frac{2\pi a}{\lambda f} u + \Phi - \Theta\right) \qquad (7)$$

Eq. (7) shows that quadratically recording the JFT produces amplitude and phase modulations. Pure amplitude modulation is obtained if $\Phi(u,v) = \Theta(u,v)$ for each u,v or if the input objects are hermitian functions (i.e., with real Fourier transforms). In these cases, the JFT is cosinusoidal with the modulation carrier frequency χ:

$$\chi = a/\lambda f \qquad (8)$$

Eq. (5) and (8) limit the range allowed for the separation distance between input objects: a should be sufficiently large to satisfy (5) and sufficiently small to keep χ below the cut-off frequency of the recording medium.

The method can be easily extended to multiple correlation among an unknown object g and a set of reference objects h_i (i = 1, ... n). The spatial extent and geometrical disposition of the input objects must be choosen to avoid overlapping of output terms. Each pair of input objects produces a modulation pattern with a spatial carrier frequency that should satisfy the conditions mentioned above.

Assuming that input objects have the same mean square value, the autocorrelation peak is greater than all cross-correlation peaks: the intensities of the correlation peaks may be taken as a measure of the similarity between the objects to be compared. Other criteria may be added to allow a more reliable identification depending on the nature of input objects[3].

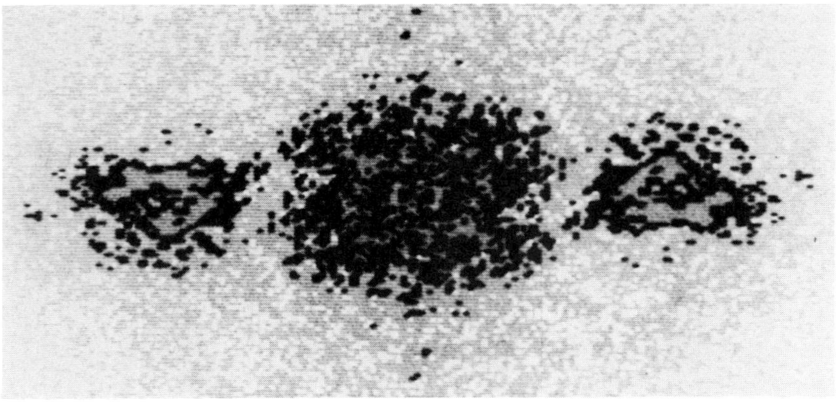

Fig. 3: Output correlation pattern observed on the back focal plane

JFT CROSS-CORRELATION OF STRAIN GENERATED PHASE OBJECTS

Phase objects can be described by:

$$f(x,y) = e^{j\Psi(x,y)} \qquad (9)$$

Therefore they do not possess the property, as real objects do, of having hermitian Fourier transforms, for which the modulus is an even function of spatial coordinates. However this property is valid for the modulus of the autocorrelation function, since autocorrelation is the Fourier transform of the spectral power density (a real function). Therefore, the autocorrelation of phase objects is symmetrical. As a consequence, the symmetry of the cross-correlations (except in particular cases) and not only the intensity of the peaks, can be considered as a measure of the similarity between objects. In optical pattern classification the symmetry of the cross-correlation between an unknown and a reference object proves often as the most effective criterion[5].

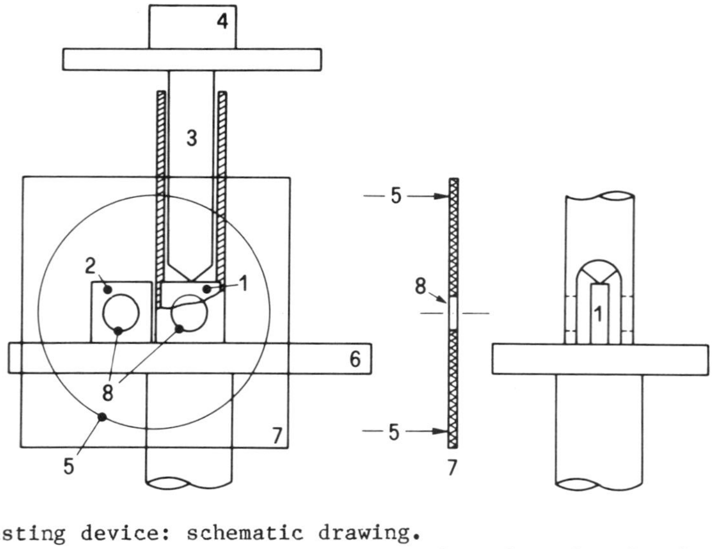

Fig. 4: Testing device: schematic drawing.
1. loaded sample - 2. unloaded sample - 3. pointed rod - 4. applied load - 5. expanded laser beam - 6. pin mounted base supported by a x-y-z translation stage.

Fig. 3 shows an example of JFT cross-correlation between two phase objects obtained by subjecting plastic samples to different strain states by means of two screws entered at different depths. The lateral zones are the cross-correlations between input objects, and are symmetrically located around the center (that is the optical axis). The central zone, which is not useful for classification purposes, is the sum of the autocorrelations of the input objects and presents the symmetry mentioned above.

JFT reconstruction can be performed either optically, as shown in fig. 2, or by means of a digital processing system, recording directly the transform with a TV camera[2,3]. The criteria to evaluate the correlation depend upon the method used to reconstruct the transform. An all-optical method is very simple and permits to use very high frequencies χ. The evaluation of cross-correlations can be obtained by means of a suitable set of photodetectors[4] and/or by visual inspection of the symmetry. Digital processing acts on line and allows the use of more sophisticated criteria[3]; however the frequency χ is generally quite lower.

The first set of tests performed on plastic samples stressed by screws permitted to recognize similar strain states, but it was not suitable for a quantitative analysis. Therefore we built the device shown schematically in fig. 4. The samples under test (2 mm thick) are illuminated by the expanded laser beam through two circular windows (diameter = 5 mm, a = 6 mm, χ = 50 lines/mm). One sample is unloaded, the other is subjected to a stress obtained by loading the upper plate and transmitting the load to the sample by means of a pointed rod. Two or three increasing load conditions were applied.

Table 1 summarizes the results we have obtained. Five pairs of samples have been tested. One of them was not mechanically worked. The other pairs were mechanically worked in order to alter the strain distribution in the material: holes were drilled in different directions (vertical in the samples of pairs 2 and 5, horizontal and parallel to the optical axis in 3, horizontal and transverse to the optical axis in 4). In the cases 1 and 2, the samples are different zones in the same bar of material; in the other cases, the samples are separate pieces. In both 2 and 5, the load is applied by means of a nail inside the hole.

In some cases the correlation area was expanded by retransforming the JFT by means of a lens with focal distance f_2 greater than f_1. The correlation areas were photographically recorded and the symmetry was evaluated by means of direct visual inspection or with the aid of a microscope (magnification of the objective = 45). Furthermore, the intensity of the correlation area was measured by a photodetector (output voltage increases when the amount of light received decreases). This procedure was useful only in two cases, as reported in Table 1, mostly because of the very small dimensions of the correlation peaks compared to the photodetector size. However, observing the symmetry permits to recognize in all cases even the smallest applied load. Figs. 5 and 6 illustrate cases 5 and 1 of Table 1 (in the latter only for two applied loads). At the left are shown the JFT's, and at the right the cross-correlation peaks (magnified in fig. 5 by f_2/f_1 = 3.6; in fig. 6 an additional magnification of 45 is used).

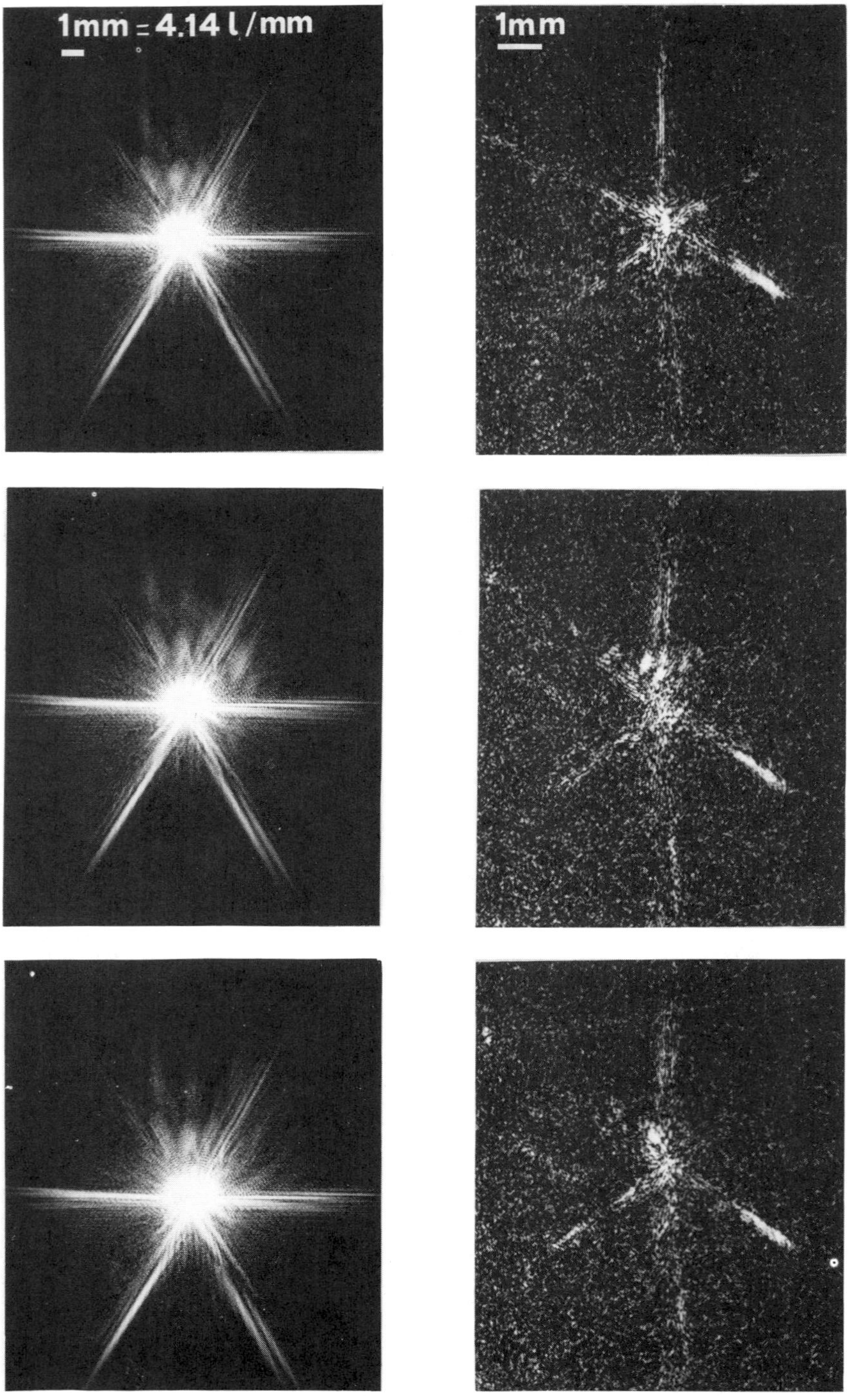

Fig. 5: JFT (at left) and enlarged central zone of the cross-correlation area for case 5 of Table 1. The applied load is zero at the top and increases from middle to bottom.

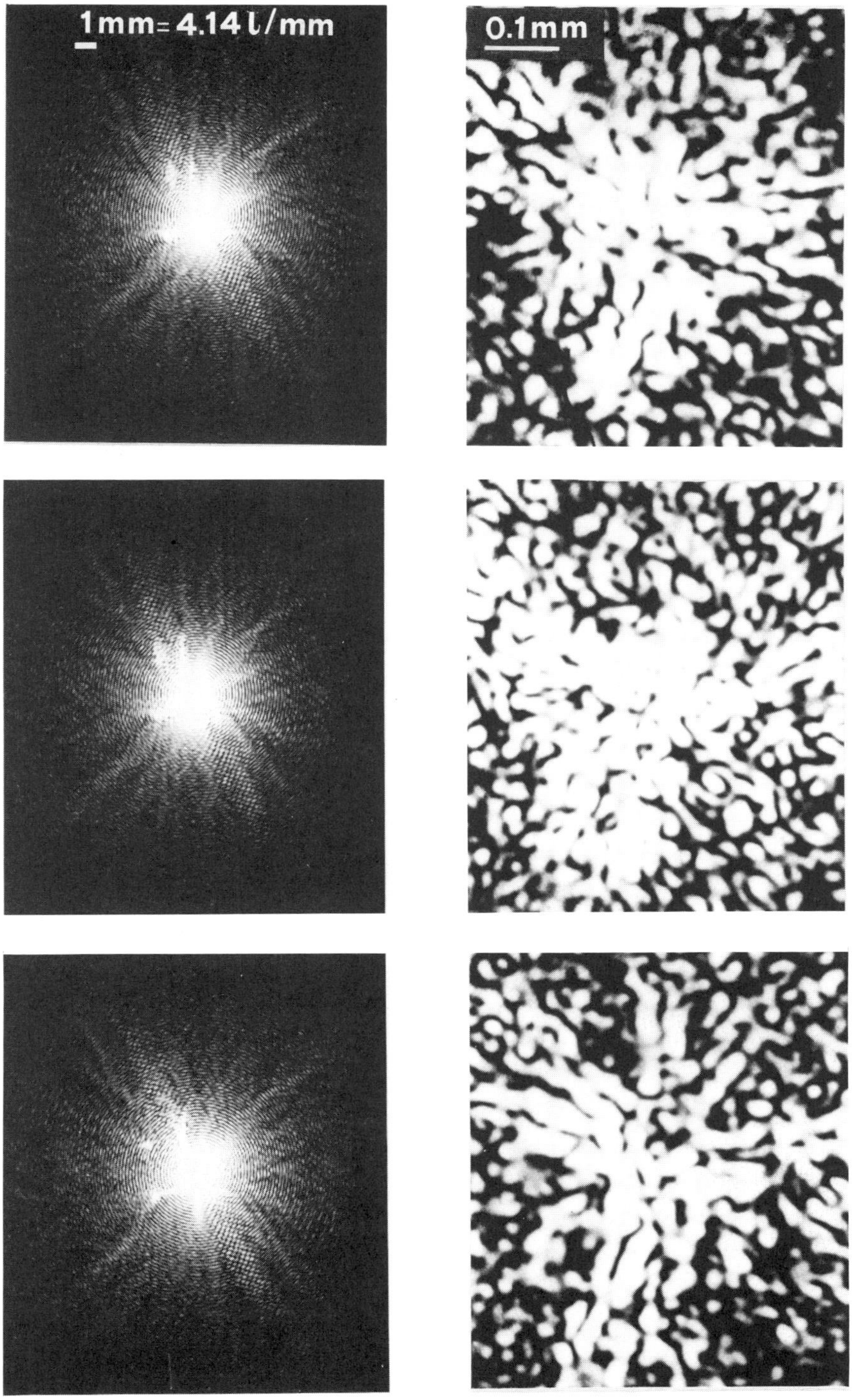

Fig. 6: JFT (at left) and cross-correlation area for case 1 of Table 1. The applied load is zero at the top and increases from middle to bottom.

TABLE 1

Sample	P (gr)	f_1 (mm)	f_2 (mm)	Symmetry (+) Vs	Mc	V (Volt)
1. 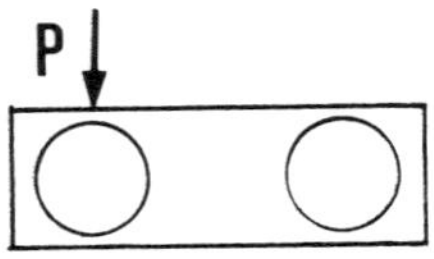	0	381	1280			
	300				*	
	2300					
	3800					
2. 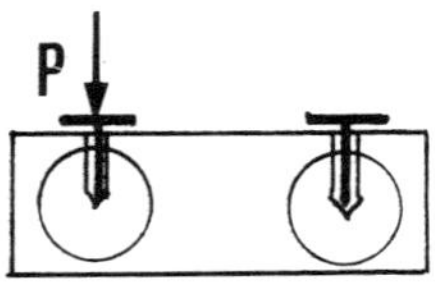	0	1280	1280			
	300			*		
	2300					
	3800					
3. 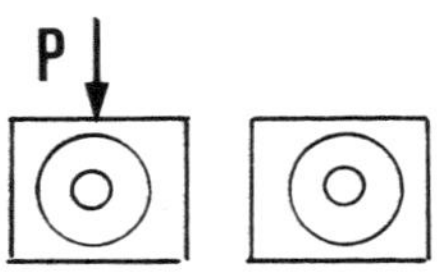	0	1280	1280			
	300				*	
	2300					
	3800					
4. 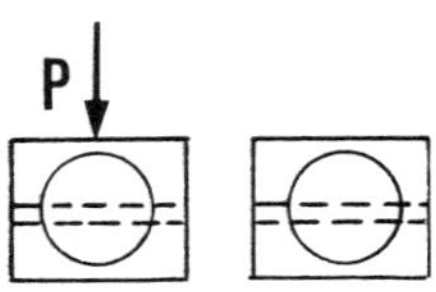	0	381	1280			7.194
	300			*		7.200
	1800					7.205
5. 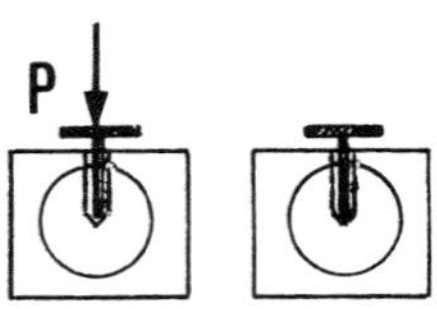	0	381	1280			7.276
	300			*		7.205
	1800					7.210

(+) The lack of symmetry becomes clearly visible starting from the load marked by *
Vs = visual inspection; Mc = microscopic inspection

JFT CROSS-CORRELATION OF RAPIDLY SELF MODIFYING PELLICLES

Multiple correlation was used: four objects (microscope cover glasses) were placed at the vertices and at the center of an equilateral triangle. In this example the cover glasses were wet by alcohol at slightly different instants. JFT was taken with 1 ms exposure time, at 10s intervals. The reconstruction of the recorded JFT's permits to follow the evaporation of the solvent.

REFERENCES

1. Y.H. Ja, Real-time nondestructive testing of phase objects using four-wave mixing with photorefractive BGO crystals, Optics and Laser Technology, vol. 17, no. 1, 36:40 (1985).
2. D. Casasent, Coherent optical pattern recognition, Proc. IEEE, vol. 67, no. 5, 813:825 (1979).
3. P. Sirotti and G. Rizzatto, Coherent optical texture recognition of digital ultrasonic images, IEEE Trans. Sonics Ultrason., vol. SU-31, no. 4, 436:440 (1984).
4. P. Sirotti and G. Rizzatto, Holographic filtering for echostructure analysis, in: Signal Processing II: Theories and Applications, H.W. Schüssler, ed., North Holland, Amsterdam (1983).
5. J.E. Rau, Detection of Difference in Real Distributions, Journal of the Optical Society of America, vol. 56, no. 11 (1966).

INTRINSIC SENSITIVITY LIMITATIONS IN CLASSICAL INTERFEROMETRY

James W. Wagner and James B. Spicer

Center for Nondestructive Evaluation
The Johns Hopkins University
Baltimore, Maryland

Introduction

Optical interferometry has long been seen as the technology with the greatest promise for providing noncontact detection for ultrasonic and acoustic emission testing. Such techniques, once fully developed, would provide couplant-free operation using only a light beam which could be scanned easily over the surface of the specimen under test. Unfortunately, there is considerable development which must take place before optical transducers can replace contact transducers for a number of testing applications. While interferometers perform quite well in the laboratory, such systems are designed usually for a specific experimental task and are not suited for general testing applications. Many of the limitations on the application of interferometric detectors are practical ones imposed, for example, by the need for surface preparation or the nature of the acoustic signal being detected. Underlying such practical limitations, there exist certain intrinsic limits to performance set by the physical properties which govern the operation of the various interferometer systems. Indeed, these intrinsic performance limits vary for the several system designs and must be understood before efforts to develop practical systems can be successful.

To evaluate and compare the ultimate performance of divers optical systems, use of the concept of signal-to-noise becomes necessary. The performance of an optical system may be expressed conveniently in terms of the system's signal-to-noise ratio. This ratio is the quotient of the desired, output signal divided by the total noise in the system. In general, maximum signal does not imply optimum system performance. Rather, system performance is maximized when the system's signal-to-noise ratio is greatest. System noise is classed broadly into two categories. One category includes noise which may be eliminated theoretically from the system. Noise such as laser output noise, detector electronics thermal noise, and random optical path fluctuations belong to this category. The second category of noise includes noise which cannot be eliminated from the system. Such noise is associated with detector quantum effects and is known as quantum noise or shot noise.[1] Only detector quantum noise poses a limit to the ultimate performance of the optical systems to be considered.[2]

For a surface-disturbance-detecting, interferometric optical system, ultimate performance implies a limit to the system's ability to detect surface changes. Such a limit to interferometer sensitivity is expressed in terms of the smallest displacement which may detected if the signal-to-noise ratio is unity. Evaluation of this sensitivity limit allows quantitative comparison of various interferometric systems.

This paper will explore the operation of the following classes of interferometric systems when used in the detection of surface disturbances: 1) path-stabilized 2) active heterodyne and 3) Sagnac. Performance of these classes with respect to various experimentally controllable parameters will be investigated en route to expressing the sensitivity limit for each class.

The class of interferometers referred to as being stabilized includes those homodyne interferometers which rely on active control of optical path lengths in order to achieve optimum signal. Such active control of the optical paths requires closed loop, electronic-mechanical/optical stabilization. Active stabilization eliminates the effects of random or systematic variations in the optical train which tend to degrade interferometer performance. Compensation for such variations may be achieved so long as the noise being rejected can be accurately distinguished from the signal; in practice this implies that the noise differs significantly in frequency content from that of the signal.

Consider the stabilized Michelson interferometer shown in figure 1. The length of the reference path with respect to the signal path is actively controlled by moving the reference mirror using a piezoelectric mount. Suppose that the sample surface supports a sinusoidal disturbance of amplitude δ and of angular frequency ω_s. The signal field is represented by the following:

$$E_s = A_s \exp j[\omega_o t - k_o L_s + 2k_o \delta \sin(\omega_s t)]$$

where A_s is the signal field amplitude, k_o is the light propagation factor , ω_o is the light frequency, and L_s is the optical length of the signal path in the absense of a surface disturbance. Similarly, the reference field is given by the following:

$$E_r = A_r \exp j[\omega_o t - k_o L_r - 2k_o \Delta]$$

where A_r is the reference field amplitude at the detector, L_r is the length of the reference path in the absense of optical path control, and Δ is half the length which the reference optical path is actively adjusted when the reference mirror is displaced by Δ. It is assumed that the quiescent path length difference $L_r - L_s$ is some multiple of the light wavelength so that subwavelength path control is explicitly given by Δ. If the polarization states of these two fields are adjusted for maximum fringe contrast then coherent interference at the detector produces the resultant field magnitude:

$$| E_{Resultant} | = A_r \left\{ 1+K^2+2\ K \cos[2k_o(\Delta+\delta \sin\omega_s t)] \right\}^{1/2}$$

where K is a field contrast factor (A_s/A_r) related to system design and loss. Making use of the Fourier-Bessel expansion [3], and using small argument approximations for the Bessel functions in the resultant field magnitude yields:

$$| E_{Resultant} | = A_r \left\{ 1+K^2+2K[\cos(2k_o\Delta) - 2k_o\delta \sin(2k_o\Delta) \sin(\omega_s t)] \right\}^{1/2} .$$

The signal of interest is that term which multiplies δ; the rest of the resultant field contributes to the detector quantum noise. The detector current is given by the following:

$$I_D = P_D \frac{\eta q}{h\nu}$$

where η is the detector quantum efficiency, ν is the light optical frequency, h is Planck's constant, and q is the carrier charge. Again, P_D is the instantaneous, optical power at the detector. In the limit of system performance, quantum noise governs system sensitivity. Detector quantum noise is related to the time average detector current as follows:

$$\langle i_q^2 \rangle = 2\ q\ \Delta\nu\ \langle I_D \rangle$$

where $\langle i_q^2 \rangle$ is the effective mean square noise current in the detector , $\Delta\nu$ is the detection electronics bandwidth, and $\langle I_D \rangle$ is the time average photodetector current.

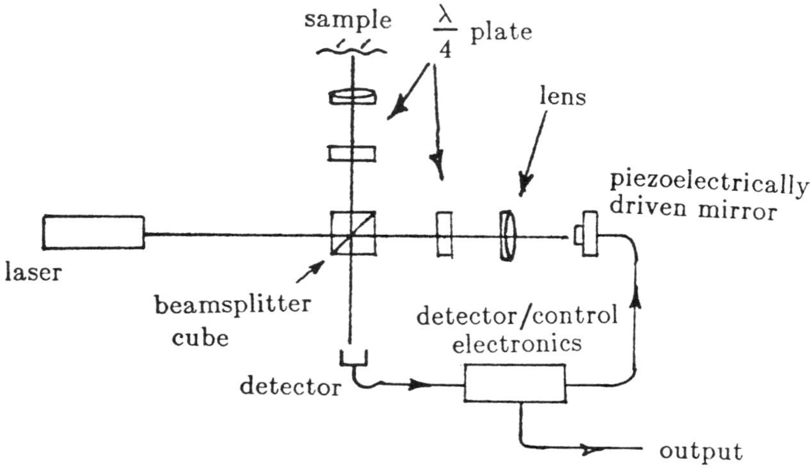

Figure 1: Stabilized Michelson Interferometer

Having defined the origins of signal and noise in the system, the performance of the system may be expressed in terms of the signal-to-noise ratio. The signal-to-noise ratio, SNR, is the quotient given by dividing the root mean square signal current by the root mean square noise current. It is understood that the signal current is taken to be that current in the detector which is functionally related to the surface displacement. Simply put, the signal-to-noise ratio is written as follows:

$$\mathrm{SNR} = \left[\frac{\langle I_s^2\rangle}{\langle i_q^2\rangle}\right]^{1/2}.$$

For the class of interferometers currently being considered, performing the required operations to obtain the signal-to-noise ratio yields the following expression:

$$\mathrm{SNR} = \left[\frac{P_o\,\eta}{4h\nu\,\Delta\nu}\right]^{1/2} \frac{2K\ \sin(2k_o\Delta)}{\left[1+K^2+2K\cos(2k_o\Delta)\right]^{1/2}} k_o\delta\,.$$

Notice that the signal-to-noise ratio is a function of Δ. Figure 2 demonstrates the variation of signal-to-noise with respect to the optical path adjustment, Δ. The curve shown is normalized to the maximum value of signal-to-noise which results if the system is assumed to be lossless and if field contrast is assumed to be unity. Note that optimal system performance occurs for a path length difference of $\approx 0.5\ \lambda$. At this path difference the signal of interest is a minimum, but so is the limiting noise. This result clearly demonstrates that optimal performance and maximum sensitivity do not occur when the signal is a maximum. Similar findings are presented by Kwaaitaal for electronically stabilized, Michelson interferometers.[2]

In practice, it is difficult to achieve field contrast of unity. For values of K not equal to one, the operating point for path length difference at maximum signal-to-noise takes on values above and below $0.5\ \lambda$. For example, if the field contrast assumes the value of $K = 0.8$ then the phase difference at maximum signal-to-noise is 2.5 radians which corresponds to a path length difference of $2\Delta = 0.4\ \lambda$ where λ is the light wavelength. Such field contrast would result if sample reflectivity were 0.64 in an otherwise lossless system.

The sensitivity limit for stabilized systems is calculated directly using the results of the signal-to-noise analysis. Demand that the signal-to-noise ratio equal unity. The surface displacement amplitude, δ , assumes its value at the detection limit and is denoted by δ_{min}. This minimum detectable amplitude may be expressed as follows:

$$\delta_{min} = [2k_o]^{-1} \left[\frac{4h\nu\,\Delta\nu}{P_o\,\eta}\right]^{1/2} \left[\frac{2^{1/2}\,[1+\cos(2k_o\Delta)]^{1/2}}{\sin(2k_o\Delta)}\right]$$

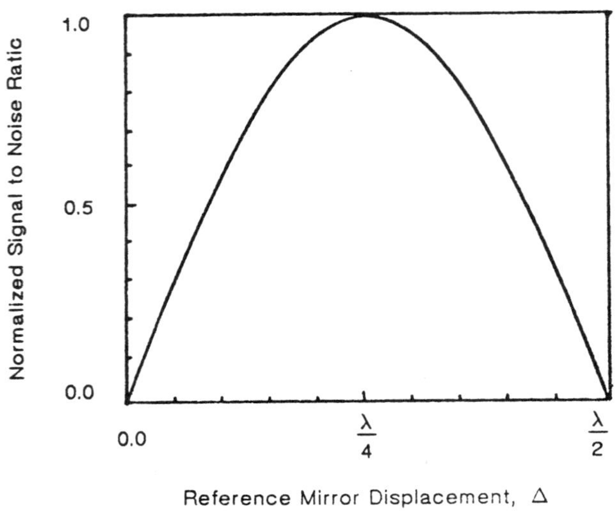

Figure 2: Normalized Signal-to-Noise Ratio for Stabilized Interferometer versus Reference Mirror Displacement

where a lossless system and unity field contrast are assumed. The expression for the minimum detectable displacement may be numerically evaluated. If laser power, P_o, is one milliwatt; detector quantum efficiency, η, is ten percent; light wavelength, λ, is 0.6328 micrometers, then the minimum detectable displacement is:

$$\delta_{min} = 5.64 \times 10^{-15} \text{ mHz}^{-1/2}.$$

Note that this displacement is given in terms of $Hz^{-1/2}$. Using such units implies that a bandwidth for the detection electronics has not been selected and that in order to obtain the minimum detectable displacement using a detection bandwidth of $\Delta\nu$ the δmin given above must be multiplied by $(\Delta\nu)^{1/2}$.

The degree to which the maximum sensitivity is achieved in practice depends on numerous parameters. The foremost consideration involves the design and execution of the electronic feedback controls. These controls must successfully hold the path difference, operating point during low frequency drift/vibration and also during large, sample surface, profile variation when the signal of interest is small. Such profile variation forces the operating point to shift to neighboring fringes where the operating point must be re-optimized in order to detect sub-fringe displacements.

Active Heterodyne Interferometers

Interferometers which belong to the active heterodyne class rely on the interference of frequency different light to carry information.[4] The interference of frequency different light produces an optical beating of the resultant field magnitude at the difference frequency. In active heterodyne systems the difference in frequency between the interfering fields (the signal and reference fields) is achieved by actively frequency shifting the light in one of the optical paths using a Bragg cell or some other frequency shifting device. The stablility and constancy of the optical beating at the detector depends on the character of the shifting device used.

Consider the heterodyne interferometer shown in figure 3. The light frequency of the reference field with respect to the signal field is actively shifted by the Bragg cell placed in the reference optical train. As before, assume the sample surface supports a sinusoidal disturbance of amplitude δ and of angular frequency ω_s. The signal field is represented by the following:

$$E_s = A_s \exp j[\omega_o t - k_o L_s + 2k_o \delta \sin(\omega_s t)]$$

where A_s is the signal field amplitude, k_o is the light propagation factor, ω_o is the light frequency, and L_s is the optical length of the signal path in the absense of a surface disturbance. Similarly, the reference field is given by the following:

$$E_r = A_r \exp j[(\omega_o + 2\omega_B)t - k_o L_r]$$

where A_r is the reference field amplitude at the detector, L_r is the length of the reference path, and ω_B is the Bragg cell frequency of operation. The magnitude of the resultant field produced by the coherent addition of these two fields is:

$$|E_{Resultant}| = A_r \left\{ 1 + K^2 + 2K \cos[2\omega_B t + 2k_o\delta \sin\omega_s t - k_o (L_s - L_r)] \right\}^{1/2}$$

where, again, K is the field contrast factor (A_s/A_r). Using the Fourier-Bessel expansion, the double angle trigonometric identity, and the small displacement approximation, the resultant field magnitude may be rewritten as follows:

$$|E_{Resultant}| = A_r \left\{ 1 + K^2 + 2K[\cos(2\omega_B t - k_o(L_s - L_r)) + k_o\delta\, [\cos((2\omega_B + \omega_s)t - k_o(L_s - L_r)) - \right.$$
$$\left. - \cos((2\omega_B - \omega_s)t - k_o(L_s - L_r))]] \right\}^{1/2}.$$

Terms which multiply δ provide the signal of interest; only terms left after the time average of the field magnitude contribute to the detector quantum noise.

Having established the character of the heterodyne signal, the performance of the system may be written in terms of the signal-to-noise ratio. Performing the previously outlined operations, the signal-to-noise ratio for the heterodyne system is as follows:

$$SNR = \left[\frac{P_o \eta}{4h\nu\, \Delta\nu} \right]^{1/2} \frac{2K}{[1+K^2]^{1/2}} k_o \delta \ .$$

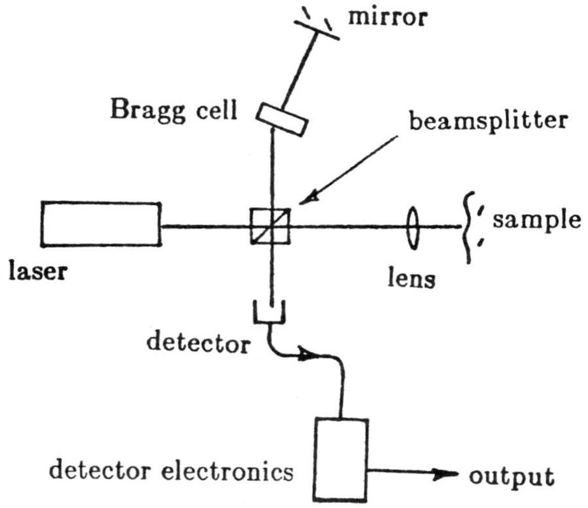

Figure 3: Heterodyne Michelson Interferometer

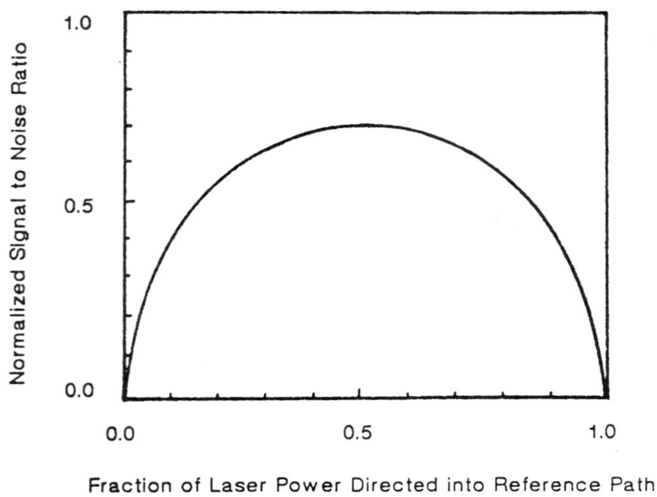

Figure 4: Normalized Signal-to-Noise Ratio versus Fraction of Laser Power Directed into Reference Path for the Heterodyne Interferometer

where all time averages were performed for the time interval $2\pi/(2\omega_B-\omega_s)$ which is assumed long in comparison to all other characteristic time intervals. Notice that the signal-to-noise ratio is completely independent of the relative lengths of the signal and the reference paths. This result implies that the interferometer performance is not affected by static, surface height changes. Figure 4 presents the functional dependence of the heterodyne signal-to-noise with respect to the fraction of the total laser power which is directed into the reference path, this corresponds to the contrast factor ranging from zero to approaching infinite values. The curve shown is for a lossless heterodyne system and is normalized to the maximum value of signal-to-noise for a path stabilized system. Note that the maximum heterodyne signal-to-noise is about 0.71 times the maximum for a stabilized system even though both sidebands are assumed to contribute to the heterodyne signal. That a difference in the maximum signal-to-noise ratio occurs can only be attributed to the intrinsic performance limits of the two systems.

As before, the sensitivity limit for heterodyne systems is calculated using the results of the signal-to-noise analysis. Assume laser power, P_o, is one milliwatt; detector quantum efficiency, η, is ten percent, and the light wavelength, λ, is 0.6328 micrometers. The minimum detectable displacement is:

$$\delta_{min} = 7.94 \times 10^{-15}\ \mathrm{mHz}^{-1/2}.$$

Again, this displacement is given in terms of $\mathrm{Hz}^{-1/2}$ and is an implicit function of the detection system bandwidth.

In practice, the degree to which the quantum noise limited sensitivity is obtained depends primarily on the design quality of the detection electronics. The method of signal demodulation considerably influences the success of the heterodyne system.[4] However, unlike the path-stabilized system, the heterodyne system does not have an optically set operating point at which the system performance is optimal. Lacking such an operating point, the heterodyne system can operate without closed loop control and can reject low frequency drift or vibration electronically while maintaining maximum sensitivity.

Sagnac Interferometers

The Sagnac class of interferometers normally is used to sense rotation rate; however, surface disturbances also may be detected using an adapted form of the Sagnac interferometer.[5] When used to detect surface disturbances, this class relies on a time -delay principle in which temporally distinct information about the surface of interest is contained in signal fields which propagate along the same optical path. The interaction of one field with the surface is time delayed relative to the interaction of another field. If the surface has changed during the delay time interval, then the interference of these fields will produce a signal related to the changes at the surface. Since the fields travel the same optical path, changes in the path may be localized to the sample surface.

Consider the Sagnac interferometer shown in figure 5. This interferometer is constructed using optical fiber as the propagation medium. The light source in the interferometer may be a laser or super radiant diode. The modulator on one side of the interferometer is a device which sinusoidally modulates the path length of the optical train.[6] In practice this device harmonically stretches the fiber by a known amount. For small surface disturbances, this modulation linearizes the output signal. It is assumed that the modulator does not significantly alter the light polarization in the fiber during the modulation cycle.[7] In general, polarization states are assumed to be properly adjusted so that maximum fringe contrast at the detector results.

For a laser of coherence length shorter than the loop's optical length, only two fields interfere. One field traverses the loop in a clockwise sense; the other, in an anti-clockwise sense. Since these two fields traverse the same optical path, in reverse fashion, the fields have equal amplitude at the detector provided system losses do not vary rapidly. The field which does not traverse the loop and the field which traverses the loop twice do not coherently interfere. These two fields contribute to the average detector photocurrent and to the quantum noise of the system.

Firstly, consider the interfering fields in the system. Again, assume the sample surface supports a sinusoidal disturbance of amplitude δ and of angular frequency ω_S. Now, require that the modulation amplitude be M and the modulation frequency be ω_M. The signal field which traverses the loop in a clockwise manner is represented as follows:

$$E_{cw} = A_{cw} \exp j\left[\omega_o t - kL_{cw} - 2k_o\delta \sin(\omega_S t) - Mk \sin(\omega_M t + \phi)\right]$$

where A_{cw} is the field amplitude, k_o is the light propagation factor in free space, k is the light propagation factor inside the fiber, ω_o is the light frequency, L_{cw} is the optical length of the path in the absense of either surface disturbance or modulation, and ϕ is a phase term. Similarly, the anti-clockwise field is given by the following:

$$E_{acw} = A_{acw} \exp j\left[\omega_o t - kL_{acw} - 2k_o\delta \sin(\omega_S(t+T)) - Mk \sin(\omega_M(t+T) + \phi)\right]$$

where A_{acw} is the field amplitude at the detector, L_{acw} is the length of the anti-clockwise path in the absense of either surface disturbance or modulation, and T is the delay time between the clockwise field arrival and the anti-clockwise field arrival at the sample surface and at the modulator. Interference at the last coupler yields a resultant field magnitude given by the following:

$$|E_{Resultant}| = 2^{1/2} A \left\{1 + \cos\left[2k_o\delta\gamma_S \sin(\omega_S t + \beta_S) + Mk\gamma_M \sin(\omega_M t + \beta_M + \phi)\right]\right\}^{1/2}.$$

where the common field amplitude $A = A_{cw} = A_{acw}$ is used and where the substitutions:

$$\beta_i = \mathrm{TAN}^{-1}\left[\frac{\sin\omega_i T}{\cos\omega_i T - 1}\right] \qquad \gamma_i = 2\sin\left(\frac{\omega_i T}{2}\right)$$

are made. It has been assumed that the quiescent path lengths for the clockwise and anti-clockwise are identical. Again, using the Fourier- Bessel expansion and the double angle trigonometric identity, the resultant field magnitude may be approximated for small displacements as follows:

$$|E_{Resultant}| = 2^{1/2} A \left\{1 + J_o(Mk\gamma_M) + 2k_o\delta\gamma_S J_1(Mk\gamma_M) \times \right.$$
$$\left. \times \left[\cos((\omega_M+\omega_S)t + \beta_M + \beta_S + \phi) - \cos((\omega_M-\omega_S)t + \beta_M - \beta_S + \phi)\right]\right\}^{1/2}.$$

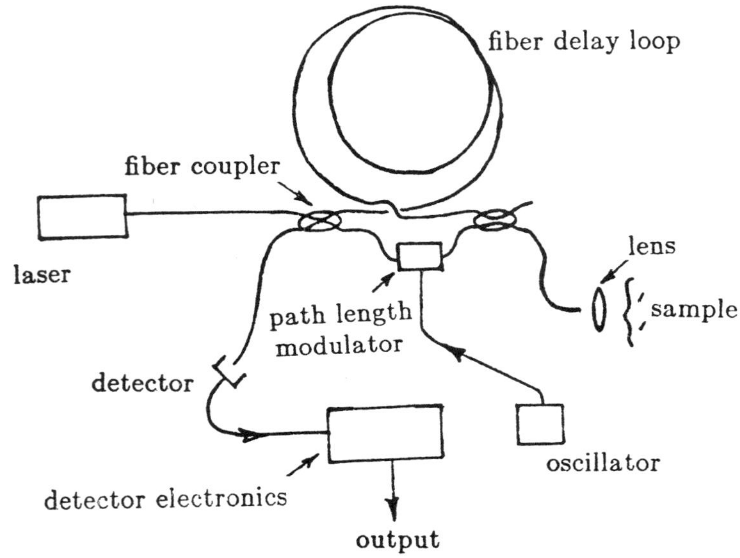

Figure 5: Sagnac Interferometer used for Surface Disturbance Detection

Having cited the terms contributing to the signal and to the noise in the Sagnac system, the signal-to-noise ratio is given as follows:

$$SNR = \left[\frac{P_o\,\eta}{h\nu\,\Delta\nu}\right]^{1/2} \frac{\varsigma\, J_1(Mk\gamma_M)}{2^{1/2}\left[1+\varsigma^2+2\varsigma\,(1+J_o(Mk\gamma_M))\right]^{1/2}} \cdot k_o\delta\gamma_s$$

where the fiber loop transmittance is given by ς. Figure 6 presents the functional variation of the Sagnac signal-to-noise ratio with respect to the modulation parameter group for zero loss in the fiber loop. The signal-to-noise ratio in this plot is normalized to the maximum signal-to-noise ratio for the path-stabilized interferometer. Notice that the signal-to-noise peaks for a modulation parameter group approximately equalling two. Recall that the modulation parameter group is as follows:

$$Mk\gamma_M = Mk\ 2\sin\left(\frac{\omega_M T}{2}\right)$$

where the appropriate substitution for γ_M has been made. If unity is given to the sine, then the modulation frequency must equal one half the reciprocal delay time. For this modulation frequency, the amplitude of modulation, M, must equal the reciprocal wavenumber, $1/k$. Qualitatively, requiring the quadrature condition indicates that the path length difference between the clockwise and anti-clockwise fields at the detector in the absense of a surface displacement is λ/π. This path difference is close to the optimum path difference predicted for a stabilized interferometer.

Unlike the stabilized and heterodyne configurations, the Sagnac configuration passively rejects low frequency noise. Passive rejection of low frequency surface displacements results from the choice of delay time. Note that the signal to noise ratio is proportional to γ_s. The appearance of this factor in the signal-to-noise ratio sinusoidally weights the frequency components of the signal. Low frequency components close to the first null in the signal-to-noise curve contribute little to the total signal. Proper choice of the delay time also allows the rejection of high frequency, systematic noise by assigning a null of the signal-to-noise curve to that noise frequency. Passive rejection of noise also entails the passive rejection of signal. The interferometer must pass a maximum of signal and a minimum of noise to be effective.

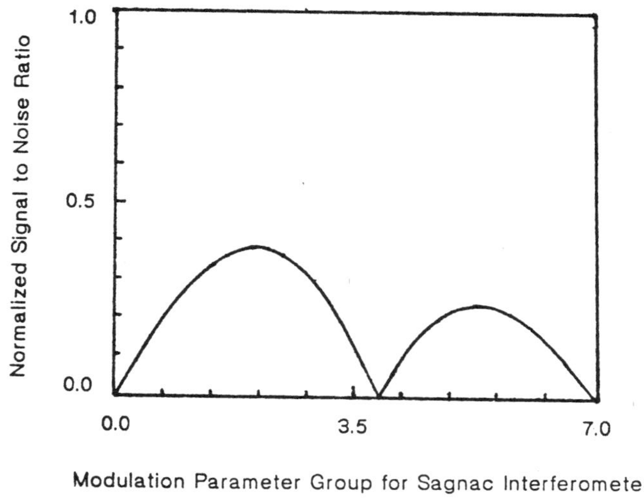

Figure 6: Normalized Signal-to-Noise Ratio for Sagnac Interferometer versus Modulation Parameter Group

As before, the sensitivity limit for Sagnac systems is calculated directly using the results of the signal-to-noise analysis. Again, assume laser power, P_o, is one milliwatt; detector quantum efficiency, η, is ten percent and light wavelength, λ, is 0.6328 micrometers. The minimum detectable displacement calculated for these values follows:

$$\delta_{min} = 1.46 \times 10^{-14}\ \mathrm{mHz}^{-1/2}.$$

Again, this displacement is given in terms of $\mathrm{Hz}^{-1/2}$ and is an implicit function of the detection system bandwidth.

The theoretical sensitivity limit of the Sagnac class is difficult to obtain in the laboratory as the assumption of zero loss is not easily achieved. Most fiber optic, interferometric systems are relatively lossy when compared to the traditional, bulk optical equivalent. The fiber based system may be compact and rugged; however, the sensitivity of a fiber system is generally lower than that of a bulk optical system.

Conclusion

Towards the end of developing practical interferometric systems for use in general testing situations, factors affecting the ultimate performance of such systems must be understood. For classical interferometry, the sensitivity limit for detection of surface disturbances repeatedly assumes a value around 10^{-15} $\mathrm{m{\cdot}Hz}^{-1/2}$ regardless of the actual system configuration. This limit is quite adequete for narrow band detection under most testing situations as losses imposed by the environment may be compensated for by the narrow detection bandwidth. Similarly, for broadband testing situations in the laboratory, this limit is satisfactory as the conditions of the test may be controlled. Losses in the optical train imposed by sample surface preparation or the nature of the displacement to be detected may be contrived so as to reach the predicted sensitivity of the system. However, for broadband detection under general testing conditions, the sensitivity afforded the classical interferometric system may not be sufficient to provide useful information about the surface disturbance under consideration. Independent of the actual interferometric system chosen for use in a given testing situation, whether it be classical or not, the intrinsic sensitivity limit of that system must be understood to ensure successful system implementation.

References

1. A. Ambrozy, "Electronic Noise," McGraw-Hill International, New York (1982).
2. Th Kwaaitaal, B. J. Luymes and G. A. van der Pijll, Noise limitations of Michelson laser interferometers, *J. Phys. D: Appl. Phys.* 13 (1980).
3. I. S. Gradshteyn and I. M. Ryzhik, "Table of Integrals, Series, and Products," Academic Press, New York (1980).
4. R. M. De La Rue, R. F. Humphryes, I. M. Mason and E. A. Ash, Acoustic-surface-wave amplitude and phase measurements using laser probes, *Proc. IEE* 119:2 (1972).
5. J. E. Bowers, R. L. Jungerman, B. T. Khuri-Yakub and G. S. Kino, An All Fiber-Optic Sensor for Surface Acoustic Wave Measurements, *J. Lightwave Tech.* LT-1:2 (1983).
6. D. A. Jackson, A. Dandridge and S. K. Sheem, Measurement of small phase shifts using a single-mode optical-fiber interferometer, *Opt. Lett.* 5:4 (1980).
7. R. Ulrich, S. C. Rashleigh and W. Eickhoff, Bending-induced birefringence in single-mode fibers, *Opt. Lett.* 5:6 (1980).

INFLUENCE OF 3-D EFFECTS ON THE CONTRAST IN PHOTOTHERMAL IMAGING

F. Lepoutre, D. Fournier and A.C. Boccara

Laboratoire d'Optique
École Supérieure de Physique et de Chimie
10, rue Vauquelin 75005, Paris, France

It has been demonstrated both experimentally and theoretically that thermal wave imaging is a very efficient quantitative tool for subsurface defect characterization. For instance, at less than one thermal diffusion length, defects equivalent to a 0.1 micron thick air slice are easily observed when using extended uniform excitation. Nevertheless, to achieve a good spatial resolution, the excitation beam is often focused on the surface.

A 3-D calculation accounting for the finite size of the excitation beam has been performed. From these calculations it appears that the smallest defects remain detectable, and the signal variation between the smallest defect and a defect of infinite thickness is much lower. For instance the phase variation of 40° which is found for a beam radius larger than 2 mm falls down to less than 10° for a radius of 100 microns for a defect located 1 mm below the surface.

Mirage effect imaging has the advantage of being noncontact. This is probably why, among all the photothermal methods, its development for NDE has been considerable these last five years. In this development the attention of the scientific community has been focused on theoretical features of the method. Indeed, experimenters have rapidly discovered that the simple 1-D interpretations are generally not valid. This paper discusses one of these 3-D particularities of the Mirage detection: the decrease of the contrast in the observation of defects parallel to the sample surface as the pump beam is focused.

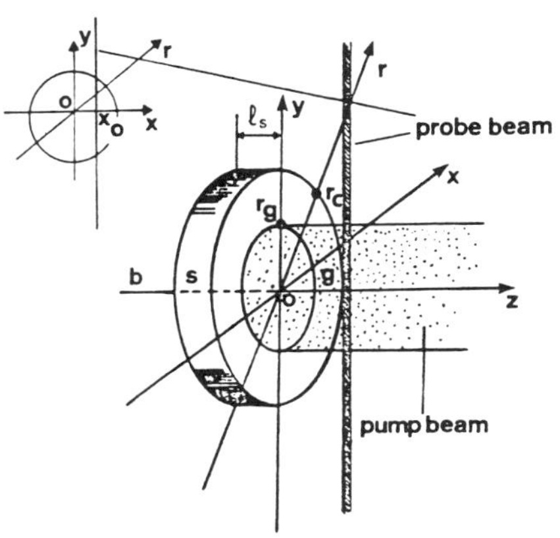

Figure 1: The cylindrical sample (s) (thickness l_s, radius r_c) is excited by a pump beam of "radius" r_g. The probe beam, parallel to Oy, crosses the fluid (g) at a distance z (normal offset), and a distance x (transverse offset) from the center of the pump beam.

We use in this paper the calculation firstly proposed by Chow[1] and McDonald[2] and recently extended to the case of Mirage detection[3]. Figure 1 defines the geometry of the problem and the notations used in the following relations. Let us recall that z is the normal offset, x_o the transverse offset, r_g the pump beam radius and r_c the sample radius. The indexes s and g will refer respectively to the sample and to the gas. We will limit in this paper our discussion to the case of the normal deflection ϕ_n with a gaussian pump beam absorbed at the sample surface.

The normal deflection ϕ_n, in a probed medium of refractive index n, is:

$$\phi_n(x_o,z) = -\frac{2}{n}\frac{\partial n}{\partial T}\sum_{m=0}^{\infty} T_m\, S_m\, \frac{\cos(\gamma_m x_o)}{\gamma_m}\, \sigma_{gm}\, \exp(-\sigma_{gm} z) \qquad (1)$$

where the variables x_o and z appear only in the cosine and the exponential respectively, γ_m is the mth root of $J_1(\gamma_m r_c) = 0$

and $$\sigma_{sm}{}^2 = \gamma_m{}^2 + j(\omega/\alpha_s) \qquad (2)$$

where ω is the pulsation and α_s the thermal diffusivity of the probed medium. S_m characterizes the pump beam distribution. For a gaussian beam of intensity I_o at r=0, S_m is given by:

$$S_m = \frac{I_o\, r_s{}^2}{2r_c{}^2\, J_o{}^2(\gamma_m\, r_c)} \qquad (3)$$

the first term (index 0) of the series being:

$$S_o = \bar{I}_o \sqrt{r^2{}_c - x_o{}^2}$$

where $\bar{I}_o$ is the intensity of the pump beam averaged on the sample surface. T_m characterizes the thermal behaviour of the sample. Let us consider the sample of Figure 2: We call R_g the thermal reflexion coefficient at the front surface and R_b the first thermal reflexion coefficient inside the sample.

R_g is given by: $$R_g = \frac{1-g}{1+g} \qquad (4)$$

where g is the ratio of the effusivities of the gas and of the sample:

$$g = \frac{k_g\sqrt{\alpha_s}}{k_s\sqrt{\alpha_g}} = \frac{k_g\ \sigma_g}{k_s\ \sigma_s} \qquad (5)$$

where k is the thermal conductivity and σ is equal to $j\frac{\omega}{\alpha}$

(α thermal diffusivity).

R_b can characterize a defect located inside the sample. For instance, for a slice of air of thickness a, R_b is equal to:

$$R_b = \frac{R_g\ (1 - \exp(-2\sigma_g a))}{1 - R_g{}^2 \exp(-2\sigma_g a)} \qquad (6)$$

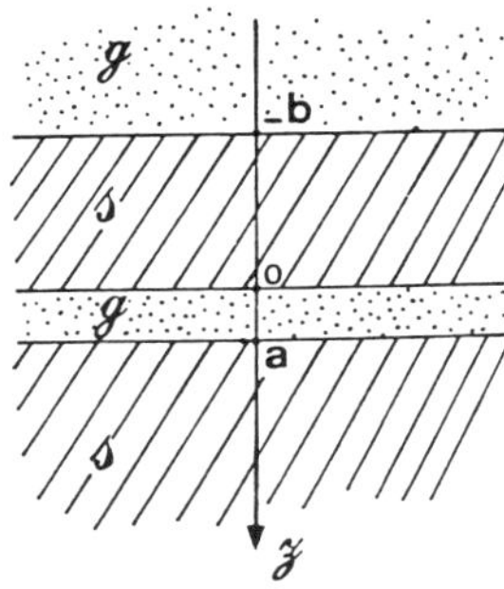

Figure 2: 1-D Model: slice of air (thickness a) at distance b below the sample surface.

The 3-D calculation of T_m leads to:

$$T_m = \frac{1}{1 + g_m} \; \frac{1}{k_s \, \sigma_{sm}} \; \frac{1 + R_{bm} \exp(-2\sigma_{smb})}{1 - R_{bm} R_{sm} \exp(-2\sigma_{smb})} \qquad (7)$$

where σ_{sm} is identical to σ_{gm} (relation (2)) for sample (index s) and g_m, R_{sm}, R_{bm} are deduced from (4), (5), (6) respectively in which σ_g and σ_s are replaced by σ_{gm} and σ_{sm}.

As mentioned in a previous paper[3] $\phi_n(x_o,z)$ may be calculated rather fastly on a microcomputer. The result shown in Figures 3-5 have been obtained from an IBM PC.

These results are presented as the phase variation ϕ of ϕ_n vs the square root of the modulation frequency f. This choice is the most convenient way to achieve the qualitative detection of subsurface defects. In such a representation a thermally thick sample would be characterized by a straight line, the slope of which being only dependent on the normal offset z. The loss of contrast appears clearly in Figure 3. This case has been calculated for a simple 1 mm thick plate of thermal diffusivity $\alpha_s = 10^{-5} m^2/s$, thermal conductivity $k_s = 50$ W/m°C surrounded by air; the normal offset z = 100 µm. Since a is infinite in this case $R_b = R_g$.

The curve $r_{g\infty}$ has been calculated by keeping only the term m=0 in equation (1), which is known to be the 1-D result. A deep minimum of about 45 degrees appears at very low frequency (10^{-2} Hz). As the pump beam is focused on the surface, the minimum appears at higher frequencies (1 Hz for r_g = 1 mm) and its depth strongly decreases. Below r = 500 µm, the minimum starts to disappear, i.e. the rear surface becomes invisible although the sample is thermally thin.

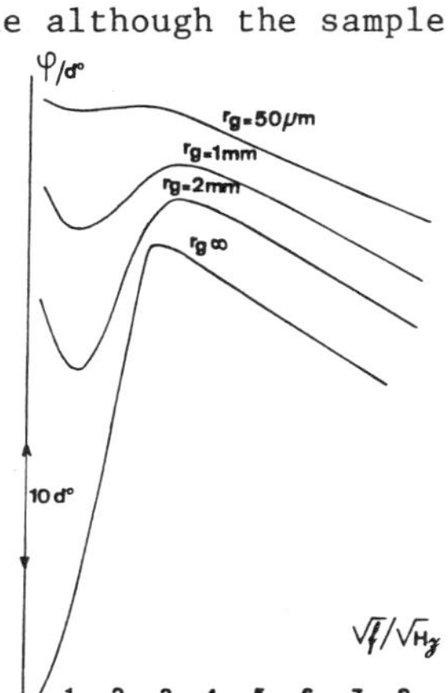

Figure 3: Case of a simple plate (a ∞). Phase ϕ of ϕ_n vs $\sqrt{f}$ for different pump beam radius r_g.
$\alpha_g = 2.10^{-5}\ m^2s^{-1}$,
$\alpha_s = 10^{-5}\ m^2s^{-1}$,
$k_g = 25.10^{-3}\ Wm^{-1}$,
$k_s = 50\ Wm^{-1}°C^{-1}$,
b = 1 mm, z= 0.1 mm

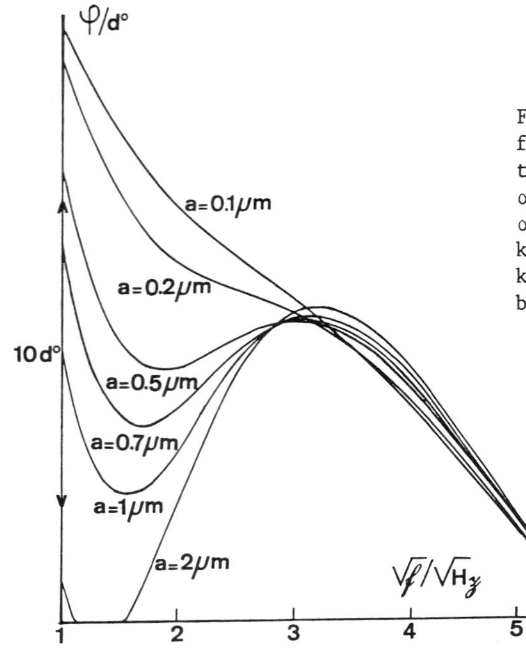

Figure 4: Case of defects of finite thickness. 1-D calculation ($r_{g\infty}$), $\alpha_g = 2.10^{-5}\ m^2s^{-1}$ $\alpha_s = 2.10^{-5}\ m^2s^{-1}$, $k_g = 25.10^{-3}\ Wm^{-1}\,°C^{-1}$, $k_s = 35\ Wm^{-1}\ °C^{-1}$, b = 0.9 mm, z = 0.2 mm.

In figures 4 and 5 we have drawn the curves obtained for a defect located at b = 900 μm below the surface of a sample of infinite thickness (see Fig. 2). Its thermal diffusivity and conductivity are respectively $\alpha_s = 10^{-5}\ m^2/s$ and $k_s = 35$ W/m°C, and the normal offset z = 200 μm. The air slice thickness α varies between 1 mm and 0.1 μm.

Figure 4 shows the result of a 1-D calculation corresponding to uniform excitation. The best sensitivity is obtained for "a" varying between 0.1 μm and 10 μm. Above 10 μm all the curves are close to the one corresponding to $r_{g\infty}$. Below 0.1 μm, they are almost straight lines. This result is kept in the 3-D curves, but (see Fig. 5) as r_g decreases the phase variations observed between the curves a = 0.1 μm and a = 10 μm fastly decrease. They are divided by a factor 2 when r_g varies from 2 mm to 1 mm and for $r_g \leqslant 500$ μm they become smaller than 3°.

It must be noticed that for all the values of r_g, the frequency at which the defect becomes detectable (i.e. when the part of the sample above the defect becomes thermally thin) is always the same. The slope of the straight line at the highest frequencies is also almost independent of r_g.

Of course samples exactly identical to the ones of Fig. 2 are not easy to realize and, in any case, it is better to work with a "true" sample. We have applied the previous discussion to the case of samples made of two coaxial tubes having the thermal parameters of Figs. 4 and 5. Between the external tube and the internal one, the contact may be imperfect, i.e. small slices of air may exist in some points and good contact in other ones.

In Figure 6 the same location on one of these samples is tested with three different pump beam radii 1.4 mm, 1 mm, 0.5 mm. With r_g = 1.4 mm and r_g = 1 mm the defect appears clearly and can be associated with z = 0.15 mm and a = 0.9 μm. The decrease of the contrast is easily verified since for r_g = 0.5 mm the defect is almost invisible. The fits of these experimental curves are not totally perfect due to the

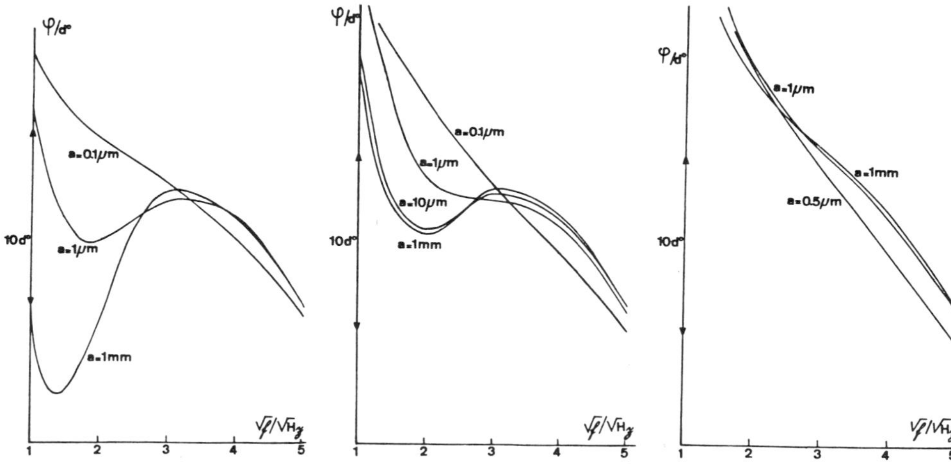

Figure 5: Case of defects of finite thickness. 3-D calculation. The pump beam radius r_g is equal to (a) 2 mm, (b) 1 mm, (c) 0.5 mm.

fact that we did not take into account the finite thickness of the internal tube in the theoretical model (see Fig. 2).

This study is not complete. The influence of some other parameters (normal and transverse offsets, distance b at which the defect is located) will be studied in a future paper. But we think that the main features are described here: one must be very careful of the size of the pump beam when detecting quantitatively subsurface defects and the pump beam radius must be at least of the order of magnitude of the distance at which the defect is located from the surface.

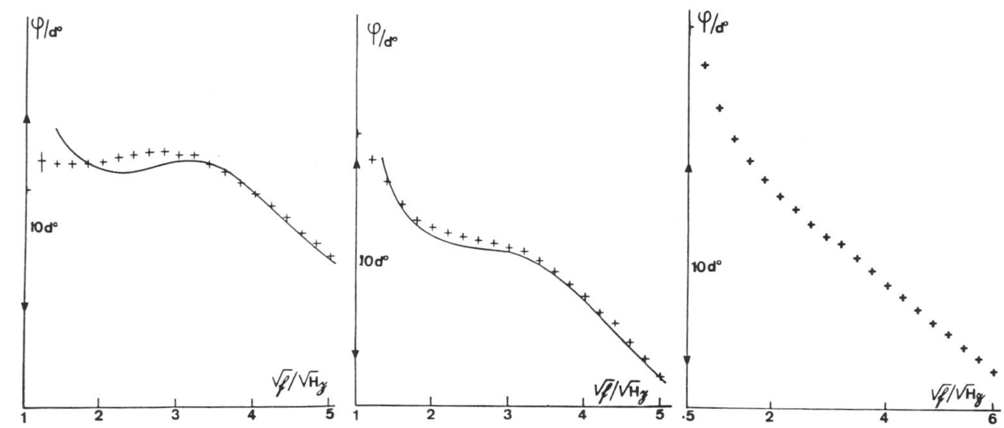

Fig. 6: (a) Experimental points with r_g = 1.4 mm, z = 0.15 mm on coaxial tubes (see text). The theoretical curve is computed with the thermal parameters of Fig. 4 and r_g = 1.4 mm, z = 0.15 mm and b = 0.9 mm. (b) Identical to (a) except r_g = 1 mm. (c) Identical to (a) except r_g = 0.5 mm and z = 0.12 mm.

REFERENCES:

1. M. Chow, J. Appl. Phys. 51, 4053 (1980).
2. F.A. McDonald, J. Appl. Phys. 52, 381 (1981).
3. F. Lepoutre, B.K. Bein and L.J. Inglehart, Can. J. Phys. (Oct. 1986).

THERMAL WAVE IMAGING FOR MATERIALS CHARACTERIZATION

J. W. Maclachlan and J. C. Murphy

Center for Nondestructive Evaluation and
Applied Physics Laboratory
The Johns Hopkins University
Laurel, MD 20707

INTRODUCTION

Novel techniques for materials characterization and nondestructive evaluation are being continually developed to meet the requirements of examining an ever-increasing range of new materials. A wide range of physical properties and processes are exploited in the current measurement technologies and it is often necessary to match the characterization technique to the materials problem at hand. A very recent addition to the repertoire of materials characterization techniques is thermal wave imaging. This field encompasses a wide range of techniques for exciting and detecting periodic temperature fields or "thermal waves" in solids. The physical probe for this materials characterization technique is heat and thermal wave imaging thus provides information about local variations in sample thermal properties such as the thermal conductivity and the heat capacity. Thermal wave imaging is a developing technique and although the contrast mechanisms can be identified in simple situations, understanding the contrast in thermal wave images for complicated sample geometries and detection schemes is very involved (1). It is readily apparent that many topics of interest in materials science lend themselves well to characterization on the basis of variations in the sample's thermal properties. The examples considered in this paper consider interfaces and show that the measurement of variations in the interruption of heat flow across these interfaces provides an effective means for characterizing the structure of the interface.

BACKGROUND

The basic principles of thermal wave imaging are illustrated in Fig. 1. A heating beam which is modulated in time strikes the sample causing a periodic temperature field to be set up in the solid. The details of this temperature field will depend on the local thermal properties of the sample as well as on the specifics of the interaction of the heating beam with the surface. For example, electrons have a range of 1-10 μm in metals for energies in the range of 5 to 30 keV, while the absorption length for photons is of the order of nanometers. The depth of penetration of the heating source relative to the depth of the thermal structures of interest can be important in determining whether these structures are detected. This has been demonstrated for thermal wave imaging of a semiconductor device using electron beam heating where increasing the interaction depth of the electrons allowed improved detection of subsurface structures (2).

Theoretical modelling of the temperature distribution in the solid involves solution of the thermal diffusion equation subject to the appropriate boundary conditions. In a simple one-dimensional model of a semi-infinite solid with surface heating the solution is of the form:

$$T(x,t) = T_s \exp(-\frac{x}{\delta}) \exp[i(\omega t - \frac{x}{\delta})] \quad (1)$$

where T_s is the surface temperature, x is the distance into the solid and δ is a parameter known as the thermal diffusion length which is given by the expression:

$$\delta = [\frac{2\kappa}{\omega\rho C_v}]^{1/2} \quad (2)$$

where κ is the thermal conductivity, ρ is the density and C_v is the specific heat capacity. Although Eq. 1 resembles the solution to a wave equation with losses (thus giving rise to the term "thermal wave"), this expression for the temperature field arises from the solution of a diffusion equation with a periodic source term. The thermal diffusion length, δ, is related to both the "thermal wavelength" and the rate of decay of the thermal waves and is thus a useful unit of measure for discussion of thermal wave measurements.

A number of direct and indirect methods are available for probing the temperature field in the solid and these techniques enable changes in the temperature field to be monitored as the heating beam is scanned over different regions of the sample. Method 1 indicated in the figure is the optical beam deflection or "mirage" technique (3,4). This measurement is based on the development of a "thermal lens" in the air layer just above the sample due to conduction of heat from the sample surface to the gas. The source of this thermal lens is the temperature-dependence of the index of refraction of air. A probe laser beam is then sent through this lens and very small variations in the amount of deflection of the beam are measured by a position sensitive detector. The deflection is given by:

$$\vec{\phi} = \int_p \frac{1}{n} \frac{dn}{dT_g} \nabla T_g \times d\vec{\ell} \quad (3)$$

where n is the gas index of refraction, T_g is the gas temperature, p is the probe beam path over the sample. Two orthogonal probe beam deflections exist, one in the plane defined by the probe ray direction and normal to the specimen surface (normal component) and the second, orthogonal to this plane (transverse component). For all relative positions of the excitation and probe beams, the normal deflection component monitors changes in specimen temperature while the transverse deflection component monitors the specimen temperature gradient in the direction orthogonal to the probe ray.

The probe beam can also be oriented in a vertical configuraton as indicated by Method 2 in the figure (5-7). The deflection signal now has two components: deflection due to passage through the thermal lens and deflection due to local variation in the sample surface topography due to thermal expansion effects. Thermoreflectance or the thermal modulation of the specimen reflectivity may also be a factor in these measurements (8). Photothermal radiometry (Method 3) is another "thermal" method for detection and is based on the enhanced IR emission of a heated object (9,10). The fourth detection method illustrated in Fig. 1 is based on the detection of the displacement generated by the periodic thermal expansion of the sample. This displacement can be monitored using a number of techniques including an attached transducer (11,12) or optical interferometry (13).

The experiments reported here incorporate combined thermal and thermoelastic detection using a system where both optical beam deflection and piezoelectric detection methods are employed. The sample mounted on a piezoelectric transducer (NBS conical transducer) is scanned beneath an argon ion laser pump beam and a helium-neon laser probe beam under computer control and line scans or images can be generated. Both the magnitude and the phase of the normal or transverse component of the optical beam deflection signal or of the piezoelectric signal can be used for imaging. A similar procedure is used for creating thermal wave images using an electron beam for heating. A scanning electron microscope has been modified to allow blanking of the electron beam over the range 0.1 to 500 kHz. The sample is mounted on a piezoelectric transducer and the magnitude or phase of the piezoelectric signal is used in conjunction with the imaging electronics in the SEM to generate both line scans and images.

SUBSURFACE HORIZONTAL BOUNDARIES

The first materials characterization problem to be presented here is the detection of subsurface horizontal boundaries. Two sample geometries

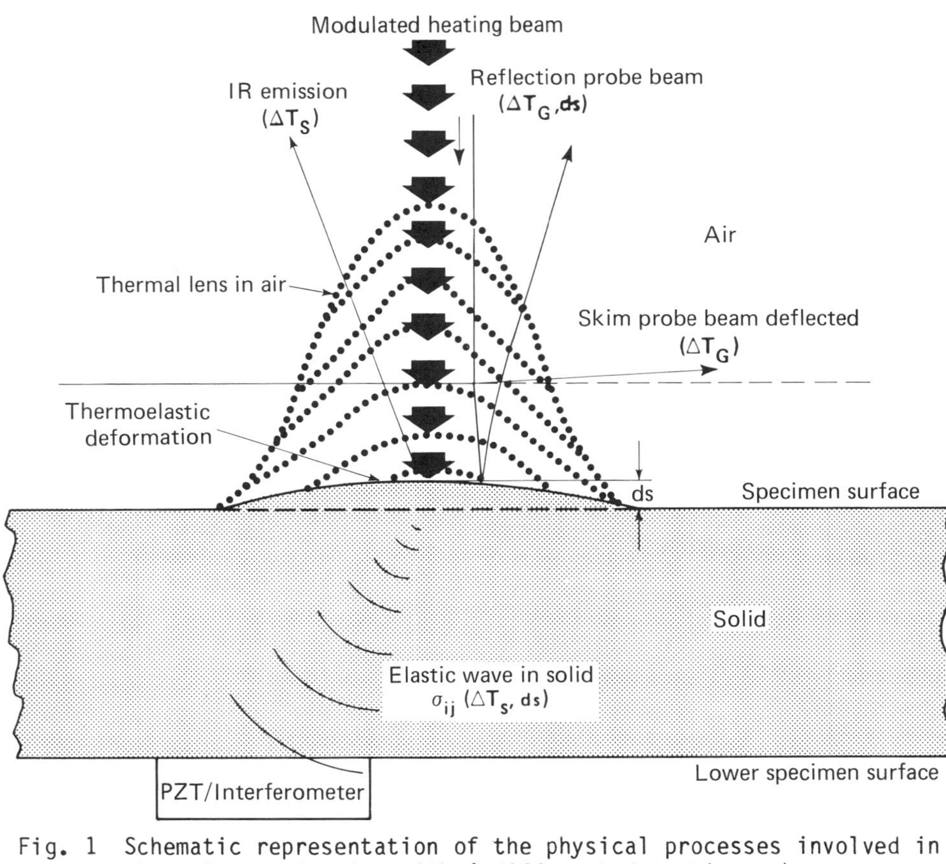

Fig. 1 Schematic representation of the physical processes involved in thermal wave imaging with 4 different detection schemes.

are considered. The buried slant slot sample is a disk of 2024 aluminum, 1 cm in diameter and 3 mm thick containing a slot halfway through the diameter of the sample at an angle of about 17° to the top surface. The buried slant hole sample is a similar disk but with a hole of diameter 1 mm drilled through the disk, also at an angle of about 17° to the top surface. Scans over the tops of these samples allow the examination of defects of ever-increasing depth in the same sample.

Figure 2(a)-(f) show both the magnitude and phase of the normal component of the optical beam deflection signal (NOBD) for the buried slant slot sample. The data is presented in image form in (a) and (b) and in a perspective plot representation in (c) and (d) to better show the func-

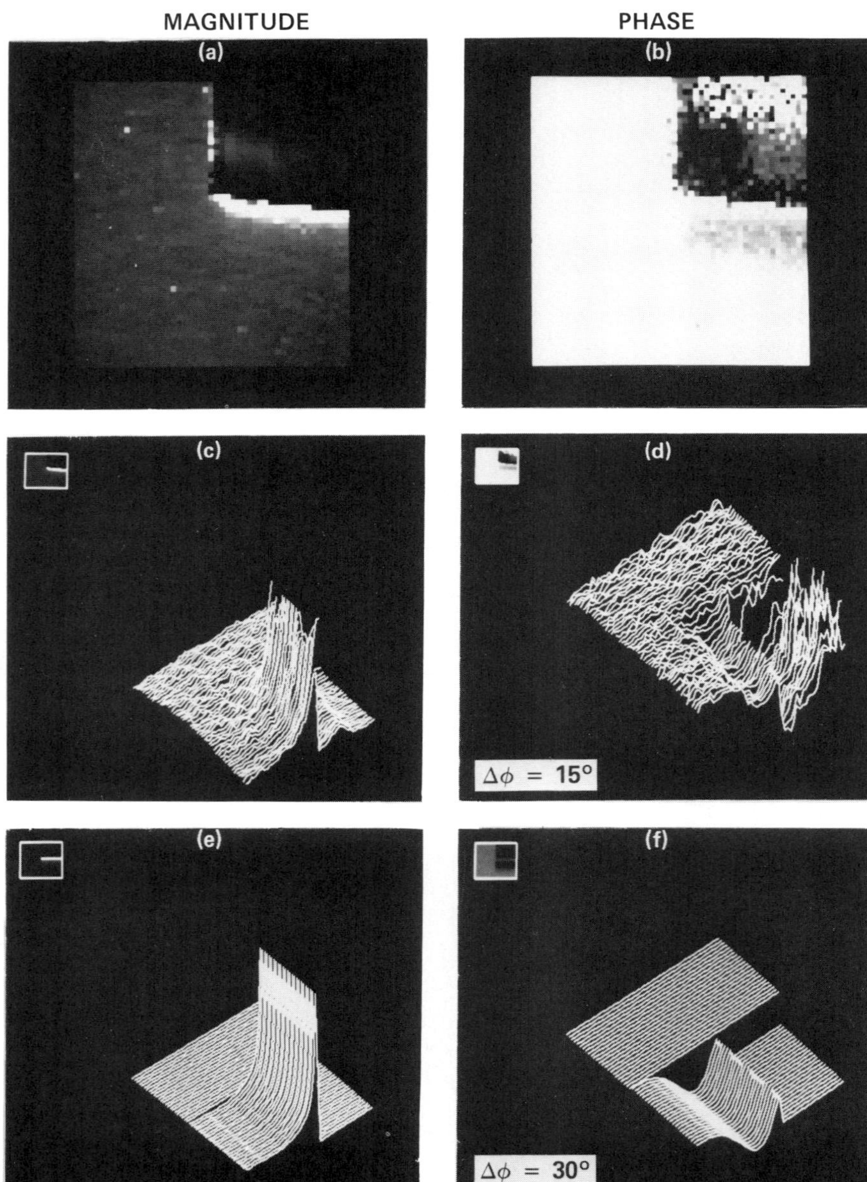

Fig. 2 Experimental (a-d) and theoretical (e-f) magnitude and phase images of buried slant slot using detection by NOBD.

tional dependence of the signal over the slot. The slot is breaking through the surface near the middle of the right hand side of the image and by the bottom of the image the slot has a depth of about 300 μm or about 1.8 thermal diffusion lengths at the modulation frequency of 500 Hz used in this experiment. The results of a one-dimensional calculation of the magnitude and phase of the surface temperature over the buried slant slot are given in (e) and (f) and show a favorable comparison with the experimental data. Although the phase gives indication of the defect to deeper depths than the magnitude, there is very little indication of the presence of the defect when it is greater than 1 to 2 thermal diffusion lengths in depth for detection by a thermal method such as optical beam deflection.

Detection of buried defects using piezoelectric detection is illustrated in Fig. 3. In this experiment the buried slant hole sample has been scanned using both electron beam (Fig. 3(a) and (b)) and laser beam (Fig. 3(c)) excitation at a modulation frequency of about 78 kHz. The results are similar for both excitation methods. The ratio of the defect depth, D to the thermal diffusion length, δ is indicated for each of the line scans in the figure and it is clear that not only does piezoelectric detection allow defects of depth up to 20 thermal diffusion lengths to be detected, but the functional dependence of the piezoelectric signal also changes as a function of depth. The source of this signal shape is under investigation but it is clearly not a scattering of the ultrasonic wave from the hole since the wavelength in aluminum at this frequency is about 70 times the diameter of the hole.

VERTICAL CRACKS

Comparison of three different detection methods, piezoelectric detection and the normal and transverse components of the optical beam deflection are shown in Fig. 4 for a 200 μm by 20 μm scan over a vertical closed crack in an aluminum alloy compact tensile specimen with all measurements made at a laser beam modulation frequency of 750 Hz. Such vertical interfaces can be very difficult to detect with conventional nondestructive testing techniques. Although the crack can be detected in both the magnitude and phase scans for piezoelectric and NOBD detection, the most telling

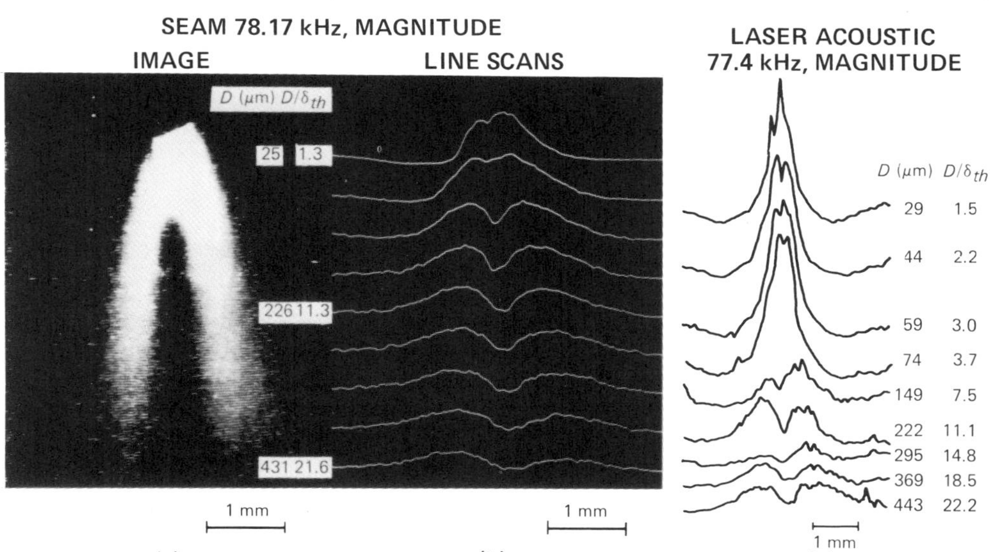

Fig. 3 Thermal wave images of buried slant hole using piezoelectric detection and excitation by (a), (b) electron beam and (c) laser beam.

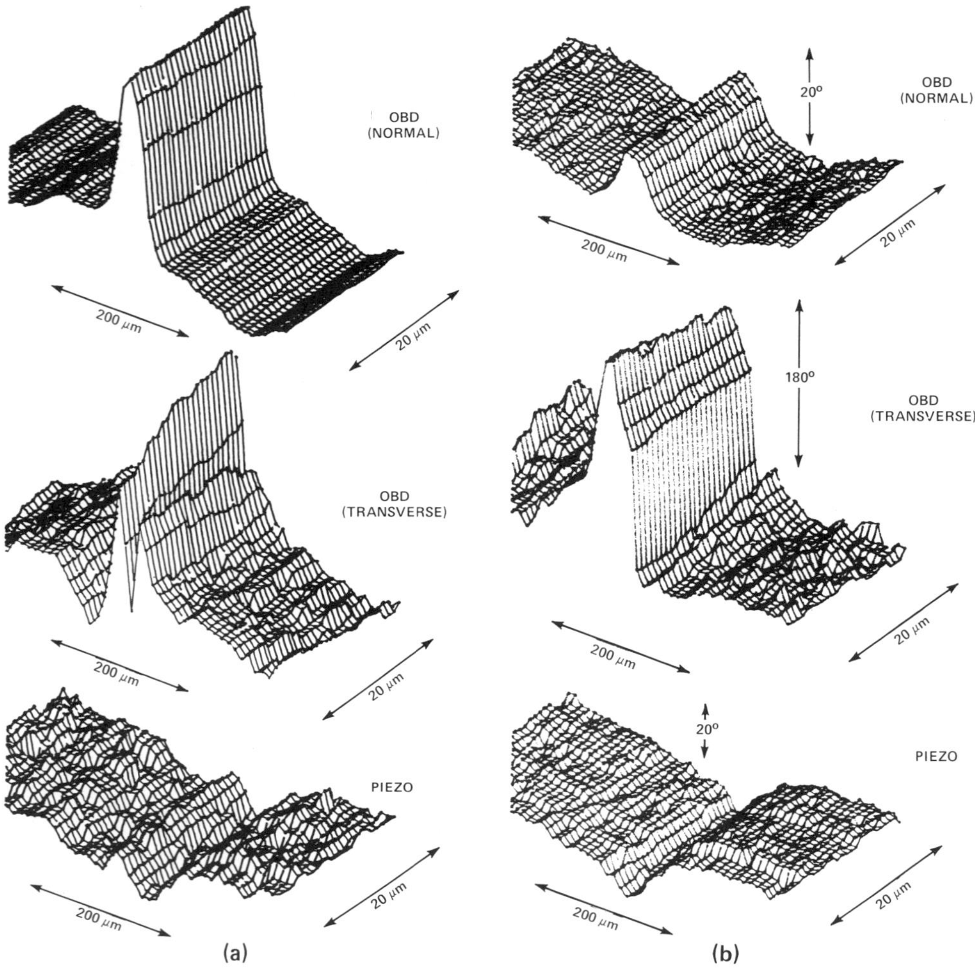

Fig. 4 A comparison of three different detection methods -- NOBD, TOBD and piezoelectric -- for thermal wave imaging of a vertical crack (a) magnitude, (b) phase.

indication of the crack presence is obtained from the double peak in the magnitude and the 180° shift in the phase of the transverse OBD signal. This is expected since the transverse signal monitors the gradient in the temperature distribution in the direction orthogonal to the probe beam direction. A vertical crack effectively interrupts this temperature gradient. The piezoelectric and NOBD signal are more dependent on the actual sample temperature and a crack in a vertical orientation does not have as large an effect on the surface temperature.

CORROSIVE ATTACK OF INCONEL 600

The final sample to be examined here is a specimen of Inconel Alloy 600 with a grain size ranging from 40 to 80 μm which has been subjected to a corrosive environment to produce intergranular attack. Such an attack results in fine cracks along grain boundaries which would be expected to strongly affect heat flow in the sample. Intact grain boundaries have been imaged using thermal wave imaging (14), but the problem in this sample is more similar to the imaging of vertical cracks as described above. Fig. 5

shows both image and perspective plot data for the magnitude and phase of the normal OBD signal at 500 Hz ($\delta \approx 50$ μm) for a scan which covers attacked material on the left and good material on the right. There is a marked difference in signal level between the good and the attacked material for both the magnitude and phase. Also, greater structure is evident in the attacked region, perhaps indicating groups of grains which are separated from other groups of grains. Other scans at higher magnification measuring the transverse OBD component revealed regions in the attacked material with 180° phase excursions, thus indicating the presence of vertical cracks.

CONCLUSIONS

We have presented examples which demonstrate that thermal wave imaging is a viable technique for materials characterization, particularly for the study of interfaces such as horizontal boundaries and vertical cracks. Just as it is often important in other materials characterization techniques to select the appropriate measurement technique for the particular materials problem, it is necessary in a thermal wave measurement to employ methods for excitation and detection of the thermal waves which best suit the material properties and sample geometry of the problem at hand. It has been shown that subsurface defects can be easily detected by optical beam deflection provided they are at a depth of a thermal diffusion length or less. For deeper defects, piezoelectric detection is required.

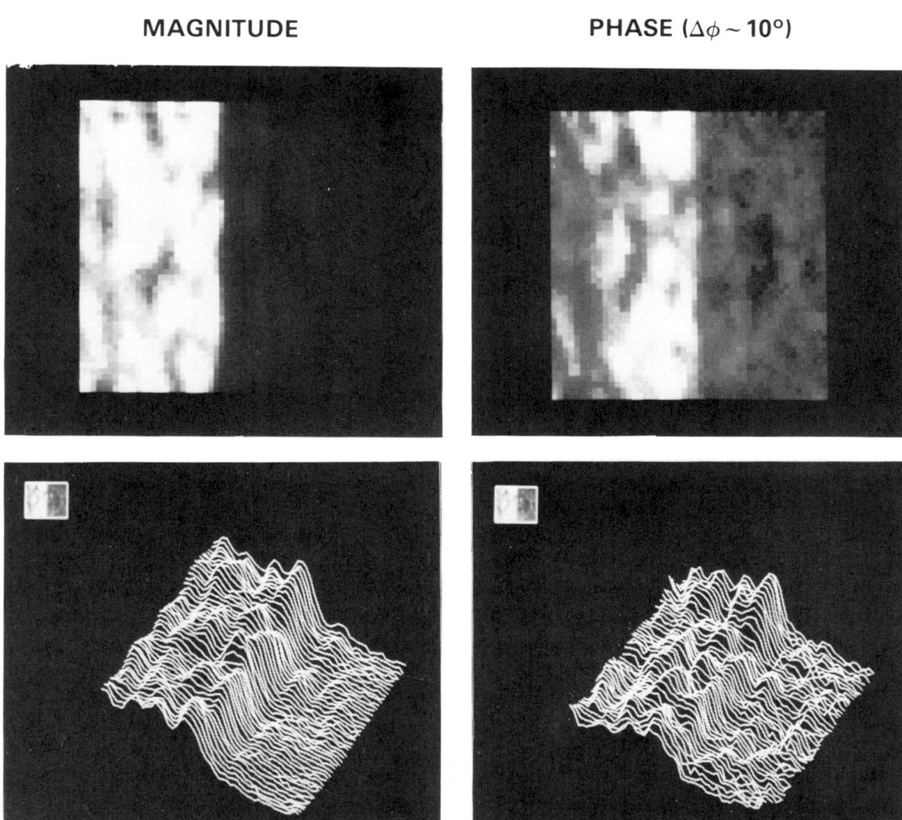

Fig. 5. Thermal wave images (magnitude and phase) of corrosive attack in Inconel 600 with detection by NOBD. Attached material is at the left of the image and good material is at the right.

ACKNOWLEDGEMENTS

We thank Ed Hackett of NSRDC, Annapolis, MD who provided us with the compact tension specimen and Jack Lareau of Combustion Engineering, Inc., Windsor, CN who prepared the Inconel 600 sample with corrosive attack. The assistance of Maureen Madey with many of the experimental measurements is also gratefully acknowledged. Funding sources for this work included The Johns Hopkins University Center for Nondestructive Evaluation and the U. S. Army Research Office and U. S. Naval Sea Systems Command under Contract No. N00024-85-C-5301. In addition, one of the authors (JWM) was supported by an AAUW International Fellowship.

REFERENCES

1. J. C. Murphy, J. W. Maclachlan, and L. C. Aamodt, Image contrast processes in thermal and thermoacoustic imaging, to appear in IEEE Trans. on Ultrasonics, Ferroelectrics and Frequency Control UFFC-33 522 (1986).
2. J. W. Maclachlan, J. C. Murphy, R. B. Givens, and F. G. Satkiewicz, Linear thermal wave imaging, in: "Proceedings of 11th World Conference on Nondestructive Testing, November 1985, Las Vegas, Nevada, Volume 1," Taylor Publishing Company, Dallas (1985).
3. A. C. Boccara, D. Fournier, and J. Badoz, Thermo-optical spectroscopy: Detection by the mirage effect, Appl. Phys. Lett. 36:136 (1980).
4. J. C. Murphy and L. C. Aamodt, Optically detected photothermal imaging, Appl. Phys. Lett. 38:196 (1981).
5. M. A. Olmstead, N. M. Amer, S. Kohn, D. Fournier, and A. C. Boccara, Photothermal displacement spectroscopy: A new optical probe for solids and surfaces, Appl. Phys. A 32:141 (1983).
6. A. Rosencwaig, J. Opsal, and D. L. Willenborg, Thin film thickness measurements with thermal waves, Appl. Phys. Lett. 43:166 (1983).
7. J. C. Murphy and L. C. Aamodt, Reflective photothermal imaging, J. Physique C6-513 (1983).
8. A. Rosencwaig, J. Opsal, W. L. Smith, and D. L. Willenborg, Detection of thermal waves through optical reflectacne, Appl. Phys. Lett. 46:1013 (1985).
9. P. E. Nordal and S. O. Kanstad, Photothermal radiometry, Physica Scripta 20:659 (1979).
10. G. Busse, Photothermal transmission probing of a metal, Infrared Phys. 20:419 (1980).
11. G. Busse and A. Rosencwaig, Subsurface imaging with photoacoustics, Appl. Phys. Lett. 36:815 (1980).
12. G. S. Cargill, Electron-acoustic microscopy, in: "Scanned Image Microscopy" E. A. Ash, ed., Academic Press, London (1980).
13. S. Ameri, E. A. Ash, V. Neuman and C. R. Petts, Photo-displacement imaging, Elec. Lett. 17:337 (1981).
14. J. W. Maclachlan, R. B. Givens, J. C. Murphy, and L. C. Aamodt, Contrast Mechanisms in Scanning Electron Acoustic Imaging of Grain Boundaries, 4th International Topical Meeting on Photoacoustic, Thermal and Related Sciences, Ville d'Esterel, Quebec, Aug. 4-8, 1985.

AUTHORS INDEX

SUBJECT INDEX